CRM Series in Mathematical Physics

Springer
New York
Berlin
Heidelberg
Barcelona
Hong Kong
London
Milan
Paris
Singapore
Tokyo

CRM Series in Mathematical Physics

Conte, The Painlevé Property: One Century Later
MacKenzie, Paranjape, and Zakrzewski, Solitons: Properties, Dynamics, Interactions, Applications
Saint-Aubin and Vinet, Algebraic Methods in Physics: A Symposium for the 60th Birthdays of Jiří Patera and Pavel Winternitz
Saint-Aubin and Vinet, Theoretical Physics at the End of the Twentieth Century: Lecture Notes of the CRM Summer School, Banff, Alberta
Semenoff and Vinet, Particles and Fields
van Diejen and Vinet, Calogero–Moser Sutherland Models

Yvan Saint-Aubin
Luc Vinet
Editors

Theoretical Physics at the End of the Twentieth Century

Lecture Notes of the CRM Summer School, Banff, Alberta

With 74 Illustrations

Springer

Yvan Saint-Aubin
Département de Mathématiques
et Statistique
Université de Montréal
C.P. 6128, Succursale Centre-ville
Montréal, Québec H3C 3J7
Canada
saint@crm.umontreal.ca

Luc Vinet
Department of Physics
and
Department of Mathematics and Statistics
McGill University
James Administration Building, Room 504
Montréal, Québec H3A 2T5
Canada
vinet@vpa.mcgill.ca

Library of Congress Cataloging-in-Publication Data
Theoretical physics at the end of the twentieth century : lecture notes of the CRM summer school, Banff, Alberta / editors Yvan Saint-Aubin, Luc Vinet.
p. cm. — (The CRM series in mathematical physics)
Includes bibliographical references.
ISBN 0-387-95311-6 (hardcover)
1. Mathematical physics — Congresses. 2. Physics — Congresses. I. Saint-Aubin, Yvan. II. Vinet, Luc. III. CRM series on mathematical physics.
QA19.2 .T44 2001
530.15—dc21 2001032817

Printed on acid-free paper.

Production managed by Yong-Soon Hwang; manufacturing supervised by Jerome Basma.
Photocomposed copy prepared from the CRM's LaTeX files.
Printed and bound by Sheridan Books, Inc., Ann Arbor, MI.
Printed in the United States of America.

9 8 7 6 5 4 3 2 1

ISBN 0-387-95311-6 SPIN 10841204

Springer-Verlag New York Berlin Heidelberg
A member of BertelsmannSpringer Science+Business Media GmbH

Series Preface

The Centre de recherches mathématiques (CRM) was created in 1968 by the Université de Montréal to promote research in the mathematical sciences. It is now a national institute that hosts several groups and holds special theme years, summer schools, workshops, and a postdoctoral program. The focus of its scientific activities ranges from pure to applied mathematics and includes statistics, theoretical computer science, mathematical methods in biology and life sciences, and mathematical and theoretical physics. The CRM also promotes collaboration between mathematicians and industry. It is subsidized by the Natural Sciences and Engineering Research Council of Canada, the Fonds FCAR of the Province de Québec, and the Canadian Institute for Advanced Research and has private endowments. Current activities, fellowships, and annual reports can be found on the CRM Web page at `www.CRM.UMontreal.CA`.

The CRM Series in Mathematical Physics includes monographs, lecture notes, and proceedings based on research pursued and events held at the Centre de recherches mathématiques.

Yvan Saint-Aubin
Montréal

Preface

When organizing a summer school at the end of the most fruitful century in the history of physics one must carefully consider the question of topics. We decided to avoid a retrospective of problems solved during the last century and we shied away from a prediction of which ones might be settled during the next. We opted for a snapshot of what theoretical physics is at the end of this century (or, should we say, the last century), namely problems actively researched in the last few years of the nineteen nineties. This choice led to twelve courses that were timely, diverse and exciting.

This book contains seven of the courses given in the summer school. They constitute up-to-date accounts of the following topics: supersymmetric Yang–Mills theory and integrable systems, branes, black holes and anti-de Sitter space, turbulence, Bose–Einstein condensation, integrability, deformed Virasoro and elliptic algebras, mesoscopic physics, QCD in extreme conditions. Young physicists will find in these texts pedagogical introductions to subjects currently active in theoretical physics, and more seasoned ones a chance to share the excitement of fields outside their immediate research interests.

Yvan Saint-Aubin
Luc Vinet
Montréal

Contents

Contributors

Alexander Altland, Teoretische Physik III, Ruhr-Universität Bochum, 44780 Bochum, Germany; `alexal@tp3.ruhr-uni-bochum.de`

Eric D'Hoker, Department of Physics and Astronomy, University of California, Los Angeles, CA 90095-1547, USA; `dhoker@physics.ucla.edu`

Michael J. Duff, Randall Laboratory, Department of Physics, University of Michigan, Ann Arbor, MI 48109-1120, USA; `mduff@umich.edu`

Krzysztof Gawędzki, Institut des Hautes Études Scientifiques, 91440 Bures-sur-Yvette, France; `bongo@ihes.fr`

Allan Griffin, Department of Physics, University of Toronto, Toronto, ON M5S 1A7, Canada; `griffin@physics.utoronto.ca`

Satoru Odake, Department of Physics, Faculty of Science, Shinshu University, Matsumoto 390-8621, Japan; `odake@azusa.shinshu-u.ac.jp`

Duong H. Phong, Departmentof Mathematics, Columbia University, New York, NY, 10027, USA; `phong@math.columbia.edu`

Ben D. Simons, Cavendish Laboratory, University of Cambridge, Madingley Road, Cambridge CB3 0HE, UK; `bds10@cus.cam.ac.uk`

Frank Wilczek, School of Natural Sciences, Institute for Advanced Study, Princeton, NJ 08540, USA; `wilczek@sns.ias.edu` (as of September 2000: Center for Theoretical Physics, Massachusetts Institute of Technology, 77 Massachusetts Ave., Cambridge, MA 02139, USA; `wilczek@mit.edu`)

1

Lectures on Supersymmetric Yang–Mills Theory and Integrable Systems

Eric D'Hoker
D.H. Phong

ABSTRACT We present a series of four self-contained lectures on the following topics;

(I) An introduction to 4–dimensional $1 \leq \mathcal{N} \leq 4$ supersymmetric Yang–Mills theory, including particle and field contents, $\mathcal{N} = 1$ and $\mathcal{N} = 2$ superfield methods and the construction of general invariant Lagrangians;

(II) A review of holomorphicity and duality in $\mathcal{N} = 2$ super-Yang–Mills, of Seiberg–Witten theory and its formulation in terms of Riemann surfaces;

(III) An introduction to mechanical Hamiltonian integrable systems, such as the Toda and Calogero–Moser systems associated with general Lie algebras; a review of the recently constructed Lax pairs with spectral parameter for twisted and untwisted elliptic Calogero–Moser systems;

(IV) A review of recent solutions of the Seiberg–Witten theory for general gauge algebra and adjoint hypermultiplet content in terms of the elliptic Calogero–Moser integrable systems.

1 Introduction

Supersymmetry (often abbreviated as susy) maps particles and fields of integer spin (bosons) into particles and fields of half odd integer spin (fermions) and vice versa. Supersymmetry was introduced in quantum field theory in [1] (for textbooks, see [2]). Supersymmetry is a continuous space-time symmetry, generated by a *supercharge* Q_α, which is a fermionic space-time spinor operator of spin $\frac{1}{2}$. As such, Q_α commutes with the usual local and global internal symmetries, such as color and flavor symmetries,

$$[Q_\alpha, \text{internal charges}] = 0. \tag{1.1}$$

As a result, in a supersymmetric theory, the spectrum is arranged in pairs (boson, fermion), where the boson and fermion have the same internal quantum numbers and the same mass.

In the present lectures, we shall concentrate on Yang–Mills theories, which are built out of particles and fields of spins at most equal to 1. Thus, we shall not deal with theories of gravity which contain also particles and fields of spins larger than 1. Before reviewing the structure of the spectrum and of the interactions of supersymmetric theories, we shall briefly explain why supersymmetry is an important and integral part of modern particle physics already today and is expected to become so even more in the near future.

1.1 Supersymmetry and the Standard Model

All the particle physics that we know of today is described to stunning accuracy by the particle contents and by the interactions of the *Standard Model.* The particle contents is given in Table 1.1; all particles have been observed, except for the Higgs.

The interactions between the spin 1 particles and quarks and leptons are dictated by the gauge invariance of the standard model, $SU(3)_c \times SU(2)_L \times U(1)_Y$. To each of the factors of the gauge group, there is an independent coupling constant, denoted by g_3, g_2 and g_1. The fine structure constant α, familiar from electromagnetism, and the weak mixing angle θ_W are combinations of the couplings g_2 and g_1.

By inspection of the particle spectrum of the Standard Model, it is immediate that there is not a single candidate for a supersymmetric pair of particles. Thus, it would appear at first sight that supersymmetry has no useful place in particle physics, since the most basic characteristic of a supersymmetric theory is not realized in the Standard Model.

However, in Nature, symmetries are often *spontaneously broken.* This means that the laws of physics are invariant under a symmetry, but the solution of the equations (such as the ground state of the theory) is not invariant under this symmetry. This situation is familiar from the Standard Model itself and occurs in all three sectors of the theory. In the $SU(3)_c$ or Quantum Chromodynamics (QCD) sector, the spontaneous breaking of

TABLE 1.1. Particle Contents of the *Standard Model*

Spin 0	physical Higgs	not yet observed
Spin $\frac{1}{2}$	leptons	$\begin{pmatrix} \nu_e \\ e^- \end{pmatrix}$ $\begin{pmatrix} \nu_\mu \\ \mu^- \end{pmatrix}$ $\begin{pmatrix} \nu_\tau \\ \tau^- \end{pmatrix}$
	quarks	$\begin{pmatrix} u \\ d \end{pmatrix}$ $\begin{pmatrix} c \\ s \end{pmatrix}$ $\begin{pmatrix} t \\ b \end{pmatrix}$
Spin 1	8 gluons g	strong interactions
	photon γ	electro-magnetic interactions
	$W^\pm$, Z	weak interactions

TABLE 1.2. Particle Contents of the *Minimal Supersymmetric Standard Model*

Non-Susy Sector	Susy Partners	Spin Susy Partners
2 Higgs (required)	Higgsinos	$\frac{1}{2}$
Leptons $(\nu_e\ e^-)\cdots$	sLeptons $(\tilde{\nu}_e\ \tilde{e}^-)\cdots$	0
Quarks $(u\ d)\cdots$	sQuarks $(\tilde{u}\ \tilde{d})\cdots$	0
Gluons g Photon γ $W^\pm$, Z	Gluinos $\tilde{g}$ photino $\tilde{\gamma}$ Winos $\widetilde{W}^\pm$, Zino $\widetilde{Z}$	$\frac{1}{2}$

approximate chiral symmetry for the up and down quarks is responsible for the comparatively small masses of the pions, as well as for the comparatively large masses of the nucleons. (This effect is discussed in detail in the Banff lectures of Frank Wilczek [3].) In the $\mathrm{SU}(2)_L \times \mathrm{U}(1)_Y$ sector, the spontaneous breaking of SU(2) is responsible for the masses of quarks, leptons, $W^\pm$ and Z particles, as well as for the emergence of unbroken electromagnetic $\mathrm{U}(1)_{\mathrm{em}}$.

Thus, by analogy with mechanisms familiar from the Standard Model, we should expect that also supersymmetry may be spontaneously broken if realized at all in Nature. In the generic mechanisms of supersymmetry breaking, the spectrum continues to be organized in pairs of bosons and fermions with equal quantum numbers, but the mass of the fermion is different from the mass of the boson of the pair.

Considering now pairs of fermions and bosons, but allowing for unequal masses, as is typical of spectra with spontaneously broken supersymmetry, we see that the Standard Model still exhibits no candidates for such pairs, and thus displays no room for supersymmetry to operate, even if spontaneously broken. To have any chance at all of realizing supersymmetry, even spontaneously broken, the particle spectrum of the Standard Model must be extended. The minimal way of doing this, via the introduction of the smallest number of supplementary particles, is called the *Minimal Supersymmetric Standard Model*. Basically, for each currently known or *non-supersymmetric* particle, we supply a hypothetical *supersymmetric partner*. There is one exception: supersymmetry requires at least two Higgs fields. The spectrum is schematically exhibited in Table 1.2.

Actually, the existence of super partners in supersymmetric theories is one of the most dramatic and most general predictions that follows from the assumption that supersymmetry has anything to do with Nature at all. Bold as the assumption of this extension of the particle spectrum may be, this is not the first time in the history of particle physics that the number of degrees of freedom associated with an elementary particle is doubled up, as schematically shown in the diagram below for the case of the electron

$$\text{Thompson electron} \implies +\text{spin} \implies +\text{positron} \implies +\text{susy partner.}$$

The precise mechanism by which supersymmetry is to be broken is not well understood at present. By analogy with QCD, the mechanism may be inherently nonperturbative, but we do not know for a fact that this is the case.

1.2 Supersymmetry and Unification of Forces

The supersymmetric extension of the Standard Model, in which a hypothetical supersymmetric partner is included for each known particle, has appeared in a rather ad hoc fashion. The full significance of supersymmetry really emerges only when the principle of supersymmetry is considered in conjunction with the principle of the unification of the strong, electromagnetic and weak forces (and most likely also gravity).

Remarkably, the Standard Model automatically contains a new scale at energies of order 10^{15} GeV. It is surprising to find such a high energy scale in the model, since the masses of the quarks, leptons and gauge bosons are on the order of or smaller than the electro-weak scale of about 250 GeV. To see how this new scale emerges, one must take into account the effect of renormalization of the three couplings g_i, $i = 1, 2, 3$. (See, for example, [4].) As the typical energy or mass scale μ of a physical process is altered, the couplings g_i vary according to the renormalization group equations, whose lowest order solutions are given by

$$\frac{4\pi}{g_i^2(\mu)} = \frac{b_i}{2\pi} \ln \frac{\mu}{\Lambda_i}, \qquad i = 1, 2, 3. \tag{1.2}$$

Here, b_i are numerical constants, dependent on the gauge group and on the matter contents, and Λ_i are integration constants; we shall not need their explicit values here. The couplings have been measured to very high accuracy at the scale of the mass of the Z particle $\mu = M_Z$, where [5]

$$g_1(M_Z) \sim 0.46 < g_2(M_Z) \sim 0.64 < g_3(M_Z) \sim 1.22. \tag{1.3}$$

These values determine the integration constants Λ_i for given b_i.

The couplings g_2 and g_3 are asymptotically free, i.e. their values decrease as the energy scale is increased because $b_2, b_3 > 0$. Now, at low energies, the strong coupling g_3 is larger than g_2, and since $b_3 > b_2$, the coupling g_3 will decrease faster than g_2. Also, the coupling g_1 is not asymptotically free and increases as the energy scale is increased because $b_1 < 0$. From these general considerations, it is not surprising that the couplings g_i will mutually intersect as the energy scale is raised.

Remarkably however, all three couplings $g_i(\mu)$ approximately intersect in one point around $M_U \sim 10^{15}$ GeV. Even more remarkably, this scale is only a few orders of magnitude below perhaps the most primary scale in physics, the Planck scale $M_P \sim 10^{19}$ GeV at which quantum gravity is supposed to become important. Adopting the philosophy that such a remarkable coincidence cannot be merely an accident and must have a fundamental physical meaning, we are led to interpret the scale M_U, where the three couplings approximately meet, as a scale where the three different forces are unified into a single one. A unified theory will be governed by a simple gauge group. The smallest such group which contains the gauge group of the Standard Model as a subgroup is SU(5), but the inclusion of a massive neutrino can be realized only in the larger SO(10), and string theory might favor a sequence all the way up to the maximal exceptional group E_8,

$$\mathrm{SU}(3)_c \times \mathrm{SU}(2)_L \times \mathrm{U}(1)_Y \subset \mathrm{SU}(5) \subset \mathrm{SO}(10) \subset \mathrm{E}_6 \subset \mathrm{E}_8. \tag{1.4}$$

Starting at the highest energy scales, just below M_P, and running down to lower energy scales, there will initially be only a single force, governed by the unified simple gauge group. The strong, electromagnetic and weak interactions will emerge as the result of a phase transition at the unification scale M_U, where the unified gauge symmetry is broken down to the $\mathrm{SU}(3)_c \times \mathrm{SU}(2)_L \times \mathrm{U}(1)_Y$ gauge group. If the breaking of the gauge symmetry occurs in successive stages, various intermediate breaking scales will also emerge.

The unification scheme is attractive for a number of additional reasons.

$\oplus$ The Abelian gauge interactions $\mathrm{U}(1)_Y$ and $\mathrm{U}(1)_{\mathrm{em}}$ are not asymptotically free and the gauge couplings g_1 and α grow indefinitely as the energy scale is increased. It is usually argued that this behavior leads to a Landau ghost, thus to a failure of unitarity at high energies and ultimately to an inconsistency of the theory. In the unification schemes where the U(1) gauge group arises from the breaking of an asymptotically free gauge theory with simple gauge group, this problem is automatically avoided, because the dynamics of the asymptotically simple group will take over at sufficiently large energies.

$\oplus$ The particle spectrum of the standard model without right handed neutrino fits economically in three family copies of the $\bar{\mathbf{5}} \oplus \mathbf{10}$ of SU(5), while the spectrum including the right handed neutrino fits precisely in three family copies of the **16** spinor representation of SO(10). In view of the recent experimental results on neutrino masses and oscillations, the SO(10) scheme is definitely the preferred one.

However, the unification scheme also has some fundamental problems.

$\ominus$ The additional gauge bosons can mediate baryon number violating transitions, and the *proton life-time* comes out around 10^{29} years, many orders of magnitude shorter than the present experimental bounds of around 10^{33} years. (See for example [6].)

$\ominus$ As the fundamental scale in the unification models is now large and on the order of 10^{15} GeV, it becomes an issue as to how the comparatively small mass scales of the electro-weak gauge bosons and the quarks and leptons come about. This is called the *hierarchy problem.* The principles of quantum field theory guarantee that these masses are free parameters, so in principle they could be adjusted by hand to be small. This requires an absurd degree of *fine tuning* of the various couplings in the theory, and is usually viewed as unnatural.

⊖ Finally, the three gauge couplings g_1, g_2 and g_3 meet only approximately in a single point. With the high precision data made available by the study of the Z-particle and its decays, we now know that strictly within the scenario of the Standard Model and its extrapolation to high energies, the three couplings will not intersect at one point.

The Supersymmetric Extension of the Standard Model, combined with the principle of unification of strong, electro-magnetic and weak forces supplies radical solutions to the above problems.

⊕ The supersymmetric extension of the SO(10) unified theory provides perfect intersection of the couplings g_1, g_2 and g_3 at $M_{\mathrm{SU}} \sim 10^{16}$ GeV. The running of the gauge couplings depends only very slightly upon the scale and mechanism of supersymmetry breaking. The result quoted here is obtained generically when the susy breaking scale is on the order of the electro-weak scale of 100 GeV.

⊕ The proton life-time for the SU(5) or SO(10) supersymmetric unified theories is now much longer (in part because the unification scale is an order of magnitude larger in the supersymmetric theories) and estimated to be on the order of 3×10^{33} years. This result is tantalizingly close to the present day bounds for this number, and if supersymmetric unified theories are viable, we may expect to see proton decay within our life-time.

⊕ The hierarchy and fine-tuning problems are dramatically reduced. The smallness of fermion masses is naturally protected by chiral symmetries. In generic nonsupersymmetric theories, the smallness of boson masses is not protected and contributions to the boson mass quadratic in the cutoff scale tend to drive boson masses towards the largest scale in the theory. In exactly supersymmetric theories, bosons and fermions are degenerate, so supersymmetry together with chiral symmetry will also protect the smallness of boson masses. In more realistic theories, where supersymmetry is broken (explicitly or spontaneously) at a scale of M_B, the smallness of boson masses will be naturally protected up to the supersymmetry breaking scale. The idea being that above this breaking scale, supersymmetry is exact for all practical purposes, and boson and fermion masses above this scale will be degenerate.

⊕ Finally, the argument of the previous point may be turned around and used to provide a rough estimate of the supersymmetry breaking scale. If supersymmetry is to protect boson masses up to the electro-weak scale, then the supersymmetry breaking scale cannot be too much larger than the electro-weak breaking scale, otherwise supersymmetry will cease to resolve the hierarchy and fine-tuning problems. Including lower bounds from present day experiments on supersymmetric particles, we obtain the challenging range

$$100\,\mathrm{GeV} \leq \text{susy breaking scale} \leq 1000\,\mathrm{GeV}\,. \tag{1.5}$$

As experiments confront energy scales larger than the susy breaking scale, one may naturally expect to find a wealth of supersymmetric partners. Remarkably, if the above scheme is borne out in Nature, the experimental discovery of supersymmetry may not be far off in the future!

1.3 Supersymmetric Yang–Mills Dynamics

In view of the remarkable confluence of circumstancial evidence pointing towards a supersymmetric extension of the Standard Model, it has become a priority in particle physics to understand better the perturbative and nonperturbative dynamics of supersymmetric gauge theories and their spontaneous breaking. Over the past few years, substantial progress has been made in the understanding of the nonperturbative dynamics of Yang–Mills theory with extended supersymmetry, namely $\mathcal{N}=2$ and $\mathcal{N}=4$. On the one hand, by generalizing Montonen–Olive [7] duality, Seiberg and Witten [8] obtained the exact low energy effective action as well as the exact spectrum of BPS states of $\mathcal{N}=2$ super Yang–Mills theory (with SU(2) gauge group) and established that the dynamics is always that of a Coulomb phase. On the other hand, by analyzing $D3$-brane configurations [9] in certain limits, Maldacena [10] arrived at an equivalence conjecture between Type IIB superstring theory on $\mathrm{AdS}_5 \times S^5$ and $\mathcal{N}=4$ superconformal Yang–Mills theory. Assuming this equivalence, further conjectures, purely on $\mathcal{N}=4$ super Yang–Mills, have emerged, suggesting an extensive degree of nonrenormalization [11] perhaps beyond that required by the $\mathcal{N}=4$ supersymmetry alone [12]. While it is unlikely that $\mathcal{N}=4$ super Yang–Mills is completely integrable in the sense of 2-dimensional models, both developments mentioned above suggest that the theory may be integrable in restricted sense. While these developments are impressive, the physically most pressing questions such as the dynamics of supersymmetry breaking remain only partially under control to this date.

The purpose of the present lectures is to present an introduction to 4-dimensional supersymmetric gauge theories, including those with extended supersymmetry, as well as to the Seiberg–Witten nonperturbative solutions. Remarkably, these solutions possess a natural and extremely useful mapping onto integrable systems in classical mechanics, a subject to which we also present a review. Finally, the map onto integrable systems, in particular onto Toda and Calogero–Moser systems, is used to give a Seiberg–Witten construction for general gauge algebras and a hypermultiplet in the adjoint representation of the gauge algebra (a theory obtained from $\mathcal{N}=4$ super Yang–Mills by adding a mass term to the hypermultiplet).

2 Supersymmetric Yang–Mills in 4 Dimensions

We begin by reviewing the particle and field contents and the construction of Lagrangians invariant under supersymmetry for spins less or equal to 1. For textbooks see [2].

2.1 Supersymmetry Algebra

Poincaré symmetry is generated by the translations $\mathcal{R}^4$ and Lorentz transformations SO(3,1), with generators P_μ and $L_{\mu\nu}$ respectively. (Here and below, $\mu, \nu = 0,1,2,3$.) The complexified Lorentz group is isomorphic to $SU(2) \times SU(2)$, and its finite-dimensional representations are usually labeled by two positive (or zero) half integers (s_+, s_-), $s_\pm \in \mathbb{Z}/2$. Scalar, 4-vector, and rank 2 symmetric tensors transform under $(0,0)$, $(\frac{1}{2},\frac{1}{2})$ and $(1,1)$ respectively, left and right chirality fermions transform under $(\frac{1}{2},0)$ and $(0,\frac{1}{2})$ respectively, and self-dual and anti-self-dual rank 2 antisymmetric tensors transform under $(1,0)$ and $(0,1)$ respectively.

Supersymmetry enlarges the Poincaré algebra by including spinor *supercharges*,

$$I = 1, \dots, \mathcal{N} \quad \begin{cases} Q^I_\alpha & \alpha = 1,2 \quad \text{left Weyl spinor} \\ \bar{Q}_{\dot{\alpha} I} = (Q^I_\alpha)^\dagger & \text{right Weyl spinor.} \end{cases} \tag{2.1}$$

Here, α is a Weyl spinor label, and $\mathcal{N}$ is the number of independent supersymmetries of the algebra. The supercharges transform as Weyl spinors of SO(3,1), and are translation invariant, so that $[P_\mu, Q^I_\alpha] = 0$. The remaining super-Lie algebra structure relations are

$$\{Q^I_\alpha, \bar{Q}_{\dot{\beta} J}\} = 2\sigma^\mu_{\alpha\dot{\beta}} P_\mu \delta^I_J \tag{2.2}$$

$$\{Q^I_\alpha, Q^J_\beta\} = 2\epsilon_{\alpha\beta} Z^{IJ}. \tag{2.3}$$

Here, we have used 2-component spinor notation, which is related to 4-component Dirac spinor notation by

$$\gamma^\mu = \begin{pmatrix} 0 & \sigma^\mu \\ \bar{\sigma}^\mu & 0 \end{pmatrix}, \quad Q^I = \begin{pmatrix} Q^I_\alpha \\ \bar{Q}^{\dot{\alpha}}_I \end{pmatrix}. \tag{2.4}$$

By construction, the generators Z^{IJ} are antisymmetric in the indices I and J, and commute with all generators of the supersymmetry algebra. For the last reason, the Z^{IJ} are usually referred to as *central charges*, and we have

$$Z^{IJ} = -Z^{JI}, \quad [Z^{IJ}, \text{anything}] = 0. \tag{2.5}$$

Note that for $\mathcal{N} = 1$, the anti-symmetry of Z implies that $Z = 0$.

The supersymmetry algebra is left invariant under a global phase rotation of all supercharges Q^I_α, forming a group $U(1)_R$. In addition, when $\mathcal{N} > 1$, the different supercharges may be rotated into one another under a unitary transformation, belonging to $\mathrm{SU}(\mathcal{N})_R$. These (automorphism) symmetries of the supersymmetry algebra are called *R-symmetries.* In quantum field theories, part or all of these R-symmetries may be broken by anomaly effects.

2.2 Massless Particle Representations

To study massless representations, we choose a Lorentz frame in which the momentum takes the form $P^\mu = (E,0,0,E)$, $E > 0$. The susy algebra relation (2.2) then reduces to

$$\{Q^I_\alpha, (Q^J_\beta)^\dagger\} = 2(\sigma^\mu P_\mu)_{\alpha\dot\beta}\delta^I_J = \begin{pmatrix} 4E & 0 \\ 0 & 0 \end{pmatrix}_{\alpha\dot\beta} \delta^I_J. \tag{2.6}$$

We consider only unitary particle representations, in which the operators Q^I_α act in a positive definite Hilbert space. The relation for $\alpha = \dot\beta = 2$ and $I = J$ implies

$$\{Q^I_2, (Q^I_2)^\dagger\} = 0 \implies Q^I_2 = 0, \ Z^{IJ} = 0. \tag{2.7}$$

The relation $Q^I_2 = 0$ follows because the left-hand side of (2.7) implies that the norm of $Q^I_2|\psi\rangle$ vanishes for any state $|\psi\rangle$ in the Hilbert space. The relation $Z^{IJ} = 0$ then follows from (2.2) for $\alpha = 2$ and $\dot\beta = 1$. The remaining supercharge operators are

- Q^I_1 which lowers helicity by $\frac{1}{2}$;
- $\bar Q^I_{\dot 1} = (Q^I_1)^\dagger$ which raises helicity by $\frac{1}{2}$.

Together, Q^I_1 and $(Q^I_1)^\dagger$, with $I = 1, \dots, \mathcal{N}$ form a representation of dimension $2^{\mathcal{N}}$ of the Clifford algebra associated with the Lie algebra $\mathrm{SO}(2\mathcal{N})$. All the states in the representation may be obtained by starting from the highest helicity state $|h\rangle$ and applying products of Q^I_1 operators for all possible values of I.

Here, we shall only be interested in CPT invariant theories, such as quantum field theories and string theories, for which the particle spectrum must be symmetric under a sign change in helicity. If the particle spectrum obtained as a Clifford representation in the above fashion is not already CPT self-conjugate, then we shall take instead the direct sum with its CPT conjugate. For helicity ≤ 1, the spectra are listed in Table 1.3. The $\mathcal{N} = 3$ and $\mathcal{N} = 4$ particle spectra then coincide, and the quantum field theories are identical.

2.3 Massive Particle Representations

For massive particle representations, we choose the rest frame with $P^\mu = (M,0,0,0)$, so that the first set of susy algebra structure relations takes

TABLE 1.3. Numbers of Massless States as a function of $\mathcal{N}$ and helicity

Helicity ≤ 1	$\mathcal{N}=1$ gauge	$\mathcal{N}=1$ chiral	$\mathcal{N}=2$ gauge	$\mathcal{N}=2$ hyper	$\mathcal{N}=3$ gauge	$\mathcal{N}=4$ gauge
1	1	0	1	0	1	1
$\frac{1}{2}$	1	1	2	2	$3+1$	4
0	0	$1+1$	$1+1$	4	$3+3$	6
$-\frac{1}{2}$	1	1	2	2	$1+3$	4
-1	1	0	1	0	1	1
Total #	2×2	2×2	2×4	8	2×8	16

the form

$$\{Q^I_\alpha, (Q^J_\beta)^\dagger\} = 2M\delta^\beta_\alpha \delta^I_J. \tag{2.8}$$

To deal with the full susy algebra, it is convenient to make use of the SU($\mathcal{N}$) R-symmetry to diagonalize in blocks of 2×2 the antisymmetric matrix $Z^{IJ} = -Z^{JI}$. To do so, we split the label I into two labels: $I=(a,i)$ where $a=1,2$ and $i=1,\dots,r$, where $\mathcal{N}=2r$ for $\mathcal{N}$ even (and we append a further single label when $\mathcal{N}$ is odd). We then have

$$Z = \mathrm{diag}(\epsilon Z_1, \dots, \epsilon Z_r, \#), \quad \epsilon^{12} = -\epsilon^{21} = 1, \tag{2.9}$$

where # equals 0 for $\mathcal{N} = 2r+1$ and # is absent for $\mathcal{N}=2r$. The Z_i, $i=1,\dots,r$ are real *central charges.* In terms of linear combinations $\mathcal{Q}^i_{\alpha\pm} \equiv \frac{1}{2}\big(Q^{1i}_\alpha \pm \sigma^0_{\alpha\dot\beta}(Q^{2i}_\beta)^\dagger\big)$, the only nonvanishing susy structure relation left is (the $\pm$ signs below are correlated)

$$\{\mathcal{Q}^i_{\alpha\pm}, (\mathcal{Q}^j_{\beta\pm})^\dagger\} = \delta^i_j \delta^\beta_\alpha (M \pm Z_i). \tag{2.10}$$

In any *unitary particle representation*, the operator on the left hand side of (2.10) must be positive, and thus we obtain the BPS (for Bogomolnyi–Prasad–Sommerfield) *bound*

$$M \geq |Z_i|, \quad i = 1,\dots, r = \left[\frac{\mathcal{N}}{2}\right]. \tag{2.11}$$

Whenever one of the values $|Z_i|$ equals M, the BPS bound is (partially) saturated and either the supercharge $\mathcal{Q}^i_{\alpha+}$ or $\mathcal{Q}^i_{\alpha-}$ must vanish. The supersymmetry representation then suffers *multiplet shortening.* More precisely, if we have $M=|Z_i|$ for $i=1,\dots,r_0$, and $M>|Z_i|$ for all other values of i, the susy algebra is effectively a Clifford algebra associated with SO$(4\mathcal{N}-2r_o)$, the corresponding representation is said to be $1/2^{r_o}$ BPS, and has dimension $2^{2\mathcal{N}-r_0}$.

2.4 Field Contents of Supersymmetric Field Theories

The analysis of the preceding two subsections has revealed that the supersymmetry particle representations for $1 \leq \mathcal{N} \leq 4$, with spin less or equal to 1, simply consist of customary spin 1 vector particles, spin $\frac{1}{2}$ fermions and spin 0 scalars. Correspondingly, *the fields in supersymmetric theories with spin less or equal to 1 are customary spin 1 gauge fields,*[1] *spin* $\frac{1}{2}$ *Weyl fermion fields and spin 0 scalar fields, but these fields are restricted to enter in multiplets of the relevant supersymmetry algebras.* Here, we shall need only the $\mathcal{N} = 2$ BPS gauge field, instead of the full $\mathcal{N} = 2$ gauge multiplet, and henceforth we shall restrict to this case.

Let $\mathcal{G}$ denote the gauge algebra, associated with a compact Lie group G. Then, two kinds of multiplets occur. For any $1 \leq \mathcal{N} \leq 4$, we have a gauge multiplet, which transforms under the adjoint representation of $\mathcal{G}$. For $\mathcal{N} = 3, 4$, this is the only possible multiplet. Furthermore, the quantum field theories with $\mathcal{N} = 3$ supersymmetry turn out to coincide with those with $\mathcal{N} = 4$ supersymmetry in view of CPT invariance. Thus, we shall limit our discussion to the $\mathcal{N} = 4$ theories, as is customarily done. For $\mathcal{N} = 1$ and $\mathcal{N} = 2$, we also have *matter multiplets*: for $\mathcal{N} = 1$, this is the *chiral multiplet*, and for $\mathcal{N} = 2$ this is the *hypermultiplet*, both of which may transform under an arbitrary (unitary, and possibly reducible) representation $\mathcal{R}$ of $\mathcal{G}$.

We shall now briefly discuss each of these field multiplets in terms of their components, which are the customary gauge field A_μ, left Weyl fermions ψ_α and λ_α and scalar fields ϕ and H. (The notation is conventional.) A Dirac fermion ψ_D is the direct sum of a left and right Weyl fermion $\psi_D = (\psi\, \bar\lambda)$, while a *Majorana* (or *real*) fermion ψ_M is a Dirac fermion with $\lambda = \psi$: $\psi_M = (\psi\, \bar\psi)$. In 4 space-time dimensions, a Majorana fermion is equivalent to a Weyl fermion.

$\mathcal{N} = 1$ Gauge Multiplet $(A_\mu\, \lambda_\alpha)$, where λ_α is the gaugino Majorana fermion;

$\mathcal{N} = 1$ Chiral Multiplet $(\psi_\alpha\, \phi)$, where ψ_α is a left Weyl fermion and ϕ a complex scalar, in the representation $\mathcal{R}$ of $\mathcal{G}$.

$\mathcal{N} = 2$ Gauge Multiplet $(A_\mu\, \lambda_{\alpha\pm}\, \phi)$, where $\lambda_{\alpha\pm}$ form a Dirac fermion, and ϕ is the complex *gauge scalar*. Under the $SU(2)_R$ symmetry of $\mathcal{N} = 2$ supersymmetry, the fields A_μ and ϕ are singlets, while the fields λ_+ and λ_- transform as a doublet.

$\mathcal{N}=2$ Hypermultiplet $(\psi_{\alpha+}\, H_\pm\, \psi_{\alpha+})$, where $\psi_{\alpha\pm}$ form a Dirac spinor and $H_\pm$ are complex scalars, transforming under the representation $\mathcal{R}$ of $\mathcal{G}$. Under the $SU(2)_R$ symmetry, $\psi_\pm$ are singlets, while H_+ and H_- transform as a doublet.

---

[1] By the principles of unitary quantum field theory, spin 1 vector particles are described by gauge fields.

$\mathcal{N} = 4$ Gauge Multiplet $(A_\mu \, \lambda_\alpha^A \, \phi^I)$, where λ_α^A, $A = 1,2,3,4$ are left Weyl fermions (equivalent to two Dirac fermions) and ϕ^I, $I = 1,\dots,6$ are real scalars (equivalent to three complex scalars). Under the $SU(4)_R$ symmetry of $\mathcal{N} = 4$ supersymmetry, the gauge field A_μ is a singlet, the fermions λ_α^A transform in the fundamental **4**, the scalars ϕ^I transform in the rank 2 antisymmetric **6**.

It is very useful to record the transformation properties and representation contents under supersymmetry *subalgebras* of the $\mathcal{N} = 2$ and $\mathcal{N} = 4$ supersymmetry algebras,

$$
\begin{aligned}
&(\mathcal{N} = 2 \text{ gauge}) = (\mathcal{N} = 1 \text{ gauge}) \oplus (\mathcal{N} = 1\text{chiral in adjoint rep. of } \mathcal{G});\\
&(\mathcal{N}{=}2 \text{ hyper}) {=} (\mathcal{N}{=}1 \text{ chiral}) \oplus (\mathcal{N}{=}1 \text{ chiral in complex conjugate rep.});\\
&(\mathcal{N} = 4 \text{ gauge}) = (\mathcal{N} = 2 \text{ gauge}) \oplus (\mathcal{N} = 2 \text{ hyper in adjoint rep. of} \mathcal{G}).
\end{aligned}
$$

We conclude by remarking that all of the above multiplets are parity self-conjugate, except for the $\mathcal{N} = 1$ chiral multiplet, since it is built out of a single left Weyl fermion.

2.5 $\mathcal{N} = 1$ Supersymmetric Lagrangians

Lagrangians invariant under supersymmetry are just customary Lagrangians of gauge, spin $\frac{1}{2}$ fermion and scalar fields (these fields are arranged in multiplets of the supersymmetry algebra, as established previously), with certain special relations amongst the coupling constants and masses. We begin with the construction of Lagrangians invariant under the smallest degree of supersymmetry, namely $\mathcal{N} = 1$. For our purposes, the Lagrangians of interest are of two restricted kinds

(1) Renormalizable $\mathcal{N} = 1$ supersymmetric gauge theories;

(2) More general *low energy effective* $\mathcal{N} = 1$ *supersymmetric theories*, with the property that any monomial term in the Lagrangian has a total of no more than two derivatives on all boson fields and no more than one derivative on all fermion fields. Such restricted Lagrangian may be viewed as describing phenomena in the limit of low energy and momenta, and are well familiar from soft pion physics.

In 4 space-time dimensions, all Lagrangians in group (1) automatically belong in group (2). Thus, we seek to construct all Lagrangians in (2).

We consider first the case of only the $\mathcal{N} = 1$ gauge multiplet $(A_\mu \, \lambda_\alpha)$, and proceed by writing down all possible gauge invariant polynomial terms of dimension 4 using minimal coupling. One finds

$$
\mathcal{L} = -\frac{1}{2g^2} \operatorname{tr} F_{\mu\nu} F^{\mu\nu} + \frac{\theta}{8\pi^2} \operatorname{tr} F_{\mu\nu} \tilde{F}^{\mu\nu} - \frac{i}{2} \operatorname{tr} \bar{\lambda} \bar{\sigma}^\mu D_\mu \lambda, \tag{2.12}
$$

where g is the gauge coupling, θ is the instanton angle, the field strength is $F_{\mu\nu} = \partial_\mu A_\nu - \partial_\nu A_\mu + i[A_\mu, A_\nu]$, $\tilde{F}_{\mu\nu} = \frac{1}{2}\epsilon_{\mu\nu\rho\kappa}F^{\rho\kappa}$ is the Poincaré dual of F, and $D_\mu = \partial_\mu\lambda + i[A_\mu, \lambda]$. Remarkably, $\mathcal{L}$ is automatically invariant under the $\mathcal{N} = 1$ supersymmetry transformations

$$\begin{aligned} \delta_\xi A_\mu &= i\bar{\xi}\bar{\sigma}_\mu\lambda - i\bar{\lambda}\bar{\sigma}_\mu\xi \\ \delta_\xi\lambda &= \sigma^{\mu\nu}F_{\mu\nu}\xi \end{aligned} \tag{2.13}$$

where ξ is a spin 1/2 valued infinitesimal supersymmetry parameter. Note that the addition in (2.12) of a Majorana mass term $m\lambda\lambda$ for λ would spoil supersymmetry.

Unfortunately, as soon as scalar fields are to be included, such as is the case when dealing with chiral multiplets, it is no longer so easy to guess supersymmetry invariant Lagrangians. For example, for a single chiral multiplet $(\psi\,\phi)$ already, the restricted case (1) of just renormalizable Lagrangians is much more involved and much less intuitive. We quote here the result and shall leave the derivation of this case as well as of the more general ones for the next subsections. We have

$$\mathcal{L} = -\partial_\mu\phi^*\partial^\mu\phi - i\bar{\psi}\bar{\sigma}^\mu\partial_\mu\psi - \left|\frac{\partial U}{\partial\phi}\right|^2 - \mathrm{Re}\left(\psi\psi\frac{\partial^2 U}{\partial\phi^2}\right) \tag{2.14}$$

where $U(\phi)$ is forced by $\mathcal{N} = 1$ supersymmetry to be a complex analytic (holomorphic) scalar function of ϕ, called the *superpotential*. Renormalizability furthermore restricts U to be a polynomial of degree no larger than 3: $U(\phi) = U_0 + U_1\phi + U_2\phi^2 + U_3\phi^3$. Here, U_i are complex coupling constants, with U_3 specifying both the scalar quartic self-coupling and Yukawa coupling, and U_2 specifying both the scalar and fermion masses.

2.6 $\mathcal{N} = 1$ *Superfield Methods*

Generally, the construction of invariants is highly facilitated by the use of tensor-type methods, in which all fields which transform linearly into one another are assembled into a single multiplet. Such methods are familiar from the construction of Lorentz invariants or gauge invariants.

The construction of field multiplets containing all fields that transform linearly into one another under supersymmetry presents a new challenge, because such multiplets must hold simultaneously bosonic and fermionic fields. To achieve such superpositions of commuting Bose fields and anti-commuting Fermi fields consistently with Lorentz invariance, requires the introduction of new anti-commuting spin 1/2 parameters or coordinates. For $\mathcal{N} = 1$ supersymmetry, we introduce a (constant) left Weyl spinor coordinate θ_α and its complex conjugate $\bar{\theta}^{\dot{\alpha}} = (\theta_\alpha)^\dagger$, satisfying

$$[x^\mu, \theta_\alpha] = [x^\mu, \bar{\theta}^{\dot{\alpha}}] = \{\theta_\alpha, \theta_\beta\} = \{\theta_\alpha, \bar{\theta}^{\dot{\beta}}\} = \{\bar{\theta}^{\dot{\alpha}}, \bar{\theta}^{\dot{\beta}}\} = 0. \tag{2.15}$$

Often, θ and $\bar\theta$ are regarded as additional space-time coordinates on the same footing as the customary coordinate x^μ, which together then parameterize *superspace.*

Superderivatives are defined by

$$D_\alpha \equiv \frac{\partial}{\partial\theta^\alpha} + i\sigma^\mu_{\alpha\dot\alpha}\bar\theta^{\dot\alpha}\partial_\mu, \quad \bar D_{\dot\alpha} \equiv -\frac{\partial}{\partial\bar\theta^{\dot\alpha}} - i\theta^\alpha\sigma^\mu_{\alpha\dot\alpha}\partial_\mu \tag{2.16}$$

where differentiation and integration of θ coordinates are defined by

$$\frac{\partial}{\partial\theta^\alpha}(1,\theta^\beta,\bar\theta^{\dot\beta}) \equiv \int d\theta^\alpha\,(1,\theta^\beta,\bar\theta^{\dot\beta}) \equiv (0,\delta_\alpha{}^\beta,0). \tag{2.17}$$

We also define Lorentz scalar differentials by

$$d^2\theta \equiv \tfrac{1}{4}d\theta_\alpha\, d\theta^\alpha, \quad d^2\bar\theta \equiv \tfrac{1}{4}d\bar\theta^{\dot\alpha}\, d\bar\theta_{\dot\alpha}, \quad d^4\theta \equiv d^2\theta\, d^2\bar\theta, \tag{2.18}$$

so that

$$\int d^2\theta\,\theta\theta = \int d^2\bar\theta\,\bar\theta\bar\theta = \int d^4\theta\,\theta\theta\bar\theta\bar\theta = 1. \tag{2.19}$$

For general notations and conventions for spinors and their contractions, see Appendix A.

A *superfield* is defined as a general function of the superspace coordinates x^μ, θ_α, $\bar\theta^{\dot\alpha}$. Since the square of each θ^α or of each $\bar\theta^{\dot\alpha}$ vanishes, superfields admit finite Taylor expansions in powers of θ and $\bar\theta$. Thus, the most general superfield $S(x,\theta,\bar\theta)$ yields the following *component expansion*

$$\begin{aligned} S(x,\theta,\bar\theta) = \phi(x) + \theta\psi(x) + \bar\theta\bar\chi(x) + \bar\theta\bar\sigma^\mu\theta A_\mu(x) + \theta\theta f(x) + \bar\theta\bar\theta g^*(x) \\ + i\theta\theta\bar\theta\bar\lambda(x) - i\bar\theta\bar\theta\theta\rho(x) + \tfrac{1}{2}\theta\theta\bar\theta\bar\theta D(x). \end{aligned} \tag{2.20}$$

Naturally, in quantum field theory, one restricts to superfields that are either bosonic or fermionic, so that the superfield has definite (anti-) commutation relations with θ and $\bar\theta$,

$$\begin{array}{ll} \text{bosonic superfield} & [S,\theta^\alpha] = [S,\bar\theta_{\dot\alpha}] = 0 \\ \text{fermionic superfield} & \{S,\theta^\alpha\} = \{S,\bar\theta_{\dot\alpha}\} = 0. \end{array} \tag{2.21}$$

Thus, if S is bosonic, the component fields ϕ, A_μ, f, g and D are bosonic as well, while the fields ψ, χ, λ and ρ are fermionic. On the other hand, if S is fermionic, the component fields ϕ, A_μ, f, g and D are fermionic as well, while the fields ψ, χ, λ and ρ are bosonic. Mathematically, the superfields belong to a $\mathbb{Z}_2$ graded algebra of functions on superspace, with the even grading associated with bosonic fields and the odd grading asssociated with fermionic fields. We shall denote the grading by $g(S)$.

Superderivatives on superfields satisfy the following graded differentiation rule

$$\begin{aligned} D_\alpha(S_1S_2) &= (D_\alpha S_1)S_2 + (-)^{g(S_1)g(S_2)}S_1(D_\alpha S_2) \\ D_{\dot\alpha}(S_1S_2) &= (D_{\dot\alpha} S_1)S_2 + (-)^{g(S_1)g(S_2)}S_1(D_{\dot\alpha} S_2), \end{aligned} \tag{2.22}$$

where $g(S_i)$ is the grading of the field S_i.

The only *elementary* superfields we shall need here are bosonic and they have the property that the zero-th order term, the field $\phi(x)$, is a Lorentz scalar. We shall also encounter a composite fermionic superfield, whose first component is a spinor, and which is associated with the gauge field strength. If S is a bosonic scalar superfield, then ϕ, f, g, D are scalars while ψ, χ λ, ρ are left-handed Weyl spinors, and A_μ is a gauge field.

On superfields, supersymmetry transformations are naturally realized in a linear way via super-differential operators (just as on ordinary fields, translations and Lorentz transformations are realized in a linear way via differential operators). The infinitesimal supersymmetry parameter is still a constant left Weyl spinor ξ, as in (2.13) and we have

$$\delta_\xi S = (\xi Q + \bar{\xi}\bar{Q})S \tag{2.23}$$

with the supercharges defined by

$$Q_\alpha = \frac{\partial}{\partial \theta^\alpha} - i\sigma^\mu_{\alpha\dot{\alpha}}\bar{\theta}^{\dot{\alpha}}\partial_\mu, \quad \bar{Q}_{\dot{\alpha}} = -\frac{\partial}{\partial \bar{\theta}^{\dot{\alpha}}} + i\theta^\alpha \sigma^\mu_{\alpha\dot{\alpha}}\partial_\mu. \tag{2.24}$$

The super-differential operators D_α and Q_α differ only by a sign change, and generate left and right actions of supersymmetry respectively. Their relevant structure relations are

$$\{Q_\alpha, \bar{Q}_{\dot{\beta}}\} = 2\sigma^\mu_{\alpha\dot{\beta}}P_\mu, \quad \{D_\alpha, \bar{D}_{\dot{\beta}}\} = -2\sigma^\mu_{\alpha\dot{\beta}}P_\mu \tag{2.25}$$

where $P_\mu = i\partial_\mu$. Since left and right actions mutually commute, all 4 components of D anti-commute with all 4 components of Q: $\{Q_\alpha, D_\beta\} = \{Q_\alpha, \bar{D}^{\dot{\beta}}\} = 0$, and their complex conjugate relations. Further useful relations are that the product of any three D's or any three Q's vanishes,

$$D_\alpha D_\beta D_\gamma = Q_\alpha Q_\beta Q_\gamma = 0 \tag{2.26}$$

as well as their complex conjugate relations.

2.7 Irreducible Superfields of $\mathcal{N} = 1$

The type of superfield introduced above is in general highly reducible, and the irreducible components may be found by imposing supersymmetric conditions on the superfield.

(a) The *Chiral Superfield* Φ is obtained by imposing the condition

$$\bar{D}_{\dot{\alpha}}\Phi = 0. \tag{2.27}$$

The *anti-chiral superfield* $\Phi^\dagger$ is obtained by imposing $D_\alpha \Phi^\dagger = 0$. These conditions are invariant under the supersymmetry transformations of (2.23) since D or $\bar{D}$ and Q or $\bar{Q}$ anti-commute. Equation (2.27) may be solved in terms of the composite coordinates

$$x_\pm^\mu = x^\mu \pm i\theta\sigma^\mu\bar{\theta}, \tag{2.28}$$

which satisfy

$$\bar{D}_{\dot{\alpha}} x_+^\mu = 0, \quad D_\alpha x_-^\mu = 0, \tag{2.29}$$

and we have (a factor of $\sqrt{2}$ has been inserted multiplying ψ to give this field standard normalization)

$$\begin{aligned} \Phi(x,\theta,\bar{\theta}) &= \phi(x_+) + \sqrt{2}\theta\psi(x_+) + \theta\theta F(x_+) \\ \Phi^\dagger(x,\theta,\bar{\theta}) &= \phi^*(x_-) + \sqrt{2}\bar{\theta}\bar{\psi}(x_-) + \bar{\theta}\bar{\theta}F^*(x_-). \end{aligned} \tag{2.30}$$

The component fields ϕ and ψ are the scalar and left Weyl spinor fields of the chiral multiplet respectively, as discussed previously. The field F has not appeared previously. The field equation for F is always algebraic, so that F is a nondynamical or *auxiliary field* of the chiral multiplet.

(b) The *Vector Superfield* is obtained by imposing the condition

$$V = V^\dagger \tag{2.31}$$

on a general superfield of the type (2.20). This condition sets $\chi = \bar{\psi}$, $g = f$ and $\rho = \lambda$ in (2.20), and requires that the fields ϕ, A_μ and D be real. It is conventional to use a specific notation for vector superfields and it is convenient to define its expansion by

$$\begin{aligned} V(x,\theta,\bar{\theta}) = v(x) &+ \theta\chi(x) + \bar{\theta}\bar{\chi}(x) + \theta\theta f(x) + \bar{\theta}\bar{\theta}f^*(x) + \bar{\theta}\bar{\sigma}^\mu\theta A_\mu(x) \\ &+ i\theta\theta\bar{\theta}\big(\bar{\lambda}(x) + \tfrac{1}{2}\bar{\sigma}^\mu\partial_\mu\chi(x)\big) - i\bar{\theta}\bar{\theta}\theta\big(\lambda(x) + \tfrac{1}{2}\sigma^\mu\partial_\mu\bar{\chi}(x)\big) \\ &+ \tfrac{1}{2}\theta\theta\bar{\theta}\bar{\theta}\big(D(x) + \tfrac{1}{2}\partial_\mu\partial^\mu v(x)\big). \end{aligned} \tag{2.32}$$

The advantage of the shifts by derivatives in the fields will become clear shortly.

The *gauge superfield* is a special case of a vector superfield. On a single (Abelian) vector superfield V, the reality condition $V^\dagger = V$ is preserved upon addition of a chiral superfield Λ and its complex conjugate $\Lambda^\dagger$, as follows,

$$V \longrightarrow V' = V + i\Lambda - i\Lambda^\dagger. \tag{2.33}$$

(Below, we shall assume that Λ has the same component field decomposition as Φ in (2.30), in order to save on some notation.) Under this transformation, the component fields λ and D of V are unchanged, v, χ and f transform in a purely algebraic way,

$$\begin{aligned} v &\longrightarrow v' = v + i\phi - i\phi^* \\ \chi &\longrightarrow \chi' = \chi + i\sqrt{2}\psi \\ f &\longrightarrow f' = f + iF \end{aligned} \tag{2.34}$$

while the field A_μ transforms as an Abelian gauge field

$$A_\mu \longrightarrow A'_\mu = A_\mu + \partial_\mu(\phi + \phi^*). \tag{2.35}$$

Thus, it is natural to view (2.33) as the superfield generalization of a gauge transformation on an Abelian gauge superfield V.

The non-Abelian generalization of the gauge field is such that V takes values in the Lie algebra $\mathcal{G}$ of the gauge group G (or equivalently transforms under the adjoint representation of $\mathcal{G}$) and that the transformation (2.33) is replaced by the following nonlinear gauge transformation law,

$$e^V \longrightarrow e^{V'} = e^{-i\Lambda^\dagger} e^V e^{i\Lambda}, \tag{2.36}$$

which again preserves the reality condition $V^\dagger = V$, assuming that Λ is a chiral superfield transforming under the adjoint representation of the gauge algebra $\mathcal{G}$.

As is clear from (2.34), (and an analogous result holds for the non-Abelian case), the component fields v, χ and f may be gauged away in an algebraic way, without implying any dynamical constraints. The gauge in which this is achieved is called the *Wess–Zumino gauge*, and is almost always imposed when performing practical calculations in the superfield formulation. What remains is the gauge superfield in *Wess–Zumino gauge*, given by

$$V(x, \theta\bar\theta) = \bar\theta\bar\sigma^\mu\theta A_\mu(x) + i\theta\theta\bar\theta\bar\lambda(x) - i\bar\theta\bar\theta\theta\lambda(x) + \tfrac{1}{2}\theta\theta\bar\theta\bar\theta D(x). \tag{2.37}$$

The component fields A_μ and λ are the gauge and gaugino fields of the gauge multiplet respectively, as discussed previously. The field D has not appeared previously and is an *auxiliary field*, just as F was an auxiliary field for the chiral multiplet.

The role of the auxiliary fields F and D in the superfield formalism is to provide a linearization of the supersymmetry transformations, as well as to allow for an off-shell realization on the fields of the supersymmetry algebra, as given in (2.23).

Working out the supersymmetry transformation (2.23) on chiral and vector superfields in terms of components, we see that the only contribution to the auxiliary fields is from the $\theta\bar\partial$ term of Q and thus takes the form of a total derivative. However, because the form (2.37) was restricted to a Wess-Zumino gauge, F and D transform by a total derivative only if F and D are themselves gauge singlets, in which case we have

$$\delta_\xi F = i\sqrt{2}\partial_\mu(\bar\xi\bar\sigma^\mu\psi) \tag{2.38}$$
$$\delta_\xi D = \partial_\mu(i\bar\xi\bar\sigma^\mu\lambda - i\bar\lambda\bar\sigma^\mu\xi). \tag{2.39}$$

(For nongauge singlet auxiliary fields, the transformation laws will involve in addition nontotal-derivative terms; we shall not need their expressions here.) These transformation properties guarantee that the F and D auxiliary fields yield supersymmetric invariant Lagrangian terms. Thus, we have two different ways of building terms for an invariant Lagrangian,

$$\text{F-terms} \qquad \mathcal{L}_F = F = \int d^2\theta\, \Phi \tag{2.40}$$

$$\text{D-terms} \qquad \mathcal{L}_D = \frac{1}{2}D = \int d^4\theta\, V. \tag{2.41}$$

The first starts from a chiral superfield Φ and the second from a vector superfield V.

2.8 General $\mathcal{N} = 1$ Susy Lagrangians via Superfields

The F and D terms used to construct invariants in the previous subsection need not be elementary fields, and may be gauge invariant composites of elementary fields instead. Allowing for this possibility, we may now derive the most general possible $\mathcal{N} = 1$ invariant Lagrangian in terms of superfields. To do so, we need the following ingredients.

(1) Any complex analytic function U depending only on left chiral superfields Φ^i (but not on their complex conjugates) is itself a left chiral superfield,

$$\bar{D}_{\dot\alpha}\Phi^i = 0 \implies \bar{D}_{\dot\alpha}U(\Phi^i) = 0. \tag{2.42}$$

Thus, for any complex analytic function U, called the *superpotential*, we may construct an invariant contribution to the Lagrangian by forming an F-term

$$\mathcal{L}_U = \int d^2\theta\, U(\Phi^i) + \text{complex conjugate.} \tag{2.43}$$

Using the component expansion of (2.30), this Lagrangian takes the component form

$$\mathcal{L}_U = \sum_i F^i \frac{\partial U}{\partial \phi^i} - \frac{1}{2}\sum_{i,j} \psi^i\psi^j \frac{\partial^2 U}{\partial\phi^i \partial\phi^j} + \text{complex conjugate.} \tag{2.44}$$

(2) Actually, the gauge field strength is a fermionic left chiral (spinor) superfield W_α, which is constructed out of the gauge superfield V by

$$W_\alpha = -\tfrac{1}{4}\bar{D}\bar{D}(e^{-V}D_\alpha e^{+V}). \tag{2.45}$$

In view of (2.26), we automatically have $\bar{D}_{\dot\beta}W_\alpha = 0$, so that W_α is chiral. Decomposing W_α in components, one finds

$$W_\alpha(x,\theta,\bar\theta) = -i\lambda_\alpha(x_+) + \theta_\alpha D(x_+) - \tfrac{i}{2}(\sigma^\mu\bar\sigma^\nu)_\alpha{}^\beta\theta_\beta F_{\mu\nu}(x_+) + \theta\theta\sigma^\mu_{\alpha\dot\beta}D_\mu\bar\lambda^{\dot\beta}(x_+). \tag{2.46}$$

The gauge field strength may be used as a chiral superfield along with elementary (scalar) chiral superfields to build up $\mathcal{N} = 1$ supersymmetric Lagrangians via F–terms. In view of our restriction to Lagrangians with no more than two derivatives on Bose fields, W can enter at most quadratically. Denoting by W^a_α the components of the gauge multiplet, with the index a running over the adjoint representation, we shall be interested in bilinears of the form

$$\begin{aligned} W^{\alpha a} W^b_\alpha = -\lambda^a\lambda^b - i\theta\lambda^a D^b - i\theta\lambda^b D^a - \tfrac{1}{2}\theta(\sigma^\mu\bar{\sigma}^\nu)(\lambda^a F^b_{\mu\nu} + \lambda^b F^a_{\mu\nu}) \\ - \theta\theta\big(i\lambda^a\sigma^\mu\partial_\mu\bar{\lambda}^b + i\lambda^b\sigma^\mu\partial_\mu\bar{\lambda}^a \\ + \tfrac{1}{4}(F^{\mu\nu a} + i\tilde{F}^{\mu\nu a})(F^b_{\mu\nu} + i\tilde{F}^b_{\mu\nu}) - D^a D^b\big) \end{aligned} \tag{2.47}$$

in Wess–Zumino gauge. It is understood that all fields on the right hand side depend upon the composite coordinate $x^\mu_+ = x^\mu + i\theta\sigma^\mu\bar{\theta}$, just as was the case in (2.46).

Elementary (scalar) chiral superfields may enter in any functional way without generating more than 2 derivatives. Thus, the most general gauge kinetic and self-interaction term is from the F-term of the gauge field strength W_α and the elementary (scalar) chiral superfields Φ^i as follows,

$$\mathcal{L}_G = \int d^2\theta\, \tau_{ab}(\Phi^i) W^a W^b + \text{complex conjugate.} \tag{2.48}$$

Here, a and b stand for the gauge index running over the adjoint representation of $\mathcal{G}$. The functions $\tau_{ab}(\Phi^i)$ are again required to be complex analytic. It is assumed that the function $\tau_{ab}(\Phi^i)$ under suitable gauge transformations of Φ^i will transform under the symmetrized square of the adjoint representation of $\mathcal{G}$.

The gauge kinetic Lagrangian may also be worked out in components, and is given by (index summation over indices i, j and a, b are suppressed for the sake of brevity)

$$\begin{aligned} \mathcal{L}_G = -\lambda^a\lambda^b\bigg(F^i\frac{\partial\tau_{ab}}{\partial\phi^i} - \frac{1}{2}\psi^i\psi^j\frac{\partial^2\tau_{ab}}{\partial\phi^i\partial\phi^j}\bigg) \\ - \frac{1}{2\sqrt{2}}\frac{\partial\tau_{ab}}{\partial\phi^i}\psi^i\bigg(-i\lambda^a D^b - i\lambda^b D^a - \frac{1}{2}(\sigma^\mu\bar{\sigma}^\nu)(\lambda^a F^b_{\mu\nu} + \lambda^b F^a_{\mu\nu})\bigg) \\ - \tau_{ab}\big(i\lambda^a\sigma^\mu\partial_\mu\bar{\lambda}^b + i\lambda^b\sigma^\mu\partial_\mu\bar{\lambda}^a \\ + \tfrac{1}{4}(F^{\mu\nu a} + i\tilde{F}^{\mu\nu a})(F^b_{\mu\nu} + i\tilde{F}^b_{\mu\nu}) - D^a D^b\big). \end{aligned} \tag{2.49}$$

(3) The left and right chiral superfields Φ^i and $(\Phi^i)^\dagger$, as well as the gauge superfield V, may be combined into a gauge invariant vector superfield $K\big(e^V\Phi^i, (\Phi^i)^\dagger\big)$, provided the gauge algebra is realized linearly on the fields Φ^i. (For the general case, including when part or all of the gauge group is realized nonlinearly, see Wess and Bagger [2]; we shall not need this

case here.) The function K is called the *Kähler potential.* Assuming that the gauge transformations Λ act on V by (2.36), the chiral superfields Φ transform as

$$\Phi \longrightarrow \Phi' = e^{-i\Lambda}\Phi, \tag{2.50}$$

so that $e^V\Phi$ transforms as Φ. An invariant Lagrangian may be constructed via a D-term,

$$\mathcal{L}_K = \int d^4\theta\, K\big(e^V\Phi^i, (\Phi^i)^\dagger\big). \tag{2.51}$$

Upon expanding $\mathcal{L}_K$ in components, one sees immediately that the leading terms are

$$\mathcal{L}_K \sim -D_\mu\phi^* D^\mu\phi - i\bar\psi\bar\sigma^\mu D_\mu\psi, \tag{2.52}$$

and thus already generates an action with two derivatives on boson fields. As a result, K must be a function only of the superfields Φ^i and $(\Phi^i)^\dagger$ and V, but not of their derivatives. The expansion of the full Kähler part Lagrangian is

$$\begin{aligned}\mathcal{L}_K = -g_{ii^*}D_\mu\phi^i D^\mu\phi^{i^*} &- ig_{ii^*}\bar\psi^{i^*}\bar\sigma^\mu D_\mu\psi^i + \tfrac{1}{4}R_{ik^*jl^*}\psi^i\psi^j\bar\psi^{k^*}\bar\psi^{l^*}\\ &+ g_{ii^*}\big(F^i - \tfrac{1}{2}\Gamma^i_{jk}\psi^j\psi^k\big)\big(F^{i^*} - \tfrac{1}{2}\Gamma^{i^*}_{j^*k^*}\psi^{j^*}\psi^{k^*}\big)\\ &-\frac{i}{2}D^a(T^a)^j{}_i\phi^i\frac{\partial K}{\partial\phi^j} + \sqrt{2}g_{ii^*}(T^a)^i{}_k\phi^k\bar\psi^{i^*}\\ &\qquad\qquad\qquad + \text{complex conjugate}.\end{aligned} \tag{2.53}$$

Here, complex conjugation is to be added only on the last line. The matrices T^a are the representation matrices of $\mathcal{G}$ in the representation under which the Φ^i transform. The covariant derivatives are given by

$$\begin{aligned}D_\mu\phi^i &= \partial_\mu\phi^i - A^a_\mu(T^a)^i{}_j\phi^j\\ D_\mu\psi^i &= \partial_\mu\psi^i - A^a_\mu(T^a)^i{}_j\psi^j + \Gamma^i_{jk}D_\mu\phi^j\psi^k\end{aligned} \tag{2.54}$$

and Γ is the usual Levi–Civita connection for the Kähler metric

$$g_{ii^*} \equiv \frac{\partial^2 K}{\partial\phi^i\partial\phi^{i^*}} \tag{2.55}$$

and $R_{ik^*jl^*}$ is its Riemann curvature tensor.

Putting together contributions from $\mathcal{L}_K$, $\mathcal{L}_U$ and $\mathcal{L}_G$, we have the most general $\mathcal{N}=1$ supersymmetric Lagrangian with the restrictions of above.

2.9 Renormalizable $\mathcal{N}=2,4$ Susy Lagrangians

We shall now derive the Lagrangians invariant under the larger supersymmetry algebras associated with $\mathcal{N}=2$ and $\mathcal{N}=4$ supersymmetry. In this subsection, we shall restrict attention to renormalizable Lagrangians, leaving the general case for the next subsection. Thus, we seek an $\mathcal{N}=2$

microscopic renormalizable super-Yang–Mills theory with gauge algebra $\mathcal{G}$, and treat $\mathcal{N}=4$ as a special case thereof. Following our general classification of supersymmetry representations, two fields are possible. The *gauge multiplet* with component fields $(A_\mu\,\lambda_\pm\,\phi)$ transforms under the adjoint representation of $\mathcal{G}$, which we simply denote by $\mathcal{G}$, and the *hypermultiplet* with component fields $(\psi_+\,H_\pm\,\psi_-)$ transforms under a representation $\mathcal{R}$ of $\mathcal{G}$.

To construct the most general renormalizable Lagrangian for these fields, we use our results on invariant $\mathcal{N}=1$ supersymmetric Lagrangians. To do so, we decompose the $\mathcal{N}=2$ gauge and hypermultiplets under an $\mathcal{N}=1$ supersymmetry,

$$\mathcal{N}=2 \text{ gauge (rep } \mathcal{G}): \quad V\oplus\Phi \qquad V\sim(A_\mu\,\lambda_+) \qquad \Phi\sim(\lambda_-\,\phi) \tag{2.56}$$

$$\mathcal{N}=2 \text{ hyper (rep } \mathcal{R}): \quad H_1\oplus H_2 \quad H_1\sim(H_+\,\psi_+) \quad H_2\sim(H_-\,\psi_-) \tag{2.57}$$

and promote the $\mathcal{N}=1$ supermultiplets V, Φ, H_1 and H_2 to $\mathcal{N}=1$ superfields by including their respective auxiliary fields. Thus, V may be viewed now as an $\mathcal{N}=1$ vector (or gauge) superfields, while Φ and H_f, $f=1,2$ may be viewed as $\mathcal{N}=1$ chiral superfields.

Following the results of §2.8, we may immediately write down the most general $\mathcal{N}=1$ and $\mathcal{G}$-gauge invariant Lagrangian for these fields,

$$\mathcal{L} = \mathrm{Re}\int d^2\theta\,\left(\tau W^a W^a + U(\Phi, H_f)\right) + \int d^4\theta\,\Big(\Phi^\dagger e^{V_\mathcal{G}}\Phi + \sum_f H_f^\dagger e^{V_\mathcal{R}} H_f\Big). \tag{2.58}$$

Here, $V_\mathcal{G}$ and $V_\mathcal{R}$ stand for the field V in the representations $\mathcal{G}$ and $\mathcal{R}$ respectively. The complex analytic function $\tau_{ab}(\Phi, H_f)$ that was allowed for a general $\mathcal{N}=1$ supersymmetric Lagrangian must be constant for renormalizability, and equal to $\tau\delta_{ab}$ by gauge invariance, for a simple algebra $\mathcal{G}$. The constant τ is related to the gauge coupling g and the instanton angle θ by

$$\tau = \frac{\theta}{2\pi} + \frac{4\pi i}{g^2}. \tag{2.59}$$

The reality of g guarantees that $\mathrm{Im}\,\tau > 0$. The superpotential $U(\Phi, H_f)$ would be a general $\mathcal{G}$-invariant function for $\mathcal{N}=1$ supersymmetry, but for $\mathcal{N}=2$ is restricted to be of the form

$$U(\Phi, H_a) = H_1^T \Phi H_2 + H_1 m H_2 \tag{2.60}$$

where m is the hypermultiplet mass matrix.

2.10 $\mathcal{N}=2$ Superfield Methods: Unconstrained Superspace

In order to realize $\mathcal{N}=2$ susy transformations linearly on complete multiplets, we seek a formulation in terms of superfields [13], just as for $\mathcal{N}=1$.

The coordinates of $\mathcal{N}=2$ *unconstrained superspace* are collectively denoted z_M, and given by

$$z_M = \{x^\mu, \theta_{\alpha i}, \bar\theta^{\dot\alpha i}, \quad i=1,2\}, \tag{2.61}$$

with $[x,\theta]=\{\theta,\theta\}=0$ as usual. The internal $SU(2)_R$ index i is raised or lowered with the help of the antisymmetric ϵ_{ij} or ϵ^{ij} tensor, where $\epsilon^{12} = -\epsilon_{12} = 1$. Unconstrained superspace is real in the sense that it is invariant under the following complex conjugation

$$(x^\mu)^\dagger = x^\mu \quad (\theta_{\alpha i})^\dagger = \bar\theta^{\dot\alpha i}. \tag{2.62}$$

Superderivatives (in the absence of central charges) are defined by

$$D^i_\alpha = \frac{\partial}{\partial\theta^\alpha_i} + i\sigma^\mu_{\alpha\dot\alpha}\bar\theta^{\dot\alpha i}\partial_\mu \quad \bar D_{\dot\alpha i} = -\frac{\partial}{\partial\bar\theta^{\dot\alpha i}} - i\theta^\alpha_i\sigma^\mu_{\alpha\dot\alpha}\partial_\mu, \tag{2.63}$$

while supercharges (in the absence of central charges) are given by

$$Q^i_\alpha = \frac{\partial}{\partial\theta^\alpha_i} - i\sigma^\mu_{\alpha\dot\alpha}\bar\theta^{\dot\alpha i}\partial_\mu \quad \bar Q_{\dot\alpha i} = -\frac{\partial}{\partial\bar\theta^{\dot\alpha i}} + i\theta^\alpha_i\sigma^\mu_{\alpha\dot\alpha}\partial_\mu. \tag{2.64}$$

Superderivatives and supercharges generate commuting left and right actions of susy, so that $\{Q^i_\alpha, D^j_\beta\} = \{Q^i_\alpha, \bar D^{\dot\beta j}\} = 0$, together with complex conjugate relations.

An *unconstrained superfield* is defined as a general function of z_M, with a Taylor expansion of the form

$$\begin{aligned} S(z_M) &= \phi(x) + \theta_i\psi^i(x) + \bar\theta^j\bar\chi_j(x) + \bar\theta^i\bar\sigma^\mu\theta_j A^j_{\mu i}(x) + \theta_{\alpha i}\theta_{j\beta}f^{\alpha\beta ij}(x) \\ &+ \bar\theta^{\dot\alpha i}\bar\theta^{\dot\beta j}g^*_{\dot\alpha\dot\beta ij}(x) + \cdots + \theta_i\theta_j\theta_k\theta_l\bar\theta^m\bar\theta^n\bar\theta^o\bar\theta^p D^{ijkl}_{mnop}(x). \end{aligned} \tag{2.65}$$

Supersymmetry transformations are linearly realized on unconstrained superfields via the super-differential operators Q and $\bar Q$ by

$$\delta_\xi S = (\xi^i Q_i + \bar\xi_i\bar Q^i)S, \tag{2.66}$$

where ξ^i and $\bar\xi_i$ are infinitesimal spinor parameters. The unconstrained superfield S, just as a general $\mathcal{N}=1$ superfield, is reducible. Thus, we seek suitable sets of $\mathcal{N}=2$ supersymmetric constraints to decompose S into irreducible components. Since Q and D mutually anticommute, we may impose on S that some of its superderivatives vanishes.

We define a $\mathcal{N}=2$ *chiral superfield* by imposing the constraints

$$\bar D_{\dot\alpha i}\Phi = 0, \quad i=1,2. \tag{2.67}$$

Notice that in order to preserve $SU(2)_R$ symmetry, we are forced to impose the above constraint for both values of i. The chirality constraints (2.67) may be solved by introducing the combinations

$$x^\mu_\pm = x^\mu \pm i\theta_i\sigma^\mu\bar\theta^i \tag{2.68}$$

which satisfy

$$\bar{D}_{\dot{\alpha}i}x_+^\mu = 0, \quad D^{\alpha i}x_-^\mu = 0. \tag{2.69}$$

Thus, the superfield Φ depends only upon x_+^μ and $\theta_{\alpha i}$ and has the component expansion

$$\begin{aligned}\Phi(x_+,\theta) = \phi(x_+) + \theta^{\alpha i}\psi_{\alpha i}(x_+) + \theta^{\alpha i}\theta^{\beta j}f_{\alpha\beta ij}(x_+)\\ + \theta^{\alpha i}\theta^j\theta^k\chi_{\alpha ijk}(x_+) + \theta\theta\theta\theta D(x_+).\end{aligned} \tag{2.70}$$

When Φ is a scalar superfield, for example, the components ϕ, $f^{[\alpha\beta](ij)}$ and D are scalars, ψ and χ are spin $\frac{1}{2}$ and $f^{(\alpha\beta)[ij]}$ is a self-dual rank 2 antisymmetric tensor field. (This chiral superfield will naturally incorporate the field strength of the $\mathcal{N} = 2$ gauge multiplet, as we shall see later). Any complex analytic function $\mathcal{F}(\Phi)$ of an $\mathcal{N} = 2$ chiral superfield is automatically also an $\mathcal{N} = 2$ chiral superfield.

The term with the highest power of θ of a (gauge invariant) $\mathcal{N} = 2$ chiral superfield transforms under $\mathcal{N} = 2$ susy by a total space-time derivative, so that it may be used to construct $\mathcal{N} = 2$ invariant Lagrangians. Introducing the chiral measure $d^4\theta \equiv d^2\theta^1\, d^2\theta^2$, the combination

$$\mathcal{L}_D = D = \int d^4\theta\, \Phi \tag{2.71}$$

is an invariant Lagrangian.

The $\mathcal{N} = 2$ chiral superfield is reducible, unlike its $\mathcal{N} = 1$ counterpart. To achieve an Abelian irreducible superfield, we may additionally impose the constraints

$$D^{\alpha i}D^j_\alpha W = \bar{D}^i_{\dot{\alpha}}\bar{D}^{\dot{\alpha}j}\overline{W}. \tag{2.72}$$

As a result we have $f_{[\alpha\beta](ij)} = \chi_{\alpha ijk} = 0$, while ϕ, ψ, $f_{(\alpha\beta)[ij]}$ and the auxiliary field D are unaffected. It is customary to represent the solution to these constraints in the form of an $\mathcal{N} = 1$ superfield expansion,

$$W(x_+,\theta) = \Phi(x_+,\theta^1) + \sqrt{2}\theta^{\alpha 2}W_\alpha(x_+,\theta^1) + \theta^2\theta^2 G(x_+,\theta^1). \tag{2.73}$$

Here, Φ is the $\mathcal{N} = 1$ chiral superfield contained in the $\mathcal{N} = 2$ gauge multiplet, while we shall denote by V the $\mathcal{N} = 1$ gauge superfield contained in the $\mathcal{N} = 2$ gauge multiplet, and denote the field strength superfield associated with V by W_α. The auxiliary field G is then a field that may be expressed in terms of Φ and V as follows,

$$\begin{aligned}G(x_+,\theta^1)\\ = -\frac{1}{2}\int d^2\bar{\theta}^1\, \Phi(x_+ - i\theta_1\sigma\bar{\theta}^1,\theta^1,\bar{\theta}^1)^\dagger e^{-2V(x_+ - i\theta_1\sigma\bar{\theta}^1,\theta^1,\bar{\theta}^1)}.\end{aligned} \tag{2.74}$$

The component field contents of W is that of an Abelian $\mathcal{N} = 2$ gauge multiplet.

The $\mathcal{N}=2$ gauge superfield for a non-Abelian gauge algebra $\mathcal{G}$ is obtained by introducing the $\mathcal{G}$-valued supergauge fields $A_{\alpha i}$ and $\bar{A}_{\dot{\alpha} i}$, and the associated gauge covariant superderivatives

$$\mathcal{D}_{\alpha i} = D_{\alpha i} + iA_{\alpha i}, \quad \overline{\mathcal{D}}_{\dot{\alpha} i} = \bar{D}_{\dot{\alpha} i} + i\bar{A}_{\dot{\alpha} i}. \tag{2.75}$$

The $\mathcal{N}=2$ gauge superfield is now governed by the Poincaré algebra, the gauge algebra $\mathcal{G}$ and the $\mathrm{SU}(2)_R$-symmetry covariant constraints

$$\overline{\mathcal{D}}_{\dot{\alpha} i} W = 0, \quad \mathcal{D}^{\alpha i}\mathcal{D}^j_\alpha W = \overline{\mathcal{D}}^i_{\dot{\alpha}}\overline{\mathcal{D}}^{\dot{\alpha} j}\overline{W}. \tag{2.76}$$

The component field contents is that of an $\mathcal{N}=2$ multiplet for gauge algebra $\mathcal{G}$.

Invariant actions for the $\mathcal{N}=2$ gauge supermultiplet may be obtained by integrating a chiral gauge invariant function of W. An $\mathcal{N}=2$ chiral superfield is obtained for any complex analytic function $\mathrm{tr}\,\mathcal{F}(W)$, called the *prepotential.* The most general gauge invariant, $\mathcal{N}=2$ supersymmetric Lagrangian is given by

$$S_{\mathcal{F}} = \int dz_M \ \mathrm{tr}\,\mathcal{F}(W) + \text{complex conjugate} \tag{2.77}$$

where the superspace measure is defined by $dz_M \equiv d^4x_+ d^4\theta$. Upon integrating out θ^2 and $\bar{\theta}^2$ from dz_M, we readily recover the $\mathcal{N}=1$ superfield form for the action $S_{\mathcal{F}}$,

$$S_{\mathcal{F}} = -\frac{1}{2}\int d^2\theta \left(\frac{\partial \mathcal{F}}{\partial \Phi^a} G^a\right) - \int d^2\theta \, \frac{\partial^2 \mathcal{F}}{\partial\Phi^a\partial\Phi^b} W^{\alpha a} W^b_\alpha \tag{2.78}$$

where the index a labels the components of the adjoint representation.

2.11 $\mathcal{N}=2$ Superfield Methods: Harmonic/Analytic Superspaces

The above construction of the gauge superfield on $\mathcal{N}=2$ unconstrained superspace does not appear to extend to the off-shell hypermultiplet. The obstacle arises from the fact that strong constraints have to be imposed on the superfield in order to retain only dynamical fields with spin less or equal to $\frac{1}{2}$. Any such set of constraints turns out to be so strong that it entails the field equations as well. Harmonic and analytic superspace methods were introduced to remedy this situation. We discuss here only the case of $\mathcal{N}=2$.

In *harmonic superspace* [14] new bosonic coordinates are introduced, in addition to the space-time coordinates z_M. From a supergroup viewpoint, additional bosonic coordinates appear naturally for any extended supersymmetry. For example, the conformal supergroups $\mathrm{SU}(2,2\mid\mathcal{N})$ have maximal bosonic subgroups $\mathrm{SU}(2,2)\times\mathrm{SU}(\mathcal{N})_R$, where the internal $\mathrm{SU}(\mathcal{N})_R$ is associated with new bosonic coordinates in superspace.

$\mathcal{N} = 2$ harmonic superspace is parameterized by x^μ, $\theta_{\alpha i}$, $\bar{\theta}^{\dot{\alpha}i}$ and the new bosonic coordinates $u_i^\pm$, $i = 1, 2$, with $u^{\pm i} = \epsilon^{ij} u_j^\pm$ and

$$\begin{pmatrix} u_1^- & u_1^+ \\ u_2^- & u_2^+ \end{pmatrix} \in \mathrm{SU}(2)_R, \qquad \begin{cases} u^{+i} u_i^- = 1, \\ u^{\pm i} u_i^\pm = 0. \end{cases} \tag{2.79}$$

The variables $u_i^\pm$ form a complex orhonormal frame, transforming under the fundamental representation of $\mathrm{SU}(2)_R$ in its index i, and under a U(1) in its $\pm$ indices. Harmonic superspace is real in the sense that it is invariant under the following complex conjugation

$$(x^\mu)^\dagger = x^\mu, \quad (\theta_{\alpha i})^\dagger = \bar{\theta}^{\dot{\alpha} i}, \quad (u^{\pm i})^\dagger = u_i^\mp. \tag{2.80}$$

Supersymmetry transformations on harmonic superspace are just the supersymmetry transformations on unconstrained superspace, implemented with the invariance of $u_i^\pm$,

$$\begin{aligned} \delta x^\mu &= i\xi^i \sigma^\mu \bar{\theta}_i - i\theta^i \sigma^\mu \bar{\xi}_i & \delta\theta_{\alpha i} &= \xi_{\alpha i} \\ \delta u_i^\pm &= 0 & \delta\bar{\theta}^i_{\dot{\alpha}} &= \bar{\xi}^i_{\dot{\alpha}}. \end{aligned} \tag{2.81}$$

An *analytic basis for harmonic superspace* with parameters x_A^μ, $\theta_\alpha^\pm$, $\bar{\theta}_{\dot{\alpha}}^\pm$, $u_i^\pm$ is obtained via the following invertible change of variables

$$\begin{aligned} x_A^\mu &\equiv x^\mu - i\theta^+ \sigma^\mu \bar{\theta}^- - i\theta^- \sigma^\mu \bar{\theta}^+ & \theta_\alpha^\pm &\equiv \theta^i_\alpha u_i^\pm \\ & & \bar{\theta}_{\dot{\alpha}}^\pm &\equiv \bar{\theta}^i_{\dot{\alpha}} u_i^\pm. \end{aligned} \tag{2.82}$$

In the analytic basis, the susy transformation laws take the form

$$\begin{aligned} \delta x_A^\mu &= -2i(\xi^i \sigma^\mu \bar{\theta}^+ + \theta^+ \sigma^\mu \bar{\xi}_i) u_i^- & \delta\theta_\alpha^\pm &= \xi^i_\alpha u_i^\pm \\ \delta u_i^\pm &= 0 & \delta\bar{\theta}_{\dot{\alpha}}^\pm &= \bar{\xi}^i_{\dot{\alpha}} u_i^\pm. \end{aligned} \tag{2.83}$$

Besides the space-time derivative $D_\mu = \partial/\partial x_A^\mu$, we have $\mathrm{SU}(2)_R$ covariant superderivatives

$$\begin{aligned} D_\alpha^+ &= \frac{\partial}{\partial\theta^{\alpha-}} & D_\alpha^- &= -\frac{\partial}{\partial\theta^{\alpha+}} + 2i\sigma^\mu_{\alpha\dot{\alpha}} \bar{\theta}^{\dot{\alpha}-} D_\mu \\ \bar{D}_{\dot{\alpha}}^+ &= \frac{\partial}{\partial\bar{\theta}^{\dot{\alpha}-}} & \bar{D}_{\dot{\alpha}}^- &= -\frac{\partial}{\partial\bar{\theta}^{\dot{\alpha}+}} - 2i\theta^{\alpha-} \sigma^\mu_{\alpha\dot{\alpha}} D_\mu \end{aligned} \tag{2.84}$$

and covariant derivatives with respect to the internal coordinates $u_i^\pm$,

$$D^{++} = u^{i+} \frac{\partial}{\partial u^{i-}} - 2i\theta^+ \sigma^\mu \bar{\theta}^+ D_\mu + \theta^{\alpha+} \frac{\partial}{\partial\theta^{\alpha-}} + \bar{\theta}^{\dot{\alpha}+} \frac{\partial}{\partial\bar{\theta}^{\dot{\alpha}-}}, \tag{2.85}$$

as well as $D^{--} = (D^{++})^\dagger$ and $D^0 = 1/2[D^{++}, D^{--}]$.

Analytic superspace is defined as the subspace of harmonic superspace parameterized by the coordinates

$$z_A = \{x_A^\mu, \theta_\alpha^+, \bar\theta_{\dot\alpha}^+\} \quad \text{and} \quad u_i^\pm \tag{2.86}$$

only (i.e. it is the subspace in which $\theta_\alpha^- = \bar\theta_{\dot\alpha}^- = 0$). All $\mathcal{N}=2$ susy transformations leave this subspace invariant. Analytic superspace is real in the sense that it has a natural complex conjugation which will be denoted by ^{c}, and acts by

$$(x_A^\mu)^c = x_A^\mu, \quad (\theta_\alpha^\pm)^c = \bar\theta_{\dot\alpha}^\pm, \quad (\bar\theta_{\dot\alpha}^\pm)^c = -\theta_\alpha^\pm, \quad (u^{i\pm})^c = -u_i^\pm. \tag{2.87}$$

The square of the operation ^{c} equals plus/minus the identity on bosonic/fermionic coordinates. (The complex conjugation † introduced previously induces a complex conjugation in the analytic basis of harmonic superspace as well $(x_A^\mu)^\dagger = x_A^\mu$, $(\theta_\alpha^\pm)^\dagger = \bar\theta_{\dot\alpha}^\mp$, $(u^{i\pm})^\dagger = \mp u_i^\mp$ which does not leave analytic superspace invariant.)

Functions on analytic superspace naturally generalize the notion of chiral superfield for $\mathcal{N}=1$. Viewed as functions on harmonic superspace, they satisfy the constraints

$$D_\alpha^+\Phi = \bar D_{\dot\alpha}^+\Phi = 0 \tag{2.88}$$

and are thus functions of the form $\Phi(z_A,u)$ with the following decomposition

$$\begin{aligned}\Phi(z_A,u) = {} & \phi(x_A,u) + \theta^+\psi(x_A,u) + \bar\theta^+\bar\chi(x_A,u) + \theta^+\theta^+ f(x_A,u) \\ & + \bar\theta^+\bar\theta^+ g(x_A,u) + \bar\theta^+\bar\sigma^\mu\theta^+ A_\mu(x_A,u) + i\theta^+\theta^+\bar\theta^+\bar\lambda(x_A,u) \\ & - i\bar\theta^+\bar\theta^+\theta^+\nu(x_A,u) + \tfrac{1}{2}\theta^+\theta^+\bar\theta^+\bar\theta^+ D(x_A,u).\end{aligned} \tag{2.89}$$

If Φ has $U(1)$ charge q, then the component field ϕ has charge q, the fields ψ and χ have charge $q-1$, f, g and A_μ have charge $q-2$, λ and ν have charge $q-3$ and D has charge $q-4$. For q even, one may impose the reality condition $(\Phi^{(q)})^c = \Phi^{(q)}$.

The dependence on the internal coordinates $u_i^\pm$ produces an infinite number of component fields which are functions of z_A. Indeed, an analytic superfield with U(1) charge $q>0$ has a decomposition of the form

$$\Phi^{(q)}(z_A,u) = \sum_{n=0}^\infty \phi^{(i_1\cdots i_{n+q}j_1\cdots j_n)}(z_A)u^+_{(i_1}\cdots u^+_{i_{n+q}}u^-_{j_1}\cdots u^-_{j_n)}. \tag{2.90}$$

Both the $\mathcal{N}=2$ gauge multiplet and the hypermultiplet may be treated as analytic superfields, with an off-shell $\mathcal{N}=2$ Lagrangian, which contains precisely the appropriate number of fields on-shell.

An invariant integration on analytic superspace may be defined in terms of the measure $dz_A \cdot du$, which may be expressed in terms of the "space-time measure" $dz_A \equiv d^4x_A d^2\theta^+ d^2\bar{\theta}^+$ and the "internal measure" du. The latter is defined by

$$\int du\; u^+_{(i_1} \cdots u^+_{i_m} u^-_{j_1} \cdots u^-_{j_n)} = \delta_{m+n,0}. \tag{2.91}$$

In order to get a U(1)-invariant Lagrangian, the integration against the measure dz_A of U(1)-charge -4 requires an analytic superfield of U(1) charge 4.

The *hypermultiplet analytic superfield* has U(1)-charge 1, and is denoted by Φ^+. A good field equation for this superfield will be such that the infinite tower of auxiliary fields associated with the u-dependence of Φ^+ collapses to just the correct number of physical fields. Thus, the field equations cannot just be built up from the superderivatives D^- and $\bar{D}^-$, but must involve derivation with respect to u. The natural candidate is the operator D^{++}, and the correct hypermultiplet free field equation is $D^{++}\Phi^+ = 0$. The most general U(1)-invariant action may be easily constructed along these lines, and one finds [14],

$$\begin{aligned} S[\Phi^+] &= \int dz_A\, du\, \{(\Phi^+)^c D^{++}\Phi^+ + U(\Phi^+, (\Phi^+)^c)\} \\ U(\Phi^+, (\Phi^+)^c) &= a(\Phi^+)^4 + b(\Phi^+)^c(\Phi^+)^3 + c((\Phi^+)^c)^2(\Phi^+)^2 \\ &\qquad + b^*((\Phi^+)^c)^3\Phi^+ + a^*((\Phi^+)^c)^4. \end{aligned} \tag{2.92}$$

The couplings all have dimensions of mass^{-2} and thus are nonrenormalizable.

The $\mathcal{N} = 2$ *gauge analytic superfield* has U(1)-charge 2, and is denoted V^{++}. It is real under the complex conjugation ^{c} that leaves analytic superspace invariant, $(V^{++})^c = V^{++}$. Gauge transformations are generated by analytic superfields Λ of U(1) charge 0, since they must act consistently on the analytic superfields V^{++} and on analytic hypermultiplets Φ^+. The appropriate transformation laws are

$$(V^{++})' = e^{i\Lambda}(V^{++} - iD^{++})e^{-i\Lambda} \quad (\Phi^+)' = e^{i\Lambda}\Phi^+. \tag{2.93}$$

Since Λ is analytic, the superderivatives $\mathcal{D}^+_\alpha = D^+_\alpha$ and $\bar{\mathcal{D}}^+_{\dot\alpha} = \bar{D}^+_{\dot\alpha}$ are automatically covariant, while the remaining covariant derivatives are constructed by

$$\mathcal{D}^{++} = D^{++} + iV^{++}, \quad \mathcal{D}^-_\alpha = D^-_\alpha + iA^-_\alpha, \quad \mathcal{D}^-_{\dot\alpha} = D^-_{\dot\alpha} + iA^-_{\dot\alpha}. \tag{2.94}$$

The superconnections A^-_α and $A^-_{\dot\alpha}$ may be expressed in terms of V^{++}, but we shall not need their explicit form here.

The gauge field strength W is an analytic superfield and was constructed in terms of the superfield V^{++} in [14],

$$W = -\tfrac{i}{4}e^{iv}\{\bar{D}^+_{\dot\alpha}\bar{D}^{\dot\alpha+}(e^{-iv}D^{--}e^{iv})\}e^{-iv}. \tag{2.95}$$

Here, v is a general superfield (i.e. neither analytic nor chiral), defined by

$$(D^{++} + iV^{++})e^{iv} = 0. \tag{2.96}$$

An explicit expression for W in terms of V^{++} was obtained in [15] and is given by

$$W(z,u) = \frac{i}{4}\bar{D}^{+}_{\dot\alpha}\bar{D}^{\dot\alpha +}\sum_{n=1}^{\infty}(-i)^n\int du_1\cdots du_n\,\frac{V^{++}(z,u_1)\cdots V^{++}(z,u_n)}{(uu_1)(u_1u_2)\cdots(u_nu)} \tag{2.97}$$

where we have defined $(u_a u_b) = u_a^{+i}u_{bi}^{+}$. Remarkably, the superfield W constructed in the analytic superspace approach turns out to be independent of u altogether, and its kinematics thus reduces to that discussed in the preceding section on unconstrained superspace.

The most general $\mathcal{N}=2$ supersymmetric and $\mathcal{G}$-gauge invariant action is thus

$$S = \mathrm{Re}\int dz_M\,du\,\mathrm{tr}\,\mathcal{F}(W;\tau) + \int dz_A\,du\,\{(\Phi^+)^c D^{++}\Phi^+ + U(\Phi^+,(\Phi^+)^c)\} \tag{2.98}$$

where the potential U was given in (2.92). The prepotential function $\mathcal{F}$ is an arbitrary complex analytic function of its arguments. In view of arguments similar to the ones given for $\mathcal{N}=1$, $\mathcal{F}$ is also complex analytic in its coupling constants τ. The renormalizable case corresponds to $\mathcal{F}(W;\tau) = i\tau W^2$ and $U=0$.

3 Seiberg–Witten Theory

Seiberg–Witten theory deals with the construction of the nonperturbative dynamics of $\mathcal{N}=2$ super-Yang–Mills theory in the limit of low energy and momenta. The prime motivation is the exploration of the fully nonperturbative dynamics of gauge theories, a problem of quantum field theory that has been outstanding for several decades now. The restriction to $\mathcal{N}=2$ theories is made because the stronger constraints of extended supersymmetry lead to simplified—yet highly nontrivial—dynamics of super-Yang–Mills theory. The dynamics of the even more restricted $\mathcal{N}=4$ theory is even simpler than that of the $\mathcal{N}=2$ theory, but the low energy approximation, obtained in Seiberg–Witten theory, becomes rather trivial in this theory.

3.1 *Wilson Effective Couplings and Actions*

Despite the large degree of supersymmetry, it is rather unlikely that the quantum field theory defined by the Lagrangian (2.58) will be exactly solvable, in the sense that all its correlation functions or even all its S-matrix elements could be obtained in explicit form. Fortunately, many of the physically most interesting questions, that have to do with the nonperturbative dynamics of the theory, are related to the properties of the vacuum and the lowest energy (and momentum) excitations above the vacuum. Thus, there already is great interest in obtaining information on the nonperturbative dynamics of these theories in the limit of low energy only. The Wilson renormalization group [16] provides the ideal tool for deriving such low energy effective theories.

- The starting point is the *bare* quantum field theory with a (energy-momentum) cutoff Λ, in which all masses, energies and momenta of all fields are restricted to be less than Λ. (It is usually safest to consider such cutoffs in the Euclidean version of the theory, where they can be imposed in a Lorentz invariant way.) The bare coupling is denoted $g(\Lambda)$.
- It is assumed that the masses, energies and momenta of physical relevance are much smaller that Λ.
- One introduces an arbitrary *renormalization scale* μ, with

$$\text{physical masses, energies, momenta} \ll \mu \ll \Lambda. \tag{3.1}$$

One considers the *Wilson effective theory*, in which only the particles and fields of masses, energies and momenta less than μ (i.e., the *light fields*) are retained as quantum degrees of freedom from the original bare theory. All particles and fields with masses, energies and momenta between the renormalization scale μ and the cutoff scale Λ (i.e., the *heavy fields*) are *integrated out.* (The separation between light and heavy modes is not usually gauge invariant. One may either work with gauge invariant degrees of freedom directly, which is always hard, or, in perturbation theory, one may replace gauge invariance with BRST invariance. Since BRST invariance is a global symmetry, separation into light and heavy modes may be achieved as usual.) The effect of this integration over the heavy fields is summarized in terms of (an infinite series of) additional terms in the Lagrangian. The resulting object is usually referred to as the *Wilson effective Lagrangian or effective action.*

For example, in a theory of only a gauge field A, with gauge field strength F and a single coupling g, the bare theory (ignoring gauge fixing and ghost terms) involves the bare field strength $F_{(\Lambda)}$ and the bare coupling $g(\Lambda)$. The Wilson effective theory involves the field strength $F_{(\mu)}$ and the effective gauge coupling $g(\mu)$, as well as an infinite series of higher dimension operators with numerical coefficients which are functions of μ and Λ,

$$\begin{aligned} \mathcal{L}_{\text{bare}} &= \frac{1}{2g(\Lambda)^2} \operatorname{tr} F_{(\Lambda)}^2 \\ \mathcal{L}_{\text{eff}} &= \frac{1}{2g(\mu)^2} \operatorname{tr} F_{(\mu)}^2 + \frac{1}{\Lambda^2} f\left(g, \frac{\mu}{\Lambda}\right) \operatorname{tr} F_{(\mu)}^3 + \cdots . \end{aligned} \tag{3.2}$$

Here, it is understood that the fields $F_{(\Lambda)}$ and $F_{(\mu)}$ contain only modes of masses and energies less than Λ and less than μ respectively.

The Wilson effective Lagrangian at scale μ essentially amounts to a Taylor series expansion in the *light fields* obtained by integrating out the heavy fields. Such a Taylor expansion is analytic by construction, and does not exhibit any infrared singularities by construction. (This may be contrasted with the effective Lagrangian defined as generating functional of connected one-particle-irreducible Feynman graphs, which does exhibit infrared singularities whenever massless particles are around.)

However, there is a generic situation in which the Wilson effective Lagrangian does exhibit singularities. This occurs when one considers not a single theory, but rather a family of theories obtained for example by varying one or several of the masses in the theory. Let m be such as mass. Assume that we begin by analyzing a theory in which the mass m is larger than the renormalization scale μ. Thus, by the rules for the construction of the Wilson effective theory, we integrate out completely all the modes associated with this mass m, and obtain a certain effective Lagrangian $\mathcal{L}_m$. No infrared singularity occurs since all the modes that were integrated out are massive. Now, consider varying m, and rendering m smaller than μ. The effective Lagrangian $\mathcal{L}_m$ still makes sense. However, as soon as $m < \mu$, $\mathcal{L}_m$ strictly speaking is no longer the Wilson effective action, since particles with mass $m < \mu$ have been completely integrated out. Nonetheless, $\mathcal{L}_m$ is a useful effective action, especially because it is analytic in m, while the true Wilson effective action will have discontinuities at $m = \mu$. As $m \to 0$, the effective action $\mathcal{L}_m$ will in general develop singularities, associated with the fact that massless modes will now have been integrated out along with massive ones. It is a fact that the only singularities in the Wilson effective action arise due to particles and fields becoming massless. Thus, the singularity structure of the effective action corresponds to the appearance of massless particles in the spectrum.

3.2 Holomorphicity and Nonrenormalization

Supersymmetric theories invariably have *holomorphic data*, such as the superpotential $U(\Phi)$ or the *gauge coupling* $\tau_{ab}(\Phi)$. Under the Wilson renormalization group procedure, which preserves supersymmetry, these holomorphic data are modified in a holomorphic way, and become the effective superpotential $U^{\mathrm{eff}}(\Phi)$ and the effective gauge coupling $\tau^{\mathrm{eff}}_{ab}(\Phi)$. The analyticity of the Wilson renormalization group procedure, pointed out in the preceding subsection, guarantees that complex analyticity will also be preserved [17, 18]. This property of supersymmetric data places severe restrictions on the renormalization group flow in supersymmetric theories, and provides one of the key ingredients for Seiberg–Witten theory.

The couplings entering these holomorphic data, such as the mass m and Yukawa coupling λ in $U(\Phi)$, or the gauge coupling τ in $\tau_{ab}(\Phi)$, are natu-

TABLE 1.4. $U(1) \times U(1)_R$ quantum numbers of chiral superfields

Field	Φ	m	λ	U and U^{eff}	$\lambda\Phi/m$
U(1)	1	−2	−3	0	0
$U(1)_R$	1	0	−1	2	0

rally complex numbers, and the dependence of U or τ_{ab} on them may be viewed as holomorphic as well. Remarkably, this holomorphicity property is also preserved under the Wilson renormalization procedure. The reason is that the complex parameters m and λ and τ may be viewed as vacuum expectation values of chiral superfields. The holomorphic dependence on these new superfields is preserved under the Wilson renormalization procedure, just as that of the dynamical chiral superfields was. The net result is that there will be highly restricted renormalization effects (or none at all) on such complex couplings. To render this discussion more concrete, we present two important examples.

(1) Absence of mass and Yukawa coupling renormalization.
Consider a renormalizable theory of a single chiral scalar superfield Φ, with superpotential $U(\Phi;\lambda,m) = m\Phi^2 + \lambda\Phi^3$. Holomorphicity considerations [18], discussed above, imply that the effective superpotential $U^{\text{eff}}(\Phi;\lambda,m)$ will be a holomorphic function of the superfield Φ and of the couplings λ and m. Now, as λ and m have both been promoted to superfields, the superpotential $U(\Phi;\lambda,m)$ has two U(1) symmetries. The first U(1) is just phase rotations of Φ, λ and m in such a way that U is invariant. The second $U(1)_R$ is the R-symmetry generator, under which the superpotential must have charge 2. The charge assignments are listed in Table 1.4. Both symmetries remain after quantization and are preserved under the Wilson renormalization group flow, so that U^{eff} has the same quantum numbers as U. From Table 1.4, it is clear that the only quantity invariant under both U(1) and $U(1)_R$ is $\lambda\Phi/m$. Since the Wilson effective Lagrangian is analytic in the light fields, U^{eff} gets only contributions of the form

$$m\Phi^2\left(\frac{\lambda\Phi}{m}\right)^n. \tag{3.3}$$

But, these are all tree level contributions and can arise in the effective superpotential U^{eff} only if they were already part of U. (One loop and higher order contributions would include nonlocal terms in the effective action, as well as contributions that are not analytic in the fields, such as is the case in the Coleman-Weinberg effective potential, which has terms of the type $\phi^4 \ln \phi^2$.) Thus, λ and m are not renormalized, a property that holds for all values of the couplings, perturbative or nonperturbative. These conclusions are borne out by explicit (super-) Feynman graph arguments in perturbation theory.

(2) Gauge Coupling Renormalization: β-functions.
The complex gauge coupling τ enters in a holomorphic way in the gauge kinetic and self-coupling terms, and we wish to derive the effects of the Wilson renormalization group procedure on this term as well [17].

$$\int d^2\theta\, \tau\, \mathrm{tr}(WW) \Longrightarrow \int d^2\theta\, \tau_{\mathrm{eff}}\, \mathrm{tr}(WW). \tag{3.4}$$

We may again use $\mathrm{U}(1)_R$ symmetry, but now, in the presence of gauge interactions, this symmetry suffers a triangle anomaly. In view of Adler–Bardeen type arguments, the perturbative contribution to the $\mathrm{U}(1)_R$ anomaly is purely 1-loop and may be evaluated directly from the triangle graph, with no need for the inclusion of radiative corrections. The result is summarized by the *Wilson β-function* for the coupling g (ignoring temporarily θ),

$$\beta_W(g) = \left.\frac{\partial g(\mu)}{\partial \ln \mu}\right|_{\Lambda, g(\Lambda)} = \big(-3C_2(\mathcal{G}) + T_2(\mathcal{R})\big)\frac{g^3}{16\pi^2}. \tag{3.5}$$

Here, we use the standard notations for the quadratic Casimir and Dynkin index in an arbitrary representation $\mathcal{R}$ of $\mathcal{G}$

$$T^a_{\mathcal{R}} T^a_{\mathcal{R}} = C_2(\mathcal{R}) I_{\mathcal{R}} \quad \mathrm{tr}(T^a_{\mathcal{R}} T^b_{\mathcal{R}}) = T_2(\mathcal{R})\delta^{ab}. \tag{3.6}$$

Integrating the equation for the coupling flow, and restoring the θ dependence in terms of a complex valued generalization of the renormalization scale μ, we have

$$\tau(\mu) - \tau(\mu') = \frac{1}{2\pi i}\big(-3C_2(\mathcal{G}) + T_2(\mathcal{R})\big)\ln\frac{\mu'}{\mu}. \tag{3.7}$$

For $3C_2(\mathcal{G}) > T_2(\mathcal{R})$, the theory is asymptotically free, while for $3C_2(\mathcal{G}) = T_2(\mathcal{G})$, the theory has vanishing β-function and will be conformally invariant provided no explicit mass parameters arise. For $3C_2(\mathcal{G}) < T_2(\mathcal{R})$, the theory becomes strongly coupled at short distances: the microscopic dynamics of such theories is not well formulated in terms of these degrees of freedom and we shall not consider this case any further here.

(3) Applications to $\mathcal{N} = 2$ and $\mathcal{N} = 4$ theories.
Let us now consider an $\mathcal{N} = 2$ super-Yang–Mills theory, where the gauge algebra is $\mathcal{G}$ and the representation of the hypermultiplet is $\mathcal{R}_H$. Then, the representation in terms of $\mathcal{N} = 1$ chiral multiplets is $\mathcal{R} = \mathcal{G} \oplus \mathcal{R}_H \oplus \mathcal{R}_H^*$, and we have $T_2(\mathcal{R}) = C_2(\mathcal{G}) + 2T_2(\mathcal{R}_H)$, so that

$$\tau(\mu) - \tau(\mu') = \frac{1}{\pi i}\big(-C_2(\mathcal{G}) + T_2(\mathcal{R}_H)\big)\ln\frac{\mu'}{\mu}. \tag{3.8}$$

Moving onto the $\mathcal{N} = 4$ theory, for which $\mathcal{R}_H = \mathcal{G}$, we see that the β function vanishes identically and the theory is UV finite. If the vacuum expectation value of the gauge scalars vanishes, this theory is exactly (super-) conformally invariant, while if the vacuum expectation value is nonzero, superconformal invariance is spontaneously broken.

3.3 Low Energy Dynamics of $\mathcal{N} = 2$ Super-Yang–Mills

The gauge multiplet part of the renormalizable $\mathcal{N} = 2$ supersymmetric Lagrangian is derived from expanding the superfields Lagrangian of (2.58) in terms of component fields and dropping the contribution of the hypermultiplet,

$$\mathcal{L} = -\frac{1}{2g^2}\operatorname{tr} FF - \frac{\theta}{16\pi^2}\operatorname{tr} F\tilde{F} - \frac{1}{2g^2}\operatorname{tr} D_\mu\phi^\dagger D^\mu\phi + \frac{1}{2g^2}\operatorname{tr}[\phi^\dagger,\phi]^2$$
$$- \frac{i}{2g^2}\operatorname{tr}\bar{\lambda}\bar{\sigma}^\mu D_\mu\lambda - \frac{i}{2g^2}\operatorname{tr}\bar{\psi}\bar{\sigma}^\mu D_\mu\psi$$
$$+ i\frac{\sqrt{2}}{g^2}\operatorname{tr}([\phi^\dagger,\psi]\lambda) - i\frac{\sqrt{2}}{g^2}\operatorname{tr}([\bar{\lambda},\phi]\bar{\psi}). \quad (3.9)$$

We note the symmetry in $\lambda \leftrightarrow \psi$, which is a manifestation of R-symmetry. To derive the low energy dynamics of this theory, we begin by identifying the ground state(s). Since $\mathcal{L}$ is supersymmetric, its Hamiltonian is positive, and the energy of any state, including the ground state is positive or zero. If we can find a state with exactly zero energy, then this state is one of the possible ground states or vacua of the theory, and supersymmetry in this vacuum will be unbroken.

At the semi-classical level, a solution of vanishing energy will require that the following quantities vanish simultaneously,

$$F_{\mu\nu} = 0, \quad D_\mu\phi = 0, \quad \operatorname{tr}([\phi^\dagger,\phi]^2) = 0, \quad (3.10)$$

so that the gauge field A_μ is pure gauge. By a gauge choice, we may pick $A_\mu = 0$, so that ϕ must be constant, and only the last equation of (3.10) remains to be solved. Since $\phi \in \mathcal{G}$, which is the Lie algebra of a compact gauge group, $\operatorname{tr}([\phi^\dagger,\phi]^2) = 0$ implies $[\phi^\dagger,\phi] = 0$. Now, ϕ is a complex field, in the adjoint representation of $\mathcal{G}$, and may thus be decomposed into its Hermitian and anti-Hermitian parts $\phi = \phi_1 + i\phi_2$, with $\phi_{1,2}^\dagger = \phi_{1,2}$ and $[\phi_1,\phi_2] = 0$. Thus, ϕ_1 and ϕ_2 may be simultaneously diagonalized, each with real eigenvalues. This implies that the original ϕ may be diagonalized with complex eigenvalues. Thus, we find not just a single zero energy state, but rather a family of zero energy states, each of which is a candidate vacuum. It is customary to denote these ground states by $|0\rangle$, but it is understood that this symbol really is parameterized by the vacuum expectation values of the gauge scalar ϕ. While these results were derived here within the semi-classical approximation, they hold true in the full quantum theory as well.

A convenient basis for the diagonal matrices is provided by the Cartan generators h_j of $\mathcal{G}$, with $j = 1,\ldots,n = \operatorname{rank}(\mathcal{G})$, which obey $[h_i,h_j] = 0$. The vacuum expectation value of the gauge scalar is then

$$\langle 0|\phi|0\rangle = \sum_{j=1}^{n} a_j h_j. \quad (3.11)$$

The complex parameters a_j, $j = 1, \ldots, n$ are called the *quantum moduli* or *quantum order parameters* of the vacuum. The semiclassical limits of these parameters are denoted by $\bar{a}_j$, $j = 1, \ldots, n$ and are called the *classical moduli* or *classical order parameters.*

For generic values of the moduli a_j, the vacuum expectation value $\langle 0|\phi|0\rangle$ will have all distinct eigenvalues. Since the field transforms under the adjoint representation of $\mathcal{G}$, the standard Higgs mechanism will break the gauge symmetry to that of the Cartan subalgebra $\mathrm{U}(1)^n$. Actually, the role of the different Cartan generators may be permuted under the set of residual (discrete) gauge transformations which form the Weyl($\mathcal{G}$) group. Thus, the detailed symmetry breaking is given by [8]

$$\mathcal{G} \longrightarrow \mathrm{U}(1)^n / \,\mathrm{Weyl}(\mathcal{G}). \tag{3.12}$$

In other words, the theory after symmetry breakdown reduces to n copies of (supersymmetric) electro-magnetism, and for this reason is often said to be in the *Coulomb phase.*

At the semi-classical level, nongeneric values of a_j may yield a larger symmetry group. It will be one of the fundamental results of Seiberg–Witten theory that in the full quantum theory, such larger residual symmetry groups do not survive quantization, so that the theory is always in the Coulomb phase. For sufficiently large gauge algebra or for sufficiently large hypermultiplet contents, novel phenomena, such as the coexistence of massless electrically charged particles with massless magnetic monopoles is possible at so-called Argyres–Douglas points or curves [19].

3.4 Particle and Field Contents

Since we are left with $n = \mathrm{rank}(\mathcal{G})$ factors of U(1), we have n different massless U(1) gauge $\mathcal{N} = 2$ supersymmetry multiplets. The fields in these multiplets may be read off from our general classification,

$$(A_{j\mu}, \lambda_{j\pm}, \phi_j) \quad \langle 0|\phi_j|0\rangle = a_j \quad j = 1, \ldots, n. \tag{3.13}$$

• For each of the $\dim(\mathcal{G}) - \mathrm{rank}(\mathcal{G})$ roots α of $\mathcal{G}$, we have a massive vector boson multiplet appearing in the spectrum as a single particle state, described by the field multiplet

$$(V_{\alpha\mu}, \lambda_{\alpha\pm}, \phi_\alpha), \quad \alpha = \text{root of } \mathcal{G}. \tag{3.14}$$

Counting the number of fields for each spin, and comparing with the $\mathcal{N} = 2$ entries of Table 1.5, we see that this massive vector boson multiplet must be a BPS multiplet! As a result, we have the BPS formula for the masses of this multiplet in terms of the central charges, which are given by $\alpha \cdot a$,

$$M_\alpha^W = |\vec{\alpha} \cdot \vec{a}| \quad \vec{a} = (a_1, \ldots, a_n). \tag{3.15}$$

TABLE 1.5. Numbers of Massive States as a function of $\mathcal{N}$ and spin

Spin ≤ 1	$\mathcal{N}=1$ gauge	$\mathcal{N}=1$ chiral	$\mathcal{N}=2$ gauge	$\mathcal{N}=2$ BPS gauge	$\mathcal{N}=2$ BPS hyper	$\mathcal{N}=4$ BPS gauge
1	1	0	1	1	0	1
$\frac{1}{2}$	2	1	4	2	2	4
0	1	2	5	1	4	6
Total #	8	4	16	8	8	16

• As the simple gauge algebra $\mathcal{G}$ is broken to a subalgebra which includes U(1) factors, general arguments show that 't Hooft–Polyakov magnetic monopoles will arise [20]. In fact, precisely one magnetic monopoles arises for each possible embedding of SU(2) into $\mathcal{G}$, i.e., for each root β of $\mathcal{G}$.

It is instructive to recall the set-up of the basic 't Hooft–Polyakov solution in the Bogomolnyi–Prasad–Sommerfield (BPS) limit [21]. The limit consists in ignoring the quartic Higgs field potential, but retaining the boundary conditions on the Higgs field that it should tend towards a constant value at spatial ∞. (This is automatically the set-up we have in $\mathcal{N}=2$ super-Yang–Mills when the gauge symmetry is broken to a U(1) factor.) Now, the magnetic monopole is a stable, static soliton solution, magnetically charged under the unbroken U(1). In the BPS limit, the energy is given by

$$
\begin{aligned}
E &= \frac{1}{g^2}\int d^3x \left(\mathrm{tr}(D_i\phi)^2 + \mathrm{tr}\, B_i^2\right) \\
&= \frac{1}{g^2}\int d^3x\ \mathrm{tr}(D_i\phi \pm B_i)^2 \mp \frac{2}{g^2}\int d^3x \partial_i(\mathrm{tr}\,\phi B_i)
\end{aligned}
\tag{3.16}
$$

where the $\pm$ signs are correlated. The second term on the rhs of (3.16) is proportional to the magnetic charge. The first term on the rhs of (3.16) is always positive or zero, and this fact guarantees that the energy is bounded from below by the magnetic charge of the monopole. This so-called *BPS bound* is really nothing else than the BPS bounds we have encountered in the study of extended supersymmetry. A BPS solution (which preserves $\mathcal{N}=1$ supersymmetry) is obtained by saturating the BPS bound upon setting the first term on the rhs of (3.16) to zero.

This construction will give us a BPS formula (at the semi-classical level) for the mass of the 't Hooft–Polyakov magnetic monopole in terms of the vacuum expectation value of the gauge scalar. Applying this construction for any SU(2) subgroup of $\mathcal{G}$, specified uniquely by a root $\vec{\beta}$ of $\mathcal{G}$, and using the fact that $\phi = \vec{a}\cdot\vec{a}$, we find the semi-classical magnetic monopole BPS mass formula

$$
M^M_\beta = |\vec{\beta}\cdot\tau\vec{a}| \quad \tau = \frac{\theta}{2\pi} + \frac{4\pi i}{g^2}. \tag{3.17}
$$

Here, we have also included the effect of the θ angle, which is to produce an electric charge on the magnetic monopole and make it into a *dyon* [22].
• In fact, a general dyon (single particle state) may be specified by two roots: one root $\vec{\alpha}$ for its electric charge as in the construction of vector boson masses; and one (co)root $\vec{\beta}$ for its magnetic charge as in the construction of the magnetic monopole masses. The full quantum BPS mass formula for such a dyon is [7, 23]

$$M^D_{(\vec{\alpha},\vec{\beta})} = |\vec{\alpha} \cdot \vec{a} + \vec{\beta} \cdot \vec{a}_D|. \tag{3.18}$$

In the semi-classical limit, the quantity $\vec{a}_D$ is given by $\vec{a}_D = \tau\vec{a}$, but at the full quantum level the expression for $\vec{a}_D$ in terms of $\vec{a}$ will be modified. One of the goals of Seiberg–Witten theory will be to predict precisely what the relation between these two quantities is. Given this information, the masses of all BPS states (including vector bosons, magnetic monopoles and dyons) will be known exactly in terms of the vacuum moduli parameters a_j. In the more constrained $\mathcal{N} = 4$ theory, the relation $a_D = \tau a$ will not be renormalized at the full quantum level.

3.5 Form of the $\mathcal{N} = 2$ Low Energy Effective Lagrangian

With the preceding subsection, we know precisely what the particle spectrum of the $\mathcal{N} = 2$ super-Yang–Mills theory will be. Upon restriction to the low energy part of the spectrum, the effects of the massive vector bosons and magnetic monoples and dyons will decouple. Only the n massless $\mathcal{N} = 2$ Abelian U(1) gauge supermultiplets will remain.[2] Thus, we are left with the problem of determining *the most general low energy effective action of n massless* U(1) *gauge supermultiplets*, with fields $(A^j_\mu, \lambda^j_\pm, \phi^j)$, as function of the vacuum moduli parameters a_j, which are the expectation values of the gauge scalars ϕ^j.

We have two methods available to do this: we could start directly from the manifestly $\mathcal{N} = 2$ supersymmetric formulation of §2.10; or start from an $\mathcal{N} = 1$ standard superfield formulation and attempt to impose the extra $\mathcal{N} = 2$ susy by hand. It turns out to be very instructive to start with the latter. To do this, we first construct the most general $\mathcal{N} = 1$ invariant Lagrangian with this field content, and we shall then enforce the supplementary constraints of $\mathcal{N} = 2$ supersymmetry [24]. In terms of $\mathcal{N} = 1$ superfields, the n multiplets decompose into r Abelian U(1) vector superfields $V_j \sim (A_{j\mu}, \lambda^j_- = \lambda^j)$ and n chiral superfields $\Phi^j \sim (\lambda^j_+ = \psi^j, \phi^j)$, which are neutral under the $\mathrm{U}(1)^n$ gauge group. The field

[2] We note here though that for special values of the vacuum moduli, one or several of the dyons may also become massless. This phenomenon and its symptoms in the Wilson effective action are central features of the theory. They will be investigated extensively in the next subsections.

strength superfields are simply

$$W_\alpha^j = -\tfrac{1}{4}\bar{D}\bar{D}D_\alpha V^j. \tag{3.19}$$

The full effective action is now gotten by putting together the superpotential part $\mathcal{L}_U$, the gauge coupling part $\mathcal{L}_G$ and the Kähler part $\mathcal{L}_K$ of §2.8. Since the chiral superfields Φ^j are neutral, the Kähler potential does not involve the gauge fields V_j. We have

$$\mathcal{L} = \int d^4\theta\, K(\Phi^j, (\Phi^j)^\dagger) + \mathrm{Re}\int d^2\theta\,\{U(\Phi^j) + \tau_{ij}(\Phi^k)W^iW^j\}. \tag{3.20}$$

Here, the superpotential $U(\Phi^j)$ and the gauge coupling $\tau_{ij}(\Phi^k)$ are complex analytic functions of the chiral superfields Φ^j.

The first step towards enforcing $\mathcal{N}=2$ supersymmetry is to expand the superfields of (3.20) in terms of components and eliminate the auxiliary fields. This step is necessary because the auxiliary fields of the $\mathcal{N}=1$ superfields do not have a natural place in or good transformation properties under $\mathcal{N}=2$ supersymmetry. The result is

$$\begin{aligned}\mathcal{L} = -g_{i\bar{j}}(D_\mu\phi^i D^\mu\bar{\phi}^{\bar{j}} + i\bar{\psi}^{\bar{j}}\bar{\sigma}^\mu D_\mu\psi^i) - \frac{1}{2}g^{i\bar{j}}\frac{\partial U}{\partial\phi^i}\frac{\partial\bar{U}}{\partial\bar{\phi}^{\bar{j}}} + \mathrm{Re}\Big\{\psi^i\psi^j\frac{\partial^2 U}{\phi^i\partial\phi^j}\Big\}\\ - \mathrm{Re}\big\{\tau_{ij}\big(\tfrac{i}{2}F^i_{\mu\nu}F^{j\mu\nu} - \tfrac{1}{2}F^i_{\mu\nu}\tilde{F}^{j\mu\nu} + \bar{\lambda}^i\bar{\sigma}^\mu D_\mu\lambda^j\big)\big\}.\end{aligned} \tag{3.21}$$

Here, the Kähler metric $g_{i\bar{j}}$ is given in terms of the Kähler potential K by

$$g_{i\bar{j}}(\phi,\bar{\phi}) = \frac{\partial^2 K(\phi,\bar{\phi})}{\partial\phi^i\partial\bar{\phi}^{\bar{j}}}. \tag{3.22}$$

The covariant derivative D_μ is taken with respect to the Levi–Civita connection Γ^i_{jk} of the metric $g_{i\bar{j}}$,

$$D_\mu\phi^i = \partial_\mu\phi^i + \Gamma^i_{jk}\phi^j\partial_\mu\phi^k. \tag{3.23}$$

No gauge field enters here, because the chiral multiplet in which ϕ^i and ψ^i belong are neutral under the gauge group.

To enforce $\mathcal{N}=2$ supersymmetry directly is still not so easy. However, as part of $\mathcal{N}=2$ supersymmetry, we know that the fields A^j_μ and ϕ^j of the $\mathcal{N}=2$ gauge supermultiplets have to transform as singlets under the $\mathrm{SU}(2)_R$ symmetry, while the fermions λ^i and ψ^i must combine into an $\mathrm{SU}(2)_R$ doublet. Thus, we may first enforce, as a necessary condition for $\mathcal{N}=2$ supersymmetry, the requirement that ψ^i and λ^i enter symmetrically.

Clearly, there is no Yukawa interaction for the field λ^i, and thus, the Yukawa interaction must also be absent for the ψ fields. Thus, $U(\phi^i)$ can be at most linear in ϕ^i. In fact, if $U(\phi^i)$ has any dependence on ϕ^i at all, then the $|\partial U|^2$ term will always be strictly positive and supersymmetry will be broken spontaneously. Discarding this possibility, U must be constant, and may hence be set to $U = 0$: the superpotential is absent!

Next, the kinetic terms of λ^i and ψ^i must be equal. This requires the relation

$$\frac{1}{2i}\left(\tau_{ij}(\phi) - \overline{\tau_{ij}(\phi)}\right) = \frac{\partial^2 K(\phi,\bar\phi)}{\partial\phi^i \partial\bar\phi^j}. \tag{3.24}$$

To solve (3.24), we make use of the holomorphicity of τ_{ij} and of its symmetry under $i \leftrightarrow j$. Considering τ_{ij} as a given holomorphic function, K is determined by a linear differential equation, which is trivially solved in terms of holomorphic functions $T_i(\phi)$,

$$K(\phi,\bar\phi) = T_i(\phi)\bar\phi^i + \overline{T_i(\phi)}\phi^i \tag{3.25}$$

with the relation

$$\frac{1}{2i}\tau_{ij}(\phi) = \frac{\partial T_j(\phi)}{\partial\phi^i} = \frac{\partial T_i(\phi)}{\partial\phi^j}. \tag{3.26}$$

The second equality in (3.26) holds in view of the fact that τ_{ij} is symmetric, and immediately implies that $T_i(\phi)$ is a pure gradient of a single holomorphic function, called the *prepotential* and usually denoted by $\mathcal{F}(\phi)$. We have then

$$T_i(\phi) = \frac{1}{2i}\frac{\partial\mathcal{F}(\phi)}{\partial\phi^i}. \tag{3.27}$$

Remarkably, the entire low energy effective action $\mathcal{L}$ has now been completely determined in terms of this single prepotential function $\mathcal{F}(\phi)$. In view of the arguments presented earlier showing holomorphicity as a function of the masses, gauge coupling an Yukawa couplings, we know that $\mathcal{F}$ is also holomorphic as a function of the gauge coupling τ and the hypermultiplet masses $m : \mathcal{F}(\phi; m_a, \tau)$. The $\mathcal{N} = 2$ low energy effective action for the r U(1) gauge multiplets is thus

$$\mathcal{L} = \mathrm{Im}(\tau_{ij})F^i_{\mu\nu}F^{j\mu\nu} + \mathrm{Re}(\tau_{ij})F^i_{\mu\nu}\tilde F^{j\mu\nu} + \mathrm{Im}(\partial_\mu\bar\phi^j\partial^\mu\phi_{Dj}) + \text{fermions} \tag{3.28}$$

with

$$\phi_{Dj} = \frac{\partial\mathcal{F}(\phi)}{\partial\phi^j} \qquad \tau_{ij}(\phi) = \frac{\partial^2\mathcal{F}(\phi)}{\partial\phi^i\partial\phi^j}. \tag{3.29}$$

Here, it is understood that the gauge scalar ϕ^j takes on a vacuum expectation value a^j, which are arbitrary complex numbers.

That the Lagrangian (3.28) is fully invariant under the $\mathcal{N} = 2$ susy follows almost trivially by obtaining it from the expansion of the manifestly $\mathcal{N} = 2$ supersymmetric Lagrangian for gauge group $\mathrm{U}(1)^n$ of (2.98) and without hypermultiplets.

3.6 Physical Properties of the Prepotential

Since $\mathcal{F}(\phi)$ only depends upon the field and not upon space-time derivatives $\partial_\mu \phi$, $\mathcal{F}$ is uniquely determined by its values on the expectation values a_j, so we consider now the complex analytic function $\mathcal{F}(a_j; m_a, \tau)$. $\mathcal{F}$ is subject to two fundamental physical properties.

- $\mathcal{F}(a_j; m_a, \tau)$ is complex analytic by $\mathcal{N}=2$ supersymmetry;
- $\tau_{ij}(a) = \partial^2 \mathcal{F}/\partial a_i \partial a_j$ obeys $\mathrm{Im}(\tau_{ij}) > 0$ since this function plays the role of a metric on the gauge kinetic terms. Unless $\mathrm{Im}(\tau_{ij}) > 0$, some gauge fields will propagate with negative probability.

As a result of the above two properties of $\mathcal{F}$, we find that $\mathcal{F}$ cannot be a single valued function of the vacuum moduli a_j (or of the gauge coupling τ). Indeed, if $\mathcal{F}$ were single-valued, then $\mathrm{Im}(\tau_{ij})$ will be a single-valued harmonic function, which by the second point above is bounded from below. But, by the maximum principle, such a harmonic function must in fact be constant. If it is not constant, then it can never be bounded from below or from above. These properties form the basis for the construction of $\mathcal{F}$ given by Seiberg and Witten [8].

Actually, we can easily explore the properties of the function $\mathcal{F}$ in the regime of large vacuum expectation values, since the effects of renormalization are then known from asymptotic freedom, and we find for say $\mathcal{G} = \mathrm{SU}(N)$,

$$\tau_{ij} = \frac{\theta_{ij}}{2\pi} + \frac{4\pi i}{g_{ij}^2} \sim \frac{i}{2\pi} \ln(a_i - a_j)^2/\mu^2, \tag{3.30}$$

as $a_j \to \infty$. Integrating up to derive $\mathcal{F}$, we find

$$\mathcal{F} \sim \frac{i}{8\pi} \sum_{i,j} (a_i - a_j)^2 \ln(a_i - a_j)^2/\mu^2. \tag{3.31}$$

The expression for $\mathcal{F}$ in this limit clearly shows that $\mathcal{F}$ is neither constant nor single-valued.

In fact, the nature of the multiple-valuedness (also called the *monodromy*) of $\mathcal{F}$ and of τ_{ij} is understood in part. A 2π rotation of any $a_i - a_j$ amounts to a shift of the θ_{ij} angle by 4π. This shift is physically immaterial since the instanton angles θ_{ij} are periodic with period 2π. However, this cannot be the source of all the monodromy of $\mathcal{F}$. For, if it were, then the expression (3.30) would continue to be the source of this monodromy down to $a_i - a_j \to 0$. But, as $a_i - a_j \to 0$, we would invariably led into a domain where $\mathrm{Im}(\tau_{ij}) < 0$, violating the positivity property $\mathrm{Im}(\tau_{ij}) > 0$ explained previously. The source of additional monodromy is much more subtle and will be explained next.

3.7 Electric-Magnetic Duality

We begin by discussing first the case of $\mathcal{G} = \mathrm{SU}(2)$ gauge group and no hypermultiplets. The fields are then the single gauge scalar ϕ, the gauge field A_μ with field strength $F_{\mu\nu}$, and the prepotential is a complex analytic function $\mathcal{F}(\phi)$ dependent on a single complex ϕ only. The effective Lagrangian is then

$$\begin{aligned} \mathcal{L} = \mathrm{Im}\{\tau(\phi)(F_{\mu\nu} + i\tilde{F}_{\mu\nu})^2\} + \partial_\mu \begin{pmatrix} \phi_D \\ \phi \end{pmatrix}^\dagger J \partial^\mu \begin{pmatrix} \phi_D \\ \phi \end{pmatrix} + \text{fermions} \\ J = \begin{pmatrix} 0 & i \\ -i & 0 \end{pmatrix}. \end{aligned} \tag{3.32}$$

The kinetic term is manifestly invariant under the linear action of an $M \in \mathrm{SL}(2,\mathcal{R}) = \mathrm{Sp}(2,\mathcal{R})$ on ϕ_D and ϕ, with

$$\begin{pmatrix} \phi_D \\ \phi \end{pmatrix} \longrightarrow M \begin{pmatrix} \phi_D \\ \phi \end{pmatrix} \qquad M^\dagger J M = J. \tag{3.33}$$

This group is generated by *translations* T_β and an inversion S, given by

$$T_\beta = \begin{pmatrix} 1 & \beta \\ 0 & 1 \end{pmatrix}, \qquad S = \begin{pmatrix} 0 & -1 \\ 1 & 0 \end{pmatrix}. \tag{3.34}$$

The action of T_β on ϕ, ϕ_D and $\tau(\phi)$ is as follows:

$$\begin{aligned} \phi_D &\longrightarrow \phi_D + \beta\phi \\ \phi &\longrightarrow \phi \\ \tau(\phi) &\longrightarrow \tau(\phi) + \beta. \end{aligned} \tag{3.35}$$

This shift is precisely of the form of a shift in the angle θ, and will be immaterial only if β is an integer multiple of 2. Thus, while the full $\mathrm{SL}(2,\mathcal{R})$ leaves the kinetic term invariant, only a subgroup $\Gamma_0(2) \subset \mathrm{SL}(2,\mathbb{Z})$ will leave the quantum mechanical physics of the problem unchanged. The action of S on ϕ, ϕ_D and $\tau(\phi)$ is as follows:

$$\begin{aligned} \phi_D &\longrightarrow -\phi \\ \phi &\longrightarrow \phi_D \\ \tau(\phi) &\longrightarrow -1/\tau(\phi). \end{aligned} \tag{3.36}$$

Considering the special case where $\theta = 0$, the transformation on the remaining coupling g in τ is by inversion: $g \to 1/g$, and thus exchanges small and large coupling regimes.

Thus far, we have implemented the action of the transformation S only on the scalar kinetic term, and it remains to implement S on the gauge field. But, the gauge part of the action does not manifestly exhibit a symmetry in which $g \to 1/g$. The key novel ingredient is the fact that upon $g \to 1/g$, the gauge field A_μ must simultaneously be replaced by its dual field $A_{D\mu}$. This transformation interchanges the roles of electric and magnetic fields, as was considered by Montonen and Olive [7].

The fact that $g \to 1/g$ is related to an electric-magnetic duality transformation may be seen intuitively from the Dirac quantization condition. If we denote by g the unit of electric charge and by g_D the unit of magnetic charge, then the Dirac condition requires $g \cdot g_D \sim 1$ in units where $\hbar = 1$. Thus, as $g \to 1/g$, we have equivalently $g \leftrightarrow g_D$, and electric and magnetic charge are being exchanged.

Electric-magnetic duality in the free Maxwell's equations $\partial_\mu \tilde{F}^{\mu\nu} = \partial_\mu F^{\mu\nu} = 0$ is a manifest symmetry, realized by $F^{\mu\nu} \leftrightarrow \tilde{F}^{\mu\nu}$. At the level of the Lagrangian, the symmetry is obscured by the fact that the Bianchi identity $\partial_\mu \tilde{F}^{\mu\nu} = 0$ has been solved for in terms of the gauge potential, thus obscuring duality symmetry.

To exhibit duality in the action, we must arrange to have both the field equations $\partial_\mu F^{\mu\nu} = 0$ and the Bianchi identities $\partial_\mu \tilde{F}^{\mu\nu} = 0$ emerge as field equations. This may be done by formulating the action in terms of an independent field strength field $F_{\mu\nu}$ and by enforcing the Bianchi identity by introducing a Lagrange multiplier field $A_{D\mu}$. We shall carry out this duality transformation only on the gauge field, but a full supersymmetric argument is possible as well. We have

$$S_G = \int \operatorname{Im} \tau(\phi)(F + i\tilde{F})^2 + 2\int A_{D\mu}\partial_\nu \tilde{F}^{\mu\nu}. \tag{3.37}$$

We consider the Lagrange multiplier $A_{D\mu}$ as a field in its own right, and introduce its field strength $F_{D\mu\nu} = \partial_\mu A_{D\nu} - \partial_\nu A_{D\mu}$. Upon integration by parts, the last term in S_G is then $\int F_{D\mu\nu}F^{\mu\nu}$, and we may complete the square in F as follows:

$$S_G = \int \operatorname{Im}\left[\tau(\phi)\left\{F + i\tilde{F} + \frac{1}{\tau(\phi)}(F_D + i\tilde{F}_D)\right\}^2 - \frac{1}{\tau(\phi)}(F_D + i\tilde{F}_D)^2\right]. \tag{3.38}$$

Integrating out the unconstrained field F by Gaussian integration, we are left with the same action S_G, but formulated entirely in terms of the dual gauge field $A_{D\mu}$, and given by

$$S_G = \int \operatorname{Im}\left[-\frac{1}{\tau(\phi)}(F_D + i\tilde{F}_D)^2\right]. \tag{3.39}$$

Thus, the S transformation in which $\tau(\phi) \to -1/\tau(\phi)$ has been realized on the gauge field as well. The precise monodromy group is generated by the transformations T_2 and S of (3.34), and this group is a subgroup $\Gamma_0(2)$ of the group $\mathrm{SL}(2, \mathbb{Z})$, which is generated by T_1 and S.

3.8 Monodromy via Elliptic Curves for SU(2) *Gauge Group*

We have now obtained a precise formulation of the monodromy problem associated with the Seiberg–Witten theory for SU(2) gauge group.

- $\binom{a_D}{a}$ transforms linearly under the $\Gamma_0(2) \subset \mathrm{SL}(2,\mathbb{Z})$ monodromy group;
- $\tau(a) = \partial a_D / \partial a$ obeys the condition $\mathrm{Im}\, \tau(a) > 0$;
- The monodromy T_2 is realized in the limit where $a \to \infty$.

Generally, such monodromy problems are very hard to solve. Here however, we have one very important hint: a complex quantity τ with positive definite imaginary part. Such quantities are also fundamental in the theory of Riemann surfaces, where they describe the moduli of Riemann surfaces and the positivity condition ensures regularity of the surface. In our case, the relevant Riemann surface is an *elliptic curve* or equivalently a *torus.*

The torus is conveniently described as an elliptic curve in $\mathbf{CP}^2$, parameterized by two complex coordinates (k, y), and given by

$$y^2 = (k - \Lambda^2)(k + \Lambda^2)(k - u). \tag{3.40}$$

Here Λ is an overall scale, and u parameterizes the modulus τ of the elliptic curve. One may view the elliptic curve as realized by a double cover of the complex plane (the two sheets corresponding to the $\pm y$ branches) with branch points $\pm\Lambda^2$, u and ∞, and branch cuts for example along $[-\Lambda^2, +\Lambda^2]$ and $[u, \infty]$.

A canonical basis for the homology 1-cycles (in which the cycles A and B intersect precisely once) is provided by the following closed curves; A: circling the branch cut $[-\Lambda^2, +\Lambda^2]$ once; and B: from branch points $+\Lambda^2$ to u circling from the upper to the lower sheets and back. This choice of homology cycles is not unique however, even when we require single intersection. Any change of homology basis by a symplectic transformation with integer coefficients will preserve canonical intersection

$$\begin{pmatrix} B \\ A \end{pmatrix} \longrightarrow M \begin{pmatrix} B \\ A \end{pmatrix}, \quad M^\dagger J M = J, \quad M \in \mathrm{SL}(2,\mathbb{Z}). \tag{3.41}$$

Notice the similarity with the transformation properties of ϕ_D and ϕ in (3.33).

There is a unique holomorphic 1-form (or *Abelian differential of the first kind*), dk/y, whose Abelian integral is

$$z(P) - z(Q) = \int_Q^P \frac{dk}{y}. \tag{3.42}$$

This integral is multiple valued, and the periods of the torus are defined by the following contour integrals

$$\omega = \oint_A \frac{dk}{y}, \qquad \omega_D = \oint_B \frac{dk}{y}. \tag{3.43}$$

The value of the matrix of periods $(\omega_D \; \omega)$ is subject to monodromy as the canonical homology basis of A and B cycles is changed, and transforms according to

$$\begin{pmatrix} \omega_D \\ \omega \end{pmatrix} \longrightarrow M \begin{pmatrix} \omega_D \\ \omega \end{pmatrix}, \qquad M^\dagger J M = J, \quad M \in \mathrm{SL}(2, \mathbb{Z}). \tag{3.44}$$

This transformation law is strongly suggestive in view of its similarity with the transformation law for ϕ_D and ϕ in (3.33). The modulus of the elliptic curve is defined by

$$\tau = \frac{\omega_D}{\omega}, \qquad \mathrm{Im}\, \tau > 0, \tag{3.45}$$

where the last condition guarantees that the curve is nonsingular.

We are now ready to identify the data of Seiberg–Witten theory and those of the family of elliptic curves parameterized by the modulus τ, or equivalently by the complex parameter u. The modulus of the curve τ is identified with the Seiberg–Witten gauge coupling $\tau(a)$, so that

$$\tau = \frac{\omega_D}{\omega} = \tau(a) = \frac{\partial a_D}{\partial a} = \left(\frac{\partial a_D}{\partial u}\right)\left(\frac{\partial a}{\partial u}\right)^{-1}. \tag{3.46}$$

These equalities are readily solved by simply identifying

$$\frac{\partial a}{\partial u} = \omega = \oint_A \frac{dk}{y} \qquad \frac{\partial a_D}{\partial u} = \omega_D = \oint_B \frac{dk}{y}. \tag{3.47}$$

Integration in u may be carried out on both equations, and we obtain the final form of the Seiberg–Witten solution,

$$a = \oint_A d\lambda, \qquad a_D = \oint_B d\lambda, \tag{3.48}$$

where the differential 1-form $d\lambda$, usually called the *Seiberg–Witten differential* is such that its u-derivative is holomorphic. Its form may be obtained explicitly by integration in u, and is unique up to the addition of any exact differential,

$$\frac{\partial d\lambda}{\partial u} = \frac{dk}{y} = \frac{dk}{\sqrt{(k^2 - \Lambda^4)(k-u)}} \Longrightarrow d\lambda = \frac{(k-u)dk}{y} + \text{exact}. \tag{3.49}$$

Notice that the Seiberg–Witten differential $d\lambda$ has a pole at $k = \infty$.

The explicit solution is obtained by using the homology A and B cycles defined above, and is given by the following parametric representation of complete elliptic integrals

$$a_D(u) = \frac{\sqrt{2}}{\pi}\int_{\Lambda^2}^{u} dk\,\frac{\sqrt{k-u}}{\sqrt{k^2-\Lambda^4}}, \quad a(u) = \frac{\sqrt{2}}{\pi}\int_{\Lambda^2}^{\Lambda^2} dk\,\frac{\sqrt{k-u}}{\sqrt{k^2-\Lambda^4}}. \tag{3.50}$$

The T_2 monodromy condition is readily verified in the limit $u \to \infty$

$$a_D(u) = \frac{\sqrt{2u}}{\pi}\int_{\Lambda^2/u}^{1} dy \frac{\sqrt{y-1}}{\sqrt{y^2-\frac{\Lambda^4}{u^2}}} \sim -i\frac{\sqrt{2u}\ln u}{\pi} \tag{3.51}$$

$$a(u) \sim \frac{\sqrt{2u}}{\pi}\int_{-\Lambda^2}^{\Lambda^2} \frac{dk}{\sqrt{\Lambda^4-k^2}} = \sqrt{2u}. \tag{3.52}$$

The function $a_D(a)$ and $\tau(a)$ are then obtained by eliminating u.

3.9 Physical Interpretation of Singularities

The elliptic curve $y^2 = (k-\Lambda^2)(k+\Lambda^2)(k-u)$ has precisely 3 singularities at $u = \pm\Lambda^2, \infty$ where the curve degenerates. These singularities translate into singular behavior of $a_D(a)$ and $\tau(a)$. For all other $u \in \mathbb{C}$, these quantities are regular.

Recall from the discussion of the Wilson effective action that singularities will arise in $\tau(a)$ and $\mathcal{F}(a)$ if and only if the mass(es) of some particle(s) that was integrated out in the Wilson remnormalization group procedure actually becomes zero. Now, we know the masses of all massive vector bosons, magnetic monopoles and dyons, because these particles belong to short BPS multiplets of $\mathcal{N}=2$ supersymmetry and so their masses obey the BPS bound

$$M = |n_2 a + n_m a_D| \tag{3.53}$$

where n_e and n_m are the electric and magnetic charges respectively. Whenever a_D/a becomes a rational number, some dyon masses will become zero. In our case, there are only three candidate values for the vacuum moduli where this can happen, and we now study those in turn, using the explicit solution of (3.50).

$\boldsymbol{u \to \infty}$ corresponds to the large a limit, where asymptotic freedom applies and we find

$$\begin{aligned} a(u) &\sim \sqrt{u} \\ a_D(u) &\sim \sqrt{u}\ln\frac{u}{\Lambda^2} \sim a\ln a. \end{aligned} \tag{3.54}$$

$\boldsymbol{u \to +\Lambda^2}$

$$\begin{aligned} a(u) &\sim \frac{i}{\pi}a_D \ln\frac{a_D}{\Lambda} \\ a_D(u) &\sim (u-\Lambda^2). \end{aligned} \tag{3.55}$$

Using the BPS mass formula, we see that at this point any particle with zero electric charge n_e will be massless: these are just the pure magnetic monopoles, with $(n_e = 0,\, n_m \neq 0)$.

$\boldsymbol{u \to -\Lambda^2}$

$$\begin{aligned}(a - a_D)(u) &\sim (u + \Lambda^2)\\ a(u) &\sim \frac{i}{\pi}(a_D - a)\ln\frac{a_D - a}{\Lambda}.\end{aligned} \tag{3.56}$$

Using the BPS mass formula, we see that at this point any particle with zero charge $n_e + n_m$ will be massless: these are the dyons, with $(n_e, n_m = -n_e)$.

The monodromy matrices at each of these singularities can be derived from the presence of the logarithm in each of the above formulas. One finds

$$D_\infty = \begin{pmatrix}1 & 2\\ 0 & 1\end{pmatrix}, \quad D_{+\Lambda^2} = \begin{pmatrix}1 & 0\\ -2 & 1\end{pmatrix}, \quad D_{-\Lambda^2} = \begin{pmatrix}-1 & 2\\ -2 & 3\end{pmatrix}. \tag{3.57}$$

As a consistency check that the above singularities are the only ones required to satisfy the monodromy conditions in a consistent way, one may check that the product of all three monodromies is proportional to the identity matrix $D_\infty D_{+\Lambda^2} D_{-\Lambda^2} = -I$.

Physically, the remarkable new phenomenon that emerges from this study of the Wilson effective action is that magnetic monopoles and dyons, which are usually encountered as heavy semi-classical soliton solutions, may actually become massless at strong coupling.

3.10 Hypergeometric Function Representation

In their original paper, Seiberg and Witten [8] were actually able to identify the additional monodromies at $u = \pm\Lambda^2$, without solving first for (a, a_D). Their arguments were as follows. Since $\mathcal{F}$ is not global, it has at least 3 singularities. As explained earlier, the appearance of a singularity in the Wilson effective action is due to a particle becoming massless. The key physical input is the realization that the particle becoming massless is a monopole. The magnetic coupling τ_D can then be evaluated in the dual formulation in terms of the dual gauge field A_D, giving the monodromy matrix $D_{+\Lambda^2}$. The remaining monodromy matrix $D_{-\Lambda^2}$ follows from the relation $D_\infty D_{+\Lambda^2} D_{-\Lambda^2} = -I$ for the homotopy group of the sphere with 3 punctures. In this way, the three monodromy matrices of (3.57) are recovered. Here we indicate briefly how the monodromy problem can be solved in terms of hypergeometric functions, if all three monodromies D_∞, $D_{+\Lambda^2}$, and $D_{-\Lambda^2}$ are already known and given by (3.57).

We look for (a_D, a) as solutions of a second order differential equation with regular singular points. For simplicity, set $\Lambda = 1$. Consider the equation

$$\left(-\frac{d^2}{dz^2}+V(z)\right)\psi(z)=0 \tag{3.58}$$

where $V(z)$ is a meromorphic function with at most double poles. Then the equation admits two independent solutions ψ_1, ψ_2 which get transformed into a linear combination of each other under analytic continuation around each of the poles of $V(z)$. In the present case, there should be 3 singularities at $z=-\Lambda^2,\Lambda^2,\infty$, and a candidate for $V(z)$ is

$$V(z)=-\frac{1}{4}\left(\frac{1-\lambda_1^2}{(z+1)^2}+\frac{1-\lambda_2^2}{(z-1)^2}-\frac{1-\lambda_1^2-\lambda_2^2+\lambda_3^2}{(z+1)(z-1)}\right). \tag{3.59}$$

As $z\to\infty$, $V(z)\sim-\frac{1}{4}(1-\lambda_3^2)/z^2$. The solutions behave asymptotically like $z^{(1\pm\lambda_3)/2}$ if $\lambda_3\neq 0$, and like $\sqrt{z}$ and $\sqrt{z}\ln z$ when $\lambda_3=0$. Comparing with the desired behavior for a and a_D, we set $\lambda_3=0$. Next, considering asymptotics as $z\to\Lambda^2$, we are led to the choice $\lambda_1=\lambda_2=1$.

Now the solutions ψ of (3.58) are related to those of the standard hypergeometric equation $x(1-x)f''+[c-(a+b+1)x]f'-abf=0$ by

$$\psi(z)=(z+1)^{(1-\lambda_1)/2}(z-1)^{(1-\lambda_2)/2}f\left(\frac{z+1}{2}\right), \tag{3.60}$$

with $a=b=-\frac{1}{2}$, $c=0$. Since a basis for the hypergeometric equation is given by

$$\begin{aligned} f_1(x)&=(1-x)^{c-a-b}F(c-a,c-b;c+1-a-b;1-x),\\ f_2(x)&=(-x)^{-a}F\left(a,a+1-c;a+1-b;\frac{1}{x}\right), \end{aligned} \tag{3.61}$$

the natural candidate for (a,a_D) is

$$a_D(u)=i\frac{u-1}{2}F\left(\frac{1}{2},\frac{1}{2};2;\frac{1-u}{2}\right) \tag{3.62}$$

$$a(u)=\sqrt{2}(u+1)^{1/2}F\left(-\frac{1}{2},\frac{1}{2};1;\frac{2}{u+1}\right). \tag{3.63}$$

Recalling that the hypergeometric function admits an integral representation

$$F(a,b;c,z)=\frac{\Gamma(c)}{\Gamma(b)\Gamma(c-b)}\int_0^1 dt\,t^{b-1}(1-t)^{c-b-1}(1-tz)^{-a} \tag{3.64}$$

we obtain the following integral representation for (a,a_D)

$$a_D(u)=i\frac{u-1}{2}\frac{\Gamma(2)}{\Gamma(\frac{1}{2})\Gamma(\frac{3}{2})}\int_0^1 dt\,t^{-1/2}(1-t)^{1/2}\left(1-t\frac{u-1}{2}\right)^{-1/2} \tag{3.65}$$

$$a(u) = \sqrt{2}(u+1)^{1/2} \frac{\Gamma(1)}{\Gamma(\frac{1}{2})\Gamma(\frac{1}{2})} \int_0^1 dt\, t^{-1/2}(1-t)^{-1/2} \times \left(1 - t\frac{u+1}{2u}\right)^{1/2}. \tag{3.66}$$

A change of variables gives back the expressions (3.50) proposed earlier for (a, a_D).

4 More General Gauge Groups, Hypermultiplets

In the previous section, we reviewed Seiberg–Witten theory for an $\mathcal{N} = 2$ super-Yang–Mills theory with gauge group SU(2) and no hypermultiplets present. The key ingredients were

- The SW curve $\Gamma(u)$ given by $y^2 = (k - \Lambda^2)(k + \Lambda^2)(k - u)$;
- The SW differential $d\lambda = (k - u)y^{-1}dk$;
- The relation between the periods of $d\lambda$, the quantum moduli a and a_D and the prepotential $\mathcal{F}$ by

$$a = \frac{1}{2\pi i}\oint_A d\lambda, \qquad a_D = \frac{1}{2\pi i}\oint_B d\lambda, \qquad a_D = \frac{\partial F}{\partial a}. \tag{4.1}$$

(Henceforth, we shall include an extra factor of $2\pi i$ in the normalization of $d\lambda$.) In this section, we shall review the known generalizations of this construction to

- higher gauge algebras SU(N), SO(N), Sp(N), and exceptional algebras;
- including hypermultiplets in various representations of the gauge algebra, with the constraint that the theory remain asymptotically free or have vanishing β-function.

A useful starting point is again the symmetries of the gauge scalar kinetic term, which we shall rewrite as follows

$$2\,\mathrm{Im}\,\partial_\mu \phi^i \partial^\mu \phi_{Di} = i\partial_\mu \begin{pmatrix}\phi_{Di}\\ \phi_i\end{pmatrix}^\dagger J_{ij} \begin{pmatrix}\phi_{Di}\\ \phi_i\end{pmatrix} \qquad J = \begin{pmatrix}0 & -I\\ I & 0\end{pmatrix} \tag{4.2}$$

where the indices $i, j = 1, \ldots, n = \mathrm{rank}(\mathcal{G})$. This part of the low energy effective Lagrangian is invariant under the continuous symmetry group $\mathrm{Sp}(2n, \mathcal{R})$. The subgroup of Abelian transformations of the type

$$T_\beta\colon \begin{pmatrix}\phi_{Di}\\ \phi_i\end{pmatrix} \longrightarrow \begin{pmatrix}I & \beta\\ 0 & I\end{pmatrix}\begin{pmatrix}\phi_{Di}\\ \phi_i\end{pmatrix} \tag{4.3}$$

plays a special role. Its effect on the gauge coupling matrix τ_{ij} is by shifts by β:

$$T_\beta: \ \mathrm{Re}(\tau_{ij}) \longrightarrow \mathrm{Re}(\tau_{ij}) + \beta_{ij}. \tag{4.4}$$

However, these transformations have the effect of shifting the instanton angles θ_{ij} by $2\pi\beta_{ij}$, and thus the entries β_{ij} must be integers. This restriction restricts the full invariance group of the gauge scalar kinetic term to the subgroup $\mathrm{Sp}(2n,\mathbb{Z})$.

The group $\mathrm{Sp}(2n,\mathbb{Z})$ may be viewed as being generated by transformations T_β, as in (4.3), with integer entries, together with the inversion $\tau \to \tau^{-1}$. This transformation exchanges weak and strong coupling, and may be interpreted as a generalized electric-magnetic duality transformation, extending the case of SU(2) gauge group.

4.1 Model of Riemann Surfaces

Consider a compact Riemann surface Γ without boundary of genus g; such a surface may be thought of as a sphere with g handles attached to it. Let $\{A_i, B_i, i = 1,\dots,g\}$ be a basis of closed 1-cycles for the homology $H_1(\Gamma,\mathbb{Z})$ of Γ, so that the intersection form # is canonical

$$\begin{aligned} \#(A_i, A_j) &= \#(B_i, B_j) = 0 \\ \#(A_i, B_j) &= -\#(B_j, A_i) = \delta_{ij}. \end{aligned} \tag{4.5}$$

Dual to this canonical homology basis are the g independent holomorphic 1-forms Ω_i, $i = 1,\dots,g$ (Abelian differentials of the 1st kind) with periods

$$\omega_{ij} = \oint_{A_i} \Omega_j, \quad \omega_{Dij} = \oint_{B_i} \Omega_j. \tag{4.6}$$

The *period matrix* τ_{ij} is defined by

$$\tau \equiv \omega_D \omega^{-1}, \quad \tau_{ij} = \sum_k \omega_{Dik} (\omega^{-1})_{kj}. \tag{4.7}$$

The *Riemann Theorems* imply two key properties of the period matrix

$$\begin{aligned} &(1)\ \ \mathrm{Im}\,\tau > 0 \text{ for nondegenerate } \Gamma \\ &(2)\ \ \tau_{ij} = \tau_{ji}. \end{aligned} \tag{4.8}$$

Finally, a change in homology basis from $\{A_i, B_i\}$ to $\{A'_i, B'_i\}$ that preserves the intersection form # is an element of $\mathrm{Sp}(2g,\mathbb{Z})$, since the intersection matrix is equivalent to the symplectic matrix in $2g$ dimensions,

$$\begin{pmatrix} B_i \\ A_i \end{pmatrix} \longrightarrow \begin{pmatrix} B'_i \\ A'_i \end{pmatrix} = M \begin{pmatrix} B_i \\ A_i \end{pmatrix}, \quad M \in \mathrm{Sp}(2g,\mathbb{Z}). \tag{4.9}$$

Thus, the monodromy group of the Abelian integrals of (4.6) is a subgroup of $\mathrm{Sp}(2g,\mathbb{Z})$.

4.2 Identifying Seiberg–Witten and Riemann Surface Data

A natural generalization of the Seiberg–Witten construction for SU(2) gauge group emerges in the following way.

• The Seiberg–Witten curve is a family of Riemann surfaces $\Gamma(u_1, \ldots, u_n)$, parameterized by n complex parameters $u_1, \ldots, u_n$, which play the role of vacuum moduli and parameterize the flat directions of $\mathcal{N} = 2$ vacua. The family of curves may contain singular curves for certain values of u_i, but these singularities should arise only when physical massless particles appear.

• The curves $\Gamma(u_i)$ must be invariant under the Weyl group Weyl($\mathcal{G}$), which is the residual gauge invariance after the gauge algebra $\mathcal{G}$ has been broken down to its Cartan subalgebra, $\mathrm{U}(1)^n$.

• The Seiberg–Witten differential $d\lambda$ is a meromorphic 1-form on Γ such that $\partial d\lambda / \partial u_i$ is a holomorphic 1-form (or Abelian differential of the first kind) on Γ. The residues at the poles of $d\lambda$ are then automatically independent of u_i, and are linear combinations of the hypermultiplet mass parameters. Physically, the fact that the residues of $d\lambda$ are independent of the u_i means that the hypermultiplet masses (entering in the $\mathcal{N} = 1$ superpotential) are not renormalized and thus are unchanged under any changes in u_i. Linearity of the residues in terms of the hypermultiplet mass parameters is required by the BPS mass formula, in which the hypermultiplet mass parameters enter linearly.

• The periods of Γ then enter in the following way. Since $\partial d\lambda / \partial u_j$ are holomorphic 1-forms, a period matrix τ_{ij} may be constructed from by integrating these 1-forms on a set of homology cycles A_i, B_i, $i = 1, \ldots, n$ as follows. We begin by defining the periods

$$\begin{aligned} \omega_{ij} &= \frac{\partial a_i}{\partial u_j} = \frac{1}{2\pi i} \oint_{A_i} \frac{\partial d\lambda}{\partial u_j} \\ \omega_{Dij} &= \frac{\partial a_{Di}}{\partial u_j} = \frac{1}{2\pi i} \oint_{B_i} \frac{\partial d\lambda}{\partial u_j} \end{aligned} \tag{4.10}$$

where we have identified the periods ω and ω_D with the relevant super-Yang–Mills quantities. The gauge coupling constant matrix τ may now be naturally connected with the Riemann surface data by

$$\tau = \frac{\partial a_D}{\partial u} \left(\frac{\partial a}{\partial u} \right)^{-1}. \tag{4.11}$$

By making this identification, we guarantee that $\mathrm{Im}(\tau) > 0$, $\tau_{ij} = \tau_{ji}$ and holomorphicity of τ_{ij} as a function of u_i. Recall that these were the key ingredients in the Seiberg–Witten construction. Upon integrating in u_k, we obtain the fundamental relations

$$a_i = \frac{1}{2\pi i} \oint_{A_i} d\lambda, \quad a_{Di} = \frac{1}{2\pi i} \oint_{B_i} d\lambda. \tag{4.12}$$

• For the simplest cases, to be explained shortly, Γ is precisely of genus $g = n$, and A_i, B_i, $i = 1, \ldots, n$, form a *basis* of $H_1(\Gamma, \mathbb{Z})$. But more generally, $n \leq g$, and A_i, B_i form a suitable subset of $H_1(\Gamma, \mathbb{Z})$.

• A crucial requirement of the Seiberg–Witten and Riemann surface correspondence is that the prepotential $\mathcal{F}$ defined by the Riemann surface data, $a_{Dk} = \partial\mathcal{F}/\partial a_k$, must have the logarithmic singularities $\mathcal{F}^{\rm pert}$ predicted by perturbative field theory in the regime of large vacuum expectation values. For $\mathcal{G} = \mathrm{SU}(N)$ and no hypermultiplet, these singularities have been described in (3.31). For general gauge algebra $\mathcal{G}$ with roots $\mathcal{R}(\mathcal{G})$ and hypermultiplet of mass m in a representation R of $\mathcal{G}$ with weights $\mathcal{W}(R)$, they are given by

$$\mathcal{F}^{\rm pert}(a) = \frac{i}{8\pi}\Bigg[\sum_{\alpha\in\mathcal{R}(\mathcal{G})}(\alpha\cdot a)^2\ln\frac{(\alpha\cdot a)^2}{\mu^2} - \sum_{\lambda\in\mathcal{W}(R)}(\lambda\cdot a+m)^2\ln\frac{(\lambda\cdot a+m)^2}{\mu^2}\Bigg]. \tag{4.13}$$

In view of the above correspondence, obtaining the Seiberg–Witten solution for any given gauge algebra and hypermultiplet representation has been reduced to obtaining the Seiberg–Witten curve $\Gamma(u_i)$, the Seiberg–Witten differential $d\lambda$ and the cycles A_i, B_i on $\Gamma(u_i)$. The monodromy problem is in general very difficult to solve, and more indirect arguments are needed to proceed. The key considerations that have been made use of in the construction for $\mathrm{SU}(N)$, $\mathrm{SO}(N)$ and $\mathrm{Sp}(N)$ gauge algebras, with either no hypermultiplets or hypermultiplets in the fundamental representation of the gauge algebra are as follows.

- Matching the singularity structure of the curve $\Gamma(u_i)$ with the appearance of massless particles in the $\mathcal{N} = 2$ super-Yang–Mills theory;
- $U(1)_R$ charge assignments;
- Decoupling limits to smaller gauge algebras and smaller hypermultiplet contents, including to the smallest of them all, namely the SU(2) gauge algebra which is known from the original Seiberg–Witten work;
- Educated guesswork.

4.3 SU(N) *Gauge Algebras, Fundamental Hypermultiplets*

We treat first the case of $\mathcal{G} = \mathrm{SU}(N)$ gauge theory (for which $n = N - 1$), and N_f hypermultiplets in the fundamental representation of $\mathrm{SU}(N)$. We shall limit ourselves to asymptotically free theories, for which the number of fundamental hypermultiplets is constrained by $N_f < 2N_c$. The curve proposed in [25] is given by

$$\Gamma(u_i) = \{(k,y); y^2 = A(k)^2 - \bar\Lambda^2 B(k)\}, \quad \bar\Lambda \equiv \Lambda^{N-N_f/2} \tag{4.14}$$

where the functions A and B are given by

$$A(k) = \prod_{i=1}^{N} (k - u_i), \quad B(k) = \prod_{\alpha=1}^{N_f} (k + m_\alpha), \tag{4.15}$$

and m_α are the hypermultiplet masses. The Seiberg–Witten differential is defined to be

$$d\lambda = k d \ln\bigl(y + A(k)\bigr). \tag{4.16}$$

We observe that even when $N_f = 0$ and $N = 2$, this curve and differential are different from the curve found originally for SU(2) in §3.8. Indeed, the Seiberg–Witten curves associated with a given gauge theory are not unique, although they may satisfy the same criteria outlined above. We shall check below that these criteria are satisfied by the curve (4.14) for the SU(N) gauge theory with N_f hypermultiplets in the fundamental representation.

When $\Lambda = 0$, the curve degenerates to two copies of the complex plane given by $y = \pm A(k)$, which have a double intersection precisely (and only) at the zeroes u_i of $A(k)$. As $\Lambda \neq 0$, the double zeros open up into branch cuts. The end points $x_k^{\pm}$ of these branch cuts are defined by

$$A(x_k^{\pm})^2 - B(x_k^{\pm}) = 0. \tag{4.17}$$

There will be precisely N branch cuts, so that the genus of the curve is $n = N - 1$. Thus, the set of cycles A_i and B_i may be chosen to be a basis of $H_1(\Gamma, \mathbb{Z})$. The Seiberg–Witten differential may be written as

$$d\lambda = \frac{k\, dk}{y} \left(A'(k) - \frac{1}{2} A(k) \frac{B'(k)}{B(k)} \right), \tag{4.18}$$

since this expression differs from the one in (4.16) by the term $\bar{\Lambda}^2\, kB'(k) \times dk/B(k)$, which has zero periods. It is easy to see that the residues of $d\lambda$ are independent of u_i, since the only poles (at finite k) appear at $k = -m_\alpha$ through the term inverse in $B(k)$. Now, around $k = -m_\alpha$, we have $y \sim A(k)$, and the behavior of $d\lambda$ is given by

$$d\lambda \sim \frac{1}{2} \frac{m_\alpha\, dk}{k + m_\alpha}. \tag{4.19}$$

Similarly, the pole at $k = \infty$ has a residue independent of u_i and thus $\partial d\lambda / \partial u_i$ is holomorphic.

There are a number of simple checks on the validity of this curve. First, by making a hypermultiplet mass, say m_{N_f} large, while keeping $m_{N_f} \bar{\Lambda}^2$ fixed, we decouple precisely the hypermultiplet with mass m_{N_f}. The limit of the corresponding curve agrees with the curve with one less hypermultiplet. Second, by letting one of the vacuum expectation values become large, one obtains a limit in which the gauge group SU(N) $\to$ SU($N-1$). Again the corresponding limits of the curves agree. Lastly, it was argued based upon the Picard-Lefschetz construction, that the strong coupling monodromies agree with the values expected from the dyon quantum numbers in the theory.

However, the true check which we need to carry out is that the family of Riemann surfaces reproduce the correct logarithmic singularities found in the perturbative regime for the prepotential $\mathcal{F}^{\rm pert}$ (cf. (4.13)). This is the regime where Λ is small compared to the gauge scalar vacuum expectation values a_i and the hypermultiplet masses m_α.

The first step is to derive the quantum order parameters $a_k = (2\pi i)^{-1} \times \oint_{A_k} d\lambda$ in terms of their classical limits u_k. Now the cycles A_k, $2 \le k \le N$, can be chosen to be contours surrounding the branch cuts between x_k^- and x_k^+, which remain at a fixed distance from u_k while the branch cuts themselves shrink to u_k as $\bar\Lambda \to 0$. This means that for Λ small compared to u_k, we may expand

$$d\lambda = k\left(\frac{A'}{A} - \frac{1}{2}\frac{B'}{B}\right) \sum_{m=0}^{\infty} \frac{\Gamma(m+\frac{1}{2})}{\Gamma(\frac{1}{2})m!} k \left(\frac{B}{A^2}\right)^m dk. \tag{4.20}$$

This reduces $d\lambda$ to a rational differential on a sphere with punctures. The method of residues applies readily to the evaluation of the periods around A_k, and we find

$$a_i = u_i + \sum_{m=1}^{\infty} \frac{\bar\Lambda^{2m}}{2^{2m}(m!)^2} \left(\frac{\partial}{\partial u_i}\right)^{2m-1} S_i(u_i;u)^m. \tag{4.21}$$

The function $S_i(k;u)$ is defined by

$$S_i(k;u) = \bar\Lambda^{-2}(k-u_i)^2 \frac{B(k)}{A(k)^2} = \frac{\prod_\alpha (k+m_\alpha)}{\prod_{l\neq i}(k-u_l)^2}. \tag{4.22}$$

The evaluation of the dual variables $a_{Dk} = (2\pi i)^{-1} \oint_{B_k} d\lambda$ is more difficult. This is because the cycles B_k correspond to paths going from x_1^- to x_k^- on each sheet, unlike the A_k cycles which can be chosen at a fixed distance from the cut x_k^-, x_k^+ cut as $\bar\Lambda \to 0$. The expansion (4.20) cannot be applied to the evaluation of the B_k periods as it stands. However, we can restore its validity by introducing a parameter ξ with $|\xi|$ small, and deform $d\lambda$ to

$$\begin{aligned} d\lambda(\xi) &= k\left(\frac{A'}{A} - \frac{1}{2}\frac{B'}{B}\right)\left(1 - \xi^2 \frac{B}{A^2}\right)^{-1/2} dk \\ &= \sum_{m=0}^{\infty} \frac{\Gamma(m+\frac{1}{2})}{\Gamma(\frac{1}{2})m!} k \left(\frac{A'}{A} - \frac{1}{2}\frac{B'}{B}\right)\left(\frac{\xi^2 B}{A^2}\right)^m dk. \end{aligned} \tag{4.23}$$

The main advantage is that for fixed small ξ, the expansion is reliable and the evaluation of the periods of $d\lambda(\xi)$ reduce to integrals of rational differentials, just as before in the evaluation of the A_k-periods (in practice, for each order ξ^{2m}, it suffices to retain terms with poles of order up to $2m$). The periods a_{Dk} of $d\lambda$ are then obtained by analytic continuation to $\xi = 1$.

It is now easy to identify the logarithmic terms in the resulting prepotential. Clearly, we have

$$d\lambda = \frac{1}{2}\sum_{\alpha=1}^{N_f} \frac{m_\alpha\, dk}{k+m_\alpha} + \sum_{l=1}^{N} a_l \frac{dk}{k-u_l} + \mathcal{O}(\Lambda) \tag{4.24}$$

since the residues of $d\lambda$ at $k + m_\alpha = 0$ are known to be $\frac{1}{2}m_\alpha$, while the periods a_i of $d\lambda$ around the A_i cycles tend to the residues of $d\lambda$ at u_i as $\Lambda \to 0$. The B_i periods of $d\lambda$ are

$$\int_{x_1^-}^{x_i^-} d\lambda = \frac{1}{2}\sum_{\alpha=1}^{N_f} m_\alpha \ln(x_i^- + m_\alpha) + \sum_{l=1}^{N} a_l \ln(x_i^- - u_l) + \mathcal{O}(\Lambda) \tag{4.25}$$

(where we have omitted, here as well as below, a similar expression with x_k^- replaced by x_1^-, arising from the lower bound of the integral). This can be rewritten as

$$\int_{x_1^-}^{x_i^-} d\lambda = a_i \ln\bar{\Lambda} + \frac{1}{2}\sum_{\alpha=1}^{N_f} (a_i + m_\alpha)\ln(x_i^- + m_\alpha) - \sum_{l=1}^{N}(a_i - a_l)\ln(x_i^- - u_l) + \mathcal{O}(\Lambda) \tag{4.26}$$

since the branch points x_i^- satisfy

$$\prod_{l=1}^{N}(x_i^- - u_l)^2 = \bar{\Lambda}^2 \prod_{j=1}^{N_f}(x_i^- + m_j). \tag{4.27}$$

Since u_i, x_i^- are analytic functions of a_j which differ from a_i by $\mathcal{O}(\bar{\Lambda})$ only, it follows that

$$\int_{k_1^-}^{k_i^-} d\lambda = a_i \ln\bar{\Lambda} + \frac{1}{2}\sum_{\alpha=1}^{N_f} (a_i + m_\alpha)\ln(a_i + m_\alpha) - \sum_{l=1}^{N}(a_i - a_l)\ln(a_i - a_l) + O(\Lambda), \tag{4.28}$$

where the terms $\mathcal{O}(\bar{\Lambda})$ are analytic functions of $\bar{\Lambda}$. Thus there are no higher powers of $\ln\bar{\Lambda}$ in the prepotential, in agreement with the nonrenormalization theorem for $\mathcal{N} = 2$ supersymmetric gauge theories. In view of the constraint $\sum_{k=1}^{N} a_k = 0$, we may consider a prepotential $\mathcal{F}(a_1, \ldots, a_N)$

which is a function of all variables a_k, whose restriction to the hyperplane $\sum_{k=1}^N a_k = 0$ is the usual prepotential. In terms of this prepotential $\mathcal{F}(a_1, \ldots, a_N)$, we have $a_{Dk} = \partial\mathcal{F}(a_1, \ldots, a_N)/\partial a_k - \partial\mathcal{F}(a_1, \ldots, a_N)/\partial a_1$. The expressions found in (4.28) identify the perturbative part of the prepotential

$$\mathcal{F}^{\text{pert}} = -\frac{1}{8\pi i}\left(\sum_{l,m=1}^{N} (a_l - a_m)^2 \ln \frac{(a_l - a_m)^2}{\Lambda^2} \right. \\ \left. - \sum_{l=1}^{N}\sum_{j=1}^{N_f} (a_l + m_j)^2 \ln \frac{(a_l + m_j)^2}{\Lambda^2}\right) \quad (4.29)$$

which is exactly the expression (4.13) required for the SU(N) gauge theory with N_f hypermultiplets of masses m_j, $1 \leq j \leq N_f$, in the fundamental representation.

The same methods can be used to determine explicitly the successive instanton corrections to the prepotential, corresponding to the successive terms $\mathcal{O}(\bar{\Lambda}^d)$ in (4.23). Clearly, the only remaining ingredient we need is a full expansion of the branch points x_k^- in terms of u_k and Λ. For this, we recast the defining equation (4.17) for the branch points as a fixed point equation

$$x_k^{\pm} = u_k \pm \bar{\Lambda} S(x_k^{\pm})^{1/2}. \quad (4.30)$$

The exact solution of this fixed point equation is given by

$$x_k^{\pm} = u_k + \sum_{m=1}^{\infty} \frac{(\pm)^m \bar{\Lambda}^m}{m!} \left(\frac{\partial}{\partial u_k}\right)^{m-1} S_k(u_k)^{m/2}. \quad (4.31)$$

It is now clear that we can evaluate the periods of the Seiberg–Witten differential to any order in $\bar{\Lambda}$. The answers will be in terms of the *classical* moduli parameters u_k. To get the correct prepotential, as we already saw in the determination of $\mathcal{F}^{\text{pert}}$, it is crucial that all expressions be recast in terms of the *quantum* moduli a_k. This can be done routinely, since the equation (4.21) allows to write a_k and u_k in terms of each other. Although the calculations can become very cumbersome beyond one instanton order, the methods are straightforward and have a certain flexibility. Originally developed in the context of hyperelliptic curves [26, 27], they have been now extended successfully by I.P. Ennes, S.G. Naculich, H. Rhedin, and H.J. Schnitzer to the study of more complicated spectral curves than (4.14). These include the ones corresponding to matter in the symmetric and the antisymmetric representation of SU(N) [28], as well as product groups [29], and other curves arising from M Theory. Another method which has been used in the determination of the prepotential from Seiberg–Witten curves is the method of Picard–Fuchs equations. These methods give exact differential equations satisfied by the periods of the Seiberg–Witten differential, but they can get complicated very quickly as the genus of the curve increases [30].

However, the prepotential obeys a simple and remarkable renormalization group type equation [31]

$$\left.\frac{\partial \mathcal{F}}{\partial \Lambda}\right|_{a_i} = \frac{2N - N_f}{2\pi i} \sum_{i=1}^{N} u_i^2, \tag{4.32}$$

which can actually be used to determine explicitly all instanton corrections, once the dependence of u_i is known in terms of the quantum moduli a_i (Simpler cases of (4.32) had been found in [32]. A broad framework for such equations had also been provided in [34], in the context of the tau function for soliton equations).

We shall limit ourselves here to exhibiting the expression for the prepotential up to 2-instanton order, and to the cases where $N_f < 2N$

$$\mathcal{F} = \mathcal{F}^{(0)} + \mathcal{F}^{(1)} + \mathcal{F}^{(2)} + \mathcal{O}(\bar{\Lambda}^6). \tag{4.33}$$

Here, $\mathcal{F}^{(0)}$ is the classical and perturbative contribution

$$\begin{aligned} 2\pi i \mathcal{F}^{(0)} = & -\frac{1}{4}\sum_{j\neq i}(a_i - a_j)^2 \ln \frac{(a_i - a_j)^2}{\Lambda^2} - (\ln 2 - 2N + N_f)\sum_i a_i^2 \\ & + \frac{1}{4}\sum_{i,\alpha}(a_i + m_\alpha)^2 \ln \frac{(a_i + m_\alpha)^2}{\Lambda^2}, \end{aligned} \tag{4.34}$$

$\mathcal{F}^{(1)}$ is the one instanton contribution, given by

$$2\pi i \mathcal{F}^{(1)} = \frac{\bar{\Lambda}^2}{4} \sum_i S_i(a_i; a), \tag{4.35}$$

and $\mathcal{F}^{(2)}$ is the two instanton contribution, given by

$$2\pi i \mathcal{F}^{(2)} = \frac{\bar{\Lambda}^4}{16}\left[\sum_{j\neq i} \frac{S_i(a_i;a)S_j(a_j;a)}{(a_i - a_j)^2} + \frac{1}{4}\sum_i S_i(a_i;a)\frac{\partial^2 S_i(a_i;a)}{\partial a_i^2}\right]. \tag{4.36}$$

Here, the function $S_i(k; a)$ is as defined in (4.22), but with u_i replaced by a_i. There is perfect agreement between the quantum field theory calculations and these results derived directly from Seiberg–Witten theory [26].

We note also that a set of linear recursion relations was very recently obtained for the prepotential, from which instanton corrections to very high order may be read off almost without calculations [35]. In a related direction, the renormalization group equation can be imbedded into a full hierarchy of equations satisfied by the prepotential [34]. These equations can also be exploited to derive instanton corrections [36]. For SU(N) theories with hypermultiplets in the fundamental representation, the prepotential has also been shown to satisfy WDVV equations [33]. Clearly, all these properties are indicative of a deeper underlying structure which is still yet to be fully uncovered. Finally, we note that the renormalization group methods may also be used to perform systematic expansions at strong coupling, for small values of the quantum moduli parameters [37]. These results have found exciting applications to twisted topological supersymmetric gauge theories as well [38]. Various reviews may be found in [39].

4.4 Classical Gauge Algebras, Fundamental Hypermultiplets

The curves $\Gamma(u_i)$ for the other classical gauge groups $\mathrm{SO}(N)$ and $\mathrm{Sp}(N)$ have also been constructed [40], and we have

$$A(k) = k^a \prod_{i=1}^{n} (k^2 - u_i^2) \quad B(k) = \Lambda^c k^b \prod_{\alpha=1}^{N_f} (k^2 - m_\alpha^2). \tag{4.37}$$

The values of the exponents a, b and c are given in Table 1.6.

For simplicity, we have here restricted for $\mathrm{Sp}(2n)$ to the case where we have at least two massless hypermultiplets. The general case was treated in [27].

In the same paper [27], it was shown that the prepotentials for the $\mathrm{SO}(N)$ and $\mathrm{Sp}(N)$ cases could be simply expressed in terms of the prepotentials for the $\mathrm{SU}(N)$ cases. One finds for the case $\mathcal{G} = \mathrm{SO}(2n+1)$,

$$\begin{aligned} &\mathcal{F}_{\mathrm{SO}(2n+1);N_f}(a_1, \ldots, a_n; m_1, \ldots, m_{N_f}; \Lambda) \\ &\quad = \mathcal{F}_{\mathrm{SU}(2n);2N_f+2}(a_1, \ldots, a_n, -a_1, \ldots, -a_n; \\ &\qquad\qquad m_1, \ldots, m_{N_f}, -m_1, \ldots, -m_{N_f}, 0, 0; \Lambda) \end{aligned} \tag{4.38}$$

for $\mathrm{Sp}(2n)$, with at least two massless hypermultiplets,

$$\begin{aligned} &\mathcal{F}_{\mathrm{Sp}(2n);N_f}(a_1, \ldots, a_n; m_1, \ldots, m_{N_f-2}, 0, 0; \Lambda) \\ &\quad = \mathcal{F}_{\mathrm{SU}(2n);2N_f-4}(a_1, \ldots, a_n, -a_1, \ldots, -a_n; \\ &\qquad\qquad m_1, \ldots, m_{N_f-2}, -m_1, \ldots, -m_{N_f-2}; \Lambda) \end{aligned} \tag{4.39}$$

TABLE 1.6. Exponents entering Seiberg–Witten curves for classical gauge algebras

$\mathcal{G}$	a	b	c
$\mathrm{SU}(n+1)$	0	0	$2n+2-2N_f$
$\mathrm{SO}(2n+1)$	0	2	$4n-2-2N_f$
$\mathrm{Sp}(2n)$	2	0	$4n+4-2N_f$
$\mathrm{SO}(2n)$	0	4	$4n-4-2N_f$

and, finally, for SO$(2n)$, we have

$$\begin{aligned}\mathcal{F}_{\mathrm{SO}(2n);N_f}&(a_1,\ldots,a_n;m_1,\ldots,m_{N_f};\Lambda)\\ &=\mathcal{F}_{\mathrm{SU}(2n);2N_f+4}(a_1,\ldots,a_n,-a_1,\ldots,-a_n;\\ &\qquad m_1,\ldots,m_{N_f},-m_1,\ldots,-m_{N_f},0,0,0,0;\Lambda).\end{aligned}\tag{4.40}$$

Again, these Seiberg–Witten expressions agree perfectly with the perturbative results known directly from quantum field theory.

5 Mechanical Integrable Systems

We consider Hamiltonian mechanical systems with a finite number of degrees of freedom, namely positions x_i and momenta p_i, $i = 1, \ldots, n$. The Hamiltonian, Poisson bracket and Hamilton equations are as always

$$\begin{aligned} &H(x_i,p_i), &\qquad \dot{x}_i = \{x_i, H\} = \frac{\partial H}{\partial p_i},\\ &\{x_i,p_j\} = \delta_{ij}, &\qquad \dot{p}_i = \{p_i, H\} = -\frac{\partial H}{\partial x_i}.\end{aligned}\tag{5.1}$$

While for the purposes of classical mechanics, these quantities and equations are customarily used with real valued x_i, p_i and H, there is absolutely no difficulty in letting x_i, p_i and H be complex. For our purposes, derived from Seiberg–Witten theory, the natural setting is complex analyticity and it will be crucial to allow for this complex generalization.

A Hamiltonian mechanical system defined in (5.1) is (*completely*) *integrable* if and only if there exist precisely n functionally independent *integrals of motion* $I_i(x,p)$, $i = 1, \ldots, n$, such that

$$\dot{I}_i = \{I_i, H\} = 0, \quad \{I_i, I_j\} = 0. \tag{5.2}$$

(More precisely, the system is integrable if the first relation holds, while it is completely integrable when both equations hold.) For a practical guide to integrable systems, see for example [41]. As a result of the existence of precisely as many integrals of motion as there are position degrees of freedom x_i, it is possible to perform a canonical transformation from the variables x_i, p_i to *action-angle variables*, I_i, ψ_i, $i = 1, \ldots, n$, for which time evolution is linear

$$I_i(t) = I_i(0), \quad \psi_i(t) = \psi_i(0) + c_i I_i t \tag{5.3}$$

for some constants c_i. In other words, for a completely integrable system, in action-angle variables, time evolution reduces to a linear flow on an n-dimensional torus, parameterized by the angle variables ψ_i.

There is no general classification of Hamiltonians $H(x_i, p_i)$ which correspond to completely integrable systems. There is also no systematic test to determine whether a given Hamiltonian $H(x_i, p_i)$ is integrable or not. Thus, deciding whether a given system is integrable is a difficult and challenging problem in mechanics.

However, there is a general correspondence that has been used very fruitfully in the study of integrable systems. This is the case whenever there exists a *Lax pair*. A Lax pair is a pair of $N \times N$ matrix-valued functions $L(x,p)$ and $M(x,p)$ which depend upon the dynamical variables x and p, such that the Lax equation $\dot{L} = [L, M]$ is equivalent to the Hamilton equations of the Hamiltonian mechanical system associated with H,

$$\dot{L} = [L, M] \Longleftrightarrow \begin{pmatrix} \dot{x}_i = \{x_i, H\} \\ \dot{p}_i = \{p_i, H\} \end{pmatrix} . \tag{5.4}$$

The dimension N of the Lax matrices is not a priori known, and there is again no systematic test or algorithm to establish the existence of a Lax pair, even less to find it explicitly. A Lax pair is not unique, since one is always free to perform a gauge transformation on L and M by

$$L^S = S^{-1} L S, \quad M^S = S^{-1} M S - S^{-1} \dot{S}. \tag{5.5}$$

As soon as a Lax pair is known to exist for a given Hamiltonian system, integrals of motion immediately follow. Indeed, traces of powers of L are found to be time independent as well as gauge invariant,

$$I_i \equiv \operatorname{tr} L^{n_i} \implies \dot{I}_i = n_i \operatorname{tr}(L^{n_i - 1}[L, M]) = 0. \tag{5.6}$$

Here, the traces of L can be defined for all positive integer values n_i, and all such I_i are integrals of motion. However, since the matrix L is of dimension N, we know from Cayley's theorem that all I_i with $n_i \geq N$ will be functionally dependent upon traces of lower powers. Thus, to get "enough" integrals of motion for complete integrability, we need $n \leq N$. For all known integrable systems, this condition appears to be fullfilled, though we are not aware of any general theorem to that effect. Amongst all traces of powers of L, we choose n functionally independent I_i, $i = 1, \ldots, n$ for certain choises of integers n_i. In particular, the original Hamiltonian will be functionally dependent upon these integrals of motion

$$H = h(I_1, \ldots, I_n) \tag{5.7}$$

since it is time-independent itself.

5.1 Lax Pairs with Spectral Parameter—Spectral Curves

A stronger form of integrability which exists in many known integrable systems, is the fact that the Lax operators L and M may depend upon an additional (complex) *spectral parameter*, usually denoted by z. The Lax operators become then matrix-valued functions $L(z)$, $M(z)$ of z. The Hamiltonian H and the whole mechanical system does not depend upon z, yet for all values of z, $L(z)$ and $M(z)$ are a Lax pair,

$$\dot{L}(z) = [L(z), M(z)] \iff \begin{pmatrix} \dot{x}_i = \{x_i, H\} \\ \dot{p}_i = \{p_i, H\} \end{pmatrix}. \tag{5.8}$$

A Lax pair $L(z)$, $M(z)$ obeying these relations is called a *Lax pair with spectral parameter*. Again, there is no systematic way to establish the existence of a Lax pair with spectral parameter, and there is no algorithm for finding it, even if the system is known to be integrable or even if a Lax pair without spectral parameter is already known explicitly.

To any Lax pair with spectral parameter, there is naturally associated a *spectral curve*, defined by

$$\Gamma = \{(k, z) \in \mathbb{C} \times \mathbb{C}; \det(kI - L(z)) = 0\}, \tag{5.9}$$

and a natural one-form $d\lambda$, defined by

$$d\lambda = k\,dz. \tag{5.10}$$

While $L(z)$ and $M(z)$ depend upon $2n$ phase space variables x_i and p_i, $i = 1, \ldots, n$, the spectral curve is time-independent and gauge invariant and depends only upon the n functionally independent integrals of motion I_i. Furthermore, if $L(z)$ is a meromorphic function of z (up to a gauge transformation, which may or may not be meromorphic), then the form $d\lambda$ is a meromorphic form on the spectral curve Γ. Depending on the periodicity properties of the Lax pair $L(z)$, $M(z)$ with respect to z, the domain of definition of z may be more properly viewed as the sphere of the torus, and a different variable than z may be more suitable in the definition 5.10 for the Seiberg–Witten differential (examples of this are discussed in subsequent sections). The curve Γ is correspondingly a branched covering of the sphere or of the torus.

5.2 The Toda Systems

The *nonperiodic Toda system* is a nonrelativistic system of $n+1$ points on a linear chain, with exponential nearest neighbor interactions (see for example [41]). The Hamiltonian is given by

$$H = \frac{1}{2}\sum_{i=1}^{n+1} p_i^2 - M^2 \sum_{i=1}^{n} e^{x_{i+1} - x_i}. \tag{5.11}$$

By translation invariance of the x_i by a common constant, it is clear that the center of mass $x_0 = \sum_i x_i$ degree of freedom may be decoupled from the system, effectively leaving a system of n coupled degrees of freedom. The simplest nonperiodic system with $n = 1$ reduces to the Liouville system.

The *periodic Toda system* is a nonrelativistic system of $n + 1$ points on a circular chain, with exponential nearest neighbor interactions, given by

$$H = \frac{1}{2} \sum_{i=1}^{n+1} p_i^2 - M^2 \sum_{i=1}^{n+1} e^{x_{i+1} - x_i}, \tag{5.12}$$

with the periodicity condition $x_{n+2} = x_1$. Again, the center of mass coordinate decouples leaving a coupled system of n degrees of freedom. The simplest periodic Toda system is for $n = 1$ and reduces to the sine-Gordon model.

Actually, both Toda systems admit a natural Lie algebraic interpretation. This may be seen by viewing the linear chain of the nonperiodic system as the Dynkin diagram for the Lie algebra $\mathrm{A}_n \sim \mathrm{SU}(n+1)$, and the circular chain of the periodic Toda system as the Dynkin diagram for the untwisted affine Lie algebra $\mathrm{A}_n^{(1)}$. The set $\mathcal{R}_*$ of *simple roots* for each Lie algebra is well known:

$$\begin{array}{lll} \mathrm{A}_n : & e_i - e_{i+1}, \quad i = 1, \ldots, n & \\ \mathrm{A}_n^{(1)} : & e_i - e_{i+1}, \quad i = 1, \ldots, n+1, & e_{r+2} = e_1. \end{array} \tag{5.13}$$

In terms of the simple roots, and the vector notation for position and momentum variables, $x = (x_1, \ldots, x_{n+1})$ and $p = (p_1, \ldots, p_{n+1})$, both Toda Hamiltonians take on a unified form,

$$H = \frac{1}{2} p^2 - M^2 \sum_{\alpha \in \mathcal{R}_*} e^{-\alpha \cdot x}. \tag{5.14}$$

Lie algebraically, the nonperiodic Toda system is associated with the finite dimensional Lie algebra A_n, while the periodic Toda system is associated with the infinite dimensional untwisted affine Lie algebra $\mathrm{A}_n^{(1)}$.

Now, for any finite-dimensional semi-simple Lie algebra $\mathcal{G}$ or one of the associated untwisted $\mathcal{G}^{(1)}$ or twisted affine Lie algebras, we may associate a Toda system based on the set of simple roots $\mathcal{R}_*$ of the Lie algebra. For a Lie algebra $\mathcal{G}$ of rank n, we introduce the position $x = (x_1, \ldots, x_n)$ and momentum $p = (p_1, \ldots, p_n)$ variables and define the Toda Hamiltonian associated with a Lie algebra $\mathcal{G}$ (which may be finite dimensional or affine) in terms of the set of simple roots $\mathcal{R}_*(\mathcal{G})$, as follows,

$$H = \frac{1}{2} p^2 - M^2 \sum_{\alpha \in \mathcal{R}_*(\mathcal{G})} e^{-\alpha \cdot x}. \tag{5.15}$$

Remarkably, this Toda system is completely integrable for any Lie algebra $\mathcal{G}$.

In fact, for any finite dimensional Lie algebra $\mathcal{G}$, or its untwisted affine extension $\mathcal{G}^{(1)}$, the Toda system admits a Lax pair [42]. To exhibit this Lax pair, we need the following ingredients,

- $h = (h_1, \dots, h_n)$ is the array of Cartan generators of $\mathcal{G}$, with $[h_i, h_j] = 0$;
- E_α is the generators of $\mathcal{G}$ associated with the root α;
- $\mathcal{R}_* = \mathcal{R}_*(\mathcal{G})$ is the set of simple roots of $\mathcal{G}$;
- α_0 is the affine root (to be included for $\mathcal{G}^{(1)}$);
- a representation ρ of $\mathcal{G}$, so that all generators h_i, E_α can be viewed as N-dimensional matrices, if ρ is of dimension N.

The Lax matrices then take on the form

$$L = p \cdot h + \sum_{\alpha \in \mathcal{R}_*} M e^{-\alpha/2 \cdot x}(E_\alpha - E_{-\alpha}) + \mu^2 e^{\alpha_0/2 \cdot x}(z E_{-\alpha_0} - z^{-1} E_{\alpha_0}), \tag{5.16}$$

$$M = -\frac{1}{2} \sum_{\alpha \in \mathcal{R}_*} M e^{-\alpha/2 \cdot x}(E_\alpha + E_{-\alpha}) + \frac{\mu^2}{2} e^{\alpha_0/2 \cdot x}(z E_{-\alpha_0} + z^{-1} E_{\alpha_0}). \tag{5.17}$$

For a nonperiodic Toda system, associated with a finite dimensional Lie algebra $\mathcal{G}$, we set $\mu = 0$ in (5.16), and any dependence on the spectral parameter z disappears. For the periodic Toda system, associated with the untwisted affine Lie algebra $\mathcal{G}^{(1)}$, we have $\mu \neq 0$, and the Lax pair has a spectral parameter z.

For non-simply laced Lie algebras $\mathcal{G}$ (i.e. for which not all roots have the same lengths), one may also define a Toda system associated with *the twisted affine Lie algebra* $(\mathcal{G}^{(1)})^\dagger$. The roots of this algebra are the duals (or co-roots) to the roots of $\mathcal{G}^{(1)}$, and the twisted affine Toda system associated with these algebras is also integrable, and admit a Lax pair with operators given by (5.16) but with roots replaced with co-roots.

For the periodic Toda systems, we have a Lax pair with spectral parameter and thus we have a spectral curve, as defined in (5.9). For the case of $\mathrm{A}_n^{(1)}$ with ρ the fundamental representation, this curve may be worked out more explicitly. Indeed, the determinant that enters (5.9) is

$$\det\bigl(kI - L(z)\bigr) = A(k) - \frac{1}{2}\left(z + \frac{\bar{\Lambda}^2}{z}\right). \tag{5.18}$$

The simplicity of the z-dependence is due to the fact that the generators E_{α_0} and $E_{-\alpha_0}$ are matrices of rank 1, so that their coefficients enter linearly in the expansion of the determinant. The remaining polynomial $A(k)$ is independent of z and is of degree $N = n+1$. Since $\operatorname{tr} L = 0$, and k multiplies the identity matrix, the form of $A(k)$ is

$$A(k) = k^N + k^{N-2}J_2 + k^{N-3}J_3 + \cdots + J_N = \prod_{i=1}^{N}(k - u_i) \tag{5.19}$$

where the parameters J_i are functions of the integrals of motion $I_i = \operatorname{tr} L^i$ only. The integrals of motion can in principle be expressed in terms of the dynamical variables x_i and p_i, but we shall not need these relations here.

By a change of variables, $y = z - A(k)$, the equation for the curve (5.9) may be evaluated explicitly using (5.19), and we get

$$\Gamma = \{(k, y) \in \mathbb{C} \times \mathbb{C}, y^2 = A(k)^2 - \bar{\Lambda}^2\}. \tag{5.20}$$

Remarkably, this curve is identical to the Seiberg–Witten curve for an $\mathcal{N} = 2$ supersymmetric SU(N) gauge theory without hypermultiplets. In fact, also the Seiberg–Witten differential admits a very simple expression, given by

$$d\lambda = k\, d\ln z = \frac{A'(k) k\, dk}{\sqrt{A(k)^2 - \bar{\Lambda}^2}}. \tag{5.21}$$

The coincidence of the spectral curves of Toda systems and the Seiberg–Witten curves of $\mathcal{N} = 2$ supersymmetric Yang–Mills theory was first noted in [43], in the case of SU(2) without matter hypermultiplets. It is natural to speculate that, for all gauge groups $\mathcal{G}$, a similar correspondence should hold between Toda systems and Seiberg–Witten curves of $\mathcal{N} = 2$ supersymmetric Yang–Mills theories without matter hypermultiplets, for all gauge groups. However, due to subtleties related to the renormalization group flow, the correct correspondence is rather between Toda systems for the dual group $(\mathcal{G}^{(1)})^\vee$ and the $\mathcal{G}$ Yang–Mills theory, as shown in [44]. We shall come back in §6 to this emergence of the dual group $(\mathcal{G}^{(1)})^\vee$, in the context of Calogero–Moser systems.

5.3 *The Calogero–Moser Systems for* SU(N)

The Calogero–Moser systems are nonrelativistic mechanical models of $n+1$ particles on a (complex line) with two body interactions, which are not limited to nearest neighbor interactions. The original Calogero–Moser system comes in three varieties [45]

The Rational Calogero–Moser System

$$H = \frac{1}{2}\sum_{i=1}^{n+1} p_i^2 - \frac{1}{2}m^2 \sum_{i \neq j}^{n+1} \frac{1}{(x_i - x_j)^2}. \tag{5.22}$$

The Trigonometric Calogero–Moser System

$$H = \frac{1}{2}\sum_{i=1}^{n+1} p_i^2 - \frac{1}{2}m^2 \sum_{i\neq j}^{n+1} \frac{1}{\sin^2(x_i - x_j)}. \tag{5.23}$$

The Elliptic Calogero–Moser System

$$H = \frac{1}{2}\sum_{i=1}^{n+1} p_i^2 - \frac{1}{2}m^2 \sum_{i\neq j}^{n+1} \wp(x_i - x_j; \omega_1, \omega_2). \tag{5.24}$$

As is clear from the above definitions, the interaction potential of the rational and trigonometric systems is a rational and trigonometric (or simply periodic) function respectively, while for the elliptic systems, the potential is the Weierstrass elliptic (doubly periodic) function defined by

$$\wp(x; \omega_1, \omega_2) \equiv \frac{1}{x^2} + \sum_{(m,n)\neq(0,0)} \left(\frac{1}{(x + 2m\omega_1 + 2n\omega_2)^2} - \frac{1}{(2m\omega_1 + 2n\omega_2)^2} \right). \tag{5.25}$$

Here, $2\omega_1$ and $2\omega_2$ are the periods, and $\wp$ may be thought of as the electric field due to an electric dipole of unit strength on a torus with sides $2\omega_1$ and $2\omega_2$. The three systems are related by the following limits: as $\omega_2 \to \infty$ with ω_1 fixed, the elliptic system tends to the trigonometric one, while as in addition $\omega_1 \to \infty$, the system tends to the rational case. In view of these limiting relations, it suffices to carry out our analysis of integrability and construction of Lax pairs (with spectral parameter) for the elliptic Calogero–Moser systems. Integrability is preserved under these limits and will smoothly carry over to the trigonometric and rational cases.

It turns out also that the elliptic Calogero–Moser system is the system of interest for Seiberg–Witten theory. An important clue is the microscopic gauge coupling τ of (2.59). In theories where this microscopic gauge coupling does not get renormalized, it can be expected to play a central role in the prepotential and hence in the Seiberg–Witten curves. Thus for theories such as super Yang–Mills with matter in the adjoint representation (so that the beta function vanishes, in view of (3.7)), the correct corresponding integrable model should involve a torus with moduli τ. This requirement is satisfied by the Calogero–Moser systems in their elliptic versions, but not in their rational or trigonometric versions. The latter versions have appeared however in many other physical contexts. We refer to [46] for some recent applications.

The Calogero–Moser systems are naturally interpreted in terms of $\mathrm{A}_n = \mathrm{SU}(n+1)$ Lie algebra data. Consider the set $\mathcal{R}$ of *all roots of* A_n; in an orthonormal basis e_i, $i = 1, \dots, n+1$, they are given by

$$\mathcal{R}(\mathrm{A}_n) = \{e_i - e_j, i \neq j = 1, \dots, n+1\}. \tag{5.26}$$

In terms of the set of all roots of A_n and the position $x = (x_1, \ldots, x_{n+1})$ and momentum $p = (p_1, \ldots, p_{n+1})$ variables, we can recast the Hamiltonian as follows

$$H = \frac{1}{2}p^2 - \frac{1}{2}m^2 \sum_{\alpha \in \mathcal{R}} \wp(\alpha \cdot x; \omega_1, \omega_2). \tag{5.27}$$

It is easy to generalize the above Hamiltonian to other Lie algebras, but we shall postpone this discussion to §6, since a great number of subtleties arise in the integrability properties of these systems.

The elliptic Calogero–Moser system for $\mathcal{G} = A_n = SU(n+1)$ admits a Lax pair *with spectral parameter* z, first derived in [47],

$$L_{ij}(z) = p_i \delta_{ij} - m(1 - \delta_{ij})\Phi(x_i - x_j; z), \tag{5.28}$$

$$M_{ij}(z) = d_i(x)\delta_{ij} + m(1 - \delta_{ij})\Phi'(x_i - x_j; z). \tag{5.29}$$

Here $d_i(x) = m \sum_{k \neq i} \wp(x_i - x_k)$, $\Phi'(x; z) = \partial_x \Phi(x; z)$, and $\Phi(x; z)$ is a Lamé function, given by

$$\Phi(x; z) = \frac{\sigma(z - x)}{\sigma(x)\sigma(z)} e^{x\zeta(z)} \tag{5.30}$$

where the Weierstrass function ζ is related to $\wp$ by $\wp(z) = -\zeta'(z)$ and $\zeta(z) = \sigma'(z)/\sigma(z)$. The Lamé function $\Phi(x; z)$ satisfies the Lamé equation,

$$\left(\frac{d^2}{dz^2} - \wp(x) \right) \Phi(x; z) = 2\wp(z)\Phi(x; z). \tag{5.31}$$

This Lax pair and the corresponding spectral curve has sparked a rich theory of elliptic solitons and spectral covers. We refer to [48, 49] for some of these developments. Complete integrability and the existence of a classical R-matrix has been investigated in [50] and [51]. For us, the above Lax pair will be of fundamental importance for a different reason: the associated spectral curve and differential $d\lambda = k\,dz$ will turn out to give exactly the Seiberg–Witten curve for the $SU(N)$ $\mathcal{N} = 2$ super-Yang–Mills theory, with a hypermultiplet in the adjoint representation.

5.4 Relation between Calogero–Moser and Toda for SU(N)

Remarkably, Toda systems may be obtained as a limit of elliptic Calogero–Moser systems [52]. Qualitatively, the limit is expected to arise as follows: in Calogero–Moser systems, all roots of the algebra are being summed over, while for Toda systems, only the simple roots enter. Thus, upon taking the limit, all interactions that are not nearest neighbor must vanish, leaving only the nearest neighbor interaction of simple roots.

The limit is taken on the parameters m, ω_2 and on the position variables x_i by letting $\omega_1 = -i\pi$ and

$$\mathrm{Re}(\omega_2) \to \infty \qquad \begin{cases} m = Me^{\delta\omega_2}, \\ x_j = X_j + 2\omega_2\delta j, \\ M, X_j, 0 < \delta \leq 1/N \text{ fixed.} \end{cases} \tag{5.32}$$

A very convenient way to investigate the limit is to expand the $\wp$-function in terms of trigonometric functions

$$\wp(x; -i\pi; \omega_2) = \frac{1}{2} \sum_{k=-\infty}^{\infty} \frac{1}{\mathrm{ch}(x + 2k\omega_2) - 1}. \tag{5.33}$$

(for a derivation of this formula, see Appendix C). Since $\wp(x; \omega_1, \omega_2)$ is even in x, we restrict to $i < j$ without loss of generality and we have

$$m^2 \wp(x_i - x_j; -i\pi, 2\omega_2) = \frac{1}{2} \sum_{k=-\infty}^{\infty} \frac{M^2 e^{2\delta\omega_2}}{\mathrm{ch}\big[X_i - X_j + 2\omega_2\big(\delta(i-j) + k\big)\big] - 1}. \tag{5.34}$$

When $\delta > 1/N$, it is easy to see that no limit will exist, which is why we have excluded this case from the outset in (5.32).

Because we have $0 < \delta \leq 1/N$ all terms in the infinite sum (5.34), except $n = 0, 1$, converge to 0. The limits of the remaining terms are

$$k = 0 \text{ term} \longrightarrow \begin{cases} M^2 \exp(X_j - X_i), & j - i = 1, \\ 0, & j - i \neq 1 \end{cases} \tag{5.35}$$

and

$$k = 1 \text{ term} \longrightarrow \begin{cases} M^2 \exp(X_N - X_1), & \delta N = 1, \\ 0, & \delta N < 1. \end{cases} \tag{5.36}$$

The contribution from the $k = 0$ term emerges for any values of δ, and it should be manifest that this limit precisely corresponds to the Toda exponential interaction between nearest neighbors, i.e. based on the simple roots of $\mathrm{A}_n = \mathrm{SU}(n+1)$. The contribution from the $k = 1$ term on the other hand is absent for all values $\delta N < 1$. In this case, the limiting Toda system is based only on the simple roots of A_n and corresponds to the nonperiodic case. When $\delta N = 1$ on the other hand, the $k = 1$ term survives and yields the Toda exponential interaction for the affine root $-\alpha_0$. In this case, the limiting Toda system is based on the simple roots of A_n as well as the affine root $-\alpha_0$, and corresponds to the periodic Toda system, associated with the untwisted affine Lie algebra $\mathrm{A}_n^{(1)}$. In summary

$$\begin{cases} \delta N < 1 & \longrightarrow \text{nonperiodic Toda for } \mathrm{A}_n \\ \delta N = 1 & \longrightarrow \text{periodic Toda for } \mathrm{A}_n^{(1)}. \end{cases} \tag{5.37}$$

5.5 Relations with KdV and KP Systems

The Toda and Calogero–Moser systems are also connected with higher dimensional integrable systems. Some remarkable connections are:

Toda Field Theory The basic fields are $\phi = (\phi_1, \dots, \phi_n)$ with $\phi_i(t,x)$ 2-dimensional fields, $\partial_\mu = (\partial_t, \partial_x)$, and Lagrangian

$$\mathcal{L} = \frac{1}{2} \partial_\mu \phi \cdot \partial^\mu \phi - M^2 \sum_{\alpha \in \mathcal{R}_*} e^{\alpha \cdot \phi}. \tag{5.38}$$

These theories are integrable as field theories and the simplest cases are Liouville theory, which enters into the study of noncritical string theory and two-dimensional quantum gravity [53], as well as sine-Gordon theory which is equivalent to the massive Thirring model of fermions [54].

Korteweg–de Vries (KdV) equation The fundamental field is a single scalar $u(t,x)$ in 2 dimensions and the field equation is

$$\partial_t u = 6u\partial_x u - \partial_x^3 u. \tag{5.39}$$

Of special interest are *rational solutions*, which are of the form

$$u(t,x) = 2 \sum_j \frac{1}{\big(x - x_j(t)\big)^2}. \tag{5.40}$$

By a theorem of Airault–McKean–Moser [55], this Ansatz for u solves the KdV equation provided the positions $x_j(t)$ are subject to certain constraints as well as to the following time-evolution equation [47]

$$\partial_t x_j = \{x_j, \operatorname{tr} L^3\} \tag{5.41}$$

where L is the Lax operator for the rational Calogero–Moser system.

Kadomzev–Petviashvili (KP) equation The fundamental field is a scalar $u(t,t',x)$ in three dimensions (one space and two times) with the field equation

$$3\partial_{t'}^2 u = \partial_x(\partial_t u - 6u\partial_x u - \partial_x^3 u). \tag{5.42}$$

The t'-independent solutions are governed by the KdV equation. It has been shown by Krichever [47] that the elliptic solutions are of the form

$$u(t,t',x) = 2 \sum_j \wp\big(x - x_j(t,t')\big) \tag{5.43}$$

provided the positions $x_j(t,t')$ obey the following time-evolution equations

$$\partial_t x_j = \{x_j, \operatorname{tr} L^3\}, \quad \partial_{t'} x_j = \{x_j, \operatorname{tr} L^2\} \tag{5.44}$$

where L is the Lax operator for the elliptic Calogero–Moser system.

So far, these properties have not yet entered directly into the connection between integrable systems and Seiberg–Witten theory. However, it is now known that the symplectic forms arising in Seiberg–Witten theory from the differential $d\lambda$ can be viewed as restrictions to the space of algebraic-geometric solutions of symplectic forms for soliton equations [56, 57]. It is likely that more connections will emerge in the near future.

5.6 *Calogero–Moser Systems for General Lie Algebras*

As Olshanetsky and Perelomov [58] realized very early on, the Hamiltonian system (5.24) is only one example of a whole series of Hamiltonian systems associated with each simple Lie algebra. More precisely, given any simple Lie algebra $\mathcal{G}$, Olshanetsky and Perelomov [58] introduced the system with Hamiltonian

$$H(x,p) = \frac{1}{2}\sum_{i=1}^{n} p_i^2 - \frac{1}{2}\sum_{\alpha\in\mathcal{R}(\mathcal{G})} m_{|\alpha|}^2 \wp(\alpha\cdot x), \tag{5.45}$$

where n is the rank of $\mathcal{G}$, $\mathcal{R}(\mathcal{G})$ denotes the set of roots of $\mathcal{G}$, and the $m_{|\alpha|}$ are mass parameters. To preserve the invariance of the Hamiltonian (5.45) under the Weyl group, the parameters $m_{|\alpha|}$ depend only on the orbit $|\alpha|$ of the root α, and not on the root α itself. (For $\mathrm{A}_{N-1} = \mathrm{SU}(N)$, it is common practice as we saw earlier to use N pairs of dynamical variables (x_i, p_i), since the roots of A_{N-1} lie conveniently on a hyperplane in $\mathbb{C}^N$. The dynamics of the system are unaffected if we shift all x_i by a constant, and the number of degrees of freedom is effectively $N-1=n$.) As in the original $\mathrm{SU}(N)$ case, the elliptic systems (5.45) admit rational and trigonometric limits. Olshanetsky and Perelomov succeeded in constructing a Lax pair for all these systems in the case of classical Lie algebras, albeit without spectral parameter [58].

Twisted Calogero–Moser Systems defined by Lie Algebras

It turns out that the Hamiltonian systems (5.45) are not the only natural extensions of the basic elliptic Calogero–Moser system. A subtlety arises for simple Lie algebras $\mathcal{G}$ which are not simply laced, i.e., algebras which admit roots of uneven length. This is the case for the algebras B_n, C_n, G_2, and F_4 in Cartan's classification. For these algebras, the following *twisted* elliptic Calogero–Moser systems were introduced by the authors in [59, 60]

$$H_{\mathcal{G}}^{\text{twisted}} = \frac{1}{2}\sum_{i=1}^{n} p_i^2 - \frac{1}{2}\sum_{\alpha\in\mathcal{R}(\mathcal{G})} m_{|\alpha|}^2 \wp_{\nu(\alpha)}(\alpha\cdot x). \tag{5.46}$$

Here the function $\nu(\alpha)$ depends only on the length of the root α. If $\mathcal{G}$ is simply laced, we set $\nu(\alpha)=1$ identically. Otherwise, for $\mathcal{G}$ non-simply laced, we set $\nu(\alpha) = 1$ when α is a long root, $\nu(\alpha) = 2$ when α is a short root and $\mathcal{G}$ is one of the algebras B_n, C_n, or F_4, and $\nu(\alpha) = 3$ when α is a short root and $\mathcal{G} = \mathrm{G}_2$. The *twisted* Weierstrass function $\wp_\nu(z)$ is defined by

$$\wp_\nu(z) = \sum_{\sigma=0}^{\nu-1} \wp\left(z + 2\omega_a \frac{\sigma}{\nu}\right), \tag{5.47}$$

where ω_a is any of the half-periods ω_1, ω_2, or $\omega_1 + \omega_2$. Thus the twisted and untwisted Calogero–Moser systems coincide for $\mathcal{G}$ simply laced. For $\mathcal{G} = \mathrm{B}_n, \mathrm{C}_n$, the twisted Calogero–Moser models introduced here are equivalent to the systems introduced by Inozemtsev [52], while for $\mathcal{G} = \mathrm{F}_4, \mathrm{G}_2$, they are new integrable systems. The original motivation for twisted Calogero–Moser systems was based on their scaling limits (which will be discussed in the next section) [59, 60].

5.7 *Scaling of Calogero–Moser to Toda for General Lie Algebras*

The key feature of the scaling limit of the SU(N) Calogero–Moser system to the SU(N) Toda system was the collapse of the sum over the entire root lattice of A_{N-1} in the Calogero–Moser Hamiltonian to the sum over only simple roots in the Toda Hamiltonian for the Kac-Moody algebra $\mathrm{A}_{N-1}^{(1)}$. Our task is to extend this mechanism to general Lie algebras. It turns out that there are two possible extensions, depending on whether we use the Coxeter number or the dual Coxeter number [60].

Scaling Limits based on the Coxeter Number

For this, we consider the following generalization of the preceding scaling limit

$$m = M q^{-\frac{1}{2}\delta}, \tag{5.48}$$

$$x = X - 2\omega_2 \delta \rho^\vee. \tag{5.49}$$

Here $x = (x_i)$, $X = (X_i)$ and $\rho^\vee$ are n-dimensional vectors. The vector x is the dynamical variable of the Calogero–Moser system. The parameters δ and $\rho^\vee$ depend on the algebra $\mathcal{G}$ and are yet to be chosen. As for M and X, they have the same interpretation as earlier, namely as respectively the mass parameter and the dynamical variables of the limiting system. Setting $\omega_1 = -i\pi$, the contribution of each root α to the Calogero–Moser potential can be expressed as

$$m^2 \wp(\alpha \cdot x) = \frac{1}{2} M^2 \sum_{k=-\infty}^{\infty} \frac{e^{2\delta\omega_2}}{\mathrm{ch}(\alpha \cdot x - 2k\omega_2) - 1}. \tag{5.50}$$

It suffices to consider positive roots α. We shall also assume that $0 \leq \delta\alpha \cdot \rho^\vee \leq 1$. The contributions of the $k = 0$ and $k = -1$ summands in (5.50) are proportional to $\exp\{2\omega_2(\delta - \delta\alpha \cdot \rho^\vee)\}$ and $\exp\{2\omega_2(\delta - 1 + \delta\alpha \cdot \rho^\vee)\}$ respectively. Thus the existence of a finite scaling limit requires that

$$\delta \leq \delta\alpha \cdot \rho^\vee \leq 1 - \delta. \tag{5.51}$$

Let α_i, $1 \leq i \leq n$ be a basis of simple roots for $\mathcal{G}$. If we want all simple roots α_i to survive in the limit, we must require that

$$\alpha_i \cdot \rho^\vee = 1, \quad 1 \leq i \leq n. \tag{5.52}$$

This condition characterizes the vector $\rho^\vee$ as the *level vector*. Next, the second condition in (5.51) can be rewritten as $\delta\{1 + \max_\alpha(\alpha \cdot \rho^\vee)\} \leq 1$. But

$$h_{\mathcal{G}} = 1 + \max_\alpha(\alpha \cdot \rho^\vee) \tag{5.53}$$

is precisely the Coxeter number of $\mathcal{G}$, and we must have $\delta \leq 1/h_{\mathcal{G}}$. Thus when $\delta < 1/h_{\mathcal{G}}$, the contributions of all the roots except for the simple roots of $\mathcal{G}$ tend to 0. On the other hand, when $\delta = 1/h_{\mathcal{G}}$, the highest root α_0 realizing the maximum over α in (5.53) survives. Since $-\alpha_0$ is the additional simple root for the affine Lie algebra $\mathcal{G}^{(1)}$, we arrive in this way at the following theorem, which was proved in [60].

Theorem 1. *Under the limit* (5.48), *with* $\delta = 1/h_{\mathcal{G}}$, *and* $\rho^\vee$ *given by the level vector, the Hamiltonian of the elliptic Calogero–Moser system for the simple Lie algebra* $\mathcal{G}$ *tends to the Hamiltonian of the Toda system for the affine Lie algebra* $\mathcal{G}^{(1)}$.

Scaling Limit based on the Dual Coxeter Number

If the Seiberg–Witten spectral curve of the $\mathcal{N} = 2$ supersymmetric gauge theory with a hypermultiplet in the adjoint representation is to be realized as the spectral curve for a Calogero–Moser system, the parameter m in the Calogero–Moser system should correspond to the mass of the hypermultiplet. In the gauge theory, the dependence of the coupling constant on the mass m is given by

$$\tau = \frac{i}{2\pi} h_{\mathcal{G}}^\vee \ln \frac{m^2}{M^2} \iff m = Mq^{-1/(2h_{\mathcal{G}}^\vee)} \tag{5.54}$$

where $h_{\mathcal{G}}^\vee$ is the quadratic Casimir of the Lie algebra $\mathcal{G}$. This shows that the correct physical limit, expressing the decoupling of the hypermultiplet as it becomes infinitely massive, is given by (5.48), but with $\delta = 1/h_{\mathcal{G}}^\vee$. To establish a closer parallel with our preceding discussion, we recall that the quadratic Casimir $h_{\mathcal{G}}^\vee$ coincides with the *dual Coxeter number* of $\mathcal{G}$, defined by

$$h_{\mathcal{G}}^\vee = 1 + \max_\alpha(\alpha^\vee \cdot \rho), \tag{5.55}$$

where $\alpha^\vee = 2\alpha/\alpha^2$ is the coroot associated to α, and

$$\rho = \frac{1}{2}\sum_{\alpha>0} \alpha \tag{5.56}$$

is the well-known Weyl vector.

For simply laced Lie algebras $\mathcal{G}$ (ADE algebras), we have $h_\mathcal{G} = h_\mathcal{G}^\vee$, and the preceding scaling limits apply. However, for non-simply laced algebras (B_n, C_n, G_2, F_4), we have $h_\mathcal{G} > h_\mathcal{G}^\vee$, and our earlier considerations show that the untwisted elliptic Calogero–Moser Hamiltonians do not tend to a finite limit under (5.48), $q \to 0$, M is kept fixed. This is why the twisted Hamiltonian systems (5.46) have to be introduced. The twisting produces precisely to an improvement in the asymptotic behavior of the potential which allows a finite, nontrivial limit. More precisely, we can write

$$m^2 \wp_\nu(x) = \frac{\nu^2}{2} \sum_{n=-\infty}^{\infty} \frac{m^2}{\operatorname{ch} \nu(x - 2n\omega_2) - 1}. \tag{5.57}$$

Setting $x = X - 2\omega_2 \delta^\vee \rho$, we obtain the following asymptotics

$$m^2 \wp_\nu(x) = \nu^2 M^2 \begin{cases} e^{-2\omega_2(\delta^\vee \alpha^\vee \cdot \rho - \delta^\vee) - \alpha^\vee \cdot X} \\ \qquad + e^{-2\omega_2(1-\delta^\vee \alpha^\vee \cdot \rho - \delta^\vee) + \alpha^\vee \cdot X}, & \text{if } \alpha \text{ is long;} \\ e^{-2\omega_2(\delta^\vee \alpha^\vee \cdot \rho - \delta^\vee) - \alpha^\vee \cdot X}, & \text{if } \alpha \text{ is short.} \end{cases} \tag{5.58}$$

This leads to the following theorem [60].

Theorem 2. *Under the limit $x = X + 2\omega_2 \rho / h_\mathcal{G}^\vee$, $m = M q^{-1/(2h_\mathcal{G}^\vee)}$, with ρ the Weyl vector and $q \to 0$, the Hamiltonian of the twisted elliptic Calogero–Moser system for the simple Lie algebra $\mathcal{G}$ tends to the Hamiltonian of the Toda system for the affine Lie algebra $(\mathcal{G}^{(1)})^\vee$.*

So far we have discussed only the scaling limits of the Hamiltonians. However, similar arguments show that the Lax pairs constructed below also have finite, nontrivial scaling limits whenever this is the case for the Hamiltonians. The spectral parameter z should scale as $e^z = Zq^{\frac{1}{2}}$, with Z fixed. The parameter Z can be identified with the loop group parameter for the resulting affine Toda system.

6 Calogero–Moser Lax Pairs for General Lie Algebras

While the Lax pair for the elliptic SU(N) Calogero–Moser system was constructed [47] shortly after the proposal of the Calogero–Moser systems was made [58], the Lax pairs for elliptic Calogero–Moser systems associated with general Lie algebras were constructed only recently [59].

The key complication encountered when passing from SU(N) to other Lie algebras is the following. For SU(N), the Lax operators L and M are $N \times N$ matrices, and thus belong to the (complexified) Lie algebra of SU(N). In other words, the existence and calculation of a Lax pair is essentially a problem in the Lie algebra of SU(N). For general Lie algebra $\mathcal{G}$ however, it will turn out that L and M cannot belong to $\mathcal{G}$ for generic values of the spectral parameter. Thus, the problem of the existence and calculation of the Lax pairs becomes a problem in a more general algebra in which $\mathcal{G}$ will have to be embedded.

Before presenting the Lax pairs with spectral parameter for general simple Lie algebras, we describe them first in a concrete and relatively simpler case, namely when $\mathcal{G}$ is one of the classical algebras B_n, C_n, or D_n, and the elliptic Calogero–Moser system is untwisted. In these cases, the operators $L(z)$ and $M(z)$ have a suggestive expression in terms of matrices.

6.1 *Lax Pairs with Spectral Parameter for Classical Lie Algebras*

Following Olshanetsky and Perelomov [58], it is convenient to consider the Lax pair for the root system $\mathcal{R}(\mathrm{BC}_n) \equiv \mathcal{R}(\mathrm{B}_n) \cup \mathcal{R}(\mathrm{C}_n)$, although this set is not strictly speaking the root system of an algebra. The expression (5.45) defines then the corresponding (untwisted) elliptic Calogero–Moser system, with mass parameters m_1, m_2, and m_4. The cases B_n, C_n, and D_n are recovered with the following choices of masses

$$\begin{aligned} \mathrm{B}_n: &\qquad m_4 = 0 \\ \mathrm{C}_n: &\qquad m_1 = 0 \\ \mathrm{D}_n: &\qquad m_1 = m_4 = 0. \end{aligned} \tag{6.1}$$

We claim that the elliptic Calogero–Moser system associated to $\mathcal{R}(\mathrm{BC}_n)$ is integrable and admits a Lax pair with spectral parameter if

$$m_1(m_1^2 - 2m_2^2 + \sqrt{2} m_2 m_4) = 0. \tag{6.2}$$

In this case, the Lax pair $L(z)$, $M(z)$ is of the form

$$L(z) = P + X, \qquad M(z) = D + Y. \tag{6.3}$$

Here the matrices P, X, D, Y are all $(2n+1) \times (2n+1)$ dimensional, with P, D diagonal matrices

$$\begin{aligned} P &= \mathrm{diag}(p_1, \ldots, p_n; -p_1, \ldots, -p_n; 0) \\ D &= \mathrm{diag}(d_1, \ldots, d_n; +d_1, \ldots, +d_n; 0) \end{aligned} \tag{6.4}$$

and X, Y of the form

$$X = \begin{pmatrix} A & B_1 & C_1 \\ B_2 & A^T & C_2 \\ C_2^T & C_1^T & 0 \end{pmatrix}, \qquad Y = \begin{pmatrix} A' & B_1' & C_1' \\ B_2' & A'^T & C_2' \\ C_2'^T & C_1'^T & 0 \end{pmatrix}. \tag{6.5}$$

The entries of the matrix D are given by

$$d_i = -2m_2 \sum_k \wp(x_k) + \frac{m_1^2}{m_2} \wp(x_i) + \sqrt{2} m_4 \wp(2x_i) + m_2 \sum_{k \neq i} [\wp(x_i - x_k) + \wp(x_i + x_k)], \tag{6.6}$$

those of the matrix X by

$$\begin{aligned} A_{ij} &= m_2(1-\delta_{ij})\Phi(x_i - x_j, z) \\ B_{1ij} &= m_2(1-\delta_{ij})\Phi(x_i + x_j, z) + \sqrt{2} m_4 \delta_{ij} \Phi(2x_i, z) \\ B_{2ij} &= m_2(1-\delta_{ij})\Phi(-x_i - x_j, z) + \sqrt{2} m_4 \delta_{ij} \Phi(-2x_i, z) \\ C_{1i} &= m_1 \Phi(x_i, z) \\ C_{2i} &= m_1 \Phi(-x_i, z) \end{aligned} \tag{6.7}$$

and those of the matrix Y are given by similar formulas, with A, B, C, and $\Phi(x,z)$ replaced respectively by A', B', C', and $\Phi'(x,z)$. We note that for $z = \omega_a$, where ω_a is one of the half periods, then the preceding Lax pair reduces to the one found by Olshanetsky and Perelomov [58].

The matrices $L(z)$, $M(z)$ can now be verified directly to be a Lax pair by calculation. However, as preparation for the general Ansatz for the Lax pair for general simple Lie algebras, we note that they arise by the following representation theoretic construction. The verification that they form a Lax pair for the elliptic BC$_n$ Calogero–Moser system can then be carried out in the general framework provided in the next section §6.2.

We embed $\mathcal{G} = \mathrm{B}_n$ into GL$(N, \mathbb{C})$, with $N = 2n+1$, by embedding the fundamental representation of B$_n$ into the fundamental representation of GL$(N, \mathbb{C})$. The weights of B$_n$ obtained by the decomposition of the adjoint representation of GL$(N, \mathbb{C})$ automatically contain all the roots of the BC$_n$ system. Let λ_I, $1 \leq I \leq N$, be the weights of the fundamental representation of B$_n$. In terms of an orthonormal basis of vectors e_i, $1 \leq i \leq n$, we have

$$\lambda_i = e_i; \ \lambda_{i+n} = -e_i \quad 1 \leq i \leq n; \quad \lambda_{2n+1} = 0. \tag{6.8}$$

Let u_I, $1 \leq I \leq N$, be the weights of the fundamental representation of GL$(N, \mathbb{C})$. They are vectors in $\mathbb{C}^N$. They can be verified to decompose orthogonally as

$$\sqrt{2} u_I = \lambda_I + v_I, \tag{6.9}$$

for vectors v_I which are orthogonal to λ_J, and satisfy $v_i = v_{i+n}$, $1 \leq i \leq n$. We can deduce the decomposition of the roots of GL$(N, \mathbb{C})$ into weights of B$_n$. There are three orbits, consisting of the weights of B$_n$ with length$^2 = 2$, which may be viewed as the roots of D_n

$$\left.\begin{aligned}\sqrt{2}(u_i - u_j) &= e_i - e_j + v_i - v_j\\ \sqrt{2}(u_{n+j} - u_{n+i}) &= e_i - e_j - v_i + v_j\\ \sqrt{2}(u_i - u_{n+j}) &= e_i + e_j + v_i - v_j\\ \sqrt{2}(u_{n+i} - u_j) &= -e_i - e_j + v_i - v_j,\end{aligned}\right\} \quad i \neq j, \tag{6.10}$$

the weights of length^2 = 4 (additional roots of C_n)

$$\begin{aligned}\sqrt{2}(u_i - u_{n+i}) &= 2e_i\\ \sqrt{2}(u_{n+i} - u_i) &= -2e_i\end{aligned} \tag{6.11}$$

and the weights of length^2 = 1 (additional roots of B_n)

$$\begin{aligned}\sqrt{2}(u_i - u_N) &= e_i + v_i - v_{2n+1}\\ \sqrt{2}(u_N - u_{n+i}) &= e_i - v_i + v_{2n+1}\\ \sqrt{2}(u_{n+i} - u_N) &= -e_i + v_i - v_{2n+1}\\ \sqrt{2}(u_N - u_i) &= -e_i - v_i + v_{2n+1}.\end{aligned} \tag{6.12}$$

Then the matrices X, Y are of the form

$$\begin{aligned}X &= \sum_{\alpha\in\mathcal{R}(\mathrm{BC}_n)} \Phi(\alpha\cdot x)\Big(\sum_{\lambda_I-\lambda_J=\alpha} C_{IJ}E_{IJ}\Big)\\ Y &= \sum_{\alpha\in\mathcal{R}(\mathrm{BC}_n)} \Phi'(\alpha\cdot x)\Big(\sum_{\lambda_I-\lambda_J=\alpha} C_{IJ}E_{IJ}\Big)\end{aligned} \tag{6.13}$$

where the matrices E_{IJ} are given by $(E_{IJ})_{LM} = \delta_{IL}\delta_{JM}$, and C_{IJ} are constants proportional to the masses m_1, m_2, and m_4.

We can now generalize this set up to the case of an arbitrary simple Lie algebra $\mathcal{G}$.

6.2 The General Ansatz

Let the rank of $\mathcal{G}$ be n, and d be its dimension. Let Λ be a representation of $\mathcal{G}$ of dimension N, of weights λ_I, $1 \le I \le N$. We fix a Cartan subalgebra $\mathcal{H}$ for $\mathrm{GL}(N,\mathbb{C})$ which contains the Cartan subalgebra $\mathcal{H}_\mathcal{G}$ of $\mathcal{G}$. A basis for $\mathcal{H}$ consists then of a basis h_i, $1 \le i \le n$, of Cartan generators for $\mathcal{G}$, together with a complementary set $\tilde{h}_i$, $n+1 \le i \le N$, of generators of $\mathrm{GL}(N,\mathbb{C})$ satisfying $[h_i,h_j] = [h_i,\tilde{h}_j] = [\tilde{h}_i,\tilde{h}_j] = 0$. By adding to the $\tilde{h}_i$ suitable linear combinations of the h_i, we may assume that the h_i and $\tilde{h}_i$ are mutually orthogonal with respect to the Cartan–Killing form $\mathrm{tr}[\mathrm{Ad}_{h_i}, \mathrm{Ad}_{\tilde{h}_j}] = 0$.

Let $u_I \in \mathbb{C}^N$ be the weights of the fundamental representation of $\mathrm{GL}(N,\mathbb{C})$. Project orthogonally the u_I's onto the λ_I's as

$$su_I = \lambda_I + v_I, \quad \lambda_I \perp v_J. \tag{6.14}$$

The coefficient s is determined by the normalizations of the various weights. Let e_i, $1 \le i \le n$, be an orthonormal basis for the weight space of $\mathcal{G}$, and set $\lambda_I = \sum_{i=1}^n \lambda_{Ii} e_i$. Then $su_I \cdot e_k = \lambda_I \cdot e_k = \lambda_{Ik}$. In particular, the vector se_k, viewed as a vector in the weight space of $\mathrm{GL}(N,\mathbb{C})$ can be expressed as $se_k = \sum_{I=1}^N (se_k \cdot u_I) u_I = \sum_{I=1}^N \lambda_{Ik} u_I$. Squaring and summing over k gives

$$s^2 = \frac{1}{n} \sum_{I=1}^N \lambda_I^2 = I_2(\Lambda) \tag{6.15}$$

where $I_2(\Lambda)$ is by definition the second Dynkin index of the representation Λ of $\mathcal{G}$. We note that

$$\alpha_{IJ} = \lambda_I - \lambda_J \tag{6.16}$$

is a weight of $\Lambda \otimes \Lambda^*$ associated to the root $u_I - u_J$ of $\mathrm{GL}(N,\mathbb{C})$. More precisely, let $E_{IJ} = u_I u_J^T$ be the usual generators for $\mathrm{GL}(N,\mathbb{C})$. Then the basic commutation relations in terms of E_{IJ}, h_i, $\tilde{h}_i$ are

$$[E_{IJ}, E_{KL}] = \delta_{JK} E_{IL} - \delta_{IL} E_{KJ} \tag{6.17}$$

$$[h, E_{IJ}] = (\lambda_I - \lambda_J) E_{IJ}, \quad [\tilde{h}, E_{IJ}] = (v_I - v_J) E_{IJ}. \tag{6.18}$$

The generators E_{II} form an alternative basis for $\mathcal{H}$. Comparing the commutators of E_{II} and $(h, \tilde{h})$ with E_{IJ}, $I \ne J$, we obtain easily the relation between E_{II} and $(h, \tilde{h})$

$$E_{II} = \frac{1}{s^2} (\lambda_I \cdot h + v_I \cdot \tilde{h}). \tag{6.19}$$

The centralizer of $\mathcal{H}_{\mathcal{G}}$ in $\mathrm{GL}(N,\mathbb{C})$ may be larger that the Cartan subalgebra $\mathcal{H}$ of $\mathrm{GL}(N,\mathbb{C})$. We denote it by $\mathcal{H} \oplus \mathrm{GL}_0$. For all simple Lie algebras in their lowest dimensional faithful representation, we have $\mathrm{GL}_0 = 0$, except in the cases of F_4 and E_8, where the dimension of GL_0 is 2 and 56 respectively.

The Lax pairs for both untwisted and twisted Calogero–Moser systems will be of the form

$$L = P + X, \quad M = D + Y, \tag{6.20}$$

where the matrices P, X, D, and Y are given by

$$X = \sum_{I \ne J} C_{IJ} \Phi_{IJ}(\alpha_{IJ}, z) E_{IJ}, \quad Y = \sum_{I \ne J} C_{IJ} \Phi'_{IJ}(\alpha_{IJ}, z) E_{IJ} \tag{6.21}$$

and by

$$P = p \cdot h, \quad D = d \cdot (h \oplus \tilde{h}) + \Delta. \tag{6.22}$$

Here Δ is an element of GL_0. The functions $\Phi_{IJ}(x,z)$ and the coefficients C_{IJ} are yet to be determined. We begin by stating the necessary and sufficient conditions for the pair $L(z)$, $M(z)$ of (6.20) to be a Lax pair for the (twisted or untwisted) Calogero–Moser systems. For this, it is convenient to introduce the following notation

$$\Phi_{IJ} = \Phi_{IJ}(\alpha_{IJ}\cdot x), \quad \Phi'_{IJ} = \Phi'_{IJ}(\alpha_{IJ}\cdot x) \tag{6.23}$$

$$\begin{aligned}\wp'_{IJ} = \Phi_{IJ}(\alpha_{IJ}\cdot x,z)\Phi'_{JI}(-\alpha_{IJ}\cdot x,z)&\\ - \Phi'_{IJ}(\alpha_{IJ}\cdot x,z)\Phi'_{JI}(-\alpha_{IJ}\cdot x,z).&\end{aligned} \tag{6.24}$$

Then the Lax equation $\dot{L}(z) = [L(z), M(z)]$ implies the Calogero–Moser system if and only if the following three identities are satisfied

$$\sum_{I\neq J} C_{IJ}C_{JI}\wp'_{IJ}\alpha_{IJ} = s^2 \sum_{\alpha\in\mathcal{R}(\mathcal{G})} m^2_{|\alpha|}\wp_{\nu(\alpha)}(\alpha\cdot x), \tag{6.25}$$

$$\sum_{I\neq J} C_{IJ}C_{JI}\wp'_{IJ}(v_I - v_J) = 0, \tag{6.26}$$

$$\begin{aligned}\sum_{K\neq I,J} C_{IK}C_{KJ}(\Phi_{IK}\Phi'_{KJ} - \Phi'_{IK}\Phi_{KJ})&\\ = sC_{IJ}\Phi_{IJ}d\cdot(u_I - u_J) + \sum_{K\neq I,J}\Delta_{IJ}C_{KJ}\Phi_{KJ}&\\ - \sum_{K\neq I,J} C_{IK}\Phi_{IK}\Delta_{KJ}.&\end{aligned} \tag{6.27}$$

In fact, applying the commutation relations (6.17) shows readily that the conditions $\dot{X} = [P,Y]$ and $\dot{x} = p$ are equivalent. The remaining terms in the equation $\dot{L} = [L,M]$ decompose into two equations, $\dot{P} = [X,Y]_{\mathcal{H}}$ and $[X,Y]_{\mathrm{GL}(N,\mathbb{C})\ominus\mathcal{H}} + [X,D] = 0$. Now the commutator $[X,Y]$ is given by

$$[X,Y] = \sum_{J\notin\{I,L\}} C_{IJ}C_{JL}(\Phi_{IJ}\Phi'_{JL} - \Phi'_{IJ}\Phi_{JL})E_{IL}. \tag{6.28}$$

The component $[X,Y]_{\mathcal{H}}$ is obtained by retaining only the terms E_{II}. Expressing E_{II} in terms of $(h,\tilde{h})$ as in (6.19), we find that the equation $\dot{P} = [X,Y]_{\mathcal{H}}$ is equivalent to

$$\dot{p}\cdot h = \frac{1}{s^2}\sum_{I\neq J} C_{IJ}C_{JI}\wp'_{IJ}(\lambda_I\cdot h + v_I\cdot\tilde{h}). \tag{6.29}$$

In view of (6.23), $\wp'_{IJ} = -\wp'_{JI}$, and we may antisymmetrize in I and J. Since h, $\tilde{h}$ are linearly independent, the resulting equation can only hold if the second identity (6.26) is satisfied. Assuming this, the equation reduces then to the Calogero–Moser equation of motion

$$\dot{p} = \frac{1}{2}\sum_{\alpha\in\mathcal{R}(\mathcal{G})} m^2_{|\alpha|}\wp'_{\nu(\alpha)}(\alpha\cdot x) \tag{6.30}$$

if the identity (6.25) is imposed. Finally, the terms in the identity (6.27) are recognized as just the coefficients of E_{IJ}, $I \neq J$, in $[X,Y]_{\mathrm{GL}(N,\mathbb{C})\ominus\mathcal{H}}$, $[d\cdot(h\oplus\tilde{h}),X]$, and $[\Delta, X]$ respectively. Thus the identity (6.27) is equivalent to the vanishing of $[X,Y]_{\mathrm{GL}(N,\mathbb{C})\ominus\mathcal{H}} + [X,D]$, completing the argument.

The following theorem was established in [59]:

Theorem 3. *A representation Λ, functions Φ_{IJ}, and coefficients C_{IJ} with a spectral parameter z satisfying* (6.25), (6.26), (6.27) *can be found for all twisted and untwisted elliptic Calogero–Moser systems associated with a simple Lie algebra $\mathcal{G}$, except possibly in the case of twisted* G_2. *In the case of* E_8, *we have to assume the existence of a ± 1-valued cocycle.*

6.3 Lax Pairs for Untwisted Calogero–Moser Systems

Returning to the case of the untwisted elliptic Calogero–Moser system for the BC_n system, we can now at the same stroke explain how the matrices (6.13) are found, and verify that they lead to a Lax pair. Given the Ansatz (6.3), the embedding of the fundamental representation of B_n into the fundamental representation of $\mathrm{GL}(N,\mathbb{C})$ ($N = 2n+1$) described in §6.1, and the choice $\Phi_{IJ}(x,z) = \Phi(x,z)$ for all I, J, the main issue reduces to the existence and determination of the constants C_{IJ}.

For each $i \neq j$, there are two distinct roots of $\mathrm{GL}(N,\mathbb{C})$ which project to $e_i - e_j$, namely $\sqrt{2}(u_i - u_j)$ and $\sqrt{2}(u_{n+j} - u_{n+i})$. The conditions (6.25) and (6.26) become

$$\begin{aligned} 2m_2^2 &= C_{ij}^2 + C_{n+j,n+i}^2 \\ 0 &= C_{ij}^2(v_i - v_j) + C_{n+j,n+i}^2(-v_i + v_j). \end{aligned} \tag{6.31}$$

Similar equations are obtained for the roots $e_i + e_j$. Using the linear independence of the vectors v_i, we find the conditions

$$\begin{aligned} m_2^2 &= C_{ij}^2 = C_{n+i,n+j}^2 = C_{n+i,j}^2 \\ 2m_4^2 &= C_{i,n+i}^2 \\ m_1^2 &= C_{iN}^2 = C_{n+i,N}^2. \end{aligned} \tag{6.32}$$

These equations can be solved by taking

$$\begin{aligned} m_2 &= C_{ij} = C_{n+i,n+j} = C_{n+i,j} \\ 2m_4 &= C_{i,n+i} \\ m_1 &= C_{iN} = C_{n+i,N}. \end{aligned} \tag{6.33}$$

With this choice of coefficients C_{IJ}, we turn to the third condition (6.27). Letting i and j take values between 1 and n, there are 6 cases to be considered:

(1) $I = i$, $J = j$;

(2) $I = i$, $J = N$;

(3) $I = i$, $J = n + j$ with $i \neq j$;

(4) $I = i$, $J = n + i$;

(5) $I = N$, $J = n + j$; and

(6) $I = n + i$, $J = n + j$.

We set $d\dot{e}_i = 0$. Then case (4) is satisfied, cases (3) and (6) give the same equations as case (1), and case (5) gives the same equation as case (2). The cases (1) and (2) themselves lead to the following equations:

$$\begin{aligned} m_2 d\cdot(v_i - v_j) &= \sum_{k\neq i,j} m_2^2[\wp(x_i - x_j) - \wp(x_k - x_j) \\ &\qquad + \wp(x_i + x_k) - \wp(x_k + x_j)] \\ &\quad + m_1^2[\wp(x_i) - \wp(x_j)] + \sqrt{2}m_2 m_4[\wp(2x_i) - \wp(2x_j)], \\ m_1 d\cdot(v_i - v_N) &= \sum_{k\neq i} m_1 m_2[\wp(x_i - x_j) + \wp(x_i + x_k) - 2\wp(x_k)] \\ &\qquad + \sqrt{2}m_1 m_4[\wp(2x_i) - \wp(x_i)]. \end{aligned} \tag{6.34}$$

We may assume that $m_2 \neq 0$, since otherwise the system decomposes trivially into a set of noninteracting one-dimensional systems. Then the most general solution of the first equation in the preceding condition is

$$d\cdot v_i = d_0 + \frac{m_1^2}{m_2}\wp(x_i) + \sqrt{2}m_4\wp(2x_i) + \sum_{k\neq i} m_2[\wp(x_i - x_k) + \wp(x_i + x_k)] \tag{6.35}$$

where d_0 is an arbitrary function of x which is independent of i. Substituting this into the second equation yields

$$\begin{aligned} m_1 d\cdot v_N = m_1 d_0 + m_1\left(-2m_2 + \sqrt{2}m_4 + \frac{m_1^2}{m_2}\right)\wp(x_i) \\ + 2m_1 m_2 \sum_k \wp(x_k). \end{aligned} \tag{6.36}$$

Since the left-hand side is independent of i, our construction works only when $m_1(m_1^2 - 2m_2^2 + \sqrt{2}m_2 m_4) = 0$, which is the condition we stated earlier in §6.1. When it is satisfied, a vector d can clearly be found. Choosing d_0 so that $d \cdot v_N = 0$, we obtain the Lax pair announced previously for the BC_n system.

We summarize now some important features of the Lax pairs we obtain in Theorem 3.

- In the case of the *untwisted* Calogero–Moser systems, we can choose $\Phi_{IJ}(x,z) = \Phi(x,z)$, $\wp_{IJ}(x) = \wp(x)$ for all $\mathcal{G}$.
- $\Delta = 0$ for all $\mathcal{G}$, except for E_8.

• For A_n, the Lax pair (6.20) corresponds to the choice of the fundamental representation for Λ. A different Lax pair can be found by taking Λ to be the antisymmetric representation.

• As we saw before, for the BC_n system, the Lax pair is obtained by imbedding B_n in $GL(N, \mathbb{C})$ with $N = 2n + 1$. When $z = \omega_a$ (half-period), the Lax pair obtained this way reduces to the Lax pair obtained by Olshanetsky and Perelomov [58].

• For the B_n and D_n systems, additional Lax pairs with spectral parameter can be found by taking Λ to be the spinor representation.

• For G_2, a first Lax pair with spectral parameter can be obtained by the above construction with Λ chosen to be the **7** of G_2. A second Lax pair with spectral parameter can be obtained by restricting the **8** of B_3 to the $\mathbf{7} \oplus \mathbf{1}$ of G_2.

• For F_4, a Lax pair can be obtained by taking Λ to be the $\mathbf{26} \oplus \mathbf{1}$ of F_4, viewed as the restriction of the **27** of E_6 to its F_4 subalgebra.

• For E_6, Λ is the **27** representation.

• For E_7, Λ is the **56** representation.

• For E_8, a Lax pair with spectral parameter can be constructed with Λ given by the **248** representation, if coefficients $c_{IJ} = \pm 1$ exist with the following cocycle conditions

$$
\begin{aligned}
c(\lambda, \lambda - \delta) c(\lambda - \delta, \mu) &= c(\lambda, \mu + \delta) c(\mu + \delta, \mu) \\
&\qquad \text{when } \delta \cdot \lambda = -\delta \cdot \mu = 1, \lambda \cdot \mu = 0 \qquad (6.37) \\
c(\lambda, \mu) c(\lambda - \delta, \mu) &= c(\lambda, \lambda - \delta) \\
&\qquad \text{when } \delta \cdot \lambda = \lambda \cdot \mu = 1, \delta \cdot \mu = 0 \qquad (6.38) \\
c(\lambda, \mu) c(\lambda, \lambda - \mu) &= -c(\lambda - \mu, -\mu) \\
&\qquad \text{when } \lambda \cdot \mu = 1. \qquad (6.39)
\end{aligned}
$$

The matrix Δ in the Lax pair is then the 8×8 matrix given by

$$
\begin{aligned}
\Delta_{ab} = &\sum_{\substack{\delta \cdot \beta_a = 1 \\ \delta \cdot \beta_b = 1}} \frac{m_2}{2} \big(c(\beta_a, \delta) c(\delta, \beta_b) \\
&\qquad + c(\beta_a, \beta_a - \delta) c(\beta_a - \delta, \beta_b) \big) \wp(\delta \cdot x) \\
&- \sum_{\substack{\delta \cdot \beta_a = 1 \\ \delta \cdot \beta_b = -1}} \frac{m_2}{2} \big(c(\beta_a, \delta) c(\delta, \beta_b) \\
&\qquad + c(\beta_a, \beta_a - \delta) c(\beta_a - \delta, \beta_b) \big) \wp(\delta \cdot x) \qquad (6.40) \\
\Delta_{aa} = &\sum_{\beta_a \cdot \delta = 1} m_2 \wp(\delta \cdot x) + 2 m_2 \wp(\beta_a \cdot x),
\end{aligned}
$$

where β_a, $1 \leq a \leq 8$, is a maximal set of 8 mutually orthogonal roots.

Explicit expressions for the constants C_{IJ} and the functions $d(x)$, and thus for the Lax pair are particularly simple when the representation Λ consists of only a single Weyl orbit of weights. This is the case when Λ is either

- the defining representation of A_n, C_n or D_n;
- any rank p totally antisymmetric representation of A_n;
- an irreducible fundamental spinor representation of B_n or D_n;
- the **27** of E_6; the **56** of E_7.

Then, the weights λ and μ of Λ provide unique labels instead of I and J, and the values of $C_{IJ} = C_{\lambda\mu}$ are given by a simple formula

$$C_{\lambda\mu} = \begin{cases} \sqrt{\frac{\alpha^2}{2}} m_{|\alpha|} & \text{when } \alpha = \lambda - \mu \text{ is a root} \\ 0 & \text{otherwise.} \end{cases} \tag{6.41}$$

The expression for the vector d may be summarized by

$$sd \cdot u_\lambda = \sum_{\substack{\lambda\cdot\delta=1 \\ \delta^2=2}} m_{|\delta|} \wp(\delta \cdot x). \tag{6.42}$$

(For C_n, the last equation has an additional term, as given in (6.6).) In each case, the number of independent couplings $m_{|\alpha|}$ equals the number of different root lengths.

We shall illustrate the methods in a particularly simple case, namely E_6. Recall that the root system of E_6 is

$$\begin{aligned} &\pm e_i \pm e_j, && 1 \le i < j \le 5 \\ &\pm \frac{1}{2}\left(\sqrt{3}e_6 + \sum_{i=1}^{5} \epsilon_i e_i\right), && \prod_{i=1}^{5} \epsilon_i = 1. \end{aligned} \tag{6.43}$$

The simple roots are chosen as in Table 1.8 in Appendix B. The weights of the **27** representation of E_6 are given by

$$\begin{aligned} &\frac{2}{\sqrt{3}} e_6 \\ &\frac{1}{2\sqrt{3}} e_6 - \frac{1}{2}\sum_{i=1}^{5} \epsilon_i e_i, && \prod_{i=1}^{5} \epsilon_i = 1 \\ &-\frac{1}{2\sqrt{3}} e_6 \pm e_i, && 1 \le i \le 5. \end{aligned} \tag{6.44}$$

The Weyl orbit structure of $\mathbf{27} \otimes \mathbf{27}^*$ is given in Table 1.7.

We shall require the following simple property of the roots α and weights λ of E_6 in the **27** representation:

$$\lambda \pm \alpha = \text{weight} \iff \lambda \cdot \alpha = \mp 1. \tag{6.45}$$

TABLE 1.7. Weyl orbit structure of $\mathbf{27} \otimes \mathbf{27}^*$ of E_6

Weyl Orbit	Multiplicity	# Weights	Length^2
[000000]	27	1	0
[000001]	6	72	2
[100010]	1	270	4

When $\lambda \pm \alpha$ is not a weight, $\lambda \cdot \alpha = 0$. To see this, we note that $|\lambda \cdot \alpha| \leq (2/\sqrt{3}) \times \sqrt{2} < 2$, since all weights λ of $\mathbf{27}$ satisfy $\lambda^2 = \frac{4}{3}$. As $\lambda \cdot \alpha$ is an integer, it can only assume the values 0, 1 or -1. If $\lambda - \alpha = \mu$ is a weight, then $\lambda \cdot \alpha - \mu \cdot \alpha = 2$. Since $\mu \cdot \alpha \in \{0, \pm 1\}$, we must have $\lambda \cdot \alpha = 1$. The argument for $\lambda + \alpha$ is similar. The reverse implication in (6.45) is a basic fact of the theory of Lie algebras.

We return now to the construction of the Lax pair. Since E_6 is simply laced, $\wp_{IJ}(x) = \wp(x)$ for all I, J. With C_{IJ} given by (6.41), the first condition (6.25) for a Lax pair reduces to

$$\sum_{\alpha \in \mathcal{R}(\mathcal{G})} \wp'(\alpha \cdot x) \sum_{\lambda_I - \lambda_J = \alpha} 1 = s^2 \sum_{\alpha \in \mathcal{R}(\mathcal{G})} \wp'(\alpha \cdot x). \tag{6.46}$$

However, it follows from (6.15) that $s^2 = \frac{1}{6} \cdot 27 \times \frac{4}{3} = 6$. On the other hand, for all roots α, the multiplicity $\sum_{\lambda_I - \lambda_J = \alpha} 1$ of the Weyl orbit is also 6. Thus the condition (6.25) is satisfied.

To obtain the second condition (6.26), it suffices to verify that

$$\sum_{\lambda_I - \lambda_J = \alpha} (v_I - v_J) = 0 \tag{6.47}$$

for each root α. Pairing the left-hand side with an arbitrary vector u_K gives

$$\sum_{\lambda_I - \lambda_J = \alpha} (v_I - v_J) \cdot u_K = \sum_{\lambda_I - \lambda_J = \alpha} \left(s(\delta_{IK} - \delta_{JK}) - \frac{1}{s} \alpha \cdot \lambda_K \right). \tag{6.48}$$

Using again the fact that s^2 and the multiplicity of the Weyl orbit are both 6, this can be rewritten as

$$\sum_{\lambda_I - \lambda_J = \alpha} (v_I - v_J) \cdot u_K = \sqrt{6} \left\{ \left(\sum_{\lambda_I - \lambda_J = \alpha} (\delta_{IK} - \lambda_{JK}) \right) - \alpha \cdot \lambda_K \right\}. \tag{6.49}$$

When $\alpha \cdot \lambda_K = 0$, $\lambda \pm \alpha$ is not a weight, and both expressions on the right hand side of (6.49) vanish. When $\alpha \cdot \lambda_K = \pm 1$, there exists a unique weight so that $\lambda_K \mp \alpha$ is a weight. Thus both expressions on the right hand side of (6.49) are ± 1, and the right hand side still vanishes. This establishes (6.26).

Recall that in the case of E_6, we choose $\Delta = 0$ and $\Phi_{IJ}(x, z) = \Phi(x, z)$ for all I, J, $I \neq J$. In view of the key identity (C.30) for $\Phi(x, z)$, the third identity (6.27) for the Lax pair can be rewritten as

$$\sqrt{6}C_{IJ}d \cdot (u_I - u_J) = m_{\sqrt{2}}^2 \Big[\sum \wp(\alpha_{IK} \cdot x) - \sum \wp(\alpha_{KJ} \cdot x) \Big], \tag{6.50}$$

where both sums on the right hand side are restricted to those $K \neq I, J$ for which $\lambda_I - \lambda_K$ and $\lambda_K - \lambda_J$ are both roots. This implies that $(\lambda_I - \lambda_K)^2 = (\lambda_J - \lambda_K)^2 = 2$, and hence $\lambda_I \cdot \lambda_K = \lambda_J \cdot \lambda_K = \frac{1}{3}$.

We consider separately the two possibilities where $(\lambda_I - \lambda_J)^2 = 4$ and $(\lambda_I - \lambda_J)^2 = 2$.

In the first case, $\lambda_I - \lambda_J$ is not a root, and $C_{IJ} = 0$. In particular, $(\lambda_I - \lambda_J)^2 = 4$, and hence $\lambda_I \cdot \lambda_J = -\frac{2}{3}$. By setting $\beta = \lambda_I - \lambda_K$, the first sum on the right of (6.50) can be viewed as the sum over all roots β with $\beta \cdot \lambda_I = 1$, $\beta \cdot \lambda_J = -1$. Indeed, in this case, $\lambda_I - \lambda_J$ is not a root, and the restriction $K \neq I, J$ is automatically satisfied. Furthermore, $\beta \cdot \lambda_I = \lambda_I \cdot \lambda_I - \lambda_K \cdot \lambda_I = 1$, $\beta \cdot \lambda_J = \lambda_I \cdot \lambda_J - \lambda_K \cdot \lambda_J = -1$. Conversely, a root β with these properties can be written as $\beta = \lambda_I - \lambda_K = \lambda_K - \lambda_J$. Similarly, by introducing $\beta = \lambda_K - \lambda_J$, the second sum on the right of (6.50) can be viewed as the sum over all roots β with $\beta \cdot \lambda_I = -1$, $\beta \cdot \lambda_J = 1$. The right hand side of (6.50) becomes

$$\sum_{\beta\cdot\lambda_I = -\beta\cdot\lambda_J = 1} \wp(\beta \cdot x) - \sum_{\beta\cdot\lambda_I = -\beta\cdot\lambda_J = -1} \wp(\beta \cdot x), \tag{6.51}$$

which vanishes since $\wp(x)$ is even. Thus the third condition for the Lax pair is satisfied when $\lambda_I - \lambda_J$ is not a root.

Assume now that $\lambda_I - \lambda_J$ is a root. Then $(\lambda_I - \lambda_J)^2 = 2$ and $\lambda_I \cdot \lambda_J = \frac{1}{3}$. Introducing $\beta = \lambda_I - \lambda_K$ and arguing as before, we can view the first sum on the right hand side of (6.50) as the sum over all roots β with $\beta \cdot \lambda_I = -1$, $\beta \cdot \lambda_J = 0$, and $\beta \neq \lambda_I - \lambda_J$. Expressing the other sum in the same way, we obtain

$$\sqrt{6}m_{\sqrt{2}}d \cdot (u_I - u_J) = m_{\sqrt{2}}^2 \Bigg(\sum_{\substack{\beta\cdot\lambda_I = -1 \\ \beta\cdot\lambda_J = 0}} \wp(\beta \cdot x) - \sum_{\substack{\beta\cdot\lambda_J = -1 \\ \beta\cdot\lambda_I = 0}} \wp(\beta \cdot x) \Bigg) \tag{6.52}$$

where we have added and subtracted the terms corresponding to $\beta = \lambda_I - \lambda_J$. Now recall that the only possible values for $\beta \cdot \lambda$ are 0 or ± 1. Thus the above right hand side can be rewritten as

$$m_{\sqrt{2}}^2 \Bigg(\sum_{\beta\cdot\lambda_I = -1} \wp(\beta \cdot x) - \sum_{\beta\cdot\lambda_J = -1} \wp(\beta \cdot x) \Bigg) \tag{6.53}$$

up to the following combination of terms

$$-m_{\sqrt{2}} \Bigg(\sum_{\substack{\beta\cdot\lambda_I = -1 \\ \beta\cdot\lambda_J = \pm 1}} \wp(\beta \cdot x) - \sum_{\substack{\beta\cdot\lambda_I = -1 \\ \beta\cdot\lambda_J = \pm 1}} \wp(\beta \cdot x) \Bigg). \tag{6.54}$$

But in this last combination, the terms with $\beta\cdot\lambda_I = \beta\cdot\lambda_J = -1$ cancel out. The remaining sums are over $\beta\cdot\lambda_I = -\beta\cdot\lambda_J = 1$ and $\beta\cdot\lambda_I = -\beta\cdot\lambda_J = 1$ respectively. Since the function $\wp(x)$ is even, the flip $\beta \to -\beta$ shows that the expression (6.54) cancels. Thus the third condition for the Lax pair is satisfied by setting $d\cdot u_I = m_{\sqrt{2}} \sum_{\beta\cdot\lambda_I=-1} \wp(\beta\cdot x)$, which is the prescription we gave earlier.

6.4 Lax Pairs for Twisted Calogero–Moser Systems

Recall that the twisted and untwisted Calogero–Moser systems differ only for non-simply laced Lie algebras, namely B_n, C_n, G_2 and F_4. These are the only algebras we discuss in this paragraph. The construction (6.20) gives then Lax pairs for all of them, with the possible exception of twisted G_2. Unlike the case of untwisted Lie algebras however, the functions Φ_{IJ} have to be chosen with care, and differ for each algebra. More specifically,

- For B_n, the Lax pair is of dimension $N = 2n$, admits two independent couplings m_1 and m_2, and

$$\Phi_{IJ}(x,z) = \begin{cases} \Phi(x,z), & \text{if } I-J \neq 0, \pm n \\ \Phi_2(\frac{1}{2}x,z), & \text{if } I-J = \pm n. \end{cases} \tag{6.55}$$

Here a new function $\Phi_2(x,z)$ is defined by

$$\Phi_2\left(\tfrac{1}{2}x,z\right) = \frac{\Phi(\frac{1}{2}x,z)\Phi(\frac{1}{2}x+\omega_1,z)}{\Phi(\omega_1,z)}. \tag{6.56}$$

- For C_n, the Lax pair is of dimension $N = 2n+2$, admits one independent coupling m_2, and

$$\Phi_{IJ}(x,z) = \Phi_2(x+\omega_{IJ},z), \tag{6.57}$$

where ω_{IJ} are given by

$$\omega_{IJ} = \begin{cases} 0, & \text{if } I \neq J = 1,2,\dots,2n+1; \\ \omega_2, & \text{if } 1 \leq I \leq 2n, J = 2n+2; \\ -\omega_2, & \text{if } 1 \leq J \leq 2n, I = 2n+2. \end{cases} \tag{6.58}$$

- For F_4, the Lax pair is of dimension $N = 24$, two independent couplings m_1 and m_2,

$$\Phi_{\lambda\mu}(x,z) = \begin{cases} \Phi(x,z), & \text{if } \lambda\cdot\mu = 0; \\ \Phi_1(x,z), & \text{if } \lambda\cdot\mu = \frac{1}{2}; \\ \Phi_2(\frac{1}{2}x,z), & \text{if } \lambda\cdot\mu = -1, \end{cases} \tag{6.59}$$

where the function $\Phi_1(x,z)$ is defined by

$$\Phi_1(x,z) = \Phi(x,z) - e^{\pi i\zeta(z)+\eta_1 z}\Phi(x+\omega_1,z). \tag{6.60}$$

Here it is more convenient to label the entries of the Lax pair directly by the weights $\lambda = \lambda_I$ and $\mu = \lambda_J$ instead of I and J.

• For G_2, candidate Lax pairs can be defined in the **6** and **8** representations of G_2, but it is still unknown whether elliptic functions $\Phi_{IJ}(x,z)$ exist which satisfy the required identities.

We illustrate the construction of Lax pairs for twisted Calogero–Moser systems by reproducing here the construction of [59] in the simplest case of B_n. Since we are dealing here with the twisted Calogero–Moser system for a non-simply laced algebra, we let $\mathcal{G} = B_n^\vee$, and choose Λ to be the fundamental representation of $B_n^\vee$, which is of dimension $N = 2n$. The weights λ of Λ are

$$\lambda_i = -\lambda_{i+n} = e_i, \quad 1 \le i \le n. \tag{6.61}$$

Via the Ansatz described in (6.25)–(6.27), the representation Λ is imbedded in $GL(N, \mathbb{C})$ with $s^2 = 2$, and

$$\sqrt{2}u_i = e_i + v_i, \ \sqrt{2}u_{n+i} = -e_i + v_i, \quad 1 \le i \le n. \tag{6.62}$$

The roots of $GL(N, \mathbb{C})$ decompose into short roots of $B_n^\vee$

$$\sqrt{2}(u_i - u_j) = e_i - e_j + v_i - v_j, \qquad i \neq j \tag{6.63}$$

$$\sqrt{2}(u_{\nu+i} - u_{\nu+j}) = e_i - e_j + v_i - v_j, \qquad i \neq j \tag{6.64}$$

$$\sqrt{2}(u_i - u_{\nu+j}) = e_i + e_j + v_i - v_j, \qquad i \neq j \tag{6.65}$$

$$\sqrt{2}(u_{\nu+i} - u_j) = -e_i - e_j + v_i - v_j, \quad i \neq j \tag{6.66}$$

and long roots of $B_n^\vee$

$$\sqrt{2}(u_i - u_{\nu+i} = 2e_i \tag{6.67}$$

$$\sqrt{2}(u_{\nu+i} - u_i) = -2e_i. \tag{6.68}$$

Let the constants C_{IJ} be given by

$$C_{IJ} = \begin{cases} m_2, & \text{if } I - J \neq 0, \pm n \\ m_1, & \text{if } I - J = \pm n. \end{cases} \tag{6.69}$$

Then the condition (6.26) in the requirements for Lax pairs is satisfied, since each short root of $B_n^\vee$ has two roots of $GL(N, \mathbb{C})$ as preimages, and they come with opposite values of $v_i - v_j$. Each long root of $B_n^\vee$ has no v-dependence at all, and thus does not enter (6.26). The condition (6.25) in the requirements for Lax pairs is an easy consequence of the functional equations (C.32) and (C.30) for the $\Phi(x,z)$ and $\Lambda(x,z)$ functions. Thus our main task is to find coefficients d_I so that the last condition (6.27) for the Lax pair be satisfied.

We set $\Delta = 0$. Using the antisymmetry of the right hand side of (6.27) under $x \to -x$ and $I \to J$ on the left-hand side, we see that $d(-x) = d(x)$.

Using this symmetry, we may restrict attention to the cases $I < J$. Two cases arise, $J = n + I$ and $J - I \neq 0, n$, which we discuss separately. Consider first the case $J = n + I$, in which (6.27) reduces to

$$\begin{aligned} & m_I \Lambda(2x_i) d \cdot 2e_i \\ & \quad = \sum_{K \neq i, n+i} m_2^2 \{\Phi(\alpha_{iK} \cdot x)\Phi'(\alpha_{K(n+i)} \cdot x) - \Phi'(\alpha_{iK} \cdot x)\Phi(\alpha_{K(n+i)} \cdot x)\} \\ & \quad = m_2^2 \Phi(2x_i) \sum_{k \neq i}^{n} \{\wp(\alpha_{ik} \cdot x) - \wp(\alpha_{k(n+i)} \cdot x) \\ & \qquad\qquad + \wp(\alpha_{i(n+k)} \cdot x) - \wp(\alpha_{(n+k)(n+i)} \cdot x)\}. \end{aligned} \tag{6.70}$$

The right hand side is easily seen to vanish, which simply requires that $d \cdot e_i = 0$. Next, we consider the case $J - I \neq 0, n$, for which (6.27) becomes

$$\begin{aligned} & m_2 \Phi(\alpha_{IJ} \cdot x) s d \cdot (u_I - u_J) \\ & = \sum_{\substack{I-K \neq 0, \pm n \\ K-J \neq 0, \pm n}} m_2^2 \{\Phi(\alpha_{IK} \cdot x)\Phi'(\alpha_{KJ} \cdot x) - \Phi'(\alpha_{IK} \cdot x)\Phi(\alpha_{KJ} \cdot x)\} \\ & \quad + \sum_{\substack{I-K = \pm n \\ K-J \neq 0, \pm n}} m_1 m_2 \{\Lambda(\alpha_{IK} \cdot x)\Phi'(\alpha_{KJ} \cdot x)\Lambda'(\alpha_{IK} \cdot x)\Phi(\alpha_{KJ} \cdot x)\} \\ & \quad + \sum_{\substack{I-K \neq 0, \pm n \\ K-J = \pm n}} m_1 m_2 \{\Phi(\alpha_{IK} \cdot x)\Lambda'(\alpha_{KJ} \cdot x)\Phi'(\alpha_{IK} \cdot x)\Lambda(\alpha_{KJ} \cdot x)\} \\ & \quad + \sum_{\substack{I-K = \pm n \\ K-J = \pm n}} m_1^2 \{\Lambda(\alpha_{IK} \cdot x)\Lambda'(\alpha_{KJ} \cdot x) - \Lambda'(\alpha_{IK} \cdot x)\Lambda(\alpha_{KJ} \cdot x)\}. \end{aligned} \tag{6.71}$$

The last sum in the preceding equation vanishes identically, because the conditions $I - K = \pm n$ and $K - J = \pm n$ imply that $I - J = 0, \pm 2n$, which is impossible since $I \neq J$. By noticing that if $I - K = \pm n$, we have that $\lambda_I = -\lambda_K$ for all I and K, we can easily make the second and third sum collapse to single terms. Thus we obtain

$$\begin{aligned} & m_2 \Phi(\alpha_{IJ} \cdot x) s d \cdot (u_I - u_J) \\ & = \sum_{\substack{I-K \neq 0, \pm n \\ K-J \neq 0, \pm n}} m_2^2 \{\Phi(\alpha_{IK} \cdot x)\Phi'(\alpha_{KJ} \cdot x) - \Phi'(\alpha_{IK} \cdot x)\Phi(\alpha_{KJ} \cdot x)\} \\ & \quad + m_1 m_2 \big\{\Lambda(2\lambda_I \cdot x)\Phi'\big(-(\lambda_I + \lambda_J) \cdot x\big) \\ & \qquad\qquad - \Lambda'(2\lambda_I \cdot x)\Phi\big(-(\lambda_I + \lambda_K) \cdot x\big)\big\} \\ & \quad + m_1 m_2 \big\{\Phi\big((\lambda_I + \lambda_J) \cdot x\big)\Lambda'(-2\lambda_J \cdot x) \\ & \qquad\qquad - -\Phi'\big((\lambda_I + \lambda_J) \cdot x\big)\Lambda(-2\lambda_J \cdot x)\big\}. \end{aligned} \tag{6.72}$$

We now make use of the relations (C.32) and (C.33) for the functions Φ and Λ to simplify the right hand side of (6.72). Omitting an overall factor of $m_2\Phi(\alpha_{IJ}\cdot x)$, (6.72) is reduced to

$$sd\cdot(u_I-u_J)=\sum_{\substack{I-K\neq 0,\pm n\\ K-J\neq 0,\pm n}}\{\wp(\alpha_{IK}\cdot x)-\wp(\alpha_{KJ}\cdot x)\} + \tfrac{1}{2}m_1\{\wp_2(\lambda_I\cdot x)-\wp_2(\lambda_J\cdot x)\}. \quad (6.73)$$

This is solved by setting

$$d\cdot v_i=\sum_{J-i\neq 0,n} m_2\wp((e_i-\lambda_j)\cdot x)+\frac{1}{2}m_1\wp_2(e_I\cdot x). \quad (6.74)$$

6.5 Scaling Limits of Lax Pairs

All the Lax pairs we constructed have finite scaling limits. This follows from the asymptotic behavior of the function $\Phi(u,z)$

$$\Phi(u,z)\to\begin{cases}+e^{-u/2}(1-Z^{-1}e^{u-\omega_2}), & \mathrm{Re}(u)\to+\infty\\ -e^{u/2}(1-Ze^{-u-\omega_2}), & \mathrm{Re}(u)\to-\infty\end{cases} \quad (6.75)$$

in the range $|\,\mathrm{Re}\,u|<2\omega_2$. This asymptotic behavior results in finite limits for the expression $C_{IJ}\Phi_{IJ}(\alpha\cdot x)$ under the scalings defined in §5.7 for the Hamiltonians. The behavior of the functions $C_{IJ}\Phi'_{IJ}(\alpha\cdot x)$ is similar, and thus both operators $L(z)$ and $M(z)$ have well defined and finite limits.

More precisely, consider first the scaling limits of the untwisted Calogero–Moser systems defined with $\delta=1/h_{\mathcal{G}}$, where $h_{\mathcal{G}}$ is the Coxeter number. This means that we let $x=X+2\omega_2\delta\rho^\vee$, $e^z=Zq^{-1/2}$ as in Theorem 1, and set

$$C_{IJ}=\begin{cases}M_{|\alpha|}e^{\delta\omega_2}c_{IJ}, & \text{when } \alpha_{IJ}=\alpha\in\mathcal{R}(\mathcal{G})\\ 0, & \text{when } \alpha_{IJ}\notin\mathcal{R}(\mathcal{G}).\end{cases} \quad (6.76)$$

Then the matrices $L(z)$, $M(z)$ converge as $\omega_2\to+\infty$ to matrices $L(Z)$, $M(Z)$ of the form (5.16), where the root system is the one associated with the affine algebra $\mathcal{G}^{(1)}$, and the generators E_α are given by

$$E_\alpha=\sum_{\alpha_{IJ}=\alpha}c_{IJ}E_{IJ}. \quad (6.77)$$

The matrices $L(Z)$, $M(Z)$ form a Lax pair for the Toda system associated with the affine algebra $\mathcal{G}^{(1)}$.

Similarly, consider now the scaling limits of the twisted Calogero–Moser systems defined with $\delta^\vee=1/h^\vee_{\mathcal{G}}$, where $h^\vee_{\mathcal{G}}$ is the dual Coxeter number. This means that we let $x=X+2\omega_2\delta^\vee\rho^\vee$, $e^z=Zq^{-1/2}$ as in Theorem 2, and set

$$C_{IJ}=\begin{cases}M_{|\alpha|}e^{\delta^\vee\omega_2}c_{IJ}, & \text{when } \alpha_{IJ}=\alpha\in\mathcal{R}(\mathcal{G})\\ 0, & \text{when } \alpha_{IJ}\notin\mathcal{R}(\mathcal{G}).\end{cases} \quad (6.78)$$

Then the matrices $L(z)$, $M(z)$ of the Lax pair for the twisted Calogero–Moser systems B_n, C_n, and F_4 all have finite limits. The entries of their limits are given as follows. For B_n and F_4, the entries of the limit of $L(z)$ and $M(z)$ are given respectively by

$$C_{IJ}\Phi_{IJ}(\alpha\cdot x,z)\to\begin{cases}\pm\kappa_{\mathcal{G}}M_{|\alpha|}c_{IJ}e^{\mp\alpha^\vee/2\cdot X}, & \text{if } l^\vee(\alpha^\vee)=\pm 1;\\ \mp\kappa_{\mathcal{G}}M_{|\alpha|}c_{IJ}e^{\pm\alpha_0^\vee/2\cdot X}, & \text{if } l^\vee(\alpha^\vee)=\pm l_0^\vee;\\ 0 & \text{otherwise}\end{cases} \tag{6.79}$$

and by

$$C_{IJ}\Phi'_{IJ}(\alpha\cdot x,z)\to\begin{cases}-\frac{1}{2}\kappa_{\mathcal{G}}M_{|\alpha|}c_{IJ}e^{\mp\alpha^\vee/2\cdot X}, & \text{if } l^\vee(\alpha^\vee)=\pm 1;\\ -\frac{1}{2}\kappa_{\mathcal{G}}M_{|\alpha|}c_{IJ}e^{\pm\alpha_0^\vee/2\cdot X}, & \text{if } l^\vee(\alpha^\vee)=\pm l_0^\vee;\\ 0 & \text{otherwise.}\end{cases} \tag{6.80}$$

Here $\kappa_{\mathcal{G}}$ is a constant depending on the Lie algebra, with $\kappa_{B_n}=1$ and $\kappa_{F_4}=2$. As for the case of C_n, the entries of the limits of the matrices $L(z)$ and $M(z)$ are given respectively by

$$C_{IJ}\Phi_{IJ}(\alpha\cdot x,z)$$
$$\to\begin{cases}\pm 2M_{|\alpha|}c_{IJ}e^{\mp\alpha^\vee/2\cdot X}, & \text{if } l^\vee(\alpha^\vee)=\pm 1;\\ \mp 2M_{|\alpha|}c_{IJ}e^{\pm\alpha_0^\vee/2\cdot X}Z^{-1/2\mp 1/2}, & \text{if } l^\vee(\alpha^\vee)=\pm l_0, I<J;\\ \mp 2M_{|\alpha|}c_{IJ}e^{\pm\alpha_0^\vee/2\cdot X}Z^{1/2\mp 1/2}, & \text{if } l^\vee(\alpha^\vee)=\pm l_0, J<I;\\ 0 & \text{otherwise,}\end{cases} \tag{6.81}$$

and by

$$C_{IJ}\Phi'_{IJ}(\alpha\cdot x,z)$$
$$\to\begin{cases}-2M_{|\alpha|}c_{IJ}e^{\mp\alpha^\vee/2\cdot X}, & \text{if } l^\vee(\alpha^\vee)=\pm 1;\\ -2M_{|\alpha|}c_{IJ}e^{\pm\alpha_0^\vee/2\cdot X}Z^{-1/2\mp 1/2}, & \text{if } l^\vee(\alpha^\vee)=\pm l_0, I<J;\\ -2M_{|\alpha|}c_{IJ}e^{\pm\alpha_0^\vee/2\cdot X}Z^{1/2\mp 1/2}, & \text{if } l^\vee(\alpha^\vee)=\pm l_0, J<I;\\ 0 & \text{otherwise.}\end{cases} \tag{6.82}$$

In each case, the resulting matrices $L(Z)$, $M(Z)$ form a Lax pair for the Toda system associated with $(\mathcal{G}^{(1)})^\vee$. In other words, starting with the Lax pair for the twisted Calogero–Moser system for B_n, C_n, and F_4, we obtain respectively a Lax pair for the Toda system associated with $(B^{(1)})^\vee = A_{2n-1}^{(2)}$, $(C_n^{(1)})^\vee = D_{n+1}^{(2)}$, and $(F_4^{(1)})^\vee = E_6^{(2)}$.

We note that recently Lax pairs of root type have been considered [61] which correspond, in the above Ansatz (5.3)–(5.5), to Λ equal to the adjoint representation of $\mathcal{G}$ and the coefficients C_{IJ} vanishing for I or J associated with zero weights. This choice yields another Lax pair for the case of E_8. However, as the authors have themselves pointed out [62], the Lax pairs they obtained do not tend to finite limits under the scalings defined in Theorems 1 and 2. Thus the corresponding spectral curves do not appear to be suitable as Seiberg–Witten curves for supersymmetric Yang–Mills theories.

7 Super-Yang–Mills and Calogero–Moser Systems

In this section, we begin by discussing the general correspondence between integrable systems and $\mathcal{N} = 2$ super Yang–Mills theory and then treat the correspondence in detail for $\mathrm{SU}(N)$ gauge group and a hypermultiplet in the adjoint representation of the gauge group. Why the effective prepotential of $\mathcal{N} = 2$ supersymmetric gauge theories can be realized by a fibration of spectral curves remains one of the most important unanswered questions in Seiberg–Witten theory. However, assuming this fact, it is easy to see, as Donagi and Witten pointed out, the emergence of integrable models.

7.1 Correspondence of Seiberg–Witten and Integrable Systems

The key ingredient is the symplectic form ω on the fibration of Jacobians (or Prym varieties) associated to the spectral curves Γ

$$\omega = \delta\Big(\sum_{i=1}^{n} d\lambda(z_i)\Big). \tag{7.1}$$

Here we have identified the Jacobian (or Prym) of Γ with the symmetric product of n copies of Γ, and the differential δ in (7.1) is taken with respect to both the z_i variables and the vacuum parameters u_i for the base of the fibration. Evidently, ω is then of the form $\omega = \omega_{ji} du_j \wedge dz_i$, with no components of the form $du_i \wedge du_j$ or $dz_i \wedge dz_j$. It follows that the Poisson brackets $\{u_k, u_l\}$ all vanish. Thus the vacuum moduli parameters of the gauge theory can be viewed as a maximal pairwise commuting set of Hamiltonians with respect to the symplectic form ω. This correspondence between $\mathcal{N} = 2$ supersymmetric gauge theories and integrable models is quite attractive in its generality. However, it should be stressed that it is probably incomplete. Indeed, in all known cases, one particular Hamiltonian amongst the infinite set of Hamiltonians seems to play a special role, namely as a beta function for the gauge theory. Also, there is at this moment no systematic rule for how to identify the integrable model corresponding to a given gauge theory.

Schematically, the aspects of the correspondence between the data that arise in both settings can be summarized as follows in Table 1.8.

In the table, τ stands for the complex Yang–Mills coupling, $\mathcal{R}$ and m stand for the hypermultiplet representation and mass respectively. The exact integrable system that arises in the correspondence will depend upon

TABLE 1.8. Data map between Seiberg–Witten theory and Integrable systems

$\mathcal{N}=2$ super-Yang–Mills Data	Integrable Systems Data
$(\mathcal{G}, \mathcal{R}, \tau, m, \Gamma, \lambda)$	Integrable Hamiltonian system associated with Lie algebra $\mathcal{G}$, $x_i, p_i \in \mathbb{C}$, $i = 1, \ldots, n = \mathrm{rank}(\mathcal{G})$
S.W. curve Γ	spectral curve $\det\big(kI - L(z)\big) = 0$
S.W. differential $d\lambda$	$d\lambda = k\,dz$
$\omega = \delta\big(\sum_{i=1}^n d\lambda(z_i)\big)$	symplectic form
vacuum moduli a_i	integrals of motion I_i
beta function of RG equation	Hamiltonian

the gauge algebra $\mathcal{G}$ and the representation $\mathcal{R}$ of the hypermultiplet. Its systematic identification is one of the major problems of the study of supersymmetric Yang–Mills theories.

One of the first cases obtained in all generality is when the gauge algebra is simple and no hypermultiplets are present [44]. We have established previously that the Seiberg–Witten curve for $\mathcal{N} = 2$ super-Yang–Mills for $\mathrm{SU}(N)$ gauge group and no hypermultiplets, arrived at by arguments of singularity analysis, R-symmetry properties and limiting behavior, is precisely of the same form as the spectral curve for the period Toda system associated with the Lie algebra $A_{N-1} = \mathrm{SU}(N)$. The correct generalization of this correspondence involves the periodic Toda system associated with the twisted affine Lie algebra $(\mathcal{G}^{(1)})^\vee$. The appearance of the dual affine algebra $(\mathcal{G}^{(1)})^\vee$ is due to the grading of the term $z+\mu/z$ in the spectral curve $\det\big(kI - L(z)\big) = 0$. When $L(z)$, $M(z)$ is the Lax pair for the $\mathcal{G}^{(1)}$ Toda system, this term will appear with the grading $h_\mathcal{G}$, if k and the Casimirs u_l of $\mathcal{G}$ are given gradings 1 and l respectively. Identifying u_l with the vacuum moduli u_l of the gauge theory, we find that the instanton generated term $z + \mu/z$ should have instead grading $h_\mathcal{G}^\vee$, where $h_\mathcal{G}^\vee$ is the dual Coxeter number. This is why the correct integrable system for pure super-Yang–Mills theory with gauge algebra $\mathcal{G}$ should be the periodic Toda system associated with $(\mathcal{G}^{(1)})^\vee$. For the classical Lie algebras $\mathcal{G}$, the spectral curves derived from the $(\mathcal{G}^{(1)})^\vee$ Toda system were found to be in precise agreement with the preceding constructions in terms of the singularity structure of the Seiberg–Witten curves, R-symmetry properties and the matching of various decoupling limits.

Thus the case of pure Yang–Mills theories, with simple gauge algebra $\mathcal{G}$ and no additional hypermultiplets, has been settled. The main problem we wish to address is the identification of the integrable models corresponding to $\mathcal{N} = 2$ supersymmetric Yang–Mills either with product gauge algebras, or with additional matter hypermultiplets. Among these, one theory is particularly attractive and has a certain universal aspect, in the sense that many other theories follow from it in suitable decoupling limits. This is the

Yang–Mills theory with simple gauge algebra $\mathcal{G}$, and one matter hypermultiplet in the adjoint representation of $\mathcal{G}$.

The first case to be treated successfully was $\mathcal{G} = \mathrm{SU}(N)$, where the integrable model was identified by Donagi and Witten [64] as the $\mathrm{SU}(N)$ Hitchin system [80]. Several authors [65] subsequently recognized the spectral curves of the $\mathrm{SU}(N)$ Hitchin system as identical to the spectral curves of the $\mathrm{SU}(N)$ elliptic Calogero–Moser system. This correspondence between $\mathrm{SU}(N)$ Seiberg–Witten theory and $\mathrm{SU}(N)$ Calogero–Moser system was however still rather rudimentary, since the vacuum parameters of the Yang–Mills theory were still obscure, and the prepotential arising from the Calogero–Moser system were not yet known to satisfy the monodromy properties required by field theory. Our purpose in this section is to address these issues in the case of $\mathcal{G} = \mathrm{SU}(N)$, and to show that for general $\mathcal{G}$, the integrable model corresponding to the Yang–Mills theory with a hypermultiplet in the adjoint representation is the twisted Calogero–Moser system associated to the Lie algebra $\mathcal{G}$.

7.2 Calogero–Moser and Seiberg–Witten Theory for SU(N)

The full correspondence between Seiberg–Witten theory for $\mathcal{N} = 2$ $\mathrm{SU}(N)$ super-Yang–Mills theory with one hypermultiplet in the adjoint representation of the gauge algebra, and the elliptic $\mathrm{SU}(N)$ Calogero–Moser systems was obtained in [66]. We describe it here in some detail.

All that we shall need here of the elliptic Calogero–Moser system is its Lax operator $L(z)$, whose $N \times N$ matrix elements are given by

$$L_{ij}(z) = p_i \delta_{ij} - m(1 - \delta_{ij}) \Phi(x_i - x_j, z). \tag{7.2}$$

Notice that the Hamiltonian is simply given in terms of L by $H(x,p) = \frac{1}{2} \operatorname{tr} L(z)^2 + C\wp(z)$ with $C = -\frac{1}{2} m^2 N(N-1)$.

The correspondence between the data of the elliptic Calogero–Moser system and those of the Seiberg–Witten theory is as follows.

- The parameter m in (7.2) is the hypermultiplet mass;
- The gauge coupling g and the θ-angle are related to the modulus of the torus (or elliptic curve) $\Sigma = \mathbb{C}/(2\omega_1\mathbb{Z} + 2\omega_2\mathbb{Z})$ by

$$\tau = \frac{\omega_2}{\omega_1} = \frac{\theta}{2\pi} + \frac{4\pi i}{g^2}; \tag{7.3}$$

- The Seiberg–Witten curve Γ is the spectral curve of the elliptic Calogero–Moser model, defined by

$$\Gamma = \{(\tilde{k}, z) \in \mathbb{C} \times \Sigma, \det\big(\tilde{k}I - L(z)\big) = 0\} \tag{7.4}$$

and the Seiberg–Witten 1-form is $d\lambda = k\, dz$. Γ is invariant under the Weyl group of $\mathrm{SU}(N)$. It can be viewed as N copies of the torus Σ, glued along suitable cuts on each copy. We use here the notation $\tilde{k}$ in the equation for the spectral curve, in order to reserve the notation k for a more convenient variable to be introduced later.

• Using the Lax equation $\dot{L} = [L, M]$, it is clear that the spectral curve is independent of time, and can be dependent only upon the constants of motion of the Calogero–Moser system, of which there are only N. These integrals of motion may be viewed as parameterized by the quantum moduli of the Seiberg–Witten system.

• Finally, $d\lambda = \tilde{k}\, dz$ is meromorphic, with a simple pole on each of the N sheets above the point $z = 0$ on the base torus. The residue at each of these poles is proportional to m, as required by the general set-up of Seiberg–Witten theory, explained in §2.

7.3 Four Fundamental Theorems

While the above mappings of the Seiberg–Witten data onto the Calogero–Moser data are certainly natural, there is no direct proof for them, and it is important to check that the results inferred from it agree with known facts from quantum field theory. To establish this, as well as a series of further predictions from the correspondence, we give four theorems (the proofs may be found in [66] for the first three theorems, and in [67] for the last one).

Theorem 4. *The spectral curve equation* $\det\big(\tilde{k}I - L(z)\big) = 0$ *can be expressed as*

$$\begin{aligned} &det\big(\tilde{k} - L(z)\big) \\ &\quad = \frac{\vartheta_1\big((z - m\partial/\partial k)/(2\omega_1) \mid \tau\big)}{\vartheta_1(z/(2\omega_1) \mid \tau)} H(k)\bigg|_{k=\tilde{k}+m\partial_z \ln \vartheta_1(z/(2\omega_1)|\tau)} = 0 \end{aligned} \tag{7.5}$$

where $H(k)$ *is a monic polynomial in* k *of degree* N*, whose zeros (or equivalently whose coefficients) correspond to the moduli of the gauge theory. If* $H(k) = \prod_{i=1}^{N}(k - k_i)$*, then*

$$\lim_{q \to 0} \frac{1}{2\pi i} \oint_{A_i} \tilde{k}\, dz = k_i - \frac{1}{2} m. \tag{7.6}$$

Here, ϑ_1 is the Jacobi ϑ-function, which admits the following simple series expansion in powers of the instanton factor $q = e^{2\pi i \tau}$

$$\vartheta_1(u \mid \tau) = \sum_{r \in 1/2 + \mathbb{Z}} q^{r^2/2} e^{2\pi i r(u+1/2)}. \tag{7.7}$$

Thus, in terms of the new variable k defined by

$$k = \tilde{k} + m\partial_z \ln \vartheta_1\left(\frac{z}{2\omega_1} \,\middle|\, \tau\right) + \frac{1}{2} m \tag{7.8}$$

the equation of the spectral curve equation can be written perturbatively in q as

$$\sum_{n\in\mathbb{Z}}(-)^n q^{n(n-1)/2} e^{nz} H(k - n\cdot m) = 0 \tag{7.9}$$

where we have set $\omega_1 = -i\pi$ without loss of generality. The series expansion (7.9) is superconvergent and sparse in the sense that it receives contributions only at integers that grow like n^2. But more important, it provides a geometric interpretation of the order parameters of the super Yang–Mills theory: at $q = 0$, the base torus Σ degenerates into a sphere with two punctures at $w = 0$ and $w = \infty$, if we introduce the new variable

$$w = e^z. \tag{7.10}$$

Each of the copies of the torus Σ making up the spectral curve Γ upstairs degenerates correspondingly into a sphere with two punctures. Now the equation (7.9) of the spectral curve Γ at $q = 0$ becomes

$$H(k) - wH(k - m) = 0. \tag{7.11}$$

Thus the punctures lying above $w = 0$ are given by the N zeroes k_i of the polynomial $H(k)$, while those lying above $w = \infty$ are given by $k_i - m$. Now the A-cycles on Γ can be chosen to be circles of a fixed radius around each of the points k_i. At the degeneration point $q = 0$, the Seiberg–Witten differential $d\lambda = \tilde{k}\, dz$ can be written as

$$d\lambda = \tilde{k}\, d\ln w = k\, d\ln\frac{H(k)}{H(k-m)} - m\, d\ln\vartheta_1\left(\frac{z}{2\pi i}\,\middle|\,\tau\right) - \frac{1}{2}m\, dz. \tag{7.12}$$

The second terms on the right does not contribute to A-periods, while the other two terms readily give

$$\frac{1}{2\pi i}\oint_{A_i} d\lambda_{|q=0} = k_i - \frac{1}{2}m, \tag{7.13}$$

as asserted in Theorem 4. This relation provides the key starting point of the correspondence between Calogero–Moser systems and supersymmetric gauge theories, by identifying the classical vacua parameters in terms of the moduli of the Calogero–Moser spectral curves.

The parametrization of spectral curves by polynomials $H(k)$ has been extended from SU(N) Calogero–Moser systems to SU(N) spin Calogero–Moser systems in [73], using methods of quantum field theory on Riemann surfaces [68]. For SU(N) Calogero–Moser systems, a different derivation from [66] has been recently obtained in [69].

The expansion (7.9) actually allows us to determine the exact value of the quantum vacua parameters a_i as power series in q. Geometrically, for $q \neq 0$, each copy of the torus making up Γ has a branch cut of length

$O(|q|^{1/2})$ which shrinks to the puncture k_i as $q \to 0$. But the contours A_i can still be chosen at a fixed, nonvanishing distance from k_i. Thus the A_i-periods of $d\lambda$ can be evaluated by residue formulas, just as in the case of the fundamental representation. More specifically, applying the methods in [26] for the perturbative solution of fixed point equations, we find

$$w = \frac{H(k)}{H(k-m)} \left[1 + \sum_{n=1}^{\infty} \frac{q^n}{n!} \frac{\partial^n}{\partial y^n} F^n(y)_{|y=1} \right] \tag{7.14}$$

where the function $F(y)$ is defined by

$$\begin{aligned} F(y) &= \sum_{n=1}^{\infty} q^{\frac{1}{2}n(n+1)} (-)^n [y^{-n} \eta_n^+(k) - y^{n+1} \eta_n^-(k-m)] \\ \eta_n^{\pm}(k) &= \frac{H(k \pm mn) H(k \mp m)}{H(k)^{n+1}}. \end{aligned} \tag{7.15}$$

We would like to stress that the expansion (7.15) is a reliable expansion only when the denominators of the expressions $\eta_n^{\pm}(k)$ and $\eta_n^{\pm}(k-m)$ are bounded away from 0. As explained earlier, this can be achieved in the evaluation of the A-periods, by fixing the A-cycles away from the points k_i. We can deduce then the full expansion in q of the Seiberg–Witten differential $d\lambda = \tilde{k} d \ln w$. Each term in the expansion is rational in k with poles at k_i and $k_i - \beta m$, and the corresponding residues are easily evaluated. For example, we obtain to order q^2

$$\begin{aligned} a_i &= k_i + q\bar{S}_i(k_i) + \tfrac{1}{4} q^2 (\bar{S}_i)'''(k_i) + \mathcal{O}(q^3) \\ \bar{S}_i(k) &\equiv \frac{H(k+m)H(k-m)}{\prod_{j \neq i}(k-k_j)^2}. \end{aligned} \tag{7.16}$$

For the super Yang–Mills theory with matter in the adjoint representation, the microscopic gauge coupling $q = e^{2\pi i \tau}$ is analogous to the renormalization scale Λ encountered earlier for the fundamental representation, and the full expansion in q of the vacua parameters a_i is required to evaluate the contributions to the prepotential $\mathcal{F}$ of instanton processes to an arbitrary order.

Theorem 5. *The prepotential of the Seiberg–Witten theory obeys a renormalization group-type equation that simply relates $\mathcal{F}$ to the Calogero–Moser Hamiltonian, expressed in terms of the quantum order parameters a_j*

$$a_j = \frac{1}{2\pi i} \oint_{A_j} d\lambda \quad \frac{\partial \mathcal{F}}{\partial \tau}\bigg|_{a_j} = H(x,p) = \frac{1}{2} \sum_{i=1}^{N} k_i^2. \tag{7.17}$$

Furthermore, in an expansion in powers of the instanton factor $q = e^{2\pi i \tau}$, the quantum order parameters a_j may be computed by residue methods in terms of the zeros of $H(k)$. (The polynomial $H(k)$ is not to be confused with the Calogero–Moser Hamiltonian $H(x,p)$; the notation is conventional.)

The microscopic gauge coupling τ corresponds geometrically to the moduli of the base torus Σ, so that the renormalization group equation (7.17) is mathematically a statement about deformations of complex structures. The main underlying equation is the following

$$\delta a_{Di} = \sum_{j=1}^{N} \oint_{A_j} \tilde{k}\Omega_i \tag{7.18}$$

where Ω_i, $1 \leq i \leq N$, is the basis of holomorphic Abelian integrals dual to the A_i basis of cycles. It can be derived by showing that the deformation $\delta d\lambda$ of the Seiberg–Witten differential satisfies a $\bar{\partial}$-equation, and solving this equation in terms of the prime form [66]. Now the Abelian differentials Ω_i can also be written as

$$\Omega_i = \frac{1}{2\pi i}\frac{\partial}{\partial a_i} d\lambda_i. \tag{7.19}$$

The equation (7.18) implies

$$\frac{\partial}{\partial \tau}\left(\frac{\partial \mathcal{F}}{\partial a_i}\right) = \frac{1}{4\pi i}\frac{\partial}{\partial a_i}\left(\sum_{j=1}^{N} \oint_{A_j} \tilde{k}^2\, dz\right) \tag{7.20}$$

from which the renormalization group equation (7.17) follows.

The general form of the prepotential $\mathcal{F}$ for the SU(N) super Yang–Mills theory with matter in the adjoint representation is

$$\mathcal{F} = \mathcal{F}^{(\mathrm{pert})} + \sum_{n=1}^{\infty} q^n \mathcal{F}^{(n)}, \tag{7.21}$$

where $\mathcal{F}^{(\mathrm{pert})}$ is the perturbative part which can be determined by standard methods of quantum field theory

$$\begin{aligned}\mathcal{F}^{(\mathrm{pert})} = \frac{\tau}{2}\sum_i a_i^2 - \frac{1}{8\pi i}\sum_{i,j} &[(a_i - a_j)^2 \ln(a_i - a_j)^2 \\ &- (a_i - a_j - m)^2 \ln(a_i - a_j - m)^2],\end{aligned} \tag{7.22}$$

and $\mathcal{F}^{(n)}$ are the terms due to instanton processes. The renormalization group equation (7.17) shows that all the coefficients $\mathcal{F}^{(n)}$ can just be read off from an expansion of the Hamiltonian $H(x,p)$ in terms of the vacua parameters a_i. Such an expansion is easily obtained by writing $H(x,p)$ in terms of the Calogero–Moser integrals of motion k_i, and inverting the relations (7.16) giving a_i in terms of k_i. For example, to two-instanton order, the results can be described as in Part (b) of the following theorem [66]:

Theorem 6. (a) *The perturbative part $\mathcal{F}^{(\mathrm{pert})}$ of the prepotential given by the Calogero–Moser spectral curves is indeed of the form (7.22);*

(b) *To two-instanton order, the instanton corrections can be expressed in terms of a single function*

$$S_i(a) = \frac{\prod_{j=1}^N [(a_i - a_j)^2 - m^2]}{\prod_{j\neq i}(a - a_j)^2}, \tag{7.23}$$

as follows

$$\begin{aligned}\mathcal{F}^{(1)} &= \frac{1}{2\pi i}\sum_i S_i(a_i) \\ \mathcal{F}^{(2)} &= \frac{1}{8\pi i}\Bigg[\sum_i S_i(a_i)\partial_i^2 S_i(a_i) \\ &\qquad + 4\sum_{i\neq j}\frac{S_i(a_i)S_j(a_j)}{(a_i - a_j)^2} - \frac{S_i(a_i)S_j(a_j)}{(a_i - a_j - m)^2}\Bigg].\end{aligned} \tag{7.24}$$

Part (a) of Theorem 6 cannot be obtained from the renormalization group equation. We discuss briefly its proof, which is important, because it is the defining criterion for the Calogero–Moser spectral curves to provide an exact solution of the four-dimensional supersymmetric gauge theory. The main problem in establishing (a) is to determine the leading terms in the periods a_{Di} of the Seiberg–Witten differential $d\lambda$. Recall that as q moves away from 0, each puncture k_i and $k_i + m$ on each sheet above the base torus Σ opens into a cut. Let $k_i^{\pm}$ be the end points of the cut near k_i. They can be identified as the solutions of the equation

$$\frac{dk}{dz} = 0. \tag{7.25}$$

This equation can be solved perturbatively in q. To $O(q)$ order, we find

$$k_i^{\pm} = k_i \pm q^{1/2}k_i^{(1)}, \quad k_i^{(1)} = 2\frac{H^{1/2}(k_i - m)H^{1/2}(k_i + m)}{\prod_{j\neq i}(k_i - k_j)}. \tag{7.26}$$

Similarly, the end points of the cut near $k_i + \beta m$ can be identified as $k_i^{\pm} + m$.
The B-cycles for the spectral curve Γ can be chosen as paths from k_i^+ to $k_i^+ + m$. Thus the dual periods a_{Di} are given by

$$a_{Di} = \frac{1}{2\pi i}\int_{k_i^+}^{k_i^+ + m} d\lambda = \frac{1}{2\pi i}\int_{k_i^+}^{k_i^+ + m} k\, d\ln w. \tag{7.27}$$

The most difficult step in the evaluation of (7.27) is how to express the differential $d\ln w$ in terms of dk. For the periods a_i, this step was easy, because the A-cycles could be kept at fixed distance from k_i. Here, the B-cycles end at k_i^+, which is at distance $q^{1/2} \to 0$ from k_i, in view of (7.26).

Thus the expansion (7.15) for w does not apply in the present case. However, it turns out that an approximation for w as a function of k can be found to order $O(q)$. It is given by

$$w = \frac{H(k)}{H(k-m)} \times \frac{1+\sqrt{1-4q\eta_1^+(k)}}{1+\sqrt{1-4q\eta_1^-(k-m)}} \tag{7.28}$$

where the functions $\eta_1^+(k)$ and $\eta_1^-(k-m)$ are defined in (7.15). Expanding in q, the differential $d\lambda$ can now be expressed in terms of rational functions of k. The same analytic continuation methods used in the case of the fundamental representation apply, giving the leading terms of a_{Di}. The perturbative part $\mathcal{F}^{(\text{pert})}$ can be read off from $a_{Di} = \partial\mathcal{F}/\partial a_i$, establishing (a).

The perturbative corrections to the prepotential are of course predicted by asymptotic freedom. The formulas in (b) for the instanton corrections $\mathcal{F}^{(1)}$ and $\mathcal{F}^{(2)}$ are new, as they have not yet been computed by direct field theory methods. Perturbative expansions of the prepotential in powers of m have also been obtained in [70].

The moduli k_i, $1 \le i \le N$, of the gauge theory are evidently integrals of motion of the system. To identify these integrals of motion, denote by S any subset of $\{1,\ldots,N\}$, and let $S^* = \{1,\ldots,N\} \setminus S$, $\wp(S) = \wp(x_i - x_j)$ when $S = \{i,j\}$. Let also p_S denote the subset of momenta p_i with $i \in S$. We have [67]:

Theorem 7. *For any K, $0 \le K \le N$, let $\sigma_K(k_1,\ldots,k_N) = \sigma_K(k)$ be the Kth symmetric polynomial of $(k_1,\ldots,k_N)$, defined by $H(u) = \sum_{K=0}^N (-)^K \times \sigma_K(k) u^{N-K}$. Then*

$$\sigma_K(k) = \sigma_K(p) + \sum_{l=1}^{[K/2]} m^{2l} \sum_{\substack{|S_i \cap S_j| = 2\delta_{ij} \\ 1 \le i,j \le l}} \sigma_{K-2l}\Big(p_{(\cup_{i=1}^l S_i)^*}\Big) \prod_{i=1}^l \left[\wp(S_i) + \frac{\eta_1}{\omega_1}\right]. \tag{7.29}$$

The proof of Theorem 7 requires some new elliptic function identities, linking combinations of the function $\Phi(x,z)$ with determinants of the $\bar\delta$-operator on the torus [68]. These identities indicate a close relationship between elliptic Calogero–Moser systems and free fermions.

7.4 Partial Decoupling of Hypermultiplet, Product Groups

The spectral curves of certain gauge theories can be easily derived from the Calogero–Moser curves by a partial decoupling of the hypermultiplet. Indeed,

- the masses of the gauge multiplet and hypermultiplet are $|a_i - a_j|$ and $|a_i - a_j + m|$. In suitable limits, some of these masses become ∞, and states with infinite mass decouple. The remaining gauge group is a subgroup of SU(N).

- When the effective coupling of a gauge subgroup is 0, the dynamics freeze and the gauge states become noninteracting.

Nontrivial decoupling limits arise when $\tau \to \infty$ and $m \to \infty$. When all a_i are finite, we obtain the pure Yang–Mills theory. When some hypermultiplets masses remain finite, the $U(1)$ factors freeze, the gauge group SU(N) is broken down to SU(N_1) $\times \cdots \times$ SU(N_p), and the remaining hypermultiplets are in, e.g., fundamental or bifundamental representations. For example, let $N = 2N_1$ be even, and set

$$k_i = v_1 + x_i, \; k_{N_1+j} = v_2 + y_j, \quad 1 \leq i, j \leq N_1, \tag{7.30}$$

with $\sum_{i=1}^{N_1} x_i = \sum_{j=1}^{N_1} y_j = 0$. (The term $v = v_1 - v_2$ is associated to the U(1) factor of the gauge group.) In the limit $m \to \infty$, $q \to 0$, with x_i, y_j, $\mu = v - m$ and $\Lambda = mq^{\frac{1}{N}}$ kept fixed, the theory reduces to a SU(N_1)$\times$SU(N_1) gauge theory, with a hypermultiplet in the bifundamental $(N_1, \bar{N}_1) \oplus (\bar{N}_1, N_1)$, and spectral curve

$$A(x) - t(-)^{N_1} B(x) - 2^{N_1} \Lambda^{N_1} \left(\frac{1}{t} - t^2 \right) = 0, \tag{7.31}$$

where $A(x) = \prod_{i=1}^{N_1}(x - x_i)$, $B(x) = \prod_{j=1}^{N}(x + \mu - y_j)$, $t = e^z$. This agrees with the curve found by Witten [71] using M Theory, and by Katz, Mayr, and Vafa [72] using geometric engineering.

The prepotential of the SU(N_1) $\times$ SU(N_1) theory can be also read off the Calogero–Moser prepotential. It is convenient to introduce $x_i^{(I)}$, $I = 1, 2$, by $x_i^{(1)} = x_i$, $x_i^{(2)} = y_i$, $1 \leq i \leq N_1$. Set

$$\begin{gathered} A_i^I = \prod_{\substack{j \neq i \\ j \in I}} (x - x_j^{(I)}), \quad B^I(x) = \prod_{\substack{j \in J \\ |I-J|=1}} (\mu \pm (x - x_j^{(J)})), \\ S_i^I(x) = \frac{B^I(x)}{A_i^I(x)^2}, \end{gathered} \tag{7.32}$$

where the $\pm$ sign in $B^I(x)$ is the same as the sign of $J - I$. Then the the first two orders of instanton corrections to the prepotential for the SU(N_1) $\times$ SU(N_1) theory are given by

$$\mathcal{F}^{(1)}_{\mathrm{SU}(N_1)\times\mathrm{SU}(N_1)} = \frac{(-2\Lambda)^{N_1}}{2\pi i} \sum_{I=1,2} \sum_{i \in I} S_i^I (x_i^{(I)})$$

$$\mathcal{F}^{(2)}_{\mathrm{SU}(N_1)\times\mathrm{SU}(N_1)} = \frac{(-2\Lambda)^{2N_1}}{8\pi i} \sum_{I=1,2} \sum_{i\in I} S_i^I\big(x_i^{(I)}\big) \frac{\partial^2 S_i^I\big(x_i^{(I)}\big)}{\partial x_i^{(I)2}} + \sum_{\substack{i\neq j\\ i,j\in I}} \frac{S_i^I\big(x_i^{(I)}\big) S_j^I\big(x_j^{(I)}\big)}{\big(x_i^{(I)} - x_j^{(I)}\big)^2}. \tag{7.33}$$

These formulas can serve as useful checks on candidates for Seiberg–Witten curves for theories with product gauge groups obtained by other methods [29, 71].

8 Calogero–Moser and Seiberg–Witten for General $\mathcal{G}$

We consider now the $\mathcal{N} = 2$ supersymmetric gauge theory for a general simple gauge algebra $\mathcal{G}$ and a hypermultiplet of mass m in the adjoint representation.

8.1 The General Case

Then, we have the following results, established in [74]:

• the Seiberg–Witten curve of the theory is given by the spectral curve $\Gamma = \big\{(k,z) \in \mathbb{C}\times\Sigma; \det\big(kI - L(z)\big) = 0\big\}$ of the *twisted* elliptic Calogero–Moser system associated to the Lie algebra $\mathcal{G}$. The Seiberg–Witten differential $d\lambda$ is given by $d\lambda = k\,dz$.

• The function $R(k,z) = \det\big(kI - L(z)\big)$ is polynomial in k and meromorphic in z. The spectral curve Γ is invariant under the Weyl group of $\mathcal{G}$. It depends on n complex moduli, which can be thought of as independent integrals of motion of the Calogero–Moser system.

• The differential $d\lambda = k\,dz$ is meromorphic on Γ, with simple poles. The position and residues of the poles are independent of the moduli. The residues are linear in the hypermultiplet mass m. (Unlike the case of $\mathrm{SU}(N)$, their exact values are difficult to determine for general $\mathcal{G}$).

• In the $m \to 0$ limit, the Calogero–Moser system reduces to a free system, the spectral curve Γ is just the product of several unglued copies of the base torus Σ, indexed by the constant eigenvalues of $L(z) = p\cdot h$. Let k_i, $1 \leq i \leq n$, be n independent eigenvalues, and A_i, B_i be the A and B cycles lifted to the corresponding sheets. For each i, we readily obtain

$$\begin{aligned} a_i &= \frac{1}{2\pi i}\oint_{A_i} d\lambda = \frac{k_i}{2\pi i}\oint_A dz = \frac{2\omega_1}{2\pi i}k_i \\ a_{Di} &= \frac{1}{2\pi i}\oint_{B_i} d\lambda = \frac{k_i}{2\pi i}\oint_B dz = \frac{2\omega_1}{2\pi i}\tau k_i. \end{aligned} \tag{8.1}$$

Thus the prepotential $\mathcal{F}$ is given by $\mathcal{F} = (\tau/2)\sum_{i=1}^n a_i^2$. This is the classical prepotential and hence the correct answer, since in the $m \to 0$ limit, the theory acquires an $\mathcal{N} = 4$ supersymmetry, and receives no quantum corrections.

• The $m \to \infty$ limit is the crucial consistency check, which motivated the introduction of the *twisted* Calogero–Moser systems in the first place [59, 60]. In view of Theorem 2 and subsequent comments, in the limit $m \to \infty$, $q \to 0$, with

$$\begin{aligned} x &= X + 2\omega_2 \frac{1}{h_{\mathcal{G}}^\vee} \rho \\ m &= M q^{-1/(2h_{\mathcal{G}}^\vee)} \end{aligned} \tag{8.2}$$

with X and M kept fixed, the Hamiltonian and spectral curve for the twisted elliptic Calogero–Moser system with Lie algebra $\mathcal{G}$ reduce to the Hamiltonian and spectral curve for the Toda system for the affine Lie algebra $(\mathcal{G}^{(1)})^\vee$. This is the correct answer. Indeed, in this limit, the gauge theory with adjoint hypermultiplet reduces to the pure Yang–Mills theory, and the Seiberg–Witten spectral curves for pure Yang–Mills with gauge algebra $\mathcal{G}$ have been shown by Martinec and Warner [44] to be the spectral curves of the Toda system for $(\mathcal{G}^{(1)})^\vee$.

• As in the known correspondences between Seiberg–Witten theory and integrable models [66], we expect the following equation

$$\frac{\partial \mathcal{F}}{\partial \tau} = H_{\mathcal{G}}^{\text{twisted}}(x, p), \tag{8.3}$$

to hold. Note that the left-hand side can be interpreted in the gauge theory as a renormalization group equation.

• For simple laced $\mathcal{G}$, the curves $R(k, z) = 0$ are modular invariant. Physically, the gauge theories for these Lie algebras are self-dual. For non-simply laced $\mathcal{G}$, the modular group is broken to the congruence subgroup $\Gamma_0(2)$ for $\mathcal{G} = \mathrm{B}_n, \mathrm{C}_n, \mathrm{F}_4$, and to $\Gamma_0(3)$ for G_2. The Hamiltonians of the twisted Calogero–Moser systems for non-simply laced $\mathcal{G}$ are also transformed under Landen transformations into the Hamiltonians of the twisted Calogero–Moser system for the dual algebra $\mathcal{G}^\vee$. It would be interesting to determine whether such transformations exist for the spectral curves or the corresponding gauge theories themselves.

Spectral curves for certain gauge theories with classical gauge algebras and matter in the adjoint representation have also been proposed in [71] and [72], based on branes and M theory. Some generalizations of the construction by Witten [71] for $\mathrm{SU}(N)$ were given in [75–77]. Relations between the dynamics of gauge theories from branes and integrable systems were proposed and analyzed in [78]. A possible role in Seiberg–Witten theory for Ruijsenaars-Schneider and related relativistic systems has been investigated in [79].

8.2 Spectral Curves for Low Rank

In the case of $\mathcal{G} = \mathrm{D}_n$, the trigonometric limit ($q \to 0$) of the spectral curve $R(k,z) = 0$ can be written down explicitly and takes a particularly simple form (see (8.16) below). Now the equation $R(k,z) = 0$ for the spectral curve is polynomial in p_j, $\wp$, g_2, g_3, and Δ, where g_2, g_3 are the usual Eisenstein series of weights 4 and 6, and $\Delta = g_2^3 - 27g_3^2$ is the discriminant. In the limit $q \to 0$, only Δ vanishes. Since the weight of Δ is 12 and the degree of the spectral curves is $2n$, the spectral curves are determined by their trigonometric limit when $n \leq 5$.

The derivation of the trigonometric limit is as follows. Let Z be the spectral parameter defined by

$$\frac{1}{Z} = \frac{1}{2}\coth\frac{z}{2} - \frac{1}{z}. \tag{8.4}$$

Then in the trigonometric limit $\tau \to i\infty$, the elliptic functions $\wp(z)$ and $\Phi(x,z)$ reduce to

$$\wp(z) \to \frac{1}{Z^2} - \frac{1}{6} \tag{8.5}$$

$$\Phi(x,z) \to \frac{1}{2}\coth\frac{x}{2} - \frac{1}{Z}. \tag{8.6}$$

Now the spectral curve Γ depends on the dynamical variables p_i, x_i, but only through n combinations $u_i = u_i(m)$. At $m = 0$, $u_i(m)$ are just the n independent Casimirs of $\mathcal{G}$, which are polynomials in p of degrees $\gamma_i + 1$. As m is deformed away from 0, the $u_i(m)$ are still recognizable by their leading p behavior. Thus we may carry out our calculations with any choice of the variables x_i. A particularly convenient choice is

$$x = \xi\rho^\vee, \quad \alpha\cdot x = \xi l(\alpha), \quad \xi \to \infty. \tag{8.7}$$

Then

$$\Phi(\alpha\cdot x, z) \to -\frac{1}{Z} + \begin{cases} +\frac{1}{2}, & \text{if } \alpha > 0 \\ -\frac{1}{2}, & \text{if } \alpha < 0. \end{cases} \tag{8.8}$$

Let $P = \mathrm{diag}(p_1, \ldots, p_n)$, $\mu = \mu^+ + \mu^-$, with $\mu^\pm$ given by

$$\mu_{ij}^+ = \begin{cases} 1, & \text{if } i < j, \\ 0, & \text{if } i \geq j, \end{cases} \qquad \mu_{ij}^- = \begin{cases} 1, & \text{if } i > j, \\ 0, & \text{if } i \leq j. \end{cases} \tag{8.9}$$

Then the function $R(k,z)$ can be expressed as

$$R(k,z) = \det\begin{pmatrix} kI - P + m\mu/Z - m(\mu^+ - \mu^-)/2 & (m/Z - m/2)\mu \\ (m/Z + m/2)\mu & kI + P + m\mu/z + m(\mu^+ - \mu^-)/2 \end{pmatrix} \tag{8.10}$$

combinations, we find

$$R(k,z) = \det\left[(kI + P - m\mu^-)(kI - P - m\mu^+) + k\left(m + 2\frac{m}{Z}\right)\mu\right]. \quad (8.11)$$

Although each of the two factors in the above determinant is triangular, the determinant is still difficult to evaluate due to the presence of the third term $k(m + m/Z)\mu$. It is here that we must introduce the rank 1 matrix $\mu + I$, and make a shift similar to the key one in the study of SU(N) Calogero–Moser systems: defining a new variable A by

$$0 = A^2 + mA + 2k\frac{m}{Z} - k^2 \quad (8.12)$$

we can write

$$R(k,z) = \det\Big[(AI + P - m\mu^-)(AI - P - m\mu^-) + \left(mA + 2k\frac{m}{Z}\right)(\mu + I)\Big]. \quad (8.13)$$

Since $\mu + I = uu^T$ with $u^T = (1, \ldots, 1)$, and $\det(M + uu^T) = \det M(1 + u^T M^{-1} u)$ for any invertible matrix M, ( ) can be expressed as

$$R(k,z) = \prod_{j=1}^{n}(A^2 - p_j^2) + \left(mA + 2\frac{m}{Z}\right)\sum_{j=1}^{n}\prod_{i=1}^{j-1}((A+m)^2 - p^2)\prod_{i=j+1}^{n}(A^2 - p_i^2). \quad (8.14)$$

In analogy with the SU(N) case, we introduce a polynomial $H(A)$ by

$$H(A) = \prod_{j=1}^{n}(A^2 - p_j^2) = \sum_{j=0}^{n}(-1)^{n-j}A^{2j}u_{2n-2j}. \quad (8.15)$$

The final expression for $R(k,z)$ is in terms of $H(A)$

$$R(k,z) = \frac{m^2 + mA - 2km/z}{m^2 + mA}H(A) + \frac{mA + 2km/Z}{m^2 + 2mA}H(A+m). \quad (8.16)$$

The similarity with the equation (7.5) for the SU(N) Calogero–Moser spectral curves is now manifest. Just as in the case of SU(N), the parameters p_j can be identified with the classical vacuum parameters of the gauge theory.

We derive now explicit formulas for $R(k,z)$ for $n \leq 5$. Let

$$R(k,z) = \sum_{j=0}^{n}(-1)^{n-j}P_{2j}u_{2n-2j}. \quad (8.17)$$

Then (8.16) implies the following recursive relation

$$0 = P_{2(j+1)} - \left(2k^2 + m^2 - 4k\frac{m}{Z}\right)P_{2j} + k^2\left(k - 2\frac{m}{Z}\right)^2 P_{2(j-1)} \tag{8.18}$$

with $P_0 = 1$, $P_2 = k^2$. This works out to

$$P_4 = k^4 - 4k^2\frac{m^2}{Z^2} + m^2k^2 \tag{8.19}$$

$$\begin{aligned}P_6 = k^6 &- 12k^4{\frac{m^2}{Z}}^2 + 16k^3\frac{m^3}{Z^3} - 4k^3\frac{m^3}{Z} - 4k^2\frac{m^4}{Z^2}\\ &+ 3k^4m^2 + k^2m^4\end{aligned} \tag{8.20}$$

$$\begin{aligned}P_8 = k^8 &- 24k^6\frac{m^2}{Z^2} + 6k^6m^2 + 64k^5\frac{m^3}{Z^3} - 16k^5\frac{m^3}{Z} - 48k^4\frac{m^4}{Z^4}\\ &- 8k^4\frac{m^4}{Z^2} + 5k^4m^4 + 32k^3\frac{m^5}{Z^3} - 8k^3\frac{m^5}{Z} - 4k^2\frac{m^6}{Z^2} + k^2m^6\end{aligned} \tag{8.21}$$

$$\begin{aligned}P_{10} = k^{10} &- 40k^8\frac{m^2}{Z^2} + 160k^7\frac{m^3}{Z^3} - 240k^6\frac{m^4}{Z^4} + 128k^5\frac{m^5}{Z^5}\\ &+ m^2\left[10k^8 - 40k^7\frac{m}{Z} + 160k^5\frac{m^3}{Z^3} - 160k^4\frac{m^4}{Z^4}\right]\\ &+ m^4\left[15k^6 - 48k^5\frac{m}{Z} + 12k^4\frac{m^2}{Z^2} + 48k^3\frac{m^3}{Z^3}\right]\\ &+ m^6\left[7k^4 - 12k^3\frac{m}{Z} - 4k^2\frac{m^2}{Z^2}\right] + m^8k^2.\end{aligned} \tag{8.22}$$

Let now $0 = R(k, z) = \sum_{j=0}^{n} Q_{2j}(k)u_{2n-2j}$ be the equation of the spectral curve for general q. Using the limit of $\wp$ and its derivatives, we may identify the functional dependence on $\wp(x)$ which gave rise to each of the terms in (8.18). The result is

$$Q_0 = 1 \tag{8.23}$$

$$Q_2 = k^2 \tag{8.24}$$

$$Q_4 = k^4 - 4k^2m^2\wp \tag{8.25}$$

$$Q_6 = k^6 - 12k^4m^2\wp - 8k^3m^3\wp' \tag{8.26}$$

$$Q_8 = k^8 - 24k^6m^2\wp - 32k^5m^3\wp' - 48k^4m^4\wp^2 + 64g_2k^2m^6\wp \tag{8.27}$$

$$\begin{aligned}Q_{10} = k^{10} &- 40k^8m^2\wp - 80k^7m^3\wp' - 240k^6m^4\wp^2 - 64k^5m^5\wp\wp'\\ &+ 704g_2k^4m^6\wp + 512g_2k^3m^7\wp' - 768k^2m^8g_3\wp.\end{aligned} \tag{8.28}$$

This completes our derivation of the spectral curves for D_n for $n \leq 5$.

8.3 Perturbative Prepotential for SO(2n)

With the parametrization (8.16) in terms of the polynomial $H(A)$, it is now easy to evaluate the logarithmic terms of the prepotential $\mathcal{F}$. We have already shifted the variable k to the variable A defined by (8.12). A complete shift of the variables (k, z) to new variables (A, u) is obtained by setting

$$e^u = \frac{k^2 - (A+m)^2}{k^2 - A^2}. \tag{8.29}$$

The equation (8.16) of the curve becomes

$$e^u = \frac{H(A+m)}{H(A)}. \tag{8.30}$$

It remains to determine the Seiberg–Witten differential $d\lambda = kdz$ in terms of A and u. Evidently, k can be solved in terms of A and u using (8.29). To write z in terms of A and u, we note that z can be expressed in terms of Z by (8.4), which can be solved in terms of A and k using (8.12). The final outcome is

$$e^z = \frac{2/Z+1}{2/Z-1} = \frac{(k-A)(k+A+m)}{(k+A)(k-A-m)} \tag{8.31}$$

$$d\lambda = -A\,du - m\,d\ln\bigl(k^2 - (A+m)^2\bigr). \tag{8.32}$$

This is now essentially the same set-up as the spectral curves and Seiberg–Witten differential for the SU(N) Calogero–Moser system. The same methods used earlier give immediately the logarithmic terms in the prepotential

$$\mathcal{F}^{\text{1-loop}} = -\frac{1}{8\pi i}\sum_{\alpha\in\mathcal{R}(D_r)} (\alpha\cdot a)^2\ln(\alpha\cdot a)^2 - (\alpha\cdot a+m)^2\ln(\alpha\cdot a+m)^2. \tag{8.33}$$

These are the logarithmic singularities expected from field theory considerations.

Acknowledgments: These notes are based on lectures delivered by E.D. at the 1999 Banff Summer School, and by D.H.P. at the meeting *Mathematics from Physics* in May 1999 at the University of Illinois, Urbana-Champaign. The authors would like to thank respectively Jacques Hurtubise, Yvan Saint-Aubin, Luc Vinet for their invitation to lecture at Banff, and Steve Bradlow, John D'Angelo, Robert Leigh, and Mike Stone for their invitation to lecture at Urbana. They would like to express their appreciation for the very warm hospitality extended to them at each place. E.D. also acknowledges the Aspen Center for Physics and the Laboratoires de Physique Théorique at École Polytechnique and at École Normale Supérieure, where part of these notes were drafted. The research of E.D. is supported in part by NSF Grants No. PHY-95-31023 and PHY-98-19686, as well as by the Centre National de Recherche Scientifique (CNRS). The work of D.H.P. is supported in part by the NSF grant DMS-98-00783.

9 References

1. Yu.A. Golfand and E.P. Lichtman, "Extension of the algebra of Poincare group generators and violation of P invariance", JETP Lett. **13** (1971), 323–326; J.-L. Gervais and B. Sakita, "Field theory interpretation of supergauges in dual models", Nucl. Phys. **B34** (1971), 632–639; D.V. Volkov and V.P. Akulov, "Possible universal neutrino interaction", JETP Lett. **16** (1972), 438–440; J. Wess and B. Zumino, "Supergauge transformations in four dimensions," Nucl. Phys. **B70** (1974), 39-50.

2. J. Wess and J. Bagger, *Supersymmetry and Supergravity*, Princeton Univ. Press, 1983; P. West, *Introduction to Supersymmetry and Supergravity*, World Scientific, 1990; S.J. Gates, M.T. Grisaru, M. Rocek, and W. Siegel, *Superspace*, Benjamin/Cummings Publ. Comp., 1983.

3. F. Wilczek, "QCD in extreme conditions," in this book.

4. M.E. Peskin and D.V. Schroeder, *An Introduction to Quantum Field Theory*, Addison-Wesley, 1995; T.P. Cheng and L.F. Li, *Gauge Theory of Elementary Particle Physics*, Oxford Univ. Press, 1984; S. Weinberg, *The Quantum Theory of Fields*, Cambridge Univ. Press, 1996.

5. WWW Reviews of Particle Physics, `http://www-pdg.lbl.gov/1999`.

6. K.S. Babu, J.C. Pati, and F. Wilczek, "Fermion masses, neutrino oscillations and proton decay in the light of SuperKamiokande," Nucl. Phys. **B566** (2000), 33–91, hep-ph/9812538.

7. C. Montonen and D. Olive, "Magnetic monopoles as gauge particles?", Phys. Lett. **B72** (1977), 117–120; P. Goddard, J. Nuyts, and D. Olive, "Gauge theories and magnetic charge," Nucl. Phys. **B125** (1977), 1–28.

8. N. Seiberg and E. Witten, "Electro-magnetic duality, monopole condensation, and confinement in $\mathcal{N} = 2$ supersymmetric Yang–Mills theory," Nucl. Phys. **B426** (1994), 19–52, hep-th/9407087; "Monopoles, duality, and chiral symmetry breaking in $\mathcal{N} = 2$ supersymmetric QCD," Nucl. Phys. **B431** (1994), 484–550, hep-th/9410167.

9. J. Polchinski, "String Theory," Cambridge Univ. Press, 1998.

10. J. Maldacena, "The large N limit of superconformal theories and supergravity," Adv. Theor. Math. Phys. **2** (1998), 231–252, hep-th/9711200; S.S. Gubser, I.R. Klebanov, and A.M. Polyakov, "Gauge theory correlators from noncritical string theory," Phys. Lett. **B428** (1998), 105–114, hep-th/9802109; E. Witten, "Anti de Sitter space and holography," Adv. Theor. Math. Phys. **2** (1998), 253–291, hep-th/9802150.

11. S. Lee, S. Minwalla, M. Rangamani, and N. Seiberg, "Three-point functions of chiral operators in $D = 4$, $\mathcal{N} = 4$ SYM at large N," Adv. Theor. Math. Phys. **2** (1998), 697–718, hep-th/9806074; E. D'Hoker, D.Z. Freedman, and W. Skiba, "Field theory tests for the correlators in the AdS/CFT correspondence," Phys. Rev. **D59** (1999), 045008 , hep-th/9807098; K. Intriligator and W. Skiba, "Bonus symmetry and the operator product expansion of $\mathcal{N} = 4$ super-Yang–Mills," Nucl. Phys. **B559** (1999), 165–183, hep-th/9905020.

12. P. Howe and P. West, "Superconformal invariants and extended supersymmetry," Phys. Lett. **B400** (1997), 307–314, hep-th/9611075; P.S. Howe, E. Sokatchev, and P. West, "3-point functions in $\mathcal{N} = 4$ Yang–Mills," Phys. Lett. **B444** (1998), 341–351, hep-th/9808162.

13. R. Grimm, M. Sohnius, and J. Wess, "Extended supersymmetry and gauge theories," Nucl. Phys. **B133** (1978), 275–284.

14. A. Galperin, E. Ivanov, S. Kalitzin, V. Ogievetsky, and E. Sokatchev, "Unconstrained $\mathcal{N} = 2$ matter, Yang–Mills and supergravity theories in harmonic superspace," Class. Quant. Grav. **1** (1984), 469–498; G.G. Hartwell, P.S. Howe, "(N, p, q) harmonic superspace," Int. J. Mod. Phys. **A10** (1995), 3901–3919; P.S. Howe, "A superspace survey," Class. Quant. Grav. **12** (1995), 1823–1880.

15. B.M. Zupnik, "The action of the supersymmetric $\mathcal{N} = 2$ gauge theory in harmonic superspace," Phys. Lett. **B183** (1987), 175–176; "Solution of the constraints of the supergauge theory in the harmonic SU(2)/ U(1)-superspace," Theor. Math. Phys. **69** (1986), 1101–1105.

16. K.G. Wilson and J. Kogut, "The renormalization group and the epsilon expansion," Phys. Rep. **12** (1974), 75–200 .

17. .A. Shifman and A.I. Vainshtein, "Solution of the anomaly puzzle in susy gauge theories and the Wilson operator expansion", Nucl. Phys. **B277** (1986), 456–486; "On holomorphic dependence and infrared effects in supersymmetric gauge theories", Nucl. Phys. **B359** (1991), 571–580; D. Amati, K. Konishi, Y. Meurice, G.C. Rossi, and G. Veneziano, "Nonperturbative aspects in supersymmetric gauge theories," Phys. Rep. **162** (1988), 169–248.

18. N. Seiberg, "Naturalness versus supersymmetric nonrenormalization theorems," Phys. Lett. **B318** (1993), 469–475; "Exact results on the space of vacua of four-dimensional SUSY gauge theories," Phys. Rev. **D49** (1994), 6857–6863.

19. P.C. Argyres and M. Douglas, "New phenomena in SU(3) supersymmetric gauge theory," Nucl. Phys. **B448** (1995), 93-126, hep-th/9505062.

20. G. 't Hooft, "Magnetic monopoles in unified gauge theories," Nucl. Phys. **B79** (1974), 276–284; A.M. Polyakov, "Particle spectrum in the quantum field theory," JETP Lett. **20** (1974), 194–195.

21. M.K. Prasad and C.M. Sommerfield, "Exact classical solution for the 't Hooft Monopole and the Julia–Zee dyon," Phys. Rev. Lett. **35** (1975), 760–762; E.B. Bogomolnyi, "The stability of classical solutions," Sov. J. Nucl. Phys. **24** (1976), 449–454.

22. E. Witten, "Dyons of charge $e\theta/2\pi$," Phys. Lett. **B86** (1979), 283–287.

23. E. Witten and D. Olive, "Supersymmetry algebras that include topological charges", Phys. Lett. **B78** (1978), 97–101.

24. B. de Wit and A. Van Proeyen, "Potentials and symmetries of general gauged $\mathcal{N} = 2$ supergravity-Yang–Mills models," Nucl. Phys. **B245** (1984), 89–117; S. Ferrara, "Calabi–Yau moduli space, special geometry and mirror symmetry," Mod. Phys. Lett. **A6** (1991), 2175–2180; A. Strominger, "Special geometry," Comm. Math. Phys. **133** (1990), 163–180.

25. A. Klemm, W. Lerche, S. Theisen, and S. Yankielowicz, "Simple singularities and $\mathcal{N} = 2$ supersymmetric gauge theories," Phys. Lett. **B344** (1995), 169–175, hep-th/9411048; P. Argyres and A. Faraggi, "The vacuum structure and spectrum of $\mathcal{N} = 2$ supersymmetric SU(N) gauge theory," Phys. Rev. Lett. **73** (1995), 3931–3934, hep-th/9411057; P. Argyres, M.R. Plesser, and A. Shapere, "The Coulomb phase of $\mathcal{N} = 2$ supersymmetric QCD," Phys. Rev. Lett. **75** (1995), 1699–1702, hep-th/9505100; M.R. Abolhasani, M. Alishahiha, and A.M. Ghezelbash, "The moduli space and monodromies of the $\mathcal{N} = 2$ supersymmetric Yang–Mills theory with any Lie gauge group," Nucl. Phys. **B480** (1996), 279–295, hep-th/9606043; M. Alishahiha, F. Ardalan, and F. Mansouri, "The moduli space of the $\mathcal{N} = 2$ supersymmetric G_2 Yang–Mills theory," Phys. Lett. **B381** (1996), 446–450, hep-th/9512005; A. Hanany and Y. Oz, "On the quantum moduli space of vacua of the $\mathcal{N} = 2$ supersymmetric SU(N_c) gauge theories," Nucl. Phys. **B452** (1995), 283–312, hep-th/9505075; A. Hanany, "On the quantum moduli space of vacua of $\mathcal{N} = 2$ supersymmetric gauge theories," Nucl. Phys. **B466** (1996), 85–100, hep-th/9509176.

26. E. D'Hoker, I.M. Krichever, and D.H. Phong, "The effective prepotential for $\mathcal{N} = 2$ supersymmetric SU(N_c) gauge theories," Nucl. Phys. **B489** (1997), 179–210, hep-th/9609041.

27. E. D'Hoker, I.M. Krichever, and D.H. Phong, "The effective prepotential of $\mathcal{N} = 2$ supersymmetric SO(N_c) and Sp(N_c) gauge theories," Nucl. Phys. **B489** (1997), 211–222, hep-th/9609145.

28. S.G. Naculich, H. Rhedin, and H.J. Schnitzer, "One-instanton tests of a Seiberg–Witten curve from M-theory: the antisymmetric representation of SU(N)," Nucl. Phys. **B533** (1988), 275–302 , hep-th/9804105; I.P. Ennes, S.G. Naculich, H. Rhedin, and H.J. Schnitzer, "One-instanton predictions of a Seiberg–Witten curve from M-theory: the symmetric representation of SU(N)," Int. J. Mod. Phys. **A14** (1999), 301–321, hep-th/9804151; "One-instanton predictions for nonhyperelliptic curves derived from M-theory," Nucl. Phys. **B536** (1988), 245–257, hep-th/9806144; "One-instanton predictions of Seiberg–Witten curves for product groups," Phys. Lett. **B452** (1999), 260–264, hep-th/9901124; "Two antisymmetric hypermultiplets in $\mathcal{N} = 2$ SU(N) gauge theory: Seiberg–Witten curve and M-theory interpretation," Nucl. Phys. **B558** (1999), 41–62, hep-th/9904078; "Tests of M-theory from $\mathcal{N} = 2$ Seiberg–Witten theory," hep-th/9911022.

29. I.P. Ennes, S.G. Naculich, H. Rhedin, and H.J. Schnitzer, "One-instanton predictions of Seiberg–Witten curves for product groups," Phys. Lett. **B452** (1999), 260–264, hep-th/9901124; U. Feichtinger, "The prepotential of $\mathcal{N} = 2$ SU(2) $\times$ SU(2) supersymmetric Yang–Mills theory with bifundamental matter," Phys. Lett. **B465** (1999), 155–162, hep-th/9908143.

30. A. Ceresole, R. D'Auria, and S. Ferrara, "On the geometry of moduli space of vacua in $\mathcal{N} = 2$ supersymmetric Yang–Mills theory," Phys. Lett. **B339** (1994), 71–76, hep-th/9408036; A. Klemm, W. Lerche, and S. Theisen, "Nonperturbative effective actions of $\mathcal{N} = 2$ supersymmetric gauge theories," Int. J. Mod. Phys. **A11** (1996), 1929–1973, hep-th/9505150; J. Isidro, A. Mukherjee, J. Nunes, and H. Schnitzer, "A new derivation of Picard–Fuchs equations for effective $\mathcal{N} = 2$ super Yang–Mills theories," Nucl. Phys. **492** (1997), 647–681, hep-th/9609116; "A note on the Picard–Fuchs equations for $\mathcal{N} = 2$ Seiberg-Witten theories," Int. J. Mod. Phys. **A13** (1998), 233–250, hep-th/9703176; "On the Picard–Fuchs equations for massive $\mathcal{N} = 2$ Seiberg-Witten theories,"Nucl. Phys. **B502** (1997), 363–382, hep-th/9704174.

31. E. D'Hoker, I.M. Krichever, and D.H. Phong, "The renormalization group equation for $\mathcal{N} = 2$ supersymmetric gauge theories," Nucl. Phys. **B494** (1997), 89–104, hep-th/9610156.

32. M. Matone, "Instantons and recursion relations in $\mathcal{N} = 2$ SUSY gauge theories," Phys. Lett. **B357** (1996), 342–348, hep-th/9506102; T. Eguchi and S.K. Yang, "Prepotentials of $\mathcal{N} = 2$ susy gauge theories and soliton equations," Mod. Phys. Lett. **A11** (1996), 131–138, hep-th/9510183; T. Nakatsu and K. Takasaki, "Whitham–Toda hierarchy and $\mathcal{N} = 2$ supersymmetric Yang–Mills theory," Mod. Phys. Lett.

A11 (1996), 157–168, hep-th/9509162; J. Sonnenschein, S. Theisen, and S. Yankielowicz, "On the relation between the holomorphic prepotential and the quantum moduli in SUSY gauge theories," Phys. Lett. **B367** (1996), 145–150, hep-th/9510129; K. Takasaki, "Whitham deformations and tau functions in $\mathcal{N} = 2$ supersymmetric gauge theories," Prog. Theor. Phys. Suppl. **135** (1999), 53–74, hep-th/9905224; J. Edelstein, M. Gomez-Reina, and J. Mas, "Instanton corrections in $\mathcal{N} = 2$ supersymmetric theories with classical gauge groups and fundamental matter hypermultiplets," Nucl. Phys. **B561** (1999), 273–292, hep-th/9904087.

33. A. Marshakov, A. Mironov, and A. Morozov, "WDVV-like equations in $\mathcal{N} = 2$ SUSY Yang–Mills theory," Phys. Lett. **B389** (1996), 43–52, hep-th/9607109; "More evidence for the WDVV equations in $\mathcal{N} =$ 2 SUSY Yang–Mills theory," Int. J. Mod. Phys. **A15** (2000), 1157–1206, hep-th/9701123; J.M. Isidro, "On the WDVV equation and M-theory," Nucl. Phys. **B539** (1999), 379–402, hep-th/9805051.

34. I.M. Krichever, "The tau function of the universal Whitham hierarchy, matrix models, and topological field theories," Comm. Pure Appl. Math. **47** (1994), 437–475; "The dispersionless Lax equations and topological minimal models," Comm. Math. Phys. **143** (1992), 415–429; B.A. Dubrovin, "Hamiltonian formalism for Whitham hierarchies and topological Landau–Ginzburg models," Comm. Math. Phys. **145** (1992), 195-207; "Integrable systems in topological field theory," Nucl. Phys. **B379** (1992), 627–689; "Geometry of 2D topological field theories," in *Integrable Systems and Quantum Groups*, eds. M. Francaviglia and S. Greco, Lecture Notes in Math., vol. 1620, Springer, 1996, 120–348.

35. G. Chan and E. D'Hoker, "Instanton recursion relations for the effective prepotential in $\mathcal{N} = 2$ super Yang–Mills," Nucl. Phys. **B564** (2000), 503–516, hep-th/9906193.

36. J.D. Edelstein and J. Mas, "Strong coupling expansion and Seiberg–Witten Whitham equations," Phys. Lett. **B452** (1999), 69–75, hep-th/9901006.

37. M. Douglas and S. Shenker, "Dynamics of SU(N) supersymmetric gauge theory," Nucl. Phys. **B447** (1995), 271–296, hep-th/9503163; E. D'Hoker and D.H. Phong, "Strong coupling expansions of SU(N) Seiberg–Witten theory," Phys. Lett. **B397** (1997), 94–103, hep-th/9701055.

38. M. Marino and G. Moore, "The Donaldson–Witten function for gauge groups of rank larger than one," Comm. Math. Phys. **199** (1998), 25–69, hep-th/9802185; M. Marino, "The uses of Whitham hierarchies," Prog. Theor. Phys. Suppl. **135** (1999), 29-52, hep-th/9905053.

39. W. Lerche, "Introduction to Seiberg–Witten theory and its stringy origins," Nucl. Phys. Proc. Suppl. **55B** (1997), 83–117, hep-th/9611190; A. Marshakov, "On integrable systems and supersymmetric gauge theories," Theor. Math. Phys. **112** (1997), 791–826, hep-th/9702083; A. Klemm, "On the geometry behind $\mathcal{N} = 2$ supersymmetric effective actions in four dimensions," Trieste 1996, *High Energy Physics and Cosmology*, 120–242, hep-th/9705131; A. Marshakov and A. Mironov, "Seiberg–Witten systems and Whitham hierarchires: a short review," hep-th/9809196; E. D'Hoker and D.H. Phong, "Seiberg–Witten theory and Integrable Systems," hep-th/9903068; C. Lozano, "Duality in topological quantum field theories," hep-th/9907123.

40. P. Argyres and A. Shapere, "The vacuum structure of $\mathcal{N} = 2$ super-QCD with classical gauge groups," Nucl. Phys. **B461** (1996), 437–459,hep-th/9509175; U.H. Danielsson and B. Sundborg, "The moduli space and monodromies of $\mathcal{N} = 2$ supersymmetric SO$(2r + 1)$ gauge theories", Phys. Lett. **B358** (1995), 273–280, hep-th/9504102; A. Brandhuber and K. Landsteiner, "On the monodromies of $\mathcal{N} = 2$ supersymmetric Yang–Mills theories with gauge group SO$(2n)$," Phys. Lett. **B358** (1995), 73–80, hep-th/9507008.

41. A.M. Perelomov, *Integrable Systems of Classical Mechanics and Lie Algebras.* I, Birkhäuser, Boston, 1990 and references therein; A.N. Leznov and M.V. Saveliev, "Group theoretic methods for integration of nonlinear dynamical systems," Birkhäuser, Boston, 1992.

42. M. Adler and P. Van Moerbeke, "The Toda lattice, Dynkin diagrams, singularities and Abelian varieties," Invent. Math. **103** (1991), 223–278; "Completely integrable systems, Euclidean Lie algebras, and curves," Advances in Math. **38** (1980), 267–317; "Linearization of Hamiltonian systems, Jacobi varieties, and representation theory," Advances in Math. **38** (1980), 318–379.

43. A. Gorskii, I.M. Krichever, A. Marshakov, A. Mironov, and A. Morozov, "Integrability and Seiberg–Witten exact solution," Phys. Lett. B355 (1995), 466–474, hep-th/9505035.

44. E. Martinec and N. Warner, "Integrable systems and supersymmetric gauge theories," Nucl. Phys. **B459** (1996), 97–112, hep-th/9509161.

45. F. Calogero, "Exactly solvable one-dimensional many-body problems," Lett. Nuovo Cim. **13** (1975), 411–416; J. Moser, "Three integrable Hamiltonian systems connected with isospectral deformations," Advances in Math. **16** (1975), 197–220; B. Sutherland, "Exact results for a quantum many-body problem in one dimension. II," Phys. Rev. **A5** (1972), 1372–1376; C. Marchioro, F. Ca-

logero, and O. Ragnisco, "Exact solution of the classical and quantal one-dimensional many-body problem with the two-body potential $V(ax) = G^2A^2/\sinh^2(ax)$," Lett. Nuovo Cim. **13** (1975), 383–387; S.N.M. Ruijsenaars, "Systems of Calogero–Moser type," in *Particles and Fields*, eds. G. Semenoff and L. Vinet, Springer, 1999, 251–352.

46. A. Polychronakos, "Lattice integrable systems of Haldane–Shastry type," Phys. Rev. Lett. **70** (1993), 2329–2331; "New integrable systems from unitary matrix models," Phys. Lett. **B277** (1992), 102–108; "Generalized statistics in one dimension," presented at Les Houches Summer School in Theoretical Physics, Session 69: Topological Aspects of Low-Dimensional Systems, Les Houches, France, 7–31 July 1998, hep-th/9902157; L. Lapointe and L. Vinet, "Exact operator solution of the Calogero-Sutherland model," Commun. Math. Phys. **178** (1996), 425–452, q-alg/9509003.

47. I.M. Krichever, "Elliptic solutions of the Kadomtsev–Petviashvili equation and integrable systems of particles," Funct. Anal. Appl. **14** (1980) 282–290.

48. A. Treibich and J.-L. Verdier, "Solitons elliptiques," The Grothendieck Festschrift, vol. III, Birkhäuser, Boston, 1990, 437–480; R. Donagi and E. Markham, "Spectral curves, algebraically integrable Hamiltonian systems, and moduli of bundles," in *Integrable Systems and Quantum Groups*, Lecture Notes in Math., vol. 1620, Springer, 1996, 1–119, alg-geom/9507017; E. Markham, "Spectral curves and integrable systems," Comp. Math. **93** (1994), 255–290; R. Donagi, "Seiberg–Witten integrable systems," in *Algebraic geometry—Santa Cruz 1995*, eds. J. Kollár, R. Lazarsfeld, and D.R. Morrison, Proc. Sympos. Pure Math., vol. 62, Part 2, Amer. Math. Soc., 1997, 3–43, alg-geom/9705010.

49. M.R. Adams, J. Harnard, and J. Hurtubise, "Coadjoint orbits, spectral curves, and Darboux coordinates," in *The Geometry of Hamiltonian Systems*, ed. T. Ratiu, Math. Sci. Res. Inst. Publ., vol. 22, Springer, 1991, 9–21; M.R. Adams, J. Harnard, and J. Hurtubise, "Darboux coordinates and Liouville–Arnold integration in loop algebras," Comm. Math. Phys. **155** (1993), 385–413; J. Hurtubise, "Integrable systems and algebraic surfaces," Duke Math. J. **83** (1996), 19–50; Erratum, Duke Math. J. **84** (1996), 815.

50. H. Braden, "A conjectured R-matrix," J. Phys. A **31** (1998), 1733–1741.

51. H.W. Braden and V.M. Buchstaber, "The general analytic solution of a functional equation of addition type," Siam J. Math. Anal. **28** (1997), 903–923.

52. I. Inozemtsev, "Lax representation with spectral parameter on a torus for integrable particle systems," Lett. Math. Phys. **17** (1989), 11–17; "The finite Toda lattices," Comm. Math. Phys. **121** (1989) 628–638.

53. E. D'Hoker and R. Jackiw, "Classical and quantum Liouville theory," Phys. Rev. **D26** (1982), 3517–3542; E. D'Hoker and P. Kurzepa, "2-D quantum gravity and Liouville theory," Mod. Phys. Lett. **A5** (1990), 1411–1421; E. D'Hoker, "Equivalence of Liouville theory and 2-D quantum gravity," Mod. Phys. Lett. **A6** (1991), 745–767.

54. M.J. Ablowitz, D.J. Kaup, A.C. Newell, and H. Segur, "Method for solving the sine-Gordon equation," Phys. Rev. Lett. **30** (1973), 1262–1264; S. Coleman, "The quantum sine-Gordon equation as the massive Thirring model," Phys. Rev. **D11** (1975), 2088–2097; E.K. Sklyanin, L.A. Takhtajan, and L.D. Faddeev, "The quantum inverse problem method. I", Theor. Math. Phys. **40** (1979), 688–706.

55. H. Airault, H. McKean, and J. Moser, "Rational and elliptic solutions of the KdV equation and integrable systems of N particles on a line," Comm. Pure Appl. Math. **30** (1977), 95–125.

56. I.M. Krichever and D.H. Phong, "On the integrable geometry of $\mathcal{N} = 2$ supersymmetric gauge theories and soliton equations," J. Differential Geometry **45** (1997), 349–389, hep-th/9604199.

57. I.M. Krichever and D.H. Phong, "Symplectic forms in the theory of solitons," in Surveys in Differential Geometry, vol. IV , eds. C.L. Terng and K. Uhlenbeck, International Press, 1998, 239–313, hep-th/9708170; I.M. Krichever, "Elliptic solutions to difference nonlinear equations and nested Bethe ansatz equations," solv-int/9804016.

58. M.A. Olshanetsky and A.M. Perelomov, "Completely integrable Hamiltonian systems connected with semisimple Lie algebras," Invent. Math. **37** (1976), 93–108; "Classical integrable finite-dimensional systems related to Lie algebras," Phys. Rep. **71 C** (1981), 313–400.

59. E. D'Hoker and D.H. Phong, "Calogero–Moser Lax pairs with spectral parameter for general Lie algebras," Nucl. Phys. **B530** (1998), 537–610, hep-th/9804124.

60. E. D'Hoker and D.H. Phong, "Calogero–Moser and Toda systems for twisted and untwisted affine Lie algebras," Nucl. Phys. **B530** (1998), 611–640, hep-th/9804125.

61. A. Bordner, E. Corrigan, and R. Sasaki, "Calogero–Moser systems. A new formulation," Prog. Theor. Phys. **100** (1998), 1107–1129, hep-th/9805106; A. Bordner, R. Sasaki, and K. Takasaki, "Calogero–Moser systems. II. Symmetries and foldings," Prog. Theor. Phys. **101** (1999), 487–518, hep-th/9809068.

62. S.P. Khastgir, R. Sasaki, and K. Takasaki, "Calogero–Moser Models. IV. Limits to Toda theory," Prog. Theor. Phys. **102** (1999), 749–776, hep-th/9907102.

63. P. Goddard and D. Olive, "Kac-Moody and Virasoro algebras in relation to quantum physics," Int. J. Mod. Phys.**A1** (1986), 303–414; V. Kac, *Infinite-Dimensional Lie Algebras*, Birkhäuser, Boston, 1983.

64. R. Donagi and E. Witten, "Supersymmetric Yang–Mills and integrable systems," Nucl. Phys. **B460** (1996), 288–334, hep-th/9510101.

65. E. Martinec, "Integrable structures in supersymmetric gauge and string theory," Phys. Lett. **B367** (1996), 91–96, hep-th/9510204; A. Gorsky and N. Nekrasov, "Elliptic Calogero–Moser system from two-dimensional current algebra," preprint ITEP-NG/1-94, hep-th/9401021; N. Nekrasov, "Holomorphic bundles and many-body systems," Comm. Math. Phys. **180** (1996), 587–604, hep-th/9503157; M. Olshanetsky, "Generalized Hitchin systems and the Knizhnik–Zamolodchikov–Bernard equation on elliptic curves," Lett. Math. Phys. **42** (1997), 59–71, hep-th/9510143.

66. E. D'Hoker and D.H. Phong, "Calogero–Moser systems in SU(N) Seiberg–Witten theory," Nucl. Phys. **B513** (1998), 405–444, hep-th/9709053.

67. E. D'Hoker and D.H. Phong, "Order parameters, free fermions, and conservation laws for Calogero–Moser systems," Asian J. Math. **2** (1998), 655–666, hep-th/9808156.

68. E. D'Hoker and D.H. Phong, "The geometry of string perturbation theory," Rev. Mod. Physics **60** (1988), 917–1065; "Loop amplitudes for the fermionic string," Nucl. Phys. **B278** (1986), 225–241; "On determinants of Laplacians on Riemann surfaces," Comm. Math. Phys. **105** (1986), 537–545.

69. K. Vaninsky, "On explicit parametrization of spectral curves for Moser–Calogero particles and its applications," Int. Math. Res. Notices (1999), 509–529.

70. J. Minahan, D. Nemeschansky, and N. Warner, "Instanton expansions for mass deformed $\mathcal{N} = 4$ super Yang–Mills theory," Nucl. Phys. **B528** (1998), 109–132, hep-th/9710146.

71. E. Witten, "Solutions of four-dimensional field theories via M-Theory," Nucl. Phys. **B500** (1997), 3–42, hep-th/9703166.

72. S. Katz, P. Mayr, and C. Vafa, "Mirror symmetry and exact solutions of 4D $\mathcal{N} = 2$ gauge theories. I.," Adv. Theor. Math. Phys. **1** (1998),

53–114, hep-th/9706110; S. Katz, A. Klemm, and C. Vafa, "Geometric engineering of quantum field theories," Nucl. Phys. **B497** (1997), 173–195, hep-th/9609239; M. Bershadsky, K. Intriligator, S. Kachru, D. R. Morrison, V. Sadov, and C. Vafa, "Geometric singularities and enhanced gauge symmetries," Nucl. Phys. **B481** (1996), 215–252, hep-th/9605200; S. Kachru and C. Vafa, "Exact results for $\mathcal{N} = 2$ compactifications of heterotic strings," Nucl. Phys. **B450** (1995), 69–89, hep-th/9505105.

73. E. D'Hoker and D. H. Phong, "Lax pairs and spectral curves for Calogero–Moser and spin Calogero–Moser systems," Regular and Chaotic Dynamics textbf3 (1998), 27–39, hep-th/9903002.

74. E. D'Hoker and D. H. Phong, "Spectral curves for super Yang–Mills with adjoint hypermultiplet for general Lie algebras," Nucl. Phys. **B534** (1998), 697–719, hep-th/9804126.

75. A. Brandhuber, J. Sonnenschein, S. Theisen, and S. Yankielowicz, "M Theory and Seiberg–Witten curves: orthogonal and symplectic groups," Nucl. Phys. **B504** (1997), 175–188, hep-th/9705232; K. Landsteiner and E. Lopez, "New curves from branes," Nucl.Phys. **B516** (1998), 273–296, hep-th/9708118; K. Landsteiner, E. Lopez, and D.A. Lowe, "$\mathcal{N} = 2$ supersymmetric gauge theories, branes and orientifolds," Nucl. Phys. **B507** (1997), 197–226, hep-th/9705199.

76. A. M. Uranga, "Towards mass deformed $\mathcal{N} = 4$ SO(N) and Sp(K) gauge theories from brane configurations," Nucl. Phys. **B526** (1998), 241–277, hep-th/9803054.

77. T. Yokono, "Orientifold four plane in brane configurations and $\mathcal{N} = 4$ USp($2N$) and SO($2N$) theory," Nucl. Phys. **B532** (1998), 210–226, hep-th/9803123; K. Landsteiner, E. Lopez, and D. Lowe, "Supersymmetric gauge theories from branes and orientifold planes," JHEP **9807** (1998), 011, hep-th/9805158.

78. A. Gorsky, "Branes and Integrability in the $\mathcal{N} = 2$ SUSY YM theory," Phys. Lett. **B410** (1997), 22–26, hep-th/9612238; A. Gorsky, S. Gukov, and A. Mironov, "Susy field theories, integrable systems and their stringy brane origin," Nucl. Phys. **B518** (1998), 689–713, hep-th/9710239; A. Cherkis and A. Kapustin, "Singular monopoles and supersymmetric gauge theories in three dimensions," Nucl. Phys. **B525** (1998), 215–234, hep-th/9711145.

79. H. Braden, A. Marshakov, A. Mironov, and A. Morozov, "The Ruijsenaars-Schneider model in the context of Seiberg–Witten theory," Nucl. Phys. **B558** (1999), 371-390, hep-th/9902205; Y. Ohta, "Instanton correction of prepotential in Ruijsenaars model associated

with $\mathcal{N} = 2$ SU(2) Seiberg–Witten," J. Math. Phys. **41** (2000), 4541–4550, hep-th/9909196; H.W. Braden, A. Marshakov, A. Mironov, and A. Morozov, "Seiberg–Witten theory for a nontrivial compactification from five to four dimensions," Phys. Lett. **B448** (1999), 195–202, hep-th/9812078; K. Takasaki, "Elliptic Calogero–Moser systems and isomonodromic deformations," J. Math. Phys. **40** (1999), 5787–5821, math.QA/9905101; A.M. Khvedelidze and D.M. Mladenov, "Euler–Calogero–Moser system from SU(2) Yang–Mills theory," Phys.Rev. **D62** (2000), 125016, hep-th/9906033; A. Gorsky and A. Mironov, "Solutions to the reflection equation and integrable systems for $\mathcal{N} = 2$ SQCD with classical groups," Nucl. Phys. **B550** (1999), 513–530, hep-th/9902030; I.M. Krichever, "Elliptic analog of the Toda lattice," Int. Math. Res. Notices (2000), 383–412, hep-th/9909224.

80. N. Hitchin, "Stable bundles and integrable systems," Duke Math. J. **54** (1987), 91–114.

81. A. Erdélyi, W. Magnus, F. Oberhettinger, and F.G. Tricomi, *Higher Transcendental Functions*, Vol. II, R.E. Krieger, 1981.

A Notations and Conventions

Vector indices run over the following values $\mu, \nu, \ldots = 0, 1, 2, 3$ and $i, j, \ldots = 1, 2, 3$. The flat Minkowski space-time metric is given by $-\eta_{00} = \eta_{11} = \eta_{22} = \eta_{33} = 1$, which is invariant under translations and under the Lorentz group SO(1,3). Three- and four-dimensional totally antisymmetric Levi–Civita symbols are defined by

$$\epsilon_{ijk} = \epsilon^{ijk}, \quad \epsilon^{123} = 1; \quad \epsilon_{\mu\nu\rho\sigma} = -\epsilon^{\mu\nu\rho\sigma}, \quad \epsilon^{0123} = 1. \tag{A.1}$$

The Poincaré dual of an antisymmetric rank 2 tensor is defined by

$$\tilde{F}_{\mu\nu} = \tfrac{1}{2}\epsilon_{\mu\nu\rho\sigma}F^{\rho\sigma} \tag{A.2}$$

and the duality operation squares to minus the identity: $\tilde{\tilde{F}} = -F$.

A.1 Spinors

The *Pauli matrices* are defined by

$$\sigma^0 = \begin{pmatrix} -1 & 0 \\ 0 & -1 \end{pmatrix}, \quad \sigma^1 = \begin{pmatrix} 0 & 1 \\ 1 & 0 \end{pmatrix}, \\ \sigma^2 = \begin{pmatrix} 0 & -i \\ i & 0 \end{pmatrix}, \quad \sigma^3 = \begin{pmatrix} 1 & 0 \\ 0 & -1 \end{pmatrix} \tag{A.3}$$

and their "conjugates" are given by $\bar\sigma^0 = \sigma^0$ and $\bar\sigma^i = -\sigma^i$. The *Clifford–Dirac* γ matrices obey the Clifford algebra relations

$$\{\gamma^\mu, \gamma^\nu\} = \gamma^\mu\gamma^\nu + \gamma^\nu\gamma^\mu = -2\eta^{\mu\nu} I \tag{A.4}$$

with γ^0 anti-Hermitian and γ^i Hermitian. The matrix γ_5 obeys

$$\{\gamma_5, \gamma^\mu\} = 0, \quad \gamma_5 = \gamma^5 = (\gamma_5)^\dagger = i\gamma^0\gamma^1\gamma^2\gamma^3, \quad (\gamma_5)^2 = I. \tag{A.5}$$

Representation matrices for the (reducible) spinor representation of the Lorentz group SO(1, 3) are defined by

$$\Sigma^{\mu\nu} = \tfrac{i}{2}[\gamma^\mu, \gamma^\nu] \quad \text{with } [\gamma_5, \Sigma^{\mu\nu}] = 0. \tag{A.6}$$

The *charge conjugation* matrix C is defined by

$$C\gamma_\mu C^{-1} = -(\gamma_\mu)^T \tag{A.7}$$

which implies the useful identities

$$\begin{gathered} C\Sigma_{\mu\nu}C^{-1} = -(\Sigma_{\mu\nu})^T, \quad C\gamma_\mu\gamma_5 C^{-1} = (\gamma_\mu\gamma_5)^T, \\ C\gamma_5 C^{-1} = (\gamma_5)^T. \end{gathered} \tag{A.8}$$

A.2 Dirac Matrices in a Weyl Basis

In a Weyl basis γ^5 is chosen to be diagonal; this basis is convenient for chiral fermions.

$$\gamma^\mu = \begin{pmatrix} 0 & \sigma^\mu \\ \bar\sigma^\mu & 0 \end{pmatrix}, \quad \gamma_5 = \begin{pmatrix} 1 & 0 \\ 0 & -1 \end{pmatrix}, \quad C = i\gamma^2\gamma^0. \tag{A.9}$$

A.3 Dirac Matrices in a Majorana Basis

In a Majorana basis all γ^μ matrices are chosen to be real. This basis is particularly convenient when dealing with Majorana fermions, or when converting from the two-component notation.

$$\begin{gathered} \gamma^0 = \begin{pmatrix} 0 & i\sigma^2 \\ i\sigma^2 & 0 \end{pmatrix}, \quad \gamma^1 = \begin{pmatrix} -\sigma^3 & 0 \\ 0 & -\sigma^3 \end{pmatrix}, \quad \gamma^2 = \begin{pmatrix} 0 & i\sigma^2 \\ -i\sigma^2 & 0 \end{pmatrix}, \\ \gamma^3 = \begin{pmatrix} \sigma^1 & 0 \\ 0 & \sigma^1 \end{pmatrix}, \quad \gamma^5 = \begin{pmatrix} \sigma^2 & 0 \\ 0 & -\sigma^2 \end{pmatrix}, \quad C = \begin{pmatrix} 0 & -i\sigma^2 \\ i\sigma^2 & 0 \end{pmatrix}. \end{gathered} \tag{A.10}$$

A.4 Two-Component Spinors

The correspondence between two-component and four-component spinors will be made in the Weyl basis, given previously. Left-handed spinors transform under the $(\frac{1}{2}, 0)$ representation of the Lorentz group and are denoted by χ_α or their transpose by χ^α. Right-handed spinors transform under the $(0, \frac{1}{2})$ representation of the Lorentz group and are denoted by $\bar{\psi}^{\dot\alpha}$ or their transpose by $\bar{\psi}_{\dot\alpha}$. One has the following relations

$$\begin{aligned} \chi^\alpha &= \epsilon^{\alpha\beta}\chi_\beta, & \chi_\alpha &= \epsilon_{\alpha\beta}\chi^\beta \\ \bar{\psi}^{\dot\alpha} &= \epsilon^{\dot\alpha\dot\beta}\bar{\psi}_{\dot\beta}, & \bar{\psi}_{\dot\alpha} &= \epsilon_{\dot\alpha\dot\beta}\bar{\psi}^{\dot\beta} \end{aligned} \tag{A.11}$$

with the following conventions for ϵ:

$$\begin{aligned} \epsilon^{12} &= \epsilon_{21} = -\epsilon^{21} = -\epsilon_{12} = 1, \quad \epsilon^{11} = \epsilon^{22} = 0, \quad \epsilon_{\alpha\beta}\epsilon^{\beta\gamma} = \delta_\alpha{}^\gamma, \\ \epsilon^{\dot1\dot2} &= \epsilon_{\dot2\dot1} = -\epsilon^{\dot2\dot1} = -\epsilon_{\dot1\dot2} = 1, \quad \epsilon^{\dot1\dot1} = \epsilon^{\dot2\dot2} = 0, \quad \epsilon_{\dot\alpha\dot\beta}\epsilon^{\dot\beta\dot\gamma} = \delta_{\dot\alpha}{}^{\dot\gamma}. \end{aligned} \tag{A.12}$$

Contraction conventions are

$$\begin{aligned} \psi\chi &= \psi^\alpha\chi_\alpha, & \bar{\psi}\bar{\chi} &= \bar{\psi}_{\dot\alpha}\bar{\chi}^{\dot\alpha}, \\ \chi\sigma^m\bar{\psi} &= \chi^\alpha\sigma^m_{\alpha\dot\alpha}\bar{\psi}^{\dot\alpha}, & \bar{\chi}\bar{\sigma}^m\psi &= \bar{\chi}_{\dot\alpha}\bar{\sigma}^{m\dot\alpha\alpha}\psi_\alpha. \end{aligned} \tag{A.13}$$

Transposition Identities

$$\begin{aligned} \psi\chi &= \chi\psi, & (\chi\psi)^\dagger &= \bar{\chi}\bar{\psi} = \bar{\psi}\bar{\chi}, \\ \chi\sigma^m\bar{\psi} &= -\bar{\psi}\bar{\sigma}^m\chi, & (\chi\sigma^m\bar{\psi})^\dagger &= \psi\sigma^m\bar{\chi} \\ \chi\sigma^m\bar{\sigma}^n\psi &= \psi\sigma^n\bar{\sigma}^m\chi, & (\chi\sigma^m\bar{\sigma}^n\psi)^\dagger &= \bar{\psi}\bar{\sigma}^n\sigma^m\bar{\chi}. \end{aligned} \tag{A.14}$$

Fierz Identity

$$(\psi\varphi)\bar{\chi}_{\dot\beta} = -\tfrac{1}{2}(\varphi\sigma^m\bar{\chi})(\psi\sigma_m)_{\dot\beta}. \tag{A.15}$$

Relating Four-Component Spinors and Two-Component Spinors in a Weyl basis

$$\begin{aligned} &\text{Left Weyl:} \quad \Psi_{\rm L} = \begin{pmatrix}\chi_\alpha\\0\end{pmatrix} && \text{Dirac:} \quad \Psi_{\rm D} = \begin{pmatrix}\chi_\alpha\\ \bar{\psi}^{\dot\alpha}\end{pmatrix} \\ &\text{Right Weyl:} \quad \Psi{\rm R} = \begin{pmatrix}0\\ \bar{\psi}^{\dot\alpha}\end{pmatrix} && \text{Majorana:} \quad \Psi_{\rm M} = \begin{pmatrix}\psi_\alpha\\ \bar{\psi}^{\dot\alpha}\end{pmatrix}. \end{aligned} \tag{A.16}$$

Transposition Identities in Four-Component Form

$$\begin{aligned} \bar{\psi}_1\Gamma\psi_2 = \bar{\psi}_2\tilde{\Gamma}\psi_1 \quad \text{with } \Gamma &= +\tilde{\Gamma} & \Gamma &= 1, \gamma_5, \gamma^m\gamma_5 \\ \Gamma &= -\tilde{\Gamma} & \Gamma &= \gamma^m, \gamma_{[\mu}\gamma_{\nu]}. \end{aligned} \tag{A.17}$$

B Lie Algebra Theory

In Figure 1.1, we give the Dynkin diagrams for the finite dimensional simple Lie algebras; for the untwisted affine Lie algebras (left column) and for the twisted affine Lie algebras (right column). The simple roots are labeled following Dynkin notation, and are given in an orthonormal basis in Table 1.9, where we also list the dimension, the Coxeter and dual Coxeter numbers (to be defined below). We list the set of all roots in Table 1.9, and of the highest roots in Table 1.10. Below we provide additional notations and definitions [63].

Let $\mathcal{G}$ be one of the finite dimensional simple Lie algebras of rank n, let α_i, and $\alpha_i^\vee \equiv 2\alpha_i/\alpha_i^2$, $i = 1, \dots, n$ be its simple roots and coroots respectively. The coroot $\alpha^\vee$ of any root is defined by $\alpha^\vee = 2\alpha/\alpha^2$. Any (co-)root admits a unique decomposition into a sum of simple (co-)roots, with integer coefficients l_i and $l_i^\vee$.

$$\alpha = \sum_{i=1}^{n} l_i \alpha_i, \qquad \alpha^\vee = \sum_{i=1}^{n} l_i^\vee \alpha_i^\vee . \tag{B.1}$$

The coefficients l_i and $l_i^\vee$ are either all positive or all negative according to whether α (or $\alpha^\vee$) is positive or negative respectively. They are related by

$$l_i^\vee = \frac{\alpha_i^2}{\alpha^2} l_i, \qquad i = 1, \dots, n. \tag{B.2}$$

The *highest root* α_0 and co-root $\alpha_0^\vee$ play special roles. The extension of the simple root system of an algebra $\mathcal{G}$ by α_0 generates the *untwisted affine Lie algebra* $\mathcal{G}^{(1)}$, while the extension of the simple coroot system of $\mathcal{G}$ by $\alpha_0^\vee$ generates the *dual affine Lie algebra* $(\mathcal{G}^{(1)})^\vee$. When $\mathcal{G}$ is non-simply laced, $(\mathcal{G}^{(1)})^\vee$ coincides with one of the *twisted affine Lie algebras*. The Dynkin diagrams of these various Lie algebras are given in Table 1.1. The decompositions of α_0 and $\alpha_0^\vee$ onto roots or coroots

$$\alpha_0 = \sum_{i=1}^{n} a_i \alpha_i, \qquad \alpha_0^\vee = \sum_{i=1}^{n} a_i^\vee \alpha_i^\vee \tag{B.3}$$

define the *marks* a_i and the *comarks* $a_i^\vee$, which are given in Table 1.5. The *Coxeter number* $h_\mathcal{G}$ and the *dual Coxeter number* $h_\mathcal{G}^\vee$ are defined by

$$h_\mathcal{G} = 1 + \sum_{i=1}^{n} a_i, \qquad h_\mathcal{G}^\vee = 1 + \sum_{i=1}^{n} a_i^\vee , \tag{B.4}$$

and their values are given in Table 1.9. For simply laced Lie algebras, for which all roots have the same length (normalized to $\alpha_i^2 = 2$), we have $a_i^\vee = a_i$ and $h_\mathcal{G} = h_\mathcal{G}^\vee$. The dual Coxeter number equals the *quadratic Casimir* operator in the adjoint representation, $h_\mathcal{G}^\vee = C_2(\mathcal{G})$.

The highest weight vectors λ_j, $j = 1, \dots, n$ of the *fundamental representations* (also called fundamental weights) of $\mathcal{G}$ are defined by

$$\alpha_i^\vee \cdot \lambda_j = \delta_{ij}. \tag{B.5}$$

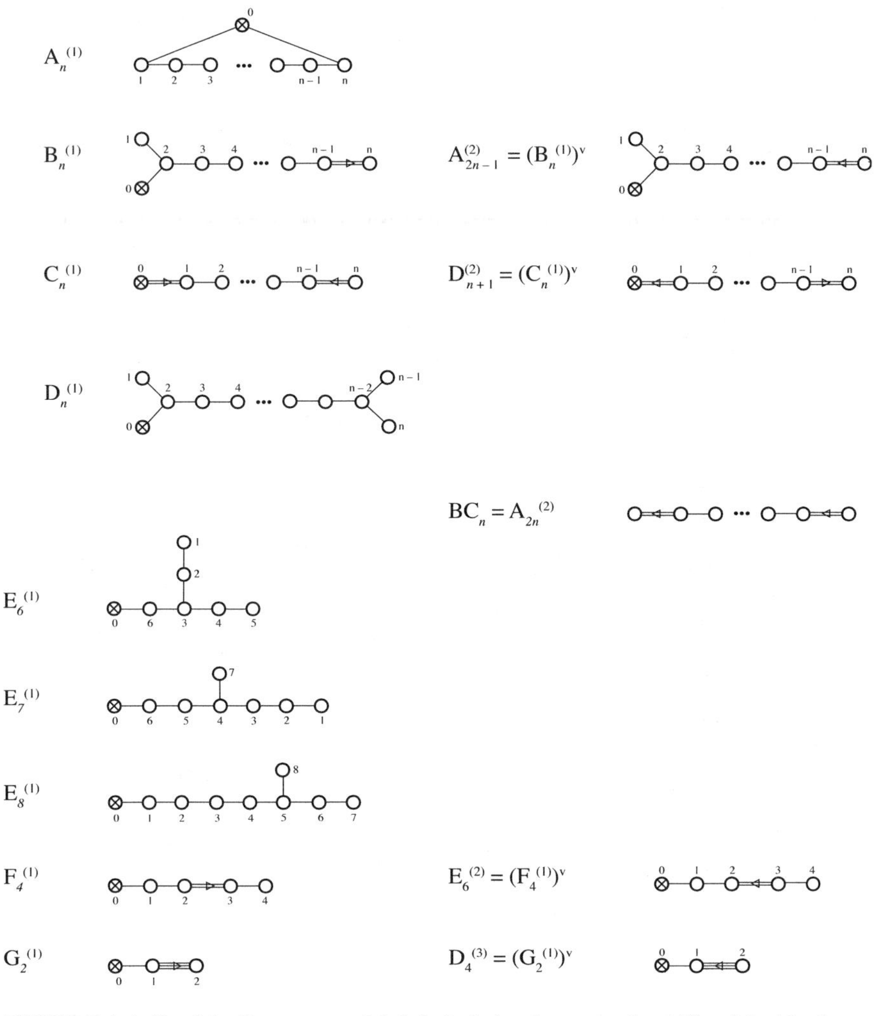

FIGURE 1.1. Dynkin diagrams, with labeled simple roots, for Affine Lie Algebras: untwisted on the left; twisted on the right. Dynkin diagram for finite-dimensional Lie algebras are obtained from unrwisted algebras by deleting the affine root, with label 0 and indicated with a cross.

TABLE 1.9. Root systems of finite dimensional simple Lie algebras

$\mathcal{G}$	all roots
A_n	$\pm(e_i - e_j)$, $1 \leq i < j \leq n+1$
B_n	$\pm(e_i - e_j)$, $\pm(e_i + e_j)$, $1 \leq i < j \leq n$; $\pm e_i$, $1 \leq i \leq n$
C_n	$\pm(e_i - e_j)$, $\pm(e_i + e_j)$, $1 \leq i < j \leq n$; $\pm 2e_i$, $1 \leq i \leq n$
D_n	$\pm(e_i - e_j)$, $\pm(e_i + e_j)$, $1 \leq i < j \leq n$
E_6	$\pm(e_i - e_j)$, $\pm(e_i + e_j)$, $1 \leq i < j \leq 5$; $\pm\frac{1}{2}(\sqrt{3}e_6 + \sum_{i=1}^5 \epsilon_i e_i)$, $\prod_i \epsilon_i = +1$
E_7	$\pm(e_i - e_j)$, $\pm(e_i + e_j)$, $1 \leq i < j \leq 6$; $\pm\frac{1}{2}(\sqrt{2}e_7 + \sum_{i=1}^6 \epsilon_i e_i)$, $\prod_i \epsilon_i = -1$
E_8	$\pm(e_i - e_j)$, $\pm(e_i + e_j)$, $1 \leq i < j \leq 8$; $\frac{1}{2}\sum_{i=1}^8 \epsilon_i e_i$, $\prod_i \epsilon_i = +1$
G_2	$\pm(e_i - e_j)$, $1 \leq i < j \leq 3$; $\pm\big(e_i - \frac{1}{3}(e_1 + e_2 + e_3)\big)$, $i = 1, 2, 3$
F_4	$\pm(e_i - e_j)$, $\pm(e_i + e_j)$, $1 \leq i < j \leq 4$; $\pm e_i$, $1 \leq i \leq 4$; $\pm\frac{1}{2}\sum_{i=1}^4 \epsilon_i e_i$

TABLE 1.10. Marks, co-marks and exponents

$\mathcal{G}$	marks (a_i)	comarks $(a_i^\vee)$	exponents γ_i
A_n	$(1,1,1,\dots,1,1)$	$(1,1,1,\dots,1,1)$	$1,2,3,\dots,n$
B_n	$(1,2,2,\dots,2,2)$	$(1,2,2,\dots,2,1)$	$1,3,5,\dots,2n-1$
C_n	$(2,2,2,\dots,2,1)$	$(1,1,1,\dots,1,1)$	$1,3,5,\dots,2n-1$
D_n	$(1,2,\dots,2,1,1)$	$(1,2,\dots,2,1,1)$	$1,3,5,\dots,2n-3,n-1$
E_6	$(1,2,3,2,1,2)$	$(1,2,3,2,1,2)$	$1,4,5,7,8,11$
E_7	$(2,3,4,3,2,1,2)$	$(2,3,4,3,2,1,2)$	$1,5,7,9,11,13,17$
E_8	$(2,3,4,5,6,4,2,3)$	$(2,3,4,5,6,4,2,3)$	$1,7,11,13,17,19,23,29$
G_2	$(2,3)$	$(2,1)$	$1,5$
F_4	$(2,3,4,2)$	$(2,3,2,1)$	$1,5,7,11$

The highest weight vector λ of any finite dimensional representation Λ of $\mathcal{G}$ is then uniquely specified by positive or zero integers q_i, $i = 1, \dots, n$, with

$$\Lambda \equiv (q_1, \dots, q_n), \qquad \lambda = \sum_{i=1}^n q_i \lambda_i. \tag{B.6}$$

The Weyl orbit of the highest weight vector of $(q_1, \dots, q_n)$ is denoted by $[q_1, \dots, q_n]$.

The *level* $l(\lambda)$ and the *co-level* $l^\vee(\alpha)$ are defined by

$$\begin{aligned} l(\lambda) &= \lambda \cdot \rho^\vee, & l(\alpha_i) &= 1, & i &= 1, \dots, n; \\ l^\vee(\lambda) &= \lambda \cdot \rho, & l^\vee(\alpha_i^\vee) &= 1, & i &= 1, \dots, n. \end{aligned} \tag{B.7}$$

Here, the *level vector* $\rho^\vee$ is related to the *Weyl vector* ρ by exchanging weights λ_i and coweights $\lambda_i^\vee = 2\lambda_i/\alpha_i^2$. Both are uniquely determined by the above normalization, and may be expressed in terms of the fundamental weights and coweights by

$$\begin{aligned} \rho &= \sum_{i=1}^{n} \lambda_i = \frac{1}{2} \sum_{\alpha \in \mathcal{R}_+(\mathcal{G})} \alpha \\ \rho^\vee &= \sum_{i=1}^{n} \lambda_i^\vee = \frac{1}{2} \sum_{\alpha^\vee \in \mathcal{R}_+(\mathcal{G})^\vee} \alpha^\vee . \end{aligned} \tag{B.8}$$

Here, we have provided the relation between the Weyl vector and the half sum of all positive roots, and its dual relation. It is clear from (A.3), (A.4) and (A.8) that the Coxeter and dual Coxeter numbers are related to the level of the highest root and the co-level of the highest coroot

$$\begin{aligned} h_\mathcal{G} &= 1 + \alpha_0 \cdot \rho^\vee = 1 + l(\alpha_0) \\ h_\mathcal{G}^\vee &= 1 + \alpha_0^\vee \cdot \rho = 1 + l^\vee(\alpha_0^\vee). \end{aligned} \tag{B.9}$$

As α (resp. $\alpha^\vee$) ranges through $\mathcal{R}(\mathcal{G})$ (resp. $\mathcal{R}(\mathcal{G})^\vee$), the maxima of $l(\alpha)$ and $l^\vee(\alpha^\vee)$ are $h_\mathcal{G} - 1$ and $h_\mathcal{G}^\vee - 1$ respectively.

The *exponents* γ_i, $i = 1, \ldots, n$ are such that the numbers $\gamma_i + 1$ are the degrees of the independent Casimir operators of the algebra $\mathcal{G}$. Their values are also listed in Table 1.5.

C Elliptic Functions

In this appendix we review some basic definitions and properties of elliptic functions on an elliptic curve of periods $2\omega_1$ and $2\omega_2$ and modulus $\tau = \omega_2/\omega_1$, $\operatorname{Im}\tau > 0$. The half periods are ω_1, ω_2 and $\omega_3 = \omega_1 + \omega_2$. For a useful source, see [81].

C.1 Basic Definitions and Properties

The Weierstrass function is defined by

$$\begin{aligned} &\wp(z; 2\omega_1, 2\omega_2) \\ &\equiv \frac{1}{z^2} + \sum_{(m_1,m_2)\neq(0,0)} \left\{ \frac{1}{(z+2\omega_1 m_1+2\omega_2 m_2)^2} - \frac{1}{(2\omega_1 m_1+2\omega_2 m_2)^2} \right\}. \end{aligned} \tag{C.1}$$

The function $\wp$ is related to the Weierstrass functions ζ and σ by

$$\wp(z; 2\omega_1, 2\omega_2) = -\frac{d}{dz}\zeta(z; 2\omega_1, 2\omega_2) = -\frac{d^2}{dz^2} \log \sigma(z; 2\omega_1, 2\omega_2). \tag{C.2}$$

These functions satisfy

$$\begin{aligned}
\wp(-z) &= \wp(z), & \wp(z+2\omega_a) &= \wp(z) \quad a=1,2,3 \\
\zeta(-z) &= -\zeta(z), & \zeta(z+2\omega_a) &= \zeta(z)+2\eta_a \\
\sigma(-z) &= \sigma(z), & \sigma(z+2\omega_a) &= -\sigma(z)e^{2\eta_a(z+2\omega_a)},
\end{aligned} \tag{C.3}$$

where $\eta_a = \zeta(\omega_a)$ and

$$\begin{aligned}
\sigma(z) &= z + \mathcal{O}(z^5) \\
\zeta(z) &= \frac{1}{z} + \mathcal{O}(z^3) \\
\wp(z) &= \frac{1}{z^2} + \mathcal{O}(z^2).
\end{aligned} \tag{C.4}$$

The function σ may be expressed in terms of the Jacobi ϑ-function

$$\sigma(z; 2\omega_1, 2\omega_2) = 2\omega_1 \exp\left(\frac{\eta_1 z^2}{2\omega_1}\right) \frac{\vartheta_1(z/(2\omega_1) \mid \tau)}{\vartheta_1'(0 \mid \tau)}, \tag{C.5}$$

which in turn is defined in terms of

$$q = e^{2\pi i \tau}, \qquad v = \frac{z}{2\omega_1} \tag{C.6}$$

by

$$\vartheta_1(u \mid \tau) = 2q^{1/4} \sin \pi u \prod_{n=1}^{\infty} (1 - q^n e^{2\pi i u})(1 - q^n e^{-2\pi i u})(1 - q^n). \tag{C.7}$$

An identity basic to the theory of θ-functions is

$$\vartheta_1'(0 \mid \tau) = \frac{2\pi}{2\omega_1} q^{1/8} \prod_{n=1}^{\infty} (1 - q^n)^3. \tag{C.8}$$

The function $\wp$ satisfies the differential equation

$$\wp'(z)^2 = 4\big(\wp(z) - \wp(\omega_1)\big)\big(\wp(z) - \wp(\omega_2)\big)\big(\wp(z) - \wp(\omega_3)\big). \tag{C.9}$$

These and additional properties of elliptic functions may be found in [81]. To evaluate scaling limits, the following formula for $\wp(z; 2\omega_1, 2\omega_2)$ [52] is more useful

$$\wp(z; 2\omega_1, 2\omega_2) = -\frac{\eta_1}{\omega_1} + \left(\frac{\pi}{2\omega_1}\right)^2 \sum_{n=-\infty}^{\infty} \frac{1}{\sinh^2[i\pi(z - 2n\omega_2)/(2\omega_1)]}, \tag{C.10}$$

where

$$\frac{\eta_1}{\omega_1} = -\frac{1}{12} + \frac{1}{2} \sum_{n=1}^{\infty} \frac{1}{\sinh^2 i\pi n \tau}. \tag{C.11}$$

Since a proof of this formula is not given in [52], we provide a proof here for the convenience of the reader. First observe that

$$(1-q^n e^{2\pi i u})(1-q^n e^{-2\pi i u}) = 4q^{2n}\sin(\pi u + n\pi\tau)\sin(\pi u - n\pi\tau). \quad \text{(C.12)}$$

It follows that

$$\vartheta_1(u \mid \tau) = 2q^{1/8}\sin(\pi u)\prod_{n=1}^{\infty}(1-q^n)^3\prod_{n=1}^{\infty}(4q^2)^n \times \prod_{n=1}^{\infty}\sin(nu+n\pi\tau)\sin(nu-n\pi\tau). \quad \text{(C.13)}$$

However,

$$\ln\prod_{n=1}^{\infty}(4q^2) = 2\ln 2\sum_{n=1}^{\infty}1 + \ln q\sum_{n=1}^{\infty}n = 2\ln 2\zeta(0) + \ln q\zeta(-1), \quad \text{(C.14)}$$

where $\zeta(s)$ is the Riemann zeta function. Since $\zeta(0) = -\frac{1}{2}$ and $\zeta(-1) = -1/12$, we find $\ln\prod_{n=1}^{\infty}(4q^2) = \frac{1}{2}q^{-1/12}$. In view of the identity (C.8), we arrive at

$$\frac{\vartheta_1(u \mid \tau)}{\vartheta_1'(0 \mid \tau)} = \frac{\omega_1}{\pi}q^{-1/12}\prod_{n=0}^{\infty}\sin\pi(u-n\tau). \quad \text{(C.15)}$$

It follows that

$$\sigma(z) = \frac{2\omega_1^2}{\pi}\exp\left(\eta_1\frac{z^2}{2\omega_1} - \frac{i\pi\tau}{6}\right)\prod_{n=-\infty}^{\infty}\sin\pi\left(\frac{z}{2\omega_1} - n\tau\right). \quad \text{(C.16)}$$

Differentiating twice, we obtain

$$\zeta(z) = \frac{\eta_1}{\omega_1}z + \frac{\pi}{2\omega_1}\sum_{n=-\infty}^{\infty}\operatorname{cotan}\pi\left(\frac{z}{2\omega_1} - n\tau\right) \quad \text{(C.17)}$$

$$\wp(z) = -\frac{\eta_1}{\omega_1} + \frac{\pi^2}{4\omega_1^2}\sum_{n=-\infty}^{\infty}\frac{1}{\sin^2\pi(z/(2\omega_1) - n\tau)}. \quad \text{(C.18)}$$

The constant η_1/ω_1 can be easily determined. If we choose $\omega_1 = -i\pi$, $\operatorname{Re}\omega_2 > 0$, and compare the expansion (C.10) with the fact that $\wp(z) = 1/z^2 + z^2 g_2/20 + O(z^4)$, we find

$$\frac{\eta_1}{\omega_1} = -\frac{1}{12} + \frac{1}{2}\sum_{n=1}^{\infty}\frac{1}{\operatorname{sh}^2 n\omega_1}. \quad \text{(C.19)}$$

C.2 Half and Third Period Functions

Elliptic functions at half and third period (which are the only ones that we shall need here, since the order of twisting is at most 3) are expressible in terms of the original periods using Landen's transformations [81]. It is convenient to make a definite choice for the period that is to be twisted; we shall choose this period to be ω_1. It is straightforward to adapt these formulas when an arbitrary period $2\omega_a$, $a = 1, \dots, 3$ is twisted.

Formulas for Twists of Order 2: Elliptic Functions with Half Periods

Henceforth, we shall reserve the notation $\wp(z)$, $\zeta(z)$ and $\sigma(z)$ for the corresponding Weierstrass functions with periods $2\omega_1$ and $2\omega_2$, as defined in §C.2. The elliptic functions at half period ω_1 are given by

$$\begin{aligned}
\wp_2(z) &= \wp(z;\omega_1, 2\omega_2) = \wp(z) + \wp(z+\omega_1) - \wp(\omega_1) \\
&= \frac{1}{\wp(\omega_1)}\left[\wp(z)\wp(z+\omega_1) - \big(\wp(\omega_1)-\wp(\omega_2)\big)\big(\wp(\omega_1) - \wp(\omega_3)\big)\right] \\
\zeta_2(z) &= \zeta(z;\omega_1, 2\omega_2) = \zeta(z) + \zeta(z+\omega_1) + z\wp(\omega_1) - \eta_1 \\
\sigma_2(z) &= \sigma(z;\omega_1, 2\omega_2) = \frac{\sigma(z)\sigma(z+\omega_1)}{\sigma(\omega_1)} e^{z^2\wp(\omega_1)/2 - z\eta_1} .
\end{aligned} \tag{C.20}$$

This gives rise to the duplication formula

$$4\wp(2z) = \wp(z) + \wp(z+\omega_1) + \wp(z+\omega_2) + \wp(z+\omega_1+\omega_2). \tag{C.21}$$

Formulas for Twists of Order 3: Elliptic Functions at Third Periods

Similarly, we have the following formulas for the third period elliptic functions

$$\begin{aligned}
\wp_3(z) &= \wp(z; 2\omega_1/3, 2\omega_2) = \wp(z) + \wp(z+2\omega_1/3) + \wp(z+4\omega_1/3) \\
&\qquad - \wp(2\omega_1/3) - \wp(4\omega_1/3) \\
\zeta_3(z) &= \zeta(z; 2\omega_1/3, 2\omega_2) = \zeta(z) + \zeta(z+2\omega_1/3) + \zeta(z+4\omega_1/3) \\
&\qquad + z\wp(2\omega_1/3) + z\wp(4\omega_1/3) - \eta_1 \\
\sigma_3(z) &= \sigma(z; 2\omega_1/3, 2\omega_2) \\
&= \frac{\sigma(z)\sigma(z+2\omega_1/3)\sigma(z+4\omega_1/3)}{\sigma(2\omega_1/3)\sigma(4\omega_1/3)} e^{z^2\wp(\omega_1)/2 - z\eta_1} .
\end{aligned} \tag{C.22}$$

This gives rise to the triplication formula

$$9\wp(3z) = \sum_{j,k=0}^{2} \wp\left(z + j\frac{2\omega_1}{3} + k\frac{2\omega_2}{3}\right). \tag{C.23}$$

All of the above formulas may be established by identifying singularities in z and establishing that the remainder must be independent of z by Liouville's theorem.

The functions at half and third periods, defined above are related to one another in a way analogous to (C.1)

$$\wp_\nu(z; 2\omega_1, 2\omega_2) = -\frac{d}{dz}\zeta_\nu(z; 2\omega_1, 2\omega_2) = -\frac{d^2}{dz^2}\log\sigma_\nu(z; 2\omega_1, 2\omega_2), \quad \text{(C.24)}$$

where $\nu = 1, 2, 3$.

C.3 The Function Φ

We define the function Φ by

$$\Phi(x, z) = \Phi(x, z; 2\omega_1, 2\omega_2) = \frac{\sigma(z-x)}{\sigma(z)\sigma(x)} e^{x\zeta(z)}, \qquad \text{(C.25)}$$

where $\sigma(z)$ and $\zeta(z)$ are the Weierstrass functions of (C.1) and (C.2). As a function of z, $\Phi(x, z)$ is periodic with periods $2\omega_1$ and $2\omega_2$, is holomorphic except for an essential singularity at $z = 0$, and has a single zero at $z = x$. As a function of x, $\Phi(x, z)$ has multiplicative monodromy, given by

$$\Phi(x + 2\omega_a, z) = \Phi(x, z)e^{2\omega_a\zeta(z) - 2\eta_a z}, \qquad \text{(C.26)}$$

is holomorphic in x except for a simple pole at $x = 0$, and has a single zero at $x = z$. Some useful asymptotics are given as follows

$$\begin{aligned} \Phi(x, z) &= \frac{1}{x} - \frac{1}{2}x\wp(z) + \mathcal{O}(x^2) \\ \Phi(x, z) &= \left\{-\frac{1}{z} + \zeta(x) + \mathcal{O}(z)\right\}e^{x\zeta(z)} \qquad z \to 0. \end{aligned} \qquad \text{(C.27)}$$

Products of the function $\Phi(x_\alpha, z)$, with $\sum_{\alpha=1}^n x_\alpha = 0$, are periodic in z, with periods $2\omega_1$ and $2\omega_2$, and meromorphic in z since the essential singularities cancel. As a result, they satisfy

$$\prod_{\alpha=1}^n \Phi(x_\alpha, z) = P_n[\wp(z); x_\alpha] + \wp'(z)Q_n[\wp(z); x_\alpha], \qquad \text{(C.28)}$$

where P_n and Q_n are polynomials of degrees $[n/2]$ and $[(n-3)/2]$ in $\wp(z)$ respectively, with x_α-dependent coefficients. The simplest case is

$$\Phi(x, z)\Phi(-x, z) = \wp(z) - \wp(x). \qquad \text{(C.29)}$$

In general, the polynomials P and Q may be determined by the fact that the r.h.s. of (C.28) has a simple zero at each point $z = x_\alpha$, and that the pole highest order in z has coefficient $(-1)^n$.

The function $\Phi(x, z)$ satisfies a fundamental differential equation,

$$\Phi(x, z)\Phi'(y, z) - \Phi(y, z)\Phi'(x, z) = \big(\wp(x) - \wp(y)\big)\Phi(x + y, z), \qquad \text{(C.30)}$$

where $\Phi'(x, z)$ denotes the derivative with respect to x of $\Phi(x, z)$.

C.4 The Functions Λ, Φ_1, Φ_2

The functions Λ and Φ_2 are defined by

$$\Lambda(2x,z)=\Phi_2(x,z)=\frac{\Phi(x,z)\Phi(x+\omega_1,z)}{\Phi(\omega_1,z)}. \tag{C.31}$$

The essential singularity in z and the monodromy in x of $\Lambda(2x,z)$ coincide with those of $\Phi(x,z)$. We shall need the following two basic differential equations,

$$\begin{aligned}\Lambda(2x,z)\Lambda'(2y,z)-\Lambda'(2x,z)\Lambda(2y,z)&\\ &=\tfrac{1}{2}\big(\wp_2(x)-\wp_2(y)\big)\Lambda(2x+2y,z),\\ \Phi_2(x,z)\Phi_2'(y,z)-\Phi_2(y,z)\Phi_2'(x,z)&\\ &=\big(\wp_2(x)-\wp_2(y)\big)\Phi_2(x+y,z),\end{aligned} \tag{C.32}$$

as well as differential equation that involves both Φ and Λ,

$$\begin{aligned}\Lambda(2x,z)\Phi'(-x-y,z)-\Lambda'(2x,z)\Phi(-x-y,z)-\Lambda(-2y,z)\Phi'(x+y,z)&\\ +\Lambda'(-2y,z)\Phi(x+y,z)=\tfrac{1}{2}\big(\wp_2(x)-\wp_2(y)\big)\Phi(x-y,z).&\end{aligned} \tag{C.33}$$

By letting $y\to x$ in (C.33), and taking into account the known zeros of Λ, we derive another useful formula

$$\Lambda(2x,z)\Lambda(-2x,z)=\wp_2(z)-\wp_2(x). \tag{C.34}$$

Actually, $\Phi_2(x,z)$ may be viewed as the function $\Phi(x,z)$ associated with a torus of periods ω_1 and $2\omega_2$.

The function $\Phi_1(x,z)$ is defined by

$$\Phi_1(x,z)=\Phi(x,z)+f(z)\Phi(x+\omega_1,z) \tag{C.35}$$

$$f(z)=-e^{\pi i\zeta(z)+\eta_1 z}. \tag{C.36}$$

It obeys the monodromy relation $\Phi_1(x+\omega_1,z)=f(z)^{-1}\Phi_1(x,z)$, as well as the following differential equations

$$\Phi_1(x,z)\Phi_1'(y,z)-\Phi_1'(x,z)\Phi_1(y,z)=\big(\wp_2(x)-\wp_2(y)\big)\Phi_1(x+y,z) \tag{C.37}$$

$$\begin{aligned}\Phi_1(x,z)\Phi'(y,z)-\Phi(y,z)\Phi_1'(x,z)&\\ =\{\wp(x+\omega_1)-\wp(y)\}\Phi_1(x+y,z)&\\ +\{\wp(x)-\wp(x+\omega_1)\}\Phi(x+y,z),&\end{aligned} \tag{C.38}$$

and

$$\begin{aligned}\Phi(2x,z)\Phi_1'(-x-y,z)-\Phi'(2x,z)\Phi_1(-x-y,z)&\\ -\Phi(-2y,z)\Phi_1'(x+y,z)+\Phi'(-2y,z)\Phi_1(x+y,z)&\\ =\big(\wp(2x)-\wp(2y)\big)\Phi_1(x-y,z),&\end{aligned} \tag{C.39}$$

$$\begin{aligned}\Lambda(2x,z)\Phi_1'(-x-y,z) - \Lambda'(2x,z)\Phi_1(-x-y,z)& \\ - \Lambda(-2y,z)\Phi_1'(x+y,z) + \Lambda'(-2y,z)\Phi_1(x+y,z)& \\ = \tfrac{1}{2}\big(\wp_2(x) - \wp_2(y)\big)\Phi_1(x-y,z).& \qquad \text{(C.40)}\end{aligned}$$

2

Lectures on Branes, Black Holes, and Anti-de Sitter Space

M.J. Duff

ABSTRACT In the light of the duality between physics in the bulk of anti-de Sitter space and a conformal field theory on the boundary, we review the M2-, D3- and M5-branes and how their near-horizon geometry yields the compactification of $D = 11$ supergravity on S^7, Type IIB supergravity on S^5 and $D = 11$ supergravity on S^4, respectively. We discuss the "Membrane at the End of the Universe" idea and its relation to the corresponding superconformal singleton theories that live on the boundary of the AdS_4, AdS_5 and AdS_7 vacua. The massless sectors of these compactifications are described by the maximally supersymmetric $D = 4$, $D = 5$ and $D = 7$ gauged supergravities. We construct the nonlinear Kaluza–Klein ansatzs describing the embeddings of the $U(1)^4$, $U(1)^3$ and $U(1)^2$ truncations of these supergravities, which admit 4-charge AdS_4, 3-charge AdS_5 and 2-charge AdS_7 black hole solutions. These enable us to embed the black hole solutions back in ten and eleven dimensions and reinterpret them as M2-, D3- and M5-branes spinning in the transverse dimensions with the black hole charges given by the angular momenta of the branes. A comprehensive Appendix lists the field equations, symmetries and transformation rules of $D = 11$ supergravity, Type IIB supergravity, and the M2-, D3- and M5-branes.

1 Introduction

1.1 Supergravity, Supermembranes and M-Theory

A vital ingredient in the quest for a unified theory embracing all physical phenomena is *supersymmetry*, a symmetry which (a) unites bosons and fermions, (b) requires the existence of gravity and (c) places an upper limit of *eleven* on the dimension of spacetime. For these reasons, in the early 1980s many physicists looked to eleven-dimensional supergravity in the hope that it might provide that elusive superunified theory. Then in 1984 superunification underwent a major paradigm-shift: eleven-dimensional supergravity was knocked off its pedestal by ten-dimensional superstrings, one-dimensional objects whose vibrational modes represent the elementary particles. Unlike eleven-dimensional supergravity, superstrings provided a perturbatively finite theory of gravity which, after compactification to four spacetime dimensions, seemed in principle capable of explaining the Standard Model of the strong, weak and electromagnetic forces including the required *chiral* representations of quarks and leptons.

Despite these major successes, however, nagging doubts persisted about superstrings. First, many of the most important questions in string theory —How do strings break supersymmetry? How do they choose the right vacuum state? How do they explain the smallness of the cosmological constant? How do they resolve the apparent paradoxes of quantum black holes?—seemed incapable of being answered within the framework of a weak coupling perturbation expansion. They seemed to call for some new, *nonperturbative*, physics. Secondly, why did there appear to be *five* different mathematically consistent superstring theories: the $E_8 \times E_8$ heterotic string, the SO(32) heterotic string, the SO(32) Type I string, the Type IIA and Type IIB strings? If one is looking for a unique *Theory of Everything*, this seems like an embarrassment of riches! Thirdly, if supersymmetry permits eleven dimensions, why do superstrings stop at ten? This question became more acute with the discoveries of the elementary *supermembrane* in 1987 and its dual partner, the solitonic *superfivebrane*, in 1992. These are supersymmetric extended objects with respectively two and five dimensions moving in an eleven-dimensional spacetime. Finally, therefore, if we are going to generalize zero-dimensional point particles to one-dimensional strings, why stop there? Why not two-dimensional membranes or more generally p-dimensional objects (inevitably dubbed *p-branes*)? Although this latter possibility was pursued by a small but dedicated group of theorists, starting in about 1986, it was largely ignored by the mainstream physics community.

Well, the year 1995 witnessed a new paradigm-shift: perturbative ten-dimensional superstrings have in their turn been superseded by a new *nonperturbative* theory called *M-theory*, which describes, amongst other things, supermembranes and superfivebranes, which subsumes the above five consistent strings theories, and which has as its low-energy limit, eleven-dimensional supergravity! According to Fields Medalist Edward Witten "M stands for magical, mystery or membrane, according to taste". New evidence in favor of this theory is appearing daily on the internet and represents the most exciting development in the subject since 1984 when the superstring revolution first burst on the scene.

$$\left.\begin{array}{r}\mathrm{E}_8 \times \mathrm{E}_8 \text{ heterotic string} \\ \mathrm{SO}(32) \text{ heterotic string} \\ \mathrm{SO}(32) \text{ Type I string} \\ \text{Type IIA string} \\ \text{Type IIB string}\end{array}\right\} \text{M theory} \tag{1.1}$$

Thus this new framework now provides the starting point for understanding a wealth of new nonperturbative phenomena, including string/string duality, Seiberg–Witten theory, quark confinement, QCD, particle physics phenomenology, quantum black holes, cosmology and, ultimately, their complete synthesis in a final theory of physics.

1.2 The Kaluza–Klein Idea

Cast your minds back to 1919. Maxwell's theory of electromagnetism was well established and Einstein had recently formulated his General Theory of Relativity. By contrast, the strong and weak interactions were not well understood. In searching for a unified theory of the fundamental forces, therefore, it was natural to attempt to merge gravity with electromagnetism. This is what Kaluza was able to do through the remarkable device of postulating an extra fifth dimension for spacetime. Consider Einstein's theory of pure gravity in five spacetime dimensions with signature $(-,+,+,+,+)$. The line element is given by

$$ds_5{}^2 = g_{MN}\, dx^M\, dx^N \tag{1.2}$$

where $M, N = 0, 1, 2, 3, 4$. Kaluza then made the $4+1$ split

$$g_{MN} = e^{\phi/\sqrt{3}} \begin{pmatrix} g_{\mu\nu} + e^{-\sqrt{3}\phi} A_\mu A_\nu & e^{-\sqrt{3}\phi} A_\mu \\ e^{-\sqrt{3}\phi} A_\nu & e^{-\sqrt{3}\phi} \end{pmatrix} \tag{1.3}$$

where $x^M = (x^\mu, y)$, $\mu = 0, 1, 2, 3$. Thus the fields $g_{\mu\nu}(x)$, $A_\mu(x)$ and $\phi(x)$ transform respectively as a tensor, a vector and a scalar under four-dimensional general coordinate transformations. All this was at the classical level, of course, but in the modern parlance of quantum field theory, they would be described as the spin 2 graviton, the spin 1 photon and the spin 0 dilaton.[1] Of course it is not enough to call A_μ by the name photon, one must demonstrate that it satisfies Maxwell's equations and here we see the Kaluza miracle at work. After making the same $4+1$ split of the five-dimensional Einstein equations $R_{MN} = 0$, we correctly recover not only the Einstein equations for $g_{\mu\nu}(x)$ but also the Maxwell equation for $A_\mu(x)$ and the massless Klein–Gordon equation for $\phi(x)$. Thus Maxwell's theory of electromagnetism is an inevitable consequence of Einstein's general theory of relativity, given that one is willing to buy the idea of a fifth dimension.

Attractive though Kaluza's idea was, it suffered from two obvious drawbacks. First, although the indices were allowed to range over $0, 1, 2, 3, 4$, for no very good reason the dependence on the extra coordinate y was suppressed. Secondly, if there is a fifth dimension why haven't we seen it? The resolution of both these problems was supplied by Oskar Klein in 1926.

[1] The dilaton was considered an embarrassment in 1919, and was (inconsistently) set equal to zero. However, it was later revived and subsequently stimulated Brans–Dicke theories of gravity. The dilaton also plays a crucial role in M-theory.

Klein insisted on treating the extra dimension seriously but assumed the fifth dimension to have circular topology so that the coordinate y is periodic, $0 \leq my \leq 2\pi$, where m has dimensions of mass. Thus the space has topology $\mathbb{R}^4 \times S^1$. It is difficult to envisage a spacetime with this topology but a simpler analogy is provided by a garden hose: at large distances it looks like a line $\mathbb{R}^1$ but closer inspection reveals that at every point on the line there is a little circle, and the topology is $\mathbb{R}^1 \times S^1$. So it was that Klein suggested that there is a little circle at each point in four-dimensional spacetime.

Let us consider Klein's proposal from a modern perspective. We start with pure gravity in five dimensions described by the action

$$I_5 = \frac{1}{2\kappa_5{}^2} \int d^5x \sqrt{-g}R. \tag{1.4}$$

I_5 is invariant under the five-dimensional general coordinate transformations

$$\delta g_{MN} = \partial_M \xi^P g_{PN} + \partial_N \xi^P g_{MP} + \xi^P \partial_P g_{MN}. \tag{1.5}$$

endinut The periodicity in y means that the fields $g_{MN}(x, y)$, $A_M(x, y)$ and $\phi(x, y)$ may be expanded in the form

$$\begin{aligned} g_{\mu\nu}(x,y) &= \sum_{n=-\infty}^{\infty} g_{\mu\nu(n)}(x) e^{inmy}, \\ A_\mu(x,y) &= \sum_{n=-\infty}^{\infty} A_{\mu(n)}(x) e^{inmy}, \\ \phi(x,y) &= \sum_{n=-\infty}^{\infty} \phi_{(n)} e^{inmy} \end{aligned} \tag{1.6}$$

with

$$g^*{}_{\mu\nu(n)}(x) = g_{\mu\nu(-n)}(x) \tag{1.7}$$

etc. So (as one now finds in all the textbooks) a Kaluza–Klein theory describes an infinite number of four-dimensional *fields*. However (as one finds in none of the textbooks) it also describes an infinite number of four-dimensional *symmetries* since we may also Fourier expand the general coordinate parameter $\xi^\mu(x, y)$ as follows

$$\begin{aligned} \xi^\mu(x,y) &= \sum_{n=-\infty}^{\infty} \xi^\mu{}_{(n)}(x) e^{inmy} \\ \xi^4(x,y) &= \sum_{n=-\infty}^{\infty} \xi^4{}_{(n)}(x) e^{inmy} \end{aligned} \tag{1.8}$$

with $\xi^{*M}{}_{(n)} = \xi^M{}_{-(n)}$.

Let us first focus on the $n = 0$ modes in (1.6), which are just Kaluza's graviton, photon and dilaton. Substituting (1.3) and (1.6) in the action (1.4), integrating over y and retaining just the $n = 0$ terms we obtain (dropping the 0 subscripts)

$$I_4 = \frac{1}{2\kappa_4^2} \int d^4x \sqrt{-g} \left[R - \frac{1}{2}\partial_\mu \phi \partial^\mu \phi - \frac{1}{4} e^{-\sqrt{3}\phi} F_{\mu\nu} F^{\mu\nu} \right] \tag{1.9}$$

where $F_{\mu\nu} = \partial_\mu A_\nu - \partial_\nu A_\mu$ and $2\pi\kappa_4^2 = m\kappa_5^2$. Newton's constant is given by $\kappa_4^2 = 8\pi G$. This form of the action explains our choice of parameterization in (1.3): we obtain the usual Einstein–Hilbert term for gravity, the conventional Maxwell kinetic term for electromagnetism and the conventional Klein–Gordon term for the dilaton. Note, however, that we have normalized ϕ and A_μ so that $1/(2\kappa_4^2)$ is common to both gravity and matter terms. From (1.5, this action is invariant under general coordinate transformations with parameter ${\xi^\mu}_0$, i.e, (again dropping the 0 subscripts)

$$\begin{aligned} \delta g_{\mu\nu} &= \partial_\mu \xi^\rho g_{\rho\nu} + \partial_\nu \xi^\rho g_{\mu\rho} + \xi^\rho \partial_\rho g_{\mu\nu}, \\ \delta A_\mu &= \partial_\mu \xi^\rho A_\rho + \xi^\rho \partial_\rho A_\mu, \\ \delta\phi &= \xi^\rho \partial_\rho \phi, \end{aligned} \tag{1.10}$$

local gauge transformations with parameter ${\xi^4}_0$

$$\delta A_\mu = \partial_\mu \xi^4 \tag{1.11}$$

and global scale transformations with parameter λ

$$\delta A_\mu = \lambda A_\mu, \quad \delta\phi = 2\lambda/\sqrt{3}. \tag{1.12}$$

The symmetry of the vacuum, determined by the VEVs

$$\langle g_{\mu\nu} \rangle = \eta_{\mu\nu}, \quad \langle A_\mu \rangle = 0, \quad \langle \phi \rangle = \phi_0 \tag{1.13}$$

is the four-dimensional Poincaré group $\times\ \mathbb{R}$. Thus, the masslessness of the graviton is due to general covariance, the masslessness of the photon to gauge invariance, but the dilaton is massless because it is the Goldstone boson associated with the spontaneous breakdown of the global scale invariance. Note that the gauge group is $\mathbb{R}$ rather than U(1) because this truncated $n = 0$ theory has lost all memory of the periodicity in y.

Now, however, let us include the $n \neq 0$ modes. An important observation is that the assumed topology of the ground state, namely $\mathbb{R}^4 \times S^1$, restricts us to general coordinate transformations periodic in y. Whereas the general covariance (1.10) and local gauge invariance (1.11) simply correspond to the $n = 0$ modes of (1.5) respectively, the global scale invariance is no longer a symmetry because it corresponds to a rescaling

$$\delta g_{MN} = -\tfrac{1}{2}\lambda g_{MN} \tag{1.14}$$

combined with a general coordinate transformation

$$\xi^4 = -\lambda y/m \tag{1.15}$$

which is now forbidden by the periodicity requirement. The field ϕ is therefore merely a pseudo-Goldstone boson.

Just as ordinary general covariance may be regarded as the local gauge symmetry corresponding to the global Poincaré algebra and local gauge invariance as the gauge symmetry corresponding to the global Abelian algebra, so the infinite parameter local transformations (1.8) correspond to an infinite-parameter global algebra with generators

$$\begin{aligned} P^{\mu}{}_n &= e^{inmy}\partial^{\mu} \\ M^{\mu\nu}{}_n &= e^{inmy}(x^{\mu}\partial^{\nu} - x^{\nu}\partial^{\mu}) \\ Q_n &= ie^{inmy}\partial/\partial(my). \end{aligned} \tag{1.16}$$

It is in fact a Kac–Moody–Virasoro generalization of the Poincaré/gauge algebra. Although this larger algebra describes a symmetry of the four-dimensional theory, the symmetry of the vacuum determined by (1.13) is only Poincaré × U(1). Thus the gauge parameters $\xi^{\mu}{}_n$ and $\xi^4{}_n$ with $n \neq 0$ each correspond to spontaneously broken generators, and it follows that for $n \neq 0$ the fields $A_{\mu n}$ and ϕ_n are the corresponding Goldstone boson fields. The gauge fields $g_{\mu\nu n}$, with two degrees of freedom, will then each acquire a mass by absorbing the two degrees of freedom of each vector Goldstone boson $A_{\mu n}$ and the one degree of freedom of each scalar Goldstone boson ϕ_n to yield a pure spin 2 massive particle with five degrees of freedom. This accords with the observation that the massive spectrum is pure spin two. Thus we find an infinite tower of charged, massive spin 2 particles with charges e_n and masses m_n given by

$$e_n = n\sqrt{2}\kappa_4 m, \quad m_n = |n|/R \tag{1.17}$$

where R is the radius of the S^1 given by $R^2m^2 = \langle g_{yy}\rangle$. Thus Klein explained (for the first time) the quantization of electric charge. (Note also that charge conjugation is just parity tranformation $y \to -y$ in the fifth dimension.) Of course, if we identify the fundamental unit of charge $e = \sqrt{2}\kappa_4 m$ with the charge on the electron, then we are forced to take m to be very large: the Planck mass 10^{19} GeV, way beyond the range of any current or forseeable accelerator. This answers the second question left unanswered by Kaluza because with m very large, the radius of the circle must be very small: the Planck size 10^{-35} *meters*, which satisfactorily accords with our everyday experience of living in four spacetime dimensions.[2]

[2] A variation on the Kaluza–Klein theme is that our universe is a 3-brane embedded in a higher dimensional spacetime [204, 205]. This is particularly compelling in the

It is interesting to note that, despite the inconsistency problems that arise in coupling a finite number of massive spin two particles to gravity and/or electromagnetism, Kaluza–Klein theory is consistent by virtue of having an *infinite* tower of such states. Any attempt to truncate to a finite nonzero number of massive modes would reintroduce the inconsistency. We also note, however, that these massive Kaluza–Klein modes have the unusual gyromagnetic ratio $g = 1$. Moreover, when we embed the theory in a superstring theory or M-theory, these Kaluza–Klein states will persist as a subset of the full string or M-theory spectrum. It is sometimes claimed that $g \neq 2$ leads to unacceptable high-energy behaviour for Compton scattering. However, although the classical value $g = 2$ is required in QED, the Standard Model, and indeed open string theory, this is not a universal rule. Tree level unitarity applies only in the energy regime $M_{\text{Planck}} > E > m/Q$ for a particle of mass m and charge Q, and this range is empty for Kaluza–Klein theory.

In summary, it seems that a five-dimensional world with one of its dimensions compactified on a circle is operationally indistinguishable from a four-dimensional world with a very particular (albeit infinite) mass spectrum.

1.3 The Field Content

Eleven is the maximum spacetime dimension in which one can formulate a consistent supersymmetric theory, as was first recognized by Nahm [37] in his classification of supersymmetry algebras. The easiest way to see this is to start in four dimensions and note that one supersymmetry relates states differing by one half unit of helicity. If we now make the reasonable assumption that there be no massless particles with spins greater than two, then we can allow up to a maximum of $N = 8$ supersymmetries taking us from helicity -2 through to helicity $+2$. Since the minimal supersymmetry generator is a Majorana spinor with four off-shell components, this means a total of 32 spinor components. A discussion of spinors and Dirac matrices in D spacetime dimensions may be found in the reprint volume of Salam and Sezgin [198]. Now in a spacetime with D dimensions and signature $(1, D-1)$, the maximum value of D admitting a 32 component spinor is $D = 11$. (Going to $D = 12$, for example, would require 64 components.) See

---

context of the Type IIB threebrane [206] since the worldvolume fields necessarily include gauge fields [58]. Thus the strong, weak and electromagnetic forces might be confined to the worldvolume of the brane while gravity propagates in the bulk. It has recently been suggested that, in such schemes, the extra dimensions might be much larger than 10^{-35} meters [207, 208] and may even be a large as a millimeter [209–211, 232]. In yet another variation, the brane occupies the boundary of AdS_5 and the extra fifth dimension is infinite [233]. Once again, the near horizon geometry of the D3-brane provides an example of this, as discussed in Section 3.2.

TABLE 2.1. Minimal spinor components and supersymmetries.

Dimension (D or d)	Minimal Spinor (M or m)	Supersymmetry (N or n)
11	32	1
10	16	2, 1
9	16	2, 1
8	16	2, 1
7	16	2, 1
6	8	4, 3, 2, 1
5	8	4, 3, 2, 1
4	4	8, ..., 1
3	2	16, ..., 1
2	1	32, ..., 1

TABLE 2.2. On-shell degrees of freedom in D dimensions. $\alpha = D/2$ if D is even, $\alpha = (D-1)/2$ if D is odd. We assume Majorana fermions and divide by two if the fermion is Majorana–Weyl. Similarly, we assume real bosons and divide by two if the tensor field strength is self-dual.

D-bein	$e_M{}^A$	$D(D-3)/2$
gravitino	Ψ_M	$2^{(\alpha-1)}(D-3)$
p-form	$A_{M_1 M_2 \cdots M_p}$	$\binom{D-2}{p}$

Table 2.1.[3] Furthermore, as we shall see later, $D = 11$ emerges naturally as the maximum dimension admitting supersymmetric extended objects, without the need for any assumptions about higher spin. Not long after Nahm's paper, Cremmer, Julia and Scherk [197] realized that supergravity not only permits up to seven extra dimensions but in fact takes its simplest and most elegant form when written in its full eleven-dimensional glory. The unique $D = 11$, $N = 1$ supermultiplet is comprised of a graviton g_{MN}, a gravitino ψ_M and 3-form gauge field A_{MNP} with 44, 128 and 84 physical degrees of freedom, respectively. For a counting of on-shell degrees of freedom in higher dimensions, see Table 2.2.

The theory may also be formulated in superspace. Ironically, however, these extra dimensions were not at first taken seriously but rather regarded merely as a useful device for deriving supergravities in four dimensions. Indeed $D = 4$, $N = 8$ supergravity was first obtained by Cremmer and Julia via the process of *dimensional reduction*, i.e. by requiring that all the fields of $D = 11$, $N = 1$ supergravity be independent of the extra seven coordinates.

---

[3] Conventions differ on how to count the supersymmetries and in later sections we follow the more usual convention [197] that $N_{\max} = 8$ in $D = 5$ and in $N_{\max} = 4$ $D = 7$.

For many years the Kaluza–Klein idea of taking extra dimensions seriously was largely forgotten but the arrival of eleven-dimensional supergravity provided the missing impetus. The kind of four-dimensional world we end up with depends on how we *compactify* these extra dimensions: maybe seven of them would allow us to give a gravitational origin, à la Kaluza–Klein, to the strong and weak forces as well as the electromagnetic. In a very influential paper, Witten [212] drew attention to the fact that in such a scheme the four-dimensional gauge group is determined by the *isometry* group of the compact manifold $\mathcal{K}$. Moreover, he proved (what to this day seems to be merely a gigantic coincidence) that seven is not only the maximum dimension of $\mathcal{K}$ permitted by supersymmetry but the minimum needed for the isometry group to coincide with the standard model gauge group $SU(3) \times SU(2) \times U(1)$.

In the early '80s there was great interest in four-dimensional N-extended supergravities for which the global $SO(N)$ is promoted to a gauge symmetry [8]. In these theories the underlying supersymmetry algebra is no longer Poincaré but rather anti-de Sitter (AdS_4) and the Lagrangian has a nonvanishing cosmological constant Λ proportional to the square of the gauge coupling constant g:

$$G\Lambda \sim -g^2 \tag{1.18}$$

where G is Newton's constant. The $N > 4$ gauged supergravities were particularly interesting since the cosmological constant Λ does not get renormalized [9] and hence the $SO(N)$ gauge symmetry has vanishing β-function.[4] The relation (1.18) suggested that there might be a Kaluza–Klein interpretation since in such theories the coupling constant of the gauge group arising from the isometries of the extra dimensions is given by

$$g^2 \sim Gm^2 \tag{1.19}$$

where m^{-1} is the size of the compact space. Moreover, there is typically a negative cosmological constant

$$\Lambda \sim -m^2. \tag{1.20}$$

Combining (1.19) and (1.20), we recover (1.18). Indeed, the maximal ($D = 4$, $N = 8$) gauged supergravity [11] was seen to correspond to the massless sector of ($D = 11$, $N = 1$) supergravity [12] compactified on an S^7 whose metric admits an $SO(8)$ isometry and 8 Killing spinors [13]. An important ingredient in these developments that had been insufficiently emphasized in earlier work on Kaluza–Klein theory was that the $AdS_4 \times S^7$ geometry was

[4]For $N \leq 4$, the *beta* function (which receives a contribution from the spin $\frac{3}{2}$ gravitinos) is positive and the pure supergravity theories are not asymptotically free. The addition of matter supermultiplets only makes the β function more positive [10] and hence gravitinos can never be confined. I am grateful to Karim Benakli, Rene Martinez Acosta and Parid Hoxha for discussions on this point.

TABLE 2.3. Compactifications and their symmetries.

Compactification	**Supergroup**	**Bosonic subgroup**
$\mathrm{AdS}_4 \times S^7$	$\mathrm{OSp}(4 \mid 8)$	$\mathrm{SO}(3,2) \times \mathrm{SO}(8)$
$\mathrm{AdS}_5 \times S^5$	$\mathrm{SU}(2,2 \mid 4)$	$\mathrm{SO}(4,2) \times \mathrm{SO}(6)$
$\mathrm{AdS}_7 \times S^4$	$\mathrm{OSp}(6,2 \mid 4)$	$\mathrm{SO}(6,2) \times \mathrm{SO}(5)$

TABLE 2.4. Superconformal groups and their singleton, doubleton and tripleton repesentations.

Supergroup	**Supermultiplet**	**Field content**
$\mathrm{OSp}(4 \mid 8)$	$(n = 8, d = 3)$ singleton	8 scalars, 8 spinors
$\mathrm{SU}(2,2 \mid 4)$	$(n = 4, d = 4)$ doubleton	1 vector, 8 spinors, 6 scalars
$\mathrm{OSp}(6,2 \mid 4)$	$\big((n_+, n_-) = (2,0), d = 6\big)$ tripleton	1 chiral 2-form, 8 spinors, 5 scalars

not fed in by hand but resulted from a *spontaneous compactification*, i.e. the vacuum state was obtained by finding a stable solution of the higher-dimensional field equations [14]. The mechanism of spontaneous compactification appropriate to the $\mathrm{AdS}_4 \times S^7$ solution of eleven-dimensional supergravity was provided by the Freund–Rubin mechanism [15] in which the 4-form field strength in spacetime $F_{\mu\nu\rho\sigma}$ $(\mu = 0,1,2,3)$ is proportional to the alternating symbol $\epsilon_{\mu\nu\rho\sigma}$ [16]:

$$F_{\mu\nu\rho\sigma} \sim \epsilon_{\mu\nu\rho\sigma}. \tag{1.21}$$

A summary of this S^7 and other X^7 compactifications of $D = 11$ supergravity down to AdS_4 may be found in [20]. By applying a similar mechanism to the 7-form dual of this field strength one could also find compactifications on $\mathrm{AdS}_7 \times S^4$ [17] whose massless sector describes gauged maximal $N = 4$, SO(5) supergravity in $D = 7$ [18, 19]. Type IIB supergravity in $D = 10$, with its self-dual 5-form field strength, also admits a Freund–Rubin compactification on $\mathrm{AdS}_5 \times S^5$ [23–25] whose massless sector describes gauged maximal $N = 8$ supergravity in $D = 5$ [26, 27].

In the three cases given above, the symmetry of the vacuum is described by the supergroups $\mathrm{OSp}(4 \mid 8)$, $\mathrm{SU}(2,2 \mid 4)$ and $\mathrm{OSp}(6,2 \mid 4)$ for the S^7, S^5 and S^4 compactifications respectively, as shown in Table 2.3. Each of these groups is known to admit the so-called singleton, doubleton or tripleton[5] supermultiplets [28] as shown in Table 2.4.

We recall that singletons are those strange representations of AdS first identified by Dirac [29] which admit no analogue in flat spacetime. They have been much studied by Fronsdal and collaborators [30, 31].

[5]Our nomenclature is based on the AdS_4, AdS_5 and AdS_7 groups having ranks 2, 3 and 4, respectively, and differs from that of Günaydin.

This Kaluza–Klein approach to $D = 11$ supergravity [200] eventually fell out of favor for three reasons. First, as emphasized by Witten [199], it is impossible to derive by the conventional Kaluza–Klein technique of compactifying on a manifold a *chiral theory* in four spacetime dimensions starting from a nonchiral theory such as eleven-dimensional supergravity. Ironically, Horava and Witten [102] were to solve this problem years later by compactifying on something that is not a manifold! Secondly, in spite of its maximal supersymmetry and other intriguing features, eleven dimensional supergravity was, after all, still a *field theory* of gravity with all the attendant problems of nonrenormalizability. For a recent discussion, see [241]. This problem also had to await the dawn of M-theory, since we now regard $D = 11$ supergravity not as a fundamental theory in its own right but the effective low-energy Lagrangian of M-theory. Thirdly, these AdS vacua necessarily have nonvanishing cosmological constant unless cancelled by fermion condensates [109], and this was deemed unacceptable at the time. However, as we shall now describe, AdS is currently undergoing a renaissance thanks to the AdS/CFT correspondence.

1.4 The AdS/CFT Correspondence

A by-product of M-theory has been the revival of anti-de Sitter space brought about by Maldacena's conjectured duality between physics in the bulk of AdS and a conformal field theory on the boundary [110]. In particular, M-theory on $\mathrm{AdS}_4 \times S^7$ is dual to a non-Abelian $(n = 8, d = 3)$ superconformal theory, Type IIB string theory on $\mathrm{AdS}_5 \times S^5$ is dual to a $d = 4$ SU(N) super Yang–Mills theory and M-theory on $\mathrm{AdS}_7 \times S^4$ is dual to a non-Abelian $\big((n_+, n_-) = (2, 0), d = 6\big)$ conformal theory. In particular, as has been spelled out most clearly in the $d = 4$ SU(N) Yang–Mills case, there is seen to be a correspondence between the Kaluza–Klein mass spectrum in the bulk and the conformal dimension of operators on the boundary [122, 124]. This duality thus holds promise not only of a deeper understanding of M-theory, but may also throw light on nonperturbative aspects of the theories that live on the boundary which can include four-dimensional gauge theories. Models of this kind, where a bulk theory with gravity is equivalent to a boundary theory without gravity, have also been advocated by 't Hooft [130] and by Susskind [131] who call them *holographic* theories. The reader may notice a striking similarity to the earlier idea of "The membrane at the end of the universe" [38, 39, 44–46], whereby the p-brane occupies the $S^1 \times S^p$ boundary of AdS_{p+2} and is described by a superconformal singleton theory and to the "membrane/supergravity bootstrap" [38, 39, 44] which conjectured that the dynamics of the supergravity in the bulk of AdS was dictated by the membrane on its boundary and vice-versa. For example, one immediately recognises that the dimensions and supersymmetries of the three conformal theories in Maldacena's duality are exactly the same as the singleton, doubleton and tripleton supermultiplets of Table 2.3. Further interconnections between the two are currently being explored. See [140, 157] and Section 5.4.

1.5 Plan of the Lectures

The first purpose of these lectures will be to explain an area of research that was very active about ten years ago, namely anti-de Sitter space, the *Membrane at the End of the Universe*, singletons, superconformal theories, the *Membrane/Supergravity Bootstrap* and all that. These topics have recently undergone a revival of interest thanks to the the AdS/CFT conjecture of Maldacena [110] which suggests a duality between physics in the bulk of AdS and a superconformal theory on its boundary. We may thus regard this earlier work as a "prequel" (AdS/CFT Episode 1?). The second purpose is to discuss some very recent work on black holes in AdS which are interesting in their own right as well as finding application in the AdS/CFT correspondence. Both these topics will first require a thorough grounding in Kaluza–Klein theory, $D = 11$ supergravity, $D = 10$ Type IIB supergravity and M2-, D3- and M5-branes.

In Section 2, we discuss the bosonic sector of $D = 11$ supergravity and those brane solutions (so called BPS branes) that preserve half of the supersymmetry: the supermembrane (M2-brane) and the superfivebrane (M5-brane). These solutions, together with the plane wave [56] and Kaluza–Klein monopole [217], which also preserve half the supersymmetry, are the progenitors of the lower dimensional BPS objects of M-theory.

In Section 3, we discuss the bosonic sector of Type IIB supergravity and the self-dual super threebrane (D3-brane) solution which also preserves one half of the supersymmetry.

These M2, D3 and M5 BPS branes may be regarded as the extremal mass = charge limit of more general two-parameter black brane solutions that exhibit event horizons. In the near-horizon limit, they tend respectively to the $AdS_4 \times S^7$, the $AdS_5 \times S^5$ and the $AdS_7 \times S^4$ vacua of the maximally supersymmetric $D = 4$, $D = 5$ and $D = 7$ gauged supergravities that correspond to the massless sector of the compactification of $D = 11$ supergravity on S^7, Type IIB supergravity on S^5 and $D = 11$ supergravity on S^4, respectively. These are also treated in Sections 2 and 3, as is the issue of consistent Kaluza–Klein truncation to the massless sector.

The zero modes of these three brane solutions are described by Green–Schwarz type covariant brane actions whose bosonic sectors are treated in Section 4. These will be crucial for Section 5, whose purpose will be to explain anti-de Sitter space, the membrane at the end of the universe, singletons, superconformal theories and the membrane/supergravity bootstrap. In particular, by focussing on the bosonic radial mode, we show in detail how the superconformal singleton action on $S^1 \times S^2$ (including the scalar "mass term") follows from the action for a spherical M2 brane on the boundary of $\mathrm{AdS}_4 \times S^7$ in the limit of large radius: the membrane at the end of the universe. (We also indicate, following [140] how this generalizes to generic branes with spherical topology but arbitrary geometry.) Then we compare and contrast these old results with the comparatively recent Maldacena conjecture. Since the AdS/CFT correspondence will be the subject matter of several other lecturers at this school, we shall not dwell on the implications of our results for this correspondence but simply present them as interesting in their own right.

As we have seen, these AdS spaces emerge as the vacua of the maximally supersymmetric $D = 4$, $D = 5$ and $D = 7$ gauged supergravities that correspond to the massless sector of the compactification of $D = 11$ supergravity on S^7, Type IIB supergravity on S^5 and $D = 11$ supergravity on S^4, respectively. Recently, therefore, black hole solutions of *gauged* supergravity have attracted a good deal of attention and these are the subject of Section 6.

In [149], for example, new anti-de Sitter black hole solutions of gauged $N = 8$, $D = 4$, SO(8) supergravity were presented. By focussing on the $\mathrm{U}(1)^4$ Cartan subgroup, nonextremal 1, 2, 3 and 4 charge solutions were found. In the extremal limit, they may preserve up to $\frac{1}{2}$, $\frac{1}{4}$, $\frac{1}{8}$ and $\frac{1}{8}$ of the supersymmetry respectively. By contrast, the magnetic solutions preserve none. Since $N = 8$, $D = 4$ supergravity is a consistent truncation of $N = 1$, $D = 11$ supergravity, resulting from the S^7 compactification, it follows that these black holes will also be solutions of this theory. In [149], it was conjectured that a subset of the extreme electric black holes preserving $\frac{1}{2}$ the supersymmetry may be identified with the S^7 Kaluza–Klein spectrum, with the non-Abelian quantum numbers provided by the fermionic zero modes.

In [203] the nonlinear Kaluza–Klein ansatzs describing the embeddings of the $\mathrm{U}(1)^4$, $\mathrm{U}(1)^3$ and $\mathrm{U}(1)^2$ truncations of these supergravities were presented, which admit 4-charge AdS_4, 3-charge AdS_5 and 2-charge AdS_7 black hole solutions. These enable us to embed the black hole solutions back in ten and eleven dimensions and reinterpret them as M2-, D3- and M5-branes spinning in the transverse dimensions with the black hole charges given by the angular momenta of the branes. It is curious that the same $\mathrm{U}(1)^4$, $\mathrm{U}(1)^3$ and $\mathrm{U}(1)^2$ black hole charges appear in these spherical compactifications as in the toroidal compactifications but for totally different reasons. Instead of arising from the intersection of different nonrotating branes, they arise from the different angular momenta of a single brane. This is indicative of deeper levels of duality yet to be uncovered.

At the end of each of these sections there are problems whose solutions may be found in Section 7.

Much of what we need to do in these lectures involves only the bosonic sectors of the supergravities and the branes. For completeness, however, a comprehensive appendix lists the complete field equations, symmetries and transformation rules of $D = 11$ supergravity, Type IIB supergravity and the M2-, D3- and M5-branes, with all the fermionic terms included.

Introductory treatments of supersymmetry and supergravity may be found in [1–5, 238], of Kaluza–Klein theories in [198, 200, 201, 212, 213], of supermembranes in [7, 38, 47, 55, 111, 214], of M-theory in [6, 101, 103, 237] and of the AdS/CFT correspondence in [157, 216, 234].

1.6 Problems 1

1. Derive the four-dimensional action (1.9) by substituting (1.3) into the five-dimensional Einstein action (1.4) and assuming all fields are independent of the fifth coordinate y.
2. Derive the particle content of $(D = 4, N = 8)$ supergravity starting from $(D = 11, N = 1)$ supergravity.

2 Eleven-Dimensional Supergravity

2.1 Bosonic Field Equations

We will frequently seek solutions to the field equations of $D = 11$ supergravity in which all fermion fields are set equal to zero. The bosonic field equations following from Appendix A are

$$R_{MN} - \frac{1}{2} g_{MN} R = \frac{1}{12}\left(F_{MPQR} F_N{}^{PQR} - \frac{1}{8} g_{MN} F_{PQRS} F^{PQRS} \right), \quad (2.1)$$

and

$$\partial_M(\sqrt{-g} F^{MUVW}) + \frac{1}{1152} \epsilon^{UVWMNOPQRST} F_{MNOP} F_{QRST} = 0 \quad (2.2)$$

or, in the language of differential forms

$$d * F + \tfrac{1}{2} F \wedge F = 0 \quad (2.3)$$

where $*$ denotes the Hodge dual.

2.2 $\mathrm{AdS}_4 \times S^7$

In this section, we examine the spontaneous compactification of $D = 11$ supergravity. We are interested in obtaining a four-dimensional theory which admits maximal spacetime symmetry. With signature $(-+++)$, this means that the vacuum should be invariant under SO(4, 1), Poincaré or SO(3, 2) according as the cosmological constant is positive, zero or negative, corresponding to de Sitter, Minkowski or anti-de Sitter space, respectively.

The first requirement of maximal symmetry is that the vev of any fermion field should vanish and accordingly we set

$$\langle \Psi_M \rangle = 0 \tag{2.4}$$

and focus our attention on vacuum solutions of the bosonic equations (2.1) and (2.2). We look for solutions of the direct product form $M_4 \times M_7$ compatible with maximal spacetime symmetry. We denote the spacetime coordinates by x^μ, $\mu = 0, 1, 2, 3$ and the internal coordinates by y^m, $m = 1, \ldots, 7$. The ansatz of Freund and Rubin [15] is to set

$$F_{\mu\nu\rho\sigma} = \frac{3a}{2}\epsilon_{\mu\nu\rho\sigma} \tag{2.5}$$

with all other components vanishing, where a is a real constant and the factor 3/2 is chosen for future convenience. For future reference, it will prove convenient to rewrite this in terms of the dual field strength

$$F_7 \equiv *F_4 + \tfrac{1}{2}A_3 \wedge F_4 = 6L^6\epsilon_7 \tag{2.6}$$

where ϵ_7 is the volume form on the internal manifold, and where $L = 2/a$. At the moment, L is just an arbitrary constant of integration, but in Section 2.4 we shall relate it to the tension of an M2 brane. Substituting into the field equations we find that (2.2) is trivially satisfied while (2.1) yields the product of a four-dimensional Einstein spacetime

$$R_{\mu\nu} = -3a^2 g_{\mu\nu} = -\frac{12}{L^2}g_{\mu\nu} \tag{2.7}$$

with signature $(- + + +)$ and seven-dimensional Einstein space

$$R_{mn} = \frac{3a^2}{2}g_{mn} = \frac{6}{L^2}g_{mn} \tag{2.8}$$

with signature $(+ + + + + + +)$.

For future reference we also record the form taken by the supercovariant derivative $\widetilde{D}_M$ (2.36) when evaluated in the Freund–Rubin background geometry. First we decompose the $D = 11$ gamma matrices Γ_A

$$\Gamma_A = (\gamma_\alpha \otimes 1, \gamma_5 \otimes \Gamma_a) \tag{2.9}$$

where

$$\{\gamma_\alpha, \gamma_\beta\} = -2\eta_{\alpha\beta} \tag{2.10}$$
$$\{\Gamma_a, \Gamma_b\} = -2\delta_{ab} \tag{2.11}$$

and where $\alpha, \beta, \ldots$, are spacetime indices for the tangent space group SO(1,3) and $a, b, \ldots$, are the extra-dimensional indices for the tangent space group SO(7). Substituting the Freund–Rubin ansatz into the covariant drivative we find that

$$\tilde{D}_\mu = D_\mu + \frac{1}{L}\gamma_\mu\gamma_5 \tag{2.12}$$

$$\tilde{D}_m = D_m - \frac{1}{2L}\Gamma_m \tag{2.13}$$

where $\gamma_\mu = e_\mu^\alpha \gamma_\alpha$ and $\Gamma_m = e_m^a \Gamma_a$.

Some comments are now in order. The constancy of a in the ansatz (2.5) is necessary to solve the field equations, but other ansatzs are possible and are discussed in [20]. The maximally spacetime symmetric solution of (2.7) is, in fact, AdS_4 since the cosmological constant

$$\Lambda = -3a^2 \tag{2.14}$$

is negative. AdS_4 can be defined as the four-dimensional hyperboloid

$$\eta_{ab} y^a y^b = -\frac{1}{a^2} \tag{2.15}$$

in $\mathbb{R}^5$ with Cartesian coordinates y^a, where

$$\eta_{ab} = \mathrm{diag}(-1,1,1,1,-1). \tag{2.16}$$

In polar coordinates $x^\mu = (t, r, \theta, \phi)$ the line element may be written

$$\begin{aligned} ds_4^2 &= g_{\mu\nu}\, dx^\mu\, dx^\nu \\ &= -(1+a^2r^2)\, dt^2 + (1+a^2r^2)^{-1}\, dr^2 + r^2(d\theta^2 + \sin^2\theta\, d\phi^2). \end{aligned} \tag{2.17}$$

It is sometimes useful to employ the change of variable

$$ar = \sinh\rho \tag{2.18}$$

for which

$$ds_4^2 = -\cosh^2\rho\, dt^2 + \frac{1}{a^2}\, d\rho^2 + \frac{1}{a^2}\sinh^2\rho(d\theta^2 + \sin^2\theta\, d\phi^2). \tag{2.19}$$

Representations of SO(3,2) are denoted $D(E_0, s)$, where E_0 is the lowest energy eigenvalue (in units of a) and s is the total angular momentum. The representation is unitary provided $E_0 \geq s + \frac{1}{2}$ for $s = 0, \frac{1}{2}$ and $E_0 \geq s+1$ for $s \geq 1$. The representations are all infinite dimensional. Of course, in addition to being the AdS group in four dimensions, SO(3,2) has the interpretation as the *conformal* group in three dimensions, where the quantum number E_0 plays the role of the conformal weight. This will prove significant in the AdS/CFT correspondence.

There are infinitely many seven-dimensional Einstein spaces M_7 satisfying (2.8) and we now turn to the question of how much unbroken supersymmetry survives. Recall from (2.4) that the vev of the gravitino has been set to zero. For a supersymmetric vacuum we require that it remain zero under a supersymmetry transformation (A.16)

$$\langle \delta \Psi_M \rangle = \langle \widetilde{D}_M \epsilon \rangle = 0. \tag{2.20}$$

To solve this in the Freund–Rubin background, we look for solutions of the form

$$\epsilon(x, y) = \epsilon(x)\eta(y) \tag{2.21}$$

where $\epsilon(x)$ is an anticommuting four-component spinor in $D = 4$ and $\eta(y)$ is a commuting eight-component spinor in $D = 7$, satisfying

$$\widetilde{D}_\mu \epsilon(x) = 0 \tag{2.22}$$

$$\widetilde{D}_m \eta(y) = 0. \tag{2.23}$$

Thus the problem of counting unbroken supersymmetries is equivalent to the problem of counting *Killing spinors*. It is not dificult to establish that AdS_4 admits the maximum number (i.e., four) as far as spacetime is concerned, and so the number N of unbroken generators of AdS supersymmetry is given by the number of Killing spinors on M_7. From (2.13) these are seen to satisfy the integrability condition [69]

$$[\widetilde{D}_m, \widetilde{D}_n] = -\tfrac{1}{4} C_{mn}{}^{ab} \Gamma_{ab} \eta = 0 \tag{2.24}$$

where $C_{mn}{}^{ab}$ is the Weyl tensor. The subgroup of Spin(7) (the double cover of the tangent space group SO(7)) generated by these linear combinations of the Spin(7) generators Γ_{ab} corresponds to the *holonomy* group $\mathcal{H}$ of the generalized connection appearing in $\tilde{D}_m$. Thus the maximum number of unbroken supersymmetries N_{max} is equal to the number of spinors left invariant by $\mathcal{H}$. This in turn is given by the number of singlets appearing in the decomposition of the 8 of Spin(7) under $\mathcal{H}$. For example, the squashed S^7 of [69, 70] has $\mathcal{H} = G_2$ and hence $N_{\text{max}} = 1$.

In the supersymmetric context, all linear irreducible representations of $N = 1$ AdS supersymmetry were classified by Heidenreich [32]. They fall into 4 classes:

1. $D(\frac{1}{2}, 0) \oplus D(1, \frac{1}{2})$
2. $D(E_0, 0) \oplus D(E_0 + \frac{1}{2}, \frac{1}{2}) \oplus D(E_0 + 1, 0)$, $E_0 \geq \frac{1}{2}$
3. $D(s + 1, s) \oplus D(s + \frac{3}{2}, s + \frac{1}{2})$, $s \geq \frac{1}{2}$
4. $D(E_0, s) \oplus D(E_0 + \frac{1}{2}, s + \frac{1}{2}) \oplus D(E_0 + \frac{1}{2}, s - \frac{1}{2}) \oplus D(E_0 + 1, s)$.

Class 1 is the singleton supermultiplet which has no analogue in Poincaré supersymmetry. Singletons are the subject of Section 2.5. Class 2 is the Wess–Zumino supermultiplet. Class 3 is the gauge supermultiplet with spins s and $s+\frac{1}{2}$ with $s \geq \frac{1}{2}$. Class 4 is the higher spin supermultiplet. The corresponding study of OSp$(4 \mid N)$ representations was neglected in the literature until their importance in Kaluza–Klein supergravity became apparent. For example, the round S^7 leads to massive $N = 8$ supermultiplets with maximum spin 2. This corresponds to an AdS type of multiplet shortening analogous to the shorteneing due to central charges in Poincaré supersymmetry [33]. Two features emerge: (1) OSp$(4 \mid N)$ multiplets may be decomposed into the OSp$(4 \mid 1)$ multiplets discussed above; (2) In the limit as $a \to 0$ and the OSp$(4|N)$ contracts to the N-extended Poincaré algebra, all short AdS multiplets become massless Poincaré multiplets.

In this section we shall focus on the maximally symmetric round S^7, for which ${C_{mn}}^{ab} = 0$ yielding the maximum $N = 8$ Killing spinors, and hence on the eleven-dimensional vacuum AdS$_4 \times S^7$. The Kaluza–Klein mechanism will give rise to an effective $D = 4$ theory with $N = 8$ supersymmetry and local SO(8) invariance, describing a massless $N = 8$ multiplet coupled to an infinite tower of massive $N = 8$ supermultiplets with masses quantized in units of L^{-1}, the inverse radius of S^7. Combining the internal SO(8) symmetry and $N = 8$ supersymmetry with the SO$(2,3)$ of the AdS$_4$ spacetime, it is readily seen that the complete symmetry is OSp$(4 \mid 8)$. It follows without any further calculation that the massless sector on S^7 is given by the familiar massless $N = 8$ supermultiplet consisting of (1 spin 2, 8 spin $\frac{3}{2}$, 28 spin 1, 56 spin $\frac{1}{2}$, 70 spin zero), but a calculation is required to see the $D = 11$ origin of these fields, which is different from that of T^7. See Table 2.8, p. 198.

2.3 Consistent Truncation to the Massless Modes

An entirely different question is whether the massive modes can be consistently truncated to yield just gauged $N = 8$ supergravity. A *consistent truncation* is defined to be one for which all solutions of the truncated theory are solutions of the original theory. It requires that the truncated fields must never appear linearly in the action, otherwise setting them to zero would result in further constraints on the massless modes over and above the equations of motion of the massless theory [158, 159, 161, 162]. Interestingly enough, for generic Kaluza–Klein theories, truncation to the massless sector is *not* consistent. To illustrate this, consider the field equations of pure gravity with a positive cosmological constant Λ in $D = 4 + k$ dimensions

$$\hat{R}_{MN} = \Lambda \hat{g}_{MN}. \tag{2.25}$$

This theory admits the ground-state solution of ($D = 4$ de Sitter spacetime) $\times$ (compact manifold M_k). In some earlier Kaluza–Klein literature, it was generally believed that the correct ansatz for the metric $\hat{g}_{MN}(x,y)$ is given by

$$\begin{aligned}\hat{g}_{\mu\nu}(x,y) &= g_{\mu\nu}(x) + A_\mu{}^i K^{mi}(y)(x) A_\nu{}^j(x) K^{nj}(y) g_{mn}(y)\\ \hat{g}_{\mu n}(x,y) &= A_\mu{}^i(x) K^{mi}(y) g_{mn}(y)\\ \hat{g}_{mn}(x,y) &= g_{mn}(y)\end{aligned} \tag{2.26}$$

where $g_{mn}(y)$ is the metric on M_k. The quantities $K^{mi}(y)$ are the Killing vectors corresponding to the isometries of this metric and i runs over the dimension of the isometry group G. The claim that this is the correct ansatz was based on the observation that substituting this ansatz into the higher-dimensional Einstein action and integrating over y, one obtains the four-dimensional Einstein–Yang–Mills action with metric $g_{\mu\nu}(x)$ and gauge potential $A_\mu{}^i(x)$.

However, the correct Kaluza–Klein ansatz must be consistent with the higher dimensional field equations and, as we shall now demonstrate, this is not in general true. For example, the four-dimensional Einstein equations read

$$R_{\mu\nu} - \tfrac{1}{2}g_{\mu\nu} + \Lambda g_{\mu\nu} = \tfrac{1}{2}\left(F_{\mu\nu}{}^i F_\nu{}^{\rho j} - \tfrac{1}{4}F_{\rho\sigma}{}^i F^{\rho\sigma j}\right) K_n{}^i K^{nj} \tag{2.27}$$

where $F_{\mu\nu}{}^i$ is the Yang–Mills field strength. The inconsistency is now apparent. The left-hand side is independent of y while the right-hand side in general depends on y via the Killing vector combination $K_n{}^i K^{nj}$. For example, when $M_k = S^k$ with its SO$(k+1)$ invariant metric.

$$K_n{}^i K^{nj} = \delta_{ij} + Y^{ij}(y) \tag{2.28}$$

where $Y^{ij}(y)$ is that harmonic of the scalar Laplacian with next to lowest nonvanishing eigenvalue $2\Lambda(k+1)/(k-1)$ belonging to the $k(k+3)/2$ dimensional representation of SO$(k+1)$.

This situation changes radically when we turn to the S^7 compactification $D = 11$ supergravity. The reason for the difference is the presence of the three-index gauge field $\hat{A}_{MNP}$ in addition to the metric $\hat{g}_{MN}$. The crucial observation is that the standard Kaluza–Klein ansatz (2.26) must be augmented by the additional ansatz [13, 161]

$$\widehat{F}_{\mu\nu pq} = -\frac{1}{2a}\epsilon_{\mu\nu\rho\sigma} F^{\rho\sigma i} \nabla_{[p} K_{q]}{}^i. \tag{2.29}$$

Substituting (2.26) and (2.29) into the $D = 11$ field equations (2.1) and (2.2) now yields the $D = 4$ Einstein equation

$$\begin{aligned}&R_{\mu\nu} - \tfrac{1}{2}g_{\mu\nu} + \Lambda g_{\mu\nu}\\ &\quad = \frac{1}{2}\left(F_{\mu\nu}{}^i F_\nu{}^{\rho j} - \frac{1}{4}F_{\rho\sigma}{}^i F^{\rho\sigma j}\right)\left(K_n{}^i K^{nj} + \frac{1}{a^2}\nabla_m K_n{}^i \nabla^m K^{ni}\right).\end{aligned} \tag{2.30}$$

The miracle of the S^7 compactification is that

$$K_n{}^i K^{nj} + \frac{1}{a^2}\nabla_m K_n{}^i \nabla^m K^{ni} = \delta^{ij} \tag{2.31}$$

and so the right-hand side of (2.30) becomes the correct energy-momentum tensor of SO(8) Yang–Mills! Indeed the round S^7 is the only M_7 solution of (2.8) known to satisfy (2.31).

2.4 The Supermembrane Solution

Historically, the equations of motion of $D = 11$ supergravity were written down in 1978 [197] and the $D = 11$ supermembrane was discovered as a fundamental object in its own right in 1987 [41, 42]. In 1990, however, it was realized [51] that the $D = 11$ supermembrane arises as a classical solution of the supergravity field equations which preserves one half of the spacetime supersymmetry. We shall now describe this solution.

We begin by making an ansatz for the $D = 11$ gauge fields $g_{MN} = e_M{}^A e_N{}^B \eta_{AB}$ and A_{MNP} corresponding to the most general three-eight split invariant under $P_3 \times \mathrm{SO}(8)$, where P_3 is the $D = 3$ Poincaré group. We split the $D = 11$ coordinates

$$x^M = (x^\mu, y^m) \tag{2.32}$$

where $\mu = 0, 1, 2$ and $m = 3, 4, \ldots, 10$, and write the line element as

$$ds^2 = e^{2A}\eta_{\mu\nu}dx^\mu dx^\nu + e^{2B}\delta_{mn}dx^m dx^n \tag{2.33}$$

and the 3-form gauge field as

$$A_{012} = e^C. \tag{2.34}$$

All other components of A_{MNP} and all components of the gravitino Ψ_M are set equal to zero. P_3 invariance requires that the arbitrary functions A, B and C depend only on y^m; SO(8) invariance then requires that this dependence be only through $y = \sqrt{\delta_{mn}y^m y^n}$.

As we shall show, the three arbitrary functions A, B, and C are reduced to one by the requirement that the field configuration (2.33) and (2.34) preserve some unbroken supersymmetry. In other words, there must exist Killing spinors ϵ satisfying

$$\widetilde{D}_M \epsilon = 0 \tag{2.35}$$

where $\widetilde{D}_M$ is the bosonic part of the supercovariant derivative appearing in the supersymmetry transformation rule of the gravitino (A.18)

$$\widetilde{D}_M = \partial_M + \frac{1}{4}\omega_M{}^{AB}\Gamma^{AB} - \frac{1}{288}(\Gamma_M{}^{PQRS} - 8\delta_M{}^P\Gamma^{QRS})F_{PQRS}. \tag{2.36}$$

We make the three-eight split

$$\Gamma^A = (\gamma_\alpha \otimes \Gamma_9, \mathbf{1} \otimes \Sigma_a) \tag{2.37}$$

where γ_α and Σ_a are the $D = 3$ and $D = 8$ Dirac matrices respectively and where

$$\Gamma_9 = \Sigma_3 \Sigma_4 \cdots \Sigma_{10} \tag{2.38}$$

so that $\Gamma_9^2 = 1$. We also decompose the spinor field as

$$\epsilon(x, y) = \zeta(x) \otimes \eta(y) \tag{2.39}$$

where ζ is a constant spinor of SO(1, 2) and η is an SO(8) spinor which may further be decomposed into chiral eigenstates via the projection operators $(1 + \Gamma_9)/2$.

In our background (2.33) and (2.34), the supercovariant derivative becomes:

$$\begin{aligned}
\widetilde{D}_\mu &= \partial_\mu - \gamma_\mu \tfrac{1}{2} \Sigma^m \partial_m A \Gamma_9 - \gamma_\mu e^{-3A} \Sigma^m \partial_m e^C, \\
\widetilde{D}_m &= \partial_m + \tfrac{1}{4} e^{-B} (\Sigma_m \Sigma^n - \Sigma^n \Sigma_m) \partial_n e^B \\
&\quad - \frac{1}{24} e^{-3A} (\Sigma_m \Sigma^n - \Sigma^n \Sigma_m) \partial_n e^C \Gamma_9 + \frac{1}{6} e^{-3A} \partial_m e^C \Gamma_9.
\end{aligned} \tag{2.40}$$

Note that the Γ_μ and Σ_m carry world indices. Hence we find that (2.35) admits two nontrivial solutions

$$(1 + \Gamma_9)\eta = 0 \tag{2.41}$$

where

$$\eta = e^{C/6} \eta_0 \tag{2.42}$$

where η_0 is a constant spinor and

$$\begin{aligned}
A &= \tfrac{1}{3} C \\
B &= -\tfrac{1}{6} C + \text{constant}.
\end{aligned} \tag{2.43}$$

In each case, (2.41) means that one half of the maximal possible supersymmetry survives.

With the substitutions (2.33), (2.34) and (2.43), the Einstein equation and the 3-form equation reduce to the single equation for one unknown:

$$\delta^{mn} \partial_m \partial_n e^{-C} = 0 \tag{2.44}$$

and hence, imposing the boundary condition that the metric be asymptotically Minkowskian, we find

$$e^{-C} = 1 + \frac{b^6}{y^6} \tag{2.45}$$

where b is a constant, at this stage arbitrary, and $y^2 = \delta_{mn} y^m y^n$. Thus the metric is given by

$$ds^2 = (1 + b^6/y^6)^{-2/3}\, dx^\mu\, dx_\mu + (1 + b^6/y^6)^{1/3}(dy^2 + y^2\, d\Omega_7^2) \qquad (2.46)$$

and the 7-form field strength by

$$F_7 \equiv *F_4 + \tfrac{1}{2} A_3 \wedge F_4 = 6b^6 \epsilon_7. \qquad (2.47)$$

Here ϵ_7 is the volume form on S^7 and Ω_7 is the volume. The mass per unit area of the membrane $\mathcal{M}_3$ is given by:

$${\kappa_{11}}^2 \mathcal{M}_3 = 3b^6 \Omega_7. \qquad (2.48)$$

This *elementary* solution is a singular solution of the supergravity equations coupled to a supermembrane source and carries a Noether "electric" charge

$$Q = \frac{1}{\sqrt{2}\kappa_{11}} \int_{S^7} F_7 = \sqrt{2}\kappa_{11} \mathcal{M}_3. \qquad (2.49)$$

Hence the solution saturates the Bogomol'nyi–Prasad–Sommerfield (BPS) bound

$$\sqrt{2}\kappa_{11} \mathcal{M}_3 \geq Q. \qquad (2.50)$$

This is a consequence of the preservation of half the supersymmetries. Of course, the equation (2.44) displays a delta-function singularity at the origin and hence requires a source term. This is provided by adding to the supergravity action (A.1) the action of the supermembrane itself [51], discussed in Section 4.1. We find, as expected, that the mass per unit volume $\mathcal{M}_3$ is just the membrane tension T_3

$$\mathcal{M}_3 = T_3. \qquad (2.51)$$

The zero modes of this solution belong to a $(d = 3, n = 8)$ supermultiplet consisting of eight scalars and eight spinors (ϕ^I, χ^I), with $I = 1, \ldots, 8$, which correspond to the eight Goldstone bosons and their superpartners associated with breaking of the eight translations transverse to the membrane worldvolume.

A straightforward generalization to exact, stable multimembrane configurations can be obtained by replacing the single membrane expression (2.45) by a linear superposition

$$e^{-C} = 1 + \sum_l \frac{b_l^6}{|\mathbf{y} - \mathbf{y}_l|^6} \qquad (2.52)$$

where $\mathbf{y}_l$ corresponds to the position of each brane, each with charge ${b_l}^6$. The ability to superpose solutions of this kind is a well-known phenomenon in soliton and instanton physics and goes by the name of the "no-static-force condition". In the present context, it means that the gravitational attractive force acting on each of the branes is exactly cancelled by an equal but repulsive force due to the 3-form. This condition is closely related to the saturation of the BPS bound and to the existence of unbroken supersymmetry. In the case that N branes with the same charge are stacked together, we have

$$e^{-C} = 1 + \frac{Nb^6}{y^6} \tag{2.53}$$

or in terms of the Schwarzschild-like coordinate r given by

$$r^6 = y^6 + Nb^6 \tag{2.54}$$

we have

$$e^{C} = 1 - \frac{Nb^6}{r^6} \tag{2.55}$$

and the solution exhibits an event horizon at $r = N^{1/6}b$. Indeed the solution may be analytically continued down to $r = 0$ where there is a curvature singularity, albeit hidden by the event horizon [65]. Of particular interest is now the near horizon limit $y \to 0$, or equivalently the large N limit, because then the metric reduces to [64–66] the $\mathrm{AdS}_4 \times S^7$ vacuum of Section 2.2 with

$$L^6 = Nb^6. \tag{2.56}$$

Thus

$$ds^2 = \frac{y^4}{L^4}\, dx^\mu\, dx_\mu + \frac{L^2}{y^2} dy^2 + L^2 d\Omega_7^2 \tag{2.57}$$

which is just $\mathrm{AdS}_4 \times S^7$ with the AdS metric written in horospherical coordinates.

2.5 $\mathrm{AdS}_7 \times S^4$

Here, we are interested in obtaining a seven-dimensional theory which admits maximal AdS spacetime symmetry. With signature $(-++++++)$, this means that the vacuum should be invariant under $\mathrm{SO}(6,2)$. We again look for solutions of the direct product form $M_7 \times M_4$ compatible with maximal spacetime symmetry. We denote the spacetime coordinates by x^μ, $(\mu = 0, \ldots, 6)$ and the internal coordinates by y^m, $(m = 1, \ldots, 4)$. The ansatz of Freund and Rubin [15] is to set

$$F_4 = 3L^2 \epsilon_4 \tag{2.58}$$

with all other components vanishing, where L is a real constant. Here ϵ_4 is the volume form on the internal manifold. As before, L is just an arbitrary constant of integration, but in Section 2.6 we shall relate it to the tension of an M5-brane. Substituting into the field equations we find that (2.2) is trivially satisfied while (2.1) yields the product of a seven dimensional Einstein spacetime

$$R_{\mu\nu} = -\frac{3}{2L^2} g_{\mu\nu} \tag{2.59}$$

with signature $(- + + + + + +)$ and four-dimensional Einstein space

$$R_{mn} = \frac{3}{L^2} g_{mn} \tag{2.60}$$

with signature $(+ + ++)$.

The maximally spacetime symmetric solution of (2.59) is AdS_7. In these lectures, we shall focus also on the maximally symmetric round S^4 solution of (2.60). The Kaluza–Klein mechanism will give rise to an effective $D = 7$ theory with $N = 4$ supersymmetry and local SO(5) invariance, describing a massless $N = 4$ multiplet coupled to an infinite tower of massive $N = 4$ supermultiplets with masses quantized in units of $2a$, the inverse radius of S^4. Combining the internal SO(5) symmetry and $N = 4$ supersymmetry with the SO(6, 2) of the AdS_7 spacetime, it is readily seen that the complete symmetry is $OSp(6, 2 \mid 4)$. It follows without any further calculation that the massless sector on S^7 is given by the massless $N = 4$ supergravity supermultiplet, and there indeed exists a consistent truncation to the massless sector [218, 219].

2.6 The Superfivebrane Solution

The superfivebrane was discovered as a soliton solution of $D = 11$ supergravity also preserving half the spacetime supersymmetry [62]. Now we make the six/five split $x^M = (x^\mu, y^m)$ where $\mu = 0, 1, 2, 3, 4, 5$ and $m = 6, \ldots, 10$ and proceed in a way similar to the supermembrane solution. In particular, we again look for solutions preserving supersymmetry but this time we look for a nonvanishing *magnetic* charge carried by F_4:

$$F_4 = 3b^3 \epsilon_4 . \tag{2.61}$$

The metric is given by

$$ds^2 = (1 + b^3/y^3)^{-1/3} \, dx^\mu \, dx_\mu + (1 + b^3/y^3)^{2/3} (dy^2 + y^2 \, d\Omega_4{}^2) \tag{2.62}$$

where the fivebrane mass per unit 5-volume $\mathcal{M}_6$, which we identify with the fivebrane tension T_6, is related to the constant b^3 by

$$b^3 = \frac{2\kappa_{11}{}^2 \mathcal{M}_6}{3\Omega_4} . \tag{2.63}$$

Here ϵ_4 is the volume form on S^4 and Ω_4 is the volume. This *solitonic* solution is a nonsingular solution of the source-free equations and carries a topological "magnetic" charge

$$P = \frac{1}{\sqrt{2}\kappa_{11}} \int_{S^4} F_4 = \sqrt{2}\kappa_{11}\mathcal{M}_6. \tag{2.64}$$

Hence the solution saturates the Bogomol'nyi bound

$$\sqrt{2}\kappa_{11}\mathcal{M}_6 \geq P. \tag{2.65}$$

Once again, this is a consequence of the preservation of half the supersymmetries. The soliton zero modes are described by the chiral antisymmetric tensor multiplet $(B^{-}{}_{\mu\nu}, \lambda^I, \phi^{[IJ]})$. Note that in addition to the five scalars corresponding to the five translational Goldstone bosons, there is also a 2-form $B^{-}{}_{\mu\nu}$ whose 3-form field strength is anti-self-dual and which describes three degrees of freedom.

The electric and magnetic charges obey a Dirac quantization rule

$$QP = 2\pi n \quad n = \text{integer}. \tag{2.66}$$

Or, in terms of the tensions,

$$2{\kappa_{11}}^2 T_3 T_6 = 2\pi n. \tag{2.67}$$

This naturally suggests a $D = 11$ membrane/fivebrane duality. Note that this reduces the three dimensionful parameters T_3, T_6 and κ_{11} down to two. Moreover, it can be shown [99] that they are not independent. To see this, we note from Appendix C that A_3 has period $2\pi/T_3$ so that F_4 is quantized according to

$$\int F_4 = \frac{2\pi n}{T_3}, \qquad n = \text{integer}. \tag{2.68}$$

Consistency of such A_3 periods with the spacetime action, (A.1), gives the relation

$$\frac{(2\pi)^2}{{\kappa_{11}}^2 T_3^3} \in 2\mathbb{Z}. \tag{2.69}$$

From (2.67), this may also be written as

$$2\pi \frac{T_6}{T_3^2} \in \mathbb{Z}. \tag{2.70}$$

Thus the tension of the singly charged fivebrane is given by

$$T_6 = \frac{1}{2\pi} T_3^2. \tag{2.71}$$

In Problem 3.4, you are asked to show that a derivation [101] based on M/IIB duality gives the same result [215].

As for the membrane, multifivebrane solutions may be obtained by superposition. If we again consider N singly charged fivebranes stacked one upon the other, we find in the near horizon, or large N, limit the $\mathrm{AdS}_7 \times S^4$ geometry with

$$L^3 = Nb^3 \tag{2.72}$$

namely

$$ds^2 = \frac{y}{L} dx^\mu dx_\mu + \frac{L^2}{y^2} dy^2 + L^2 d\Omega_4{}^2. \tag{2.73}$$

2.7 Problems 2

1. Derive the bosonic field equations (2.1) and (2.2) by varying the $D = 11$ supergravity action (A.1).
2. Prove that the $D = 11$ superfivebrane preserves one half of the supersymmetry.

3 Type IIB Supergravity

3.1 Bosonic Field Equations

Next we consider Type IIB supergravity in $D = 10$ [21, 22] which also describes 128 + 128 degrees of freedom, and corresponds to the field-theory limit of the Type IIB superstring. The spectrum of the supergravity theory consists of a complex scalar B, a complex spinor λ, a complex 2-form A_{MN}, a complex Weyl gravitino ψ_M, a real graviton $e_M{}^R$ and a real 4-form A_{MNPQ} whose 5-form field strength F_{MNPQR} obeys a self-duality condition. Owing to this self-duality, there exists no covariant action principle and it is therefore simplest to work directly with the field equations. Once again, it is convenient to isolate just the bosonic field equations following from Appendix B which read

$$D^M P_M = \frac{1}{24} \kappa_{10}{}^2 G_{MNP} G^{MNP} \tag{3.1}$$

$$D^P G_{MNP} = P^P G^*{}_{MNP} - \tfrac{2}{3} i \kappa_{10} F_{MNPQR} G^{*PQR} \tag{3.2}$$

$$\begin{aligned} R_{MP} - \tfrac{1}{2} g_{MP} R = {} & P_M P^*{}_P + P^*{}_M P_P - g_{MP} P^R P^*{}_R \\ & + \tfrac{1}{6} \kappa_{10}{}^2 F_{R_1 \cdots R_4 M} F^{R_1 \cdots R_4}{}_P \\ & + \tfrac{1}{8} \kappa_{10}{}^2 (G_M{}^{RS} G^*{}_{PRS} + G^*{}_M{}^{RS} G_{PRS}) \\ & - \frac{1}{24} \kappa_{10}{}^2 g_{MP} G^{RST} G^*{}_{RST} \end{aligned} \tag{3.3}$$

$$F_{MNPQR} = *F_{MNPQR} \tag{3.4}$$

where

$$P_M = f^2 \partial_M B \tag{3.5}$$
$$G_{MNP} = f(F_{MNP} - BF^*{}_{MNP}) \tag{3.6}$$
$$F_{MNP} = 3\partial_{[M} A_{NP]} \tag{3.7}$$
$$f = (1 - B^* B)^{-1/2} \tag{3.8}$$
$$F_{MNPQR} = 5\partial_{[M} A_{NPQR]} - \tfrac{5}{4}\kappa_{10} \Im(A_{[MN} F^*{}_{PQR]}). \tag{3.9}$$

Our notation is that X^* is the complex conjugate of X while $*X$ is the Hodge dual of X.

3.2 $\mathrm{AdS}_5 \times S^5$

In this section, we want to obtain a five-dimensional theory which admits maximal AdS spacetime symmetry. With signature $(-++++)$, this means that the vacuum should be invariant under SO(4, 2). We again look for solutions of the direct product form $M_5 \times M_5'$ compatible with maximal spacetime symmetry. We denote the spacetime coordinates by x^μ, $(\mu = 0, \ldots, 4)$ and the internal coordinates by y^m, $(m = 1, \ldots, 5)$. The ansatz of Freund and Rubin [15] is to set

$$F_5 = G_5 + *G_5 \tag{3.10}$$

and

$$G_5 = 4L^4 \epsilon_5 \tag{3.11}$$

with all other components vanishing, where L is a real constant. Here ϵ_5 is the volume form on the internal manifold. As before, L is just an arbitrary constant of integration, but in Section 3.3, we shall relate it to the tension of an D3-brane. Substituting into the field equations we find that (2.2) is trivially satisfied while (2.1) yields the product of a five-dimensional Einstein spacetime

$$R_{\mu\nu} = -\frac{4}{L^2} g_{\mu\nu} \tag{3.12}$$

with signature $(-++++)$ and five-dimensional Einstein space

$$R_{mn} = \frac{4}{L^2} g_{mn} \tag{3.13}$$

with signature $(+++++)$.

The maximally spacetime symmetric solution of (3.12) is AdS_5. In these lectures, we shall focus also on the maximally symmetric round S^5 solution of (3.13). The Kaluza–Klein mechanism will give rise to an effective $D = 5$ theory with $N = 8$ supersymmetry and local SO(6) invariance, describing a massless $N = 8$ multiplet coupled to an infinite tower of massive $N = 8$

supermultiplets with masses quantized in units of a, the inverse radius of S^5. Combining the internal SO(6) symmetry and $N = 8$ supersymmetry with the SO(2, 4) of the AdS_5 spacetime, it is readily seen that the complete symmetry is SU(2, 2|4). It follows without any further calculation that the massless sector on S^5 is given by the massless $N = 8$ supergravity supermultiplet. To date, however, the proof of a complete consistent truncation to the massless sector is still lacking.

3.3 The Self-Dual Superthreebrane Solution

The resulting field equations admit as a solution the self-dual threebrane [57] which in the extremal limit preserves half the supersymmetry [58] just like the extremal 2-brane and 5-brane of $D = 11$ supergravity. We need keep only the graviton, dilaton and 4-form and make the four/six split $x^M = (x^\mu, y^m)$ where $\mu = 0, 1, 2, 3$ and $m = 5, \ldots, 10$. The solution is given by

$$ds^2 = (1 + b^4/y^4)^{-1/2} \, dx^\mu \, dx_\mu + (1 + b^4/y^4)^{1/2}(dy^2 + y^2 \, d\Omega_5{}^2) \quad (3.14)$$

$$e^{2\phi} = e^{2\phi_0} = \text{constant} \quad (3.15)$$

$$F_5 = G_5 + *G_5 \quad (3.16)$$

where

$$G_5 = 4b^4 \epsilon_5. \quad (3.17)$$

Here we have employed the string frame metric $g_{MN}(\text{string}) = e^{\phi/2} \times g_{MN}(\text{Einstein})$. The threebrane mass per unit volume $\mathcal{M}_4$ is then related to the tension T_4 by [7]

$$\mathcal{M}_4 = e^{-\phi_0} T_4 \quad (3.18)$$

and related to the constant k_4 by

$$b^4 = \frac{2\kappa_{10}{}^2 \mathcal{M}_4}{\Omega_5}. \quad (3.19)$$

It is perhaps worth saying a few more words about the self-duality of the superthreebrane. By virtue of the self-duality condition (3.4), the electric Noether charge

$$Q = \frac{1}{\sqrt{2}\kappa_{10}} \int_{S^5} *F_5 \quad (3.20)$$

coincides with the topological magnetic charge

$$P = \frac{1}{\sqrt{2}\kappa_{10}} \int_{S^5} F_5 \quad (3.21)$$

so

$$Q = P \quad (3.22)$$

and hence

$$2\kappa_{10}^2 \mathcal{M}_4^2 = 2\pi n. \tag{3.23}$$

(Note that such a condition is possible only in theories allowing a real self-duality condition, i.e. in $D = 2$ modulo 4 dimensions, assuming Minkowski signature. The $D = 6$ self-dual string of [72] is another example.)

We can also count bosonic and fermionic zero modes. We know that one half of the supersymmetries are broken, hence we have 16 fermionic zero modes. Regrouping these 16 fermionic zero modes, we get four Majorana spinors in $d = 4$. Hence the $d = 4$ worldvolume supersymmetry is $N = 4$. Worldvolume supersymmetry implies that the number of fermionic and bosonic on-shell degrees of freedom must be equal, so we need a total of eight bosonic zero modes. There are the usual six bosonic translation zero modes, but we are still short of two. The two extra zero modes come from the excitation of the complex antisymmetric field strength G_{MNP} and correspond to a real vector field on the worldvolume [58]. Together with the other zero modes, these fields make up the $d = 4$, $N = 4$ gauge supermultiplet $(A_\mu, \lambda^I, \phi^{[IJ]})$.

As for the membrane and fivebrane, multithreebrane solutions may be obtained by superposition. If we again consider N singly charged threebranes stacked one upon the other, we find in the near horizon, or large N, limit the $\mathrm{AdS}_7 \times S^4$ geometry with

$$L^4 = Nb^4 \tag{3.24}$$

namely

$$ds^2 = \frac{y^2}{L^2}\, dx^\mu\, dx_\mu + \frac{L^2}{y^2}\, dy^2 + L^2\, d\Omega_5{}^2. \tag{3.25}$$

3.4 Problems 3

1. Show, using the duality between M-theory on T^2 and Type IIB theory on S^1, that the M-theory membrane tension T_3 and the M-theory fivebrane tension T_6 are related by

$$T_6 = \frac{1}{2\pi} T_3^2. \tag{3.26}$$

2. Write down the dictionary that relates the fields of M-theory on T^2 to IIB on S^1.

4 The M2-Brane, D3-Brane and M5-Brane

4.1 The M2-Brane

The $8 + 8$ zero modes of the M2-brane are described by a supersymmetric $2 + 1$ dimensional worldvolume action. However, this obscures the underlying $D = 11$ spacetime supersymmetry. It is possible to construct a covariant action with the scalars and spinors being given by the 11 bosonic coordinates X^M and the 32 fermionic coordinates θ^α of a $D = 11$ superspace [41, 42]. Indeed, in the case of the M2 brane this came first. The $8 + 8$ physical degrees of freedom, and the worldvolume supersymmetry then emerge by going to a physical gauge, as discussed in Section 5.2. We begin with the bosonic sector of the $d = 3$ worldvolume of the $D = 11$ supermembrane which follows from Appendix C:

$$S_3 = T_3 \int d^3\xi \Big[-\frac{1}{2}\sqrt{-\gamma}\gamma^{ij}\partial_i x^M \partial_j x^N g_{MN}(x) + \frac{1}{2}\sqrt{-\gamma} + \frac{1}{3!}\epsilon^{ijk}\partial_i x^M \partial_j x^N \partial_k x^P A_{MNP}(x) \Big], \quad (4.1)$$

where T_3 is the membrane tension, ξ^i $(i = 0, 1, 2)$ are the worldvolume coordinates, γ^{ij} is the worldvolume metric and $x^M(\xi)$ are the spacetime coordinates $(M = 0, 1, \ldots, 10)$. Kappa symmetry, discussed in Appendix C, demands that the background metric g_{MN} and background 3-form potential A_{MNP} obey the classical field equations of $D = 11$ supergravity (2.1) and (2.2).

Varying (4.1) with respect to x^M yields

$$\partial_i(\sqrt{-\gamma}\gamma^{ij}\partial_j x^N g_{MN}) - \frac{1}{2}\sqrt{-\gamma}\gamma^{ij}\partial_i x^N \partial_j x^P \partial_M g_{NP} + \frac{1}{3!}\epsilon^{ijk}\partial_i x^N \partial_j x^P \partial_k x^Q F_{MNPQ} = 0 \quad (4.2)$$

while varying with respect to γ_{ij} yields the embedding equation

$$\gamma_{ij} = \partial_i x^M \partial_j x^N g_{MN}. \quad (4.3)$$

An important issue is the existence of membrane configurations preserving some supersymmetry [44, 45]. We usually seek supersymmetric membrane vacuum states for which both the spacetime gravitino Ψ_M and the fermionic coordinates $\theta(\xi)$ are zero, so that the membrane action reduces to (4.1). In this case, the criterion for surviving supersymmetry is that the vacuum expectation values of Ψ_M and $\theta(\xi)$ remain zero under some appropriate combination of κ symmetry and supersymmetry transformations. We shall work to linear order in fermions, which is as high as we need go to investigate the supersymmetry of a purely bosonic background. From Appendices A and C, these transformations rules are

$$\delta\Psi_M = \widetilde{D}_M \epsilon(x) \quad (4.4)$$

$$\delta\theta = (1 + \Gamma)\kappa(\xi) + \epsilon(x) \quad (4.5)$$

where ϵ and κ are the supersymmetry and κ symmetry parameters, respectively. Here $\widetilde{D}_M$ is the $D = 11$ supercovariant derivative (2.36) and Γ is given by

$$\Gamma = \frac{1}{3!}\sqrt{-\gamma}\epsilon^{ijk}\partial_i x^M \partial_j x^N \partial_k x^P \Gamma_{MNP}. \tag{4.6}$$

As a consequence of the embedding equation (4.3), Γ satisfies $\Gamma^2 = 1$. Since Γ is tracefree, this implies that $(1 \pm \Gamma)/2$ are projection operators with 16 zero eigenvalues. Irrespective of the background geometry the κ symmetry may therefore be used to set 16 of the 32 components of θ to zero. A convenient choice is

$$(1 + \Gamma)\theta = 0. \tag{4.7}$$

Acting with (4.5), we see that around a purely bosonic background, (4.7) is preserved if $(1 + \Gamma)\delta\theta = 0$, i.e., combined κ and ϵ transformations for which

$$2(1 + \Gamma)\kappa + (1 + \Gamma)\epsilon = 0 \tag{4.8}$$

which implies that $\delta\theta = (1 - \Gamma)\epsilon/2$. Thus a bosonic vacuum has residual supersymmetries corresponding to solutions of [44, 45]

$$\tilde{D}_M \epsilon = 0, \quad \Gamma\epsilon(x) = \epsilon(x). \tag{4.9}$$

We shall make use of this in seeking the membrane at the end of the universe in Section 5.4.

4.2 The M5-Brane

The bosonic M5 worldvolume equations of motion follow from Appendix D. We work with equations of motion rather than an action because of the appearance of a self-dual worldvolume 3-form field strength h_{abc}.

The suitable pullbacks of the spacetime 3-form potential, and the induced metric are

$$\begin{aligned} A_{ijk} &= \partial_i x^M \partial_j x^N \partial_k x^P A_{PNM} \\ \gamma_{ij} &= \partial_i x^M E_M{}^A \partial_j x^N E_N{}^B \eta_{AB}. \end{aligned} \tag{4.10}$$

ξ^i $(i = 0, \ldots, 5)$ are the worldvolume coordinates, γ^{ij} is the worldvolume metric and $x^M(\xi)$ are the spacetime coordinates $(M = 0, 1, \ldots, 10)$. We introduce the worldvolume 2-form A_{ij} and corresponding worldvolume 3-form:

$$\mathcal{F}_{ijk} \equiv 3\partial_{[i}A_{jk]} - A_{ijk}. \tag{4.11}$$

The field equation for A_{ij} is

$$G^{mn}\nabla_m \mathcal{F}_{npq} = Q^{-1}[4Y - 2(mY + Ym) + mYm]_{pq}, \tag{4.12}$$

and the field equations for the x^M are

$$\begin{aligned} &G^{mn}\nabla_m \mathcal{E}_n{}^C \\ &= \frac{Q}{\sqrt{-g}}\epsilon^{m_1\cdots m_6}\left(\frac{1}{6!}F^A_{m_1\cdots m_6} + \frac{1}{(3!)^2}F^A{}_{m_1 m_2 m_3}\mathcal{F}_{m_4 m_5 m_6}\right)P_A{}^C. \end{aligned} \tag{4.13}$$

Several definitions are in order. To begin with,

$$\begin{aligned} m_a{}^b &= \delta_a{}^b - 2k_a{}^b, \quad k_a{}^b \equiv h_{acd}h^{bcd}, \quad Q \equiv (1 - \tfrac{2}{3}\operatorname{tr} k^2), \\ Y_{ab} &\equiv [4 * F - 2(m * F + *Fm) + m * Fm]_{ab}, \\ P_A{}^C &\equiv \delta_A{}^C - \mathcal{E}_A{}^m\mathcal{E}_m{}^C, \quad *F^{ab} := \frac{1}{4!\sqrt{-g}}\epsilon^{abcdef}F_{cdef}. \end{aligned} \tag{4.14}$$

The fields $\mathcal{F}_3$, F_4 and its Hodge dual F_7 are given by

$$\mathcal{F}_3 = dA_2 - A_3, \quad F_4 = dA_3, \quad F_7 = dA_6 + \tfrac{1}{2}A_3 \wedge F_4. \tag{4.15}$$

The target space indices on F_4 and F_7 have been converted to worldvolume indices with factors of $\mathcal{E}_m{}^A$ defined as

$$\mathcal{E}_m{}^A(x) \equiv \partial_m x^M E_M{}^A. \tag{4.16}$$

The metric

$$\gamma_{mn}(x) \equiv \mathcal{E}_m{}^A\mathcal{E}_n{}^B\eta_{AB} = e_m{}^a e_n{}^b \eta_{ab} \tag{4.17}$$

is the standard induced metric with determinant γ, and G^{mn} is another metric defined as

$$G^{mn} \equiv (m^2)^{ab} e_a{}^m e_b{}^n. \tag{4.18}$$

Let us note that the connection in the covariant derivative ∇_m occurring in (4.13) is the Levi–Civita connection for the induced metric γ_{mn}.

A key relation between h_{abc} and $\mathcal{F}_{abc}$ follows from the Bianchi identity $d\mathcal{F}_3 = -F_4$,and is given by

$$h_{abc} = \tfrac{1}{4}m_a{}^d\mathcal{F}_{bcd}. \tag{4.19}$$

We conclude this section by elucidating the consequences of the central equation (4.19). To this end, we first note the useful identities

$$\begin{aligned} h_{abe}h^{cde} &= \delta_{[a}^{[c}k_{b]}^{d]} \\ k_{ac}k_b^c &= \tfrac{1}{6}\eta_{ab}\operatorname{tr} k^2 \\ k_a{}^d h_{bcd} &= k_{[a}{}^d h_{bc]d} \end{aligned} \tag{4.20}$$

which are consequences of the linear self-duality of h_{abc}. Taking the Hodge dual of (4.19) one finds $*\mathcal{F}_{abc} = -\mathcal{F}_{abc} + 2Q^{-1}m_a{}^d\mathcal{F}_{bcd}$. Using the identity $m^2 = 2m - Q$, we readily find the nonlinear self-duality equation

$$*\mathcal{F}_{mnp} = Q^{-1}G_m{}^q\mathcal{F}_{npq}. \tag{4.21}$$

This equation can be expressed solely in terms of $\mathcal{F}_3$. To do this, we first insert (4.19) into (4.20), which yields the identities

$$\begin{aligned} \mathcal{F}_{abe}\mathcal{F}^{cde} &= 2\delta_{[a}^{[c}X_{b]}^{d]} + \tfrac{1}{2}K^{-2}X_{[a}{}^c X_{b]}{}^d + 2(K^2-1)\delta_{[a}^c\delta_{b]}^d \\ X_{ac}X_b^c &= 4K^2(K^2-1)\eta_{ab} \\ X_a{}^d\mathcal{F}_{bcd} &= X_{[a}{}^d\mathcal{F}_{bc]d} \end{aligned} \tag{4.22}$$

where we have defined

$$K \equiv \sqrt{1 + \frac{1}{24}\mathcal{F}^{abc}\mathcal{F}_{abc}}$$
$$X_{ab} \equiv \tfrac{1}{2}K * \mathcal{F}_a{}^{cd}\mathcal{F}_{bcd}. \tag{4.23}$$

Next we derive the identities

$$Q(K+1) \equiv 2$$
$$X_{ab} \equiv \frac{1}{2}\mathcal{F}_{acd}\mathcal{F}_b{}^{cd} - \frac{1}{12}\eta_{ab}\mathcal{F}_{cde}\mathcal{F}^{cde} = 4K(1+K)k_{ab}. \tag{4.24}$$

We can now express (4.21) entirely in terms of $\mathcal{F}_3$ by deriving the identity

$$Q^{-1}G_{mn} = K\eta_{mn} - \tfrac{1}{2}K^{-1}X_{mn}. \tag{4.25}$$

Another way of writing (4.21) is

$$\mathcal{F}^-_{abc} = \tfrac{1}{2}(1+K)^{-2}\mathcal{F}^+_{ade}\mathcal{F}^{+def}\mathcal{F}^+_{fbc}, \tag{4.26}$$

where K is a root of the quartic equation

$$(K+1)^3(K-1) = \frac{1}{24}\mathcal{F}^{+abc}\mathcal{F}^+_{ade}\mathcal{F}^{+def}\mathcal{F}^+_{fbc}. \tag{4.27}$$

4.3 The D3-Brane

From Appendix E, the bosonic sector of the D3-brane coupled the background of Type IIB supergravity is given by the action

$$S_4 = -T_4\int d^4\xi e^{-\phi}\sqrt{-\det(\gamma_{ij}+\mathcal{F}_{ij})}$$
$$+ T_4\Big[\int A_4 + \int A_2\wedge\mathcal{F} + \frac{1}{2!}\int A_0\wedge\mathcal{F}\wedge\mathcal{F}\Big] \tag{4.28}$$

where $\mathcal{F}_{ij}$ are the components of a modified 2-form field strength

$$\mathcal{F} = F - B, \tag{4.29}$$

where $F = dV$ is the usual field strength 2-form of the Born–Infeld field V and B is the pullback to the worldvolume of the NS-NS Type IIB 2-form potential B, and where A_0, A_2 and A_4 are the R-R forms. We use the same letter for superspace forms and their pullbacks to the worldvolume.

4.4 Problems 4

1. Show that the equations for the bosonic sector of the Type IIA superstring in $D = 10$ follow from those of the supermembrane in $D = 11$ by assuming an $M_{10}\times S^1$ topology, wrapping the membrane around the S^1 and taking the small radius limit [44].

5 ADS/CFT: The Membrane at the End of the Universe

5.1 Singletons Live on the Boundary

As emphasized by Fronsdal et al. [30, 31], singletons are best thought of as living not in the $(d+1)$-dimensional bulk of the AdS_{d+1} spacetime but rather on the d-dimensional $S^1 \times S^{d-1}$ boundary where the AdS group $SO(d-1,2)$ plays the role of the *conformal* group. Remaining for the moment with our 4-dimensional example, consider a scalar field $\Phi(t,r,\theta,\phi)$ on AdS_4 with metric (2.33), described by the action

$$S_{\rm bulk} = \int_{AdS_4} d^4x\, \sqrt{-g}\tfrac{1}{2}\Phi[g^{\mu\nu}\nabla_\mu\nabla_\nu - M^2]\Phi. \tag{5.1}$$

Note that this differs from the conventional Klein–Gordon action by a boundary term. Since the scalar Laplacian on AdS_4 has eigenvalues $E_0(E_0-3)a^2$, the critical value of M^2 for a singleton with $(E_0,s)=(\frac{1}{2},0)$ is

$$M^2 = \frac{5}{4}a^2. \tag{5.2}$$

In this case, one can show with some effort [30, 31] that as $r \to \infty$,

$$\Phi(t,r,\theta,\phi) \to r^{-1/2}\phi(t,\theta,\phi) \tag{5.3}$$

and hence that the radial dependence drops out:

$$S_{\rm boundary} = \int_{S^1\times S^2} d^3\xi\, \sqrt{-h}\big[-\tfrac{1}{2}h^{ij}\nabla_i\phi\nabla_j\phi - \tfrac{1}{8}a^2\phi^2\big]. \tag{5.4}$$

Here we are integrating over a 3-manifold with $S^1 \times S^2$ topology and with metric

$$h^{ij}\, d\xi^i\, d\xi^j = -dt^2 + \frac{1}{a^2}(d\theta^2 + \sin^2\theta\, d\phi^2). \tag{5.5}$$

This 3-manifold is sometimes referred to as the *boundary* of AdS_4 but note that the metric h_{ij} is not obtained by taking the $r \to \infty$ limit of $g_{\mu\nu}$ but rather the $r \to \infty$ limit of the conformally rescaled metric $\Omega^2 g_{\mu\nu}$ where $\Omega = 1/ar$. The radius of the S^2 is a^{-1}, not infinity. Most particle physicists are familiar with the conformal group in flat Minkowski space. It is the group of coordinate transformations which leave invariant the Minkowski lightcone. In the case of three-dimensional Minkowski space, M_3, it is $SO(3,2)$. In the present context, however, the spacetime is curved with topology $S^1 \times S^2$, but still admits $SO(3,2)$ as its conformal group,[6]

[6] One sometimes finds the statement in the physics literature that the only compact spaces admitting conformal Killing vectors are those isomorphic to spheres. By a theorem of Yano and Nagano [71], this is true for *Einstein spaces*, but $S^1 \times S^2$ is not Einstein.

i.e., as the group which leaves invariant the three-dimensional lightcone $h_{ij}\, d\xi^i\, d\xi^j = 0$. The failure to discriminate between these different kinds of conformal invariance is, we believe, a source of confusion in the singleton literature. In particular, the ϕ^2 "mass" term appearing in the action (5.4) would be incompatible with conformal invariance if the action were on M_3 but is essential for conformal invariance on $S^1 \times S^2$. Moreover, the coefficient $-a^2/8$ is uniquely fixed [39]. In Section 5.4 we shall derive the singleton action (5.4) including the scalar mass terms, starting from the membrane action (4.1).

So although singleton actions of the form (5.1) and their superpartners appeared in the Kaluza–Klein harmonic expansions on $\mathrm{AdS}_4 \times S^7$ [34–36], they could be gauged away everywhere except on the boundary where the above OSp(4|8) corresponds to the superconformal group [37]. One finds an $(n = 8, d = 3)$ supermultiplet with 8 scalars ϕ^A and 8 spinors $\chi^{\dot{A}}$, where the indices A and $\dot{A}$ range over 1 to 8 and denote the 8_s and 8_c representations of SO(8), respectively. The OSp(4|8) action is a generalization of (5.4) and is given by [39]

$$S_{\text{singleton}} = \int_{S^1 \times S^2} d^3\xi \sqrt{-h}\big[-\tfrac{1}{2}h^{ij}\nabla_i\phi^A\nabla_j\phi^A - \tfrac{1}{8}a^2\phi^A\phi^A + \tfrac{i}{2}\bar{\chi}^{\dot{A}}(1-\gamma)\gamma^i D_i\chi^{\dot{A}}\big] \tag{5.6}$$

where $\gamma = -\gamma_0\gamma_1\gamma_2$ and where D_i is the covariant derivative appropriate to the $S^1 \times S^2$ background.

In the case of $\mathrm{AdS}_5 \times S^5$ one finds a $(n = 4, d = 4)$ supermultiplet with 1 vector A_i, $(i = 0, 1, 2, 3)$, a complex spinor ${\lambda^a}_+$, $(a = 1, 2, 3, 4)$, obeying $\gamma_5 {\lambda^a}_+ = {\lambda^a}_+$ and 6 real scalars ϕ^{ab}, obeying $\phi^{ab} = -\phi^{ab}$, $\phi^{ab} = \epsilon^{abcd}\phi_{cd}/2$. The corresponding action for the doubletons of SU(2, 2|4) is [40]

$$S_{\text{doubleton}} = \int_{S^1 \times S^3} \big[-\tfrac{1}{4}F_{ij}F^{ij} - \tfrac{1}{4}a^2\phi_{ab}\phi^{ab} - \tfrac{1}{4}\partial_i\phi_{ab}\partial^i\phi^{ab} + i\bar{\lambda}_{+a}\gamma^i D_i{\lambda^a}_+\big] \tag{5.7}$$

where $F_{ij} = 2\partial_{[i}A_{j]}$. However, in contrast to the singletons, we know of no derivation of this doubleton action on the boundary starting from an action in the bulk analogous to (5.4).

In the case of $\mathrm{AdS}_7 \times S^4$ one finds a $\big((n_+, n_-) = (2, 0), d = 6\big)$ supermultiplet with a 2-form B_{ij}, $(i = 0, 1, \ldots, 5)$, whose field strength is self-dual, 8 spinors ${\lambda^A}_+$, $(A = 1, 2, 3, 4)$, obeying $\gamma^7{\lambda^A}_+ = {\lambda^A}_+$ and 5 scalars ϕ^a, $(a = 1, 2, \ldots, 5)$. The OSp(6, 2|4) tripleton covariant field equations on $S^1 \times S^5$ are [40]:

$$\begin{aligned}(\nabla^i\nabla_i - 4a^2)\phi^a &= 0\\ \gamma^i D_i{\lambda^A}_+ &= 0\\ H^{ijk} &= \frac{1}{3!}\sqrt{-h}\epsilon^{ijklmn}H_{lmn}\end{aligned} \tag{5.8}$$

where $H_{ijk} = 3\partial_{[i}B_{jk]}$. Once again, we know of no derivation of these tripleton field equations on the boundary starting from equations in the bulk.

5.2 The Membrane as a Singleton: The Membrane/Supergravity Bootstrap

Being defined over the boundary of AdS_4, the $OSp(4 \mid 8)$ singleton action (5.6) is a *three dimensional* theory with signature $(-,+,+)$ describing 8 scalars and 8 spinors. With the discovery of the eleven-dimensional supermembrane [41, 42], it was noted that 8 scalars and 8 spinors on a three-dimensional worldvolume with signature $(-,+,+)$ is just what is obtained after gauge-fixing the supermembrane action! Moreover, kappa-symmetry of this supermembrane action forces the background fields to obey the field equations of $(N=1, D=11)$ supergravity. It was therefore suggested [38] that on the $AdS_4 \times S^7$ supergravity background, the supermembrane whose worldvolume occupies the $S^1 \times S^2$ boundary of the AdS_4 could be regarded as the singleton of $OSp(4|8)$. Noting that these singletons also appear in the Kaluza–Klein harmonic expansion of this supergravity background, this further suggested a form of bootstrap [38] in which the supergravity gives rise to the membrane on the boundary which in turn yields the supergravity in the bulk. This conjecture received further support with the subsequent discovery of the "membrane at the end of the universe" [38, 39, 44–46] to be discussed in Section 5.4, and the realisation [51] that the eleven-dimensional supermembrane emerges as a solution of the $D = 11$ supergravity field equations.

The possibility of a similar 3-brane/supergravity bootstrap arising for the $SU(2,2 \mid 4)$ doubletons on $AdS_5 \times S^5$ and a similar 5-brane/supergravity bootstrap arising for the $OSp(6,2 \mid 4)$ tripletons on $AdS_7 \times S^4$ was also considered [38]. Ironically, however, it was (erroneously as we now know) rejected since the only supermembranes that were known at the time [52] had worldvolume theories described by *scalar* supermultiplets, whereas the doubletons and tripletons required vector and tensor supermultiplets, respectively. See Section 5.3.

Nevertheless, since everything seemed to fit nicely for the $(d=3, D=11)$ slot on the brane-scan of supersymmetric extended objects with worldvolume dimension d, there followed a good deal of activity relating other super p-branes in other dimensions to singletons and superconformal field theories [39, 40, 43–45, 47–50]. In particular, it was pointed out [39, 40, 48] that there was a one-to one-correspondence between the 12 points on the brane-scan as it was then known [52] and the 12 superconformal groups in Nahm's classification [37] admitting singleton representations, as shown in Table 2.5. To understand these 12 points on the brane-scan, we recall the matching of physical bose and fermi degrees of freedom on the worldvolume. As the p-brane moves through spacetime, its trajectory is described by the functions $X^M(\xi)$ where X^M are the spacetime coordinates ($M = 0, 1, \ldots, D-1$) and ξ^i are the worldvolume coordinates ($i = 0, 1, \ldots, d-1$). It is often convenient to make the so-called *static gauge choice* by making the $D = d + (D-d)$ split

$$X^M(\xi) = \big(X^\mu(\xi), Y^m(\xi)\big), \tag{5.9}$$

where $\mu = 0, 1, \ldots, d-1$ and $m = d, \ldots, D-1$, and then setting

$$X^\mu(\xi) = \xi^\mu. \tag{5.10}$$

Thus the only physical worldvolume degrees of freedom are given by the $(D-d)Y^m(\xi)$. So the number of on-shell bosonic degrees of freedom is

$$N_B = D - d. \tag{5.11}$$

To describe the super p-brane we augment the D bosonic coordinates $X^M(\xi)$ with anticommuting fermionic coordinates $\theta^\alpha(\xi)$. Depending on D, this spinor could be Dirac, Weyl, Majorana or Majorana–Weyl. The fermionic kappa symmetry means that half of the spinor degrees of freedom are redundant and may be eliminated by a physical gauge choice. The net result is that the theory exhibits a *d-dimensional worldvolume supersymmetry* where the number of fermionic generators is exactly half of the generators in the original spacetime supersymmetry. This partial breaking of supersymmetry is a key idea. Let M be the number of real components of the minimal spinor and N the number of supersymmetries in D spacetime dimensions and let m and n be the corresponding quantities in d worldvolume dimensions. Let us first consider $d > 2$. Since kappa symmetry always halves the number of fermionic degrees of freedom and going on-shell halves it again, the number of on-shell fermionic degrees of freedom is

$$N_F = \tfrac{1}{2}mn = \tfrac{1}{4}MN. \tag{5.12}$$

Worldvolume supersymmetry demands $N_B = N_F$ and hence

$$D - d = \tfrac{1}{2}mn = \tfrac{1}{4}MN. \tag{5.13}$$

A list of dimensions, number of real dimensions of the minimal spinor and possible supersymmetries is given in Table 2.1, from which we see that there are only 8 solutions of (5.13) all with $N = 1$, which exactly match the singletons shown in Table 2.5. We note in particular that $D_{\max} = 11$ since $M \geq 64$ for $D \geq 12$ and hence (5.13) cannot be satisfied. Similarly $d_{\max} = 6$ since $m \geq 16$ for $d \geq 7$. The case $d = 2$ is special because of the ability to treat left and right moving modes independently. If we require the sum of both left and right moving bosons and fermions to be equal, then we again find the condition (5.13). This provides a further 4 solutions

TABLE 2.5. The brane scan of superconformal groups admitting singletons.

$D \uparrow$							
11	·			$\mathbf{OSp(4\,\vert\,8)}$			
10	·		$[\mathrm{OSp}(2\,\vert\,8)]^2$				$\mathrm{OSp}(6,2\,\vert\,2)$
9	·					$F(4)$	
8	·				$\mathrm{SU}(2,2\,\vert\,2)$		
7	·			$\mathrm{OSp}(4\,\vert\,4)$			
6	·		$[\mathrm{OSp}(2\,\vert\,4)]^2$		$\mathrm{SU}(2,2\,\vert\,1)$		
5	·			$\mathrm{OSp}(4\,\vert\,2)$			
4	·		$[\mathrm{OSp}(2\,\vert\,2)]^2$	$\mathrm{OSp}(4\,\vert\,1)$			
3	·		$[\mathrm{OSp}(2\,\vert\,1)]^2$				
2	·						
1	·						
0	·	·	·	·	·	·	·
	0	1	2	3	4	5	6 $d \rightarrow$

all with $N = 2$, corresponding to Type II superstrings in $D = 3, 4, 6$ and 10 (or 8 solutions in all if we treat Type IIA and Type IIB separately). Both the gauge-fixed Type IIA and Type IIB superstrings will display $(8, 8)$ supersymmetry on the worldsheet. If we require only left (or right) matching, then (5.13) is replaced by

$$D - 2 = n = \tfrac{1}{2} MN, \tag{5.14}$$

which allows another 4 solutions in $D = 3, 4, 6$ and 10, all with $N = 1$. The gauge-fixed theory will display $(8, 0)$ worldsheet supersymmetry. The heterotic string falls into this category.

The number of dimensions transverse to the brane, $D - d$, equals the number of scalars in the singleton supermultiplet. (The two factors appearing in the $d = 2$ case is simply a reflection of the ability of strings to have right and left movers. For brevity, we have written the Type II assignments in Table 2.5, but more generally we could have $\mathrm{OSp}(2\,|\,p) \times \mathrm{OSp}(2\,|\,q)$ where p and q are the number of left and right supersymmetries [53].) Note that the $d = 6$ upper limit on the worldvolume dimension is consistent with the requirement of renormalizability [47].

Note, however, that the $(d = 3, D = 11)$, $\mathrm{OSp}(4\,|\,8)$ slot (written in boldface) occupies a privileged position in that the corresponding $D = 11$ supergravity theory admits the $\mathrm{AdS}_4 \times S^7$ solution with $\mathrm{OSp}(4\,|\,8)$ symmetry, whereas the other supergravities do not admit solutions with the superconformal group as a symmetry. For example, $D = 10$ supergravity admits an $\mathrm{AdS}_3 \times S^7$ solution [54, 65], but it does not have the full $[\mathrm{OSp}(2\,|\,8)]^2$ symmetry because the dilaton is nontrivial and acts as a conformal Killing vector on the AdS_3. This is slightly mysterious, since the bulk theory has less symmetry than the boundary theory. We shall return to this in Sections 5.5 and 5.7.

5.3 Doubletons and Tripletons Revisited

These early works focussed on scalar supermultiplets because these were the only p-branes known in 1988 [52]. However, with the discovery in 1990 of Type II p-brane solitons [57–61], vector and tensor multiplets were also seen to play a role. In particular, the worldvolume fields of the self-dual Type IIB superthreebrane were shown to be described by an $(n = 4, d = 4)$ gauge theory [58], which on the boundary of AdS_5 is just the doubleton supermultiplet of the superconformal group $\mathrm{SU}(2,2\,|\,4)$! Thus one can after all entertain a 3-brane-doubleton-supergravity bootstrap similar to the membrane-singleton-supergravity bootstrap of Section 5.2, and we may now draw the doubleton brane scan of Table 2.6. Once again, the restriction to $d = 4$ is consistent with renormalizability. Note, however, that the $(d = 4, D = 10)$, $\mathrm{SU}(2,2\,|\,4)$ slot (written in boldface) occupies a privileged position in that the corresponding $D = 10$ Type IIB supergravity admits the $\mathrm{AdS}_5 \times S^5$ solution with $\mathrm{SU}(2,2\,|\,4)$ symmetry, whereas the other supergravities do not admit solutions with the superconformal group as a symmetry since, as discussed in Section 5.5, the dilaton is again nontrivial.

Similarly, with the discovery of the M-theory fivebrane [62], it was realized [64] that the zero modes are described by an $\big((n_+, n_-) = (2,0), d = 6\big)$ multiplet with a chiral 2-form, 8 spinors and 5 scalars, which on the boundary of AdS_7 is just the tripleton supermultiplet of the superconformal group $\mathrm{OSp}(6,2\,|\,4)$! (These zero modes are the same as those of the Type IIA fivebrane, found previously in [59, 60]). Thus one can after all also entertain a 5-brane-tripleton-supergravity bootstrap similar to the membrane-singleton-supergravity bootstrap of Section 5.2. Thus we may now draw the tripleton brane scan of Table 2.7. Note once again, however, that the $(d = 6, D = 11)$, $\mathrm{OSp}(6,2\,|\,4)$ slot (written in boldface) occupies a privileged position in that the corresponding $D = 11$ supergravity admits the $\mathrm{AdS}_7 \times S^4$ solution with $\mathrm{OSp}(6,2\,|\,4)$ symmetry, whereas the other supergravities do not admit solutions with the superconformal group as a symmetry since, as discussed in Section 5.5, the dilaton is again nontrivial.

With the inclusion of branes with vector and tensor supermultiplets on their worldvolume, another curiosity arises. Whereas the singleton brane scan of Table 2.5 exhausts all the scalar branes and the tripleton brane scan of Table 2.7 exhausts all the tensor branes, the doubleton brane scan of Table 2.6 is only a subset of all the vector branes [6, 55]. The Type IIB 3-brane is special because gauge theories are conformal only in $d = 4$.

TABLE 2.6. The brane scan of superconformal groups admitting doubletons.

$D \uparrow$ \ $d \rightarrow$	0	1	2	3	4	5	6
11	·						
10	·				**SU(2, 2 | 4)**		
9	·						
8	·				SU(2, 2 | 2)		
7	·						
6	·						
5	·						
4	·				SU(2, 2 | 1)		
3	·						
2	·						
1	·						
0	·	·	·	·	·	·	·

TABLE 2.7. The brane scan of superconformal groups admitting tripletons.

$D \uparrow$ \ $d \rightarrow$	0	1	2	3	4	5	6
11	·						**OSp(6, 2 | 4)**
10	·						
9	·						
8	·						
7	·						OSp(6, 2 | 2)
6	·						
5	·						
4	·						
3	·						
2	·						
1	·						
0	·	·	·	·	·	·	·

5.4 The Membrane at the End of the Universe

As further evidence of the membrane/supergravity bootstrap idea, solutions of the $D = 11$ supermembrane equations (4.2) and (4.3) were sought for which the spacetime is $\text{AdS}_4 \times M^7$ and for which the supermembrane occupies the $S^1 \times S^2$ boundary of the AdS_4. As we shall now recall, the BPS condition was achieved only as $r \to \infty$, hence the name *the Membrane at the End of the Universe* [44–46, 78].

We substitute (2.19) and (2.5) into (4.2) and (4.3) and look for solutions of the form

$$t = \xi^0, \quad \theta = \xi^1, \quad \phi = \xi^2 \tag{5.15}$$

so that the membrane is embedded in the AdS_4 as

$$ds_4^{\,2} = -\cosh^2\rho(d\xi^0)^2 + \frac{1}{a^2}d\rho^2 + \frac{1}{a^2}\sinh^2\rho\big((d\xi^1)^2 + \sin^2\xi^1(d\xi^2)^2\big). \tag{5.16}$$

In order to show that this configuration does indeed satisfy the requirements of unbroken supersymmetry (4.9), we first exhibit a spinor ϵ satisfying $\widetilde{D}_M\epsilon = 0$ everywhere and then show how it satisfies $\Gamma\epsilon = \epsilon$ as $r \to \infty$. It is not difficult to show that

$$\Gamma = \gamma \otimes I \tag{5.17}$$

where $\gamma \equiv \gamma_{012}$. First we look for spinors of the form (2.39), then the general solution for $\epsilon(x)$ is

$$\epsilon = \sqrt{2}(\sinh\rho/2 + \gamma\cosh\rho/2)\zeta(t, \theta, \phi) \tag{5.18}$$

where ζ satisfies

$$(1+\gamma)\big(\nabla_i + \tfrac{1}{2}a\gamma_i\gamma_3\big)\zeta = 0 \tag{5.19}$$

where ∇_i is the covariant derivative on the $S^1 \times S^2$ boundary of AdS_4 with metric (5.5). One can show [44] that this equation has four solutions, implying the well-known result that AdS_4 has four Killing spinors.

Since $r = \infty$ corresponds to $\rho = \infty$, it follows from (5.18), (5.17) that ϵ becomes an eigenstate of γ and hence $\Gamma\epsilon = \epsilon$ "at the end of the universe". Thus we have shown that the membrane at the end of the universe is supersymmetric whenever M_7 admits Killing spinors. As we have seen in Section 2.2, the number N of such spinors depends on the Weyl holonomy of M_7 and lies between 0 and 8. It remains to show that the action for such a membrane is indeed the $\mathrm{OSp}(4 \mid N)$ singleton action (5.6). To simplify matters we shall now demonstrate this for the bosonic radial mode $r(\xi)$. Substituting the $\mathrm{AdS}_4 \times M_7$ solution and the brane ansatz (5.15) into the membrane action (4.1) we find

$$S = T_3 \int_{S^1\times S^2} d^3\xi\,\left[-\sqrt{-\det(g_{ij} + a^{-2}\partial_i\rho\partial_j\rho)} + \sinh^3\rho\right] \tag{5.20}$$

where from (4.3) and (2.19)

$$g_{ij} = \begin{pmatrix} -\cosh^2\rho & 0 & 0 \\ 0 & (\sinh^2\rho)/a^2 & 0 \\ 0 & 0 & (\sinh^2\rho\sin^2\theta)/a^2 \end{pmatrix}. \tag{5.21}$$

Hence

$$S = T_3 \int_{S^1 \times S^2} d^3\xi \sqrt{-h} \Big[-\frac{1}{2a^2} \cosh \rho h^{ij} \partial_i \rho \partial_j \rho - \cosh \rho \sinh^2 \rho + \sinh^3 \rho \Big]. \quad (5.22)$$

Since we are interested in the $r \to \infty$ limit, we consider only large ρ, for which

$$S = \frac{T_3}{8} \int_{S^1 \times S^2} d^3\xi \sqrt{-h} \Big[-\frac{2}{a^2} e^\rho h^{ij} \partial_i \rho \partial_j \rho - 2e^\rho \Big]. \quad (5.23)$$

So, bearing in mind that $T_3 \sim a^3$ and making the change of variable

$$e^\rho \sim \frac{1}{a}\phi^2 \quad (5.24)$$

we find

$$S = \int_{S^1 \times S^2} d^3\xi \sqrt{-h} \Big[-\frac{1}{2} h^{ij} \nabla_i \phi \nabla_j \phi - \frac{1}{8} a^2 \phi^2 \Big] \quad (5.25)$$

which is just the singleton action (5.4), including the scalar mass terms necessary for conformal invariance on $S^1 \times S^2$.

Following [140], this result may be generalized to arbitrary $(d-1)$-branes occupying the conformal boundary M of an arbitrary Einstein $(d+1)$-dimensional manifold W. We shall adapt the notation and the signature used in [140] for our convenience. The boundary M has a natural conformal structure but not a natural metric. Let h_{ij} be an arbitrary metric on the boundary in its conformal class. Here the ξ^i, $i = 0, \ldots, d-1$ are an arbitrary set of local coordinates on the boundary. There is then a unique way [196] to extend the ξ^i to coordinates on W near the boundary, adding an additional coordinate ρ that tends to infinity on the boundary, such that the metric in a neighborhood of the boundary is

$$ds^2 = \frac{1}{a^2} \left(d\rho^2 + \frac{1}{4} e^{2\rho} h_{ij} d\xi^i d\xi^j - P_{ij} d\xi^i d\xi^j + O(e^{-2\rho}) \right) \quad (5.26)$$

where

$$P_{ij} = \frac{2(D-1)R_{ij} - g_{ij}R}{2(D-1)(D-2)}, \quad (5.27)$$

and R_{ij} is the Ricci tensor of M, which implies

$$g^{ij} P_{ij} = \frac{R}{2(d-1)}. \quad (5.28)$$

We find

$$S = \frac{T_d}{2^d} \int_M d^d\xi \sqrt{-h} \Big[-\frac{2}{a^2} e^\rho h^{ij} \partial_i \rho \partial_j \rho - \frac{2}{(d-1)(d-2)} e^\rho R \Big] \quad (5.29)$$

we recognise the curvature term as that required for Weyl invariance of the action [39].

5.5 Near Horizon Geometry and p-Brane Aristocracy

More recently, AdS has emerged as the near-horizon geometry of black p-brane solutions [7, 64–66] in D dimensions. The dual brane, with worldvolume dimension $\tilde{d} = D - d - 2$, interpolates between D-dimensional Minkowski space M_D and $\mathrm{AdS}_{\tilde{d}+1} \times S^{d+1}$ (or $M_{\tilde{d}+1} \times S^3$ if $d = 2$). To see this, we recall that such branes arise generically as solitons of the following action [72]:

$$I = \frac{1}{2\kappa_D{}^2} \int d^D x \sqrt{-g} \left[R - \frac{1}{2} (\partial \phi)^2 - \frac{1}{2(d+1)!} e^{-\alpha \phi} F_{d+1}{}^2 \right] \tag{5.30}$$

where F_{d+1} is the field strength of a d-form potential A_d and α is the constant

$$\alpha^2 = 4 - \frac{2d\tilde{d}}{d+\tilde{d}}. \tag{5.31}$$

Written in terms of the $(d-1)$-brane sigma-model metric $e^{-\alpha/d\phi} g_{MN}$, the solutions are [7, 72]

$$\begin{aligned} ds^2 &= H^{(2-d)/d}\, dx^\mu\, dx_\mu + H^{2/d} (dy^2 + y^2 d\Omega_{d+1}{}^2) \\ e^{2\phi} &= H^\alpha \\ F_{d+1} &= dL^d \epsilon_{d+1} \end{aligned} \tag{5.32}$$

where

$$H = 1 + \frac{L^d}{y^d}. \tag{5.33}$$

For a stack of N singly charged branes $L^d = Nb^d$ and the near horizon, or large N, geometry corresponds to

$$ds^2 \sim \frac{y}{L}^{2-d} dx^\mu dx_\mu + \frac{L^2}{y^2} dy^2 + L^2 d\Omega^2_{d+1}. \tag{5.34}$$

Or, defining the new coordinate

$$y = L e^{\zeta/L} \tag{5.35}$$

we get

$$\begin{aligned} ds^2 &\sim e^{(2-d)\zeta/L}\, dx^\mu\, dx_\mu + d\zeta^2 + L^2\, d\Omega_{d+1}{}^2 \\ \phi &\sim \frac{d\alpha}{2L} \zeta \\ F_{d+1} &\sim dL^d \epsilon_{d+1}. \end{aligned} \tag{5.36}$$

Thus for $d \neq 2$ the near-horizon geometry is $\mathrm{AdS}_{\tilde{d}+1} \times S^{d+1}$. Note, however, that the gradient of the dilaton is generically nonzero and plays the role of a conformal Killing vector on $\mathrm{AdS}_{\tilde{d}+1}$. Consequently, there is no enhancement of symmetry in the near-horizon limit. The unbroken supersymmetry remains one-half and the bosonic symmetry remains $P_{\tilde{d}} \times \mathrm{SO}(d+2)$. (If $d = 2$, then (5.36) reduces to

$$\begin{aligned} ds^2 &\sim dx^\mu\, dx_\mu + d\zeta^2 + L^2\, d\Omega_3{}^2 \\ \phi &\sim \frac{\alpha}{L}\zeta \\ F_3 &\sim 2L^2\epsilon_3 \end{aligned} \tag{5.37}$$

which is $M_{\tilde{d}+1} \times S^3$, with a linear dilaton vacuum. The bosonic symmetry remains $P_{\tilde{d}} \times \mathrm{SO}(4)$.)

Of particular interest are the ($\alpha = 0$) subset of solitons for which the dilaton is zero or constant: the *nondilatonic p-branes*. From (5.31) we see that for branes with one kind of charge there are only 3 cases:

$$\begin{aligned} D = 11: &\quad d = 6, \tilde{d} = 3 \\ D = 10: &\quad d = 4, \tilde{d} = 4 \\ D = 11: &\quad d = 3, \tilde{d} = 6 \end{aligned}$$

which are precisely the three cases that occupied privileged positions on the singleton, doubleton and tripleton brane-scans of Tables 2.5, 2.6 and 2.7. Then the near-horizon geometry coincides with the $\mathrm{AdS}_{\tilde{d}+1} \times S^{d+1}$ nondilatonic maximally symmetric compactifications of the corresponding supergravities. The supersymmetry doubles and the bosonic symmetry is also enhanced to $\mathrm{SO}(\tilde{d}, 2) \times \mathrm{SO}(d+2)$. Thus the total symmetry is given by the conformal supergroups $\mathrm{OSp}(4|8)$, $\mathrm{SU}(2,2|4)$ and $\mathrm{OSp}(6,2|4)$, respectively.

For bound states of branes with M kinds of charge, the constant α gets replaced by [73, 74, 76]

$$\alpha^2 = \frac{4}{M} - \frac{2d\tilde{d}}{d+\tilde{d}}. \tag{5.38}$$

A nondilatonic solution ($\alpha = 0$) occurs for $M = 2$:

$$D = 6: \quad d = 2, \tilde{d} = 2$$

which is just the dyonic string [75], of which the self-dual string [72] is a special case, whose near-horizon geometry is $\mathrm{AdS}_3 \times S^3$. For $M = 3$ we have

$$D = 5: \quad d = 2, \tilde{d} = 1$$

which is the 3-charge black hole [80], whose near-horizon geometry is $\mathrm{AdS}_2 \times S^3$, and

$$D = 5: \quad d = 1, \tilde{d} = 2$$

which is the 3-charge string [80] whose near-horizon geometry is $\mathrm{AdS}_3 \times S^2$. For $M = 4$ we have

$$D = 4: \quad d = 1, \tilde{d} = 1$$

which is the 4-charge black hole [81, 82], of which the Reissner-Nordström solution is a special case [76], and whose near-horizon geometry is $AdS_2 \times S^2$ [77].

Thus we see that not all branes are created equal. A *p-brane aristocracy* obtains whose members are those branes whose near-horizon geometries have as their symmetry the conformal supergroups. As an example of a plebian brane we can consider the ten-dimensional superstring:

$$D = 10: \quad d = 6, \tilde{d} = 2$$

whose near-horizon geometry is $AdS_3 \times S^7$ but with a nontrivial dilaton of Section 5.2 which does not have the conformal group $[OSp(2|8)]^2$ as its symmetry, even though this group appears in the $(D = 10, \tilde{d} = 2)$ slot on the singleton brane-scan of Table 2.5. In which case, of course, one may ask what role do these singletons play. We shall return to this in Section 5.7.

The original membrane at the end of the universe was a spherical brane embedded in the AdS_4 geometry as given in 5.4. Alternatively, one could take as the membrane at the end of the universe to be the flat near-horizon membrane, which is embedded as

$$ds^2 = e^{4\zeta/L}\left(-(d\xi^0)^2 + (d\xi^1)^2 + (d\xi^2)^2\right) + d\zeta^2 \tag{5.39}$$

and has M_3 topology. It is still possible to associate an $OSp(4 \mid 8)$ action but this time it is defined over M_3 and has no scalar mass terms [78, 79]. One can continue to call these fields "singletons", of course, if by singleton one simply means anything transforming according to the $D(\frac{1}{2}, 0)$ and $D(1, \frac{1}{2})$ representations of $SO(3, 2)$. A comparison of these two approaches is discussed in some detail in [78].

5.6 Supermembranes with Fewer Supersymmetries. Skew-Whiffing

So far we have focussed attention on compactifications to $AdS_{\tilde{d}+1}$ on round spheres S^{d+1} which have maximal supersymmetry, but the supergravity equations admit infinitely many other compactifications on Einstein spaces X^{d+1} which have fewer supersymmetries [20]. Indeed generic X^{d+1} have no supersymmetries at all.[7] We note in this connection the *skew-whiffing theorem* [20], which states that for every $AdS_{\tilde{d}+1}$ compactification preserving supersymmetry, there exists one with no supersymmetry simply obtained by reversing the orientation of X^{d+1} (or, equivalently, reversing the sign of F_{d+1}). The only exceptions are when X^{d+1} are round spheres which preserve the maximum supersymmetry for either orientation. A corollary

[7]Thus in the early eighties, the most highly prized solutions were those with many supersymmetries. Nowadays, bragging rights seem to go to those which have none!

is that other *symmetric spaces*, which necessarily admit an orientation-reversing isometry, can have no supersymmeties. Examples are provided by products of round spheres.

The question naturally arises as to whether these compactifications with fewer supersymmetries also arise as near-horizon geometries of p-brane solitons. The answer is yes and the soliton solutions are easy to construct [67, 68]. One simply makes the replacement

$$d\Omega_{d+1}{}^2 \to d\widehat{\Omega}_{d+1}{}^2 \tag{5.40}$$

in (5.32), where $d\widehat{\Omega}_{d+1}{}^2$ is the metric on an arbitrary Einstein space X^{d+1} with the same scalar curvature as the round S^{d+1}. The space need only be Einstein; it need not be homogeneous [67]. (There also exist brane solutions on Ricci flat X^{d+1} [67] but we shall not discuss them here.) Note, however, that these non-round-spherical solutions do not tend to $(D-d)$-dimensional Minkowski space as $r \to \infty$. Instead the metric on the $(D-\tilde{d})$-dimensional space transverse to the brane is asymptotic to a generalized cone

$$ds_{D-\tilde{d}}{}^2 = dr^2 + r^2\, d\widehat{\Omega}_{d+1}{}^2 \tag{5.41}$$

and $(D-d)$-dimensional translational invariance is absent except when X^{d+1} is a round sphere. The number of supersymmetries preserved by these p-branes is determined by the number of Killing spinors on X^{d+1}.

To illustrate these ideas let us focus on the eleven-dimensional supermembrane. The usual supermembrane interpolates between M_{11} and $\mathrm{AdS}_4\times$ round S^7, has symmetry $P_3\times \mathrm{SO}(8)$ and preserves $\frac{1}{2}$ of the spacetime supersymmetries for either orientation of the round S^7. Replacing the round S^7 by generic Einstein spaces X^7 leads to membranes with symmetry $P_3 \times G$, where G is the isometry group of X^7. For example, $G = \mathrm{SO}(5)\times\mathrm{SO}(3)$ for the squashed S^7 [69, 70]. For one orientation of X^7, they preserve $N/16$ spacetime supersymmetries where $1 \le N \le 8$ is the number of Killing spinors on X^7; for the opposite orientation they preserve no supersymmetries since then X^7 has no Killing spinors. For example, $N = 1$ for the left-squashed S^7 owing to its G_2 holonomy [20, 69, 70], whereas $N = 0$ for the right-squashed S^7. However, all these solutions satisfy the same Bogomol'nyi bound between the mass and charge as the usual supermembrane [67]. Of course, skew-whiffing is not the only way to obtain vacua with less than maximal supersymmetry. A summary of known X^7, their supersymmetries and stability properties is given in [20]. Note, however, that skew-whiffed vacua are automatically stable at the classical level since skew-whiffing affects only the spin $\frac{3}{2}$, $\frac{1}{2}$ and 0^- towers in the Kaluza–Klein spectrum, whereas the criterion for classical stability involves only the 0^+ tower [20].

5.7 The Maldacena Conjecture

The year 1998 marks a revolution in anti-de Sitter space brought about by Maldacena's conjectured duality between physics in the bulk of AdS and a conformal field theory on the boundary [110]. In particular, M-theory on $AdS_4 \times S^7$ is dual to a non-Abelian $(n = 8, d = 3)$ superconformal theory, Type IIB string theory on $AdS_5 \times S^5$ is dual to a $d = 4$ SU(N) super Yang–Mills theory and M-theory on $AdS_7 \times S^4$ is dual to a non-Abelian $\big((n_+, n_-) = (2, 0), d = 6\big)$ conformal theory. In particular, as has been spelled out most clearly in the $d = 4$ SU(N) Yang–Mills case, there is seen to be a correspondence between the Kaluza–Klein mass spectrum in the bulk and the conformal dimension of operators on the boundary [122, 124].

One immediately recognises that the dimensions and supersymmetries of these three conformal theories are exactly the same as the singleton, doubleton and tripleton supermultiplets of Section 2.5. Moreover, both the old and new AdS/CFT correspondences are *holographic* in the sense of [130, 131]. Following Maldacena's conjecture [110], therefore, a number of papers appeared reviving the old singleton-AdS-membrane-superconformal field theory connections [78, 84, 117–129, 155] and applying them to this new duality context. What are the differences?

The first difference is that the branes were spherical as opposed to flat, since the "boundary" in question for the singletons and p-branes at the end of the universe was the $S^1 \times S^p$ with finite radius $1/a$ for the S^p. As such, superconformal invariance requires mass terms for the scalars, as discussed in Section 5.1. The branes in the Maldacena conjecture, on the other hand, are the flat Minkowski space branes embedded in AdS in the way dictated by the near-horizon geometry.

Another curious difference is that, with the exception of the three aristocratic branes, all the slots on the three brane-scans of superconformal field theories corresponded to bulk supergravities whose brane solutions are dilatonic, and hence have a symmetry smaller than the boundary theory. It seems that the branes at the end of the universe do not care about the dilaton because the $\rho =$ constant surfaces in (5.16) (or the $\zeta =$ constant surfaces in (5.39)) posess the full superconformal symmetry even though the bulk AdS solution does not. In other words, they admit the maximal set of conformal Killing vectors even though the bulk admits less than the maximal set of Killing vectors. This contrasts with the new AdS/CFT conjecture where a nonconformal supergravity solution in the bulk [65] is deemed to be dual to nonconformal field theory on the boundary [117]. It is not obvious at the moment whether this difference is real or apparent and it would be interesting to pursue the matter further.

Thirdly, although the *membrane/supergravity bootstrap* idea, that physics in the bulk of AdS should be dictated by the membrane on the boundary and vice versa, might nowadays be called "holographic", it was not motivated by considerations of entropy. Rather, in analogy with string theory where there are self-consistency requirements between the spacetime and worldsheet dynamics, it was expected that a similar phenomenon should hold for membranes and the membrane on the boundary of AdS was simply a special case.

Fourthly, attention was focussed on *free* superconformal theories on the boundary as opposed to the *interacting* theories currently under consideration. For example, although the worldvolume fields of the Type IIB 3-brane were known to be described by an $(n = 4, d = 4)$ gauge theory [58], we now know that this brane admits the interpretation of a Dirichlet brane [111] and that the superposition of N such branes yields a *non-Abelian* $\mathrm{SU}(N)$ gauge theory [112]. These observations are crucial to the new duality conjecture [110]. For earlier related work on coincident threebranes and $n = 4$ super Yang Mills, see [113–116].

Let us consider the solution for N coincident 3-branes corresponding to N units of 5-form flux [57, 58]. As we have already seen in Section 3.2, $L = N^{1/4}b$ and the near horizon limit may equally well be regarded as the large N limit. We find $\mathrm{AdS}_5 \times S^5$, but with an AdS radius proportional to $N^{1/4}$. The philosophy is that Type IIB supergravity is a good approximation for large N and that Type IIB stringy excitations correspond to operators whose dimensions diverge for $N \to \infty$. This makes contact with the whole industry of large N QCD. These large N, non-Abelian features were absent in the considerations of a 3-brane/supergravity bootstrap discussed in Section 5.3, as was the precise correspondence between the Kaluza–Klein mass spectrum in the bulk and the conformal dimension of operators on the boundary [122, 124]. Nevertheless, as we hope these lectures show, there are sufficently many similarities between the current bulk/boundary duality and the old Membrane at the End of the Universe idea to merit further comparisons.

5.8 Problems 5

1. Repeat the analysis of Section 5.5 for *rotating* branes.

6 Anti-de Sitter Black Holes

6.1 Introduction

Anti-de Sitter black hole solutions of gauged extended supergravities [143] are currently attracting a good deal of attention [144–154] due, in large part, to the AdS/CFT correspondence. As we have seen, these gauged extended supergravities arise as the massless modes of various Kaluza–Klein compactifications of both $D = 11$ and $D = 10$ supergravities. The three examples studied in these lectures are gauged $D = 4$, $N = 8$ SO(8) supergravity [11, 156] arising from $D = 11$ supergravity on S^7 [13, 20] whose black hole solutions are discussed in [149]; gauged $D = 5$, $N = 8$ SO(6) supergravity [18, 27] arising from Type IIB supergravity on S^5 [23–25] whose black hole solutions are discussed in [144, 148]; and gauged $D = 7$, $N = 4$ SO(5) supergravity [18, 19] arising from $D = 11$ supergravity on S^4 [17] whose black hole solutions are given in Section 6.8 and in [63, 151].[8] In the absence of the black holes, these three AdS compactifications are singled out as arising from the near-horizon geometry of the extremal nonrotating M2-, D3- and M5-branes [64–66, 157]. One of our goals will be to embed these known lower-dimensional black hole solutions into ten or eleven dimensions, thus allowing a higher dimensional interpretation in terms of *rotating* M2-, D3- and M5-branes.

Since these gauged supergravity theories may be obtained by consistently truncating the massive modes of the full Kaluza–Klein theories, it follows that all solutions of the lower-dimensional theories will also be solutions of the higher-dimensional ones [158, 159]. In principle, therefore, once we know the Kaluza–Klein ansatz for the massless sector, it ought to be straightforward to read off the higher dimensional solutions. In practice, however, this is a formidable task. The correct massless ansatz for the S^7 compactification took many years to finalize [160, 162], and is still highly implicit, while for the S^5 compactifications the complete massless ansatzs are still unknown. The S^4 case has recently been given in its entirety in [218, 219]. For our present purposes, it suffices to consider truncations of the gauged supergravities to include only gauge fields in the Cartan subalgebras of the full gauge groups, namely $U(1)^4$, $U(1)^3$ and $U(1)^2$ for the S^7, S^5 and S^4 compactifications, respectively. These truncated theories will admit respectively the 4-charge AdS_4, 3-charge AdS_5 and 2-charge AdS_7 black hole solutions.

The simplest of the three is perhaps the $D = 5$, $N = 8$ maximal gauged supergravity, for which there is a consistent $N = 2$ (i.e., minimal) truncation to supergravity coupled to two Abelian vector multiplets. This has the bosonic field content of a graviton, three U(1) gauge fields and two scalars. In these lectures we obtain the complete nonlinear Kaluza–Klein ansatz for the compactification of $D = 10$ Type IIB supergravity on S^5, truncated to the $U(1)^3$ Cartan subgroup of SO(6).

In four dimensions there is a consistent truncation of gauged $N = 8$ maximal supergravity to gauged $N = 2$ supergravity coupled to three vector multiplets. The bosonic sector consists of a graviton, four vectors and three complex scalars, whose real and imaginary parts correspond to three

[8]BPS black holes arising in the $SU(2) \times SU(2)$ version of gauged $N = 4$ supergravity in $D = 4$, which is the massless sector of the $S^3 \times S^3$ compactification of $N = 1$ supergravity in $D = 10$, were discussed in [147]. These solutions are not asymptotically AdS.

"axions" and three "dilatons".[9] The inclusion of the axions is necessary for providing a consistent truncation; the full bosonic Lagrangian in this case is obtained in Appendix F. This truncation corresponds to the $U(1)^4$ Cartan subgroup of the non-Abelian SO(8), for which there exist AdS black hole solutions with four electric charges [149]. While one would ideally wish to obtain a complete Kaluza–Klein ansatz for the $N = 2$ truncation, in practice the complexity arising from the inclusion of the axions is considerable. Thus in these lectures we omit the axions in the Kaluza–Klein reduction. This is of course sufficient for the embedding of the electric black hole solutions in $D = 11$ as they do not involve the axions.

Finally, in seven dimensions, maximal $N = 4$ gauged SO(5) supergravity admits a consistent truncation to $N = 2$ supergravity, comprising the metric, a 2-form potential, three vectors and a dilaton, coupled to a vector multiplet comprising a vector and three scalars. We obtain the Kaluza–Klein ansatz for an S^4 reduction of $D = 11$ supergravity, including two $U(1)^2$ gauge fields and two dilatonic scalars. This is sufficient for the consideration of the embedding of the $D = 7$ black holes in $D = 11$.

Having obtained the explicit Kaluza–Klein reduction ansatzs, this allows an investigation of the embedding of the various AdS black holes of $D = 4$, $D = 5$ and $D = 7$ in the respective higher-dimensional supergravities. An important point here is that one must know the exact Kaluza–Klein reduction ansatz for the reduction of the supergravity theory itself, and not just for a specific solution, in order to show that the metric, gauge fields and scalar fields of the lower-dimensional solution are indeed precisely embeddable in the higher-dimensional theory. It is worth remarking, in this regard, that it is the scalar fields that present most of the subtleties and complexities in the implementation of the reduction procedure.

Having embedded these black holes in ten or eleven dimensions, an interesting question then arises as to their higher-dimensional interpretation. It was noted some time ago [45], in the context of a "test" membrane moving in a fixed $AdS_4 \times S^7$, that a 4-dimensional BPS state (whose AdS energy is equal to its electric charge) admits the eleven-dimensional interpretation of an M2-brane [41, 42, 51] that is rotating in the extra dimensions. Moreover, the electric charge is equal to the spin.

---

[9] Interestingly enough, the ungauged version of this theory, obtained by switching off the gauge coupling and performing some dualisations, appears in the T^2 compactification of $D = 6$, $N = 1$ string theory. The four vectors are the two Kaluza–Klein and two winding gauge fields, while the three complex scalars S, T and U correspond to the axion-dilaton, the Kähler form and complex structure of the torus. This STU system plays a crucial role in four-dimensional string/string/string triality [163]. The black hole solutions of this theory [163, 164], and their embedding in ungauged $N = 8$ supergravity [165], [166] arising from the T^7 compactification of M-theory as intersections [167, 168] are also well known.

Recently there has been an upsurge of activities on the study of rotating p-branes [150–152, 169–178]. In particular, in [150] AdS Reissner–Nordström black holes, (i.e. the charged black holes without scalars) of AdS supergravity in $D = 4$ and $D = 5$ were studied, and shown to be related to the rotating solutions of M-/string theory. In [151] the near-extreme spinning D3-brane with one angular momentum was shown to reproduce the metric and the gauge fields of the $k = 0^+$ limit of $D = 5$ gauged supergravity black holes [148], with the anticipation that the result would generalize to multiple angular momenta. However, the identification of the scalar fields was not given. In addition, in [151, 152], the equivalence of the thermodynamics of the near-extreme spinning branes and the corresponding large black holes of $D = 4, 5, 7$ gauged supergravity was given. While incomplete, these works provided some initial stages in the investigation of the sphere compactifications of M-/string theory.

Unlike black holes that are asymptotically Minkowskian, for which the horizons are always spherical, it is known that AdS black holes can also admit horizons of more general topology. Following the embedding procedure described above, we demonstrate that the AdS_4 black holes with toroidal horizon can indeed be interpreted as the near-horizon structures of an M2 brane rotating in the extra dimensions. The four charges corresponding to the $U(1)^4$ Cartan subgroup are just the four angular momenta. Similarly, the 5-dimensional charged black hole with toroidal horizon corresponds to a rotating D3-brane and the 7-dimensional charged black hole with toroidal horizon to a rotating M5-brane. In each case, the event horizon coincides with the worldvolume of the brane.[10] Additionally, one may use the Kaluza–Klein ansatz to obtain the higher-dimensional interpretation of AdS black holes with horizons of other topologies. We conjecture that these correspond to the near-horizon limits of rotating p-branes whose world-volumes have these topologies. (In fact the rotating "test" membrane in [45] had S^2 topology.)

In these lectures we also obtain the general rotating p-brane solutions in arbitrary dimensions, supported by a single $(p + 2)$-form charge, and discuss their sphere reductions. These rotating p-branes are easily constructed, merely by performing standard diagonal dimensional oxidations of the general rotating black holes that were constructed in [169].

6.2 S^5 Reduction of Type IIB Supergravity

The S^5 reduction [23–25] of Type IIB supergravity gives rise to $N = 8$, $D = 5$ gauged supergravity, with SO(6) Yang–Mills gauge group [18, 27]. The complete details of this reduction, as with any sphere reduction, would be of great complexity, and in fact no example has ever been fully worked

[10] This is a concrete realisation of the "Membrane Paradigm" [179].

out. For our present purposes, however, it suffices to consider the truncation of the five-dimensional theory to $N = 2$ supersymmetry. In this truncation, which is of course a consistent one, the gauge group is reduced down to the $U(1) \times U(1) \times U(1)$ Cartan subgroup of SO(6). The bosonic sector of the theory comprises these three gauge bosons, the metric, and two scalar fields. (The consistency of the truncation to this field content can be seen by considering the S^1 reduction of ungauged minimal nonchiral supergravity in $D = 6$, whose bosonic fields $(g_{\mu\nu}, A_{(2)}, \phi)$ reduce to give precisely the field content we are considering here in $D = 5$. After gauging, one would obtain the $U(1)^3$ gauged theory.)

Even to construct the S^5 reduction ansatz for this truncated $N = 2$ theory is somewhat nontrivial, owing to the presence of the scalar fields. It is most conveniently expressed in terms of the parameterisation of sphere metrics given in [180].

We find that the ansatz for the reduction of the ten-dimensional metric is

$$ds_{10}{}^2 = \sqrt{\tilde{\Delta}} ds_5{}^2 + \frac{1}{g^2 \sqrt{\tilde{\Delta}}} \sum_{i=1}^{3} X_i^{-1} \left(d\mu_i^2 + \mu_i^2 (d\phi_i + g A^i)^2 \right), \tag{6.1}$$

where the two scalars are parameterised in terms of the three quantities X_i, which are subject to the constraint $X_1, X_2, X_3 = 1$. They can be parameterised in terms of two dilatons φ_1 and φ_2 as

$$X_i = e^{-\vec{a}_i/2 \cdot \vec{\varphi}}, \tag{6.2}$$

where $\vec{a}_i$ satisfy the dot products

$$M_{ij} \equiv \vec{a}_i \cdot \vec{a}_j = 4\delta_{ij} - \tfrac{4}{3}. \tag{6.3}$$

A convenient choice is

$$\vec{a}_1 = \left(\frac{2}{\sqrt{6}}, \sqrt{2} \right), \quad \vec{a}_2 = \left(\frac{2}{\sqrt{6}}, -\sqrt{2} \right), \quad \vec{a}_3 = \left(-\frac{4}{\sqrt{6}}, 0 \right). \tag{6.4}$$

The three quantities μ_i are subject to the constraint $\sum_i \mu_i^2 = 1$, and the metric on the unit round 5-sphere can be written in terms of these as

$$d\Omega_5^2 = \sum_i (d\mu_i^2 + \mu_i^2 \, d\phi_i^2). \tag{6.5}$$

The μ_i can be parameterised in terms of angles on a 2-sphere, for example as

$$\mu_1 = \sin\theta, \quad \mu_2 = \cos\theta \sin\psi, \quad \mu_3 = \cos\theta \cos\psi. \tag{6.6}$$

Note that $\tilde{\Delta}$ is given by

$$\tilde{\Delta} = \sum_{i=1}^{3} X_i \mu_i^2, \tag{6.7}$$

and is therefore expressed purely in terms of the scalar fields, and the coordinates on the compactifying 5-sphere. The constant g in (6.1) is the inverse of the radius of the compactifying 5-sphere, and is equal to the gauge coupling constant. We find that the ansatz for the reduction of the 5-form field strength is $F_{(5)} = G_{(5)} + {*G_{(5)}}$, where

$$\begin{aligned} G_{(5)} = 2g \sum_i (X_i^2 \mu_i^2 - \tilde{\Delta} X_i) \epsilon_{(5)} - \frac{1}{2g} \sum_i X_i^{-1} \bar{*} dX_i \wedge d(\mu_i^2) \\ + \frac{1}{2g^2} \sum_i X_i^{-2} d(\mu_i^2) \wedge (d\phi_i + g A^i_{(1)}) \wedge \bar{*} F^i_{(2)}. \quad (6.8) \end{aligned}$$

Here, $F^i_{(2)} = dA^i_{(1)}$, $\epsilon_{(5)}$ is the volume form of the 5-dimensional metric $ds_5{}^2$, and $\bar{*}$ denotes the Hodge dual with respect to the five-dimensional metric $ds_5{}^2$.

Substituting these ansatzs into the equations for motion for the Type IIB theory, we obtain five-dimensional equations of motion that can be derived from the Lagrangian[11]

$$\begin{aligned} e^{-1}\mathcal{L}_5 = R - \frac{1}{2}(\partial\varphi_1)^2 - \frac{1}{2}(\partial\varphi_2)^2 + 4g^2 \sum_i X_i^{-1} \\ - \frac{1}{4} \sum_i X_i^{-2} (F^i_{(2)})^2 + \frac{1}{4} \epsilon^{\mu\nu\rho\sigma\lambda} F^1_{\mu\nu} F^2_{\rho\sigma} A^3_\lambda. \quad (6.9) \end{aligned}$$

(The other bosonic fields of the Type IIB theory are set to zero in this U$(1)^3$ truncated reduction.) Note that the ten-dimensional Bianchi identity $dF_{(5)} = 0$ gives rise to the equations of motion for the scalars and gauge fields in five dimensions.

Thus we have established that the reduction ansatzs (6.1) and (6.8) describe the exact embedding of the five-dimensional $N = 2$ gauged U$(1)^3$ supergravity into Type IIB supergravity.

The bosonic Lagrangian (6.9) can be further truncated down to smaller sectors. For example, we can consistently set $\varphi_2 = 0$, implying that $X_1 = X_2 = X_3^{-1/2}$, provided that $F^1_{(2)} = F^2_{(2)} = F_{(2)}/\sqrt{2}$. The Lagrangian then becomes

$$\begin{aligned} e^{-1}\mathcal{L}_5 = R - \frac{1}{2}(\partial\varphi_1)^2 + 4g^2(2e^{\varphi_1/\sqrt{6}} + e^{-2\varphi_1/\sqrt{6}}) - \frac{1}{4} e^{2\varphi_1\sqrt{6}} (F_{(2)})^2 \\ - \frac{1}{4} e^{-4\varphi_1\sqrt{6}} (F^3_{(2)})^2 + \frac{1}{8} \epsilon^{\mu\nu\rho\sigma\lambda} F_{\mu\nu} F_{\rho\sigma} A^3_\lambda. \quad (6.10) \end{aligned}$$

[11] We shall make some more detailed comments on certain general features of these spherical Kaluza–Klein reductions in Section 6.5, where we consider the S^7 reduction of $D = 11$ supergravity.

It is also possible to set both scalars to zero, implying that $X_i = 1$, provided that $F^i_{(2)} = F_{(2)}/\sqrt{3}$. The Lagrangian is then given by

$$e^{-1}\mathcal{L}_5 = R + 12g^2 - \frac{1}{4}F^2_{(2)} + \frac{1}{12\sqrt{3}}\epsilon^{\mu\nu\rho\sigma\lambda}F_{\mu\nu}F_{\rho\sigma}A_\lambda. \tag{6.11}$$

The embedding of the truncated Lagrangian (6.11) in $D = 10$ dimensions was discussed in [150].

6.3 $D = 5$ AdS Black Holes

The Lagrangian (6.9) admits a three-charge AdS black hole solution, given by [148]

$$\begin{aligned} ds_5{}^2 &= -(H_1H_2H_3)^{-2/3}f\,dt^2 + (H_1H_2H_3)^{1/3}(f^{-1}\,dr^2 + r^2\,d\Omega^2_{3,k}), \\ X_i &= H_i^{-1}(H_1H_2H_3)^{1/3}, \quad A^i_{(1)} = \sqrt{k}(1 - H_i^{-1})\coth\beta_i\,dt, \end{aligned} \tag{6.12}$$

and

$$f = k - \frac{\mu}{r^2} + g^2r^2(H_1H_2H_3), \quad H_i = 1 + \frac{\mu\sinh^2\beta_i}{kr^2}. \tag{6.13}$$

Here k can be 1, 0 or -1, corresponding to the foliating surfaces of the transverse space being S^3, T^3 or H^3, with unit metric $d\Omega^2_{3,k}$, where H^3 denotes the hyperbolic 3-space of constant negative curvature. In the case of $k = 0$, one first needs to make the rescaling [151] $\sinh^2\beta_i \to k\sinh^2\beta_i$, followed by sending k to zero. The gauge potential for the $k = 0$ case is then given by

$$A^i_{(1)} = \frac{1 - H_i^{-1}}{\sinh\beta_i}dt. \tag{6.14}$$

6.4 Rotating D3-Brane

In this section, we show that the $k = 0$ three-charge AdS black hole of the $N = 2$ gauged supergravity in $D = 5$ given in (6.12) can be embedded in $D = 10$ as a solution that is precisely the decoupling limit of the rotating D3-brane. The higher-dimensional solutions corresponding to five-dimensional AdS black holes with $k = 1$ and $k = -1$ can also be easily obtained, by substituting the five-dimensional solutions into the S^5 reduction ansatzs.

There can be three angular momenta, ℓ_i, $i = 1, 2, 3$, in the rotating D3-brane. The generic single-charge rotating p-branes can be obtained by dimensional oxidation of the generic single-charge rotating black holes constructed in [169]. See problems of Section 6.15. We find that the metric of the rotating D3-brane is given by[12]

$$\begin{aligned} ds_{10}^{\;2} &= H^{-1/2}\left(-\left(1-\frac{2m}{r^4\Delta}\right)dt^2+dx_1^2+dx_2^2+dx_3^2\right)\\ &\quad + H^{1/2}\Bigg[\frac{\Delta\, dr^2}{H_1H_2H_3-2mr^{-4}}+r^2\sum_{i=1}^{3}H_i(d\mu_i^2+\mu_i^2\,d\phi_i^2)\\ &\quad -\frac{4m\cosh\alpha}{r^4H\Delta}\,dt\left(\sum_{i=1}^{3}\ell_i\mu_i^2\,d\phi_i\right)+\frac{2m}{r^4H\Delta}\left(\sum_{i=1}^{3}\ell_i\mu_i^2\,d\phi_i\right)^2\Bigg], \end{aligned} \tag{6.15}$$

where the functions Δ, H, and H_i are given by

$$\begin{aligned} \Delta &= H_1H_2H_3\sum_{i=1}^{3}\frac{\mu_i^2}{H_i}, \qquad H = 1+\frac{2m\sinh^2\alpha}{r^4\Delta},\\ H_i &= 1+\frac{\ell_i^2}{r^2}, \qquad i=1,2,3. \end{aligned} \tag{6.16}$$

The rotating D3-brane is supported by the self-dual 5-form field strength $F_{(5)}$ of the Type IIB theory. It is given by $F_{(5)} = G_{(5)} + *G_{(5)}$, where $G_{(5)} = dB_{(4)}$ and

$$B_{(4)} = \frac{1-H^{-1}}{\sinh\alpha}\left(-\cosh\alpha\, dt+\sum_{i=1}^{3}\ell_i\mu_i^2\,d\phi_i\right)\wedge d^3x. \tag{6.17}$$

As is well known, the nonrotating D3-brane has a "decoupling limit" where the spacetime of the D3-brane becomes a product space $M_5\times S^5$. If the D3-brane is extremal, M_5 is a five-dimensional anti-de Sitter spacetime. More generally, when the D3-brane is nonextremal, M_5 is the Carter-Novotny-Horsky metric [181], which can thus be viewed as a "nonextremal" generalisation of AdS$_5$. A similar limit also exists for the rotating D3-brane, and can be achieved by making the rescalings

$$\begin{aligned} &m\to\epsilon^4 m, \quad \sinh\alpha\to\epsilon^{-2}\sinh\alpha,\\ &r\to\epsilon r, \quad x^\mu\to\epsilon^{-1}x^\mu, \quad \ell_i\to\epsilon\ell_i, \end{aligned} \tag{6.18}$$

and then sending $\epsilon\to 0$. (Note that when this limit is taken, we also have $\cosh\alpha\to\epsilon^{-2}\sinh\alpha$.) This has the effect that the last term in (6.15) is set to zero and that

$$H = 1+\frac{2m\sinh^2\alpha}{r^4\Delta}\;\to\;\frac{2m\sinh^2\alpha}{r^4\Delta}. \tag{6.19}$$

---

[12]This metric agrees with previously obtained results [175, 176], after correcting some typographical errors.

In this limit, the metric (6.15) becomes

$$ds_{10}^2 = \sqrt{\tilde{\Delta}}[-(H_1H_2H_3)^{-2/3} f\, dt^2 + (H_1H_2H_3)^{1/3}(f^{-1}\, dr^2 + r^2\, d\vec{y}\cdot d\vec{y})] \\ + \frac{1}{g^2\sqrt{\tilde{\Delta}}} \sum_{i=1}^{3} X_i^{-1}\big(d\mu_i^2 + \mu_i^2(d\phi_i + gA^i)^2\big), \quad (6.20)$$

where

$$\vec{y} = g\vec{x}, \quad g^2 = \frac{1}{\sqrt{2m}\sinh\alpha}, \qquad \mu = 2mg^2. \quad (6.21)$$

The metric (6.20) precisely matches the dimensional reduction ansatz (6.1), with the lower dimensional fields given by

$$ds_5^2 = -(H_1H_2H_3)^{-2/3} f\, dt^2 + (H_1H_2H_3)^{1/3}(f^{-1}\, dr^2 + r^2\, d\vec{y}\cdot d\vec{y}), \\ X_i = H_i^{-1}(H_1H_2H_3)^{1/3}, \quad A^i_{(1)} = \frac{1-H_i^{-1}}{g\ell_i \sinh\alpha}\, dt, \quad (6.22)$$

where

$$f = -\frac{\mu}{r^2} + g^2r^2H_1H_2H_3, \quad g^2 = \frac{1}{\sqrt{2m}\sinh\alpha}, \qquad \mu = 2mg^2. \quad (6.23)$$

To complete the story, we note that the 5-form field strength in the decoupling limit is given by $F_{(5)} = G_{(5)} + {*G_{(5)}}$, where $G_{(5)} = dB_{(4)}$ and

$$B_{(4)} = -g^4r^4\Delta\, dt\wedge d^3x + \frac{1}{\sinh\alpha}\Big(\sum_{i=1}^{3} \ell_i\mu_i^2\, d\phi_i\Big)\wedge d^3x. \quad (6.24)$$

This gives precisely the field strength in the dimensional reduction ansatz (6.8).

Thus we see that the solution (6.22) is precisely the $k = 0$ three-charge AdS black hole given in the previous subsection, after reparameterising the angular momenta $\ell_i^2 = \mu\sinh^2\beta_i$. This shows that the embedding of the three-charge AdS $k = 0$ black hole in gauged $N = 2$ supergravity in five dimensions gives a ten-dimensional solution that is precisely the decoupling limit of the rotating D3-brane. Single-charge AdS black holes coming from the reduction of the metric of a rotating D3-brane with one angular momentum were obtained in [151], however without the explicit embedding of the scalar fields. The connection between the thermodynamics of AdS black holes and rotating p-branes was discussed in [151, 152].

It is also straightforward to oxidise the $k = 1$ and $k = -1$ AdS black holes back to $D = 10$ Type IIB. The metric is the same form as (6.20) with $d\vec{y}\cdot d\vec{y}$ replaced by the unit metric for S^3 or H^3 respectively. The 5-form field strength follows by substituting the five-dimensional fields into (6.8).

6.5 S^7 Reduction of $D = 11$ Supergravity

The S^7 reduction of eleven-dimensional supergravity gives rise to SO(8) gauged $N = 8$ supergravity in four dimensions. One may again consider a consistent truncation to $N = 2$, for which the bosonic sector comprises the metric, four commuting U(1) gauge potentials, three dilatons and three axions. (That this is a consistent truncation can be seen by reducing minimal nonchiral six-dimensional supergravity on T^2, for which the reduction of $(g_{\mu\nu}, A_{(2)}, \phi)$ will give precisely the field content we are considering. After gauging, this would give the $U(1)^4$ gauged theory. See Appendix F for an extended discussion of this.) We have not yet determined the complete reduction ansatz for the entire truncated theory where the axions are included, but we can give the exact ansatz in the case where one sets the axions to zero. This will not, of course, be a consistent truncation, since the U(1) gauge fields will provide source terms of the form $\epsilon^{\mu\nu\rho\sigma} F_{\mu\nu} F_{\rho\sigma}$ for the axions. Nevertheless, one can use the axion-free ansatz for discussing the exact embedding of four-dimensional solutions for which the axions are zero. The full $N = 2$ four-dimensional theory, including the axions, is obtained in Appendix F.

The reduction ansatz for the eleven-dimensional metric is

$$ds_{11}{}^2 = \tilde{\Delta}^{2/3}\, ds_4{}^2 + g^{-2}\tilde{\Delta}^{-1/3} \sum_i X_i^{-1}\big(d\mu_i^2 + \mu_i^2 (d\phi_i + g A_{(1)}^i)^2\big), \tag{6.25}$$

where $\tilde{\Delta} = \sum_{i=1}^4 X_i \mu_i^2$. The four quantities μ_i satisfy $\sum_i \mu_i^2 = 1$. They can be parameterised in terms of angles on the 3-sphere as

$$\begin{aligned} \mu_1 &= \sin\theta, & \mu_2 &= \cos\theta \sin\varphi, \\ \mu_3 &= \cos\theta\cos\varphi\sin\psi, & \mu_4 &= \cos\theta\cos\varphi\cos\psi. \end{aligned} \tag{6.26}$$

The four X_i, which satisfy $X_1 X_2 X_3 X_4 = 1$, can be parameterised in terms of three dilatonic scalars $\vec{\varphi} = (\varphi_1, \varphi_2, \varphi_3)$:

$$X_i = e^{-(\vec{a}_i \cdot \vec{\varphi})/2}, \tag{6.27}$$

where the $\vec{a}_i$ satisfy the dot products

$$M_{ij} \equiv \vec{a}_i \cdot \vec{a}_j = 4\delta_{ij} - 1. \tag{6.28}$$

A convenient choice, corresponding to the combinations of (F.11), is

$$\begin{aligned} \vec{a}_1 &= (1,1,1), & \vec{a}_2 &= (1,-1,-1), \\ \vec{a}_3 &= (-1,1,-1), & \vec{a}_4 &= (-1,-1,1). \end{aligned} \tag{6.29}$$

The reduction ansatz for the 4-form field strength is

$$\begin{aligned} F_{(4)} = 2g \sum_i (X_i^2 \mu_i^2 - \tilde{\Delta} X_i)\epsilon_{(4)} &+ \frac{1}{2g}\sum_i X_i^{-1} \bar{*} dX_i \wedge d(\mu_i^2) \\ &- \frac{1}{2g^2}\sum_i X_i^{-2}\, d(\mu_i^2) \wedge (d\phi_i + g A_{(1)}^i) \wedge \bar{*} F_{(2)}^i. \end{aligned} \tag{6.30}$$

Here, $\bar{*}$ denotes the Hodge dual with respect to the four-dimensional metric ds_4^2, and $\epsilon_{(4)}$ denotes its volume form.[13]

It is of interest to note that the eleven-dimensional Bianchi identity $dF_{(4)} = 0$ already gives rise to the four-dimensional equations of motion for the scalars and gauge potentials, namely

$$\begin{aligned} d\bar{*}d\log(X_i) &= \frac{1}{4}\sum_j M_{ij} X_j^{-2} \bar{*}F_{(2)}^j \wedge F_{(2)}^j + g^2 \sum_{j,k} M_{ij} X_j X_k \\ &\qquad - g^2 \sum_j M_{ij} X_j^2, \\ d(X_i^{-2} \bar{*}F_{(2)}^i) &= 0. \end{aligned} \tag{6.31}$$

It is straightforward to see that these equations of motion can be obtained from the four-dimensional Lagrangian

$$\begin{aligned} e^{-1}\mathcal{L}_4 = R - \frac{1}{2}(\partial\vec{\varphi})^2 + 8g^2(\cosh\varphi_1 + \cosh\varphi_2 + \cosh\varphi_3) \\ - \frac{1}{4}\sum_{i=1}^{4} e^{\vec{a}_i\cdot\vec{\varphi}}(F_{(2)}^i)^2. \end{aligned} \tag{6.32}$$

One might think that it would be possible to obtain the four-dimensional Lagrangian by substituting the ansatzs (6.25) and (6.30) into the eleven-dimensional Lagrangian. In fact this is not the case, and one must work at the level of the eleven-dimensional equations of motion. One way of understanding this is from the fact that the ansatz for $F_{(4)}$ does not *identically* satisfy the Bianchi identity. Rather, as we have seen, it satisfies it modulo the use of the four-dimensional equations of motion for the scalars and gauge fields. In other words, the ansatz is made on the eleven-dimensional 4-form $F_{(4)}$ rather than on the fundamental potential $A_{(3)}$ itself. Consequently, it would not be correct to insert the ansatz for $F_{(4)}$ into the Lagrangian.

We may further illustrate this point by showing, as an example, how the scalar potential arises in the four-dimensional Einstein equation. This comes from considering the eleven-dimensional Einstein equation,

$$\hat{R}_{AB} - \frac{1}{2}\hat{R}\eta_{AB} = \frac{1}{12}\left(F_{AB}^2 - \frac{1}{8}F^2\eta_{AB}\right), \tag{6.33}$$

with vielbein indices A, B ranging just over the four-dimensional spacetime directions α, β. From the ansatz (6.25), the relevant terms in the eleven-

[13] If the ansatzs (6.25) and (6.30) are linearised around an $\mathrm{AdS}_4 \times S^7$ background, they can be seen to be in agreement with previous results that were derived at the linear level [13]. The full nonlinear metric ansatz (6.25) should be in agreement with the appropriate specialisation of the ansatz given in [182].

dimensional Ricci tensor and Ricci scalar are given by

$$\hat{R}_{\alpha\beta} = \frac{4g^2}{3\tilde{\Delta}^{8/3}}\Big[-\Big(\sum_i X_i^2\mu_i^2\Big)^2 + \tilde{\Delta}\sum_i X_i^2\mu_i^2\sum_j X_j + \tilde{\Delta}\sum_i X_i^3\mu_i^2 - \tilde{\Delta}^2\sum_i X_i^2\Big]\eta_{\alpha\beta} + \tilde{\Delta}^{-2/3}R_{\alpha\beta} + \cdots, \qquad (6.34)$$

$$\hat{R} = \frac{2g^2}{3\tilde{\Delta}^{8/3}}\Big[-\Big(\sum_i X_i^2\mu_i^2\Big)^2 - 2\tilde{\Delta}\sum_i X_i^2\mu_i^2\sum_j X_j + 4\tilde{\Delta}\sum_i X_i^3\mu_i^2 + 6\tilde{\Delta}^2\Big(\sum_i X_i\Big)^2 - 7\tilde{\Delta}^2\sum_i X_i^2\Big] + \tilde{\Delta}^{-2/3}R + \cdots, \qquad (6.35)$$

where $R_{\alpha\beta}$ and R are the four-dimensional Ricci tensor and scalar, and the ellipsis indicate that terms not involving purely the undifferentiated scalars have been omitted for the purposes of the present illustrative discussion. From the ansatz (6.30) for the 4-form, the eleven-dimensional energy-momentum tensor vielbein components in the four-dimensional spacetime directions are given by

$$\frac{1}{12}\Big(F_{\alpha\beta}^2 - \frac{1}{8}F^2\eta_{\alpha\beta}\Big) = -g^2\tilde{\Delta}^{-8/3}\Big(\sum_i (X_i^2\mu_i^2 - \tilde{\Delta}X_i)\Big)^2\eta_{\alpha\beta}. \qquad (6.36)$$

Substituting (6.34), (6.35) and (6.36) into (6.33), we find that all the angular dependence coming from the μ_i variables cancels, and that the scalar potential terms in the four-dimensional Einstein equation are given by

$$R_{\alpha\beta} - \tfrac{1}{2}R\eta_{\alpha\beta} = -\tfrac{1}{2}g^2V\eta_{\alpha\beta}, \qquad (6.37)$$

with V given by

$$V = -4\sum_{i<j} X_iX_j = -8(\cosh\varphi_1 + \cosh\varphi_2 + \cosh\varphi_3). \qquad (6.38)$$

Since (6.37) derives from the Lagrangian $R - g^2, V$, we see that we have precisely produced the hoped-for potential terms of the gauged supergravity Lagrangian (6.32). This sample calculation also serves to illustrate that the angular dependence coming from the μ_i variables would not have cancelled if we had merely substituted the ansatzs (6.25) and (6.30) into the eleven-dimensional Lagrangian. It also shows that the cancellation of the μ_i dependence in the higher-dimensional equations of motion depends crucially on "conspiracies" between the contributions from the metric and the 4-form field strength. This is quite different from the situation in toroidal reductions, where each term in the higher-dimensional theory reduces consistently by itself, without the need for any such conspiracies. Note, furthermore, that the required conspiracies needed for the success of the spherical reduction depend on the 4-form field strength occurring with precisely the correct coefficient relative to the Einstein–Hilbert term. This normalisation is not a free parameter, but is governed by the strength of the FFA term in the eleven-dimensional theory. Thus ultimately the consistency of the spherical reduction ansatz can be traced back to the supersymmetry of the eleven-dimensional theory.

Note that the Lagrangian (6.32) can be further truncated, to pure Einstein–Maxwell with a cosmological constant, by setting all the field strengths equal, $F^i_{(2)} = \frac{1}{2}F_{(2)}$, and setting all the scalars to zero:

$$e^{-1}\mathcal{L}_4 = R - \tfrac{1}{4}(F_{(2)})^2 + 24g^2. \tag{6.39}$$

The embedding of this theory into $D = 11$ supergravity was obtained in [159]. The ansatz for the metric and field strength for the embedding in [159] was given in terms of a decomposition of the 7-sphere as a U(1) bundle over CP^3. This is identical, after a transformation of coordinates, to the Einstein–Maxwell embedding given in [150]. In the same spirit, the S^5 reduction to five-dimensional Einstein–Maxwell can be described using the method presented in [159], with S^5 viewed as a U(1) bundle over CP^2. (An analogous consistent embedding of four-dimensional Einstein–Yang–Mills with an SU(2) gauge group, and a cosmological constant, in $D = 11$ supergravity was obtained in [183]. This involves a decomposition of S^7 as an SU(2) bundle over S^4.)

6.6 $D = 4$ AdS Black Holes

The $D = 4$, $N = 2$ gauged supergravity coupled to three vector multiplets admits 4-charge AdS black hole solutions, given by [149, 153]

$$\begin{aligned} ds_4^2 &= -(H_1H_2H_3H_4)^{-1/2} f\, dt^2 \\ &\qquad\qquad + (H_1H_2H_3H_4)^{1/2}(f^{-1}dr^2 + r^2 d\Omega_{2,k}^2), \\ X_i &= H_i^{-1}(H_1H_2H_3H_4)^{1/4}, \quad A^i_{(1)} = \sqrt{k}(1 - H_i^{-1}) \coth\beta_i\, dt, \end{aligned} \tag{6.40}$$

and

$$f = k - \frac{\mu}{r} + 4g^2r^2(H_1H_2H_3H_4), \quad H_i = 1 + \frac{\mu \sinh^2\beta_i}{kr}. \tag{6.41}$$

Here, k can be 1, 0 or -1, corresponding to the cases where the foliations in the transverse space have the metric $d\Omega_{2,k}^2$ on the unit S^2, T^2 or H^2, where H^2 is the unit hyperbolic 2-space of constant negative curvature. In the case of $k = 0$, one must first make the rescaling $\sinh^2\beta_i \to k\sinh^2\beta_i$ before sending k to zero. The gauge potential for the $k = 0$ case is then given by

$$A^i_{(1)} = \frac{1 - H_i^{-1}}{\sinh\beta_i}\, dt. \tag{6.42}$$

6.7 Rotating M2-Brane

There are four angular momenta, ℓ_i, $i = 1,2,3,4$, in the rotating M2-brane. The solution can be obtained by oxidising the $D = 9$ rotating black hole [169]. After the oxidation, we find that the metric of the rotating M2-brane is given by

$$\begin{aligned} ds_{11}{}^2 &= H^{-2/3}\Big(-\Big(1-\frac{2m}{r^6\Delta}\Big)\,dt^2 + dx_1^2 + dx_2^2\Big) \\ &\quad + H^{1/3}\Bigg[\frac{\Delta\,dr^2}{H_1H_2H_3H_4 - \frac{2m}{r^6}} + r^2\sum_{i=1}^{4} H_i(d\mu_i^2 + \mu_i^2\,d\phi_i^2) \\ &\quad - \frac{4m\cosh\alpha}{r^6 H\Delta}dt\Big(\sum_{i=1}^{4}\ell_i\mu_i^2\,d\phi_i\Big) + \frac{2m}{r^4 H\Delta}\Big(\sum_{i=1}^{4}\ell_i\mu_i^2\,d\phi_i\Big)^2\Bigg], \end{aligned} \tag{6.43}$$

where the functions Δ, H and H_i are given by

$$\begin{aligned} \Delta &= H_1H_2H_3H_4\sum_{i=1}^{4}\frac{\mu_i^2}{H_i}, \qquad H = 1 + \frac{2m\sinh^2\alpha}{r^6\Delta}, \\ H_i &= 1 + \frac{\ell_i^2}{r^2}, \qquad i = 1,2,3,4. \end{aligned} \tag{6.44}$$

The 3-form gauge potential is given by

$$A_{(3)} = \frac{1-H^{-1}}{\sinh\alpha}(-\cosh\alpha\,dt + \ell_i\mu_i^2\,d\phi_i)\wedge d^2x. \tag{6.45}$$

Following the previous D3-brane example, we consider the decoupling limit, which is obtained by making the rescaling

$$\begin{aligned} m &\to \epsilon^6 m, \quad \sinh\alpha \to \epsilon^{-3}\sinh\alpha, \\ r &\to \epsilon r, \quad x^\mu \to \epsilon^{-2}x^\mu, \quad \ell_i \to \epsilon\ell_i \end{aligned} \tag{6.46}$$

and then sending $\epsilon \to 0$. This has the effect that the last term in (6.43) is set to zero and that the 1 in function H (6.44) is removed. In this limit, the rotating M2-brane (6.43) becomes

$$\begin{aligned} ds_{11}{}^2 &= \tilde{\Delta}^{2/3}[-(H_1H_2H_3H_4)^{-1/2} f\,dt^2 \\ &\quad + (H_1H_2H_3H_4)^{1/2}(f^{-1}\,d\rho^2 + \rho^2 d\vec{y}\cdot d\vec{y})] \\ &\quad + g^{-2}\tilde{\Delta}^{-1/3}\sum_{i=1}^{4} X_i^{-1}\big(d\mu_i^2 + \mu_i^2(d\phi_i + gA^i)^2\big), \end{aligned} \tag{6.47}$$

where

$$\begin{aligned} \rho &= \tfrac{1}{2}gr^2, \quad \vec{y} = 2g\vec{x}, \quad f = -\frac{\mu}{\rho} + 4g^2\rho^2 H_1H_2H_3H_4, \\ g^2 &= (2m\sinh^2\alpha)^{-1/3}, \quad \mu = mg^5, \quad \tilde{\Delta} = \sum_i X_i\mu_i^2. \end{aligned} \tag{6.48}$$

This is precisely of the form of the metric ansatz in the dimensional reduction given by (6.25). The lower dimensional fields are given by

$$\begin{aligned} ds_4^2 &= -(H_1 H_2 H_3 H_4)^{-1/2} f\, dt^2 \\ &\qquad + (H_1 H_2 H_3 H_4)^{1/2} (f^{-1} d\rho^2 + \rho^2 d\vec{x} \cdot d\vec{x}) \\ X_i &= H_i^{-1} (H_1 H_2 H_3 H_4)^{1/4}, \quad A^i = \frac{1 - H_i^{-1}}{g\ell_i \sinh\alpha} dt. \end{aligned} \tag{6.49}$$

In the decoupling limit, the gauge potential $A_{(3)}$ given in (6.45) for the rotating M2-brane becomes, after a gauge transformation,

$$A_{(3)} = -g^6 r^6 \Delta\, dt \wedge d^2x + \frac{1}{\sinh\alpha} \sum_i \ell_i \mu_i^2\, d\phi_i \wedge d^2x. \tag{6.50}$$

We find that its field strength $F_{(4)} = dA_{(3)}$ is also of the form given in (6.30) for the dimensional reduction ansatz. Thus we have established an exact embedding of four-dimensional nonextremal 4-charge AdS black holes into eleven-dimensional supergravity, and furthermore, that they become precisely the decoupling limit of the rotating M2-branes. It should, of course, be emphasised that the four-dimensional AdS black holes that we are considering at this point have T^2 rather than S^2 horizons, corresponding to $k = 0$ in (6.40) and (6.41).

It is also straightforward to oxidise the $k = 1$ and $k = -1$ AdS black hole solutions back to $D = 11$, by substituting the four-dimensional fields into the ansatzs (6.25) and (6.30).

6.8 S^4 reduction of $D = 11$ Supergravity

The Kaluza–Klein reduction of eleven-dimensional supergravity on S^4 gives rise to $N = 4$ gauged SO(5) supergravity in seven dimensions. In a similar manner to the S^5 and S^7 reductions that we discussed previously, we may consider an $N = 2$ truncation of this seven-dimensional theory. As described in the introduction, the truncated theory comprises $N = 2$ supergravity coupled to a vector multiplet, comprising the metric, 2-form potential, four vector potentials and four scalars in total. For our present purposes, we shall focus on a further truncation where only the metric, two gauge potentials (which are associated with the $U(1) \times U(1)$ Cartan subgroup of SO(5)) and two scalars are retained. This is not in general a consistent truncation, but, as in the case of the neglect of the axions in the S^7 reduction, it is consistent for a subset of solutions where the truncated fields are not excited by the ones that are retained. In particular, solutions of the $N = 2$ theory for which $F^1_{(2)} \wedge F^2_{(2)} = 0$, such as the AdS black holes, will also be solutions of this truncated theory.

We find that we can obtain this truncated theory by making the following Kaluza–Klein S^4-reduction ansatz:

$$ds_{11}{}^2 = \tilde{\Delta}^{1/3}\, ds_7{}^2 + g^{-2}\tilde{\Delta}^{-2/3}\Big(X_0^{-1}\, d\mu_0^2 + \sum_{i=1}^{2} X_i^{-1}\big(d\mu_i^2 + \mu_i^2(d\phi_i + gA_{(1)}^i)^2\big)\Big), \quad (6.51)$$

$$\begin{aligned} * F_{(4)} &= 2g \sum_{\alpha=0}^{2}(X_\alpha^2\mu_\alpha^2 - \tilde{\Delta}X_\alpha)\epsilon_{(7)} + g\tilde{\Delta}X_0\epsilon_{(7)} \\ &+ \frac{1}{2g}\sum_{\alpha=0}^{2} X_\alpha^{-1}\bar{*}dX_\alpha \wedge d(\mu_\alpha^2) \\ &+ \frac{1}{2g^2}\sum_{i=1}^{2} X_i^{-2}\, d(\mu_i^2)\wedge(d\phi_i + gA_{(1)}^i)\wedge\bar{*}F_{(2)}^i, \end{aligned} \quad (6.52)$$

where we have defined the auxiliary variable $X_0 \equiv (X_1X_2)^{-2}$. Here, $\bar{*}$ denotes the Hodge dual with respect to the seven-dimensional metric $ds_7{}^2$, $\epsilon_{(7)}$ denotes its volume form, and $*$ denotes the Hodge dualisation in the eleven-dimensional metric. The quantity $\tilde{\Delta}$ is given by

$$\tilde{\Delta} = \sum_{\alpha=0}^{2} X_\alpha\mu_\alpha^2, \quad (6.53)$$

where μ_0, μ_1 and μ_2 satisfy $\mu_0^2+\mu_1^2+\mu_2^2 = 1$. The two scalar fields X_i can be parameterised in terms of two canonically-normalised dilatons $\vec{\varphi} = (\varphi_1, \varphi_2)$ by writing

$$X_i = e^{-(\vec{a}_i\cdot\vec{\varphi})/2}, \quad (6.54)$$

where the dilaton vectors satisfy the relations $\vec{a}_i\cdot\vec{a}_j = 4\delta_{ij} - \frac{8}{5}$. A convenient parameterisation is given by

$$\vec{a}_1 = \Big(\sqrt{2}, \sqrt{\tfrac{2}{5}}\Big), \quad \vec{a}_2 = \Big(-\sqrt{2}, \sqrt{\tfrac{2}{5}}\Big). \quad (6.55)$$

Note that the two X_i are independent here, unlike in the cases of the three X_i in $D = 5$ or the four X_i in $D = 4$, which satisfied $\prod_i X_i = 1$. The auxiliary variable X_0 that we have introduced in order to make the expressions more symmetrical can be written as $X_0 = e^{-(\vec{a}_0\cdot\vec{\varphi})/2}$, where $\vec{a}_0 = -2(\vec{a}_1 + \vec{a}_2) = (0, -4\sqrt{2/5})$.

After substituting into the eleven-dimensional equations of motion, one obtains seven-dimensional equations that can be derived from the Lagrangian

$$e^{-1}\mathcal{L}_7 = R - \frac{1}{2}(\partial\vec{\varphi})^2 - g^2V - \frac{1}{4}\sum_{i=1}^{2} e^{\vec{a}_i\cdot\vec{\varphi}}(F_{(2)}^i)^2, \quad (6.56)$$

where the potential V is given by

$$V = -4X_1X_2 - 2X_1^{-1}X_2^{-2} - 2X_2^{-1}X_1^{-2} + \tfrac{1}{2}(X_1X_2)^{-4}. \quad (6.57)$$

This potential has a more complicated structure than those in the $D = 5$ and $D = 4$ gauged theories, and in particular it has not only a maximum at $X_1 = X_2 = 1$, but also a saddle point at $X_1 = X_2 = 2^{-1/5}$ [184]. Note that by making use of the auxiliary variable $X_0 = (X_1 X_2)^{-2}$, the potential can be re-expressed as

$$V = -4X_1X_2 - 2X_0X_1 - 2X_0X_2 + \tfrac{1}{2}X_0^2. \tag{6.58}$$

It is interesting to note that the Lagrangian (6.56) can be further consistently truncated, by setting $X_1 = X_2 = X$, and $F^1_{(2)} = F^2_{(2)} = F_{(2)}/\sqrt{2}$. This implies that the dilatonic scalar φ_1 is set to zero, in terms of the parameterisation defined by (6.55). This gives

$$e^{-1}\mathcal{L}_7 = R - \tfrac{1}{2}(\partial\varphi_2)^2 + g^2(4X^2 + 4X^{-3} - \tfrac{1}{2}X^{-8}) - \tfrac{1}{4}X^{-2}(F_{(2)})^2, \tag{6.59}$$

where

$$X = e^{-\varphi_2/\sqrt{10}}. \tag{6.60}$$

This scalar potential was used in [185] to construct supersymmetric domain wall solutions.

6.9 $D = 7$ AdS Black Holes

This Lagrangian (6.56) admits 2-charge AdS black-hole solutions, given by

$$\begin{aligned} ds_7^2 &= -(H_1H_2)^{-4/5} f\, dt^2 + (H_1H_2)^{1/5}(f^{-1}\, dr^2 + r^2\, d\Omega_{5,k}{}^2), \\ f &= k - \frac{\mu}{r^4} + \frac{1}{4}g^2r^2H_1H_2, \quad X_i = (H_1H_2)^{2/5}H_i^{-1}, \\ A^i_{(1)} &= \sqrt{k}\coth\beta_i(1 - H_i^{-1})\, dt, \quad H_i = 1 + \frac{\mu\sinh^2\beta_i}{r^4}, \end{aligned} \tag{6.61}$$

where $d\Omega_{5,k}{}^2$ is the metric on a unit S^5, T^5 or H^5 according to whether $k = 1, 0$ or -1. As in the previous cases we discussed, the $k = 0$ solution is obtained by first rescaling $\sinh^2\beta_i \to k\sinh^2\beta_i$ before setting $k = 0$. The metric of the $D = 7$ AdS black hole was obtained in [151], by isolating the spacetime direction of the rotating M5-brane metric.

6.10 Rotating M5-Brane

There are two angular momenta, ℓ_1 and ℓ_2, in the rotating M5-brane [170, 177]. Its metric is given by

$$\begin{aligned} &{ds_{11}}^2 \\ &= H^{-1/3}\left(-\left(1-\frac{2m}{r^3\Delta}\right)dt^2+dx_1^2+\cdots+dx_5^2\right) \\ &\quad + H^{2/3}\left[\frac{\Delta\, dr^2}{H_1H_2-\frac{2m}{r^3}}+r^2\left(d\mu_0^2+\sum_{i=1}^{2}H_i(d\mu_i^2+\mu_i^2\,d\phi_i^2)\right)\right. \\ &\qquad \left. -\frac{4m\cosh\alpha}{r^3H\Delta}\,dt\left(\sum_{i=1}^{2}\ell_i\mu_i^2\,d\phi_i\right)^2+\frac{2m}{r^3H\Delta}\left(\sum_{i=1}^{2}\ell_i\mu_i^2\,d\phi_i\right)^2\right], \end{aligned} \tag{6.62}$$

where Δ, H and H_i are given by

$$\begin{gathered} \Delta = H_1H_2\left(\mu_0^2+\frac{\mu_1^2}{H_1}+\frac{\mu_2^2}{H_2}\right), \qquad H = 1+\frac{2m\sinh^2\alpha}{r^3\Delta}, \\ H_1 = 1+\frac{\ell_1^2}{r^2}, \qquad H_2 = 1+\frac{\ell_2^2}{r^2}. \end{gathered} \tag{6.63}$$

The three quantities μ_0, μ_1 and μ_2 satisfy $\mu_0^2+\mu_1^2+\mu_2^2=1$. The 4-form field strength is given by $F_4 = *dA_6$, where

$$A_6 = \frac{1-H^{-1}}{\sinh\alpha}(\cosh\alpha\,dt+\ell_1\mu_1^2\,d\phi_1+\ell_2\mu_2^2\,d\phi_2)\wedge d^5x. \tag{6.64}$$

The decoupling limit is defined by

$$\begin{gathered} m\to\epsilon^3 m, \qquad \sinh\alpha\to\epsilon^{-3/2}\sinh\alpha, \\ r\to\epsilon r, \qquad x^\mu\to\epsilon^{-1/2}x^\mu, \qquad \ell_i\to\epsilon\ell_i, \end{gathered} \tag{6.65}$$

with $\epsilon\to 0$. In this limit, the metric becomes

$$\begin{aligned} &{ds_{11}}^2 \\ &= \tilde\Delta^{1/3}[-(H_1H_2)^{-4/5}f\,dt^2+(H_1H_2)^{1/5}(f^{-1}\,d\rho^2+\rho^2\,d\vec y\cdot d\vec y)] \\ &\quad + g^{-2}\tilde\Delta^{-2/3}\left((X_1X_2)^2\,d\mu_0^2+\sum_{i=1}^{2}X_i^{-1}\left(d\mu_i^2+\mu_i^2(d\phi_i+gA_{(1)}^i)^2\right)\right), \end{aligned} \tag{6.66}$$

where

$$\begin{gathered} \rho^2 = 4rg^{-1}, \quad \vec y = \tfrac{1}{2}g\vec x, \quad \tilde\Delta = (X_1X_2)^{-2}\mu_0^2+X_1\mu_1^2+X_2\mu_2^2, \\ g^2 = (2m\sinh^2\alpha)^{-2/3}, \quad \mu = 32mg^{-1}. \end{gathered} \tag{6.67}$$

The metric (6.66) fits precisely the dimensional reduction ansatz given in (6.51). The lower dimensional fields are given by

$$\begin{aligned} ds_7^2 &= -(H_1H_2)^{-4/5} f\, dt^2 + (H_1H_2)^{1/5}(f^{-1}\, d\rho^2 + \rho^2\, d\vec{y}\cdot d\vec{y}), \\ X_i &= (H_1H_2)^{2/5} H_i^{-1}, \quad f = -\frac{\mu}{\rho^4} + \frac{1}{4} g^2 \rho^2 H_1 H_2, \\ A^i_{(1)} &= \frac{1 - H_i^{-1}}{g\ell_i \sinh\alpha}\, dt. \end{aligned} \tag{6.68}$$

This is precisely the $k = 0$ AdS_7 black hole obtained in the previous section, with the angular momenta reparameterised as $\ell_i = \mu g^2 \sinh^2 \beta_i / 16$. This establishes that the 2-charge $k = 0$ AdS black hole in $D = 7$ can be reinterpreted as the decoupling limit of the rotating M2-brane. (Of course in this example, one can *only* discuss the embedding when the scalar fields are included, since there is no choice of charge parameters for which the scalar fields vanish in the seven-dimensional black holes. This contrasts with the cases of the rotating D3-branes and M5-branes, where the special choice of setting all the charges equal allows the discussion of a simplified ansatz where the scalars are omitted.)

6.11 *Charge as Angular Momentum*

To those students of Kaluza–Klein theory who are used to the idea that electric charge is *momentum* in an extra dimension, it may come as a bit of a shock to learn that the electric charges of the AdS black holes correspond to *angular momenta* in extra dimensions. My understanding of this is as follows and it relies crucially on the interpretation of $AdS_{\tilde{d}+1} \times S^{d+1}$ vacua as the near-horizon limit of $(\tilde{d}-1)$-brane geometries.

Momentum and angular momentum correspond to generators of the Poincaré group or AdS group, and hence to spacetimes that are asymptotically Minkowski or AdS. If the spacetime is asymptotically M^D, for example, it makes sense to talk about angular momentum in D dimensions. On the other hand, if the spacetime is asymptotically $AdS_{\tilde{d}+1} \times S^{d+1}$, it makes sense to talk about angular momentum in $(\tilde{d}+1)$ dimensions but not in $D = (\tilde{d} + d + 2)$ dimensions. Of course, there is an $SO(d+2)$ symmetry but this is just the isometry group of S^{d+1}; I see no reason (yet) to call this angular momentum.[14]

The ability to call $SO(d+2)$ angular momentum depends crucially on the fact that the $AdS_{\tilde{d}+1}$ vacua are near horizon geometries of $(\tilde{d}-1)$-branes and that these brane geometries tend asymptotically to $M^{\tilde{d}+d+2}$. We can now legitimately refer to S^{d+1} as the rotation group in the d dimensions

[14]Somewhat confusingly, some authors refer to the black holes as "spinning in the S^{d+1} directions" but this is, in my opinion, an abuse of language.

transverse to the branes, and if we allow these branes to spin it will describe their angular momentum.

(Having said this, one should probably not refer to the electric charge of ordinary five-dimensional Kaluza–Klein as "momentum in the fifth direction" either. After all, the symmetry group of $M^4 \times S^1$ is $P_4 \times \mathrm{U}(1)$ and not P_5.)

6.12 Magnetic Black Holes

We have seen that the $N = 8$ gauged supergravity naturally admits a four-electric-charge black hole solution. In fact it turns out that this solution is easily generalized to give magnetically charged black holes; although the full theory involves non-Abelian SO(8) gauge fields, the $\mathrm{U}(1)^4$ truncation of (6.32) gives rise to bosonic equations of motion that are symmetric under the electric-magnetic duality

$$F^i \to X_i^{-2} * F^i, \quad X_i \to X_i^{-1}. \tag{6.69}$$

The resulting four magnetic charge solution in the case $k = 1$ has the form

$$\begin{aligned} ds^2 &= -(H_1H_2H_3H_4)^{-1/2} f dt^2 \\ &\qquad + (H_1H_2H_3H_4)^{1/2}(f^{-1}dr^2 + r^2 d\Omega^2), \\ X_i &= H_i(H_1H_2H_3H_4)^{-1/4} \\ H_i &= 1 + \frac{\mu \sinh^2 \beta_i}{r}, \quad f = 1 - \frac{\mu}{r} + 4g^2r^2(H_1H_2H_3H_4), \\ F^i_{\theta\phi} &= \mu \cosh \beta_i \sinh \beta_i \sin \theta. \end{aligned} \tag{6.70}$$

While the extremal limit is once again reached by taking $\mu \to 0$ and $\beta_i \to \infty$ with $P_\alpha \equiv \mu \sinh^2 \beta_i$ fixed, the resulting extremal black hole is in fact not supersymmetric whenever $g \neq 0$! In the case of the magnetic Reissner–Nordström black hole, this phenomenon was previously found in [91]. (Note, however, that it is possible to obtain magnetic black holes that do preserve some supersymmetry if one allows for event horizons with nonspherical topologies [194].) To see that (6.70) admits no Killing spinors, we note that while the scalar potential is symmetric under $X_i \to X_i^{-1}$, the scalar related terms in the supersymmetry variations are not. In particular, focusing on $\delta\chi$, we find for example

$$\begin{aligned} \delta(2\chi^{(3)i_{(1)}}) = \frac{1}{\sqrt{2}}\epsilon^{ij}\gamma^r \Big\{ &\partial_r \log \frac{H_1H_3}{H_2H_4}[\delta^{jk} - i\eta f^{-1/2}\gamma_{\bar{0}}\gamma^5\epsilon_{jk}] \\ &+ \partial_r\big((H_1 + H_3) - (H_2 + H_4)\big)[\sqrt{2}grf^{-1/2}\gamma_{\bar{r}}\delta^{jk}]\Big\}\epsilon_{k_{(1)}} \end{aligned} \tag{6.71}$$

(where $\eta \equiv \eta_1 = \eta_2 = \eta_3 = \eta_4$), indicating explicitly that the g-dependent term on the last line has a different structure than the others. This is in contrast to the electric solution. Additionally, note that the matrices $[i\gamma_{\overline{0}}\gamma^5\epsilon_{ij}]$ and $[\gamma_{\overline{r}}\delta^{ij}]$ now commute, while previously, for the electric black hole, they had anticommuted in the absence of γ^5.

For both of the above reasons, we see that whenever $g \neq 0$ none of the supersymmetry variations vanish, and hence the magnetic solution is non-BPS (regardless of the choice of signs of the magnetic charges). In the $g \to 0$ limit, on the other hand, the last line of (6.71) drops out, and we are left with

$$\delta(2\chi^{3i_{(1)}}) = \sqrt{2}\epsilon^{ij}\gamma^r\partial_r \log \frac{H_1H_3}{H_2H_4}\tilde{P}_\eta\epsilon_{j_{(1)}}, \tag{6.72}$$

where $\tilde{P}^{ij}_\eta = \frac{1}{2}[\delta^{ij} - i\eta\gamma_{\overline{0}}\gamma^5\epsilon_{ij}]$ is the projection appropriate to a magnetically charged solution.

Just as for the electrically charged solutions, we may embed the magnetically charged solutions back in eleven dimensions. Although the electrically charged black holes then have the interpretation as the near-horizon limit of a rotating M2-brane, we do not have any such simple interpretation in the magnetic case.

6.13 Kaluza–Klein States as Black Holes

For *ungauged* $N = 8$ supergravity, the supersymmetry algebra admits 4 central charges Z_1, Z_2, Z_3, Z_4. States fall into 5 categories according as they are annihilated by $4 \geq q \geq 0$ supersymmetry generators. q also counts the number of Z's that obey the bound $M = Z_{\max}$. Non-rotating black holes (in the sense of vanishing bosonic Kerr angular momentum L) belong to superspin $L = 0$ supermultiplets [163, 189]. Starting with a spin $J = 0$ member, the rest of the black hole multiplet may then be filled out using the fermionic zero-modes [190, 191]. The spin will run from $J = 0$ up to $J = (8 - q)/2$. For *gauged* supergravity, the algebra is different with no central charges but the same multiplet shortening phenomenon still occurs [192, 200]. So we can be confident that the above black holes preserving 4, 2, 1, 0 supersymmetries will belong to supermultiplets with maximum spins 2, 3, 7/2, 4.[15] Unfortunately, as far as we know, the analogue of the $M = Z_{\max}$ condition has never been been spelled out in the literature. It is presumably some relation between the AdS quantum numbers (E_0, s) and the SO(8) Casimirs.

It seems entirely consistent, therefore, to identify a subset of the maximum spin 2 black hole supermultiplets with the S^7 Kaluza–Klein spectrum, in analogy with the black hole Kaluza–Klein correspondence of ungauged supergravity [193, 200]. The subset in question will correspond to electric black holes whose mass is quantized in units of the inverse S^7

---

[15]The maximum spin 5/2 solutions are (mysteriously?) absent just as for ungauged supergravity.

radius. However, this raises the puzzle of how the black holes carrying only U(1) charges can be identified with the Kaluza–Klein particles carrying nontrivial SO(8) representations. Although we have not demonstrated this explicitly, it seems reasonable to suppose that it is the fermion zero modes that provide the nontrivial SO(8) quantum numbers just as they provide the nontrivial spin. The fact that these nonAbelian charges arise from *fermionic hair* also nicely circumvents the usual no-hair theorems of classical relativity. In this connection, it would be interesting to repeat the gyromagnetic ratio calculations of [191] and verify that the fermionic hair again yields a gyromagnetic ratio equal to 1, as demanded by Kaluza–Klein reasoning.

It is furthermore tempting, in analogy with the ungauged case, to identify the 2, 3 and 4 charge solutions as 2, 3 and 4-particle bound states of the singly charged solution [163, 195]. However, although the quantum number assignments are consistent with this, we do not have multi-center solutions in the AdS case. Such a bound state interpretation would, of course, lead to states of arbitrarily high spin.

Another difference between the S^7 and the T^7 compactifications is that the $g \rightarrow 0$ limit of the gauged supergravity does not directly coincide with the massless sector of the T^7 compactification. They differ by various dualizations. Thus it was possible, for example, to find 4-charge solutions with all charges electric as opposed to the 2-electric and 2-magnetic charges of the ungauged theory. Moreover, whereas 2 charges are Kaluza–Klein modes and 2 are winding modes in the Type IIA string theory context, there is no T or U-duality associated with the S^7 compactification.

One might also generalize the purely electric and purely magnetic solutions discussed here to dyonic black hole solutions of gauged $N = 8$ supergravity. Neither magnetic nor dyonic black holes have any Kaluza–Klein interpretation and do not appear in the spectrum of the S^7 compactification of $D = 11$ supergravity. It would be interesting to provide their M-theory interpretation and to determine their role in the AdS/CFT correspondence.

6.14 Recent Developments

Since these lectures were delivered, several papers have appeared on related topics:

- Explicit Kaluza–Klein ansatzs for deriving various gauged supergravities have recently been given: $N = 4$, $D = 7$ gauged supergravity from $D = 11$ supergravity [218, 219]; $N = 1$, $D = 7$ gauged supergravity from $D = 11$ supergravity [220]; $N = 4$, $\mathrm{SU}(2) \times \mathrm{U}(1)$, $D = 5$ gauged supergravity from $D = 10$ Type IIB supergravity [221]; $N = 4$, SO(4), $D = 4$ gauged supergravity from $D = 11$ supergravity [222]. The complete ansatz for $N = 8$, SO(6), $D = 5$ gauged supergravity from $D = 10$ Type IIB supergravity remains unsolved.

• The CFT duals of AdS black holes rotating in spacetime and carrying electric charge as a consequence of the rotation of the corresponding branes in the transverse dimensions, as discussed in Section 6, are treated in [223] and [240]. These papers also exploit the appearance of the $N = 8$ superconformal singleton action [39] as the membrane on the boundary, discussed in Section 5.

• The thermodynamics of spinning branes and their field theory duals are discussed in [242], while the phase structure of noncommutative field theories and spinning brane bound states are discussed in [243].

• Unitary supermultiplets of $\mathrm{Osp}(6, 2|4)$, discussed in Section 5, and the AdS_7/CFT_6 correspondence are treated in [224].

• An application of the correspondence between AdS_4 and a $d = 3$ conformal field theory involving the squashed S^7 [20, 69] discussed in Section 2.2 is given in [239].

• The *Membrane at the End of the Universe* finds application in [235], which resolves some puzzles in the AdS/CFT correspondence.

• A review of M-theory cosmology may be found in [236].

6.15 Problems 6

1. Calculate the Ricci tensor for the general Kaluza–Klein ansatz for odd sphere S^{2k-1} reductions of the D-dimensional metric.

2. Compute the near-horizon limit of the rotating p-brane carrying a single charge given in 7.5.

7 Solutions to Problems

7.1 Solutions 1

1. In fact we will solve a slightly more general problem by starting with the D-dimensional Einstein action

$$I_D = \frac{1}{2\kappa_D{}^2} \int d^D x \sqrt{-\hat{g}} \hat{R} \tag{7.1}$$

which after integration by parts may be written

$$I_D = \frac{1}{2\kappa_D{}^2} \int d^D x \sqrt{-\hat{g}} (\Omega_{ABC}\Omega^{ABC} - 2\Omega_{ABC}\Omega^{CAB} - 4\Omega_{CA}{}^{A}\Omega^{C}{}_{B}{}^{B}) \tag{7.2}$$

where Ω_{MNP} are the anholonomy coefficients (A.14). We write the vielbein as

$$e_M{}^A = \begin{pmatrix} e_\mu{}^\alpha & e_\mu{}^a \\ e_m{}^\alpha & e_m{}^a \end{pmatrix} \tag{7.3}$$

so that the metric takes the form

$$\hat{g}_{\hat{\mu}\hat{\nu}} = \begin{pmatrix} g_{\mu\nu} + g_{mn}A_\mu{}^m A_\nu{}^n & g_{mn}A_\mu{}^m \\ g_{mn}A_\nu{}^n & g_{mn} \end{pmatrix} \tag{7.4}$$

where

$$A_\mu{}^m = e^m{}_a e_\mu{}^a. \tag{7.5}$$

Note that

$$\det g_{MN} = \det g_{\mu\nu} \det g_{mn} \equiv \det g_{\mu\nu}\Delta. \tag{7.6}$$

If we choose a gauge $e_m{}^\alpha = 0$, the only nonvanishing components are $\Omega_{\alpha\beta\gamma}$,

$$\begin{aligned} \Omega_{\beta\alpha c} &= e_{mc}e^\mu{}_\alpha e^\nu{}_\beta F_{\mu\nu}{}^m \\ \Omega_{\alpha bc} &= -\Omega_{b\alpha c} = -e^m{}_b e^\mu{}_\alpha \partial_\mu e_{mc} \end{aligned} \tag{7.7}$$

where

$$F_{\mu\nu}{}^m = \partial_\mu A_\nu{}^m - \partial_\nu A_\mu{}^m. \tag{7.8}$$

Here we have assumed that all fields are independent of the extra coordinates y^m. This yields the four-dimensional action

$$I_4 = \frac{1}{2\kappa_4{}^2}\int d^4x\,\sqrt{-g}\sqrt{\Delta}\big[R - \tfrac{1}{4}g_{ij}F_{\mu\nu}{}^i F_{\rho\sigma}{}^j g^{\mu\rho}g^{\nu\sigma} + \tfrac{1}{4}g^{\rho\sigma}(g^{ik}g^{jl} - g^{il}g^{jk})\partial_\rho g_{ik}\partial_\sigma g_{jl}\big]. \tag{7.9}$$

We can eliminate $\sqrt{\Delta}$ by a Weyl rescaling

$$e_\mu{}^a \to \Delta^{1/4}e_\mu{}^a \tag{7.10}$$

to obtain

$$I_4 = \frac{1}{2\kappa_4{}^2}\int d^4x\sqrt{-g}\big[R - \tfrac{1}{4}\sqrt{\Delta}g_{ij}F_{\mu\nu}{}^i F_{\rho\sigma}{}^j g^{\mu\rho}g^{\nu\sigma} + \tfrac{1}{4}g^{\mu\nu}\partial_\mu g_{ij}\partial_\nu g^{ij} - \tfrac{1}{8}g^{\mu\nu}\partial_\mu \ln\Delta\partial_\nu \ln\Delta\big]. \tag{7.11}$$

In the special case of one extra dimension, $g_{11} = e^{-\sqrt{3}\phi}$, $\Delta = e^{-\phi/\sqrt{3}}$, we obtain the required result.

Note that, of the original D-dimensional diffeomorphism group with parameters $\xi^M(x,y)$, the surviving symmetry is the 4-dimensional diffeomorphism group with parameters $\xi^\mu(x)$, an Abelian $\mathrm{U}(1)^k$ gauge invariance with parameter $\xi^m(x)$ and a global $\mathrm{SL}(n,\mathbb{R})$ invariance with constant parameter $a^m{}_n$ with $a^m{}_m = 0$:

$$\begin{aligned} \xi^\mu &= \xi^\mu(x) \\ \xi^m &= \xi^m(x) + a^m{}_n y^n. \end{aligned} \tag{7.12}$$

The choice of gauge $e_m{}^\alpha = 0$ does not restrict these invariances, since

$$\delta e_m{}^\alpha = \xi^\mu\partial_\mu e_m{}^\alpha + a_m{}^n e_n{}^\alpha = 0. \tag{7.13}$$

TABLE 2.8. The fields of $(D = 4, N = 8)$ supergravity coming from $D = 11$, $N = 1$ on T^7 and S^7.

$D = 11$	$D = 4$	field	spin	T^7	S^7
g_{MN}	$g_{\mu\nu}$	tensor	2	1	1
	$g_{\mu n}$	vectors	1	7	28
	g_{mn}	scalars	0	28	35
$\Psi_M{}^\alpha$	$\Psi_\mu{}^{\alpha'\alpha''}$	vector-spinors	$\frac{3}{2}$	8	8
	$\Psi_m{}^{\alpha'\alpha''}$	spinors	$\frac{1}{2}$	56	56
A_{MNP}	$A_{\mu\nu\rho}$	3-form	–	1	0
	$A_{\mu\nu p}$	2-forms	0	7	0
	$A_{\mu np}$	1-forms	1	21	0
	A_{mnp}	0-forms	0	35	35

2. We make a $4 + 7$ split of the world indices:

$$x^M = (x^\mu, y^m) \tag{7.14}$$

where $\mu = 0, 1, 2, 3$ and $m = 4, 5, 6, 7, 8, 9, 10$ and write the SO(1, 10) spinor index as

$$\alpha = (\alpha', \alpha'') \tag{7.15}$$

where $\alpha' = 1, \ldots, 4$ is a spinor index of SO(1, 3) and $\alpha'' = 1, \ldots, 8$ is a spinor index of SO(7). Then the fields split according to the T^7 column of Table 2.8.

The spin assigments are obvious except for the 3-form and 2-form gauge fields. A 3-form is dual to a nonpropagating auxilary field c:

$$\sqrt{-g} g^{\mu\alpha} g^{\nu\beta} g^{\rho\gamma} g^{\sigma\delta} \partial_\alpha A_{\beta\gamma\delta} = \epsilon^{\mu\nu\rho\sigma} c \tag{7.16}$$

and c =constant as a result of the Bianchi identity $ddA \equiv 0$. Similarly, a 2-form gauge field is dual to a scalar field ϕ:

$$\sqrt{-g} g^{\mu\alpha} g^{\nu\beta} g^{\rho\gamma} \partial_\alpha A_{\beta\gamma} = \epsilon^{\mu\nu\rho\sigma} \partial_\sigma \phi \tag{7.17}$$

and $\partial^2 \phi = 0$ as a result of the Bianchi identity.

To summarize: the particle content of $N = 8$, $D = 4$ supergravity is (1 spin 2, 8 spin 3/2, 28 spin 1, 56 spin 1/2, 70 spin zero).

7.2 Solutions 2

1. The Einstein equation follows from varying with respect to g_{MN} and the 3-form equation follows by varying with respect to A_{MNP}. We shall need the identities:

$$\begin{aligned} \delta(\sqrt{-g} g^{MN} R_{MN}) &= (\delta\sqrt{-g}) g^{MN} R_{MN} \\ &\quad + \sqrt{-g}(\delta g^{MN}) R_{MN} \sqrt{-g} g^{MN} (\delta R_{MN}) \end{aligned} \tag{7.18}$$

and

$$\delta\sqrt{-g} = \tfrac{1}{2}\sqrt{-g}g^{MN}\delta g_{MN} \tag{7.19}$$
$$\delta g^{MN} = -g^{MP}g^{NQ}\delta g_{PQ} \tag{7.20}$$
$$\sqrt{-g}g^{MN}\delta R_{MN} = \partial_N(\sqrt{-g}g^{MN}\delta\Gamma_{ML}{}^{L}) - \partial_L(\sqrt{-g}g^{MN}\delta\Gamma_{MN}{}^{L}). \tag{7.21}$$

Since the last expression is a total divergence we can drop the final term in (7.18). Similarly

$$\delta(\sqrt{-g}g^{MR}g^{NS}g^{PT}g^{QU}F_{MNPQ}F_{RSTU}) \\ = \sqrt{-g}(4F_M{}^{PQR}F_{NPQR} - \frac{1}{2}g_{MN}F^{PQRS}F_{PQRS}). \tag{7.22}$$

Bearing in mind that the AFF term is independent of the metric, we obtain the Einstein equation (2.1).
A straightforward variation of the action with respect to A_{MNP} then yields the 3-form equation (2.34).

2. We begin by making an ansatz for the $D = 11$ gauge fields $g_{MN} = e_M{}^A e_N{}^B \eta_{AB}$ and A_{MNP} corresponding to the most general six-five split invariant under $P_6 \times \mathrm{SO}(5)$, where P_6 is the $D = 6$ Poincaré group. We split the $D = 11$ coordinates

$$x^M = (x^\mu, y^m) \tag{7.23}$$

where $\mu = 0, 1, 2, 3, 4, 5$ and $m = 6, 7, 8, 9, 10$, and write the line element as

$$ds^2 = e^{2A}\eta_{\mu\nu}\, dx^\mu\, dx^\nu + e^{2B}\delta_{mn}\, dx^m\, dx^n \tag{7.24}$$

and the 4-form gauge field strength as

$$F_{mnpq} = \epsilon_{mnpqr}\partial^r e^C. \tag{7.25}$$

All other components of F_{MNPQ} and all components of the gravitino Ψ_M are set equal to zero. P_6 invariance requires that the arbitrary functions A, B and C depend only on y^m; SO(5) invariance then requires that this dependence be only through $y = \sqrt{\delta_{mn}y^m y^n}$.
As we shall show, the three arbitrary functions A, B, and C are reduced to one by the requirement that the field configuration (7.24) and (7.25) preserve some unbroken supersymmetry. In other words, there must exist Killing spinors ϵ satisfying

$$\widetilde{D}_M\epsilon = 0 \tag{7.26}$$

where $\widetilde{D}_M$ is the bosonic part of the supercovariant derivative appearing in the supersymmetry transformation rule of the gravitino (A.18).
We make the six-five split

$$\Gamma^A = (\gamma_\alpha \otimes \mathbf{1}, \gamma_7 \otimes \Sigma_a) \tag{7.27}$$

where γ_α and Σ_a are the $D = 6$ and $D = 5$ Dirac matrices respectively and where

$$\gamma_7 = \gamma_0 \cdots \gamma_5 \tag{7.28}$$

so that $\gamma_7^2 = 1$. We also decompose the spinor field as

$$\epsilon(x, y) = \zeta(x) \otimes \eta(y) \tag{7.29}$$

where ζ is a constant spinor of SO(1,5) which may further be decomposed into chiral eigenstates via the projection operators $(1 \pm \gamma_7)/2$ and η is an SO(5) spinor.

In our background (7.24) and (7.25), the supercovariant derivative becomes:

$$\begin{aligned} \widetilde{D}_\mu &= \partial_\mu - \frac{1}{2}\gamma^\mu \gamma_7 \Sigma^m \partial_m e^{2A} - \frac{1}{12}\gamma_\mu \Sigma^m \partial_m e^C, \\ \widetilde{D}_m &= \partial_m + \tfrac{1}{2}\Sigma_m \Sigma^n \partial_n e^B - \tfrac{1}{6}\partial_m e^C \gamma_7 - \tfrac{1}{6}(\Sigma_m \Sigma^n - \Sigma_n \Sigma^m)\partial_m e^C. \end{aligned} \tag{7.30}$$

Hence we find that (2.35) admits nontrivial solutions

$$(1 + \gamma_7)\eta = 0 \tag{7.31}$$

where

$$\eta = e^{-C/6}\eta_0 \tag{7.32}$$

where η_0 is a constant spinor and

$$\begin{aligned} A &= \tfrac{1}{3}C \\ B &= -\tfrac{2}{3}C + \text{constant}. \end{aligned} \tag{7.33}$$

In each case, (7.31) means that one half of the maximal possible supersymmetry survives.

With the substitutions (7.24), (7.25) and (7.33), the Einstein equation and the 4-form equation reduce to the single equation for one unknown:

$$\delta^{mn}\partial_m \partial_n e^{-C} = 0 \tag{7.34}$$

and hence, imposing the boundary condition that the metric be asymptotically Minkowskian, we find

$$e^{-C} = 1 + \frac{b^3}{y^3} \tag{7.35}$$

where k_6 is a constant, at this stage arbitrary. Thus the metric is given by

$$ds^2 = (1 + b^3/y^3)^{-1/3}\, dx^\mu\, dx_\mu + (1 + b^3/y^3)^{2/3}(dy^2 + y^2\, d\Omega_4^2) \tag{7.36}$$

and the four-form field strength by

$$F_4 = 3b^3\epsilon_4. \tag{7.37}$$

Here ϵ_4 is the volume form on S^4 and Ω_4 is the volume.

7.3 Solutions 3

1. We use the duality between M-theory compactified on a 2-torus T^2 of area A (measured with the M theory metric g^M) and the Type IIB theory compactified on a circle S^1 of radius R (measured with the IIB metric $g^B = \beta^2 g^M$) [101]. Thus p-brane tensions of dimension $d = p+1$ will be related to each other by $T_d^M = \beta^d T_d^B$.
The $D = 9$ particle spectrum will involve Kaluza–Klein $M0$-branes of mass $T_I^M(KK) = 2\pi A^{-1/2}$ which are identified with winding states of mass $T_1^B(W) = 2\pi R T_2^B$, coming from wrapping the IIB string with tension T_2^B around S^1, so

$$2\pi A^{-1/2} = \beta 2\pi R T_2^B. \tag{7.38}$$

There will also be winding $M0$-branes of mass AT_3^M coming from wrapping the M2-brane with tension T_3^M around T^2, which are identified with the IIB Kaluza–Klein states of mass R^{-1}, so

$$AT_3^M = \beta R^{-1}. \tag{7.39}$$

Hence

$$A = \beta^4 \left(\frac{T_2^B}{T_3^M} \right)^2 \tag{7.40}$$

and

$$R = \beta^{-3} \frac{T_3^M}{(T_2^B)^2}. \tag{7.41}$$

Next we identify the M2-brane in $D = 9$ with the D3-brane with tension T_4^B wrapped around S^1:

$$T_3^M = \beta^3 2\pi R T_4^B \tag{7.42}$$

and the D3-brane in $D = 9$ with the M5-brane with tension T_6^M wrapped around T^2:

$$AT_6^M = \beta^4 T_4^B. \tag{7.43}$$

Eliminating R we find

$$T_4^B = \frac{1}{2\pi} (T_2^B)^2 \tag{7.44}$$

and eliminating A we find

$$\frac{T_6^M}{(T_3^M)^2} = \frac{T_4^B}{(T_2^B)^2} \tag{7.45}$$

and hence that

$$T_6^M = \frac{1}{2\pi} (T_3^M)^2 \tag{7.46}$$

in agreement with the purely $D = 11$ result (2.71).

2. The Lagrangian of $D = 9$ $N = 2$ supergravity as low-energy limit of Type IIA string compactified on a circle can be obtained from dimensional reduction of Type IIA supergravity in $D = 10$, which itself can be obtained from dimensional reduction of $D = 11$ supergravity. Using the notation adopted in [165], the bosonic sector of the theory contains the vielbeins, a dilaton ϕ together with a second dilatonic scalar φ (which measures the size of the compactifying circle), one 4-form field strength $\widetilde{F}_4 = dA_3$, two 3-forms $\widetilde{F}_3^{(i)} = dA_2^{(i)}$, three 2-forms $\widetilde{F}_2^{(12)} = dA_1^{(12)}$ and $\tilde{\mathcal{F}}_2^{(i)} = d\mathcal{A}_1^{(i)}$ and one 1-form $\tilde{\mathcal{F}}_1^{(12)} = d\mathcal{A}_0^{(12)}$. The full bosonic Lagrangian is given by

$$\begin{aligned} e^{-1}\mathcal{L}_{\text{IIA}} = R &- \frac{1}{2}(\partial\phi)^2 - \frac{1}{2}(\partial\varphi)^2 - \frac{1}{2}\left(\mathcal{F}_1^{(12)}\right)^2 e^{-3\phi/2-\sqrt{7}\varphi/2} \\ &- \frac{1}{48}(F_4)^2 e^{-\phi/2-3\varphi/(2\sqrt{7})} - \frac{1}{12}\left(F_3^{(1)}\right)^2 e^{\phi-\varphi/sqrt7} \\ &- \frac{1}{12}\left(F_3^{(2)}\right)^2 e^{-\phi/2+5\varphi/(2\sqrt{7})} - \frac{1}{4}\left(F_2^{(12)}\right)^2 e^{\phi+3\varphi/\sqrt{7}} \\ &- \frac{1}{4}\left(\mathcal{F}_2^{(1)}\right)^2 e^{-3\phi/2-\varphi/(2\sqrt{7})} - \frac{1}{4}\left(\mathcal{F}_2^{(2)}\right)^2 e^{-4\varphi/\sqrt{7}} \\ &- \frac{1}{2}\widetilde{F}_4 \wedge \widetilde{F}_4 \wedge A_1^{(12)} - \widetilde{F}_3^{(1)} \wedge \widetilde{F}_3^{(2)} \wedge A_3. \end{aligned} \tag{7.47}$$

Here we are using the notation that field strengths without tildes include the various Chern-Simon modifications, whilst field strengths written with tildes do not include the mofications. Thus we have

$$\begin{gathered} F_4 = \widetilde{F}_4 - \widetilde{F}_3^{(1)} \wedge \mathcal{A}_1^{(1)} - \widetilde{F}_3^{(2)} \wedge \mathcal{A}_1^{(2)} - \tfrac{1}{2}\widetilde{F}_2^{(12)} \wedge \mathcal{A}_1^{(1)} \wedge \mathcal{A}_1^{(2)}, \\ F_3^{(1)} = \widetilde{F}_3^{(1)} - \widetilde{F}_2^{(12)} \wedge \mathcal{A}_1^{(2)}, \\ F_3^{(2)} = \widetilde{F}_3^{(2)} + F_2^{(12)} \wedge \mathcal{A}_1^{(1)} - \mathcal{A}_0^{(12)}\left(\widetilde{F}_3^{(1)} - F_2^{(12)} \wedge \mathcal{A}_1^{(2)}\right), \\ F_2^{(12)} = \widetilde{F}_2^{(12)}, \quad \mathcal{F}_2^{(1)} = \mathcal{F}_2^{(1)} + \mathcal{A}_0^{(12)}\mathcal{F}_1^{(2)}, \quad \mathcal{F}_2^{(2)} = \tilde{\mathcal{F}}_2^{(2)}, \\ \mathcal{F}_1^{(12)} = \tilde{\mathcal{F}}_1^{(12)}. \end{gathered} \tag{7.48}$$

The Lagrangian of the $D = 9$ supergravity as low-energy limit of Type IIB string compactified on a circle can be obtained from dimensional reduction of Type IIB supergravity in $D = 10$. It is given by

$$\begin{aligned} e^{-1}\mathcal{L}_{\text{IIB}} = R &- \frac{1}{2}(\partial\phi)^2 - \frac{1}{2}(\partial\varphi)^2 - \frac{1}{2}e^{-2\phi}(\partial\chi)^2 \\ &- \frac{1}{48}e^{2\varphi/\sqrt{7}}F_4^2 - \frac{1}{12}e^{\phi-\varphi/\sqrt{7}}\left(F_3^{(\text{NS})}\right)^2 \\ &- \frac{1}{2}e^{-\phi-\varphi/\sqrt{7}}\left(F_3^{(\text{R})}\right)^2 - \frac{1}{4}e^{-4\varphi/\sqrt{7}}(\mathcal{F}_2)^2 \\ &- \frac{1}{4}e^{-\phi+3\varphi/\sqrt{7}}\left(F_2^{(\text{R})}\right)^2 - \frac{1}{4}e^{\phi+3\varphi/\sqrt{7}}\left(F_2^{(\text{NS})}\right)^2 \\ &- \frac{1}{2}\widetilde{F}_4 \wedge \widetilde{F}_4 \wedge \mathcal{A}_1 - \widetilde{F}_3^{(\text{NS})} \wedge \widetilde{F}_3^{(\text{R})} \wedge A_3. \end{aligned} \tag{7.49}$$

TABLE 2.9. Gauge potentials of Type II theories in $D = 10$ and $D = 9$.

	IIA			IIB	
	$D = 10$	$D = 9$	T-duality	$D = 9$	$D = 10$
R-R fields	A_3	A_3	$\longleftrightarrow$	A_3	B_4
		$A_2^{(1)}$	$\longleftrightarrow$	A_2^{R}	A_2^{R}
	$\mathcal{A}_1^{(1)}$	$\mathcal{A}_1^{(1)}$	$\longleftrightarrow$	A_1^{R}	
		$\mathcal{A}_0^{(12)}$	$\longleftrightarrow$	χ	χ
NS-NS fields	$G_{\mu\nu}$	$\mathcal{A}_1^{(2)}$	$\longleftrightarrow$	A_1^{NS}	A_2^{NS}
	$A_2^{(1)}$	$A_2^{(1)}$	$\longleftrightarrow$	A_2^{NS}	
		$A_1^{(12)}$	$\longleftrightarrow$	$\mathcal{A}_1$	$G_{\mu\nu}$

Note that in $D = 10$, there are two 2-form antisymmetric tensors, one of which is the NS-NS field $A_2^{(\mathrm{NS})}$, and the other is the R-R field $A_2^{(\mathrm{R})}$. The dimensional reduction of these two 2-form gauge potentials also gives rise to two vector fields in $D = 9$, denoted by $A_1^{(\mathrm{NS})}$ and $A_1^{(\mathrm{R})}$ respectively.
The $D = 10$ IIA string and IIB string are related by a perturbative T-duality, in that Type IIA string compactified on a circle with radius R is equivalent to Type IIB string compactified on a circle with radius $1/R$. At the level of their low-energy effective actions, it implies that there is only one $D = 9$ $N = 2$ supergravity. The Lagrangians (7.47) and (7.49) are related to each other [86] by local field redefinitions. The relations between the gauge potentials of these two nine-dimensional theories (including the axions) are summarized in Table 2.9.
The relation between the dilatonic scalars of the two nine-dimensional theories is given by

$$\begin{pmatrix} \phi \\ \varphi \end{pmatrix}_{\mathrm{IIA}} = \begin{pmatrix} \frac{3}{4} & -\frac{\sqrt{7}}{4} \\ -\frac{\sqrt{7}}{4} & -\frac{3}{4} \end{pmatrix} \begin{pmatrix} \phi \\ \varphi \end{pmatrix}_{\mathrm{IIB}} . \tag{7.50}$$

The dimensional reduction of the ten-dimensional string metric to $D = 9$ is given by

$$\begin{aligned} ds_{\mathrm{str}}{}^2 &= e^{-\phi/2}\, ds_{10}{}^2 \\ &= e^{-\phi/2}(e^{\varphi/(2\sqrt{7})}\, ds_9{}^2 + e^{-\sqrt{7}\varphi/2}(dz_2 + \mathcal{A})^2), \end{aligned}$$

where $ds_{10}{}^2$ and $ds_9{}^2$ are the Einstein-frame metrics in $D = 10$ and $D = 9$. The radius of the compactifying circle, measured using the ten-dimensional string metric, is therefore given by $R = e^{-\phi/4-\sqrt{7}\varphi/4}$. It follows from (7.50) that the radii R_{IIA} and R_{IIB} of the compactifying circles, measured using their respective ten-dimensional string metrics, are related by $R_{\mathrm{IIA}} = 1/R_{\mathrm{IIB}}$.

7.4 Solutions 4

1. We begin with the bosonic sector of the $d = 3$ worldvolume of the $D = 11$ supermembrane (4.1). To see how a double worldvolume/spacetime compactification of the $D = 11$ supermembrane theory on S^1 leads to the Type IIA string in $D = 10$, let us denote all $(d = 3, D = 11)$ quantities by a hat and all $(d = 2, D = 10)$ quantities without. We then make a ten-one split of the spacetime coordinates

$$\hat{x}^{\widehat{M}} = (x^M, y), \quad M = 0, 1, \ldots, 9, \tag{7.51}$$

and a two-one split of the worldvolume coordinates

$$\hat{\xi}^{\hat{\imath}} = (\xi^i, \xi^2), \quad i = 0, 1, \tag{7.52}$$

in order to make the partial gauge choice

$$\xi^2 = y, \tag{7.53}$$

which identifies the eleventh dimension of spacetime with the third dimension of the worldvolume. The dimensional reduction is then effected by taking the background fields $\hat{g}_{\widehat{M}\widehat{N}}$ and $\hat{A}_{\widehat{M}\widehat{N}\widehat{P}}$ to be independent of y. The string backgrounds of dilaton ϕ, string σ-model metric g_{MN}, 1-form A_M, 2-form B_{MN} and 3-form A_{MNP} are given by

$$\begin{aligned} \hat{g}_{MN} &= e^{-2\phi/3} \begin{pmatrix} g_{MN} + e^{\Phi} A_M A_N & e^{2\phi} A_M \\ e^{2\phi} A_N & e^{\Phi} \end{pmatrix} \\ \hat{A}_{MNP} &= A_{MNP} \\ \hat{A}_{MNY} &= A_{MN}. \end{aligned} \tag{7.54}$$

The choice of dilaton prefactor, $e^{-2\phi/3}$, is dictated by the requirement that g_{MN} be the $D = 10$ string σ-model metric. (To obtain the $D = 10$ fivebrane σ-model metric, the prefactor is unity because the reduction is then spacetime only and not simultaneous worldvolume/spacetime. This explains the remarkable “coincidence” between $\hat{g}_{MN}$ and the $D = 10$ fivebrane σ-model metric.)
Varying the supermembrane action (4.1) with respect to the metric $\widehat{\gamma}_{\hat{\imath}\hat{\jmath}}$ yields the embedding equation

$$\widehat{\gamma}_{\hat{\imath}\hat{\jmath}} = \partial_{\hat{\imath}} \hat{x}^{\widehat{M}} \partial_{\hat{\jmath}} \hat{x}^{\widehat{N}} \hat{g}_{\widehat{M}\widehat{N}}(\hat{x}) \tag{7.55}$$

while varying with respect to $\hat{x}^{\widehat{M}}$ yields the equation of motion

$$\begin{aligned} \partial_{\hat{\imath}}(\sqrt{-\widehat{\gamma}}\widehat{\gamma}^{\hat{\imath}\hat{\jmath}} \partial_{\hat{\jmath}} \hat{x}^{\widehat{M}}) &+ \sqrt{-\widehat{\gamma}}\widehat{\gamma}^{\hat{\imath}\hat{\jmath}} \partial_{\hat{\imath}} \hat{x}^{\widehat{N}} \partial_{\hat{\jmath}} \hat{x}^{\widehat{P}} \widehat{\Gamma}_{\widehat{N}\widehat{P}}{}^{\widehat{M}} \\ &= \tfrac{1}{6} \epsilon^{\hat{\imath}\hat{\jmath}\hat{k}} \partial_{\hat{\imath}} \hat{x}^{\widehat{N}} \partial_{\hat{\jmath}} \hat{x}^{\widehat{P}} \partial_{\hat{k}} \hat{x}^{\widehat{Q}} \widehat{F}^{\widehat{M}}{}_{\widehat{N}\widehat{P}\widehat{Q}} \end{aligned} \tag{7.56}$$

where $\widehat{F}_{\widehat{M}\widehat{N}\widehat{P}\widehat{Q}}$ is the field strength of $\hat{A}_{\widehat{M}\widehat{N}\widehat{P}}$,

$$\widehat{F}_{\widehat{M}\widehat{N}\widehat{P}\widehat{Q}} = 4\partial_{[\widehat{M}}\hat{A}_{\widehat{N}\widehat{P}\widehat{Q}]}. \tag{7.57}$$

Having made the two-one split of the worldvolume coordinates and the ten-one split of the spacetime coordinates, and having made the partial gauge choice, the double dimensional reduction is then effected by demanding that

$$\frac{\partial x^M}{\partial \xi^2} = 0 \tag{7.58}$$

and

$$\frac{\partial \hat{g}_{\widehat{M}\widehat{N}}}{\partial y} - 0 = \frac{\partial \hat{A}_{\widehat{M}\widehat{N}\widehat{P}}}{\partial y} = 0. \tag{7.59}$$

From (7.54), the induced metric on the worldvolume is now given by

$$\widehat{\gamma}_{\hat{\imath}\hat{\jmath}} = e^{-2\phi/3}\begin{pmatrix} \gamma_{ij} + e^2\phi A_i A_j & e^{2\phi}A_i \\ e^{2\phi}A_j & e^{2\phi} \end{pmatrix} \tag{7.60}$$

where

$$\begin{aligned} \gamma_{ij} &= \partial_i x^M \partial_j x^N g_{MN} \\ A_i &= \partial_i x^M A_M. \end{aligned} \tag{7.61}$$

Note that

$$\sqrt{-\widehat{\gamma}} = \sqrt{-\gamma}. \tag{7.62}$$

Substituting these expressions into the field equations (7.56) yields in the case $\widehat{M} = M$

$$\partial_i(\sqrt{-\gamma}\gamma^{ij}\partial_j x^M) + \sqrt{-\gamma}\gamma^{ij}\partial_i x^N \partial_j x^P \Gamma_{NP}{}^M = \tfrac{1}{2}\epsilon^{ij}\partial_i x^N \partial_j x^P F^M{}_{NP} \tag{7.63}$$

where F_{MNP} is the field strength of A_{MNP},

$$F_{MNP} = 3\partial_{[M}A_{NP]}. \tag{7.64}$$

In the case $\widehat{M} = y$, (7.56) is an identity as it must be for consistency. But (7.63) is just the ten-dimensional string equation of motion derivable from the action

$$S_2 = T_2 \int d^2\xi \left[-\tfrac{1}{2}\sqrt{-\gamma}\gamma^{ij}\partial_i x^M \partial_j x^N g_{MN}(x) - \tfrac{1}{2!}\epsilon^{ij}\partial_i x^M \partial_j x^N A_{MN}(x)\right]. \tag{7.65}$$

One may repeat the procedure in superspace to obtain

$$S_2 = T_2 \int d^2\xi \left[-\tfrac{1}{2}\sqrt{-\gamma}\gamma^{ij}E_i^a E_j^b \eta_{ab} + \tfrac{1}{2!}\epsilon^{ij}\partial_i X^M \partial_j X^N A_{MN}(Z)\right] \tag{7.66}$$

which is just the action of the Type IIA superstring. Note that the ten-dimensional bosonic R-R fields A_{MNP}, A_M and have decoupled in (7.65). They have not disappeared from the theory, however, since their coupling still survives in the R-R sector of (7.66).

7.5 Solutions 5

1. Rotating p-branes in arbitrary dimensions, supported by a single $(p+2)$-form charge, are all straightforwardly obtained by diagonally oxidising the rotating black holes constructed in [169]. There are two cases arising, depending on whether $\tilde{d}$ is even or odd.
In this case $\tilde{d}+2 = 2N$, there are N angular momenta ℓ_i, with $i = 1, 2, \ldots, N$. We find that the metric of the rotating $(n-2)$-brane solution to the equations is

$$\begin{aligned} ds_D{}^2 = & H^{-\tilde{d}/(D-2)}\left(-\left(1-\frac{2m}{r^{\tilde{d}}\Delta}\right)dt^2 + d\vec{x}\cdot d\vec{x}\right) \\ & + H^{d/(D-2)}\left[\frac{\Delta\, dr^2}{H_1\cdots H_N - 2mr^{-\tilde{d}}} + r^2\sum_{i=1}^{N} H_i(d\mu_i^2 + \mu_i^2\, d\phi_i^2)\right. \\ & \left. -\frac{4m\cosh\beta}{r^{\tilde{d}}H\Delta}dt\left(\sum_{i=1}^{N}\ell_i\mu_i^2\, d\phi_i\right) + \frac{2m}{r^{\tilde{d}}H\Delta}\left(\sum_{i=1}^{N}\ell_i\mu_i^2\, d\phi_i\right)^2\right], \end{aligned} \tag{7.67}$$

where the functions Δ, H and H_i are given by

$$\begin{aligned} \Delta = H_1\cdots H_N\sum_{i=1}^{N}\frac{\mu_i^2}{H_i}, \quad & H = 1 + \frac{2m\sinh^2\beta}{r^{\tilde{d}}\Delta}, \\ H_i = 1 + \frac{\ell_i^2}{r^2}, \quad & i = 1, 2, \ldots, N. \end{aligned} \tag{7.68}$$

The dilaton ϕ and gauge potential $A_{(n-1)}$ are given by

$$\begin{aligned} e^{2\phi/a} &= H, \\ A_{(n-1)} &= \frac{1-H^{-1}}{\sinh\beta}\left(\cosh\beta\, dt + \sum_{i=1}^{N}\ell_i\mu_i^2\, d\phi_i\right)\wedge d^{n-2}x. \end{aligned} \tag{7.69}$$

The N quantities μ_i, as usual, are subject to the constraint $\sum_i \mu_i^2 = 1$. One can parameterise the μ_i in terms of $(N-1)$ unconstrained angles. A common choice is

$$\begin{aligned} \mu_i &= \sin\psi_i\prod_{j=1}^{i-1}\cos\psi_j, \quad i \le N-1, \\ \mu_N &= \prod_{j=1}^{N-1}\cos\psi_j. \end{aligned} \tag{7.70}$$

Note that $\prod_{j=1}^{n}\cos\psi_j$ is defined to be equal to 1 if $n \le 0$.
In the case $\tilde{d}+2 = 2N+1$, the solution has the same form as above, but with the range of the index i extended to include 0. However, there is no angular momentum parameter or azimuthal coordinate associated with the extra index value, and so $\ell_0 = 0$ and $\phi_0 = 0$. Otherwise, all the formulæ given above are generalised simply by extending the summation to span the range $0 \le i \le N$. Of course $H_0 = 1$ as a consequence of $\ell_0 = 0$.

7.6 Solutions 6

1. The general Kaluza–Klein ansatz for odd sphere S^{2k-1} reductions of the D-dimensional metric may be expressed in the form

$$ds_D{}^2 = \tilde{\Delta}^a\, ds_d{}^2 + \tilde{\Delta}^{-b}\sum_{i=1}^{k} X_i^{-1}\left(d\mu_i^2 + \mu_i^2\left(d\phi_i + A^i_{(1)}\right)^2\right), \tag{7.71}$$

where we have set the radius of S^{2k-1} to unity. There are $k-1$ scalar degrees of freedom parameterised by the k quantities X_i satisfying the constraint $\prod_{i=1}^{k} X_i = 1$. This form of the line element encompasses both the S^5 reduction of Type IIB supergravity and the S^7 reduction of eleven dimensional supergravity. As defined previously, $\tilde{\Delta} = \sum_{i=1}^{k} X_i\mu_i^2$ and $\sum_{i=1}^{k}\mu_i^2 = 1$.
In the absence of the gauge fields, this metric has a block diagonal form, with the blocks corresponding to the d-dimensional spacetime, the $k-1$ direction cosines μ_i and the k azimuthal rotation angles ϕ_i. The main difficulty in computing the curvature of (7.71) lies in the fact that the μ_i's are constrained. Nevertheless, we may perform an asymmetric choice of using the first $k-1$ of them as actual coordinates, while expressing μ_k as the constrained quantity $\mu_k = (1-\sum_{i=1}^{k-1}\mu_i^2)^{1/2}$.
Since numerous terms are involved in the computation, it is imperative to clarify our notation. We denote the lower-dimensional spacetime indices by $\mu, \nu, \dots = 0, 1, \dots, d-1$, the direction cosine indices by $\alpha, \beta, \gamma, \dots = 1, 2, \dots, k-1$ and the azimuthal indices by $i, j, \dots = 1, 2, \dots, k$. Note that for instance implicit sums over α always run over $k-1$ values, while sums over i always run over the full k values.
Thus (with vanishing gauge fields) the D-dimensional metric may be expressed in the form

$$G_{MN} = \mathrm{diag}[\tilde{\Delta}^a g_{\mu\nu}, \tilde{\Delta}^{-b}\hat{g}_{ij}, \tilde{\Delta}^{-b}\tilde{g}_{\alpha\beta}], \tag{7.72}$$

where $\hat{g}_{ij} = X_i^{-1}\mu_i^2\delta_{ij}$ is diagonal and

$$\begin{aligned}\tilde{g}_{\alpha\beta} &= X_\alpha^{-1}\delta_{\alpha\beta} + X_k^{-1}\hat{\mu}_\alpha\hat{\mu}_\beta,\\ \tilde{g}^{\alpha\beta} &= X_\alpha\delta_{\alpha\beta} - \tilde{\Delta}^{-1}X_\alpha X_\beta\mu_\alpha\mu_\beta,\end{aligned} \tag{7.73}$$

with $\hat{\mu}_\alpha \equiv \mu_\alpha/\mu_k$. Note that $\det\tilde{g}_{\alpha\beta} = \tilde{\Delta}/\mu_k^2$. As the μ_α themselves are coordinates, this allows the use of expressions such as $\partial_\alpha\mu_k = -\hat{\mu}_\alpha$ and

$\partial_\alpha \hat{\mu}_\beta = \mu_k^{-1}(\delta_{\alpha\beta} + \hat{\mu}_\alpha \hat{\mu}_\beta)$. In addition, all $\alpha, \beta, \ldots,$ indices are raised and lowered with the metric $\tilde{g}_{\alpha\beta}$.

Using this specific form of the metric $\tilde{g}_{\alpha\beta}$ and the fact that $\det \hat{g}_{ij} = \prod_{i=1}^{k} \mu_i^2$, we find $\det G_{MN} = \det g_{\mu\nu} \tilde{\Delta}^{\kappa+2a} \prod_{i=1}^{k-1} \mu_i^2$ where the product provides the measure over the internal S^{2k-1}. Here $\kappa = a(d-2) - b(2k-1) + 1$ so that $\sqrt{-G}R \sim \sqrt{-g}\tilde{\Delta}^{\kappa/2}$. Hence one expects $\kappa = 0$ in order to prevent any $\tilde{\Delta}$ dependence from appearing in front of the lower-dimensional Einstein term. Indeed we see that κ vanishes for both the S^5 and the S^7 reductions considered in the text.

We have only computed selected components of the full D-dimensional Ricci tensor which are of interest in the Kaluza–Klein reduction. While we have used an asymmetric parameterisation of the direction cosines, the final results are symmetric in all k of the μ_i's. For the lower-dimensional components of R_{MN} we find

$$
\begin{aligned}
R_{\mu\nu} = {} & R_{\mu\nu}^{(d)} - \frac{1}{2}\kappa \tilde{\Delta}^{-1}\partial_\mu \partial_\nu \tilde{\Delta} \\
& + \frac{1}{4}\big((a+2)\kappa - (a+b)b(2k-1) + a + 2b\big)\tilde{\Delta}^{-2}\partial_\mu \tilde{\Delta}\partial_\nu \tilde{\Delta} \\
& + \frac{1}{4}(\partial_\mu \tilde{g}^{\alpha\beta}\partial_\nu \tilde{g}_{\alpha\beta} - X_i^{-2}\partial_\mu X_i \partial_\nu X_i) \\
& + g_{\mu\nu}\left[-\frac{a}{2}\tilde{\Delta}^{-1}\nabla^2\tilde{\Delta} - \frac{a}{4}(\kappa-2)\tilde{\Delta}^{-2}\partial^\rho \tilde{\Delta}\partial_\rho \tilde{\Delta}\right] \\
& + g_{\mu\nu}\tilde{\Delta}^{a+b}\left[-\frac{a}{4}(\kappa + 2a + 2b - 3)\tilde{\Delta}^{-2}\partial^\alpha \tilde{\Delta}\partial_\alpha \tilde{\Delta}\right. \\
& \qquad \left. - \frac{a}{2}(\tilde{\Delta}^{-1}\tilde{\nabla}^2\tilde{\Delta} + \tilde{\Delta}^{-1}\mu_i^{-1}\partial^\alpha \mu_i \partial_\alpha \tilde{\Delta})\right],
\end{aligned} \tag{7.74}
$$

where $R_{\mu\nu}^{(d)}$ denotes the Ricci tensor of the d-dimensional spacetime metric $g_{\mu\nu}$. We have also determined the internal Ricci components R_{ij} and $R_{\alpha\beta}$ necessary for computing the D-dimensional scalar curvature. For the former we find

$$
\begin{aligned}
R_{ij} = {} & \hat{g}_{ij}\tilde{\Delta}^{-a-b}\left[\frac{1}{4}\kappa\tilde{\Delta}^{-1}X_i^{-1}\partial^\rho \tilde{\Delta}\partial_\rho X_i + \frac{b}{2}\tilde{\Delta}^{-1}\nabla^2\tilde{\Delta}\right. \\
& \qquad \left. + \frac{b}{4}(\kappa-2)\tilde{\Delta}^{-2}\partial^\rho\tilde{\Delta}\partial_\rho\tilde{\Delta}\right] \\
& + \hat{g}_{ij}\left[-\frac{1}{2}(\kappa + 2a + 2b - 1)\tilde{\Delta}^{-1}\mu_i^{-1}\partial^\alpha \mu_i \partial_\alpha \tilde{\Delta}\right. \\
& \qquad - \mu_i^{-1}\tilde{\nabla}^2\mu_i + \tilde{\Delta}^{-1}\Big(X_i \sum X - X_i^2\Big) \\
& \qquad + \frac{b}{4}(\kappa + 2a + 2b - 3)\tilde{\Delta}^{-2}\partial^\alpha\tilde{\Delta}\partial_\alpha\tilde{\Delta} \\
& \qquad \left. + \frac{b}{2}\tilde{\Delta}^{-1}\tilde{\nabla}^2\tilde{\Delta} + \frac{b}{2}\tilde{\Delta}^{-1}\mu_l^{-1}\partial^\alpha\mu_l\partial_\alpha\tilde{\Delta}\right]
\end{aligned} \tag{7.75}
$$

(no sum on i), while for the latter we have

$$
\begin{aligned}
R_{\alpha\beta} = {} & \tilde{\Delta}^{-a-b}\left[\frac{1}{2}\tilde{g}^{\gamma\delta}\partial^{\rho}\tilde{g}_{\alpha\gamma}\partial_{\rho}\tilde{g}_{\beta\delta} - \frac{1}{4}\kappa\tilde{\Delta}^{-1}\partial^{\rho}\tilde{g}_{\alpha\beta}\partial_{\rho}\tilde{\Delta} - \frac{1}{2}\nabla^{2}\tilde{g}_{\alpha\beta}\right] \\
& + \tilde{g}_{\alpha\beta}\tilde{\Delta}^{-a-b}\left[\frac{b}{4}(\kappa-2)\tilde{\Delta}^{-2}\partial^{\rho}\tilde{\Delta}\partial_{\rho}\tilde{\Delta} + \frac{b}{2}\tilde{\Delta}^{-1}\nabla^{2}\tilde{\Delta}\right] \\
& + \widetilde{R}_{\alpha\beta} - \frac{1}{2}(\kappa+2a+2b-1)\tilde{\Delta}^{-1}\widetilde{\nabla}_{\alpha}\widetilde{\nabla}_{\beta}\tilde{\Delta} - \mu_i^{-1}\widetilde{\nabla}_{\alpha}\widetilde{\nabla}_{\beta}\mu_i \\
& - \frac{1}{4}\big((b-2)\kappa + a(a+b)d + (b-2)(2a+2b-1)\big)\tilde{\Delta}^{-2}\partial_{\alpha}\tilde{\Delta}\partial_{\beta}\tilde{\Delta} \\
& + \tilde{g}_{\alpha\beta}\left[\frac{b}{2}\tilde{\Delta}^{-1}\widetilde{\nabla}^{2}\tilde{\Delta} + \frac{b}{4}(\kappa+2a+2b-3)\tilde{\Delta}^{-2}\partial^{\gamma}\tilde{\Delta}\partial_{\gamma}\tilde{\Delta}\right. \\
& \qquad\qquad \left. + \frac{b}{2}\tilde{\Delta}^{-1}\mu_i^{-1}\partial^{\gamma}\mu_i\partial_{\gamma}\tilde{\Delta}\right]. \qquad (7.76)
\end{aligned}
$$

Note that $\widetilde{R}_{\alpha\beta}$ as well as the covariant derivatives $\widetilde{\nabla}_{\alpha}$ are defined with respect to the $(k-1)$-dimensional metric $d\tilde{s}^2 = \sum_{i=1}^{k} X_i^{-1} d\mu_i^2$. While these expressions are rather unwieldy, they simplify considerably in both the S^5 and the S^7 reductions, as many of the coefficients take on simple values. Finally, by taking the trace of the above, we find the expression for the D-dimensional curvature scalar

$$
\begin{aligned}
R = {} & \tilde{\Delta}^{-a}\big[R^{(d)} - (\kappa+a)\nabla^{\rho}(\tilde{\Delta}^{-1}\nabla_{\rho}\tilde{\Delta}) + \tfrac{1}{4}(\partial^{\rho}\tilde{g}_{\alpha\beta}\partial_{\rho}\tilde{g}^{\alpha\beta} - X_i^{-2}\partial^{\rho}X_i\partial_{\rho}X_i) \\
& \quad + \tfrac{1}{4}\big(-(\kappa+a)(\kappa-1) + 2b - (a+b)b(2k-1)\big)\tilde{\Delta}^{-2}\partial^{\rho}\tilde{\Delta}\partial_{\rho}\tilde{\Delta}\big] \\
& + \tilde{\Delta}^{b}\Big[\widetilde{R} - (\kappa+2a+b-1)(\tilde{\Delta}^{-1}\widetilde{\nabla}^{2}\tilde{\Delta} + \tilde{\Delta}^{-1}\mu_i^{-1}\partial^{\alpha}\mu_i\partial_{\alpha}\tilde{\Delta}) \\
& \quad + \tfrac{1}{4}\big(-(\kappa+2a+b-1)(\kappa+2a+2b-5) - a(a+b)d\big)\tilde{\Delta}^{-2}\partial^{\alpha}\tilde{\Delta}\partial_{\alpha}\tilde{\Delta} \\
& \quad - 2\mu_i^{-1}\widetilde{\nabla}^{2}\mu_i + \tilde{\Delta}^{-1}\Big(\sum X\Big)^{2} - \tilde{\Delta}^{-1}\sum X^{2}\Big]. \qquad (7.77)
\end{aligned}
$$

To make contact with the Kaluza–Klein reductions, we note that explicit computation of $\widetilde{R}$ yields

$$
\begin{aligned}
\widetilde{R} = \tilde{\Delta}^{-1}\Big[& 2\tilde{\Delta}^{-1}\sum X^{3}\mu^{2} - 2\tilde{\Delta}^{-1}\sum X\sum X^{2}\mu^{2} \\
& + \Big(\sum X\Big)^{2} - \sum X^{2}\Big], \qquad (7.78)
\end{aligned}
$$

where we have followed a shorthand notation of removing indices so that, $\sum X^3\mu^2 \equiv \sum_{i=1}^{k} X_i^3\mu_i^2$, for example. Note that for the special case of $k=3$, corresponding to S^5, not all of the above quantities are independent. As a result we find that this expression simplifies to yield $\widetilde{R}_{(k=3)} = 2\tilde{\Delta}^{-2}X_1X_2X_3 = 2\tilde{\Delta}^{-2}$. Additionally, we often find the following identities useful:

$$\begin{aligned} \partial^\alpha \tilde{\Delta}\partial_\alpha \tilde{\Delta} &= -4\Big[\tilde{\Delta}^{-1}\Big(\sum X^2\mu^2\Big)^2 - \sum X^3\mu^2\Big], \\ \widetilde{\nabla}^2\tilde{\Delta} &= 2\Big[\tilde{\Delta}^{-2}\Big(\sum X^2\mu^2\Big)^2 - \tilde{\Delta}^{-1}\sum X^3\mu^2 \\ &\qquad - \tilde{\Delta}^{-1}\sum X\sum X^2\mu^2 + \sum X^2\Big], \qquad (7.79) \\ \mu_i^{-1}\partial^\alpha\mu_i\partial_\alpha\tilde{\Delta} &= -2\Big[\tilde{\Delta}^{-1}\sum X\sum X^2\mu^2 - \sum X^2\Big], \\ \mu_i^{-1}\widetilde{\nabla}^2\mu_i &= \tilde{\Delta}^{-1}\Big[\tilde{\Delta}^{-1}\sum X\sum X^2\mu^2 - \Big(\sum X\Big)^2\Big]. \end{aligned}$$

The S^5 reduction of Type IIB supergravity discussed in Section 6.2 corresponds to the choice of $d=5$ and $a=b=\frac{1}{2}$. In this case we obtain

$$\begin{aligned} \tilde{\Delta}^{1/2}R^{(D=10)}_{(k=3)} &= R^{(5)} - \tfrac{1}{2}\nabla^\rho(\tilde{\Delta}^{-1}\nabla_\rho\tilde{\Delta}) \\ &\quad - \tfrac{1}{4}(-\partial^\rho\tilde{g}_{\alpha\beta}\partial_\rho\tilde{g}^{\alpha\beta} + X_i^{-2}\partial^\rho X_i\partial_\rho X_i + \tilde{\Delta}^{-2}\partial^\rho\tilde{\Delta}\partial_\rho\tilde{\Delta}) \\ &\quad + 2\Big(\tilde{\Delta}^{-1} + 3\sum X^{-1}\Big), \qquad (7.80) \end{aligned}$$

where we have used the simplified expression for $\widetilde{R}_{(k=3)}$ given above. On the other hand, the S^7 reduction of eleven dimensional supergravity, given by the line element (6.25), corresponds to the choice of $d=4$ and $a=\frac{2}{3}$, $b=\frac{1}{3}$. The eleven-dimensional curvature scalar is

$$\begin{aligned} \tilde{\Delta}^{2/3}R^{(D=11)}_{(k=4)} &= R^{(4)} - \tfrac{2}{3}\nabla^\rho(\tilde{\Delta}^{-1}\nabla_\rho\tilde{\Delta}) \\ &\quad - \tfrac{1}{4}(-\partial^\rho\tilde{g}_{\alpha\beta}\partial_\rho\tilde{g}^{\alpha\beta} + X_i^{-2}\partial^\rho X_i\partial_\rho X_i + \tilde{\Delta}^{-2}\partial^\rho\tilde{\Delta}\partial_\rho\tilde{\Delta}) \\ &\quad - \tfrac{2}{3}\tilde{\Delta}^{-2}\Big(\sum X^2\mu^2\Big)^2 + \tfrac{8}{3}\tilde{\Delta}^{-1}\sum X^3\mu^2 \\ &\quad - \tfrac{4}{3}\tilde{\Delta}^{-1}\sum X\sum X^2\mu^2 + 4\Big(\sum X\Big)^2 - \frac{14}{3}\sum X^2. \qquad (7.81) \end{aligned}$$

The last two lines involve undifferentiated scalars, and is used in (6.35). Curiously, the scalar kinetic terms in both cases have an identical structure save for a total derivative, and take on a standard Kaluza–Klein appearance (since $X_i^{-2}\partial^\rho X_i\partial_\rho X_i = -\partial^\rho\hat{g}_{ij}\partial_\rho\hat{g}^{ij}$). Finally note that the implicitly defined term $\partial^\rho\tilde{g}_{\alpha\beta}\partial_\rho\tilde{g}^{\alpha\beta}$ may be evaluated to give

$$\begin{aligned} -\partial^\rho\tilde{g}_{\alpha\beta}\partial_\rho\tilde{g}^{\alpha\beta} &= X_i^{-2}\partial^\rho X_i\partial_\rho X_i + \tilde{\Delta}^{-2}\partial^\rho\tilde{\Delta}\partial_\rho\tilde{\Delta} \\ &\quad - 2\tilde{\Delta}^{-1}X_i^{-1}\partial^\rho X_i\partial_\rho X_i\mu_i^2. \qquad (7.82) \end{aligned}$$

2. The general expression for a rotating p-brane carrying a single charge is given in 7.5. Following the procedure in the previous sections, we may

take the limit of large p-brane charge, by performing the rescalings

$$\begin{aligned} m &\to \epsilon^{\tilde{d}} m, \quad \sinh\beta \to \epsilon^{-\tilde{d}/2} \sinh\beta, \\ r &\to \epsilon r, \quad x^\mu \to \epsilon^{1-\tilde{d}/2} x^\mu, \quad \ell_i \to \epsilon \ell_i, \end{aligned} \tag{7.83}$$

and then sending ϵ to zero. We find that the metric becomes $ds^2 = \epsilon^{\alpha^2/2}\, d\tilde{s}^2$, where α is given by (5.31) and the metric $d\tilde{s}^2$ is given by

$$\begin{aligned} d\tilde{s}_D^2 = \tilde{\Delta}^{\tilde{d}/(D-2)} e^{\tilde{d}\varphi} \Big[&-(H_1 \cdots H_N)^{-(d-2)/(d-1)} f\, dt^2 + (H_1 \cdots H_N)^{1/(d-1)} \\ &\times \Big(\Big(\frac{\tilde{d}}{d} g\rho\Big)^{(D-2)\alpha^2/(2\tilde{d})} f^{-1}\, d\rho^2 + \rho^2\, d\vec{y}\cdot d\vec{y} \Big) \Big] \\ &+ g^{-2} \tilde{\Delta}^{-d/(D-2)} e^{-d\varphi} \sum_{i=1}^{N} X_i^{-1} \big(d\mu_i^2 + \mu_i^2 (d\phi_i + g A^i)^2 \big), \end{aligned} \tag{7.84}$$

where

$$\begin{aligned} g\rho &= (d/\tilde{d})(gr)^{\tilde{d}/d}, \quad \vec{y} = g(\tilde{d}/d)\vec{x}, \\ g^{-\tilde{d}} &= 2m \sinh^2\beta, \quad \mu = 2m(d/\tilde{d})^{d-2} g^{2+\tilde{d}-d}, \end{aligned} \tag{7.85}$$

and

$$\begin{aligned} f &= -\frac{\mu}{\rho^{d-2}} + (\tilde{d}/d)^2 g^2 \rho^2 (H_1 \cdots H_N), \\ X_i &= (H_1 \cdots H_N)^{d/[(d-1)\tilde{d}]} H_i^{-1}, \quad \tilde{\Delta} = \frac{\sum_i X_i \mu_i^2}{(X_1 \cdots X_N)^2}, \\ e^{-2\tilde{d}\varphi/\alpha^2} &= (\tilde{d}/d) g\rho, \quad A^i = \frac{1 - H_i^{-1}}{g\ell_i \sinh\beta}\, dt. \end{aligned} \tag{7.86}$$

It follows that the $(d+1)$-dimensional metric becomes

$$\begin{aligned} ds_{d+1}{}^2 = &-(H_1 \cdots H_N)^{-(d-2)/(d-1)} f\, dt^2 \\ &+ (H_1 \cdots H_N)^{1/(d-1)} \big(e^{-(D-2)\varphi} f^{-1}\, d\rho^2 + \rho^2\, d\vec{y}\cdot d\vec{y} \big). \end{aligned} \tag{7.87}$$

The Einstein-frame metric is given by $ds_{\rm E}{}^2 = e^{-(D-2)\varphi/(d-1)}\, ds_{d+1}{}^2$. This is the metric of an N-charge black hole in a domain-wall background. In the case when $a = 0$, the domain wall specialises to AdS_{d+1}.

Acknowledgments: Much of the recent work on AdS black holes described in these lectures was carried out in collaboration with my colleagues Miriam Cvetic, Parid Hoxha, Jim Liu, Hong Lü, Jianxin Lu, Rene Martinez Acosta, Chris Pope, Hisham Sati and Tuan Tran. In writing these lectures, I would especially like to thank Jim Liu and Jianxin Lu for their help and Arthur

Greenspoon for a careful reading of the manuscript. Conversations with Ergin Sezgin on singletons and related topics are also gratefully acknowledged. I would also like to thank the organizers of the Banff Summer School, Yvan Saint-Aubin and Luc Vinet, for their hospitality. Research was supported in part by NSF Grant PHY-9722090.

8 References

1. J. Wess and J. Bagger, *Supersymmetry and supergravity*, Princeton Univ. Press, Princeton, NJ, 1983.

2. S.J. Gates, Jr., M.T. Grisaru, M. Rocek, and W. Siegel, *Superspace. One thousand and one lessons in supersymmetry*, Benjamin/Cummings, Reading, MA, 1983.

3. P.P. Srivastava, *Supersymmetry, superfields and supergravity*, Adam Hilger, Bristol, 1986.

4. P. West, *Introduction to supersymmetry and supergravity*, World Scientific, Singapore, 1986.

5. D. Bailin and A. Love, *Supersymmetric gauge field theory and string theory*, Institute of Physics, Bristol, 1994.

6. M.J. Duff, *M-theory (the theory formerly known as strings)*, Int. J. Mod. Phys. **A11** (1996), 5623.

7. M.J. Duff, R.R. Khuri, and J.X. Lu, *String solitons*, Phys. Rep. **259** (1995), 213, hep-th/9412184

8. D.Z. Freedman and A. Das, *Gauge internal symmetry in extended supergravity*, Phys. Lett. **B74** (1977), 333.

9. S.M. Christensen, M.J. Duff, G.W. Gibbons, and M. Rocek, *Vanishing one-loop β-function in gauged $N > 4$ supergravity*, Phys. Rev. Lett. **45** (1980), 161.

10. M.J. Duff, *Ultraviolet divergences in extended supergravity*, in Proceedings of the 1981 Trieste conference "Supergravity 81," eds. S. Ferarra and J.G. Taylor, Cambridge Univ. Press, Cambridge–New York, 1982.

11. B. de Wit and H. Nicolai, *$N = 8$ supergravity*, Nucl. Phys. **B208** (1982), 323.

12. E. Cremmer, B. Julia and J. Scherk, *Supergravity theory in eleven dimensions*, Phys. Lett. **B76** (1978), 409.

13. M.J. Duff and C.N. Pope, *Kaluza–Klein supergravity and the seven sphere*, in Proceedings of the 1982 Trieste conference "Supersymmetry and Supergravity 82", eds. S. Ferrara, J.G. Taylor, and P. van Nieuwenhuizen, World Scientific, Singapore, 1983.

14. E. Cremmer and J. Scherk, *Spontaneous compactification of extra space dimensions*, Nucl. Phys. **B118** (1977), 61.

15. P.G.O. Freund and M.A. Rubin, *Dynamics of dimensional reduction*, Phys. Lett. **B97** (1980), 233.

16. M.J. Duff and P. van Nieuwenhuizen, *Quantum inequivalence of different field representations*, Phys. Lett. **B94** (1980), 179.

17. K. Pilch, P. van Nieuwenhuizen, and P.K. Townsend *Compactification of $d = 11$ supergravity on S^4 (or $11 = 7 + 4$, too)*, Nucl. Phys. **B242** (1984), 377.

18. M. Pernici, K. Pilch, and P. van Nieuwenhuizen, *Gauged maximally extended supergravity in seven dimensions*, Phys. Lett. **B143** (1984), 103.

19. P.K. Townsend and P. van Nieuwenhuizen, *Gauged seven-dimensional supergravity*, Phys. Lett. **B125** (1983), 41.

20. M.J. Duff, B.E.W. Nilsson, and C.N. Pope, *Kaluza–Klein supergravity*, Phys. Rep. **130** (1986), 1.

21. J.H. Schwarz, *Covariant field equations of chiral $N = 2$ $D = 10$ supergravity*, Nucl. Phys. **B226** (1983), 269.

22. P. Howe and P.C. West, *The complete $N = 2$, $d = 10$ supergravity*, Nucl. Phys. **B238** (1984), 181.

23. J.H. Schwarz, *Spontaneous compactification of extended supergravity in ten dimensions*, Physica **A124** (1984), 543.

24. M. Günaydin and N. Marcus, *The spectrum of the S^5 compactification of the $N = 2$, $D = 10$ supergravity and the unitary supermultiplet*, Class. Quant. Grav. **2** (1985), L11.

25. H.J. Kim, L.J. Romans, and P. van Nieuwenhuizen, *Mass spectrum of chiral ten-dimensional $N = 2$ supergravity on S^5*, Phys. Rev. **D32** (1985), 389.

26. M. Pernici, K. Pilch, and P. van Nieuwenhuizen, *Gauged $N = 8$, $d = 5$ supergravity*, Nucl. Phys. **B259** (1985), 460.

27. M. Günaydin, L.J. Romans, and N.P. Warner, *Gauged $N = 8$ supergravity in five dimensions*, Phys. Lett. **B154** (1985), 268.

28. M. Günaydin, *Singleton and doubleton supermultiplets of space-time supergroups and infinite spin superalgebras*, in Proceedings of the 1989 Trieste Conference "Supermembranes and Physics in 2 + 1 Dimensions", eds. M.J. Duff, C.N. Pope, and E. Sezgin, World Scientific, Singapore 1990.

29. P.A.M. Dirac, *A remarkable representation of the* 3+2 *de Sitter group*, J. Math. Phys. **4** (1963), 901.

30. C. Fronsdal, *Dirac supermultiplet*, Phys. Rev. **D26** (1982), 1988.

31. M. Flato and C. Fronsdal, *Quantum field theory of singletons. The rac*, J. Math. Phys. **22** (1981), 1100.

32. W. Heidenreich, *All linear, unitary, irreducible representations of de Sitter supersymmetry with positive energy*, Phys. Lett. **B110** (1982), 461.

33. D.Z. Freedman and H. Nicolai, *Multiplet shortening in* OSp(4|8), Nucl. Phys. **B237** (1984), 342.

34. E. Sezgin, *The spectrum of the eleven dimensional supergravity compactified on the round seven sphere*, Trieste preprint, 1983, in Supergravity in Diverse Dimensions, vol. 2, eds A. Salam and E. Sezgin, World Scientific, Singapore, 1989, 1367; *The spectrum of $D = 11$ supergravity via harmonic expansions on $S^4 \times S^7$*, Fortschr. Phys. **34** (1986), 217.

35. H. Nicolai and E. Sezgin, *Singleton representations of* $\mathrm{OSp}(N, 4)$, Phys. Lett. **B143** (1984), 389.

36. M. Günaydin, L.J. Romans, and N.P. Warner, *Spectrum generating algebras in Kaluza–Klein theories*, Phys. Lett. **B146** (1984), 401.

37. W. Nahm, *Supersymmetries and their representations*, Nucl. Phys. **B135** (1978), 149.

38. M.J. Duff, *Supermembranes: The first fifteen weeks*, Class. Quant. Grav. **5** (1988), 189.

39. M.P. Blencowe and M.J. Duff, *Supersingletons*, Phys. Lett. **B203** (1988), 229.

40. H. Nicolai, E. Sezgin, and Y. Tanii, *Conformally invariant supersymmetric field theories on $S^p \times S^1$ and super p-branes*, Nucl. Phys. **B305** (1988), 483.

41. E. Bergshoeff, E. Sezgin, and P. Townsend, *Supermembranes and eleven-dimensional supergravity*, Phys. Lett. **B189** (1987), 75.

42. E. Bergshoeff, E. Sezgin, and P. Townsend, *Properties of the eleven-dimensional supermembrane theory*, Ann. Physics **185** (1988), 330.

43. E. Bergshoeff, E. Sezgin, and Y. Tanii, *Stress tensor commutators and Schwinger terms in singleton theories*, Int. J. Mod. Phys. **A5** (1990), 3599.

44. E. Bergshoeff, M.J. Duff, C.N. Pope, and E. Sezgin, *Supersymmetric supermembrane vacua and singletons*, Phys. Lett. **B199** (1988), 69.

45. E. Bergshoeff, M.J. Duff, C.N. Pope, and E. Sezgin, *Compactifications of the eleven-dimensional supermembrane*, Phys. Lett. **B224** (1989), 71.

46. M.J. Duff and C. Sutton, *The membrane at the end of the universe*, New. Sci. **118** (1988), 67.

47. M.J. Duff, *Classical and quantum supermembranes*, Class. Quant. Grav. **6** (1989), 1577.

48. M.P. Blencowe and M.J. Duff, *Supermembranes and the signature of spacetime*, Nucl. Phys. **B310** (1988), 387.

49. E. Bergshoeff, A. Salam, E. Sezgin, and Y. Tanii, *$N = 8$ supersingleton quantum field theory*, Nucl. Phys. **B305** (1988), 497.

50. M.J. Duff, C.N. Pope, and E. Sezgin, *A stable supermembrane vacuum with a discrete spectrum*, Phys. Lett. **B225** (1989), 319.

51. M.J. Duff and K.S. Stelle, *Multi-membrane solutions of $D = 11$ supergravity*, Phys. Lett. **B253** (1991), 113.

52. A. Achucarro, J. Evans, P. Townsend, and D. Wiltshire, *Super p-branes*, Phys. Lett. **B198** (1987), 441.

53. M. Günaydin, B.E.W. Nilsson, G. Sierra, and P.K. Townsend, *Singletons and superstrings*, Phys. Lett. **B176** (1986), 45.

54. M.J. Duff, P.K. Townsend, and P. van Nieuwenhuizen, *Spontaneous compactification of supergravity on the three-sphere*, Phys. Lett. **B122** (1983), 232.

55. M.J. Duff, *Supermembranes*, lectures given at the T.A.S.I. Summer School, University of Colorado, Boulder, June 1996, the Topical Meeting, Imperial College, London, July 1996 and the 26th British Universities Summer School in Theoretical Elementary Particle Physics, University of Swansea, September 1996. Fields, strings and duality Eds. C. Efthimiou and B. Greene. World Scientific, Singapore, 1997, 219

56. C.M. Hull, *Duality and the signature of spacetime*, JHEP **9811** (1998) 017, hep-th/9807127.

57. G.T. Horowitz and A. Strominger, *Black strings and p-branes*, Nucl. Phys. **B360** (1991), 197.

58. M.J. Duff and J.X. Lu, *The self-dual Type IIB superthreebrane*, Phys. Lett. **B273** (1991), 409.

59. C.G. Callan, J.A. Harvey, and A. Strominger, *World sheet approach to heterotic instantons and solitons*, Nucl. Phys. **B359** (1991), 611.

60. C.G. Callan, J.A. Harvey, and A. Strominger, *Worldbrane actions for string solitons*, Nucl. Phys. **B367** (1991), 60.

61. M.J. Duff and J.X. Lu, *Type II p-branes: the brane scan revisited*, Nucl. Phys. **B390** (1993), 276.

62. R. Gueven, *Black p-brane solutions of $D = 11$ supergravity theory*, Phys. Lett. **B276** (1992), 49.

63. J.T. Liu and R. Minasian, *Black holes and membranes in* AdS_7, Phys. Lett. **B457** (1999), 39, hep-th/9903269.

64. G.W. Gibbons and P.K. Townsend, *Vacuum interpolation in supergravity via super p-branes*, Phys. Rev. Lett. **71** (1993), 3754.

65. M.J. Duff, G.W. Gibbons, and P.K. Townsend, *Macroscopic superstrings as interpolating solitons*, Phys. Lett. B. **332** (1994), 321.

66. G.W. Gibbons, G.T. Horowitz, and P.K. Townsend, *Higher-dimensional resolution of dilatonic black hole singularities*, Class. Quant. Grav. **12** (1995), 297.

67. M.J. Duff, H. Lü, C.N. Pope, and E. Sezgin, *Supermembranes with fewer supersymmetries*, Phys. Lett. **B371** (1996), 206, hep-th/9511162.

68. L. Castellani, A. Ceresole, R. D'Auria, S. Ferrara, P. Fré, and M. Trigiante, *G/H M-branes and* AdS_{p+2} *geometries*, Nucl. Phys. **B527** (1998) 142, hep-th/9803039.

69. M.A. Awada, M.J. Duff, and C.N. Pope, *$N = 8$ supergravity breaks down to $N = 1$*, Phys. Rev. Lett. **50** (1983), 294.

70. M.J. Duff, B.E.W. Nilsson, and C.N. Pope, *Spontaneous supersymmetry breaking via the squashed seven sphere*, Phys. Rev. Lett. **50** (1983), 2043.

71. K. Yano and T. Nagano , *Einstein spaces admitting a one-parameter group of conformal transformations*, Ann. Math. **69** (1959), 451.

72. M.J. Duff and J.X. Lu, *Black and super p-branes in diverse dimensions*, Nucl. Phys. **B416** (1994), 301.

73. M.J. Duff and J. Rahmfeld, *Bound states of black holes and other p-branes*, Nucl. Phys. **B481** (1996), 332.

74. N. Khviengia, Z. Khviengia, H. Lü, and C.N. Pope, *Interecting M-branes and bound states*, Phys. Lett. **B388** (1996), 21.

75. M.J. Duff, S. Ferrara, R. Khuri, and J. Rahmfeld, *Supersymmetry and dual string solitons*, Phys. Lett. **B356** (1995), 479.

76. M.J. Duff and J. Rahmfeld, *Massive string states as extreme black holes*, Phys. Lett. **B345** (1995), 441.

77. D.Z. Freedman and G.W. Gibbons, *Electrovac ground state in gauged* $SU(2) \times SU(2)$ *supergravity*, Nucl. Phys. **B233**, (1984), 24.

78. P. Claus, R. Kallosh, J. Kumar, P.K. Townsend, and A. Van Proeyen, *Supergravity and the large N limit of theories with sixteen supercharges*, JHEP **9806** (1998), 004, hep-th/9801206.

79. G. Dall'Agata, D. Fabbri, C. Fraser, P. Fré, P. Termonia, and M. Trigiante, *The* OSp(4|8) *singleton action from the supermembrane*, Nucl. Phys. **B542** (1999), 157, hep-th/9807115.

80. A.A. Tseytlin, *Extreme dyonic black holes in string theory*, Mod. Phys. Lett. **A11** (1996), 689, hep-th/9601177.

81. M. Cvetic and A.A. Tseytlin, *General class of BPS saturated black holes as exact superstring solutions*, Phys. Lett. **B366** (1996), 95, hep-th/9510097.

82. M. Cvetic and A.A. Tseytlin, *Solitonic strings and BPS saturated dyonic black holes*, Phys. Rev. **D53** (1996), 5619, hep-th/9512031.

83. M.J. Duff, H. Lü, and C.N. Pope, *Supersymmetry without supersymmetry*, Phys. Lett. **B409** (1997), 136, hep-th/9704186.

84. S. Kachru and E. Silverstein, *4d conformal field theories and strings on orbifolds*, Phys. Rev. Lett. 80 (1998), 4855, hep-th/9802183.

85. M.J. Duff, B.E.W. Nilsson, and C.N. Pope, *Compactification of* $D = 11$ *supergravity on* $K3 \times T^3$, Phys. Lett. **B129** (1983), 39.

86. M.J. Duff, H. Lü, and C.N. Pope, $\mathrm{AdS}_5 \times S^5$ *untwisted*, Nucl. Phys. **B532** (1998), 181, hep-th/9803061.

87. M.J. Duff, H. Lü, and C.N. Pope, $\mathrm{AdS}_3 \times S^3$ *(un)twisted and squashed*, Nucl.,Phys. **B544** (1999), 145, hep th/9807173.

88. E. Halyo, *Supergravity on* $\mathrm{AdS}_{5/4} \times$ *Hopf fibrations and conformal field theories*, Mod. Phys. Lett. **A15** (2000) 397, hep-th/9803193.

89. D.N. Page and C.N. Pope, *Which compactifications of $D = 11$ supergravity are stable?*, Phys. Lett. **B144** (1984), 346.

90. Y. Oz and J. Terning, *Orbifolds of* $\mathrm{AdS}_5 \times S^5$ *and 4-D conformal field theories*, Nucl. Phys. **B532** (1998) 163, hep-th/9803167.

91. L. Romans, *New compactifications of chiral $N = 2$, $d = 10$ supergravity*, Phys. Lett. **B153** (1985), 392.

92. I. Klebanov and E. Witten, *Superconformal field theory on three-branes at a Calabi–Yau singularity*, Nucl. Phys. **B536** (1998) 199, hep-th/9807080.

93. L.J. Romans, *Gauged $N = 4$ supergravities in five dimensions and their magnetovac backgrounds*, Nucl. Phys. **B267** (1986), 443.

94. M.J. Duff, P. Howe, T. Inami, and K.S. Stelle, *Superstrings in $D = 10$ from supermembranes in $D = 11$*, Phys. Lett. **B191** (1987), 70.

95. M.J. Duff and J.X. Lu, *Duality rotations in membrane theory*, Nucl. Phys. **B347** (1990), 394.

96. C.M. Hull and P.K. Townsend, *Unity of superstring dualities*, Nucl. Phys. **B438** (1995), 109.

97. P.K. Townsend, *The eleven-dimensional supermembrane revisited*, Phys. Lett. **B350** (1995), 184.

98. E. Witten, *String theory dynamics in various dimensions*, Nucl. Phys. **B443** (1995), 85.

99. M.J. Duff, J.T. Liu, and R. Minasian, *Eleven-dimensional origin of string/string duality: A one loop test*, Nucl. Phys. **B452** (1995), 261.

100. K. Becker, M. Becker, and A. Strominger, *Fivebranes, membranes and nonperturbative string theory*, Nucl. Phys. **B456** (1995), 130.

101. J.H. Schwarz, *The power of M-theory*, Phys. Lett. **B360** (1995), 13.

102. P. Horava and E. Witten, *Heterotic and Type I string dynamics from eleven dimensions*, Nucl. Phys. **B460** (1996), 506.

103. P. Townsend, *D-Branes From M-Branes*, Phys. Lett. **B373** (1996), 68.

104. O. Aharony, J. Sonnenschein, and S. Yankielowicz, *Interactions of strings and D-branes from M-theory*, Nucl. Phys. **B474** (1996), 309.

105. T. Banks, W. Fischler, S.H. Shenker, and L. Susskind, *M-theory as a matrix model: a conjecture*, Phys. Rev. **D55** (1997), 5112.

106. R. Slansky, *Group theory for unified model building*, Phys. Rep. **79** (1981), 1.

107. B.E.W. Nilsson and C.N. Pope, *Hopf fibration of* $D = 11$ *supergravity*, Class. Quant. Grav. **1** (1984), 499.

108. E. Witten, *Strong coupling and the cosmological constant*, Modern Phys. Lett. **A10** (1995), 2153.

109. M.J. Duff and C. Orzalesi, *The cosmological constant in spontaneously compactified* $D = 11$ *supergravity*, Phys. Lett. **B122** (1983), 37.

110. J. Maldacena, *The large* N *limit of superconformal field theories and supergravity*, Int. J. Theor. Phys. **38** (1999), 1113, hep-th/9711200.

111. J. Polchinski, *Dirichlet branes and Ramond–Ramond charges*, Phys. Rev. Lett. **75** (1995), 4724

112. E. Witten, *Bound states of strings and p-branes*, Nucl. Phys. **B460** (1996), 335, hep-th/9510135.

113. S.S. Gubser, I.R. Klebanov, and A. W. Peet, *Entropy and temperature of black 3-branes*, Phys. Rev. **D54** (1996), 3915, hep-th/9602135.

114. I.R. Klebanov, *World volume approach to absorption by nondilatonic branes*, Nucl. Phys. **B496** (1997), 231, hep-th/9702076.

115. S.S. Gubser, I.R. Klebanov, and A.A. Tseytlin, *String theory and classical absorption by three-branes*, Nucl. Phys. **B499** (1997), 217, hep-th/9703040.

116. S.S. Gubser and I.R. Klebanov, *Absorption by branes and Schwinger terms in the world volume theory*, Phys. Lett. **B413** (1997), 41, hep-th/9708005.

117. N. Itzhak, J.M. Maldacena, J. Sonnenschein, and S. Yankielowicz, *Supergravity and the large* N *limit of theories with sixteen supercharges*, Phys. Rev. **D58** (1998), 046004, hep-th/9802042.

118. S. Ferrara and C. Fronsdal, *Conformal Maxwell theory as a singleton field theory on* AdS_5*, IIB three-branes and duality*, Class. Quant. Grav. **15** (1998), 2153, hep-th/9712239.

119. R. Kallosh, J. Kumar, and A. Rajaraman, *Special conformal symmetry of worldvolume actions*, Phys. Rev. **D57** (1998), 6452, hep-th/9712073.

120. H.J. Boonstra, B. Peters, and K. Skenderis, *Branes and anti-de Sitter space-time*, Fortsch. Phys. **47** (1999), 109, hep-th/9801076.

121. M. Günaydin and D. Minic, *Singletons, doubletons and M-theory*, Nucl. Phys. **B523** (1998), 145, hep-th/9802047.

122. S.S. Gubser, I.R. Klebanov, and A.M. Polyakov, *Gauge theory correlators from noncritical string theory*, Phys. Lett. **B428** (1998), 105, hep-th/9802109.

123. G.T. Horowitz and H. Ooguri, *Spectrum of large N gauge theory from supergravity*, Phys. Rev. Lett. **80** (1998) 4116, hep-th/9802116.

124. E. Witten, *Anti-de Sitter space and holography*, Adv. Theor. Math. Phys. **2** (1998), 253, hep-th/9802150.

125. M. Berkooz, *A supergravity dual of a $(1,0)$ field theory in six dimensions*, Phys. Lett. **B437** (1998), 315, hep-th/9802195.

126. S. Ferrara, C. Fronsdal, and A. Zaffaroni, *On $N = 8$ supergravity on AdS_5 and $N = 4$ superconformal Yang–Mills theory*, Nucl. Phys. **B532** (1998), 153, hep-th/9802203.

127. S.-J. Rey and J. Lee, *Macroscopic strings as heavy quarks in large N gauge theory and anti-de Sitter supergravity*, hep-th/9803001.

128. J. Maldacena, *Wilson loops in large N field theories*, Phys. Rev. Lett. **80** (1998), 4859, hep-th/9803002.

129. A. Lawrence, N. Nekrasov, and C. Vafa, *On conformal field theories in four dimensions*, Nucl. Phys. **B533** (1998), 199, hep-th/9803015.

130. G. 't Hooft, *Dimensional reduction in quantum gravity*, gr-qc/9310026.

131. L. Susskind, *The world as a hologram*, J. Math. Phys. **36** (1995), 6377, hep-th/9409089.

132. S.W. Hawking and C.N. Pope, *Generalised spin structures in quantum gravity*, Phys. Lett. **B73** (1978), 42.

133. C.N. Pope, *Eigenfunctions and spin^c structures in CP^2*, Phys. Lett. **B97** (1980), 417.

134. M.J. Duff, H. Lü, and C.N. Pope, *Heterotic phase transitions and singularities of the gauge dyonic string*, hep-th/9603037.

135. M.J. Duff, J.T. Liu, H. Lü, and C.N. Pope, *Gauge dyonic strings and their global limit*, Nucl. Phys. **B529** (1998) 137, hep-th/9711089.

136. H.J. Boonstra, B. Peeters, and K. Skenderis, *Duality and asymptotic geometries*, Phys. Lett. **B411** (1997), 59, hep-th/9706192.

137. H.J. Boonstra, B. Peeters, and K. Skenderis, *Brane intersections, anti-de Sitter spacetimes and dual superconformal theories*, Nucl.Phys. B533 (1998), 127, hep-th/9803231.

138. G.T. Horowitz, J. Maldacena, and A. Strominger, *Nonextremal black hole microstates and U duality*, Phys. Lett. **B383** (1996), 151, hep-th/9603109.

139. G.T. Horowitz and A. Strominger, *Counting states of near extremal black holes*, Phys. Rev. Lett. **77** (1996), 2368, hep-th/9602051.

140. N. Seiberg and E. Witten, *The D1/D5 system and singular CFT*, JHEP **9904** (1999), 017, hep-th/9903224.

141. J. Maldacena, J. Michelson, and A. Strominger, *Anti-de Sitter fragmentation*, JHEP **9902** (1999), 011, hep-th/9812073.

142. J. Maldacena and A. Strominger, AdS_3 *black holes and a stringy exclusion principle*, JHEP **9812** (1998) 005, hep-th/9804085.

143. L.J. Romans, *Supersymmetric, cold and lukewarm black holes in cosmological Einstein–Maxwell theory*, Nucl. Phys. **B383** (1992), 395; *Black holes in cosmological Einstein–Maxwell theory*, in Proceedings of the 1992 Trieste summer school "High Energy Physics and Cosmology", eds. E. Gava, K.S. Narain, S. Randjbar-Daemi, E Sezgin, and Q. Shafi, World Scientific, Singapore, 1993, 416.

144. K. Behrndt, A.H. Chamseddine, and W.A. Sabra, *BPS black holes in $N = 2$ five-dimensional AdS supergravity*, Phys. Lett. **B442** (1998), 97, hep-th/9807187.

145. D. Birmingham, *Topological black holes in anti-de Sitter space*, Class. Quant. Grav. **16** (1999), 1197, hep-th/9808032.

146. M.M. Caldarelli and D. Klemm, *Supersymmetry of anti-de Sitter black holes*, Nucl. Phys. **B545** (1999), 434, hep-th/9808097.

147. D. Klemm, *BPS black holes in gauged $N = 4$, $D = 4$ supergravity*, Nucl. Phys. B545 (1999) **461**, hep-th/9810090.

148. K. Behrndt, M. Cvetic, and W.A. Sabra, *Non-extreme black holes five-dimensional $N = 2$ AdS supergravity*, Nucl. Phys. **B553** (1999) 317, hep-th/9810227.

149. M.J. Duff and J.T. Liu, *Anti-de Sitter black holes in gauged $N = 8$ supergravity*, Nucl. Phys. **B554** (1999), 237, hep-th/9901149.

150. A. Chamblin, R. Emparan, C.V. Johnson, and R.C. Myers, *Charged AdS black holes and catastrophic holography*, Phys. Rev. **D60** (1999), 064018, hep-th/9902170.

151. M. Cvetic and S.S. Gubser, *Phases of R charged black holes, spinning branes and strongly coupled gauge theories*, JHEP **9904** (1999), 024, hep-th/9902195.

152. M. Cvetic and S.S. Gubser, *Thermodynamic stability and phases of general spinning branes*, JHEP **9907** (1999), 010, hep-th/9903132.

153. W. Sabra, *Anti-de Sitter black holes in $N = 2$ gauged supergravity*, Phys. Lett. **B458** (1999) 36, hep-th/9903143.

154. M.M. Caldarelli and D. Klemm, *M-theory and stringy corrections to anti-de Sitter black holes and conformal field theories*, Nucl. Phys. **B555** (1999), 157; hep-th/9903078.

155. E. Witten, *Anti-de Sitter space, thermal phase transition, and confinement in gauge theories*, Adv. Theor. Math. Phys. **2** (1998), 505, hep-th/9803131.

156. B. de Wit and H. Nicolai, *$N = 8$ supergravity with local* $\mathrm{SO}(8)\times\mathrm{SU}(8)$ *invariance*, Phys. Lett. **B108** (1982), 285.

157. M.J. Duff, *Anti-de Sitter space, branes, singletons, superconformal field theories and all that*, Int. J. Mod. Phys. **A14** (1999), 815, hep-th/9808100.

158. M.J. Duff and C.N. Pope, *Consistent truncations in Kaluza–Klein theories*, Nucl. Phys. **B255** (1985), 355.

159. C.N. Pope, *Consistency of truncations in Kaluza–Klein*, published in the Proceedings of the 1984 Santa Fe meeting.

160. B. de Wit, H. Nicolai, and N.P. Warner, *The embedding of gauged $N = 8$ supergravity into $D = 11$ supergravity*, Nucl. Phys. **B255** (1985), 29.

161. M.J. Duff, B.E.W. Nilsson, C.N. Pope, and N. Warner, *On the consistency of the Kaluza–Klein ansatz*, Phys. Lett. **B149** (1984), 90.

162. B. de Wit and H. Nicolai, *The consistency of the S^7 truncation in $D = 11$ supergravity*, Nucl. Phys. **B281** (1987), 211.

163. M.J. Duff, J.T. Liu, and J. Rahmfeld, *Four-dimensional string-string-string triality*, Nucl. Phys. **B459** (1996), 125, hep-th/9508094.

164. M. Cvetic and D. Youm, *Dyonic BPS saturated black holes of heterotic string on a six torus*, Phys. Rev. **D53** (1996), 584, hep-th/9507090.

165. H. Lü and C.N. Pope, *p-brane solitons in maximal supergravities*, Nucl. Phys. **B465** (1996), 127, hep-th/9512012.

166. R.R. Khuri and T. Ortin, *Supersymmetric black holes in $N = 8$ supergravity*, Nucl. Phys. **B467** (1996), 355.

167. M. Cvetic and A. A. Tseytlin, *General class of BPS saturated dyonic black holes as exact superstring solutions*, Phys. Lett. **B366** (1996), 95, hep-th/9510097.

168. A.A. Tseytlin, *Harmonic superpositions of M-branes*, Nucl. Phys. **B475** (1996), 149, hep-th/9604035.

169. M. Cvetic and D. Youm, *Near BPS saturated rotating electrically charged black holes as string states*, Nucl. Phys. **B477** (1996), 449, hep-th/9605051.

170. M. Cvetic and D. Youm, *Rotating intersecting M-branes*, Nucl. Phys. **B499** (1997), 253, hep-th/9612229.

171. S. Gubser, *Thermodynamics of spinning D3-branes*, Nucl. Phys. **B551** (1999), 667, hep-th/9810225.

172. C. Csaki, Y. Oz, J. Russo, and J. Terning, *Large N QCD from rotating branes*, Phys. Rev. **D59** (1999), 065012, hep-th/9810186.

173. P. Kraus, F. Larsen, and S.P. Trivedi, *The Coulomb branch of gauge theory from rotating branes*, Adv. Theor. Math. Phys. **3** (1999), 13, hep-th/9901056.

174. R. Cai and K. Soh, *Critical behavior in the rotating D-branes*, Mod. Phys. Lett. **A14** (1999), 1895, hep-th/9812121.

175. P. Kraus, F. Larsen, and S.P. Trivedi, *The Coulomb branch of gauge theory from rotating branes*, JHEP **9903** (1999), 003, hep-th/9811120.

176. J.G. Russo and K. Sfetsos, *Rotating D3-branes and QCD in three dimensions*, Adv. Theor. Math. Phys. **3** (1999), 131, hep-th/9901056.

177. C. Csaki, J. Russo, K. Sfetsos, and J. Terning, *Supergravity models for 3 + 1 dimensional QCD*, Phys. Rev. **D60** (1999), 044001, hep-th/9902067.

178. K. Sfetsos, *Rotating NS5-brane solution and its exact string theoretical description*, Fortsch. Phys. **48** (2000), 199, hep-th/9903201.

179. K.S. Thorne, R.H. Price and D.A. Macdonald (eds.), *Black holes: the membrane paradigm*, Yale Univ. Press, New Haven, CT, 1986.

180. R.C. Myers and M.J. Perry, *Black holes in higher dimensional space-times*, Ann. Phys. **172** (1986), 304.

181. M. Cvetic, H. Lü, and C.N. Pope, *Spacetimes of boosted p-branes, and CFT in infinite momentum frame*, Nucl.Phys. **B545** (1999), 309, hep-th/9810123.

182. B.E.W. Nilsson, *On the embedding of* $d = 4$, $N = 8$ *gauged supergravity in* $d = 11$, $N = 1$ *supergravity*, Phys. Lett. **B155** (1985), 54.

183. C.N. Pope, *The embedding of the Einstein–Yang–Mills equations in* $D = 11$ *supergravity*, Class. Quant. Grav. **2** (1985), L77.

184. M. Pernici, K. Pilch, and P. van Nieuwenhuizen, *Noncompact gaugings and critical points of maximal supergravity in seven dimensions*, Nucl. Phys. **B249** (1985), 381.

185. H. Lü, C.N. Pope, E. Sezgin, and K.S. Stelle, *Dilatonic p-brane solitons*, Phys. Lett. **B371** (1996), 46, hep-th/9511203.

186. E. Cremmer and B. Julia, *The* $N = 8$ *supergravity theory. I. The Lagrangian*, Phys. Lett. **B80** (1978), 48.

187. E. Cremmer and B. Julia, *The* SO(8) *supergravity*, Nucl. Phys. **B159** (1979), 141.

188. E. Cremmer, B. Julia, H. Lü, and C.N. Pope, *Dualisation of dualities*, Nucl. Phys. **B523** (1998), 73, hep-th/9710119.

189. M.J. Duff and J. Rahmfeld, *Bound states of black holes and other p-branes*, Nucl. Phys. **B481** (1996), 332.

190. P.C. Aichelburg and F. Embacher, *Exact superpartners of* $N = 2$ *supergravity solitons*, Phys. Rev. **D34** (1986), 3006.

191. M.J. Duff, J.T. Liu, and J. Rahmfeld, *Dipole moments of black holes and string states*, Nucl. Phys. **B494** (1997), 161.

192. D.Z. Freedman and H. Nicolai, *Multiplet shortening in* $\mathrm{OSp}(N, 4)$, Nucl. Phys. **B237** (1984), 342.

193. M.J. Duff and J. Rahmfeld, *Massive string states as extreme black holes*, Phys. Lett. **B345** (1995), 441.

194. M.M. Caldarelli and D. Klemm, *Supersymmetry of anti-de Sitter black holes*, Nucl. Phys. **B545** (1999), 434, hep-th/9808097.

195. J. Rahmfeld, *Extremal black holes as bound states*, Phys. Lett. **B372** (1996), 198.

196. C.R. Graham and R. Lee, *Einstein metrics with prescribed conformal infinity on the ball*, Adv. Math. **87** (1991), 186.

197. E. Cremmer, *Symmetries in extended supergravity*, in Proceedings of the 1981 Trieste conference "Supergravity 81," eds. S. Ferarra and J.G. Taylor, Cambridge Univ. Press, Cambridge–New York, 1982.

198. A. Salam and E. Sezgin, *Supergravities in diverse dimensions*, World Scientific, Singapore, 1989.

199. E. Witten, *Fermion quantum numbers in Kaluza–Klein theory*, in Proceedings of the "Shelter Island II" conference (1983), eds. R. Jackiw, N.N. Khuri, S. Weinberg, and E. Witten M.I.T. Press, Cambridge, MA–London, 1985, 227.

200. M.J. Duff, B.E.W. Nilsson, and C.N. Pope, *Kaluza–Klein supergravity*, Phys. Rep. **130** (1986), 1.

201. M.J. Duff, *Kaluza–Klein theory in perspective*, in the Proceedings of the Nobel Symposium "Oskar Klein Centenary," Stockholm, September 1994, ed. U. Lindstrom, World Scientific, Singapore 1995, hep-th/9410046.

202. P.K. Townsend, *M(embrane) theory on T^9*, in the "Strings 97" Proceedings, eds. F.A. Bais, E.A. Bergshoeff, B. de Wit, R. Dijkgraaf, A.N. Schellekens, E.P. Verlinde, and H.L. Verlinde, Nucl. Phys. **B68** Proc. Suppl. (1998), 11, hep-th/9708054.

203. M. Cvetic, M.J. Duff, P. Hoxha, J.T. Liu, H. Lu, J.X. Lü, R. Martinez-Acosta, C.N. Pope, H.Sati, and T.A. Tran, *Embedding AdS black holes in ten and eleven dimensions*, Nucl. Phys. **B558** (1999), 96, hep-th/9903214.

204. V.A. Rubakov and M.E. Shaposhnikov, *Do we live inside a domain wall?*, Phys. Lett. **B125** (1983), 136.

205. G.W. Gibbons and D.L. Wiltshire, *Space-time as a membrane in higher dimensions*, Nucl. Phys. **B287** (1987), 717.

206. G.T. Horowitz and A. Strominger, *Black strings and p-branes*, Nucl. Phys. **B360** (1991), 197.

207. I. Antoniadis, *A possible new dimension at a few TeV*, Phys. Lett. **B246** (1990), 377.

208. E. Witten, *Strong coupling expansion of Calabi-Yau compactification*, Nucl. Phys. **B471** (1996), 135.

209. J. Lykken, *Weak scale superstrings*, Phys. Rev. **D54** (1996), 3693.

210. N. Arkani-Hamed, S. Dimopoulos, and G. Dvali, *The hierarchy problem and new dimensions at a millimeter*, Phys. Lett. **B429** (1998), 263.

211. K.R. Dienes, E. Dudas, and T. Gherghetta, *Extra space-time dimensions and unification*, Phys. Lett. **B436** (1998), 55.

212. E. Witten, *Search for a realistic Kaluza–Klein theory*, Nucl. Phys. **B186** (1981), 412–428.

213. L. Castellani, R. D'Auria, and P. Fré, *Supergravity and superstrings: a geometric perspective* (in 3 volumes), World Scientific Press, Singapore, 1991.

214. K.S. Stelle, *An introduction to p-branes*, In Seoul/Sokcho 1997 "Dualities in Gauge and String Theories," eds. Y.M. Cho and S.Nam, World Scientific, Singapore, 1998, 39.

215. S.P. de Alwis, *A note on brane tension and M-theory*, Phys.Lett. B388 (1996) 291, hep-th/9607011.

216. O. Aharony, S. Gubser, J. Maldacena, H. Ooguri, and Y. Oz, *Large N field theories , string theory and gravity*, Phys. Rep. **323** (2000), 183, hep-th/9905111.

217. S.K. Han and I.G. Koh, *$N = 4$ supersymmetry remaining in Kaluza–Klein monopole background in $D = 11$ supergravity*, Phys. Rev. **D31** (1985), 2503.

218. H. Nastase, D. Vaman, and P. van Nieuwenhuizen, *Consistent nonlinear KK reduction of 11-D supergravity on* $\mathrm{AdS}_7 \times S^4$ *and selfduality in odd dimensions*, Phys. Lett. **B469** (1999) 96–102, hep-th/9905075.

219. H. Nastase, D. Vaman, and P. van Nieuwenhuizen, *Consistency of the* $\mathrm{AdS}_7 \times S^4$ *reduction and the origin of selfduality in odd dimensions*, Nucl. Phys. **B581** (2000), 179-239, hep-th/9911238.

220. H. Lü and C.N. Pope, *Exact embedding of $N = 1$, $D = 7$ gauged supergravity in $D = 11$*, Phys. Lett. **B467** (1999), 67, hep-th/9906168.

221. H. Lü, C.N. Pope, and T.A. Tran, *Five-dimensional $N = 4$,* $\mathrm{SU}(2) \times \mathrm{U}(1)$ *gauged supergravity from type IIB*, Phys. Lett. **B475** (2000), 261, hep-th/9909203.

222. M. Cvetic, H. Lü, and C.N. Pope, *Four-dimensional $N = 4$* $\mathrm{SO}(4)$ *gauged supergravity from $D = 11$*, Nucl. Phys. **B574** (2000), 761, hep-th/9910252.

223. S.W. Hawking and H.S. Reall, *Charged and rotating AdS black holes and their CFT duals*, Phys. Rev. **D61** (2000) 024014, hep-th/9908109.

224. M. Günaydin and S. Takemae, *Unitary supermultiplets of* OSp(8^* | 4) *and their* AdS_7/CFT_6 *duality*, Nucl. Phys. **B578** (2000), 405–448, hep-th/9910110.

225. P.S. Howe and E. Sezgin, *Superbranes*, Phys. Lett. **B390** (1997), 133–142, hep-th/9607227.

226. E. Sezgin and P. Sundell, *Aspects of the M5-brane*, hep-th/9902171.

227. P.S. Howe, E. Sezgin and P.C. West, *Covariant field equations of the M-theory five-brane*, Phys. Lett. **399B** (1997), 49.

228. P.S. Howe and E. Sezgin, $D = 11$, $p = 5$, Phys. Lett. **B394** (1997), 62.

229. I. Bandos, K. Lechner, A. Nurmagambetov, P. Pasti, D. Sorokin, and M. Tonin, *Covariant action for the superfivebrane of M-theory*, Phys. Rev. Lett. **78** (1997), 4332.

230. I. Bandos, K. Lechner, A. Nurmagambetov, P. Pasti, D. Soroki, and M. Tonin, *On the equivalence of different formulations of the M theory five-brane*, Phys. Lett. **B408** (1997), 135, hep-th/9703127.

231. E. Bergshoeff and P. K. Townsend, *Super D-branes*, Nucl. Phys. **B490** (1997), 145.

232. L. Randall and R. Sundrum, *A large mass hierarchy from a small extra dimension*, Phys. Rev. Lett. **83** (1999), 3370, hep-ph/9905221.

233. L. Randall and R. Sundrum, *An alternative to compactification*, Phys. Rev. Lett. **83** (1999), 4690, hep-th/9906064.

234. J.L. Petersen, *Introduction to the Maldacena conjecture on AdS/CFT*, Int. J. Mod. Phys. **A14** (1999), 3597, hep-th/9902131.

235. E. Witten and S.T. Yau, *Connectedness of the boundary in the AdS/CFT correspondence*, hep-th/9910245.

236. T. Banks, *M theory and cosmology*, hep-th/9911067

237. M.J. Duff, *The world in eleven dimensions: supergravity, supermembranes and M-theory*, a reprint volume with commentaries, I.O.P. Publishing, Bristol, 1999.

238. P. van Nieuwenhuizen, *Supergravity*, Phys. Rep. **68** (1981), 189.

239. C. Ahn and S-J. Rey, *Three-dimensional CFTs and RG flow from squashing M2-brane horizon*, Nucl. Phys. **B565** (2000), 210–214, hep-th/9908110

240. K. Landsteiner and E. Lopez, *The thermodynamic potentials of Kerr-AdS black holes and their CFT duals*, JHEP **9912** (1999) 020, hep-th/9911124.

241. S. Deser and D. Seminara, *Counterterms/M-theory corrections to D = 11 supergravity*, Phys. Rev. Lett. **82** (1999), 2435.

242. T. Harmark and N.A. Obers, *Thermodynamics of spinning branes and their dual field theories*, JHEP **0001** (2000) 008, hep-th/9910036.

243. T. Harmark and N.A. Obers, *Phase structure of noncommutative field spinning branes and their dual field theories*, JHEP **0003** (2000) 024, hep-th/9911169.

A The Lagrangian, Symmetries and Transformation Rules of $D = 11$ Supergravity

The unique supermultiplet in $D = 11$ consists of a graviton described by the elfbein $e_M{}^A$, a gravitino described by the Rarita-Schwinger vector-spinor Ψ_M and the 3-form gauge field A_{MNP}. Here letters at the beginning of the alphabet are local Lorentz indices and letters from the middle of the alphabet are spacetime world indices. The Lagrangian is given by

$$\begin{aligned}
\mathcal{L}_{11} = {} & \frac{1}{2\kappa_{11}{}^2} e\left[R - \frac{1}{2!\,4!} F^{MNPQ} F_{MNPQ}\right] \\
& - \frac{1}{12\kappa_{11}{}^2} \frac{1}{3!\,4!^2} \epsilon^{M_1 \cdots M_{11}} A_{M_1 M_2 M_3} F_{M_4 M_5 M_6 M_7} F_{M_8 M_9 M_{10} M_{11}} \\
& + \frac{1}{2\kappa_{11}{}^2} e\left[-2i\bar{\Psi}_M \Gamma^{MNP} D_N\left(\frac{\omega+\hat{\omega}}{2}\right)\Psi_P\right. \\
& \qquad + \frac{i}{96}(\bar{\Psi}_M \Gamma^{MNPQRS}\Psi_N + 12\bar{\Psi}^P \Gamma^{QR}\Psi_S) \\
& \qquad \left. \times (F_{PQRS} + \widehat{F}_{PQRS})\right]
\end{aligned} \tag{A.1}$$

where

$$\bar{\Psi} = \Psi^\dagger \Gamma^0 \tag{A.2}$$

and Ψ is a Majorana spinor obeying

$$\bar{\Psi} = \Psi^T C^{-1} \tag{A.3}$$

where C is the charge conjugation matrix obeying

$$C^{-1}\Gamma_A C = -\Gamma_A{}^T \tag{A.4}$$

and Γ_A are the $D = 11$ Dirac matrices

$$\{\Gamma^A, \Gamma^B\} = 2\eta^{AB} \tag{A.5}$$

$$\eta^{AB} = \text{diag}(-, +, +, \cdots +). \tag{A.6}$$

Furthermore,

$$e = \det e_M{}^A \tag{A.7}$$

$$F_{MNPQ} = 4\partial_{[M}A_{NPQ]} \tag{A.8}$$

$$\widehat{F}_{MNPQ} = F_{MNPQ} - 3\bar{\Psi}_{[M}\Gamma_{NP}\Psi_{Q]} \tag{A.9}$$

$$D_M(\omega)\Psi_N = \partial_M\Psi_N + \tfrac{1}{4}\omega_{NAB}\Gamma^{AB} \tag{A.10}$$

$$\hat{\omega}_{NAB} = \omega_{NAB} - \tfrac{i}{4}\bar{\Psi}_P\Gamma_{NAB}{}^{PQ}\Psi_Q \tag{A.11}$$

$$\begin{aligned}\omega_{NAB} = \omega^0{}_{NAB}(e) + \tfrac{i}{4}[&\bar{\Psi}_P\Gamma_{NAB}{}^{PQ}\Psi_Q \\ &- 2(\bar{\Psi}_N\Gamma_B\Psi_A - \bar{\Psi}_N\Gamma_A\Psi_B + \bar{\Psi}_B\Gamma_N\Psi_A)]\end{aligned} \tag{A.12}$$

$$\omega^0{}_{NAB}(e) = e^N{}_A e^P{}_B(\Omega_{PMN} + \Omega_{MNP} - \Omega_{NPM}) \tag{A.13}$$

where Ω_{MNP} are the anholonomy coefficients

$$[\partial_A, \partial_B] = [e^M{}_A\partial_M, e^N{}_B\partial_N] = \Omega_{AB}{}^C\partial_C \tag{A.14}$$

and $e^M{}_A$ is the inverse of $e_M{}^A$:

$$e_M{}^A e^{NB}\eta_{AB} = \delta_M{}^N. \tag{A.15}$$

The action is invariant under the following symmetry transformations:
a) $D = 11$ general coordinate transformations with parameter ξ^M:

$$\begin{aligned}\delta e_M{}^A &= e_N{}^A\partial_M\xi^N + \xi^N\partial_N e_M A \\ \delta\Psi_M &= \Psi_N\partial_M\xi^N + \xi^N\partial_N\Psi_M \\ \delta A_{MNP} &= 3A_{Q[MN}\partial_{P]}\xi^Q + \xi^Q\partial_Q A_{MNP}.\end{aligned} \tag{A.16}$$

b) Local SO(1,10) Lorentz transformations with parameter $\alpha_{AB} = -\alpha_{BA}$:

$$\begin{aligned}\delta e_M{}^A &= -e_M{}^B\alpha_B{}^A, \\ \delta\Psi_M &= -\tfrac{1}{4}\alpha_{AB}\Gamma^{AB}\Psi_M, \\ \delta A_{MNP} &= 0.\end{aligned} \tag{A.17}$$

c) $N = 1$ supersymmetry transformations withanticommuting parameter ϵ:

$$\begin{aligned}\delta e_M{}^A &= i\bar{\epsilon}\Gamma^A\Psi_M, \\ \delta\Psi_M &= D_M(\hat{\omega})\epsilon - \frac{1}{288}(\Gamma_M{}^{PQRS} - 8\delta_M{}^P\Gamma^{QRS})\widehat{F}_{PQRS}\epsilon, \\ \delta A_{MNP} &= 3i\bar{\epsilon}\Gamma_{[MN}\Psi_{P]}.\end{aligned} \tag{A.18}$$

d) Abelian gauge transformations with parameter $\Lambda_{MN} = -\Lambda_{NM}$:

$$\begin{aligned} \delta e_M{}^A &= 0, \\ \delta \Psi_M &= 0, \\ \delta A_{MNP} &= \partial_{[M}\Lambda_{NP]}. \end{aligned} \tag{A.19}$$

e) an odd number of space or time reflections together with

$$A_{MNP} \to -A_{MNP}. \tag{A.20}$$

B The Field Equations, Symmetries and Transformation Rules of Type IIB Supergravity

Next we consider Type IIB supergravity in $D = 10$ [21, 22] which also describes $128 + 128$ degrees of freedom, and corresponds to the field-theory limit of the Type IIB superstring. The spectrum of the supergravity theory consists of a complex scalar A, a complex spinor λ, a complex 2-form A_{MN}, a complex Weyl gravitino ψ_M, a real graviton $e_M{}^R$ and a real 4-form A_{MNPQ} whose 5-form field strength F_{MNPQR} obeys a self-duality condition. The complex Weyl spinors ψ_M and λ have opposite handedness

$$\gamma_{11}\psi_M = -\psi_M \tag{B.1}$$

$$\gamma_{11}\lambda = \lambda. \tag{B.2}$$

Owing to the self-duality of F_{MNPQR}, there exists no covariant action principle. It is therefore simplest to work directly with the field equations, which are:

$$D^M P_M = \frac{1}{24}\kappa_{10}{}^2 G_{MNP}G^{MNP} + O(\psi^2) \tag{B.3}$$

$$\gamma^M D_M \lambda = \frac{1}{240} i\kappa_{10}\gamma^{P_1\cdots P_5}\lambda F_{P_1\cdots P_5} + O(\psi^2) \tag{B.4}$$

$$D^P G_{MNP} = P^P G^*{}_{MNP} - \frac{2}{3} i\kappa_{10} F_{MNPQR} G^{*PQR} + O(\psi^2) \tag{B.5}$$

$$\begin{aligned} \gamma^{MNP} D_M \psi_P = {} & -i\frac{1}{2}\gamma^P\gamma^M\lambda^* P_P \\ & - \frac{1}{48} i\kappa_{10}\gamma^{NPQ}\gamma^M \lambda G^*{}_{NPQ} + O(\psi^3) \end{aligned} \tag{B.6}$$

$$\begin{aligned} R_{MP} - \frac{1}{2} g_{MP} R = {} & P_M P^*{}_P + P^*{}_M P_P - g_{MP} P^R P^*{}_R \\ & + \frac{1}{6}\kappa_{10}{}^2 F_{R_1\cdots R_4 M} F^{R_1\cdots R_4}{}_P \\ & + \frac{1}{8}\kappa_{10}{}^2 (G_M{}^{RS} G^*{}_{PRS} + G^*{}_M{}^{RS} G_{PRS}) \\ & - \frac{1}{24}\kappa_{10}{}^2 g_{MP} G^{RST} G^*{}_{RST} + O(\psi^2) \end{aligned} \tag{B.7}$$

$$F_{MNPQR} = *F_{MNPQR} \tag{B.8}$$

where

$$\begin{aligned}
P_M &= f^2 \partial_M B & \text{(B.9)}\\
G_{MNP} &= f(F_{MNP} - BF^*{}_{MNP}) & \text{(B.10)}\\
F_{MNP} &= 3\partial_{[M} A_{NP]} & \text{(B.11)}\\
f &= (1 - B^* B)^{-1/2} & \text{(B.12)}\\
F_{MNPQR} &= 5\partial_{[M} A_{NPQR]} - \tfrac{5}{4}\kappa_{10}\Im(A_{[MN}F^*{}_{PQR]}) & \text{(B.13)}
\end{aligned}$$

and where the supercovariant derivative D_M involves the composite U(1) gauge field

$$Q_M = f^2 \Im(B\partial_M B^*). \tag{B.14}$$

Our notation is that X^* is the complex conjugate of X while $*X$ is the Hodge dual of X.

We also have the identities

$$\begin{aligned}
D_{[M} P_{P]} &= 0 & \text{(B.15)}\\
D_{[M} G_{NPQ]} &= -P_{[M} G^{*NPQ]} & \text{(B.16)}\\
\partial_{[P} Q_{M]} &= -iP_{[P} P^*{}_{M]} & \text{(B.17)}\\
\partial_{[N} F_{M_1 \cdots M_5]} &= \frac{5}{12} i\kappa G_{[NM_1M_2} G^*{}_{M_3M_4M_5]}. & \text{(B.18)}
\end{aligned}$$

The symmetries are $D = 10$ general covariance, local SO(1,9) Lorentz transformations, local supersymmetry with complex Weyl spinor parameter ϵ

$$\gamma_{11}\epsilon = -\epsilon \tag{B.19}$$

and a global SL(2, $\mathbb{R}$) invariance. The supersymmetry transformation rules are

$$\begin{aligned}
\delta B &= \kappa_{10} f^{-2} \bar{\epsilon}^* \lambda & \text{(B.20)}\\
\delta A_{MN} &= f(\bar{\epsilon}\gamma_{MN}\lambda + 4i\bar{\epsilon}^*\gamma_{[M}\psi_{N]} + B\bar{\epsilon}^*\gamma_{MN}\lambda + 4iB\bar{\epsilon}\gamma_{[M}\psi^*{}_{N]}) & \text{(B.21)}\\
\delta\lambda &= \frac{i}{\kappa_{10}}\gamma^M \epsilon^* \widehat{P}_M - \frac{1}{24} i\gamma^{MNP}\epsilon\widehat{G}_{MNP} & \text{(B.22)}
\end{aligned}$$

$$
\begin{aligned}
\delta\psi_M = {} & \frac{1}{\kappa_{10}} D_M\epsilon + \frac{1}{480} i\gamma^{P_1\cdots P_5}\gamma_M \epsilon \widehat{F}_{P_1\cdots P_5} \\
& + \frac{1}{96}(\gamma_M{}^{NPQ}\widehat{G}_{NPQ} - 9\gamma^{PQ}\widehat{G}_{MPQ})\epsilon^* \\
& - \frac{7}{16}\kappa_{10}\left(\gamma_P\lambda\bar{\psi}_M\gamma^P\epsilon^* - \frac{1}{1680}\gamma_{P_1\cdots P_5}\lambda\bar{\psi}_M\gamma^{P_1\cdots P_5}\epsilon^*\right) \\
& + \frac{1}{32} i\kappa_{10}\Bigg[\left(\frac{9}{4}\gamma_M\gamma^P + 3\gamma^P\gamma_M\right)\epsilon\bar{\lambda}\gamma_P\lambda \\
& \qquad - \left(\frac{1}{24}\gamma_P\gamma^{P_1P_2P_3} + \frac{1}{6}\gamma^{P_1P_2P_3}\gamma_M\right)\epsilon\bar{\lambda}\gamma_{P_1P_2P_3}\lambda \\
& \qquad + \frac{1}{960}\gamma_M\gamma^{P_1\cdots P_5}\epsilon\bar{\lambda}\gamma_{P_1\cdots P_5}\lambda\Bigg]
\end{aligned} \tag{B.23}
$$

$$\delta e_M{}^R = -2\kappa_{10}\Im(\bar{\epsilon}\gamma^R\psi_N) \tag{B.24}$$

$$\delta A_{MNPQ} = 2\Re(\hat{\epsilon}\gamma_{MNP}\psi_{Q]}) - \tfrac{3}{8}\kappa_{10}(A_{[MN}\delta A^*{}_{PQ]} - A^*{}_{[MN}\delta A_{PQ]} \tag{B.25}$$

where

$$\widehat{P}_M = P_M - \kappa_{10}{}^2\bar{\psi}^*{}_M\lambda \tag{B.26}$$

$$\widehat{G}_{MNP} = G_{MNP} - 3\kappa_{10}\bar{\psi}_{[M}\gamma_{NP]}\lambda - 6i\kappa_{10}\bar{\psi}^*{}_{[M}\gamma_N\psi_{P]} \tag{B.27}$$

and

$$\widehat{F}_{MNPQR} = F_{MNPQR} - 5\kappa_{10}\bar{\psi}_{[M}\gamma_{NPQ}\psi_{R]} - \frac{1}{16}\kappa_{10}\bar{\lambda}\gamma_{MNPQR}\lambda \tag{B.28}$$

and where

$$D_M\epsilon = \left(\partial_M + \tfrac{1}{4}\omega_M{}^{RS}\gamma_{RS} - \tfrac{1}{2}iQ_M\right)\epsilon \tag{B.29}$$

where

$$\omega_{MNP} = \Omega_{NPM} + \Omega_{NMP} + \Omega_{MPN} \tag{B.30}$$

and

$$\Omega_{MNP} = e^R{}_P\partial_{[M}e_{N]R} + \kappa_{10}{}^2\Im(\bar{\psi}_M\gamma_P\gamma_N). \tag{B.31}$$

The fields transform under $SL(2,\mathbb{R})$ with parameters α and γ as

$$\delta B = \alpha + 2i\gamma B - \alpha^* B^2 \tag{B.32}$$

$$\delta\lambda = \tfrac{3}{2}i[\gamma + \Im(\alpha B^*)]\lambda \tag{B.33}$$

$$\delta\psi_M = \tfrac{1}{2}i[\gamma + \Im(\alpha B^*)]\psi_M \tag{B.34}$$

$$\delta A_{MN} = i\gamma A_{MN} + \alpha A^*{}_{MN} \tag{B.35}$$

$$\delta e_M{}^R = 0 \tag{B.36}$$

$$\delta F_{MNPQR} = 0. \tag{B.37}$$

C The Lagrangian, Symmetries and Transformation Rules of the M2-Brane

Let us introduce the coordinates Z^M of a curved superspace

$$Z^M = (x^\mu, \theta^\alpha) \tag{C.1}$$

and the supervielbein $E_M{}^A(Z)$ where $M = \mu, \alpha$ are world indices and $A = a, \alpha$ are tangent space indices. We also define the pull-back

$$E_i^A = \partial_i Z^M E_M{}^A. \tag{C.2}$$

We also need the super d-form $A_{CBA}(Z)$. Then the supermembrane action is [41, 42]

$$S = T_2 \int d^3\xi \left[-\tfrac{1}{2}\sqrt{-\gamma}\gamma^{ij} E_i{}^a E_j{}^b \eta_{ab} + \tfrac{1}{2}\sqrt{-\gamma} + \tfrac{1}{3!}\epsilon^{i_1\cdots i_d} E_i{}^A E_j{}^B E_k{}^C A_{CBA} \right]. \tag{C.3}$$

Note that there is a kinetic term, a worldvolume cosmological term, and a Wess-Zumino term. The target-space symmetries are superdiffeomorphisms, Lorentz invariance and d-form gauge invariance. The worldvolume symmetries are ordinary diffeomorphisms and kappa invariance which we now examine in more detail. The transformation rules are

$$\delta Z^M E^a{}_M = 0, \quad \delta Z^M E^\alpha{}_M = \kappa^\beta (1+\Gamma)^\alpha{}_\beta \tag{C.4}$$

where $\kappa^\beta(\xi)$ is an anticommuting spacetime spinor but worldvolume scalar, and where

$$\Gamma^\alpha{}_\beta = \frac{1}{3!\sqrt{-\gamma}} \epsilon^{ijk} E_i^a E_j^b \cdots E_k^c \Gamma_{abc}. \tag{C.5}$$

Here Γ_a are the Dirac matrices in spacetime and

$$\Gamma_{a_1..a_d} = \Gamma_{[a_1\cdots a_d]}. \tag{C.6}$$

The matrix Γ of (C.5) is traceless and satisfies

$$\Gamma^2 = 1 \tag{C.7}$$

when the equations of motion are satisfied and hence the matrices $(1 \pm \Gamma)/2$ act as projection operators. The transformation rule (C.4) therefore permits us to gauge away one half of the fermion degrees of freedom. As described below, this gives rise to a matching of physical boson and fermion degrees of freedom on the worldvolume.

The kappa symmetry is achieved only if certain constraints on the antisymmetric tensor field strength $F_{MNP..Q}$(Z) and the supertorsion are satisfied. One can show that the constraints on the background fields $E_M{}^A$ and A_{MNP} are nothing but the equations of motion of eleven-dimensional supergravity.

D The Field Equations, Symmetries and Transformation Rules of the M5-Brane

The M5-brane equations are elegantly derived from the superembedding formalism which treats the brane as a supermanifold M embedded in a larger spacetime supermanifold $\underline{M}$. In this subsection, therefore, we use the notations and conventions of [225, 226]. In particular, we denote by $z^{\underline{M}} = (x^{\underline{m}}, \theta^{\underline{\mu}})$ the local coordinates on $\underline{M}$, and $A = (a, \alpha)$ is the target tangent space index. We use the ununderlined version of these indices to label the corresponding quantities on the worldvolume. The embedded submanifold M, with local coordinates y^M, is given as $z^{\underline{M}}(y)$.

The suitable pullbacks of the super 3-form and the induced metric are

$$\begin{aligned} A_{ijk} &= \partial_i z^{\underline{M}} \partial_j z^{\underline{N}} \partial_k z^{\underline{P}} A_{\underline{PNM}}, \\ \gamma_{ij} &= (\partial_i z^{\underline{M}} E_{\underline{M}}{}^{\underline{a}})(\partial_j z^{\underline{N}} E_{\underline{N}}{}^{\underline{b}})\eta_{\underline{ab}}, \end{aligned} \tag{D.1}$$

where $\eta_{\underline{ab}}$ is the Minkowski metric in eleven dimensions and $E_{\underline{M}}{}^{\underline{A}}$ is the target space supervielbein. We define the basis one-forms $E^{\underline{A}} = d\xi^i \partial_i z^{\underline{M}} E_{\underline{M}}{}^{\underline{A}}$ and $E^A = d\sigma^r\, \partial_r y^M E_M{}^A$, where $E_M{}^A$ is the supervielbein on M_5. The embedding matrix $E_A{}^{\underline{A}}$ plays an important role in the description of the model, and it is defined as

$$E_A{}^{\underline{A}} \equiv E_A{}^M \partial_M z^{\underline{M}} E_{\underline{M}}{}^{\underline{A}}. \tag{D.2}$$

Here, we give the nonlinear field equations of the superfivebrane equations, up to second order fermionic terms, that follow from the superembedding condition $E_\alpha{}^{\underline{a}} = 0$, which are proposed to arise equally well from the $\mathcal{F}$-constraint $\mathcal{F}_{\alpha BC} = 0$, where we have introduced the super 2-form A_2 and the following super 3-form in M_5:

$$\mathcal{F}_3 \equiv dA_2 - f_5^* A_3, \tag{D.3}$$

and where f_5^* is the pullback associated with the embedding map $f_5 \colon M_5 \hookrightarrow \underline{M}$. The details of the procedures can be found in [226, 227]. A key point is the emergence of a super 3-form h in world superspace. This form arises in the following component of the embedding matrix

$$E_\alpha{}^{\underline{\alpha}} = u_\alpha{}^{\underline{\alpha}} + h_\alpha{}^{\beta'} u_{\beta'}{}^{\underline{\alpha}}, \tag{D.4}$$

where, upon the splitting of the indices to exhibit the $USp(4)$ $\mathbb{R}$-symmetry group indices $i = 1, \ldots, 4$, we have

$$h_\alpha{}^{\beta'} \to h_{\alpha i \beta}{}^{j} = \tfrac{1}{6}\delta_i^j (\gamma^{abc})_{\alpha\beta} h_{abc}, \tag{D.5}$$

where h_{abc} is a self-dual field defined on M. The pair $(u_\alpha{}^{\underline{\alpha}}, u_{\alpha'}{}^{\underline{\alpha}})$ make up an element of the group Spin(1, 10).

The superembedding formalism was shown to give the following complete M5-brane equations of motion:

$$\begin{aligned} E_a{}^{\underline{\alpha}} E_{\underline{\alpha}}{}^{\beta'} (\Gamma^a)_{\beta'}{}^{\alpha} &= 0, \\ \eta^{ab} \nabla_a E_b{}^{\underline{a}} E_{\underline{a}}{}^{b'} &= -\tfrac{1}{8} (\Gamma^{b'a})_{\gamma'}{}^{\beta} Z_{a\beta}{}^{\gamma'}, \\ \widehat{\nabla}^c h_{abc} &= -\frac{1}{32} (\Gamma^c \Gamma_{ab})_{\gamma'}{}^{\beta} Z_{c\beta}{}^{\gamma'}, \end{aligned} \tag{D.6}$$

where

$$Z_{a\beta}{}^{\gamma'} = E_\beta{}^{\underline{\beta}} \big(E_a{}^{\underline{a}} T_{\underline{a}\underline{\beta}}{}^{\underline{\gamma}} - E_a{}^{\underline{\delta}} E_{\underline{\beta}}{}^{\gamma} (\nabla_\gamma E_{\underline{\delta}}{}^{\delta'}) E_{\delta'}{}^{\underline{\gamma}} \big) E_{\underline{\gamma}}{}^{\gamma'}. \tag{D.7}$$

Recall that the inverse of the pair $(E_A{}^{\underline{A}}, E_{A'}{}^{\underline{A}})$ is denoted by $(E_{\underline{A}}{}^{A}, E_{\underline{A}}{}^{A'})$ and that $A = (a, \alpha)$ label the tangential directions while $A' = (a', \alpha')$ label the normal directions to the M5-brane worldvolume.

The nonvanishing parts of the target space torsion components $T_{\underline{a}\underline{\beta}}{}^{\underline{\gamma}}$ are given by

$$\begin{aligned} T_{\underline{\alpha}\underline{\beta}}{}^{\underline{c}} &= -i(\Gamma^{\underline{c}})_{\underline{\alpha}\underline{\beta}}, \\ T_{\underline{a}\underline{\beta}}{}^{\underline{\gamma}} &= -\frac{1}{36} (\Gamma^{\underline{bcd}})_{\underline{\beta}}{}^{\underline{\gamma}} F_{\underline{abcd}} - \frac{1}{288} (\Gamma_{\underline{abcde}})_{\underline{\beta}}{}^{\underline{\gamma}} F^{\underline{bcde}}, \end{aligned} \tag{D.8}$$

and $T_{\underline{a}\underline{b}}{}^{\underline{\gamma}}$. The only other nonvanishing components of F_4 are

$$F_{\underline{ab}\underline{\gamma}\underline{\delta}} = -i(\Gamma_{\underline{ab}})_{\underline{\gamma}\underline{\delta}}. \tag{D.9}$$

The covariant derivative $\widehat{\nabla}$ has an additional, composite SO(5, 1) connection of the form $(\nabla u)u^{-1}$ as explained in more detail in [227].

The M5-brane equations of motion (D.6) live in superspace [225, 228]. The component (i.e., Green–Schwarz) form of these equations has also been worked out [227]. Up to fermionic bilinears, the final result is:

$$\begin{aligned} &\mathcal{E}_a (1 - \Gamma) \gamma^b m_b{}^a = 0, \\ &G^{mn} \nabla_m \mathcal{F}_{npq} = Q^{-1} [4Y - 2(mY + Ym) + mYm]_{pq}, \\ &G^{mn} \nabla_m \mathcal{E}_n{}^{\underline{c}} \\ &\quad = \frac{Q}{\sqrt{-g}} \epsilon^{m_1 \cdots m_6} \left(\frac{1}{6!} F^{\underline{a}}{}_{m_1 \cdots m_6} + \frac{1}{(3!)^2} F^{\underline{a}}{}_{m_1 m_2 m_3} \mathcal{F}_{m_4 m_5 m_6} \right) P_{\underline{a}}{}^{\underline{c}}. \end{aligned} \tag{D.10}$$

Several definitions are in order. To begin with,

$$\begin{aligned} m_a{}^b \equiv \delta_a{}^b - 2k_a{}^b, \quad k_a{}^b \equiv h_{acd} h^{bcd}, \quad Q \equiv (1 - \tfrac{2}{3} \operatorname{tr} k^2), \\ Y_{ab} \equiv [4 * F - 2(m * F + *Fm) + m * Fm]_{ab}, \\ P_{\underline{a}}{}^{\underline{c}} \equiv \delta_{\underline{a}}{}^{\underline{c}} - \mathcal{E}_{\underline{a}}{}^{m} \mathcal{E}_m{}^{\underline{c}}, \quad *F^{ab} := \frac{1}{4! \sqrt{-g}} \epsilon^{abcdef} F_{cdef}, \end{aligned} \tag{D.11}$$

The fields $\mathcal{F}_{abc}$, $F_{\underline{a}_1 \cdots \underline{a}_4}$ and its Hodge dual $F_{\underline{a}_1 \cdots \underline{a}_7}$ are the purely bosonic components of the superforms

$$\mathcal{F}_3 = dA_2 - \underline{C}_3, \quad F_4 = dA_3, \quad F_7 = dA_6 + \tfrac{1}{2} A_3 \wedge F_4. \tag{D.12}$$

The remaining nonvanishing component of F_7 is

$$F_{\underline{\alpha\beta abcde}} = -i(\Gamma_{\underline{abcde}})_{\underline{\alpha\beta}}. \tag{D.13}$$

The target space indices on F_4 and F_7 have been converted to worldvolume indices with factors of $\mathcal{E}_m{}^{\underline{a}}$ which are the supersymmetric line elements defined as

$$\begin{aligned} \mathcal{E}_m{}^{\underline{a}}(x) &\equiv \partial_m z^{\underline{M}} E_{\underline{M}}{}^{\underline{a}} \qquad \text{at } \theta = 0, \\ \mathcal{E}_m{}^{\underline{\alpha}}(x) &\equiv \partial_m z^{\underline{M}} E_{\underline{M}}{}^{\underline{\alpha}} \qquad \text{at } \theta = 0. \end{aligned} \tag{D.14}$$

The metric

$$\gamma_{mn}(x) \equiv \mathcal{E}_m{}^{\underline{a}}\mathcal{E}_n{}^{\underline{b}}\eta_{\underline{ab}} = e_m{}^a e_n{}^b \eta_{ab} \tag{D.15}$$

is the standard GS induced metric with determinant g, and G^{mn} is another metric defined as

$$G^{mn} \equiv (m^2)^{ab} e_a{}^m e_b{}^n. \tag{D.16}$$

Let us note that the connection in the covariant derivative ∇_m occurring in (4.13) is the Levi-Civita connection for the induced metric g_{mn} up to fermionic bilinears.

A key relation between h_{abc} and $\mathcal{F}_{abc}$ follows from the dimension-0 components of the Bianchi identity $d\mathcal{F}_3 = -\underline{F}_4$, and is given by

$$h_{abc} = \tfrac{1}{4} m_a{}^d \mathcal{F}_{bcd}. \tag{D.17}$$

The matrix Γ is the $\theta = 0$ component of the matrix $\Gamma_{(5)}$ introduced above and it is given by

$$\Gamma = -\bar{\Gamma} + \tfrac{1}{3} h^{mnp}\Gamma_{mnp} = -[\exp(-\tfrac{1}{3}\Gamma^{mnp}h_{mnp})]\bar{\Gamma}, \tag{D.18}$$

where

$$\begin{aligned} \overline{\gamma} &\equiv \frac{1}{6!\sqrt{-g}} \epsilon^{m_1\cdots m_6}\Gamma_{m_1\cdots m_6}, \\ \Gamma_m \equiv \mathcal{E}_m{}^{\underline{a}}\Gamma_{\underline{a}}, &\quad \Gamma^b \equiv \Gamma^m e_m{}^b, \quad e_m{}^b \equiv E_m{}^a m_a{}^b. \end{aligned} \tag{D.19}$$

The κ-symmetry transformation rules are

$$\begin{aligned} \delta_\kappa z^{\underline{a}} &= 0, \\ \delta_\kappa z^{\underline{\alpha}} &= \kappa^{\underline{\gamma}}\tfrac{1}{2}(1+\Gamma)_{\underline{\gamma}}{}^{\underline{\alpha}}, \\ \delta_\kappa h_{abc} &= -\frac{i}{16} m_{d[a}\mathcal{E}^d(1-\Gamma)\Gamma_{bc]}\kappa, \end{aligned} \tag{D.20}$$

where Γ is given by (D.18). The κ-symmetry transformations are the fermionic diffeomorphisms of the M5-brane worldvolume with parameter $\kappa^\alpha = \kappa^{\underline{\alpha}} E_{\underline{\alpha}}{}^\alpha$. It follows that

$$\delta_\kappa \mathcal{F}_3 = \{d, i_\kappa\}\mathcal{F}_3 = -i_\kappa \underline{F}_4, \tag{D.21}$$

which can also be verified by direct computation by combining (4.19) and (D.20).

The equations of motion (4.13) have been shown [229] to be equivalent to those which follow from an action with auxiliary scalar field [230].

We conclude this section by elucidating the consequences of the central equation (4.19). To this end, we first note the useful identities

$$\begin{aligned} h_{abe}h^{cde} &= \delta_{[a}^{[c}k_{b]}^{d]}, \\ k_{ac}k_b^{\ c} &= \tfrac{1}{6}\eta_{ab}\,\mathrm{tr}\,k^2, \\ k_a^{\ d}h_{bcd} &= k_{[a}^{\ \ d}h_{bc]d}, \end{aligned} \tag{D.22}$$

which are consequences of the linear self-duality of h_{abc}. Taking the Hodge dual of (4.19) one finds $*\mathcal{F}_{abc} = -\mathcal{F}_{abc} + 2Q^{-1}m_a^{\ d}\mathcal{F}_{bcd}$. Using the identity $m^2 = 2m - Q$, we readily find the nonlinear self-duality equation

$$*\mathcal{F}_{mnp} = Q^{-1}G_m^{\ q}\mathcal{F}_{npq}. \tag{D.23}$$

This equation can be expressed solely in terms of $\mathcal{F}_3$. To do this, we first insert (4.19) into (4.20), which yields the identities

$$\begin{aligned} \mathcal{F}_{abe}\mathcal{F}^{cde} &= 2\delta_{[a}^{[c}X_{b]}^{d]} + \tfrac{1}{2}K^{-2}X_{[a}^{\ \ c}X_{b]}^{\ \ d} + 2(K^2-1)\delta_{[a}^{c}\delta_{b]}^{d}, \\ X_{ac}X_b^{\ c} &= 4K^2(K^2-1)\eta_{ab}, \\ X_a^{\ d}\mathcal{F}_{bcd} &= X_{[a}^{\ \ d}\mathcal{F}_{bc]d}, \end{aligned} \tag{D.24}$$

where we have defined

$$K \equiv \sqrt{1 + \frac{1}{24}\mathcal{F}^{abc}\mathcal{F}_{abc}}, \tag{D.25}$$

$$X_{ab} \equiv \tfrac{1}{2}K * \mathcal{F}_a^{\ cd}\mathcal{F}_{bcd}. \tag{D.26}$$

Next we derive the identities

$$\begin{aligned} &Q(K+1) = 2, \\ &X_{ab} = \frac{1}{2}\mathcal{F}_{acd}\mathcal{F}_b^{\ cd} - \frac{1}{12}\eta_{ab}\mathcal{F}_{cde}\mathcal{F}^{cde} = 4K(1+K)k_{ab}. \end{aligned} \tag{D.27}$$

We can now express (4.21) entirely in terms of $\mathcal{F}_3$ by deriving the identity

$$Q^{-1}G_{mn} = K\eta_{mn} - \tfrac{1}{2}K^{-1}X_{mn}. \tag{D.28}$$

Another way of writing (4.21) is

$$\mathcal{F}^-_{abc} = \tfrac{1}{2}(1+K)^{-2}\mathcal{F}^+_{ade}\mathcal{F}^{+def}\mathcal{F}^+_{fbc}, \tag{D.29}$$

where K is a root of the quartic equation

$$(K+1)^3(K-1) = \frac{1}{24}\mathcal{F}^{+abc}\mathcal{F}^+_{ade}\mathcal{F}^{+def}\mathcal{F}^+_{fbc}. \tag{D.30}$$

E The Lagrangian, Symmetries and Transformation Rules of the D3-Brane

Here we present a Lorentz invariant and supersymmetric worldvolume action for all Type II Dirichlet p-branes, $p \leq 9$, in a general Type II supergravity background [231]. The super Dirichlet p-brane action, in a general $N = 2$ supergravity background (for the string-frame metric) takes the form

$$S_d = S_{d\mathrm{DBI}} + S_{d\mathrm{WZ}} \tag{E.1}$$

where

$$S_{d\mathrm{DBI}} = -T_d \int d^d\,\xi e^{-\phi}\sqrt{-\det(\gamma_{ij} + \mathcal{F}_{ij})} \tag{E.2}$$

is a Dirac–Born–Infeld type action and I_{WZ} is a Wess–Zumino (WZ) type action to be discussed below; $\mathcal{F}_{ij}$ are the components of a 'modified' 2-form field strength

$$\mathcal{F} = F - B, \tag{E.3}$$

where $F = dV$ is the usual field strength 2-form of the BI field V and B is the pullback to the worldvolume of a 2-form potential B on superspace, whose leading component in a θ-expansion is the 2-form potential of Neveu–Schwarz/Neveu–Schwarz (NS-NS) origin in Type II superstring theory. We use the same letter for superspace forms and their pullbacks to the worldvolume since it should be clear from the context which is meant. Superspace forms may be expanded in the coordinate basis of 1-forms dZ^M or on the inertial frame basis $E^A = dZ^M E_M{}^A$, where $E_M{}^A$ is the supervielbein. The basis E^A decomposes under the action of the Lorentz group into a Lorentz vector E^a and a Lorentz spinor. The latter is a 32-component Majorana spinor for IIA superspace and a pair of chiral Majorana spinors for IIB superspace. Thus

$$(E^a, E^{\alpha I}), \quad I = 1, 2 \tag{E.4}$$

for Type IIB. We allow α to run from 1 to 32 but include a chiral projector as appropriate. The world volume metric g_{ij} appearing in (E.2) is defined in the standard way as

$$\gamma_{ij} = E_i^a E_j^b \eta_{ab} \tag{E.5}$$

where η is the $D = 10$ Minkowski metric and

$$E_i^A = \partial_i Z^M E_M{}^A. \tag{E.6}$$

Thus I_{DBI} is a straightforward extension to superspace of the corresponding term in the bosonic action. The same is true for the WZ term. We introduce a R-R potential A as a formal sum of r-form superspace potentials $A^{(r)}$, i.e.

$$A = \sum_{r=0}^{10} A^{(r)}. \tag{E.7}$$

The even potentials are those of IIB supergravity while the odd ones are those of IIA supergravity. In the bosonic case one could omit the 10-form gauge potential $F^{(10)}$ on the grounds that its 11-form field strength is identically zero. But an 11-form field strength on *superspace* is not identically zero; in fact we shall see that it is nonzero even in a Minkowski background, a fact that is crucial to the κ-symmetry of the super 9-brane action.

The WZ term can now be written as

$$S_{d\mathrm{WZ}} = T_d \int_W A e^{\mathcal{F}} + m I_{\mathrm{CS}} \tag{E.8}$$

where, in the first term, the product is understood to be the exterior product of forms and the form of appropriate degree is chosen in the 'form-expansion' of the integrand, i.e., $(p+1)$ for a Dirichlet p-brane. The I_{CS} term is a $(p+1)$-form Chern–Simons (CS) action that is present (for odd p) in a *massive* IIA background; its coefficient m is the IIA mass parameter. This WZ term is formally the same as the known *bosonic* Dirichlet p-brane WZ action, but here the forms $C^{(r)}$ and B are taken to be forms on *superspace*, e.g.,

$$A^{(r)} = \frac{1}{r!} dZ^{M_1} \cdots dZ^{M_r} A_{M_r \cdots M_1}. \tag{E.9}$$

This illustrates the standard normalization and the 'reverse order' convention for components of superspace forms. This convention goes hand in hand with the convention for exterior differentiation of superspace forms in which the exterior derivative acts 'from the right'. Thus,

$$dA^{(r)} = \frac{1}{r!} dZ^{M_1} \cdots dZ^{M_r} dZ^N \partial_N A_{M_r \cdots M_1}. \tag{E.10}$$

The field strength for the R-R field A is

$$F(A) = dA - HA + me^B \tag{E.11}$$

where m is the mass parameter of the IIA theory. $F(A)$ can be written as the formal sum

$$F(A) = \sum_{n=1}^{11} F^{(n)}. \tag{E.12}$$

Note that the top form is an 11-form because we included a 10-form $A^{(10)}$ in the definition of A. The field strengths $F^{(n)}$ will be subject to superspace constraints, to be given below for bosonic backgrounds, in addition to the constraint relating the bosonic components of $F^{(n)}$ to the Hodge dual of the bosonic components of $F^{(10-n)}$.

Each term in (E.1) is separately invariant under the global IIA or IIB super-Poincaré group as well as under $(p+1)$-dimensional general coordinate transformations. However, local kappa symmetry is achieved by a subtle conspiracy between them, just as in the case of super p-branes with scalar supermultiplets.

In these lectures, we are primarily interested in the D3-brane and so may set $p = 3$ in the above formulae.

F $D = 4$, $N = 2$ Gauged Supergravity

The SO(8) gauged $N = 8$ supergravity in four dimensions was obtained in [11, 156] by gauging an SO(8) subgroup of the global E_7 symmetry group of [186, 187]. To avoid some of the complications of non-Abelian gauge fields, one may consider a truncation of this model to $N = 2$, for which the bosonic sector comprises the metric, four commuting U(1) gauge potentials, three dilatons and three axions. In the absence of axions, this truncation was obtained in [149] by working in the symmetric gauge for the 56-bein and incorporating three real scalars. As was noted, there is a straightforward generalization of the scalar ansatz to allow for complex scalars. Taking into account the E_7 self-duality condition $\overline{\phi}^{ijkl} = \phi_{ijkl} = \frac{1}{4!}\epsilon_{ijklmnpq}\phi^{mnpq}$, the scalar ansatz of [149] may be generalized as:

$$
\begin{aligned}
\overline{\phi}^{ijkl} &= \phi_{ijkl} \\
&= \sqrt{2}[\Phi^{(1)}\epsilon^{(12)} + \Phi^{(2)}\epsilon^{(13)} + \Phi^{(3)}\epsilon^{(14)} + \overline{\Phi}^{(1)}\epsilon^{(34)} \\
&\qquad + \overline{\Phi}^{(2)}\epsilon^{(24)} + \overline{\Phi}^{(3)}\epsilon^{(23)}]_{ijkl}, \qquad \text{(F.1)}
\end{aligned}
$$

where we follow the notation and conventions of [149] (including the definition of SO(8) index pairs). Note that the three complex scalars may be parameterised in terms of their magnitudes and phases as $\Phi^{(i)} = \phi^{(i)}e^{i\theta^{(i)}}$.

Here, we shall consider the full $N = 2$ truncation, where the three axions are included as well as the other fields. In fact the structure of the potential is little changed. We find that the Lagrangian including the axions may be written in the form

$$
\begin{aligned}
e^{-1}\mathcal{L}_4 = R &- \frac{1}{2}\sum_i\left((\partial\phi^{(i)})^2 + \sinh^2\phi^{(i)}(\partial\theta^{(i)})^2\right) \\
&- \frac{1}{2}(F^{(\alpha)+}_{\mu\nu}\mathcal{M}_{\alpha\beta}F^{(\beta)+\mu\nu} + \text{h.c.}) - g^2V, \qquad \text{(F.2)}
\end{aligned}
$$

where the potential is given simply by

$$
V = -8(\cosh\phi^{(1)} + \cosh\phi^{(2)} + \cosh\phi^{(3)}). \qquad \text{(F.3)}
$$

The complex symmetric scalar matrix $\mathcal{M}$ is quite complicated, and incorporates all three complex scalars $\Phi^{(\alpha)}$ in a symmetric manner; this is presented below.

In terms of the $N = 2$ truncation, the three complex scalars each parameterise an $\mathrm{SL}(2;\mathbb{R})/\mathrm{SO}(2)$ coset. This may be made explicit by performing the change of variables $(\phi^{(i)}, \theta^{(i)}) \to (\varphi_i, \chi_i)$:

$$\begin{aligned} \cosh\phi^{(i)} &= \cosh\varphi_i + \tfrac{1}{2}\chi_i^2 e^{\varphi_i}, \\ \cos\theta^{(i)}\sinh\phi^{(i)} &= \sinh\varphi_i - \tfrac{1}{2}\chi_i^2 e^{\varphi_i}, \\ \sin\theta^{(i)}\sinh\phi^{(i)} &= \chi_i e^{\varphi_i}. \end{aligned} \tag{F.4}$$

Defining the dilaton-axion combinations

$$A_i = 1 + \chi_i^2 e^{2\varphi_i}, \tag{F.5}$$

as well as

$$\begin{aligned} B_1 &= \chi_2\chi_3 e^{\varphi_2+\varphi_3} + i\chi_1 e^{\varphi_1}, \\ B_2 &= \chi_1\chi_3 e^{\varphi_1+\varphi_3} + i\chi_2 e^{\varphi_2}, \\ B_3 &= \chi_1\chi_2 e^{\varphi_1+\varphi_2} + i\chi_3 e^{\varphi_3}, \end{aligned} \tag{F.6}$$

we finally obtain the bosonic Lagrangian

$$\begin{aligned} e^{-1}\mathcal{L}_4 = R - \frac{1}{2}\sum_i \left((\partial\varphi_i)^2 + e^{2\varphi_i}(\partial\chi_i)^2\right) \\ - \frac{1}{2}(F_{\mu\nu}^{(\alpha)+}\mathcal{M}_{\alpha\beta}F^{(\beta)+\mu\nu} + \text{h.c.}) - g^2 V. \end{aligned} \tag{F.7}$$

The potential V is now given by

$$V = -8\sum_i (\cosh\varphi_i + \tfrac{1}{2}\chi_i^2 e^{\varphi_i}), \tag{F.8}$$

and the scalar matrix is

$$\mathcal{M} = \frac{1}{D}\begin{bmatrix} e^{-\lambda_1} & e^{\varphi_1}B_1 & e^{\varphi_2}B_2 & e^{\varphi_3}B_3 \\ e^{\varphi_1}B_1 & e^{-\lambda_2}A_2A_3 & -e^{-\varphi_3}A_3B_3 & -e^{-\varphi_2}A_2B_2 \\ e^{\varphi_2}B_2 & -e^{-\varphi_3}A_3B_3 & e^{-\lambda_3}A_1A_3 & -e^{-\varphi_1}A_1B_1 \\ e^{\varphi_3}B_3 & -e^{-\varphi_2}A_2B_2 & -e^{-\varphi_1}A_1B_1 & e^{-\lambda_4}A_1A_2 \end{bmatrix}, \tag{F.9}$$

where

$$D = 1 + \chi_1^2 e^{2\varphi_1} + \chi_2^2 e^{2\varphi_2} + \chi_3^2 e^{2\varphi_3} - 2i\chi_1\chi_2\chi_3 e^{\varphi_1+\varphi_2+\varphi_3}. \tag{F.10}$$

The scalar combinations $\{\lambda\}$ are defined as in [149]:

$$\begin{aligned} \lambda_1 &= -\varphi_1 - \varphi_2 - \varphi_3, \\ \lambda_2 &= -\varphi_1 + \varphi_2 + \varphi_3, \\ \lambda_3 &= \varphi_1 - \varphi_2 + \varphi_3, \\ \lambda_4 &= \varphi_1 + \varphi_2 - \varphi_3. \end{aligned} \tag{F.11}$$

While this $N = 2$ truncation of the $N = 8$ theory essentially treats all four U(1) gauge fields equally, it was noted that one can make contact with the theory obtained by reduction of a closed string on T^2 through dualisation of two of the gauge fields. To be specific, we dualise $F^{(2)}_{\mu\nu}$ and $F^{(4)}_{\mu\nu}$, which singles out the dilaton-axion pair $S = \chi_2 + ie^{-\varphi_2}$. After an additional field redefinition $S \to -1/S$, we obtain the bosonic Lagrangian

$$e^{-1}\mathcal{L}_4^{\text{dualized}} = R - \tfrac{1}{2}(\partial\varphi_2)^2 - \tfrac{1}{2}e^{2\varphi_2}(\partial\chi_2)^2 + \tfrac{1}{8}\operatorname{Tr}(\partial ML\partial ML) - g^2V \\ - \tfrac{1}{4}e^{-\varphi_2}F^T(LML)F - \tfrac{1}{4}\chi_2 F^T L{*}F, \quad \text{(F.12)}$$

where the potential is still given by (F.8). The scalar matrix M is given in terms of the $\mathrm{SL}(2;\mathbb{R}) \times \mathrm{SL}(2;\mathbb{R})$ vielbein

$$\mathcal{V} = e^{\varphi_3/2}\begin{bmatrix} 1 & -\chi_3 \\ 0 & e^{-\varphi_3}\end{bmatrix} \otimes e^{\varphi_1/2}\begin{bmatrix} 1 & -\chi_1 \\ 0 & e^{-\varphi_1}\end{bmatrix}, \quad \text{(F.13)}$$

by $M = \mathcal{V}^T\mathcal{V}$, and the gauge fields have been arranged in the particular order

$$F_{\mu\nu} = \begin{bmatrix} F^{(3)}_{\mu\nu} & \widetilde{F}^{(4)}_{\mu\nu} & \widetilde{F}^{(2)}_{\mu\nu} & -F^{(1)}_{\mu\nu}\end{bmatrix}^T. \quad \text{(F.14)}$$

Finally, $L = \sigma^2 \otimes \sigma^2$ satisfies $L^2 = I_4$, where σ^2 is the standard Pauli matrix. It is worth mentioning that the pure scalar Lagrangian can be expressed as

$$e^{-1}\mathcal{L}_{\text{scalar}} = \sum_{i=1}^{3}[-\tfrac{1}{2}\operatorname{tr}\partial\mathcal{M}_i\partial\mathcal{M}_i^{-1} + 4g^2\operatorname{tr}\mathcal{M}_i], \quad \text{(F.15)}$$

where $\mathcal{M}_i = \mathcal{V}_i^T\mathcal{V}_i$ and $\mathcal{V}_i$ is given by

$$\mathcal{V}_i = e^{\varphi_i/2}\begin{bmatrix} 1 & -\chi_i \\ 0 & e^{-\varphi_i}\end{bmatrix}. \quad \text{(F.16)}$$

We see that, save for the potential, the dualized Lagrangian is indeed of the form obtained from T^2 compactification from six dimensions. In this case, two of the $\mathrm{SL}(2;\mathbb{R})$'s now correspond to T-dualities while the third corresponds to S-duality. Note that the initial choice of which two field strengths to dualise has determined which of the three dilaton-axion pairs (φ_i, χ_i) is to be identified with the strong-weak coupling $\mathrm{SL}(2;\mathbb{R})$.

Having shown that the bosonic Lagrangian is considerably simplified by dualising to the field variables that arise in the T^2 reduction, we may re-express the result (F.12) in the more explicit notation of [165, 188]. Thus the bosonic sector of the gauged $U(1)^4$ theory may be written as

$$\begin{aligned} e^{-1}\mathcal{L}_4 = R &- \tfrac{1}{2}(\partial\vec{\varphi})^2 - \tfrac{1}{2}e^{-\vec{a}\cdot\vec{\varphi}}(\partial\chi)^2 - \tfrac{1}{2}e^{\vec{a}_{12}\cdot\vec{\varphi}}(F_{(1)12})^2 \\ &- \tfrac{1}{2}e^{\vec{b}_{12}\cdot\vec{\varphi}}(\mathcal{F}^1_{(1)2})^2 - g^2 V \\ &- \tfrac{1}{4}\sum_{i=1}^{2}(e^{\vec{a}_i\cdot\vec{\varphi}}(F_{(2)i})^2 + e^{\vec{b}_i\cdot\vec{\varphi}}(\mathcal{F}^i_{(2)})^2) \\ &- \tfrac{1}{2}\chi\epsilon^{\mu\nu\rho\sigma}F_{\mu\nu i}\mathcal{F}^i_{\rho\sigma}, \end{aligned}$$

where the field strengths are given by

$$\begin{aligned} &F_{(2)1} = dA_{(1)1} + dA_{(0)12}\mathcal{A}^2_{(1)}, \quad \mathcal{F}^1_{(2)} = d\mathcal{A}^1_{(2)} - d\mathcal{A}^1_{(0)2}\mathcal{A}^2_{(1)}, \\ &F_{(2)2} = dA_{(1)2} - \mathcal{A}^1_{(0)2}dA_{(1)1} - dA_{(0)12}\mathcal{A}^1_{(1)}, \quad \mathcal{F}^2_{(2)} = d\mathcal{A}^2_{(1)}. \end{aligned} \tag{F.17}$$

Here χ, $A_{(0)12}$ and $\mathcal{A}^1_{(0)2}$ are the axions χ_2, χ_1 and χ_3, and the potential is given by (F.8).

The inclusion of the potential term in the gauged supergravity theory breaks all three $SL(2;\mathbb{R})$ symmetries to $O(2)$, acting as $\tau \to (a\tau + b)/(c\tau + d)$, where

$$\begin{pmatrix} a & b \\ c & d \end{pmatrix} = \begin{pmatrix} \cos\alpha/2 & \sin\alpha/2 \\ -\sin\alpha/2 & \cos\alpha/2 \end{pmatrix}. \tag{F.18}$$

In terms of the original (ϕ, θ) scalar variables in (F.2), this $O(2)$ subgroup corresponds to $\theta \to \theta + \alpha$. The $O(2)$ symmetry is, however, sufficient for generating dyonic solutions. Nevertheless, we note that the fermionic sector and in particular the supersymmetry transformations are not invariant under this symmetry of the bosonic sector. One manifestation of this particular situation is the fact that magnetic black holes of this theory are not supersymmetric, even though they may be extremal. Furthermore, in a related note, while it is clear that the dualisation procedure performed above runs into difficulty in the full $N = 8$ theory with non-Abelian SO(8) gauging, the fermionic sector does not admit such a straightforward dualisation, even in the $N = 2$ Abelian truncation. This is easily seen by the fact that the gravitini are necessarily charged under the gauge fields and hence couple to the bare gauge potentials themselves.

3

Easy Turbulence

Krzysztof Gawędzki

It seems a safe bet that the understanding of developed turbulence, a long standing challenge for theoretical and mathematical physics, will enter into the third millennium as an unsolved problem. This is an introductory course to the subject. We discuss

in Lecture 1: the Navier Stokes equations, existence of solutions, statistical description, energy balance and cascade picture;

in Lecture 2: the Kolmogorov theory of three-dimensional turbulence versus intermittency, the Kraichnan–Batchelor theory of two-dimensional turbulence;

in Lecture 3: the Richardson dispersion law and the breakdown of the Lagrangian flow;

in Lecture 4: direct and inverse cascades and intermittency in the Kraichnan model of passive advection.

Lecture 1

Theoretical physics pursues two goals. On one side, it searches for fundamental laws of nature. On the other side, it studies the theoretical and phenomenological consequences of the laws already found. Hydrodynamics represents the second case. Its fundamental laws, in the form of the Navier-Stokes (NS) equation or its variations, were formulated in the first half of the nineteenth century by Navier (1823) and Stokes (1843) as a modification of the even older Euler equation (1755). The NS equation describes the temporal evolution of a *velocity* field $\mathbf{v}(t, \mathbf{x})$ in gasses or liquids. It takes the form

$$\partial_t \mathbf{v} + (\mathbf{v} \cdot \boldsymbol{\nabla})\mathbf{v} - \nu \boldsymbol{\nabla}^2 \mathbf{v} = \frac{1}{\rho}(\mathbf{f} - \boldsymbol{\nabla} p), \tag{1.1}$$

where ν is the viscosity of the fluid ($\cong 1.5 \times 10^{-5}$ m^2/sec for air and $\cong 10^{-6}$ m^2/sec for water), ρ is the fluid density, $\mathbf{f}$ is the external (intensive) force and p is the pressure. In most physical applications, the dimensionality of the space is 3 or 2, but the equations make sense in a general dimension d. The Euler equation without the $\nu\nabla^2\mathbf{v}$ term is really the $\mathbf{F} = m\mathbf{a}$ (or rather $\mathbf{a} = m^{-1}\mathbf{F}$) relation for the volume element of the fluid. The $\nu\nabla^2\mathbf{v}$ term in the NS equation represents the friction forces. The relation (1.1) has to be supplemented with the continuity equation $\partial_t\rho + \nabla\cdot(\rho\mathbf{v}) = 0$, and an equation of state relating ρ and p. In most applications, one may assume that the fluid is *incompressible*, i.e. that ρ is constant and that $\mathbf{v}$ is divergence free:[1]

$$\nabla\cdot\mathbf{v} = 0. \tag{1.2}$$

It follows then by taking the divergence of both sides of Eq. (1.1) that $\nabla^2 p = -\nabla\cdot(\mathbf{v}\cdot\nabla)\mathbf{v}$ so that pressure may be calculated for given velocity field. It may be also eliminated by applying to the NS equation the transverse projector which leaves the divergenceless $\mathbf{v}$ unchanged. The incompressible Euler equation has a nice infinite-dimensional geometric interpretation: it describes the geodesic flow on the group of volume preserving diffeomorphisms.[2] In one space dimension and without the incompressibility constraint and the pressure term, Eq. (1.1) becomes the Burgers equation, one of the simplest nonlinear equations.

The incompressible Euler and NS equations are examples of nonlinear partial differential evolution equations. After a century and a half of studies, they still pose major open problems as far as the control of their solutions is concerned. The most interesting questions touch on the short-distance (ultra-violet) behavior. Suppose that we start from smooth initial data and the force $\mathbf{f}$ is smooth. For simplicity, let us assume compact support of both (we may also consider the compact space or the case with boundary conditions). It is known that the smooth solutions of the so posed initial value problem are unique and exist for short time. Do they exist for all times? The answer is positive in 2 dimensions for both Euler and NS equations but in 3 dimensions the answer is not known. It is usually expected to be positive in the NS case. The opinions about the Euler case (no blowup versus finite-time blowup for special smooth initial conditions) are more divided and fluctuate in time.

In an important 1933 paper on the NS equation, Leray has introduced the notion of weak solutions of the equation. A vector field $\mathbf{v}(t,x)$ locally in L^2 is a weak solution if it satisfies the equations in the distributional sense, i.e. if for any smooth vector field $\mathbf{u}$ without divergence and any smooth function φ, both with compact supports,

$$\int\left[(\partial_t u^i + \nu\Delta u^i + (\partial_j u^i)v^j)v^i + u^i f^i\right] = 0 \quad\text{and}\quad \int(\partial_i\varphi)v^i = 0.$$

[1] We shall absorb in this case the constant $1/\rho$ into p and $\mathbf{f}$.

[2] Recall that the Euler top is related to the geodesic flow on the group SO(3).

Leray showed by compactness arguments existence of global weak solutions of the 3-dimensional NS equations with additional properties (e.g., with space derivatives locally square integrable). The weak solutions are not unique (there are weak solutions of the 2-dimensional Euler equation with compact support).

The NS equation is invariant under rescalings. Let

$$\begin{aligned}\tilde{\mathbf{v}}(t,\mathbf{x}) &= \tau s^{-1}\mathbf{v}(\tau t, s\mathbf{x}),\\ \tilde{\mathbf{f}}(t,\mathbf{x}) &= \tau^2 s^{-1}\mathbf{f}(\tau t, s\mathbf{x}),\\ \tilde{p}(t,\mathbf{x}) &= \tau^2 s^{-2}p(\tau t, s\mathbf{x}),\\ \tilde{\nu} &= \tau s^{-2}\nu.\end{aligned}$$

If $\mathbf{v}$ and p solve the NS equation with viscosity ν and force $\mathbf{f}$ then $\tilde{\mathbf{v}}$ and $\tilde{p}$ give a solution for viscosity $\tilde{\nu}$ and force $\tilde{\mathbf{f}}$. It is convenient to introduce the dimensionless version of the (inverse) viscosity, the Reynolds number

$$\mathrm{Re} = \frac{L\delta_L v}{\nu},$$

where $\delta_L v$ is a characteristic size of velocity differences over scale L of the order of the size of the system. Note the scale-dependent nature of the concept. For the flow in a pipe of radius L, we may take $\delta_L v$ as velocity in the middle of the pipe (the velocity vanishes on the wall of the pipe). The following are the basic phenomenological observations about hydrodynamics. If $\mathrm{Re} \ll 1$, one encounters regular (*laminar*) flows. For Re between ~ 1 and $\sim 10^2$, one observes complicated phenomena depending on the precise situation. For $\mathrm{Re} \gg 10^2$, very irregular (*turbulent*) flows set in. They show for high Re (*developed turbulence*) a certain degree of similarity for different circumstances.

Somewhat simplifying, one could say that for the laminar flows the nonlinear term $(\mathbf{v}\cdot\boldsymbol{\nabla})\mathbf{v}$ of the NS equation plays a smaller role. This is a relatively well understood regime, also rigorously. Following Gallavotti's article cited at the end of the lecture, define the running Reynolds number

$$\mathrm{Re}_r = \frac{r^2}{\nu}\left(\frac{1}{|B_r|}\int_{B_r}|\boldsymbol{\nabla}\mathbf{v}|^2\right)^{1/2}$$

where $B_r = \{(s,\mathbf{y}) \mid |s-t| < r^2/\nu, |\mathbf{y}-\mathbf{x}| < r\}$ is a neighborhood of the space-time point $(t,\mathbf{x})$. Note that we may rewrite

$$\mathrm{Re}_r = \frac{r\delta_r v}{\nu}$$

where $\delta_r v$, the mean velocity difference on scale r, is calculated by multiplying the mean square gradient of v over B_r by r. The best regularity result about the weak solutions of the NS equation is due to Caffarelli–Kohn–Nirenberg and says that there exists $\epsilon > 0$ such that if $\mathrm{Re}_r \leq \epsilon$

then the solution is smooth in the $B_{\epsilon r}$ neighborhood of $(t, \mathbf{x})$. This implies that, for a weak solution, the Hausdorff dimension of the set of singularities is ≤ 1. Note the spirit of the result in line with the phenomenological characterization of laminar flows.

One expects that in the regime of intermediate Re between ~ 1 and $\sim 10^2$ only a finite number of modes of fluid play an important role. These modes may be effectively described by ordinary differential equations to which the theory of dynamical systems (bifurcations, strange attractors, chaology) may be applied. Indeed, the dynamical system theory has been used with much success to describe specific situations, as the flow between rotating cylinders, for example. It is not clear, however, if the dynamical system ideas may be useful to describe the fully developed turbulence.

The importance of the NS equations is far from being limited to the mathematical questions. It extends to meteorology, aeronautics and maritime engineering, to mention only three of the most important domains of practical applications. The regime of large Re, where the nonlinear term of the NS equation becomes very important dominates in practical situations (in medium size river, Re $\sim 10^7$). One has to admit that the interests of engineers and theoretical physicists are somewhat different. The first ones are interested mainly in flows around obstacles (e.g., an airplane wing) whereas the second ones show a tendency to concentrate on flows far from boundaries where the simplifying assumptions of homogeneity and isotropy seem in place. Nevertheless, a good understanding of such idealized flows would certainly have practical consequences.

For high Reynolds numbers, it is reasonable to attempt a statistical description of complicated turbulent flows. In the theoretical approach, the statistics may be generated by considering random initial data or/and random forcing. Since some degree of universality is observed in this situation independently on the way the flow is excited, one often assumes that the force $\mathbf{f}$ is a random Gaussian field with mean zero and the covariance

$$\langle f^i(t, \mathbf{x}) f^j(s, \mathbf{y}) \rangle = \delta(t - s) \chi^{ij} \left(\frac{\mathbf{x} - \mathbf{y}}{L} \right) \tag{1.3}$$

with $\partial_i \chi^{ij} = 0$ and the *injection scale* L regulating the decay of χ, i.e., the scale on which the force acts. $\mathbf{v}(t, \mathbf{x})$ becomes then a random field and the NS equation takes, schematically, the form

$$\partial_t \Phi = -F(\Phi) + \eta. \tag{1.4}$$

Many other dynamical problems in physics may be put in such a form with $F(\Phi)$ being a nonlinear functional of local densities $\Phi(t, \mathbf{x})$ of physical quantities and η a random noise. The case of the NS equation should be contrasted with another example of Eq. (1.4), provided by the Langevin equation describing the approach to equilibrium in systems of statistical mechanics or field theory. In the latter case, the nonlinearity is of the gradient type:

$$F(\Phi) = \frac{\delta S(\Phi)}{\delta \Phi}$$

with $S(\Phi)$ a local functional, e.g., $S(\Phi) = \frac{1}{2}\int(\nabla\Phi)^2 + \frac{1}{7}2m^2\int\Phi^2 + \lambda\int\Phi^4$ in the Φ^4 field theory. The noise is to be taken Gaussian:

$$\langle \eta(t, \mathbf{x})\eta(s, \mathbf{y})\rangle = \delta(t - s)L^{-d}\chi\Big(\frac{\mathbf{x} - \mathbf{y}}{L}\Big). \tag{1.5}$$

The covariance $L^{-d}\chi(\mathbf{x}/L)$ regulates the theory on short distances $\lesssim L$ and is close to the delta-function $\delta(\mathbf{x})$ for small L. On the contrary, in the case of the NS equations we are mostly interested in forces acting on large distances $\sim L$ (e.g., the convective forces in the atmosphere) so that the force covariance $\chi^{ij}(\mathbf{x}/L)$ becomes close to a constant in the position space, i.e., to a multiple of the delta-function in the wavenumber space. Such regime in field theory would correspond to distances shorter than the ultraviolet cutoff, with the behavior strongly dependent on the detailed form of the cutoff. Another difference is that in Eq. (1.1) the nonlinear term, dominant for high Reynolds numbers, is not of the gradient type. Finally, the presence of pressure renders it also nonlocal, which is another complication. All these differences make the case of the NS equation quite different from that of the Langevin equation describing, in the stationary regime, equilibrium states. They make the NS problem, strongly coupled for high Re, resistant to the methods employed successfully in the study of equilibrium states like perturbative approaches or the renormalization group.

The main characteristic of the stationary regimes of the randomly forced NS equations is the presence of nonvanishing fluxes of conserved quantities, forbidden in equilibrium states. By integrating the scalar product of the incompressible NS equation with $\mathbf{v}$ over the space and assuming that the flow velocity vanishes sufficiently fast at large distances (or at the boundary), one deduces the energy balance:

$$\partial_t \frac{1}{2}\int \mathbf{v}^2 = -\nu\int(\boldsymbol{\nabla}\mathbf{v})^2 + \int \mathbf{f}\cdot\mathbf{v}. \tag{1.6}$$

The equation states that the rate of change of fluid energy is equal to the energy injection rate $\int \mathbf{f}\cdot\mathbf{v}$ (work of the external forces per unit time) minus the energy dissipation per unit time $\nu\int(\boldsymbol{\nabla}\mathbf{v})^2$ due to the viscous friction. In a stationary state, the mean overall energy of the fluid is constant in time so that the energy balance equation (1.6) implies that

$$\int \langle \nu(\boldsymbol{\nabla}\mathbf{v})^2\rangle = \int \langle \mathbf{v}\cdot\mathbf{f}\rangle,$$

where $\langle\ \rangle$ denotes the ensemble average, or, that in the homogeneous state,

$$\bar{\epsilon} \equiv \langle \nu(\boldsymbol{\nabla}\mathbf{v})^2\rangle = \langle \mathbf{v}\cdot\mathbf{f}\rangle \equiv \bar{\varphi} \tag{1.7}$$

where $\bar{\epsilon}$ denotes the mean dissipation rate and $\bar{\varphi}$ the mean injection rate of energy, both with the dimension length^2/time^3. In the situation where the energy injection is a large distance process (e.g., in the atmospheric turbulence or shear flows) one expects that for high Re a *scale separation* occurs, with the energy dissipation taking place on much smaller distances. Pictorially, energy is transmitted to the fluid by the excitation of large eddies on scale L which subsequently break to smaller scale eddies and so on. This way the injected energy is passed to shorter and shorter scales without substantial loss, until the viscous scale η is reached where the friction dissipates energy. Such an *energy cascade*, described first by Richardson in 1922, is characterized by the integral scale L, the viscous scale η and the energy dissipation rate $\bar{\epsilon}$. The scale ratio L/η should grow with the Reynolds number. The interval of distance scales r satisfying $L \gg r \gg \eta$ is called the *inertial range.*

The cascade picture may be formulated in more quantitative terms by introducing the quantities

$$\bar{\epsilon}_{\leq K} = \nu \int_{|\mathbf{k}|\leq K} \Big(\int \mathrm{e}^{-i\mathbf{k}\cdot\mathbf{x}} \langle \boldsymbol{\nabla}\mathbf{v}(\mathbf{x}) \cdot \boldsymbol{\nabla}\mathbf{v}(\mathbf{0}) \rangle \, d\mathbf{x} \Big) \, d\!\!\!{}^-\mathbf{k},$$

$$\bar{\varphi}_{\leq K} = \int_{|\mathbf{k}|\leq K} \Big(\int \mathrm{e}^{-i\mathbf{k}\cdot\mathbf{x}} \langle \mathbf{v}(\mathbf{x}) \cdot \mathbf{f}(\mathbf{0}) \rangle \, d\mathbf{x} \Big) \, d\!\!\!{}^-\mathbf{k}$$

interpreted as the mean dissipation and mean injection rate in wavenumbers $\mathbf{k}$ with $|\mathbf{k}| \leq K$ ($d\!\!\!{}^-\mathbf{k} \equiv d\mathbf{k}/(2\pi)^d$). The injection of energy limited to distances $\gtrsim L$ means that, as a function of K, $\bar{\varphi}_{\leq K}$ is close to $\bar{\epsilon}$ everywhere except for $K \lesssim 1/L$ where it falls to zero with $K \to 0$. E.g., for the time-decorrelated force with an appropriate interpretation of Eq. (1.1) as a stochastic differential equation, the relation $\langle \mathbf{v}(t,\mathbf{x}) \mathbf{f}(t,\mathbf{0}) \rangle = \frac{1}{2} \operatorname{tr} \chi(\mathbf{x}/L)$ holds so that $\bar{\varphi}_{\leq K} = \frac{1}{2} \int_{|\mathbf{k}|\leq K} \big(\int \mathrm{e}^{-i\mathbf{k}\cdot\mathbf{x}} \operatorname{tr} \chi(\mathbf{x}/L) \, d\mathbf{x} \big) d\!\!\!{}^-\mathbf{k}$ and such behavior of $\bar{\varphi}_{\leq K}$ follows since χ is close to a delta function in the wavenumber space. The cascade picture should imply that the mean dissipation rate $\bar{\epsilon}_{\leq K}$ is negligable for $K \ll 1/\eta$ and then grows to $\bar{\epsilon}$. The difference

$$\bar{\varphi}_{\leq K} - \bar{\epsilon}_{\leq K} \equiv \bar{\pi}_K$$

has the interpretation of the energy flux out of the wavenumbers $\mathbf{k}$ with $|\mathbf{k}| \leq K$. This flux should be approximately constant and equal to $\bar{\epsilon}$ in the inertial range $1/L \ll K \ll 1/\eta$, see Fig. 3.1.

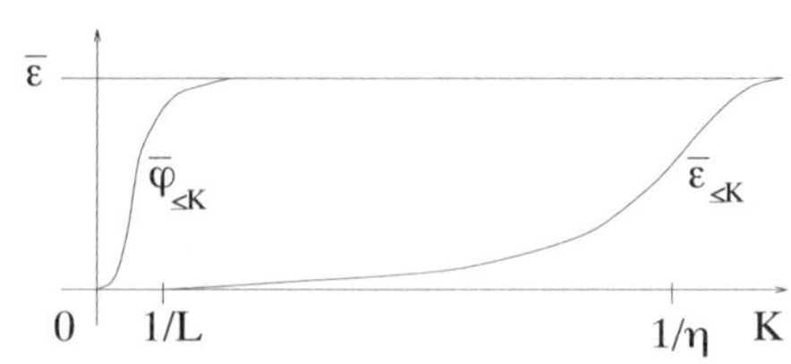

FIGURE 3.1.

References

For the original formulation of the NS equations see:

1. C.L.M.H. Navier, *Mémoire sur le lois du mouvement des fluides.* Mém. Acad. Roy. Sci. **6** (1823), 389–440.

2. G.G. Stokes, *On some cases of fluid motion.* Trans. Camb. Phil. Soc. **8** (1843), 105.

For the Burgers equation see:

3. J.M. Burgers, *The Nonlinear Diffusion Equation.* D. Reidel, Dordrecht, 1974.

A brief introduction to mathematical problems of the incompressible Navier–Stokes equations, from which we have borrowed the discussion of the regularity results, is

4. G. Gallavotti, *Some rigorous results about 3D Navier–Stokes*, Lecture Notes, "Turbulence in spatially ordered systems," Les Houches 1992, mp_arc/92-109.

For a textbook on mathematical theory of NS equations, see:

5. R. Temam, *Navier–Stokes Equations.* North-Holland, Amsterdam, 1984.

The dynamical system approach to turbulence is described, e.g., in:

6. D. Ruelle, *The turbulent fluid as a dynamical system*, in *New Perspectives in Turbulence*, ed. L. Sirovich, Springer, Berlin 1991, 123–138.

7. T. Bohr, M.H. Jensen, G. Paladin, and A. Vulpiani, *Dynamical System Approach to Turbulence.* Cambridge University Press, Cambridge 1998.

The mathematical approach to the statistics of solutions is discussed in the textbook

8. M.J. Vishik and A.V. Fursikov, *Mathematical Problems of Statistical Hydrodynamics.* Kluwer, Dordrecht 1988.

The classical textbooks on fluid dynamics are:

9. G.K. Batchelor, *Introduction to Fluid Dynamics.* Cambridge Univ. Press, Cambridge 1967.

10. A.S. Monin and A.M. Yaglom, *Statistical Fluid Mechanics I & II.* MIT Press, Cambridge MA 1971 & 1975.

For the original description of the turbulent cascade see:

11. L.F. Richardson, *Weather Prediction by Numerical Process.* Cambridge Univ. Press, Cambridge 1922

Lecture 2

In 1941, A.N. Kolmogorov proposed a scaling theory of the developed turbulence which has deeply marked the ensuing turbulence research and remains a reference point for most of the modern research on the subject. It is based on the exact statistical relations for the turbulent velocities obtained with general assumptions and going back to the 1938 work of von Kármán and Horwarth. The NS equation implies that

$$\mathbf{v}(t+\Delta t) = \mathbf{v}(t)+[-(\mathbf{v}\cdot\boldsymbol{\nabla}\mathbf{v})\mathbf{v}+\nu\boldsymbol{\nabla}^2\mathbf{v}-\boldsymbol{\nabla}p]|_t\Delta t+\int_t^{t+\Delta t}\mathbf{f}(s)\,ds+\mathcal{O}\big((\Delta t)^2\big)$$

(the term involving the white noise in time $\mathbf{f}$ is $\mathcal{O}(t^{1/2})$). Consequently, we obtain for the statistical expectation of the scalar product of two velocities at the same time but general points:

$$\begin{aligned}\langle\mathbf{v}(t+\Delta t,\mathbf{x})\cdot\mathbf{v}(t+\Delta t,\mathbf{y})\rangle &= \langle\mathbf{v}(t,\mathbf{x})\cdot\mathbf{v}(t,\mathbf{y})\rangle\\ &+\Big[-\langle(\mathbf{v}\cdot\boldsymbol{\nabla})\mathbf{v}(\mathbf{x})\cdot\mathbf{v}(\mathbf{y})\rangle-\langle\mathbf{v}(\mathbf{x})\cdot(\mathbf{v}\cdot\boldsymbol{\nabla})\mathbf{v}(\mathbf{y})\rangle\\ &\qquad+\nu\langle\boldsymbol{\nabla}^2\mathbf{v}(\mathbf{x})\cdot\mathbf{v}(\mathbf{y})\rangle+\nu\langle\mathbf{v}(\mathbf{x})\cdot\boldsymbol{\nabla}^2\mathbf{v}(\mathbf{y})\rangle+\operatorname{tr}\chi\Big(\frac{\mathbf{x}-\mathbf{y}}{L}\Big)\Big]\Big|_t\Delta t\\ &+\mathcal{O}\big((\Delta t^2)\big).\end{aligned}$$

We have dropped the pressure terms assuming the homogeneity (i.e., translation invariance) of the statistical state. The second line terms may be rewritten as

$$\tfrac{1}{2}\boldsymbol{\nabla}_{\mathbf{x}}\cdot\big\langle\big(\mathbf{v}(\mathbf{x})-\mathbf{v}(\mathbf{y})\big)\big(\mathbf{v}(\mathbf{x})-\mathbf{v}(\mathbf{y})\big)^2\big\rangle$$

and the next two ones as

$$-2\nu\langle\boldsymbol{\nabla}\mathbf{v}(\mathbf{x})\cdot\boldsymbol{\nabla}\mathbf{v}(\mathbf{y})\rangle.$$

Equating the $\mathcal{O}(\Delta t)$ terms, we obtain

$$\begin{aligned}\partial_t\langle\mathbf{v}(\mathbf{x})\cdot\mathbf{v}(\mathbf{y})\rangle = \frac{1}{2}\boldsymbol{\nabla}_{\mathbf{x}}\cdot\big\langle\big(\mathbf{v}(\mathbf{x})-\mathbf{v}(\mathbf{y})\big)\big(\mathbf{v}(\mathbf{x})-\mathbf{v}(\mathbf{y})\big)^2\big\rangle\\ -2\nu\langle\boldsymbol{\nabla}\mathbf{v}(\mathbf{x})\cdot\boldsymbol{\nabla}\mathbf{v}(\mathbf{y})\rangle+\operatorname{tr}\chi\Big(\frac{\mathbf{x}-\mathbf{y}}{L}\Big),\end{aligned}\tag{2.8}$$

which is the basic relation between the 2-point and the 3-point correlation functions of velocity.

In three (and more) dimensions, we expect the stabilization of the correlation functions for long times. In the stationary regime, the time derivatives of the equal-time functions vanish and we infer that

$$\begin{aligned}-\frac{1}{4}\boldsymbol{\nabla}_{\mathbf{x}}\cdot\big\langle\big(\mathbf{v}(\mathbf{x})-\mathbf{v}(\mathbf{y})\big)\big(\mathbf{v}(\mathbf{x})-\mathbf{v}(\mathbf{y})\big)^2\big\rangle+\nu\langle\boldsymbol{\nabla}\mathbf{v}(\mathbf{x})\cdot\boldsymbol{\nabla}\mathbf{v}(\mathbf{y})\rangle\\ =\frac{1}{2}\operatorname{tr}\chi\Big(\frac{\mathbf{x}-\mathbf{y}}{L}\Big).\end{aligned}\tag{2.9}$$

Taking first the limit $\mathbf{y} \to \mathbf{x}$ for positive ν and assuming that the presence of the latter smoothes out the behavior of $\big\langle \big(\mathbf{v}(\mathbf{x}) - v(\mathbf{y})\big)\big(\mathbf{v}(\mathbf{x}) - \mathbf{v}(\mathbf{y})\big)^2 \big\rangle$ so that the first term on the left hand side vanishes in the limit, we obtain

$$\bar{\epsilon} = \tfrac{1}{2}\,\mathrm{tr}\,\chi(\mathbf{0})$$

which is nothing else but the energy balance equation (1.7) for the case of the time decorrelated force injecting energy at the mean rate $\bar{\varphi} = \frac{1}{2}\,\mathrm{tr}\,\chi(\mathbf{0})$.

To deduce further implications of Eq. (2.9), we take its invicid limit $\nu \to 0$ keeping the points separate. This gives

$$-\frac{1}{4}\nabla_{\mathbf{x}} \cdot \big\langle \big(\mathbf{v}(\mathbf{x}) - \mathbf{v}(\mathbf{y})\big)\big(\mathbf{v}(\mathbf{x}) - \mathbf{v}(\mathbf{y})\big)^2 \big\rangle\big|_{\nu=0} = \frac{1}{2}\,\mathrm{tr}\,\chi\Big(\frac{\mathbf{x}-\mathbf{y}}{L}\Big). \tag{2.10}$$

The assumption that the force acts only on distances $\gtrsim L$ means that, for $|\mathbf{x}-\mathbf{y}| \ll L$, $\chi(\mathbf{x}-\mathbf{y}/L) = \chi(\mathbf{0})$, approximately, so that Eq. (2.10) implies that in the inertial range,

$$-\tfrac{1}{4}\nabla_{\mathbf{x}} \cdot \big\langle \big(\mathbf{v}(\mathbf{x}) - \mathbf{v}(\mathbf{y})\big)^2 \big(\mathbf{v}(\mathbf{x}) - \mathbf{v}(\mathbf{y})\big) \big\rangle = \bar{\epsilon}.$$

Assuming the isotropy (i.e. the rotation invariance), this implies the relation

$$\begin{aligned}\big\langle \big(v^i(\mathbf{x}) - v^i(\mathbf{y})\big)\big(v^j(\mathbf{x}) - v^j(\mathbf{y})\big)\big(v^k(\mathbf{x}) - v^k(\mathbf{y})\big) \big\rangle \\ = -\frac{4\bar{\epsilon}}{d(d+2)}(\delta^{ij}r^k + \delta^{ik}r^j + \delta^{jk}r^i),\end{aligned} \tag{2.11}$$

where $\mathbf{r} = \mathbf{x}-\mathbf{y}$. In other words, the 3-point function of equal-time velocity difference is in the inertial range linear in the point separation. In particular, for the so called longitudinal 3-point structure function, we obtain

$$S_3^{\|}(r) \equiv \Big\langle \Big(\big(\mathbf{v}(\mathbf{x}) - \mathbf{v}(\mathbf{y})\big) \cdot \frac{\mathbf{r}}{r}\Big)^3 \Big\rangle = -\frac{12}{d(d+2)}\bar{\epsilon} r \tag{2.12}$$

known, for $d = 3$, as the *Kolmogorov four-fifths law.*

One may deduce a stronger version of the above relation which takes the form of the operator product expansion for the $\nu \to 0$ limit of the dissipation operator $\epsilon = \nu(\nabla \mathbf{v})^2$:

$$\epsilon(\mathbf{x}) = -\frac{1}{4}\lim_{\mathbf{y}\to\mathbf{x}} \nabla_{\mathbf{x}} \cdot \big[\big(\mathbf{v}(\mathbf{x}) - \mathbf{v}(\mathbf{y})\big)\big(\mathbf{v}(\mathbf{x}) - \mathbf{v}(\mathbf{y})\big)^2\big]\big|_{\nu=0} \tag{2.13}$$

which should hold inside expectations in the $\nu \to 0$ limit. As noticed recently by Duchon and Robert, the relation (2.13) holds for all weak solutions of the Euler equation which are limits of strong solutions of the NS equation. Relation (2.13) is often called a *dissipative anomaly*: the dissipation ϵ whose definition involves a factor of ν does not vanish when $\nu \to 0$.

In his 1941 paper, Kolmogorov went one step further by postulating the universal character of the turbulent cascade in the inertial range with the equal-time correlators of velocity differences over inertial range distances given by universal functions of the latter and of the dissipation rate $\bar{\epsilon}$. In particular this implies that the velocity structure functions $S_n^{||}(r) = \left\langle \left((\mathbf{v}(\mathbf{x}) - \mathbf{v}(\mathbf{y})) \cdot \mathbf{r}/r \right)^n \right\rangle$ are determined by dimensional reasons up to universal constants:

$$S_n^{||}(r) = C_n \bar{\epsilon}^{n/3} r^{n/3}. \tag{2.14}$$

Indeed, the right hand side is the only function of $\bar{\epsilon}$ and r with the dimension (length/time)^n. The physical content of the Kolmogorov theory is that the typical velocity v_r of size r eddies behaves as $\bar{\epsilon}^{1/3} r^{1/3}$. During the *eddy turnover time* $\tau_r = r/v_r \propto \bar{\epsilon}^{-1/3} r^{2/3}$, these eddies transfer their energy with the density $\frac{1}{2} v_r^2 \propto \bar{\epsilon}^{2/3} r^{2/3}$ to the shorter scale resulting in the constant energy flux $\propto \bar{\epsilon}$. For the scale-dependent Reynolds number one obtains then $\mathrm{Re}_r \propto \bar{\epsilon}^{1/3} r^{4/3}/\nu$. In particular, $\mathrm{Re} = \mathrm{Re}_L \propto \bar{\epsilon}^{1/3} L^{4/3}/\nu$ and $1 = \mathrm{Re}_\eta \propto \bar{\epsilon}^{1/3} \eta^{4/3}/\nu$. Hence $\eta/L \propto R^{-3/4}$ and it decreases with Re.

As we have seen, the $n = 3$ relation (2.14) coincides with the Kolmogorov four-fifths law. The structure functions are measured, more or less directly, in atmospheric or ocean flows, in water jets, in aerodynamic tunnels or in subtle experiments with helium gas in between rotating cylinders or plates. They are also accessible in numerical simulations. One extracts then the scaling exponents assuming the behavior

$$S_n^{||}(x) \propto r^{\zeta_n}.$$

ζ_3 agrees well with the theoretical prediction $\zeta_3 = 1$. Here are some other exponents obtained from wind tunnel data

$$\zeta_2 = .70\ (.67), \quad \zeta_4 = 1.28\ (1.33), \quad \zeta_5 = 1.53\ (1.67),$$
$$\zeta_6 = 1.77\ (2), \quad \zeta_7 = 2.01\ (2.33)$$

with the Kolmogorov values in the parenthesis for comparison. The discrepancy is quite pronounced (its direction for the even functions is determined by the Hölder inequality implying that ζ_n is a concave function of n). One of the main open problems in the theory of fully developed turbulence is to explain, starting from the first principles (i.e., from the NS equation), the breakdown of the Kolmogorov theory leading to the anomalous structure-function exponents which indicate that the distribution of $\mathbf{v}(t, \mathbf{x})$ in the inertial range is rather different from Gaussian. The discrepancy may be measured by the *skewness* $S_3/S_2^{3/2}$ or the *flatness* S_4/S_2^2 which grow with diminishing distance instead of being equal to their Gaussian values 0 and 3. Hence the domination of the short scales by the large deviations of the velocity differences indicating an enhanced short-distance activity: the phenomenon called *intermittency*. The intuitive explanation of intermittency which was advanced is that only a part of the fluid modes (temporal or/and spatial) participates in the turbulent cascade, with the proportion of active modes decreasing with diminishing scale. This forces the active short-distance modes to transfer more energy and, consequently, to be more excited. This picture led to multiple (multi)fractal models of the cascade reviewed in the book by Frisch cited at the end of this lecture. Such models, although interesting phenomenologically, are not based on the NS equation and allow to obtain essentially arbitrary spectra of exponents. They do not really explain the mechanism of the breakdown of the normal scaling in realistic flows.

For the *energy spectrum*

$$\bar{e}_K \equiv \frac{1}{2}\frac{d}{dK}\int_{|\mathbf{k}|\leq K}\Big(\int \langle \mathbf{v}(\mathbf{x})\cdot\mathbf{v}(0)\rangle \mathrm{e}^{-i\mathbf{k}\cdot\mathbf{x}}\, d\mathbf{x}\Big)\, d\!\!^{-}\mathbf{k},$$

the Kolmogorov theory predicts

$$\bar{e}_K \propto \bar{\epsilon}^{2/3} K^{-5/3} \tag{2.15}$$

for $1/L \ll K \ll 1/\eta$ just by the dimensional count. The experimental data seem to confirm this behavior (with the possibility of a slight discrepancy consistent with the value of ζ_2 cited above). For the mean dissipation rate, we obtain: $\bar{\epsilon}_{\leq K} = 2\nu\int_0^K K'^2 \bar{e}_{K'}\, dK' \propto \nu\bar{\epsilon}^{2/3}K^{4/3}$ confirming that $\bar{\epsilon}_{\leq K}/\bar{\epsilon} = \mathrm{Re}_{K^{-1}}^{-1} \ll \mathrm{Re}_{\eta}^{-1} = 1$ in the inertial range. Deep in the dissipative regime $K \gg 1/\eta$, $\bar{e}_K$ falls off much faster than in the inertial interval.

An important object in the turbulence theory is the *vorticity* field that measures the strength and the orientation of eddies. In three dimensions it is a (pseudo-)vector field $\boldsymbol{\omega} = \boldsymbol{\nabla}\times\mathbf{v}$ and it satisfies, for the incompressible flow, the equation

$$\partial_t\boldsymbol{\omega} + (\mathbf{v}\cdot\boldsymbol{\nabla})\boldsymbol{\omega} - (\boldsymbol{\omega}\cdot\boldsymbol{\nabla})\mathbf{v} - \nu\boldsymbol{\nabla}^2\boldsymbol{\omega} = \boldsymbol{\nabla}\times\mathbf{f}. \tag{2.16}$$

Note that $(\mathbf{v}\cdot\boldsymbol{\nabla})\boldsymbol{\omega} - (\boldsymbol{\omega}\cdot\boldsymbol{\nabla})\mathbf{v}$ is the commutator of two vector fields, so that, at vanishing ν and $\mathbf{f}$, the equation (2.16) implies that the vorticity is transported by velocity field as a vector. This involves stretching of ω by the velocity *strain* $\frac{1}{2}(\partial_i v^j + \partial_j v^i)$, an important mechanism for the energy transfer between scales. In two dimensions, however, the vorticity reduces to a (pseudo-)scalar field $\omega = \epsilon^{ij}\partial_i v^j$ whose evolution is governed by the equation

$$\partial_t\omega + (\mathbf{v}\cdot\boldsymbol{\nabla})\omega - \nu\boldsymbol{\nabla}^2\omega = \epsilon^{ij}\partial_i f^j. \tag{2.17}$$

Here, at vanishing ν and $\mathbf{f}$, the vorticity is simply transported by the velocity field along the *Lagrangian trajectories* $\mathbf{x}(t)$ of the (imaginary) fluid particles s.t. $d\mathbf{x}/dt = \mathbf{v}(t,\mathbf{x})$. In particular, the two-dimensional flow, unlike the three-dimensional one, conserves, besides energy, the *enstrophy* $\Psi = \frac{1}{2}\int\omega^2$ (as well as the integrals of higher powers of vorticity).

In the seminal 1968 paper, R.H. Kraichnan has realized that the conservation of enstrophy implies a very different cascade picture in two dimensions, as compared to the three-dimensional one, see also the paper of Batchelor of 1969. First of all, the *enstrophy spectrum* is related to the energy spectrum:

$$\bar{\psi}_K \equiv \frac{1}{2}\frac{d}{dK}\int_{|\mathbf{k}|\leq K}\Big(\int \langle\omega(\mathbf{x})\omega(\mathbf{0})\rangle \mathrm{e}^{-i\mathbf{k}\cdot\mathbf{x}}\, d\mathbf{x}\Big) d\!\!{}^-\mathbf{k} = K^2 \bar{e}_K .$$

Similary, the mean enstrophy flux out of wavenumbers $\mathbf{k}$ with $|\mathbf{k}| \leq K$, if local in the momentum space, would have to be equal K^2 times the energy flux $\bar{\pi}_K$ so that the constancy of both fluxes is impossible. Kraichnan reasoned that, in such a situation, it will be the enstrophy flux which is constant on scales smaller than the injection scale so that the small scales will support a *direct enstrophy cascade* towards smaller and smaller distances, with all the enstrophy dissipation taking place on the shortest scales. The energy flux towards small scales will then be damped and, as a result, the energy will be, instead, transferred to scales longer than the injection scale L in an *inverse energy cascade* process. If not impaired by boundaries or large scale friction, this process would lead to the pumping of energy into the constant mode at the rate equal to $\frac{1}{2}\,\mathrm{tr}\,\chi(\mathbf{0})$.

If we assume that $\langle \mathbf{v}(t,\mathbf{x}) \cdot \mathbf{v}(t,\mathbf{y})\rangle - t\,\mathrm{tr}\,\chi(\mathbf{0})$ stabilizes at long times, then Eq. (2.8) will lead to the relation

$$\frac{1}{4}\boldsymbol{\nabla}_{\mathbf{x}} \cdot \big\langle \big(\mathbf{v}(\mathbf{x}) - \mathbf{v}(\mathbf{y})\big)\big(\mathbf{v}(\mathbf{x}) - \mathbf{v}(\mathbf{y})\big)^2\big\rangle - \nu\langle \boldsymbol{\nabla}\mathbf{v}(\mathbf{x}) \cdot \boldsymbol{\nabla}\mathbf{v}(\mathbf{y})\rangle = \frac{1}{2}\,\mathrm{tr}\Big[\chi(\mathbf{0}) - \chi\Big(\frac{\mathbf{x}-\mathbf{y}}{L}\Big)\Big]. \tag{2.18}$$

In particular, in the limit $\nu \to 0$ and for $|\mathbf{x} - \mathbf{y}| \gg L$,

$$\tfrac{1}{4}\boldsymbol{\nabla}_{\mathbf{x}} \cdot \big\langle \big(\mathbf{v}(\mathbf{x}) - \mathbf{v}(\mathbf{y})\big)\big(\mathbf{v}(\mathbf{x}) - \mathbf{v}(\mathbf{y})\big)^2\big\rangle = \tfrac{1}{2}\,\mathrm{tr}\,\chi(\mathbf{0})$$

implying the relation similar to the two-dimensional version of Eq. (2.11) but with the inverted sign. In particular, we obtain

$$S_3^{\|}(r) = \tfrac{3}{2}\bar{\epsilon} r \quad \text{for } r \gg L,$$

a *three-halfs law.*

The inverse cascade of the two-dimensional turbulence has been recently under an intensive study. Both experimental and numerical data confirm the above prediction and indicate that in this regime all structure functions, although non-Gaussian, scale with the Kolmogorov exponents $\zeta_n = n/3$ indicating that the *inverse cascade is not intermittent.* This implies for the energy spectrum, the behavior (2.15) for $K \ll 1/L$.

For $|\mathbf{x} - \mathbf{y}| \ll L$, Eq. (2.18) reduces in the $\nu \to 0$ limit to the relation

$$\frac{1}{4}\boldsymbol{\nabla}_{\mathbf{x}} \cdot \big\langle \big(\mathbf{v}(\mathbf{x}) - \mathbf{v}(\mathbf{y})\big)\big(\mathbf{v}(\mathbf{x}) - \mathbf{v}(\mathbf{y})\big)^2\big\rangle = \frac{1}{2}\,\mathrm{tr}\Big[\chi(\mathbf{0}) - \chi\Big(\frac{\mathbf{x}-\mathbf{y}}{L}\Big)\Big] = \mathcal{O}(r^2) \tag{2.19}$$

implying the scaling

$$S_3^{\|}(r) \sim r^3 \quad \text{for } r \ll L,$$

i.e. in the direct cascade regime. It is possible to infer in two dimensions another exact relation for the 3-point functions by proceeding from Eq. (2.17) the same way that we did before. One obtains

$$\begin{aligned}\frac{1}{2}\partial_t\langle\omega(t,\mathbf{x})\omega(t,\mathbf{y})\rangle &= \frac{1}{4}\boldsymbol{\nabla}_{\mathbf{x}}\cdot\big\langle\big(\mathbf{v}(t,\mathbf{x})-\mathbf{v}(t,\mathbf{y})\big)\big(\omega(t,\mathbf{x})-\omega(t,\mathbf{y})\big)^2\big\rangle \\ &\quad - \nu\langle\boldsymbol{\nabla}\omega(t,\mathbf{x})\cdot\boldsymbol{\nabla}\omega(t,\mathbf{y})\rangle - \frac{1}{2}\boldsymbol{\nabla}_{\mathbf{x}}^2\,\mathrm{tr}\,\chi\Big(\frac{\mathbf{x}-\mathbf{y}}{L}\Big). \qquad (2.20)\end{aligned}$$

Our previous assumption about the stabilization of the velocity 2-point function modulo a growing constant implies that the 2-point function of the vorticity reaches a stationary regime. We deduce from this, as before for the three-dimensional velocities, that the mean dissipation rate of vorticity

$$\bar{\epsilon}_\omega \equiv \nu\langle(\boldsymbol{\nabla}\omega)^2\rangle = -\tfrac{1}{2}\boldsymbol{\nabla}_{\mathbf{x}}^2\,\mathrm{tr}\,\chi(\mathbf{0})$$

is ν-independent and that, in the limit $\nu \to 0$,

$$\begin{aligned}-\frac{1}{4}\boldsymbol{\nabla}_{\mathbf{x}}\cdot\big\langle\big(\mathbf{v}(t,\mathbf{x})-\mathbf{v}(t,\mathbf{y})\big)\big(\omega(t,\mathbf{x})-\omega(t,\mathbf{y})\big)^2\big\rangle \\ = -\frac{1}{2}\boldsymbol{\nabla}_{\mathbf{x}}^2\,\mathrm{tr}\,\chi\Big(\frac{\mathbf{x}-\mathbf{y}}{L}\Big). \qquad (2.21)\end{aligned}$$

As noticed in a recent paper of Bernard, see the bibliography, the relations (2.19) and (2.21)), both consistent with the $\mathcal{O}(r^3)$ behavior of the 3-point functions of velocity differences in the direct cascade regime, allow to find exactly its asymptotic form. If we assume next that the n-point functions of the velocity differences scale accordingly, i.e., with the power n, then, for $n = 2$, we infer the Kraichnanmhyp Batchelor energy spectrum in the direct cascade

$$\bar{e}_K \sim K^{-3} \quad \text{for } K \gg \frac{1}{L}, \qquad (2.22)$$

confirmed by experimental observations. Intermittency, if existent in two-dimensional turbulence, seems much smaller then in three dimensions, especially in the inverse cascade. There are, however, theoretical predictions of logarithmic corrections to the power-law scaling in the direct cascade, not yet accessible to experimental or numerical verification.

Summarizing

In three dimensions, in the inertial range of the direct (short distance) energy cascade, the 3-point velocity structure function scales linearly in the distance. The energy spectrum is close to $\propto K^{-5/3}$, with the anomalous scaling of higher-point structure functions signaling intermittency. In two dimensions, in the direct (short distance) enstrophy cascade, the 3-point structure function scales as the 3rd power of the distance and the energy spectrum is close to $\propto K^{-3}$. In the (long distance) inverse energy cascade, one observes the Kolmogorov scaling, similarly as in the three dimensional direct cascade, but with reduced intermittency. Assuming the above spectra, one infers that both in two and in three dimensions, the mean enstrophy density given by $\int K^2 \bar{e}_K \, dK$ diverges in the ultraviolet in the invicid limit $\nu \to 0$. Accordingly, one should expect that the stationary fully turbulent state is carried by weak solutions of the Euler equation with divergent enstrophy. Hence the importance of studying such solutions which, in three dimensions, dissipate energy on short distances by a mechanism which is the topic of the next lecture.

References

The first exact statistical relations in turbulent flows were obtained in:

1. T. von Kármán and L. Horwarth, *On the statistical theory of isotropic turbulence*, Proc. Roy. Soc. London **A164** (1938), 192–215.

Kolmogorov's original work is

2. A. N. Kolmogorov, *The local structure of turbulence in incompressible viscous fluid for very large Reynolds' numbers*, C.R. Acad. Sci. URSS **30** (1941), 301–305.

A nice textbook reviewing the Kolmogorov theory of turbulence and much more, from which we have borrowed the discussion of the energy cascade and of the four-fifths law, is:

3. U. Frisch, *Turbulence: the Legacy of A.N. Kolmogorov*, Cambridge Univ. Press, Cambridge, 1995.

The dissipative anomaly in weak solutions of the NS and Euler equations has been discussed in:

4. J. Duchon and R. Robert, *Inertial energy dissipation for weak solutions of incompressible Euler and Navier–Stokes equations*, Nonlinearity **13** (2000), 249–255.

For the data about the anomalous scaling in three-dimensional turbulence see:

5. R. Benzi, S. Ciliberto, C. Baudet, and G. Ruiz Chavaria, *On the scaling of three dimensional homogeneous and isotropic turbulence.* Physica **D80** (1995), 385–398.

The original works on the two-dimensional turbulence are:

6. R.H. Kraichnan, *Inertial ranges in two-dimensional turbulence*, Phys. Fluids **10** (1967), 1417–1423.

7. G.K. Batchelor, *Computation of the energy spectrum in homogeneous two-dimensional turbulence*, Phys. Fluids Suppl. II **12** (1969), 233–239.

Our discussion of the 3-point functions in two-dimensional turbulence follows:

8. D. Bernard, *On the three point velocity correlation functions in 2d forced turbulence*, chao-dyn/9902010.

For recent experimental results on, respectively, the inverse and the direct cascades, see:

9. J. Paret and P. Tabeling, *Intermittency in the 2D inverse cascade of energy: experimental observations*, Phys. Fluids **10** (1998), 3126–3136.

10. J. Paret, M.C. Jullien, and P. Tabeling, *Vorticity statistics in the two-dimensional enstrophy cascade*, Phys. Rev. Lett. **83** (1999), 3418–3421, physics/9904044.

For a recent numerical work on the inverse cascade see:

11. G. Boffetta, A. Celani, and M. Vergassola, *Inverse cascade in two-dimensional turbulence: deviations from Gaussianity*, chao-dyn/9906016.

Lecture 3

One of the first quantitative laws of fully developed turbulence has been formulated by L.F. Richardson in 1926. Based on observations, e.g., of the movements of balloon probes released in the atmosphere, Richardson noted that the mean rate of growth of the square of the separation between two such probes is proportional to the four-thirds power of the distance, instead of being distance-independent, as in the Brownian diffusion. In other words, the Richardson dispersion law states that

$$\frac{d\rho^2}{dt} \propto \rho^{4/3}, \tag{3.23}$$

where $\rho(t)$ is the time t distance between two Lagrangian trajectories satisfying the ordinary differential equation (ODE)

$$\frac{d\mathbf{x}}{dt} = \mathbf{v}(t, \mathbf{x}). \tag{3.24}$$

Of course, the Richardson law is compatible with the Kolmogorov theory that implies that the velocity differences $|\mathbf{v}(t,\mathbf{x}) - \mathbf{v}(t,\mathbf{y})|$ scale as $\rho^{1/3}$ for $\rho = |\mathbf{x}-\mathbf{y}|$ in the inertial range, but it deserves a more detailed discussion.

Let us note that Eq. (3.23) is solved by

$$\rho^{2/3} = \rho_0^{2/3} + Ct$$

which, in the limit when the initial (i.e., time zero) separation goes to zero, reduces to the relation

$$\rho^2 \propto t^3. \tag{3.25}$$

Of course, some care is needed since the law (3.23) was observed for separations in the inertial range, so for $\rho \gg \eta$, where η is the viscous scale. In the limit $\mathrm{Re} \to \infty$, however, η tends to zero and the growth (3.25) should hold for arbitrarily short separations. In the turbulent atmosphere, for example, the viscous scale η is of the order of millimeter so that neglecting it when we look at scales of meters or even kilometers is a reasonable approximation. We infer this way that at $\mathrm{Re} = \infty$,

infinitesimally close Lagrangian trajectories separate in finite time.

Of course, ρ is the average separation, with the mean taken over different velocity fields of the statistical turbulent ensemble. One expects, however, that the trajectory separation is a self-averaging quantity so that this type of behavior holds already in a fixed typical velocity realization if we average over initial positions and times of release of two very close trajectories. What is this strange behavior of the trajectories? We are not used to such behaviors of the solutions of the ODE's. For example, in the integrable dynamical systems the distance between two trajectories

$$\rho \simeq \mathcal{O}(1).$$

In the dissipative systems,

$$\rho \simeq \mathcal{O}(\mathrm{e}^{-\lambda t}).$$

In the chaotic dynamical systems,

$$\rho \simeq \mathcal{O}(\mathrm{e}^{\lambda t}), \tag{3.26}$$

with the Lyapunov exponent $\lambda > 0$. Even the last case, where the nearby trajectories separate fast, is quite different from the behavior (3.25). Indeed, for the exponential separation (3.26), infinitesimally close trajectories keep shadowing each other and never separate to a finite distance. Definitely, chaos and fully developed turbulence are very different phenomena.

In fact the *explosive separation* (3.25) has a quite dramatic consequence: it means that, when $\mathrm{Re} \to \infty$, the very concept of a Lagrangian trajectory determined by its initial position in a fixed velocity realization breaks down. This breakdown of the Lagrangian flow is related to a breakdown of the theorem about the existence and unicity of solutions of the ODE (3.24). The theorem requires $\mathbf{v}(t, \mathbf{x})$ to be Lipschitz in $\mathbf{x}$, i.e., $|\mathbf{v}(t, \mathbf{x}) - \mathbf{v}(t, \mathbf{y})| \sim |\mathbf{x} - \mathbf{y}|$, whereas at $\mathrm{Re} = \infty$ the velocities are only Hölder continuous: $|\mathbf{v}(t, \mathbf{x}) - \mathbf{v}(t, \mathbf{y})| \sim |\mathbf{x} - \mathbf{y}|^\alpha$ with the exponent $\alpha < 1$ ($\alpha \simeq \frac{1}{3}$). We expect that for such velocities one may still maintain a probabilistic description of Lagrangian trajectories with such objects as the probability distribution function (PDF) $P^{t,s}(\mathbf{x}, \mathbf{y}|\mathbf{v})$ of the time s position $\mathbf{y}$ of the trajectory starting at time t at point $\mathbf{x}$ still making sense. $P^{t,s}(\mathbf{x}, \mathbf{y}|\mathbf{v})$ will rather be diffuse, however, than concentrated at one $\mathbf{y}$. In other words, we expect that, at $\mathrm{Re} = \infty$, the Lagrangian trajectories become stochastic already in a fixed realization of the velocity field, providing a mechanism for the energy dissipation in weak solutions of the Euler equation.

This important idea which seems to be a direct consequence in the limit of high Reynolds numbers of the Richardson dispersion law or of the Kolmogorov scaling of velocity differences has been rarely stressed in the past. It has appeared in a study of weak solutions of the Euler equations (in a somewhat different form) and in a recent analytic study of the Lagrangian trajectories in a simple statistical ensemble of velocities with the spatial Hölder continuity of the typical realizations built in. We shall spend the rest of this lecture by reviewing the latter results with the aim to substantiate the preceding discussion.

Following R.H. Kraichnan who initiated in 1968 the study of transport properties of velocities decorrelated in time, see next lecture, we shall consider a Gaussian ensemble of velocities with mean zero and 2-point function given by

$$\langle v^\alpha(t, \mathbf{x}) v^\beta(s, \mathbf{y}) \rangle = \delta(t - s)\big(D_0 \delta^{\alpha\beta} - d^{\alpha\beta}(\mathbf{x} - \mathbf{y})\big) \tag{3.27}$$

with D_0 a constant and $d^{\alpha\beta}(\mathbf{x}) \propto r^\xi$ for small $r \equiv |\mathbf{x}|$. $0 < \xi < 2$ will be the parameter of the ensemble. The constant D_0 drops out in the correlations of the velocity differences. E.g.,

$$\begin{aligned}\big\langle \big(v^\alpha(t, \mathbf{x}) - v^\alpha(t, \mathbf{0})\big)\big(v^\beta(s, \mathbf{x}) - v^\beta(s, \mathbf{0})\big)\big\rangle \\ = \delta(t - s)\big(d^{\alpha\beta}(\mathbf{x}) + d^{\beta\alpha}(\mathbf{x})\big).\end{aligned} \tag{3.28}$$

One may take

$$\begin{aligned}D_0 - d^{\alpha\beta}(\mathbf{x}) \\ = \int \frac{e^{i\mathbf{k}\cdot\mathbf{x}}}{(\mathbf{k}^2 + L^{-2})^{(d+\xi)/2}} \left(A \frac{k^\alpha k^\beta}{\mathbf{k}^2} + B\left(\delta^{\alpha\beta} - \frac{k^\alpha k^\beta}{\mathbf{k}^2}\right)\right) đ\mathbf{k}\end{aligned} \tag{3.29}$$

with the infrared cutoff L. For $A = 0$, the typical velocities are incompressible: $\nabla \cdot \mathbf{v} = 0$, whereas for $B = 0$ one obtains potential flows: $\mathbf{v} = \nabla \phi$. It is convenient to characterize the resulting velocity ensemble by, besides ξ, the *compressibility degree*

$$\wp = \frac{\langle(\boldsymbol{\nabla}\cdot\mathbf{v})^2\rangle}{\langle(\boldsymbol{\nabla}\mathbf{v})^2\rangle} = \frac{1}{1+(d-1)B/A}$$

contained between 0 and 1. $\wp = 0$ corresponds to the incompressible case whereas $\wp = 1$ to the potential one.

The above ensemble is not very realistic in its assumption of temporal velocity decorrelation. Recall, however, that the eddy turnover time τ_r is predicted by the Kolmogorov theory to behave as $\sim r^{2/3}$ so that the time correlation of the short scale eddies in small. The Kraichnan ensemble builds in the scaling behavior of the velocities in space with the Hölder continuity $|\mathbf{v}(t,\mathbf{x}) - \mathbf{v}(t,\mathbf{y})| \sim |\mathbf{x}-\mathbf{y}|^{\xi/2}$ of typical realizations, up to logarithmic corrections. To compare to the realistic ensemble of turbulent velocities, note that, for small r, the right hand side of Eq. (3.28) behaves as

$$\text{const.}\ \delta\left(\frac{t-s}{r^{2-\xi}}\right) r^{2\xi-2}.$$

We infer then that the velocity v_r of size r eddies scales as $\sim r^{\xi-1}$ corresponding to the eddy turnover time $\tau_r = r/v_r \sim r^{2-\xi}$. Within this limited comparison, $\xi = 4/3$ gives the Kolmogorov scaling of velocities. Of course, the Kraichnan ensemble, as Gaussian, is not intermittent. Also its odd-point correlation functions vanish. We shall see in next lecture, however, that it leads to intermittency in transport phenomena.

In the velocity fields of the Kraichnan ensemble, we shall consider the Lagrangian trajectories, perturbed first by a small noise, i.e., satisfying the stochastic differential equation:

$$\frac{d\mathbf{x}}{ds} = \mathbf{v}(s,\mathbf{x}) + \sqrt{2\kappa}\,\frac{d\boldsymbol{\beta}(s)}{ds}, \tag{3.30}$$

where $\boldsymbol{\beta}(s)$ is a d-dimensional Brownian motion and $\kappa > 0$. For a smooth velocity $\mathbf{v}$, solutions of Eq. (3.30) form a Markov process which may be characterized by the transition probabilities

$$P^{t,s}(\mathbf{x},\mathbf{y}|\mathbf{v}) = \overline{\delta\big(\mathbf{y} - \mathbf{x}_{t,\mathbf{x}}(s)\big)}, \tag{3.31}$$

where $\mathbf{x}_{t,\mathbf{x}}(s)$ denotes the solution of the differential equation (3.30) (with s as the running time) passing at time t by $\mathbf{x}$ and the overbar stands for the averaging with respect to the Brownian motion $\boldsymbol{\beta}$. The transition probabilities satisfy the linear equation

$$\big(\partial_t + \mathbf{v}(t,\mathbf{x})\cdot\boldsymbol{\nabla}_{\mathbf{x}} - \kappa\boldsymbol{\nabla}^2_{\mathbf{x}}\big)P^{t,s}(\mathbf{x},\mathbf{y}|\mathbf{v}) = 0,$$

with the initial condition $P^{t,t}(\mathbf{x},\mathbf{y}|\mathbf{v}) = \delta(\mathbf{x}-\mathbf{y})$, so that, in the operator language,

$$P^{t,s}(\mathbf{v}) = \mathcal{T}e^{\int_s^t[-\mathbf{v}(\sigma)\cdot\boldsymbol{\nabla}+\kappa\boldsymbol{\nabla}^2]\,d\sigma},$$

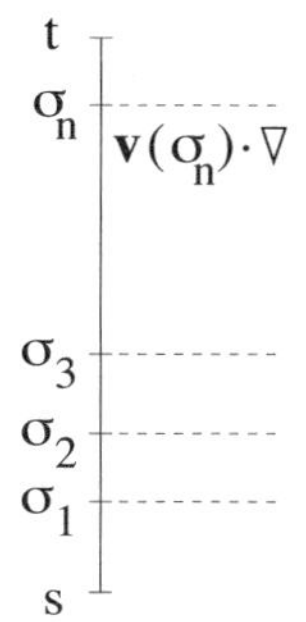

FIGURE 3.2.

where $\mathcal{T}$ denotes the time ordering (we have assumed $t \geq s$ here, the case $t < s$ is treated similarly).

As noticed by Y. Le Jan and O. Raimond, the transition probabilities $P^{t,s}(\mathbf{v})$ still make sense for velocities $\mathbf{v}$ of the Kraichnan ensemble, despite the poor regularity properties of the latter. These authors have rewritten the above expression for $P^{t,s}(\mathbf{v})$ in the Wick-ordered form:

$$\begin{aligned} P^{t,s}(\mathbf{v}) &= :\mathcal{T} e^{\int_s^t [-\mathbf{v}(\sigma)\cdot\boldsymbol{\nabla} + (\kappa + D_0/2)\boldsymbol{\nabla}^2]\, d\sigma}: \\ &= \sum_{n=0}^{\infty} (-1)^n \int_s^t d\sigma_n \cdots \int_s^{\sigma_2} d\sigma_1 \\ &\qquad :e^{(\kappa+D_0/2)(t-\sigma_n)\boldsymbol{\nabla}^2}\mathbf{v}(\sigma_n)\cdot\boldsymbol{\nabla} e^{(\kappa+D_0/2)(\sigma_n-\sigma_{n-1})\boldsymbol{\nabla}^2}\cdots \\ &\qquad\quad e^{(\kappa+D_0/2)(\sigma_2-\sigma_1)\boldsymbol{\nabla}^2}\mathbf{v}(\sigma_1)\cdot\boldsymbol{\nabla} e^{(\kappa+D_0/2)(\sigma_1-s)\boldsymbol{\nabla}^2}:, \end{aligned} \tag{3.32}$$

which is an infinite sum of Wick ordered monomials of $\mathbf{v}$ that may be represented by the diagrams of Fig. 3.2.

The homogeneous Wick polynomials of different degrees are orthogonal in the L^2 scalar product w.r.t. the Gaussian measure of the velocity process. By establishing the bound

$$\langle |P^{t,s}(\mathbf{v})g|^2\rangle \leq e^{D_0(t-s)\boldsymbol{\nabla}^2/2}|g|^2$$

where g are functions on $\mathbb{R}^d$, one shows then that the series giving $P^{s,t}(\mathbf{v})f$ converges in the space of square-integrable functionals of $\mathbf{v}$, so also for almost all (a.a.) $\mathbf{v}$, as long as g is bounded. It defines for a.a. velocities the transition probabilities $P^{t,s}(\mathbf{x},\mathbf{y}|\mathbf{v})$ of a Markov process which are continuous as functions of $\kappa \geq 0$. Note that

$$P_1^{t,s}(\mathbf{x},\mathbf{y}) \equiv \langle P^{t,s}(\mathbf{v})(\mathbf{x},\mathbf{y})\rangle = e^{(\kappa+D_0/2)|t-s|\boldsymbol{\nabla}^2}(\mathbf{x},\mathbf{y}). \tag{3.33}$$

The essential question that we want to address is about the nature of the Markov process obtained in the limit $\kappa \to 0$. Are the limiting transition probabilities concentrated at single points $\mathbf{y}$ leading to deterministic

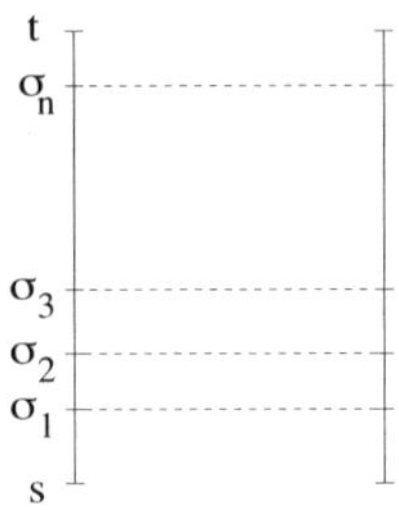

FIGURE 3.3.

Lagrangian trajectories determined, in a fixed velocity realization, by the initial position or, on the contrary, do they stay diffuse? A way to study this question is to examine the joint PDF (probability distribution function) of the equal-time values of two solutions of Eq. (3.30) averaged over the velocity ensemble:

$$P_2^{t,s}(\mathbf{x}_1,\mathbf{x}_2;\mathbf{y}_1,\mathbf{y}_2) = \langle P^{t,s}(\mathbf{x}_1,\mathbf{y}_1|\mathbf{v})P^{t,s}(\mathbf{x}_2,\mathbf{y}_2|\mathbf{v})\rangle.$$

The average on the right hand side is given by the sum of terms described by the diagrams of Fig. 3 with the broken-line propagators corresponding to the spatial part $\big(D_0 - d(\cdot)\big)$ of the velocity 2-point functions (3.27). The whole sum becomes the perturbative expansion for the heat kernel of the second order differential operator:

$$P_2^{t,s}(\underline{\mathbf{x}};\underline{\mathbf{y}}) = \mathrm{e}^{-|t-s|M_2}(\underline{\mathbf{x}};\underline{\mathbf{y}}), \tag{3.34}$$

where

$$\begin{aligned} M_2 &= -(\kappa+\tfrac{1}{2}D_0)(\boldsymbol{\nabla}^2_{\mathbf{x}_1}+\boldsymbol{\nabla}^2_{\mathbf{x}_2}) - \big(D_0\delta^{\alpha\beta} - d^{\alpha\beta}(\mathbf{x}_1-\mathbf{x}_2)\big)\nabla_{x_1^\alpha}\nabla_{x_2^\beta} \\ &= d^{\alpha\beta}(\mathbf{x}_1-\mathbf{x}_2)\nabla_{x_1^\alpha}\nabla_{x_2^\beta} - \kappa(\boldsymbol{\nabla}^2_{\mathbf{x}_1}+\boldsymbol{\nabla}^2_{\mathbf{x}_2}) + \tfrac{1}{2}D_0(\boldsymbol{\nabla}_{\mathbf{x}_1}+\boldsymbol{\nabla}_{\mathbf{x}_2})^2. \end{aligned} \tag{3.35}$$

The last term drops out in the translation-invariant sector. In other words, two solutions of Eq. (3.30) undergo, in their relative motion, an effective diffusion with the diffusion coefficient dependent on their relative position.

The PDF $P_2^{t,s}(r;\rho)$ of the distance ρ between the time s positions of two solutions, given their time t distance r, is expressed by the heat kernel of the operator M_2 restricted to the translation and rotation invariant sector. The latter becomes an explicit second order differential operator in the radial variable:

$$M_2^{\mathrm{inv}} = -C\,r^{\xi-a}\partial_r\,r^a\partial_r - 2\,\kappa\,r^{-d+1}\partial_r\,r^{d-1}\partial_r \tag{3.36}$$

with the exponent $a = a(\xi,\wp)$ (for simplicity, we give the formula after the removal of the infrared cutoff L in Eq. (3.29)). The operator M_2^{inv} may be

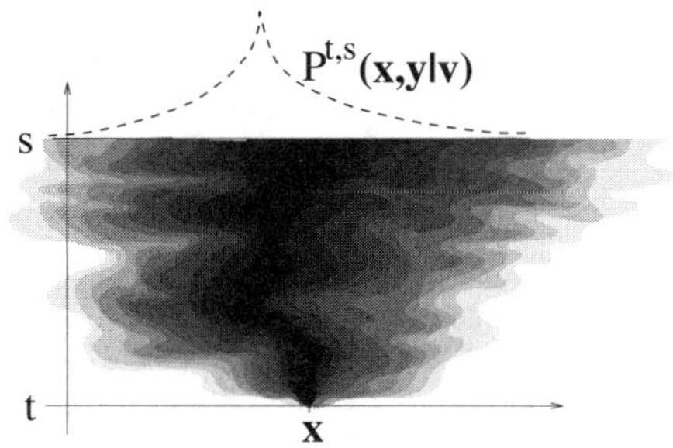

FIGURE 3.4.

transformed by a change of variables and a similarity transformation to a Schrödinger operator on the half-line. In particular,

$$\lim_{\kappa\to 0} M_2^{inv} = C'\, u^c \left[-\partial_u^2 + \frac{b^2 - 1/4}{u^2} \right] u^{-c}, \tag{3.37}$$

where $u = r^{2-\xi/2}$ and $b = b(\xi, \wp)$, $c = c(\xi, \wp)$. The limit $\kappa \to 0$ of the PDF $P_2^{t,s}(r, \rho)$ can be explicitly controlled. Two different regimes appear in this limit.

3.1 Weakly Compressible Regime

For weak compressibility $\wp < d/\xi^2$, which corresponds to $b < 1$, the distance PDF $P_2^{t,s}(r, \rho)$ is an integrable function of ρ and stays such in the limit $r \to 0$:

$$\lim_{r\to 0}\lim_{\kappa\to 0} P_2^{t,s}(r;\rho)d\rho \propto \left(\frac{\rho^{2-\xi}}{|t-s|}\right)^{1-b} \mathrm{e}^{-\rho^{2-\xi}/4C'|t-s|}\,\frac{d\rho}{\rho}. \tag{3.38}$$

This behavior excludes concentration of the transition probabilities $P^{t,s}(\mathbf{x}, \mathbf{y}|\mathbf{v})$ at single points $\mathbf{y}$. In particular, we obtain the Richardson dispersion law in the form

$$\lim_{r\to 0}\lim_{\kappa\to 0} \int \rho^2 P_2^{0,t}(r, \rho)\, d\rho \propto t^{\frac{2}{2-\xi}}$$

indicating an explosive separation of the Lagrangian trajectories and reproducing, for $\xi = \frac{4}{3}$, the mean growth (3.25). As we see, the trajectories in a fixed typical realization of the velocity fields are not determined by the initial position but rather form a Markov process with diffuse transition probabilities $P^{t,s}(\mathbf{x}, \mathbf{y}|\mathbf{v})$, as predicted above, see Fig. 3.4.

3.2 Strongly Compressible Regime

For strong compressibility $\wp \geq \frac{d}{\xi^2}$, i.e. for $b \geq 1$, one observes a different behavior due to the strong repulsive singularity at $u = 0$ in the operator

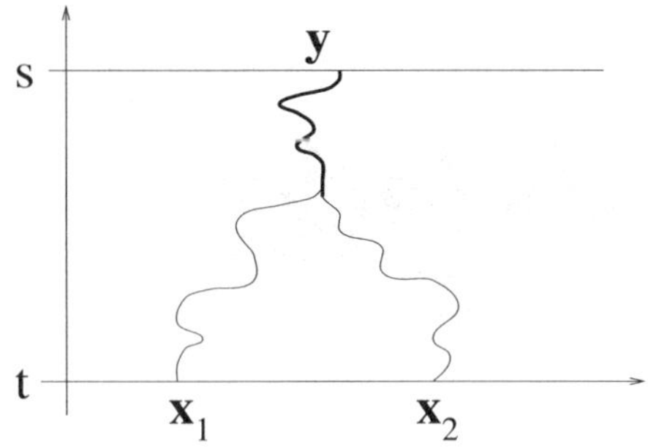

FIGURE 3.5.

(3.37):

$$\lim_{\kappa\to 0} P_2^{t,s}(r;\rho)d\rho = p^{t,s}(r)\delta(\rho)d\rho + \text{regular}$$

with the coefficient $p^{t,s}(r)$ of the delta-function converging to 1 when $r \to 0$ or $|t-s| \to \infty$. In particular,

$$\lim_{r\to 0}\lim_{\kappa\to 0} P_2^{t,s}(r;\rho)d\rho = \delta(\rho)d\rho$$

in this regime implying the concentration of the transition probabilities $P^{t,s}(\mathbf{x},\mathbf{y}|\mathbf{v})$ at single points $\mathbf{y}$ and the existence, in a fixed typical velocity realization, of Lagrangian trajectories determined by their initial positions. The presence of the singular term in $P_2^{t,s}(r;\rho)$ for $\kappa = 0$ indicates, however, an *implosive collapse* of distinct Lagrangian trajectories with a positive probability which grows in time, see Fig. 3.5.

We infer that in Hölder-continuous velocity fields, there is a competition of the tendency of two Lagrangian trajectories to separate or to collapse explosively. For weak compressibility, this is the first tendency that wins. The trapping effects increase, however, with the degree of compressibly and lead, in the Kraichnan ensemble, to a sharp transition in the behavior of the trajectories at $\wp = d/\xi^2$.

References

The Richardson dispersion law was formulated in:

1. L.F. Richardson, *Atmospheric diffusion shown on a distance-neighbour graph*, Proc. Roy. Soc. London **A110** (1926), 709–737.

Weak solutions of Euler equations with Lagrangian trajectories forming a stochastic process were discussed in:

2. Y. Brenier, *A homogenized model for vortex sheet*, Arch. Rational Mech. Anal. **138** (1997), 319–363.

3. A. Shnirelman, *Weak solutions with decreasing energy of incompressible Euler equations*, Commun. Math. Phys. **210** (2000), 541–603.

The introduction of velocities decorrelated in time in turbulence studies goes back to:

4. R.H. Kraichnan, *Small-scale structure of a scalar field convected by turbulence*, Phys. Fluids **11** (1968), 945–963.

The construction of the transition probabilities $P^{t,s}(\mathbf{x},\mathbf{y}|\mathbf{v})$ was described in:

5. Y. Le Jan and O. Raimond, *Solution statistiques fortes des équations différentielles stochastiques*, C.R. Acad. Sci. **327** (1998), 893–89.

The two regimes of behavior of the Lagrangian trajectories in the Kraichnan ensemble of velocities were analyzed in:

6. K. Gawędzki and M. Vergassola, *Phase transition in the passive scalar advection*, Physica **D138** (2000), 63–90, cond-mat/9811399.

Lecture 4

What are the consequences on the hydrodynamical properties of the flows of the observed dramatic behaviors of Lagrangian trajectories, violating the Newton–Leibniz paradigm about existence and uniqueness of solutions of the ODE's? The answer to this question may be the clue to a consistent theory of developed turbulence. Here, we shall content ourselves with discussing the transport properties in the flows induced by the velocities of the Kraichnan ensemble. We shall see that these properties differ drastically for weak and strong compressibility as a result of different behaviors of the Lagrangian trajectories. For concreteness, we shall look at the passive transport of the scalar quantity $\theta(t,\mathbf{x})$ (called tracer) whose evolution is described by the advection-diffusion equation

$$\partial_t\theta + (\mathbf{v}\cdot\boldsymbol{\nabla})\theta - \kappa\boldsymbol{\nabla}^2\theta = f, \tag{4.39}$$

where $f(t,\mathbf{x})$ denotes now a scalar source. In the incompressible case, θ may also describe the temperature field or the density of pollutant. The passivity of the advection means that the back reaction of θ on the velocity is ignored.

When the velocity field is sufficiently smooth, it is easy to solve the above linear equation for θ. For $\kappa = 0$ and $f = 0$, the scalar is simply carried by the Lagrangian flow

$$\theta(t,\mathbf{x}) = \theta\big(s,\mathbf{x}_{t,\mathbf{x}}(s)\big),$$

where $\mathbf{x}_{t,\mathbf{x}}(s)$ is the Lagrangian trajectory passing at time t by point $\mathbf{x}$. Note that the forward evolution of θ corresponds to the backward Lagrangian flow. In the presence of the source f, the scalar is also created or depleted along the trajectory:

$$\theta(t,\mathbf{x}) = \theta\big(s,\mathbf{x}_{t,\mathbf{x}}(s)\big) + \int_s^t f\big(\sigma,\mathbf{x}_{t,\mathbf{x}}(\sigma)\big)\,d\sigma.$$

Finally, when $\kappa \neq 0$, $\mathbf{x}_{t,\mathbf{x}}(s)$ should be taken as a solution of the equation (3.30) for the Lagrangian trajectories perturbed by the Brownian motion and the above formulae should be averaged over the latter. Thus

$$\begin{aligned}\theta(t,\mathbf{x}) &= \int \overline{\delta\big(\mathbf{y}-\mathbf{x}_{t,\mathbf{x}}(s)\big)}\theta(s,\mathbf{y})\,d\mathbf{y} \\ &\qquad + \int_s^t \left(\int \overline{\delta\big(\mathbf{y}-\mathbf{x}_{t,\mathbf{x}}(\sigma)\big)} f(\sigma,\mathbf{y})\,d\mathbf{y}\right) d\sigma \\ &= \int P^{t,s}(\mathbf{x},\mathbf{y}|\mathbf{v})\theta(s,\mathbf{y})\,d\mathbf{y} \\ &\qquad + \int_s^t \left(\int P^{t,\sigma}(\mathbf{x},\mathbf{y}|\mathbf{v}) f(\sigma,\mathbf{y})\,d\mathbf{y}\right) d\sigma, \end{aligned} \tag{4.40}$$

see Eq. (3.31). The right hand side still makes sense for a.a. velocities of the Kraichnan ensemble and defines a weak solution of the linear differential equation (4.39), i.e., the one satisfying the equation in the distributional sense.[3]

First, let us assume that we are given a random distribution of the scalar at time s, independent of the (later) velocities, and we wish to study its distribution at the later time t. In free decay, i.e. in the absence of the source f, by taking averages over the initial distribution and over the velocities, we obtain for the 1-point function of θ:

$$\begin{aligned}\langle\theta(t,\mathbf{x})\rangle &= \int P_1^{t,s}(\mathbf{x},\mathbf{y})\langle\theta(s,\mathbf{y})\rangle\,d\mathbf{y} \\ &= \int \mathrm{e}^{-(s-t)(\kappa+D_0/2)\boldsymbol{\nabla}^2}(\mathbf{x},\mathbf{y})\langle\theta(s,\mathbf{y})\rangle\,d\mathbf{y},\end{aligned}$$

see Eq. (3.33). In other words, the 1-point function decays diffusively. Note that the initial diffusivity κ is increased by the *eddy diffusivity* $\frac{1}{2}D_0$. Similarly, for the 2-point function,

$$\langle\theta(t,\mathbf{x}_1)\theta(t,\mathbf{x}_2)\rangle = \int P_2^{t,s}(\mathbf{x}_1,\mathbf{x}_2;\mathbf{y}_1,\mathbf{y}_2)\langle\theta(s,\mathbf{y}_1)\theta(s,\mathbf{y}_2)\rangle\,d\mathbf{y}_1\,d\mathbf{y}_2 \tag{4.41}$$

[3] Eq. (4.39) is an infinite-dimensional stochastic differential equation due to the white temporal dependence of $\mathbf{v}$ and it should be treated according to the Stratonovich prescription; the weakness of the solution is, however, due to its poor spatial regularity resulting from the nondifferentiability in space of typical velocities and the related stochastic character of the Lagrangian trajectories.

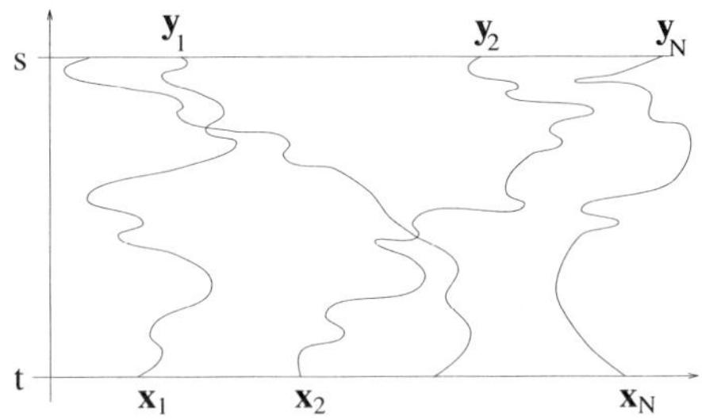

FIGURE 3.6.

and, for the N-point one,

$$\Big\langle \prod_{n=1}^{N} \theta(t,\mathbf{x}_n) \Big\rangle = \int P_N^{t,s}(\underline{\mathbf{x}};\underline{\mathbf{y}}) \Big\langle \prod_{n=1}^{N} \theta(s,\mathbf{y}_n) \Big\rangle \, d\underline{\mathbf{y}},$$

where

$$P_N^{t,s}(\underline{\mathbf{x}};\underline{\mathbf{y}}) = \Big\langle \prod_{n=1}^{N} P^{t,s}(\mathbf{x}_n,\mathbf{y}_n|\mathbf{v}) \Big\rangle = P_N^{s,t}(\underline{\mathbf{x}};\underline{\mathbf{y}})$$

is the joint PDF of the time s positions $\underline{\mathbf{y}}$ of the Lagrangian trajectories, given their time t positions $\underline{\mathbf{x}}$, see Fig. 3.6 (the last equality follows from the stationarity and time-reflection invariance of the Kraichnan ensemble). It is easy to see that the PDF's $P_N^{t,s}(\underline{\mathbf{x}};\underline{\mathbf{y}})$ are again given by the heat kernels of second order differential operators:

$$P_N^{t,s}(\underline{\mathbf{x}};\underline{\mathbf{y}}) = \mathrm{e}^{-|t-s|M_N}(\underline{\mathbf{x}};\underline{\mathbf{y}}),$$

where

$$M_N = \sum_{n<m} d^{\alpha\beta}(\mathbf{x}_n - \mathbf{x}_m)\nabla_{x_n^\alpha}\nabla_{x_m^\beta} - \kappa\sum_{n=1}^{N} \boldsymbol{\nabla}_{\mathbf{x}_n}^2 + \frac{1}{2}D_0\Big(\sum_n \boldsymbol{\nabla}_{\mathbf{x}_n}\Big)^2.$$

Assuming a homogeneous and isotropic initial distribution of θ, we may replace the PDF $P_2^{t,s}(\mathbf{x}_1,\mathbf{x}_2;\mathbf{y}_1,\mathbf{y}_2)$ in Eq. (4.41) by its translation- and rotation-invariant version $P_2^{t,s}(r,\rho)$. In particular, taking the limit $r \to 0$, we infer that, for $\kappa = 0$ and in the weakly compressible case $\wp < d/\xi^2$, the mean scalar energy density

$$\bar{e}_\theta(t) \equiv \Big\langle \frac{1}{2}\theta(t)^2 \Big\rangle = \frac{1}{2}\int P_2^{t,s}(0,\rho)\langle\theta(s,\mathbf{y})\theta(s,\mathbf{0})\rangle \, d\rho < \bar{e}_\theta(s) \qquad (4.42)$$

if the 2-point function $\langle\theta(s,\mathbf{y})\theta(s,\mathbf{0})\rangle$ decays for large $\rho = |\mathbf{y}|$. Indeed, the latter is bounded by its value at $\rho = 0$ so that the relation (4.42) follows since $P_2^{t,s}(0,\rho)$ is a strictly positive probability density, see Eq. (3.38). Hence the mean energy density of the scalar decreases with time (i.e., is dissipated) even for $\kappa = 0$. On the contrary, in the strongly compressible case $\wp \geq d/\xi^2$, the limit PDF $P_2^{t,s}(0, \rho) = \delta(\rho)$ and the mean energy density is conserved in the unforced evolution at $\kappa = 0$:

$$\bar{e}_\theta(t) = \bar{e}_\theta(s).$$

In the presence of the source f which, as before, we shall assume random Gaussian, independent of velocities and the initial distribution of θ, with mean zero and covariance

$$\langle f(t, \mathbf{x}) f(s, \mathbf{y}) \rangle = \delta(t - s) \chi\left(\frac{|\mathbf{x} - \mathbf{y}|}{L}\right), \tag{4.43}$$

the 1-point function of the scalar diffuses as before and for the 2-point function, we obtain from Eq. (4.40)

$$\begin{aligned} \langle \theta(t, \mathbf{x}_1) \theta(t, \mathbf{x}_2) \rangle = & \int P_2^{t,s}(\underline{\mathbf{x}}; \underline{\mathbf{y}}) \langle \theta(s, \mathbf{y}_1) \theta(s, \mathbf{y}_2) \rangle \, d\underline{\mathbf{y}} \\ & + \int_s^t \left(\int P_2^{t,\sigma}(\underline{\mathbf{x}}; \underline{\mathbf{y}}) \chi\left(\frac{|\mathbf{y}_1 - \mathbf{y}_2|}{L}\right) d\underline{\mathbf{y}} \right) d\sigma. \end{aligned} \tag{4.44}$$

The 2-point function solves now the equation

$$\begin{aligned} \partial_t \langle \theta(\mathbf{x}_1) \theta(\mathbf{x}_2) \rangle &= -M_2 \langle \theta(\mathbf{x}_1) \theta(\mathbf{x}_2) \rangle + \chi\left(\frac{|\mathbf{x}_1 - \mathbf{x}_2|}{L}\right) \\ &= -d^{\alpha\beta}(\mathbf{x}_1 - \mathbf{x}_2) \langle \nabla_\alpha \theta(\mathbf{x}_1) \nabla_\beta \theta(\mathbf{x}_2) \rangle \\ &\qquad - 2\kappa \langle \boldsymbol{\nabla} \theta(\mathbf{x}_1) \cdot \boldsymbol{\nabla} \theta(\mathbf{x}_2) \rangle + \chi\left(\frac{|\mathbf{x}_1 - \mathbf{x}_2|}{L}\right), \end{aligned} \tag{4.45}$$

see Eqs. (3.34) and (3.35). This is the passive scalar counterpart of the NS relation (2.8) that we discussed in Lecture 2. In the weakly compressible regime $\wp < d/\xi^2$, the scalar 2-point function reaches a stationary state. Arguing as before for the NS case, we infer in this state the energy balance

$$\bar{\epsilon}_\theta = \tfrac{1}{2} \chi(0)$$

for the scalar mean dissipation rate $\bar{\epsilon}_\theta = \langle \kappa (\boldsymbol{\nabla} \theta)^2 \rangle$ and the relation

$$\frac{1}{2} d^{\alpha\beta}(\mathbf{x}_1 - \mathbf{x}_2) \nabla_{x_1^\alpha} \nabla_{x_2^\beta} \langle \theta(\mathbf{x}_1) \theta(\mathbf{x}_2) \rangle = \frac{1}{2} \chi\left(\frac{|\mathbf{x}_1 - \mathbf{x}_2|}{L}\right).$$

The latter may be easily solved for the stationary 2-point function of the scalar giving

$$\langle \theta(\mathbf{x}) \theta(\mathbf{0}) \rangle = A_2(\chi) L^{2-\xi} - \text{const.}\, \bar{\epsilon}_\theta r^{2-\xi} + \mathcal{O}(L^{-2}), \tag{4.46}$$

where $r \equiv |\mathbf{x}|$ or, for the scalar 2-point structure function,

$$S_2(r) \equiv \left\langle \left(\theta(\mathbf{x}) - \theta(0) \right)^2 \right\rangle \propto \bar{\epsilon}_\theta r^{2-\xi} \tag{4.47}$$

for $\kappa = 0$ and $r \ll L$. This is an analogue of the Kolmogorov $\frac{4}{5}$ law. It may be viewed as resulting from the operator product expansion for the $\kappa \to 0$ limit of the dissipation operator $\epsilon_\theta = \kappa(\boldsymbol{\nabla}\theta)^2$

$$\epsilon_\theta(\mathbf{x}) = \lim_{\mathbf{y}\to\mathbf{x}} d^{\alpha\beta}(\mathbf{x}-\mathbf{y})\nabla_\alpha\theta(\mathbf{x})\nabla_\beta\theta(\mathbf{y})|_{\kappa=0} \tag{4.48}$$

valid inside expectations in the limit $\kappa \to 0$ and also in fixed realizations of θ. Eq. (4.48) expresses the dissipative anomaly in the weakly compressible regime of the Kraichnan model, analogous to the dissipative anomaly (2.13) for the Navier–Stokes case. One may also check directly the approximate constancy of the scalar energy flux towards large wavenumbers, establishing the existence of a direct scalar energy cascade.

In the strongly compressible regime $\wp > d/\xi^2$, the behavior of the scalar 2 point function (4.44) in the limit $\kappa \to 0$ is quite different. Now the 2-point function does not stabilize but has a constant contribution growing linearly in time. The dissipation rate vanishes and scalar energy is pumped into the constant mode at a constant rate equal to the injection rate $\frac{1}{2}\chi(0)$. This signals the presence of an inverse cascade of scalar energy towards small wavenumbers. The 2-point structure function of the scalar stabilizes, however, and its stationary limit satisfies the equation

$$M_2\langle(\theta(\mathbf{x}) - \theta(\mathbf{0})^2)\rangle = 2\Big(\chi(0) - \chi\Big(\frac{|\mathbf{x}|}{L}\Big)\Big)$$

from which one infers that

$$S_2(r) \propto r^{2-\xi}$$

for $r \gg L$.

As for intermittency of the scalar statistics, the two regimes also show very different behaviors. The question here is whether the higher structure functions of the scalar $S_N(r) \equiv \big\langle\big(\theta(\mathbf{x}) - \theta(0)\big)^N\big\rangle$ scale with powers $N(2-\xi)/2$ as the dimensional analysis would suggest, in analogy to the Kolmogorov theory. Although, by assumption, in the Kraichnan model there is no intermittency in the statistics of the velocity differences, the numerical studies of the incompressible model indicate strong intermittency of the scalar differences signaled by anomalous values of the scaling exponents. Unlike in the NS case, we have now some analytic understanding of this phenomenon.

First, it is not difficult to see that the higher equal-time correlators of the scalar satisfy the evolution equations generalizing Eq. (4.45):

$$\partial_t\Big\langle\prod_{n=1}^N \theta(\mathbf{x}_n)\Big\rangle = -M_N\Big\langle\prod_{n=1}^N \theta(\mathbf{x}_n)\Big\rangle + \sum_{p<q}\Big\langle\prod_{n\neq p,q}\theta(\mathbf{x}_n)\Big\rangle\chi\Big(\frac{|\mathbf{x}_p - \mathbf{x}_q|}{L}\Big). \tag{4.49}$$

In the weakly compressible regime, the correlation functions stabilize for long time and, besides, we expect that the limits $t \to \infty$ and $\kappa \to 0$ commute. The stationary equal time correlations satisfy a similar equation but with the vanishing left hand side. By inverting operators M_N, we may then compute the stationary N-point functions of the scalar recursively, a rare situation, indeed, since in most hydrodynamical problems the evolution equations for the correlation functions, called the Hopf equations, do not close. A semi-rigorous analysis shows that, for small ξ in the limit $\kappa \to 0$ and on short distances or for large injection scale L,

$$\left\langle \prod_{n=1}^{N} \theta(\mathbf{x}_n) \right\rangle = A_N(\chi) L^{N(2-\xi)/2-\zeta_N} \varphi_N(\underline{\mathbf{x}}) + \mathcal{O}(L^{-2+\mathcal{O}(\xi)}) + \cdots,$$

where $\varphi_N(\underline{\mathbf{x}})$ are scaling zero modes of the operators M_N, i.e.

$$M_N \varphi_N(\underline{\mathbf{x}}) = 0, \quad \varphi_N(\lambda \underline{\mathbf{x}}) = \lambda^{\zeta_N} \varphi_N(\underline{\mathbf{x}})$$

with

$$\zeta_N = \frac{N}{2}(2-\xi) - \frac{N(N-2)(1+2\wp)}{2(d+2)} \xi + \mathcal{O}(\xi^2).$$

The coefficients $A_N(\chi)$ are nonuniversal amplitudes and the dots denote terms that do not depend of, at least, one variable and, as such, do not contribute to the correlations of scalar differences. In particular, the structure functions

$$S_N(r) \propto L^{N(2-\xi)/2-\zeta_N} r^{\zeta_N}.$$

This *zero mode dominance* of the stationary higher-point functions of the scalar (note that such zero modes drop out in the stationary version of Eq. (4.49)) has been exhibited by the perturbative analysis of the Green functions of operators M_N around $\xi = 0$. It has been confirmed by perturbative analyses in powers of the inverse dimension and of $(2-\xi)$ and by numerical results, see Fig. 3.7 representing the values of the simulations by Frisch-Mazzino-Vergassola of the 4-point function anomalous exponent in the three-dimensional incompressible Kraichnan model.

What is the physical meaning of the zero modes of the operators M_N that dominate the short-distance asymptotics of the N-point functions of scalar differences? They are *slow modes* of the effective diffusion of Lagrangian trajectories with generators M_N. Indeed, for generic scaling function $\psi_N(\underline{\mathbf{x}})$ of scaling dimension σ_N, viewed as a function of time t positions of the Lagrangian trajectories, the effective time evolution is

$$\langle \psi_N \rangle_t \equiv \int P_N^{0,t}(\underline{\mathbf{x}}, \underline{\mathbf{y}}) \psi_N(\underline{\mathbf{y}}) \, d\mathbf{y} \sim t^{\sigma_N/(2-\xi)} \tag{4.50}$$

for large t, i.e., it exhibits a *super-diffusive* growth. But if $\psi_N = \phi_N$ is a zero mode of M_N then the above expectation is conserved in time (such statistically conserved modes are accompanied by descendent ones for which the time growth is slower than (4.50)).

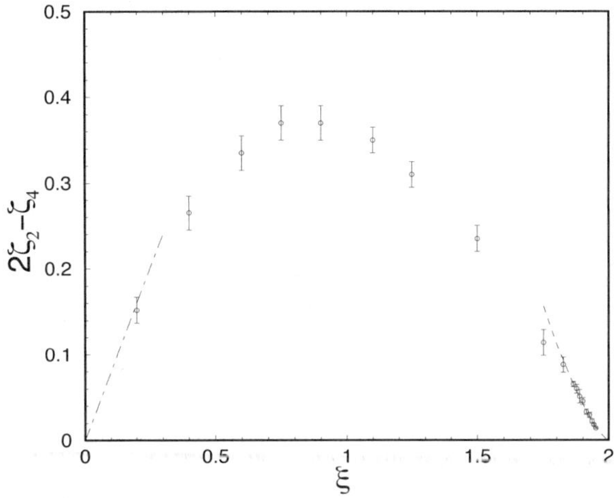

FIGURE 3.7.

In the strongly compressible phase with the inverse cascade of scalar energy, the behavior of the higher structure functions is different. In fact, only the lower ones stabilize, but the ones that do, scale normally on large distances. In this regime one can find exactly the stationary form of the PDF of the scalar difference:

$$\left\langle \delta\left(\vartheta - \frac{\theta(\mathbf{x})-\theta(\mathbf{0})}{r^{(2-\xi)/2}}\right)\right\rangle \propto [\chi(0) + C'\vartheta^2]^{-b-1/2}$$

for $r \gg L$. Its scaling form indicates that there is no intermittency in the inverse cascade of the scalar (the non-Gaussianity is scale-independent). Its poor decay at infinity corresponds to the fact that only lower structure functions reach a stationary regime.

As we see, the transport of a scalar tracer by velocities distributed according to the Kraichnan ensemble shows two different phases characterized by different direction of the scalar energy cascades and different degrees of intermittency. The phase transition occurs at the value $\wp = d/\xi^2$ of the compressibility degree, where the behavior of the Lagrangian trajectories changes drastically from the explosive separation to the implosive collapse. These two phases are quite reminiscent of the behavior of the three dimensional versus two-dimensional developed turbulence. That suggests that one should put more stress on the Lagrangian methods in studying the latter, especially on the properties of the Lagrangian flow in the weak solutions of the Euler equation. Of course the NS or the Euler equation, unlike the scalar advection one, are nonlinear. They describe velocity fields that are not only carried by their own Lagrangian trajectories but also stretched and there are nonlocal effects due to pressure. Some of these effects, however, may be studied already in various models of passive advections (passive vectors, linearized NS equation, etc.). It seems that the study of easy models of turbulence has a potential to teach us important lessons that we have to master to stand a chance of solving the fully-fledged problem of developed turbulence.

References

For the early works on passive scalar advection see:

1. A.M. Obukhov, *Structure of the temperature field in a turbulent flow*, Izv. Akad. Nauk SSSR, Geogr. Geofiz. **13** (1949), 58–69.

2. S. Corrsin, *On the spectrum of isotropic temperature fluctuations in an isotropic turbulence*, J. Appl. Phys. **22** (1951), 469–473.

That the scalar 2-point function may be found exactly if velocities are time decorrelated was realized in Kraichnan's 1968 paper cited at the end of the previous lecture.

The presence of scalar intermittency in the incompressible Kraichnan model model was first suggested in:

3. R.H. Kraichnan, *Anomalous scaling of a randomly advected passive scalar*, Phys. Rev. Lett. **72** (1994), 1016–1019.

The zero-mode dominance picture of the scalar intermittency has been proposed in

4. B. Shraiman and E. Siggia, *Anomalous scaling of a passive scalar in turbulent flow*, C.R. Acad. Sci. **321** (1995), 279–284.

5. K. Gawędzki and A. Kupiainen, *Anomalous scaling of the passive scalar*, Phys. Rev. Lett. **75** (1995), 3834–3837.

6. M. Chertkov, G. Falkovich, I. Kolokolov and V. Lebedev, *Normal and anomalous scaling of the fourth-order correlation function of a randomly advected scalar*, Phys. Rev. **E52** (1995), 4924–4941.

The numerical data of Fig. 3.7 come from:

7. U. Frisch, A. Mazzino, and M. Vergassola, *Intermittency in passive scalar advection.*, Phys. Rev. Lett. **80** (1998), 5532–5535.

The phase transition with increase of compressibility in the $\xi = 2$ limiting case of the Kraichnan model was observed in:

8. M. Chertkov, I. Kolokolov and M. Vergassola, *Inverse versus direct cascades in turbulent advection*, Phys. Rev. Lett. **80** (1998), 512–515.

The analysis of the phase transition for $\xi < 2$ was done in the 1988 preprint of M. Vergassola and the present author cited at the end of previous lecture.

For some more details on the Kraichnan model discussed in an informal way, see:

9. K. Gawędzki, *Intermittency of passive advection*, in "Advances in Turbulence VII," ed. U. Frisch, Kluwer Acad. Publ. 1998, 493–502.

4

BEC and the New World of Coherent Matter Waves

Allan Griffin

After decades of effort to produce an atomic Bose condensate, this was finally achieved in June, 1995 in a laser-cooled magnetically trapped gas of ^{87}Rb atoms by Eric Cornell and Carl Wieman at JILA (University of Colorado and N.I.S.T.). As of July, 1999 (4 years later), there are now over 20 experimental groups around the world who can routinely produce and study such atomic Bose condensates. In addition, over 1000 theoretical papers have been published!

I will give three *introductory* lectures on this new Quantum Phase of Matter, which I think will continue to be a growth point of fundamental physics research for the next decade and will also be the source of new technologies based on using this source of *coherent matter waves.* A brief sketch of my three lectures is as follows:

1. A brief history of BEC studies:

- Before 1995, going back to pioneering work of Einstein (1925) and Fritz London (1938).
- Introduce the concept of a *Bose macroscopic wavefunction* $\Phi(\mathbf{r}, t)$ describing a Bose condensate and the key Gross–Pitaevskii equation of motion for $\Phi(\mathbf{r}, t)$.
- Review some of the results obtained over the last 4 years—and indicate why trapped Bose gases are so interesting, possibly even more than superfluid ^{4}He and BCS superconductors, which also involve Bose condensation.

2. Dynamics of a pure condensate ($T \ll T_{\mathrm{BEC}}$):

- Time-dependent Gross–Pitaevskii (1961) equation of motion.
- Crucial effect of weak interatomic interactions.
- Collective oscillations of the condensate.
- The Bose condensate as a *classical* quantum object!

3. Dynamics of coupled condensate and noncondensate components (two-fluid hydrodynamics):

- Derivation of a *quantum* Boltzmann equation for the noncondensate excited atoms and a generalized GP equation for the condensate.
- More complex behaviour than in superfluid ^{4}He.
- Comparison with the well-known Landau two-fluid theory (1941).

Further references which are relevant for these lectures are:

1. BEC Homepage (maintained by a BEC theorist, Mark Edwards): `http://amo.phy.gasou.edu/bec.html/`. This widely-used homepage has played a crucial role in BEC research.

2. A recent article [1] by Dalfovo, Giorgini, Pitaevskii and Stringari is an authoritative review of recent theory on atomic Bose condensates and expands on the material I cover in Sections 1 and 2.

3. Articles in a book [2], *Bose-Einstein condensation in atomic gases*, in Proceedings of the International School of Physics "Enrico Fermi", eds. M. Inguscio, S. Stringari and C. Wieman. This contains many review articles on current BEC research. In particular, I call attention to (all these can be downloaded from the LANL website under cond-mat):

- W. Ketterle, D. S. Durfee and D. M. Stamper-Kurn, "Making, probing and understanding Bose-Einstein condensates"—a 100 page review of recent experiments [3].
- A. Griffin "A brief history of our understanding of BEC: From Bose to Beliaev" [4].
- A. Griffin, "Theory of excitations of the condensate and noncondensate at finite temperature" [4].
- A. L. Fetter, "Theory of a dilute low-temperature trapped Bose condensate." A very detailed analysis at $T = 0$.

4. A long article by Zaremba, Nikuni and Griffin [5] on the nonequilibrium dynamics of trapped Bose gases at finite temperatures. This expands on the material covered in Section 3 of these lectures.

1 An Overview of Past and Recent Work

1.1 Some History before 1980

Einstein predicted that a noninteracting gas of atoms (Bosons) would undergo a *phase transition* at low temperatures, when a *macroscopic* $(0(N))$ number of atoms occupy the lowest energy level (in a uniform ideal Bose gas, this is the the zero momentum single-particle state). His work was inspired by a novel derivation of the Planck distribution for photons by Bose in 1924. The basic physics of this phase transition is worked out in every text in statistical mechanics [6]. The simplest way of estimating $T_{\rm BEC}$ is to note that the transition occurs when

$$\lambda_T \gtrsim d = \text{average distance between atoms} \sim \frac{1}{n^{1/3}}, \tag{1.1}$$

where λ_T is the thermal de Broglie wavelength of a gas of atoms at temperature T,

$$\lambda_T \equiv \left(\frac{2\pi\hbar^2}{mk_BT}\right)^{1/2}. \tag{1.2}$$

The criterion in (1.1) is equivalent to $n\lambda_T^3 \gtrsim 1$, while a more careful analysis [6] gives $n\lambda_T^3 = \zeta(\frac{3}{2}) = 2.612$. One sees that $\lambda_T \to$ large as $T \to 0$. When $\lambda_T \gtrsim d$, all the atoms become *correlated* and the gas exhibits new collective behaviour (even in *absence* of interactions). Using (1.1), one finds that

$$k_BT_{\text{BEC}} \sim \frac{2\pi\hbar^2}{m} n^{2/3}. \tag{1.3}$$

Below T_{BEC}, the number of atoms in the $\mathbf{p} = 0$ single-particle state increases and is given by the well-known formula

$$\frac{N_c(T)}{N} = \left[1 - \left(\frac{T}{T_{\text{BEC}}}\right)^{3/2}\right]. \tag{1.4}$$

At $T = 0$, all the atoms in an ideal gas are in this $\mathbf{p} = 0$ state (this state is the Bose condensate in noninteracting 3D gas).

Nothing much happened until 1938. Then the neglected work of Einstein was re-discovered and developed by Fritz London [7], who suggested that it might be the basis of an explanation for the strange effects noticed in liquid ^{4}He at $T_c \sim 2.17K$. London's suggestion was based on the fact that the ^{4}He atom was a "composite" Boson ($S = 0$) and the formula in (1.3)) gives $T_{\text{BEC}} \sim 3K$ if we used the density for liquid ^{4}He. L. Tisza used London's idea and suggested (somehow!) that the condensate atoms act in a coherent way—a *new collective degree of freedom moving without friction.* This "picture" led to a rudimentary two-fluid model that could explain experiments showing superfluidity (especially by Kapitza as well as by Allen and Meisner) as a counterflow of the superfluid and the normal fluid. In a dilute, weakly interacting Bose gas, these two components are the condensate and noncondensate, respectively.

Until the 1960's, the theory of BEC in interacting systems was dominated by efforts to use it to understand superfluid ^{4}He. The London–Tisza scenario was essentially correct, but to formulate it properly needed field-theoretic many body techniques and the concept of broken-symmetry, which were only developed in the period 1957–1965. In this period, a large amount of work was done on a *toy* problem, a dilute weakly interacting Bose gas, since a Bose liquid like superfluid ^{4}He was too difficult to deal with theoretically. These early studies [8] are the foundation of our current understanding of trapped atomic gases.

A phenomenological theory of superfluid ^{4}He was introduced by Landau in 1941, based on the idea of quasiparticles (phonon-roton spectrum) and a two-fluid superfluid hydrodynamics [9]. This brilliant theory has been very successful and is the basis of modern descriptions of superfluid ^{4}He [10]. However, it made no explicit mention of BEC or even the fact that ^{4}He atoms obeyed Bose statistics. Only in the 1960's did it become clear that Landau formulation had its *microscopic* basis in the existence of a condensate macroscopic wavefunction,

$$\Phi(\mathbf{r}, t) = \langle \hat{\psi}(\mathbf{r}) \rangle = \sqrt{N_c(\mathbf{r}, t)} e^{i\theta(\mathbf{r},t)}, \tag{1.5}$$

where $\hat{\psi}(\mathbf{r})$ is the quantum field operator (see Section 2). This concept was first formally introduced by Beliaev in 1957 [11], extending the pioneering work of Bogoliubov in 1947 [12]. The superfluid motion is associated with the gradient of the phase of this two-component order parameter,

$$e^{i\theta} = e^{i(\theta_0 + \mathbf{r} \cdot \nabla\theta)}; \mathbf{k}_s \equiv \nabla\theta \equiv \frac{m\mathbf{v}_s}{\hbar}. \tag{1.6}$$

The essential relation between superfluidity and $\Phi(\mathbf{r}, t)$ is simply and elegantly described in the classic monograph by Nozières and Pines [13]. For further discussion of the development of our current understanding of Bose condensates, see the review article by Griffin in [2].

It might be useful to make a brief digression here on the BCS theory of superconductors based on formation of Cooper pairs (total spin $S = 0$), which was developed in 1957. In the BCS theory, superconductors exhibit the same kind of *macroscopic* quantum behaviour as superfluid ^{4}He, as first argued in the late 1930's by F. London. In the Gor'kov version of the BCS theory, the spin-singlet Cooper pair order parameter

$$\Phi(\mathbf{r}, t) = \langle \hat{\psi}_\uparrow(\mathbf{r}) \hat{\psi}_\downarrow(\mathbf{r}) \rangle \tag{1.7}$$

is the equivalent of

$$\Phi(\mathbf{r}, t) = \langle \hat{\psi}(\mathbf{r}) \rangle \tag{1.8}$$

in Bose superfluids. The essential equivalence of these two systems was obscured by the complexity of the original many-particle BCS-wavefunction, due to the large spatial size of overlapping Cooper pairs. As a result, Cooper pairs *form* and *condense* into a coherent state at the *same* temperature. It was only in the 1980's that theorists (Leggett, Nozières and others [14]) realized that for small Cooper pairs (tightly bound and hence high T_c), the BCS theory smoothly goes over to theory of a weakly interacting Bose gas of nonoverlapping Cooper pairs. Recently a gas of ^{40}K atoms (a composite fermion) have been laser-cooled at JILA to slightly below the Fermi temperature [15]. There are ways to make the interaction attractive (see later) and then one might look for the formation of Cooper pairs [16] of ultra-cold Fermi atoms!

1.2 More Recent Developments (1980–1995)

Since the 1970's, there has been increasing interest by experimentalists to find a "pure" form of BEC, namely, in a low temperature gas. The two early candidates for Bosons were excitons (electron-hole pairs) in semiconductors and spin-polarized hydrogen atoms (see articles by Greytak and by Wolfe et al. in the book mentioned in [14]). A gas of $H_\uparrow$ atoms was predicted to be stable as a gas even at $T = 0$. This is because the atoms cannot combine since there is no bound state of the interatomic potential between two spin-polarized H atoms. Thus, one cannot form liquid or solid phase. Many people got interested in BEC in $H_\uparrow$ gas, including theorists [17].

Several of the key ideas that led to success in alkali atoms in 1995 grew out of the pioneering work on $H_\uparrow$ gas in the 1980's. However 3-body interactions become increasingly important at higher densities and these cause spin flips, allowing formation of H-molecules. High densities were needed since cooling was by cryogenic methods, which could reach $\sim 10^{-4}$K but no lower. BEC in $H_\uparrow$ gas was finally produced at MIT in June, 1998, after almost 20 years of work [18]. Unfortunately, the alkali atom condensates are much easier to create and study, and appear to be more interesting gases.

Since the early 1990's, attention has focussed on the alkali atoms: Li, Na, K, Rb, Cs. The atoms are Bosons, with an even number of neutrons. The strategy was to use laser-cooling to get to very low temperatures, where the low density gas would Bose-condense. The essential idea behind laser cooling is that when an atom absorbs a photon, it slows down. In the summer of 1995, BEC was announced by three groups led by:

- C. Wieman and E. Cornell (JILA), using ^{87}Rb atoms [19].
- W. Ketterle (MIT), using ^{23}Na atoms [20].
- R. Hulet (Rice University), using ^{7}Li atoms [21].

Parenthetically, it is now the general feeling that the original Rice data, interesting as it was, did not give an unambiguous signature of a (very small, since the interaction is attractive) condensate in ^{7}Li gas [22].

Alkali atoms are perfect for BEC studies. They have a magnetic moment, and hence can be trapped by magnetic fields. They essentially have a "one-electron" structure. They are thus simple atoms, and have been well studied by atomic physicists. One can easily selectively flip the "spin" of higher energy trapped atoms. These "hot" atoms are then quickly ejected from the magnetic trap and the remaining atoms quickly thermalize to a lower temperature. This "evaporative cooling" is very efficient and quickly brings one into the temperature region required for BEC.

It is useful to mention a few experimental facts about the magnetic traps currently in use. As it turns out, these traps are well described as a harmonic potential

$$V_{\rm ex}(\mathbf{r}) = \frac{1}{2}m\omega_0^2 r^2 \qquad \text{(isotropic)} \tag{1.9}$$

$$= \frac{1}{2}m(\omega_{0x}^2 x^2 + \omega_{0y}^2 y^2 + \omega_{0z}^2 z^2) \qquad \text{(anisotropic)}. \tag{1.10}$$

Most current traps are either:

$$\begin{array}{ll} \text{pancakes,} & \omega_{0z} \gg \omega_{0x} (= \omega_{0y}) \\ \text{cigars,} & \omega_{0z} \ll \omega_{0x} (= \omega_{0y}) \end{array} \tag{1.11}$$

and the trap frequencies are of the order $\omega_0 \sim 2\pi \times 100$Hz. In 1995, the first condensates were small $\sim 10^3$ atoms and $T_{\rm BEC} \sim 100$nK. However in 1999, the condensates can be quite large $\sim 10^8$ atoms at $T_{\rm BEC} \sim \mu$K. These have a size $\sim$ many microns, which can be easily seen optically. When the condensates are small, the trap is turned off and cloud allowed to expand, and then measured by optical methods. The results are simple to analyze if gas is noninteracting. However, more analysis is needed to include the effects of interactions during expansion.

Early reports on atomic condensates discussed the system as an *ideal* Bose gas. It was soon realized that even in these very dilute gases, the interactions played a crucial role. Indeed, since the 1960's, it has been understood that an interacting Bose gas is quite different from an ideal Bose gas. In particular, interactions stabilize (or "lock") the phase of the condensate and allow *coherent* properties to emerge. (We recall that a free Bose gas has a condensate but is not a superfluid). Of course, interactions are also crucial for cooling. After hot atoms are removed by rf-induced spin-flips, it is important that remaining atoms can quickly re-thermalize through collisions.

However, it is useful to first consider an *ideal* Bose gas in a trap, to illustrate some characteristic features. For atoms in an external potential, we have

$$N = \sum_i f_0(\epsilon_i), \tag{1.12}$$

where the Bose distribution is

$$f_0(\epsilon_i) = \frac{1}{e^{\beta(\epsilon_i - \mu)} - 1}. \tag{1.13}$$

In a harmonic trap, the energy levels are:

$$\epsilon_i = \epsilon_{n_x n_y n_z} = \left(n_x + n_y + n_z + \tfrac{3}{2}\right)\hbar\omega_0, \tag{1.14}$$

with $n_x, n_y, n_z = 0, 1, 2, 3 \ldots$. The condensate is described by the ground state single-particle wavefunction

$$\phi_0(\mathbf{r}) \sim e^{-r^2/2a_{\rm HO}^2}, \tag{1.15}$$

where the SH oscillator length is $a_{\rm HO} \equiv (\hbar/[m\omega_0])^{1/2} \sim 1\mu m$ in current traps. Clearly $a_{\rm HO}$ gives the "size" of condensate $n_{c0}(\mathbf{r}) = |\Phi_0(\mathbf{r})|^2$ in a trapped gas. For the noncondensate density $\tilde{n}_0(\mathbf{r})$ (often called the "thermal cloud"), we can use the *semiclassical limit*, since the thermal energy ($k_B T$) is much larger than the spacing between the SH energy levels ($\hbar\omega_0$). Then we have

$$\begin{aligned} \tilde{n}_0(\mathbf{r}) &\sim e^{-V_{\rm ex}(\mathbf{r})/k_B T} \\ &= e^{-r^2/2R_T^2}, \end{aligned} \tag{1.16}$$

where

$$R_T = \sqrt{\frac{k_B T}{m\omega_0^2}} = a_{\rm HO}\left(\frac{kT}{\hbar\omega_0}\right)^{1/2} \gg a_{\rm HO}. \tag{1.17}$$

Thus we see that the size of the thermal cloud R_T is much larger than the condensate. The signature for condensate is this *sharp* high density peak at the centre of the trap, which suddenly starts to grow out of the broad thermal distribution at the predicted transition temperature $T_{\rm BEC}$. As $T \to 0$ (effectively $T \lesssim 0.4 T_{\rm BEC}$), the thermal cloud steadily disappears as all atoms go into the ground state $\phi_0(\mathbf{r})$ given by (1.15), which is the macroscopic wavefunction for a noninteracting trapped Bose gas. The temperature of the gas is measured from the temperature dependence of the tail of the thermal distribution given by (1.16).

It is easy to calculate the transition temperature for atoms in an harmonic trap. Separating out the condensate contribution in (1.15), we have

$$N = N_c + \sum_{i \neq 0} f^0(\epsilon_i); N_c = \int d\mathbf{r} |\phi_0(\mathbf{r})|^2. \tag{1.18}$$

Making a change of variable $\beta\hbar\omega_0 n_x \equiv \bar{n}_x$, and using the continuum approximation, we have

$$\begin{aligned} N - N_c &\simeq \left(\frac{k_B T}{\hbar\omega_0}\right)^3 \int_0^\infty d\bar{n}_x \int_0^\infty d\bar{n}_y \int_0^\infty d\bar{n}_z \frac{1}{e^{(\bar{n}_x+\bar{n}_y+\bar{n}_z)} - 1} \\ &= \zeta(3)\left(\frac{k_B T}{\hbar\omega_0}\right)^3. \end{aligned} \tag{1.19}$$

We note that the chemical potential $\mu_0 = \frac{3}{2}\hbar\omega_0$, but $\frac{\hbar\omega_0}{k_B T} \ll 1$ and hence the zero point energy has been neglected in (1.19)). Since $N_c = 0$ at $T_{\rm BEC}$, we have

$$k_B T_{\rm BEC} \simeq 0.94 (N^{1/3}) \hbar\omega_0, \quad \text{with } \frac{N_c(T)}{N} = \left[1 - \left(\frac{T}{T_{\rm BEC}}\right)^3\right]. \tag{1.20}$$

One can improve on these simple estimates for $T_{\rm BEC}$ and $N_c(T)$, but the corrections are only a few percent at best [1].

Interactions make a dilute, weakly-interacting Bose condensed gas into a full nontrivial many body problem, even though the system is very dilute and the interactions are weak. In a dilute gas, we need only consider binary collisions. The real interatomic potential $v(\mathbf{r})$ has a hard core with a radius of a few Angstroms and a weak, long-range attractive tail. In a dilute, very cold gas, we can approximate $v(\mathbf{r})$ using the s-wave scattering length approximation effectively replacing $v(\mathbf{r})$ by a pseudopotential [23]

$$v(\mathbf{r}) \Rightarrow \frac{4\pi h^2}{m} a\delta(\mathbf{r}) \equiv g\delta(\mathbf{r}). \tag{1.21}$$

We require $a \ll$ average distance between atoms, or $na^3 \ll 1$, which is very well satisfied in these gases. For alkali atoms, $v(\mathbf{r})$ almost has a bound state of two atoms. This quasi-bound state is very sensitive to the long range part of the potential, and thus the value of the $s-$wave scattering length a can be very large. Current values for atoms used in BEC experiments are:

$$\begin{aligned} {}^{87}\mathrm{Rb}: &\quad a = 58\,\text{Å} \\ {}^{23}\mathrm{Na}: &\quad a = 28\,\text{Å} \\ {}^{7}\mathrm{Li}: &\quad a = -14\,\text{Å}. \end{aligned} \tag{1.22}$$

One can adjust the energy of the quasi-bound state and, as a result, change the value of s-wave scattering length a with a small magnetic field. Near a so-called *Feshbach resonance*, one can even change the interaction sign, going from repulsive ($a > 0$) to attractive ($a < 0$). There is a lot of current work [24] trying to exploit this ability to change the interaction strength and sign by "turning a knob", perhaps even making $a \to 0$!

The key equation for the macroscopic wavefunction for a $T = 0$ condensate was written down and discussed by Pitaevskii [25] and Gross [26] in 1961:

$$i\hbar\frac{\partial\Phi(\mathbf{r},t)}{\partial t} = \left[-\frac{\hbar^2\nabla^2}{2m} + V_{\mathrm{ex}}(\mathbf{r}) + gn_c(\mathbf{r},t)\right]\Phi(\mathbf{r},t), \tag{1.23}$$

where $n_c = |\Phi|^2$. This equation describes the condensate atoms moving in dynamic self-consistent Hartree field produced by the condensate,

$$V_H(\mathbf{r},t) = \int d\mathbf{r}' v(\mathbf{r}-\mathbf{r}')n_c(\mathbf{r}',t) = gn_c(\mathbf{r},t). \tag{1.24}$$

The nonlinear GP equation (1.23) will be the subject of Section 2. Hundreds of papers have been written on it in the last four years. For $T \lesssim 0.4T_{\mathrm{BEC}}$, it describes both the static properties and the dynamic fluctuations (linear and nonlinear) very well, usually within a few percent [1, 2].

To complete this brief introduction, I mention several other research topics that make trapped Bose-condensed gases so exciting:

1. Alkali atoms have several different atomic hyperfine states. Apart from Cs, the alkali atoms have a nuclear spin $I = \frac{3}{2}$ and an electron spin $S = 1/2$.

Thus the total spin operator $\mathbf{F} = \mathbf{I} + \mathbf{S}$ has values of 1 and 2, leading to 8 possible atomic states. One usually works with one of these atomic hyperfine states—trapped in a magnetic well. However, purely optical traps (MIT) using the dipole force of a laser beam can be used to trap low energy atoms in different hyperfine states. Thus one can now also deal with gases with several different states, i.e., a multicomponent Bose gas. Moreover, one can induce transitions between different hyperfine states. One sees that spin is a new degree of freedom in such multicomponent Bose-condensed gases [1], [2].

2. Independent of the special "coherent" features of a Bose condensate, these trapped gases give us a source of high density, very cold atoms. Lene Hau [27] has used this high density to slow down the speed of light to that of slow car ($\sim$ 40 km/hour) using self-induced transparency. One can also switch on an optical lattice (produced by intersecting laser beams) on a trapped Bose gas [28]. Turning off the magnetic trap, the low energy atoms will occupy the potential minima of this periodic lattice. This could not be done with high energy atoms since the dipole-induced potentials of the optical lattice are very weak. With ultra-cold trapped fermions, one may also be able to produce a Hubbard model, of the kind extensively studied in connection with the cuprate-oxide high temperature superconductors.

We agree with the opinion of Pitaevskii [29] that the discovery of BEC in alkali gases "can be considered as one of the most beautiful results of experimental physics in our century". The next two sections will flesh out this qualitative overview with some theoretical calculations on the collective oscillations at $T \ll T_{\mathrm{BEC}}$ (pure condensate) and at finite temperatures $T \sim T_{\mathrm{BEC}}$ (mixture of condensate and noncondensate) of these strange quantum "wisps of matter".

2 Dynamics of the Pure Condensate

The theory of interacting Bose-condensed fluids is most usefully discussed using quantum field operators. This procedure was formalized by Beliaev (1957) and developed by Bogoliubov [30], Gavoret and Nozières [31], Martin and Hohenberg [32], and others in the 1960's [33]. We recall:

$$\begin{aligned} \hat{\psi}^{+}(\mathbf{r}) &= \text{creates atom at } \mathbf{r} \\ \hat{\psi}(\mathbf{r}) &= \text{destroys atom at } \mathbf{r}. \end{aligned} \tag{2.1}$$

These fields satisfy the usual Bose commutation relations, such as

$$[\hat{\psi}(\mathbf{r}), \hat{\psi}^{+}(\mathbf{r}')] = \delta(\mathbf{r} - \mathbf{r}'). \tag{2.2}$$

All observables can be written in terms of these quantum field operators, such as the interaction energy

$$\widehat{V}_{\rm ext} = \frac{1}{2}\int d\mathbf{r}\int d\mathbf{r}'\,\hat{\psi}^+(\mathbf{r}')\hat{\psi}^+(\mathbf{r})v(\mathbf{r}-\mathbf{r}')\hat{\psi}(\mathbf{r}')\hat{\psi}(\mathbf{r}) = \frac{1}{2}g\int d\mathbf{r}\,\hat{\psi}^+(\mathbf{r})\hat{\psi}^+(\mathbf{r})\hat{\psi}(\mathbf{r})\hat{\psi}(\mathbf{r}). \tag{2.3}$$

The crucial idea due to Bogoliubov (1947) and later generalized by Beliaev is to separate out the condensate part

$$\hat{\psi}(\mathbf{r}) = \langle\hat{\psi}(\mathbf{r})\rangle + \tilde{\psi}(\mathbf{r}), \tag{2.4}$$

where

$$\langle\hat{\psi}(\mathbf{r})\rangle \equiv \Phi(\mathbf{r}) = \text{Bose macroscopic wavefunction.} \tag{2.5}$$

This quantity plays the role of the order parameter for the superfluid phase transition:

$$\begin{aligned}\Phi(\mathbf{r}) &= 0, \qquad T > T_c,\\ &\neq 0, \qquad T < T_c.\end{aligned} \tag{2.6}$$

We note that $\Phi(\mathbf{r}) \equiv \sqrt{N_c}e^{i\theta}$ is a 2-component order parameter. Clearly, $\Phi(\mathbf{r})$ is not simply related to the many-particle wavefunction $\Psi(\mathbf{r}_1, \mathbf{r}_2, \ldots, \mathbf{r}_N)$. The thermal average in $\langle\hat{\psi}(\mathbf{r})\rangle$ involves a small symmetry-breaking perturbation to allow Φ to be finite,

$$\widehat{H}_{\rm SB} = \int d\mathbf{r}\,[\eta(\mathbf{r})\hat{\psi}^+(\mathbf{r}) + \eta^*(\mathbf{r})\hat{\psi}(\mathbf{r})]. \tag{2.7}$$

It is useful to make a few comments on the physics behind $\Phi(\mathbf{r}, t)$. $\Phi(\mathbf{r}, t)$ is a *coherent* state, with a "clamped" value of phase—rather than a Fock-state of fixed N, with no well-defined phase. $\Phi(\mathbf{r}, t)$ acts like a *classical* field, since quantum fluctuations are negligible when N_c is large. Probably P.W. Anderson deserves the greatest credit for understanding (in the period 1958–1963) the new physics behind working with a broken-symmetry state $\Phi(\mathbf{r}, t)$, both in BCS superconductors and in superfluid ^{4}He [34]. It captures the physics of the new phase of matter (such as the occurence of the Josephson effect) and the associated superfluidity. The symmetry-breaking perturbation allows $\langle\hat{\psi}\rangle$ to be finite. More precisely, it allows the system to internally set up off-diagonal symmetry-breaking fields, which persist even when the external symmetry-breaking perturbation in (2.7) is set to zero at the end ($\eta \to 0$). The same sort of physics is behind the BCS theory of superconductors.

The exact Heisenberg equation of motion for the field operator is

$$i\hbar\frac{\partial\hat{\psi}(\mathbf{r},t)}{\partial t} = \left[-\frac{\hbar^2\nabla^2}{2m} + V_{\rm ex}(\mathbf{r}) + \delta V(\mathbf{r},t)\right]\hat{\psi}(\mathbf{r},t) + \eta(\mathbf{r}) + g\hat{\psi}^+(\mathbf{r},t)\hat{\psi}(\mathbf{r},t)\hat{\psi}(\mathbf{r},t), \tag{2.8}$$

where $\delta V(\mathbf{r},t)$ is a small time-dependent driving potential. This gives an exact equation of motion for $\Phi(\mathbf{r},t) \equiv \langle\hat{\psi}(\mathbf{r},t)\rangle$,

$$i\hbar\frac{\partial\Phi(\mathbf{r},t)}{\partial t} = \left[-\frac{\hbar^2\nabla^2}{2m} + V_{\text{ex}}(\mathbf{r}) + \delta V(\mathbf{r},t)\right]\Phi(\mathbf{r},t) \\ + \eta(\mathbf{r}) + g\langle\hat{\psi}^+(\mathbf{r},t)\hat{\psi}(\mathbf{r},t)\hat{\psi}(\mathbf{r},t)\rangle, \quad (2.9)$$

with

$$\hat{\psi}^+\hat{\psi}\hat{\psi} = |\Phi|^2\Phi + 2|\Phi|^2\tilde{\psi} + \Phi^2\tilde{\psi}^+ + \Phi^*\tilde{\psi}\tilde{\psi} + 2\Phi\tilde{\psi}^+\tilde{\psi} + \tilde{\psi}^+\tilde{\psi}\tilde{\psi}. \quad (2.10)$$

Taking the symmetry-breaking average, one finds

$$\langle\hat{\psi}^+\hat{\psi}\hat{\psi}\rangle = n_c\Phi + \tilde{m}\Phi^* + 2\tilde{n}\Phi + \langle\tilde{\psi}^+\tilde{\psi}\tilde{\psi}\rangle, \quad (2.11)$$

where

$$n_c(\mathbf{r},t) \equiv |\Phi(\mathbf{r},t)|^2 = \text{condensate density}$$
$$\tilde{n}(\mathbf{r},t) \equiv \langle\tilde{\psi}^+(\mathbf{r},t)\tilde{\psi}(\mathbf{r},t)\rangle = \text{noncondensate density}$$
$$\tilde{m}(\mathbf{r},t) \equiv \langle\tilde{\psi}(\mathbf{r},t)\tilde{\psi}(\mathbf{r},t)\rangle = \text{off-diagonal (anomalous) density.}$$

Here we have separated out the *condensate* and *noncondensate* parts

$$\hat{\psi} \Longrightarrow \langle\hat{\psi}\rangle + \tilde{\psi} = \Phi + \tilde{\psi}. \quad (2.12)$$

In general, the equation (2.9) for $\Phi(\mathbf{r},t)$ is not closed—it is coupled to the dynamics of the noncondensate. However, in this section we limit ourselves to $T \ll T_{\text{BEC}}$, where we can assume the noncondensate fraction is negligible, leaving

$$i\hbar\frac{\partial\Phi(\mathbf{r},t)}{\partial t} = \left[-\frac{\hbar^2\nabla_r^2}{2m} + V_{\text{ex}}(\mathbf{r}) + g|\Phi(\mathbf{r},t)|^2\right]\Phi(\mathbf{r},t). \quad (2.13)$$

This is the famous time-dependent Gross-Pitaevskii equation for the condensate macroscopic wavefunction. It gives a complete description of the dynamics of a coherent matter wave at $T = 0$.

2.1 Static Condensate

We first consider the time-dependent stationary GP equation, which has the solution

$$\langle\tilde{\psi}(\mathbf{r},t)\rangle \equiv \Phi(\mathbf{r},t) = \Phi_0(\mathbf{r})e^{-i\mu t/\hbar}, \quad (2.14)$$

where μ is the chemical potential. The physics behind this can be seen from

$$\langle N-1|\hat{\psi}(\mathbf{r},t)|N\rangle = e^{iE_{N-1}t/\hbar}\langle N-1|\hat{\psi}(\mathbf{r})|N\rangle e^{-iE_Nt/\hbar} \\ = \langle N-1|\sqrt{N}|N-1\rangle e^{-i(E_N-E_{N-1})t/\hbar} \\ = \sqrt{N}e^{-i\mu t/\hbar}. \quad (2.15)$$

Using (2.14) in (2.13) gives

$$i\hbar\left(-\frac{i\mu}{\hbar}\right)\Phi_0(\mathbf{r}) = \left[-\frac{\hbar^2\nabla^2}{2m} + V_{\mathrm{ex}}(\mathbf{r}) + g|\Phi_0(\mathbf{r})|^2\right]\Phi_0(\mathbf{r}). \tag{2.16}$$

The *static* GP equation for the static condensate wavefunction $\Phi_0(\mathbf{r})$ is thus

$$\left[-\frac{\hbar^2\nabla^2}{2m} - \mu + V_{\mathrm{ex}}(\mathbf{r}) + g|\Phi_0(\mathbf{r})|^2\right]\Phi_0(\mathbf{r}) = 0. \tag{2.17}$$

A simple approximation in solving (2.17) is to ignore the kinetic energy of the condensate, i.e., neglect the $-\hbar^2\nabla^2/(2m)$ term. This is called the "Thomas–Fermi" approximation (TF) in the recent Bose gas literature [1]. In this TF approximation, the static GP (2.17) equation for $\Phi_0(\mathbf{r})$ reduces to

$$[V_{\mathrm{ex}}(\mathbf{r}) + g|\Phi_0(\mathbf{r})|^2] = \mu, \tag{2.18}$$

which is easly inverted to give the condensate density profile

$$\begin{aligned} n_{c0}(\mathbf{r}) &= \frac{1}{g}[\mu - V_{\mathrm{ex}}(\mathbf{r})] \\ &= \frac{1}{g}\left[\mu - \frac{1}{2}m\omega_0^2 r^2\right] > 0. \end{aligned} \tag{2.19}$$

Clearly in the TF approximation, the "size" of the condensate is R_{TF}, where

$$\mu = \tfrac{1}{2}m\omega_0^2 R_{\mathrm{TF}}^2. \tag{2.20}$$

One finds μ from the condition $\int d\mathbf{r}n_c(\mathbf{r}) = N_c = N$, or

$$N_c = 4\pi\int_0^{R_{\mathrm{TF}}} dr r^2 \frac{1}{g}\left[\mu - \frac{1}{2}m\omega_0^2 r^2\right]. \tag{2.21}$$

This gives

$$\mu = \hbar\omega_0\left[15\frac{Na}{a_{\mathrm{HO}}}\right]^{2/5}; \quad a_{\mathrm{HO}} \equiv (\hbar/m\omega_0)^{1/2}. \tag{2.22}$$

We note that SH oscillator length a_{HO} is the size of the ground state wavefunction (1.15) of an atom in a parabolic potential. Combining (2.21) and (2.20) gives

$$\begin{aligned} R_{\mathrm{TF}} &= a_{\mathrm{HO}}\left(15\frac{Na}{a_{\mathrm{HO}}}\right)^{1/5} \\ &\gg a_{\mathrm{HO}}, \quad \text{if } \frac{Na}{a_{\mathrm{HO}}} \gg 1. \end{aligned} \tag{2.23}$$

We thus find the surprising result that interactions (while weak) spread out the ideal gas condensate ($R_{\mathrm{TF}} \gg a_{\mathrm{HO}}$) and decrease the density of the condensate at centre of trap. The TF approximation for $n_{c0}(\mathbf{r})$ is very good for large N, except for a small region near the edge of condensate ($\simeq R_{\rm TF}$). Experimental data confirms these GP predictions for the $n_{c0}(\mathbf{r})$ condensate profile, emphasizing that the condensate is *not* simply the ground state wavefunction of the harmonic trap potential.

2.2 Dynamics of the Condensate (Collective Modes)

If we linearize around the static equilibrium value of the condensate

$$\Phi(\mathbf{r},t) = e^{-i\mu t/\hbar}[\Phi_0(\mathbf{r}) + \delta\Phi(\mathbf{r},t)], \tag{2.24}$$

where $\delta\Phi \ll \Phi_0$, we see that

$$i\hbar\frac{\partial\Phi}{\partial t} = \left[-\frac{\hbar\nabla^2}{2m} + V_{\rm ex}(\mathbf{r}) + g[|\Phi_0|^2 + \Phi_0^*\delta\Phi + \Phi_0\delta\Phi^*][\Phi_0 + \delta\Phi]e^{-i\mu t/\hbar}\right] \tag{2.25}$$

which gives

$$\begin{aligned} i\hbar\frac{\partial\delta\Phi(\mathbf{r},t)}{\partial t} = \left[-\frac{\hbar\nabla^2}{2m} + V_{\rm ext}(\mathbf{r}) + 2g|\Phi_0|^2 - \mu\right]\delta\Phi(\mathbf{r},t) \\ + g\Phi_0^2\delta\Phi^*(\mathbf{r},t). \end{aligned} \tag{2.26}$$

We also have a similar equation of motion for $\delta\Phi^*(\mathbf{r},t)$. Solving these two coupled equations with the ansatz

$$\delta\Phi(\mathbf{r},t) = u(\mathbf{r})e^{-i\omega t} + v(\mathbf{r})e^{i\omega t}, \tag{2.27}$$

we find two coupled "Bogoliubov equations" for the amplitudes u and v [35, 36]:

$$\begin{aligned} \left[-\frac{\hbar^2\nabla^2}{2m} + V_{\rm ex}(\mathbf{r}) - \mu + 2gn_{c0}(\mathbf{r})\right]u(\mathbf{r}) + gn_{c0}(\mathbf{r})v(\mathbf{r}) = E_i u(\mathbf{r}) \\ \left[-\frac{\hbar^2\nabla^2}{2m} + V_{\rm ex}(\mathbf{r}) - \mu + 2gn_{c0}(\mathbf{r})\right]v(\mathbf{r}) + gn_{c0}(\mathbf{r})u(\mathbf{r}) = -E_i v(\mathbf{r}). \end{aligned} \tag{2.28}$$

Here $E_i \equiv \hbar\omega$ are the excitation energies of the condensate.

The equations in (2.28) have been solved numerically by several groups and the observed oscillations are in good agreement with these predictions [1, 36]. As an illustration of the physics, it is useful to solve (2.28) for a uniform Bose gas. In this case we have

$$\begin{aligned} u(\mathbf{r}) &= ue^{i\mathbf{k}\cdot\mathbf{r}} \\ v(\mathbf{r}) &= ve^{i\mathbf{k}\cdot\mathbf{r}}, \end{aligned} \tag{2.29}$$

which gives

$$(\hbar\omega)^2 = \left[\frac{\hbar^2k^2}{2m} - \mu + 2gn_{c0}\right]^2 - [gn_{c0}]^2$$
$$= \epsilon_k^2 + 2gn_{c0}\epsilon_k. \tag{2.30}$$

This is the famous Bogoliubov spectrum at $T = 0$ [12], [35]. Here we have used $\mu_0 = gn_{c0}$ discussed earlier (see (2.18)). One finds a phonon region at long wavelengths

$$\hbar\omega_k = \hbar v_B k; \quad v_B \equiv \left(\frac{gn_{c0}}{m}\right)^{1/2}. \tag{2.31}$$

The cross-over from particle-like to this collective phonon region occurs at k_c, where

$$\frac{\hbar^2k_c^2}{2m} = 2gn_{c0} \to k_c = \sqrt{4mn_{c0}g/\hbar}. \tag{2.32}$$

This shows how the interactions changes the qualitative nature of low energy exciations in a Bose-condensed gas. This feature can be shown to stabilize superfluid motion against dissipation [13].

These oscillations of the condensate can be understood as excitations involving the *noncondensate*. Using (2.12), the Hamiltonian is given by [37, 38]

$$\begin{aligned}\hat{H} - \mu\hat{N} = &\int d\mathbf{r}\,\tilde{\psi}^+(\mathbf{r})\left[-\frac{\hbar^2\nabla^2}{2m} + V_{\rm ex}(\mathbf{r}) - \mu\right]\tilde{\psi}(\mathbf{r})\\ &+ \int d\mathbf{r}\,\Phi_0^*(\mathbf{r})\left[-\frac{\hbar^2\nabla^2}{2m} + V_{\rm ex}(\mathbf{r}) - \mu\right]\Phi_0(\mathbf{r})\\ &+ 2g\int d\mathbf{r}\,|\Phi_0(\mathbf{r})|^2\tilde{\psi}^+(\mathbf{r})\tilde{\psi}(\mathbf{r})\\ &+ \frac{1}{2}g\int d\mathbf{r}\,\Phi_0^2(\mathbf{r})\tilde{\psi}^+(\mathbf{r})\tilde{\psi}^+(\mathbf{r})\\ &+ \frac{1}{2}g\int d\mathbf{r}\,\Phi_0^{*2}(\mathbf{r})\tilde{\psi}(\mathbf{r})\tilde{\psi}(\mathbf{r}).\end{aligned} \tag{2.33}$$

We can diagonalize the quadratic part of the Hamiltonian, using

$$\tilde{\psi}(\mathbf{r}) = \sum_i [u_i(\mathbf{r})\hat{\alpha}_i + v_i^*(\mathbf{r})\alpha_i^+], \tag{2.34}$$

where $[\hat{\alpha}_i, \hat{\alpha}_j^+] = \delta_{ij}$ (Boson quasiparticles). Thus one finds

$$\hat{H} - \mu\hat{N} = \text{const.} + \sum_i \hbar\omega_i\hat{\alpha}_i^+\hat{\alpha}_i. \tag{2.35}$$

This transformation shows how the noncondensate part of Hamiltonian can be reduced to a system of *noninteracting* quasiparticles with a spectrum identical to the condensate fluctuations. This equivalence is easy to understand. The condensate fluctuations

$$\delta\Phi \equiv \langle\hat{\psi}(\mathbf{r})\rangle - \Phi_0 \tag{2.36}$$

can be calculated to first order in the symmetry-breaking perturbation (2.7),

$$H_{sb} = \int d\mathbf{r}\,[\eta\hat{\psi}^+ + \eta^*\hat{\psi}]. \tag{2.37}$$

Then standard linear response theory [4] gives (schematically)

$$\begin{aligned}\delta\Phi &\sim \int \langle[\hat{\psi}, H_{sb}]\rangle \\ &\sim \int \langle[\tilde{\psi}, \tilde{\psi}^+]\rangle\eta + \langle[\tilde{\psi}, \tilde{\psi}]\rangle\eta^*.\end{aligned} \tag{2.38}$$

This shows that the single-particle Green's functions of the noncondensate fields have the same spectrum as $\delta\Phi$. This identity of the spectrum of density fluctuations and single-particle excitations is a characteristic signature of all Bose-condensed systems which persists at finite temperatures [4, 8].

One interesting collective oscillation is the dipole mode corresponding to *rigid* oscillation of the centre of mass of the static condensate profile, and predicted to have the trap frequency ω_0. This mode is described by

$$n_c(\mathbf{r}, t) = n_{c0}\big(\mathbf{r} - \boldsymbol{\eta}(t)\big), \quad \dot{\boldsymbol{\eta}}(t) = \mathbf{v}_c, \tag{2.39}$$

where the time-dependent centre of mass satisfies

$$\frac{\partial^2\boldsymbol{\eta}(t)}{\partial t^2} = -\omega_0^2\boldsymbol{\eta}(t). \tag{2.40}$$

This mode at frequency ω_0 is special feature of a parabolic trap and is called the Kohn mode for in the case of interacting fermions [39]. This "sloshing mode" is used in BEC experiments to measure the natural frequency ω_0 of the trap and it exists at finite temperatures as well (see Section 3).

2.3 Quantum Hydrodynamic Formulation

One often rewrites the time-dependent GP equation using the amplitude and phase variables [1]

$$\Phi(\mathbf{r}, t) = \sqrt{n_c}e^{i\theta}. \tag{2.41}$$

Inserting this into the GP equation (2.13) and separating out the real and imaginary parts of the equation gives:

$$\frac{\partial n_c(\mathbf{r}, t)}{\partial t} + \boldsymbol{\nabla}\cdot n_c(\mathbf{r}, t)\mathbf{v}_c(\mathbf{r}, t) = 0, \qquad \text{continuity equation} \tag{2.42}$$

$$\hbar\frac{\partial\theta(\mathbf{r}, t)}{\partial t} = -\Big[\mu_c(\mathbf{r}, t) + \frac{1}{2}mv_c^2(\mathbf{r}, t)\Big], \qquad \text{Josephson equation.} \tag{2.43}$$

Here the gradient of the phase is related to the superfluid velocity by

$$m\mathbf{v}_c(\mathbf{r},t) \equiv \hbar\boldsymbol{\nabla}\theta(\mathbf{r},t) \tag{2.44}$$

and the condensate chemical potential is

$$\mu_c(\mathbf{r},t) \equiv -\frac{\hbar^2\nabla^2\sqrt{n_c}}{2m\sqrt{n_c}} + V_{ex}(\mathbf{r}) + gn_c(\mathbf{r},t). \tag{2.45}$$

Taking the gradient of (2.43) gives

$$m\left(\frac{\partial\mathbf{v}_c}{\partial t} + \frac{1}{2}\boldsymbol{\nabla}\mathbf{v}_c^2\right) = -\boldsymbol{\nabla}\mu_c. \tag{2.46}$$

The equations in (2.42) and (2.46) "look" like those in classical hydrodynamic theories. They show that the condensate can be described in terms of coherent motions involving two variables:

$$n_c(\mathbf{r},t), \mathbf{v}_c(\mathbf{r},t). \tag{2.47}$$

The Landau 2-fluid equations [9, 10] reduce to these same equations at $T = 0$ (where $\rho_s = \rho$, $\rho_n = 0$), namely

$$\begin{aligned}\frac{\partial n_c}{\partial t} + \boldsymbol{\nabla}\cdot n_c\mathbf{v}_c &= 0\\ m\left(\frac{\partial\mathbf{v}_c}{\partial t} + \frac{1}{2}\nabla\mathbf{v}_c^2\right) &= -\boldsymbol{\nabla}\mu.\end{aligned} \tag{2.48}$$

We will find these equations useful in Section 3, where they complement the hydrodynamic equations describing the *noncondensate* in the collision-dominated region.

This approach also allows a simple theory developed by Stringari [40] for linearized collective modes when we use the TF approximation. Taking the time-derivative of (2.42) gives ($\mathbf{v}_{c0} = 0$)

$$\frac{\partial^2\delta n_c}{\partial t^2} = -\boldsymbol{\nabla}\cdot\left[n_{c0}\left(\frac{\partial\delta\mathbf{v}_c}{\partial t}\right)biggr\right]. \tag{2.49}$$

Using (2.46), we have

$$\begin{aligned}\frac{\partial\delta\mathbf{v}_c}{\partial t} &= -\frac{1}{m}\boldsymbol{\nabla}[V_{\mathrm{ex}}(\mathbf{r}) + gn_{c0}(\mathbf{r}) + g\delta n_c(\mathbf{r},t)]\\ &= -\frac{g}{m}\boldsymbol{\nabla}\delta n_c(\mathbf{r},t).\end{aligned}$$

Combining this last result with (2.49), we obtain the very useful Stringari equation of motion [40]

$$\frac{\partial^2\delta n_c}{\partial t^2} = \boldsymbol{\nabla}\cdot\left\{\left[\mu - \frac{1}{2}m\omega_0^2 r^2\right]\boldsymbol{\nabla}\delta n_c\right\}. \tag{2.50}$$

This describes the collective oscillations of the condensate in terms of a single differential equation. As one example, the breathing mode of the condensate has a frequency $\hbar\omega = \sqrt{5}\hbar\omega_0$. This example points out that in the TF limit (large N_c), the frequencies are independent of the interaction strength and the size of the condensate N_c.

We also note that using (2.50), (2.50) can be equally well rewritten in terms of the superfluid velocity $\mathbf{v}_c(\mathbf{r}, t)$ defined in (2.44). This emphasizes that the condensate fluctuations are directly related to the existence of phase fluctuations. Their existence may thus be viewed as "evidence" of superfluidity, the latter being always a consequence of the phase coherence of the macroscopic wavefunction given by (2.41) [13].

The great thing about the collective oscillations of a condensate in a trapped gas is you can "see" them. A beautiful example from MIT is shown in Fig. 2 of [1]. As Ketterle has remarked, these condensates are *robust*– one can kick them, shake them and these "wisps" of Bose-condensed matter keep their integrity.

2.4 Interference of Coherent Matter Waves

In the pioneering matter wave interference experiments done at MIT using a de-tuned cigar-shaped trap [41], one first destroys the condensate at the centre using laser beam. Then the confining trap is turned off and the two condensates are allowed to expand and interfere. One observes nice interference fringes at the mid-point, as expected. Using

$$\begin{aligned} \Phi_{\text{system}} &= \Phi_A(\mathbf{r}, t) + \Phi_B(\mathbf{r}, t) \\ &= \sqrt{N_A} e^{i\theta_A} + \sqrt{N_B} e^{i\theta_B}, \end{aligned} \tag{2.51}$$

the density is given by

$$\begin{aligned} n(\mathbf{r}, t) &= |\Phi_{\text{system}}|^2 \\ &= N_A + N_B + 2\sqrt{N_A N_B} \cos \Delta\theta, \end{aligned} \tag{2.52}$$

where $\Delta\theta = \theta_A - \theta_B$. In the region of interference, the density is low and hence interaction effects are *small* (i.e., the gn_c term is small in the GP equation). Asymptotically, the solution of GP equation gives $\theta(\mathbf{r}, t) = \frac{mr^2}{2\hbar t}$. This implies [1]

$$\Delta\theta = \frac{m(z + d/2)^2}{2\hbar t} - \frac{m(z - d/2)^2}{2\hbar t} = \frac{mzd}{\hbar t} \equiv \frac{2\pi z}{\lambda(t)}, \tag{2.53}$$

where $\lambda(t) = 2\pi\hbar t/md$. This wavelength is in good agreement with experimental observations. (See Fig. 2 of [41]).

A condensate described by $\Phi(\mathbf{r}, t)$ may be viewed as a "classical" matter wave, as recently emphasized by Pitaevskii and Stringari [42]. This is quite *different* from ordinary quantum de Broglie waves, since one can ignore quantum fluctuations (large N_c). It is also quite different than ordinary (classical) macroscopic objects and electromagnetic waves, since $\Phi(\mathbf{r},t)$ is described by the GP equation (2.13) which involves Planck's constant $\hbar$. Thus $\Phi(\mathbf{r},t)$ is a classical object which is described by a quantum equation!! This promises to be a challenge for the quantum theory of measurement.

One can describe a two-component Bose gas (see Section 1) using coupled equations:

$$\begin{aligned} i\hbar\frac{\partial\Phi_1}{\partial t} &= \left[-\frac{\hbar^2\nabla^2}{2m} + V_1(\mathbf{r}) + g_1|\Phi_1|^2 + g_{12}|\Phi_2|^2\right]\Phi_1 \\ i\hbar\frac{\partial\Phi_2}{\partial t} &= \left[-\frac{\hbar^2\nabla^2}{2m} + V_2(\mathbf{r}) + g_2|\Phi_2|^2 + g_{12}|\Phi_1|^2\right]\Phi_2. \end{aligned} \tag{2.54}$$

There are two *coupled* GP equations for *two* macroscopic wave functions. Extensive studies [43] have been made at JILA using the two atomic hyperfine states of ^{87}Rb:

$$|F=1, m_F=-1\rangle, \quad |F=2, m_F=1\rangle. \tag{2.55}$$

In particular, one can study interesting interference effects between these coupled wavefunctions. Recent work at JILA has used such two-component Bose fluids to produce the long sought-for vortex state in one of the components [44].

3 Coupled Dynamics of the Condensate and Noncondensate

In this section, we switch our attention from $T=0$ (i.e., $T \lesssim 0.4\,T_{\text{BEC}}$) to finite temperatures, where N_c and $\tilde{N}$ are comparable in size. We first consider how the GP equation of motion for $\Phi(\mathbf{r},t)$ is modified. As a first step, we could use [4, 45],

$$i\hbar\frac{\partial\Phi(\mathbf{r},t)}{\partial t} = \Big[-\frac{\hbar^2\nabla^2}{2m} + V_{ex}(\mathbf{r}) + gn_c(\mathbf{r},t) + 2g\tilde{n}(\mathbf{r},t)\Big]\Phi(\mathbf{r},t). \tag{3.1}$$

The last term takes into account that the condensate moves in the dynamic Hartree–Fock (HF) field produced by noncondensate atoms. Immediately, one sees this generalized GP equation is no longer closed. It requires a theory of the noncondensate fluctuations, as described by $\tilde{n}(\mathbf{r},t) = \tilde{n}_0(\mathbf{r}) + \delta\tilde{n}(\mathbf{r},t)$.

A simpler version of (3.1) is to treat the effect of the noncondensate as a *static* HF field [38], [46]:

$$2g\tilde{n}(\mathbf{r},t) \simeq 2g\tilde{n}_0(\mathbf{r}). \tag{3.2}$$

1. This corresponds to treating the condensate moving in a static HF field of the noncondensate.

2. $\tilde{n}_0(\mathbf{r})$ can be calculated (self-consistently) using the fluctuations of Φ, as discussed in Section 2. One finds that the depletion of the condensate at $T = 0$ is only a few percent.

3. This procedure gives reasonable results for the thermodynamic properties at finite temperatures, as discussed in the recent literature [1]. Within the Thomas–Fermi approximation (good for $N \gtrsim 10^3$ atoms), the linearized version of (3.1) using (3.2) leads to the same Stringari equation at finite T as the $T = 0$ result in (2.50). Since the solutions of (2.50) do not depend on the magnitude of the condensate, one concludes that the collective modes of the condensate will show no temperature dependence, even though the condensate is being thermally depleted. This prediction does not appear to agree with experimental results when $T \gtrsim 0.6\,T_{\mathrm{BEC}}$. This suggests that the *dynamics* of the noncondensate has to be included.

We now go on to determining $\tilde{n}(\mathbf{r}, t)$ directly by deriving a *quantum* Boltzmann equation for the single-particle distribution function of excited atoms $f(\mathbf{p}, \mathbf{r}, t)$ and then use:

$$\tilde{n}(\mathbf{r}, t) \equiv \int \frac{d\mathbf{p}}{(2\pi)^3} f(\mathbf{p}, \mathbf{r}, t). \tag{3.3}$$

This procedure generalizes the approach of Boltzmann (1880's) for a classical gas, including the effect of binary collisions. Clearly one must make some approximations! One wants, initially, to find a *useful* kinetic equation that builds in *just* enough physics. Here I will discuss such a quantum Boltzmann equation for a trapped Bose-condensed gas at finite temperatures, which has been extensively discussed by Zaremba, Nikuni and the author [5], [47]. It is only valid in the so-called semi-classical limit, where it is sufficient to work with $f(\mathbf{p}, \mathbf{r}, t)$. The conditions are

$$k_B T \gg gn, \quad k_B T \gg \hbar\omega_0. \tag{3.4}$$

In this domain, one also can assume that the important thermal excitations can be approximated by simple Hartree–Fock particle-like spectrum:

$$\tilde{\varepsilon}(\mathbf{r}, t) = \frac{p^2}{2m} + 2g[n_c(\mathbf{r}, t) + \tilde{n}(\mathbf{r}, t)] + V_{\mathrm{ex}}(\mathbf{r}) \equiv \frac{p^2}{2m} + U(\mathbf{r}, t). \tag{3.5}$$

Clearly the resulting kinetic equation is not valid at very low temperatures, where the thermal excitations are described by a Bogoliubov-type spectrum.

We simply write down our quantum kinetic equation [5],

$$\frac{\partial f(\mathbf{p}, \mathbf{r}, t)}{\partial t} + \frac{\mathbf{p}}{m} \cdot \boldsymbol{\nabla}_r f(\mathbf{p}, \mathbf{r}, t) - \boldsymbol{\nabla}_r U(\mathbf{r}, t) \cdot \boldsymbol{\nabla}_p f(\mathbf{p}, \mathbf{r}, t) = C_{22}[f] + C_{12}[f]. \tag{3.6}$$

The right-hand side describes how binary collisions effect the value of the single-particle distribution function $f(\mathbf{p},\mathbf{r},t)$. The effect of collisions between excited atoms in the noncondensate is described by:

$$C_{22}[f] = \frac{2g^2}{(2\pi)^5\hbar^7}\int d\mathbf{p}_2\int d\mathbf{p}_3\int d\mathbf{p}_4\,\delta(\mathbf{p}+\mathbf{p}_2-\mathbf{p}_3-\mathbf{p}_4)\delta(\tilde{\varepsilon}_p+\tilde{\varepsilon}_{p_2}-\tilde{\varepsilon}_{p_3}-\tilde{\varepsilon}_{p_4})$$
$$\times\,[(1+f)(1+f_2)f_3f_4 - ff_2(1+f_3)(1+f_4)]. \quad (3.7)$$

This collision integral was discussed in detail in 1933 by Uehling and Uhlenbeck for $T > T_{\mathrm{BEC}}$ [48]. We recall that creating a Boson gives a factor $(1+f)$ and destroying a Boson gives f. In the classical high temperature limit, $f \ll 1$ and the collision integral C_{22} considerably simplifies.

Where does $f(\mathbf{p},\mathbf{r},t)$ come from in a microscopic derivation of (3.6)? Basically we calculate the nonequilibrium real-time single-particle Green's functions of the noncondensate field operators (using the Kadanoff-Baym formalism [49]). This gives (schematically)

$$g_1(1,1') \sim \langle\tilde{\psi}^+(1)\tilde{\psi}(1')\rangle, \quad (3.8)$$

where $1 \equiv \mathbf{r}_1, t_1$; $1' = \mathbf{r}'_1, t'_1$. We then express this as $g_1(\mathbf{r},t;\mathbf{R},T)$, where the *relative* and *centre of mass* coordinates are

$$\begin{aligned} \mathbf{r} &= \mathbf{r}_1 - \mathbf{r}'_1, & \mathbf{R} &= \tfrac{1}{2}(\mathbf{r}_1 + \mathbf{r}'_1), \\ t &= t_1 - t'_1, & T &= \tfrac{1}{2}(t_1 + t'_1). \end{aligned} \quad (3.9)$$

Finally we Fourier transform $g_1(\mathbf{r},t;\mathbf{R},T)$ to find $g_1(\mathbf{p},\omega;R,T)$, which gives the number of atoms at $\mathbf{R},T$ with momentum $\mathbf{p}$ and energy $\hbar\omega$. The single-particle Wigner distribution function is given by

$$f(\mathbf{p},\mathbf{R},T) \equiv \int_{-\infty}^{\infty} d\omega\, g_1(\mathbf{p},\omega;\mathbf{R},T). \quad (3.10)$$

The Wigner distribution function $f(\mathbf{p},\mathbf{R},T)$ is the quantum generalization the classical single-particle distribution function [49]. These remarks should indicate how we can go from an equation of motion for the single-particle Green's function (within a given self-energy approximation) to a kinetic equation for $f(\mathbf{p},\mathbf{R},T)$. We refer to the classic account given (for non-Bose-condensed gases) in the book by Kadanoff and Baym [49] for further details. This powerful approach was generalized to uniform Bose-condensed gases by Kane and Kadanoff [50], and has been extended to trapped gases in recent work [51].

In addition to C_{22} collisions, we also have collisions which involve *one* condensate atom:

$$C_{12}[f] = \frac{2g^2}{(2\pi)^2\hbar^4}\int d\mathbf{p}_1 \int d\mathbf{p}_2 \int d\mathbf{p}_3\, \delta(m\mathbf{v}_c + \mathbf{p}_1 - \mathbf{p}_2 - \mathbf{p}_3) \\ \times \delta(\varepsilon_c + \tilde{\varepsilon}_{p1} - \tilde{\varepsilon}_{p2} - \tilde{\varepsilon}_{p3}) \\ \times [\delta(\mathbf{p}-\mathbf{p}_1) - \delta(\mathbf{p}-\mathbf{p}_2) - \delta(\mathbf{p}-\mathbf{p}_3)] \\ \times [n_c(1+f_1)f_2 f_3 - n_c f_1(1+f_2)(1+f_3)]. \quad (3.11)$$

Here the condensate atom has

$$\begin{aligned} &\text{energy:} && \varepsilon_c = \mu_c + \tfrac{1}{2}mv_c^2; \quad \mu_c = V_{\text{ex}} + gn_c + 2g\tilde{n} \\ &\text{momentum:} && \mathbf{p}_c = m\mathbf{v}_c. \end{aligned} \quad (3.12)$$

We note the *key* difference between C_{12} and C_{22} collisions:

- C_{22} and C_{12} conserve energy and momentum in collisions.
- C_{12} does *not* (but C_{22} does) conserve the number of condensate atoms. C_{12} describes how atoms are "kicked" in and out of condensate.

It turns out the generalized GP equation (3.1) is also modified by a term related to $C_{12}[f]$. This makes sense, since the C_{12} collisions modify the condensate wavefunction $\Phi(\mathbf{r},t)$. One finds the new GP equation is given by (see also [52]):

$$i\hbar\frac{\partial\Phi(\mathbf{r},t)}{\partial t} \\ = \left[-\frac{\hbar^2\nabla^2}{2m} + V_{\text{ex}}(\mathbf{r}) + gn_c(\mathbf{r},t) + 2g\tilde{n}(\mathbf{r},t) - iR(\mathbf{r},t)\right]\Phi(\mathbf{r},t), \quad (3.13)$$

where

$$R(\mathbf{r},t) \equiv \int \frac{d\mathbf{p}}{(2\pi)^3}\frac{C_{12}[f(\mathbf{p},\mathbf{r},t)]}{2n_c(\mathbf{r},t)}. \quad (3.14)$$

More precisely, the dissipative iR term in (3.13) arises from a three field correlation function in the exact equation of motion [see (2.9) and (2.11)], and is given by [5]

$$\int \frac{d\mathbf{p}}{(2\pi)^3} C_{12}[f] = \frac{2g}{\hbar}\sqrt{n_c}\,\text{Im}\langle\tilde{\psi}^+\tilde{\psi}\tilde{\psi}\rangle. \quad (3.15)$$

We have to solve for $f(\mathbf{p},\mathbf{r},t)$ and $\Phi(\mathbf{r},t)$, treating $C_{12}[f]$ very carefully. We see that there will be an exchange of atoms between the $\tilde{n}(\mathbf{r},t)$ and $n_c(\mathbf{r},t)$ components through the C_{12} collisions. We can use these coupled equations for a variety of problems. In these lectures, we will consider the collective oscillations of the combined system composed of condensate and noncondensate. It is useful to introduce *two* regimes to describe collective modes in interacting systems [4, 53]:

I. Collisionless (produced by *mean fields*)

$$\omega\tau_R \gg 1 \quad \text{or} \quad T \ll \tau_R \left(\omega \equiv \frac{2\pi}{T}\right).$$

II. Hydrodynamic (produced by *collisions*)

$$\omega\tau_R \ll 1 \quad \text{or} \quad T \gg \tau_R,$$

where τ_R is some appropriate relaxation time. What should we use for τ_R? For a classical gas, this is the collision time [6]

$$\frac{1}{\tau_c} = \tilde{n}\sigma\bar{v} \tag{3.16}$$

where

$\sigma = 8\pi a^2$ (for Bose particles); a = s-wave scattering length.

$\bar{v} \simeq$ average velocity of atoms $\sim \sqrt{\frac{k_B T}{m}}$.

$\tilde{n}$ = density of excited atoms.

Even for a Bose-condensed gas, taking $\tau_R \sim \tau_c$ is a reasonable first estimate [47]. To get into the interesting hydrodynamic region ($\omega\tau_R \ll 1$), we need small τ_R, i.e., a large density $\tilde{n}$ or a large collision cross-section σ (perhaps using a Feshbach resonance, as discussed in Section 1).

Let us look at the kinetic equation (3.6), writing it in the schematic form:

$$\hat{\mathcal{L}}f = C_{22}[f] + C_{12}[f]. \tag{3.17}$$

In the collisionless region, we need only solve $\hat{\mathcal{L}}f = 0$. In contrast, in the hydrodynamic region, the collisions are so *strong* they produce *local* equilibrium [6]. That is, they force f to satisfy $C_{22}[f] = 0$. The unique solution $\tilde{f}$ of this equation is well-known to be given by

$$\tilde{f}(\mathbf{p},\mathbf{r},t) = \frac{1}{e^{\beta[(\mathbf{p}-m\mathbf{v}_n)^2/(2m)+U(\mathbf{r},t)-\tilde{\mu}(r,t)]} - 1}, \tag{3.18}$$

where $\mathbf{v}_n$ is the average local velocity and $\tilde{\mu}$ is the local chemical potential of the thermal atoms. This local equilibrium Bose distribution involves the local variables β, $\mathbf{v}_n$, $\tilde{\mu}$ and U, all of which depend on $(\mathbf{r},t)$.

Why must $\tilde{f}$ have the form in (3.18)? To satisfy $C_{22}[f_1] = 0$, we must have [see (3.7)]

$$(1+f_1)(1+f_2)f_3f_4 - f_1f_2(1+f_3)(1+f_4) = 0, \tag{3.19}$$

and this requires that f be given by the Bose distribution. We have used the fact that

$$f(x) \equiv \frac{1}{e^x - 1} = -[f(-x)+1] \tag{3.20}$$

and that

$$\left.\begin{aligned}\mathbf{p}_1+\mathbf{p}_2&=\mathbf{p}_3+\mathbf{p}_4\\ \tilde{\varepsilon}_{p_1}+\tilde{\varepsilon}_{p_2}&=\tilde{\varepsilon}_{p_3}+\tilde{\varepsilon}_{p_4}\end{aligned}\right\}\text{ energy and momentum conservation,} \tag{3.21}$$

where $\tilde{\varepsilon}_p = p^2/(2m) + U(\mathbf{r}, t)$. As an aside, using a kinetic equation is the most physical way of deriving the equilibrium Bose distribution. The standard approach in statistical mechanics texts based on calculating a partition function does not bring out the reason *why*

$$f_{B,F} = \frac{1}{e^{\beta(\epsilon-\mu)} \mp 1}. \tag{3.22}$$

However, while the fact that $\tilde{f}$ is given by the local equilibrium Bose distribution in (3.18) ensures that $C_{22}[\tilde{f}] = 0$, one finds that $C_{12}[\tilde{f}] \neq 0$. More precisely, we find from (3.11)

$$\begin{aligned}&[(1+\tilde{f}_1)\tilde{f}_2\tilde{f}_3 - \tilde{f}_1(1+\tilde{f}_2)(1+\tilde{f}_3)]\\ &\qquad\propto [e^{-\beta[\tilde{\mu}-\mu_c-m(\mathbf{v}_n-\mathbf{v}_c)^2/2]} - 1](1+\tilde{f}_1)\tilde{f}_2\tilde{f}_3.\end{aligned} \tag{3.23}$$

The expression in the square bracket only vanishes if the condensate and noncondensate are in *diffusive* equilibrium, which requires that

$$\tilde{\mu} = \mu_c + \tfrac{1}{2}m(\mathbf{v}_n - \mathbf{v}_c)^2. \tag{3.24}$$

When we *perturb* the system, this may not be true, i.e., the two components may be out of diffusive equilibrium.

We can now derive hydrodynamic equations for noncondensate by taking moments of Boltzmann equation, the standard procedure used in classical gases [6]. The first moment gives a continuity equation with a source term:

$$\int d\mathbf{p}\,\{\mathcal{L}\tilde{f} = C_{12}[\tilde{f}]\} \rightarrow \frac{\partial \tilde{n}}{\partial t} = -\boldsymbol{\nabla}\cdot(\tilde{n}\mathbf{v}_n) + \Gamma_{12}[\tilde{f}], \tag{3.25}$$

where

$$\begin{aligned}\tilde{n} &\equiv \int \frac{d\mathbf{p}}{(2\pi)^3}\tilde{f}(\mathbf{p},\mathbf{r},t)\\ \tilde{n}\mathbf{v}_n &\equiv \int \frac{d\mathbf{p}}{(2\pi)^3}\frac{\mathbf{p}}{m}\tilde{f}(\mathbf{p},\mathbf{r},t)\\ \Gamma_{12}[\tilde{f}] &\equiv \int \frac{d\mathbf{p}}{(2\pi)^3}C_{12}[\tilde{f}].\end{aligned} \tag{3.26}$$

More explicitly, we find

$$\begin{aligned}\Gamma_{12}[\tilde{f}] &= \frac{2g^2 n_c}{(2\pi)^5\hbar^7}[e^{-\beta[\tilde{\mu}-\mu_c-m(\mathbf{v}_n-\mathbf{v}_c)^2/2]} - 1]\\ &\quad\times \int d\mathbf{p}_1 \int d\mathbf{p}_2 \int d\mathbf{p}_3\, \delta(m\mathbf{v}_c + \mathbf{p}_1 - \mathbf{p}_2 - \mathbf{p}_3)\\ &\qquad\qquad \times \delta(\varepsilon_c + \tilde{\varepsilon}_1 - \tilde{\varepsilon}_2 - \tilde{\varepsilon}_3)(1+\tilde{f}_1)\tilde{f}_2\tilde{f}_3\\ &\equiv [e^{-\beta[\tilde{\mu}-\mu_c-m(\mathbf{v}_n-\mathbf{v}_c)^2/2-1]} - 1]\frac{n_c}{\tau_{12}}.\end{aligned} \tag{3.27}$$

We note that τ_{12} is a collision time [47] which describes the C_{12} collisions between the condensate and noncondensate atoms. Combining (3.25) with the continuity equation which results from (3.13),

$$\frac{\partial n_c}{\partial t} = -\boldsymbol{\nabla}\cdot(n_c\mathbf{v}_c) - \Gamma_{12}[\tilde{f}], \tag{3.28}$$

we see that the source term Γ_{12} cancels out to give

$$\frac{\partial(n_c+\tilde{n})}{\partial t} = -\boldsymbol{\nabla}\cdot(n_c\mathbf{v}_c + \tilde{n}\mathbf{v}_n). \tag{3.29}$$

Thus our theory gives the exact continuity equation for the total local density $n = n_c + \tilde{n}$.

Similarly, one finds

$$\begin{aligned}\int d\mathbf{p}\,\mathbf{p}\{\hat{\mathcal{L}}\tilde{f} = C_{12}[\tilde{f}]\} &\to m\tilde{n}\left(\frac{\partial \mathbf{v}_n}{\partial t} + \frac{1}{2}\boldsymbol{\nabla}\mathbf{v}_n^2\right)\\ &= -\boldsymbol{\nabla}\tilde{P}(\mathbf{r},t) - \tilde{n}\nabla U(\mathbf{r},t) - m(\mathbf{v}_n - \mathbf{v}_c)\Gamma_{12}[\tilde{f}],\end{aligned} \tag{3.30}$$

where the kinetic pressure is given by

$$\tilde{P}(\mathbf{r},t) = \frac{m}{3}\int\frac{d\mathbf{p}}{(2\pi)^3}(\mathbf{p} - m\mathbf{v}_n)^2\tilde{f}(\mathbf{p},\mathbf{r},t). \tag{3.31}$$

The second moment gives

$$\begin{aligned}\int d\mathbf{p}\,p^2\{\mathcal{L}\tilde{f} = C_{12}[\tilde{f}]\} &\to \frac{\partial\tilde{P}}{\partial t} + \boldsymbol{\nabla}\cdot(\tilde{P}\mathbf{v}_n)\\ &= -\tfrac{2}{3}\tilde{P}\boldsymbol{\nabla}\cdot\mathbf{v}_n + \tfrac{2}{3}\left[\mu_c + \tfrac{1}{2}m(\mathbf{v}_n - \mathbf{v}_c)^2 - U\right]\Gamma_{12}[\tilde{f}].\end{aligned}$$

The detailed derivation of these results is not important here [5]. The main thing is that the hydrodynamic equations (3.25), (3.30) and (3) can be shown to describe the noncondensate in terms of three new "coarse-grained" variables:

$$\tilde{n}(\mathbf{r},t), \mathbf{v}_n(\mathbf{r},t) \quad \text{and} \quad \tilde{P}(\mathbf{r},t).$$

These are coupled to the two additional variables which describe the condensate:

$$n_c(\mathbf{r},t), \quad \mathbf{v}_c(\mathbf{r},t).$$

We note that the two condensate equations of motion given by (2.46) and (3.28) are always "hydrodynamic" in form. In contrast, it is only in the collision-dominated region that the noncondensate dynamics can be described in terms of a few collective variables. We thus have 5 variables and 5 equations, which form a closed system. Both components exhibit *coupled, coherent* collective motions. This is the essence of two-fluid superfluid behaviour [9, 10], a new unexplored frontier in trapped Bose gases.

What is *new* about the two-fluid hydrodynamic equations derived above is the role of the source term $\Gamma_{12}[\tilde{f}]$. In a linearized theory expanded around the *static* equilibrium Bose distribution $\tilde{f}_0$ (where $\Gamma_{12}[\tilde{f}_0]$ vanishes), one finds [5, 47]

$$\Gamma_{12}[\tilde{f}] = \delta\Gamma_{12}[\tilde{f}] = -\frac{\beta_0 n_{0c}}{\tau_{12}^0}\delta\mu_{\text{diff}}, \tag{3.32}$$

where $\mu_{\text{diff}} \equiv \tilde{\mu} - \mu_c$. We find an equation of motion of the kind

$$\frac{\partial \delta\mu_{\text{diff}}}{\partial t} = -\frac{\delta\mu_{\text{diff}}}{\tau_\mu} + \cdots, \tag{3.33}$$

where (see Eq. (87) in [5]).

$$\frac{1}{\tau_\mu} \equiv \left(\frac{gn_{c0}}{k_B T}\right)\frac{1}{\sigma}\frac{1}{\tau_{12}^0}. \tag{3.34}$$

Here σ involves various static equilibrium thermodynamic functions. The new relaxation time τ_μ (which we can calculate!) determines *how fast* $\tilde{\mu} \to \mu_c$, i.e., how fast we reach diffusive equilibrium between the condensate and noncondensate. We can have

$$\left.\begin{array}{l}\omega\tau_{22} \ll 1 \\ \omega\tau_{12} \ll 1\end{array}\right\} \text{required for hydrodynamics} \tag{3.35}$$

but simultaneously

$$\omega\tau_\mu \gg 1 \tag{3.36}$$

near T_{BEC}, where $n_{c0} \to 0$. Our hydrodynamic equations predict the existence of a new relaxational mode [5, 47]

$$\omega \simeq -i/\tau_\mu. \tag{3.37}$$

This mode is not included in the standard Landau 2-fluid equations (where ρ_s and ρ_n are assumed to be always in local equilibrium with each other).

In a uniform gas, the two-fluid hydrodynamic equations give two normal mode solutions [5, 54]:

- First sound (oscillation of the noncondensate mainly)

$$\omega = u_1 k, \quad u_1^2 \simeq \frac{5}{3}\frac{\tilde{P}_0}{m\tilde{n}_0} \sim \frac{kT}{m}. \tag{3.38}$$

- Second sound (oscillation of the condensate mainly)

$$\omega = u_2 k, \quad u_2^2 \simeq \frac{gn_{c0}}{m}. \tag{3.39}$$

We note the second sound mode is the hydrodynamic version of famous $T = 0$ Bogoliubov phonon mode discussed in Section 2. It is the "soft mode" at T_{BEC}. This second sound mode couples to the new relaxational mode given in (3.37) and is damped as a result, the maximum damping occuring when $\omega\tau_\mu = 1$.

In a trapped gas, we can work out the spectrum of hydrodynamic oscillations ($\sim e^{-i\omega t}$). Both the condensate and noncondensate components have the same frequency. The most interesting one is the dipole mode, described by

$$\begin{aligned} \tilde{n}(\mathbf{r},t) &= \tilde{n}_0\big(\mathbf{r} - \boldsymbol{\eta}_n(t)\big), \qquad & \dot{\boldsymbol{\eta}}_n(t) &= \mathbf{v}_n \\ n_c(\mathbf{r},t) &= n_{c0}\big(\mathbf{r} - \boldsymbol{\eta}_c(t)\big), \qquad & \dot{\boldsymbol{\eta}}_c(t) &= \mathbf{v}_c. \end{aligned} \tag{3.40}$$

One finds there are two modes of this kind [45], [5]:

- In-phase (or Kohn) mode, where $\boldsymbol{\eta}_n = \boldsymbol{\eta}_c$ and $\omega = \omega_0$ (trap frequency). It is the finite temperature version of the sloshing mode described by (2.39) and (2.40). We note that this mode is generic (occuring in both the hydrodynamic and collisionless limit) and is not damped [5].

- Out-of-phase dipole mode, with $\boldsymbol{\eta}_n \neq \boldsymbol{\eta}_c$ and in *opposite* directions. The frequency of this mode is different from the trap frequency. This out-of-phase mode is of special interest since it is the analogue of the out-of-phase second sound mode in superfluid ^{4}He.

We conclude this section with some remarks:

1. The *specific calculation* sketched above is also of interest in the general field of nonequilibrium statistical physics. It describes the detailed dynamics of a system with a two-component order parameter self-consistently coupled to a gas of excitations based on a fully microscopic theory.

2. More work is needed to extend our analysis to *low* but finite temperatures and also into the critical region *very close* to T_{BEC}. In both cases, our simple Hartree–Fock particle-like thermal excitation spectrum (3.5) is no longer valid.

3. The *classical* kinetic theory of gases has been a rich subject in mathematical physics in the twentieth century, with well-known contributions by people like Boltzmann, Hilbert, Enskog, Chapman, Uhlenbeck and Burnett. These new equations of motion for a Bose-condensed gas promise to yield a lot of new physics in the next century—and surprises, as our work in this Section has already shown.

Acknowledgments: The work in Section 3 is part of an on-going collaboration with Eugene Zaremba (Queen's University) and Tetsuro Nukuni (Tokyo Institute of Technology). I would like to thank Yvan Saint-Aubin for his strong desire that BEC be represented at this CRM Summer School and insisting that I could fit it into my post-sabbatical schedule. My work is supported by a research grant from NSERC. I would also like to thank JILA (University of Colorado) for support during the time when I prepared these lectures.

4 References

1. F. Dalfovo, S. Giorgini, L.P. Pitaevskii, and S. Stringari, *Rev. Mod. Phys.* **71**, 463 (1999).
2. *Bose–Einstein Condensation in Atomic Gases*, Proceedings of the International School of Physics "Enrico Fermi", eds. M.I. Inguscio, S. Stringari, and C.E. Wieman (IOS Press, Amsterdam, 1999).
3. W. Ketterle et al., in [2].
4. A. Griffin, in [2].
5. E. Zaremba, T. Nikuni, and A. Griffin, *J. Low Temp. Phys.* **116**, 277 (1999).
6. K. Huang, *Statistical Mechanics*, 2nd. edition (John Wiley, New York, 1987).
7. F. London, *Nature* **141**, 643 (1938); see also the historical overview by Griffin in [2].
8. A. Griffin, *Excitations in a Bose-Condensed Liquid* (Cambridge Univ. Press, 1993).
9. L.D. Landau, *J. Phys. U.S.S.R.* **5**, 71 (1941).
10. I.M. Khalatnikov, *An Introduction to the Theory of Superfluidity* (W.A. Benjamin, New York, 1965).
11. S.T. Beliaev, *Sov. Phys.—JETP* **7**, 289 (1958).
12. N.N. Bogoliubov, *J. Phys. U.S.S.R.* **11**, 23 (1947).
13. P. Nozières and D. Pines, *The Theory of Quantum Liquids*, Vol. II: *Superfluid Bose Liquids* (Addison-Wesley, Redwood City, 1990).
14. For a review, see M. Randeria in *Bose–Einstein Condensation*, eds. A. Griffin, D. W. Snoke, and S. Stringari (Cambridge Univ. Press, New York, 1995). This book contains review articles on Bose condensation in a variety of systems, and was published just before the experimental discovery of BEC in laser-cooled atomic gases in 1995.
15. B. De Marco and D.S. Jin, *Science* **285**, 1703 (1999).

16. For a review of the theory of Cooper pairs in trapped Fermi gases, see the article by H.T.C. Stoof and M. Houbiers in [2].

17. See the review article by T.J. Greytak in [14].

18. See article in Physics Today, p. 17, October, 1998.

19. M.H. Anderson, J.R. Ensher, M.R. Matthews, C.E. Wieman, and E.A. Cornell, *Science* **269**, 198 (1995); see also Physics Today, p. 17, August, 1995.

20. K.B. Davis, M.-O. Mewes, M.R. Andrews, N.J. van Druten, D.S. Durfee, D.M. Kurn, and W. Ketterle, *Phys. Rev. Lett.* **75**, 3969 (1995).

21. C.C. Bradley, C.A. Sackett, J.J. Tollett, and R.G. Hulet, *Phys. Rev. Lett.* **75** 1687 (1995).

22. C.C. Bradley, C.A. Sackett, and R.G. Hulet, *Phys. Rev. Lett.* **78**, 985 (1997).

23. See the review article by J. Dalibard in [2].

24. S. Inouye, M.R. Andrews, J. Stenger, H.-J. Miesner, D.M. Stamper-Kurn, and W. Ketterle, *Nature* **392**, 151 (1998).

25. L.P. Pitaevskii, *Sov. Phys.-JETP* **13**, 451 (1961).

26. E.P. Gross, *Nuovo Cimento* **20**, 454 (1961).

27. See the article in Physics Today, p. 17, July, 1999.

28. D. Jaksch, C. Bruder, J.I. Cirac, C.W. Gardiner, and P. Zoller, *Phys. Rev. Lett.* **81**, 3108 (1998).

29. L.P. Pitaevskii, in a review article in *Recent Progress in Many-Body Theories*, edited by D. Nielson and R. Bishop (World Scientific, Singapore, 1998), p. 3.

30. N.N. Bogoliubov, in *Lectures on Quantum Statistics,* Vol. 2 (Gordon and Breach, New York, 1970), p. 148.

31. J. Gavoret and P. Nozières, *Ann. Phys. (N.Y.)* **28**, 349, (1964).

32. P.C. Hohenberg and P. C. Martin, *Ann. Phys. (N.Y.)* **34**, 291, (1965).

33. For a recent review of the field-theoretic treatment of Bose-condensed fluids, see the book in [14]. The second order Beliaev self-energy approximation is discussed in Hua Shi and A. Griffin, *Physics Reports* **304**, 1 (1998).

34. See, for example, P.W. Anderson, *Rev. Mod. Phys.* **38**, 298 (1966); see also numerous reprinted articles in P.W. Anderson, *A Career in Theoretical Physics* (World Scientific, Singapore, 1994).

35. See Chapter 14 of A.L. Fetter and J.D. Walecka, *Quantum Theory of Many-Particle Systems* (McGraw-Hill, New York, 1971).

36. M. Edwards, P.A. Ruprecht, K. Burnett, R.J. Dodd, and C.W. Clark, *Phys. Rev. Lett.* **77**, 1671 (1996).

37. A.L. Fetter, *Ann. Phys. (N.Y.)* **70**, 67 (1972).

38. A. Griffin, *Phys. Rev.* **B53**, 9341 (1996).

39. See, for example, J.F. Dobson, *Phys. Rev. Lett.* **73**, 7244 (1994).

40. S. Stringari, *Phys. Rev. Lett.* **77**, 2360 (1996).

41. M.R. Andrews, C.G. Townsend, H.-J. Miesner, D.S. Durfee, D.M. Kurn, and W. Ketterle, *Science*, **275**, 637 (1997).

42. L. Pitaevskii and S. Stringari, cond-mat/9905409.

43. See, for example, E.A. Cornell, D.S. Hall, M.R. Matthews, and C.E. Wieman, *J. Low Temp. Phys.* **113**, 151 (1998).

44. M.R. Matthews, B.P. Anderson, P.C. Haljan, D.S. Hall, C.E. Wieman, and E.A. Cornell, *Phys. Rev. Lett.* **83**, 2948 (1999); J.E. Williams and M. Holland, *Nature* **401**, 568 (1999).

45. E. Zaremba, T. Nikuni, and A. Griffin, *Phys. Rev.* **A57**, 4695 (1998).

46. D.A.W. Hutchinson, E. Zaremba, and A. Griffin, *Phys. Rev. Lett.* **78**, 1842 (1997).

47. T. Nikuni, E. Zaremba, and A. Griffin, *Phys. Rev. Lett.* **83**, 10 (1999).

48. E.A. Uehling and G.E. Uhlenbeck, *Phys. Rev.* **43**, 552 (1933).

49. L.P. Kadanoff and G. Baym, *Quantum Statistical Mechanics* (W.A. Benjamin, New York, 1962).

50. J.W. Kane and L.P. Kadanoff, *J. Math. Phys.* **6**, 1902 (1965).

51. M. Imamovic-Tomasovic and A. Griffin, *Phys. Rev.* **A60**, 494 (1999); to be published.

52. H.T.C. Stoof, *J. Low. Temp. Phys.* **114**, 11 (1999).

53. See Chapter 1 of D. Pines and P. Nozières, *The Theory of Quantum Liquids*, Vol. I (W.A. Benjamin, New York, 1966).

54. A. Griffin and E. Zaremba, *Phys. Rev.* **A56**, 4826 (1997).

5

Beyond CFT: Deformed Virasoro and Elliptic Algebras

Satoru Odake

ABSTRACT In this lecture we discuss "beyond CFT" from a symmetry point of view. After reviewing the Virasoro algebra, we introduce the deformed Virasoro algebras and elliptic algebras. These algebras appear in solvable lattice models and we study them by the free field approach.

1 Introduction

The conformal field theory (CFT) is a theory which is invariant under the conformal transformation. CFT in 2 dimensions [1] can be applied to the string theory as a worldsheet theory and statistical critical phenomena in 2-dimensional space. This theory has made a remarkable progress contacting with various branches of mathematics [2, 3]. The main reason is that in two-dimensional space (or $1+1$-dimensional spacetime), the group of conformal transformations is infinite-dimensional. Its algebra is known as the Virasoro algebra in the field theory realization, and this symmetry is very powerful. By using its detailed representation theory one can determine the spectrum and even calculate correlation functions. Statistical critical phenomena in 2-dimensional space are understood systematically by CFT and the list of critical exponents is obtained by the representation theory of the Virasoro algebra. In the first superstring revolution (middle 80's), the string theory was regarded as CFT on a worldsheet, and the knowledge obtained in the study of CFT developed string theories very much. Recent progress in string theory (the second superstring revolution (middle 90's)) is based on spacetime symmetry consideration, e.g., duality, D-brane, and also AdS/CFT correspondence, but importance of worldsheet symmetries remains unchanged.

Quantum field theory and critical phenomena are the systems with infinite degrees of freedom. Consequently they are difficult to treat. However we can sometimes solve some models, so-called solvable models. Here we loosely use the word "solvable" if some physical quantities of that model can be calculated exactly, for example, 2D Ising model, XYZ spin chain, solvable lattice model, CFT, 4D super Yang-Mills theory, etc. Although solvable models themselves are interesting for us, a main purpose for physicists in studying solvable models is to create new ideas and concepts and to develop them through the study of solvable models. Physics should explain the real world. The matter in the real world is very complicated and we need some approximations to study it. An approximation is not a bad thing if it grasps the essence of the problem. Physicists have developed a variety of concepts, approximation methods, calculation techniques, etc., through the study of solvable models, in order to apply them to the real world.

Symmetry is one of the main ideas of modern physics. If the model has some symmetry, its analysis becomes much simpler by using the representation theory of the symmetry algebra. By reversing this direction, symmetry is also used for a model building. For example, the general relativity and the gauge theory are constructed by imposing the invariance under the general coordinate transformation and the gauge transformation respectively. From the symmetry point of view, "solvable" in the system with infinite degrees of freedom is stated as the following "equation":

$$\frac{\text{System of } \textbf{infinite} \text{ degrees of freedom}}{\textbf{Infinite}\text{-dimensional symmetry}} = \text{System can be described by } \textbf{finite} \text{ degrees of freedom.} \tag{1.1}$$

This is the reason why we are interested in infinite-dimensional symmetries.

Common feature of many solvable models is the factorization of the scattering S matrix [4], in other words the Yang–Baxter equation [5–7]. Solutions of the Yang–Baxter equation are related to the Lie algebras [8–10], and three types of solutions are known; rational, trigonometric and elliptic. Associated for each type of solution (R matrix), algebras are defined [11],

$$\begin{aligned}
&\text{rational} && \longrightarrow \text{Yangian,}\\
&\text{trigonometric} && \longrightarrow \text{quantum group (quantum algebra),}\\
&\text{elliptic} && \longrightarrow \text{elliptic quantum group (elliptic algebra).}
\end{aligned}$$

This elliptic algebra is one of the topics in this lecture.

Since CFT is invariant under the scale transformation, CFT has no scale, in other words, it is a massless theory. If we add to CFT the perturbation which breaks the conformal symmetry, then the theory becomes massive. General massive theories are very difficult. So we restrict ourselves to its subset, massive integrable models (MIM). If we perturb CFT in a "good" manner (for example $(1,3)$ or $(1,2)$ perturbation [12–15]), infinitely many conserved quantities survive. In the terminology of statistical mechanics, CFT corresponds to on-critical theory, and perturbation corresponds to off-critical procedure, and a lattice analog of MIM is a solvable lattice model. CFT is controlled by the Virasoro symmetry, but MIM is massive, therefore there is no Virasoro symmetry.

CFT	on-critical	massless	Virasoro
↓ good perturbation	↓ off-critical	↓	↓
massive integrable model	solvable lattice model	massive	no Virasoro

A natural question arises:

> What symmetry ensures the integrability or infinitely many conserved quantities of MIM or solvable lattice model?

We would like to answer this question. This is our main motivation for recent study.

In some cases the Yangian or the quantum group symmetry plays an important role. Kyoto group investigated the XXZ spin chain and clarified its symmetry, the quantum affine Lie algebra $U_q(\widehat{\mathfrak{sl}}_2)$ [16]. They studied XXZ spin chain from the representation theory point of view and developed vertex operator calculation technique. But naively one expected some deformation of the Virasoro algebra. Such algebras, deformed Virasoro algebra (DVA) and deformed W_N algebras (DWA), were constructed in different points of view [17–20], using a correspondence of singular vectors and multivariable orthogonal symmetric polynomials or using the Wakimoto realization at the critical level. Later it was shown that this deformed Virasoro algebra appears in the Andrews–Baxter–Forester (ABF) model as a symmetry [21]. This DVA corresponds to $A_1^{(1)}$ algebra. DVA corresponding to $A_2^{(2)}$ was obtained in [22].

Another possibility is elliptic quantum groups (elliptic algebras). Corresponding to the two types of elliptic solutions of the Yang–Baxter equation, there are two type of elliptic quantum groups [23, 24]. These two elliptic quantum groups have a common structure [25]; They are quasi-Hopf algebras [11]. Along this line, explicit formulas for the twistors were presented and the vertex type algebra $\mathcal{A}_{q,p}(\widehat{\mathfrak{sl}}_n)$ and the face type algebra $\mathcal{B}_{q,\lambda}(\mathfrak{g})$ were defined in [26].

In this lecture we would like to (i) introduce the deformed Virasoro algebras and elliptic algebras, (ii) present free field approach and vertex operator calculation technique for the solvable lattice models. Contents of this lecture is presented in previous pages. In Section 2 we review the Virasoro algebra which is needed to understand the deformed case. The deformed Virasoro algebra of type $A_1^{(1)}$ is defined and its properties are given in Section 3. In Section 4 we review solvable lattice models and introduce the elliptic quantum groups. Section 5 is devoted to an application of these ideas to the ABF model. ABF model in regime III corresponds to the $(1,3)$-perturbation of the minimal unitary CFT. Vertex operators are bosonized and local height probabilities (LHP's) are calculated. Another deformed Virasoro algebra DVA ($A_2^{(2)}$) is defined in Section 6 and the dilute A_L models are studied by free field approach. Dilute A_L model in regime 2^+ corresponds to the $(1,2)$-perturbation of the minimal unitary CFT. In Section 7 we mention other topics that are not treated in this lecture. Appendix A is a summary of notations and formulas used throughout this lecture.

2 Conformal Field Theory and Virasoro Algebra

2.1 Conformal Field Theory

The conformal field theory (CFT) is a theory which is invariant under the conformal transformation. In two-dimensional space (or $1+1$-dimensional spacetime) the conformal transformation is any holomorphic map $z \mapsto w = w(z)$ where z is a complex coordinate of the space. Therefore the conformal group is infinite dimensional.

An important property of the CFT (bulk theory) is the factorization into a holomorphic part (z, left mover) and an antiholomorphic part ($\bar{z}$, right mover). We can treat them independently. Usually we treat z part (chiral part) only. To get final physical quantities, however, we have to glue chiral and antichiral parts with appropriate physical conditions.

As a quantum field theory, infinitesimal conformal transformation,

$$l_n\colon z \mapsto z + \epsilon z^{n+1}, \tag{2.1}$$

is generated by the chiral part of the energy momentum tensor, $L(z)$. The algebra generated by this $L(z)$ is called the Virasoro algebra. Invariance under the conformal transformation imposes that correlation functions satisfy the conformal Ward identity. Correlation functions are severely controlled by this infinite dimensional Virasoro symmetry.

General review of the CFT is not the aim of this lecture. Since many good review and books are available now, for various topics of the CFT, see [1–3].

2.2 Virasoro Algebra

In this and next subsections we review some properties of the Virasoro algebra in order to compare them with the deformed one in the next section.

2.2.1 Definition and Consistency

Definition

The Virasoro algebra is a Lie algebra over $\mathbb{C}$ generated by L_n ($n \in \mathbb{Z}$) and c, and their relation is

$$[L_n, L_m] = (n-m)L_{n+m} + \frac{c}{12}(n^3-n)\delta_{n+m,0}, \quad [L_n, c] = 0. \tag{2.2}$$

In terms of the Virasoro current $L(z) = \sum_{n\in\mathbb{Z}} L_n z^{-n-2}$, this relation is equivalent to the following operator product expansion (OPE),

$$L(z)L(w) = \frac{c}{2(z-w)^4} + \frac{2L(w)}{(z-w)^2} + \frac{\partial L(w)}{z-w} + \text{reg.} \tag{2.3}$$

As a formal power series this can be written as

$$[L(z), L(w)] = \frac{1}{z^4}\delta'''\left(\frac{w}{z}\right)\frac{c}{12} + \frac{1}{z^2}\delta'\left(\frac{w}{z}\right)2L(w) + \frac{1}{z^1}\delta\left(\frac{w}{z}\right)\partial L(w), \tag{2.4}$$

or

$$[z^2L(z), w^2L(w)] = \left(\frac{w}{z}\right)^2\delta'''\left(\frac{w}{z}\right)\frac{c}{12} + \frac{w}{z}\delta'\left(\frac{w}{z}\right)(z^2L(z)+w^2L(w)), \tag{2.5}$$

where $\delta(z) = \sum_{n\in\mathbb{Z}} z^n$ (see Appendix A.2). Meaning of (2.5) is that the coefficient of $z^n w^m$ in (2.5) gives (2.2).

Consistency

Mathematically the Virasoro algebra is a one-dimensional central extension of diff(S^1), the Lie algebra of diffeomorphism group of a circle S^1. Vector fields on S^1 form a Lie algebra,

$$[l_n, l_m] = (n-m)l_{n+m}, \tag{2.6}$$

where $l_n = -z^{n+1}d/dz = ie^{in\theta}d/d\theta$ ($z = e^{i\theta}$ is a coordinate of S^1). Let us consider its central extension,

$$[L_n, L_m] = (n-m)L_{n+m} + cf(n,m), \tag{2.7}$$

where c is a central element and $f(n,m)$ is a number. From antisymmetry of bracket [,] and the Jacobi identity, $f(n,m)$ should satisfy

$$f(n,m) = -f(m,n), \tag{2.8}$$
$$(n-m)f(l,n+m) + (m-l)f(n,m+l) + (l-n)f(m,l+n) = 0. \tag{2.9}$$

Its nontrivial solution has the following form,

$$f(n,m) = \text{const}\cdot n^3\delta_{n+m,0}. \tag{2.10}$$

Proof. From (2.9) with $l=0$, we have $f(n,m) = (n-m)f(n+m,0)/(n+m)$ for $n+m \neq 0$, which satisfies eqs. (2.8), (2.9) with $n+m+l \neq 0$. By setting $L'_n = L_n + cf(n,0)/n$ ($n \neq 0$) and $L'_0 = L_0$, (2.7) becomes

$$[L'_n, L'_m] = (n-m)L'_{n+m} + c\delta_{n+m,0}f(n,-n).$$

Hence we are enough to determine $a_n = f(n,-n)$. Eq. (2.9) with $l = -n-m$ implies

$$-(n-m)a_{n+m} + (2m+n)a_n - (2n+m)a_m = 0.$$

For $m = 1$, we obtain

$$a_n = \tfrac{1}{6}(n^3 - n)a_2 + \big(n - \tfrac{1}{6}(n^3 - n)\big)a_1,$$

and this solution satisfies eqs. (2.8), (2.9) for all cases. The term proportional to n in a_n, say bn, can be deleted by $L''_n = L'_n + cb\delta_{n,0}/2$. □

The central term in Virasoro algebra is chosen so that it vanishes for $n = 1, 0, -1$.

2.2.2 Representation Theory

We consider the highest representation of the Virasoro algebra. The highest weight state $|h\rangle$ ($h \in \mathbb{C}$) is characterized by

$$L_n|h\rangle = 0 \quad (n > 0), \quad L_0|h\rangle = h|h\rangle, \tag{2.11}$$

and the Verma module is

$$M = \bigoplus_{l\geq 0} \bigoplus_{n_1\geq\cdots\geq n_l>0} \mathbb{C}L_{-n_1}\ldots L_{-n_l}|h\rangle. \tag{2.12}$$

Descendant $L_{-n_1}\ldots L_{-n_l}|h\rangle$ is also an eigenstate of L_0 with eigenvalue $h + \sum_{i=1}^{l} n_i$. $\sum_{i=1}^{l} n_i$ is called a level of the state. Since L_0 corresponds to energy (exactly speaking Hamiltonian is $L_0 + \bar{L}_0$), this representation has energy bounded below. The highest weight state $|h\rangle$ is created by a primary field with conformal weight h, $\phi_h(z)$, from the vacuum $|\mathbf{0}\rangle$ ($L_n|\mathbf{0}\rangle = 0$ for $n \geq -1$); $|h\rangle = \lim_{z\to 0}\phi_h(z)|\mathbf{0}\rangle$.

At level N there are $p(N)$ independent states, $L_{-n_1}\ldots L_{-n_l}|\lambda\rangle$ ($n_1 \geq \cdots \geq n_l > 0$, $\sum_{i=1}^{l} n_i = N$). Here $p(N)$ is the number of partition and its generating function is given by

$$\sum_{N=0}^{\infty} p(N)y^N = \prod_{n=1}^{\infty} \frac{1}{1-y^n}. \tag{2.13}$$

Let us number these states by the reverse lexicographic ordering for the l-tuple $(n_1,\ldots,n_l)$, i.e., $|h;N,1\rangle = L_{-N}|\lambda\rangle$, $|h;N,2\rangle = L_{-N+1}L_{-1}|\lambda\rangle$, …, $|h;N,p(N)\rangle = L_{-1}^{N}|\lambda\rangle$.

The first problem of representation theory is whether M is irreducible or not, namely M has Virasoro invariant subspaces or not. If M has a singular vector at level N, $|\chi\rangle$, which is defined by

$$L_n|\chi\rangle = 0 \ (n > 0), \quad L_0|\chi\rangle = (h+N)|\chi\rangle, \tag{2.14}$$

then $|\chi\rangle$ generates an invariant subspace and we have to quotient out it from M in order to get an irreducible module $\mathcal{L}$. Existence of singular vectors can be detected by considering an "inner product" (bilinear form). Let us introduce the dual module M^* on which the Virasoro algebra acts as $L_n^\dagger = L_{-n}$. M^* is generated by $\langle h|$ which satisfies $\langle h|L_n = 0$ $(n < 0)$, $\langle h|L_0 = h\langle h|$ and $\langle h|h\rangle = 1$. At level N there are $p(N)$ states $\langle h; N, 1| = \langle h|L_N$, $\langle h; N, 2| = \langle h|L_1 L_{N-1}, \ldots, \langle h; N, p(N)| = \langle h|L_1^N$. A state $|\psi\rangle \in M$ is called a null state if it is orthogonal to all states, i.e., $\langle\psi'|\psi\rangle = 0$ for $\forall\langle\psi'| \in M^*$. Descendants of $|\chi\rangle$ are null states, because (i) $\langle h|\chi\rangle = 0$ (consider $\langle h|L_0|\chi\rangle$) and (ii) $\langle h|\ldots L_{m_2}L_{m_1} \cdot L_{-n_1}L_{-n_2}\ldots|\chi\rangle = \sum(\ldots)\langle h|\chi\rangle = 0$. Physical meaning of quotienting out invariant subspaces is projecting out null states which decouple from all the states.

We give an example of inner product at level 1, 2, 3.

$$
\begin{aligned}
&\langle h|L_1 L_{-1}|h\rangle = 2h, \\
&\begin{pmatrix} \langle h|L_2 L_{-2}|h\rangle & \langle h|L_2 L_{-1}^2|h\rangle \\ \langle h|L_1^2 L_{-2}|h\rangle & \langle h|L_1^2 L_{-1}^2|h\rangle \end{pmatrix} = \begin{pmatrix} 4h + \frac{1}{2}c & 6h \\ 6h & 4h(2h+1) \end{pmatrix}, \\
&\begin{pmatrix} \langle h|L_3 L_{-3}|h\rangle & \langle h|L_3 L_{-2}L_{-1}|h\rangle & \langle h|L_3 L_{-1}^3|h\rangle \\ \langle h|L_1 L_2 L_{-3}|h\rangle & \langle h|L_1 L_2 L_{-2}L_{-1}|h\rangle & \langle h|L_1 L_2 L_{-1}^3|h\rangle \\ \langle h|L_1^3 L_{-3}|h\rangle & \langle h|L_1^3 L_{-2}L_{-1}|h\rangle & \langle h|L_1^3 L_{-1}^3|h\rangle \end{pmatrix} \\
&\qquad\qquad = \begin{pmatrix} 6h + 2c & 10h & 24h \\ 10h & h(8h + 8 + c) & 12h(3h+1) \\ 24h & 12h(3h+1) & 24h(h+1)(2h+1) \end{pmatrix}
\end{aligned}
$$

A determinant of this matrix is called the Kac determinant and its zeros indicate existence of null states. The Kac determinant at level 2 is $2h\big(16h^2 + 2(c-5)h + c\big)$ which vanishes for $h = 0$ and $h = h_\pm = \big(5 - c \pm \sqrt{(1-c)(c-25)}\big)/16$. For $h = 0$, $L_{-1}^2|h\rangle$ is a null state which is a descendant of the singular vector at level 1 $L_{-1}|h\rangle$. For $h = h_\pm$, there exists a new singular vector,

$$
|\chi_\pm\rangle = \left(L_{-2} - \frac{3}{2(2h_\pm + 1)}L_{-1}^2\right)|h_\pm\rangle. \tag{2.15}
$$

At the general level N, the Kac determinant is given by [27]

$$
\det(\langle h; N, i|h; N, j\rangle)_{1\le i,j\le p(N)} = \prod_{\substack{l,k\ge 1 \\ lk\le N}} \big(2lk(h - h_{l,k})\big)^{p(N-lk)}. \tag{2.16}
$$

Here we have parameterized the central charge c and conformal weight $h_{l,k}$ by a parameter β,

$$
c = 1 - 6\alpha_0^2, \tag{2.17}
$$

$$
h_{l,k} = \frac{1}{4}\left(\left(\sqrt{\beta}l - \frac{1}{\sqrt{\beta}}k\right)^2 - \alpha_0^2\right), \tag{2.18}
$$

where α_0 is

$$\alpha_0 = \sqrt{\beta} - \frac{1}{\sqrt{\beta}}. \tag{2.19}$$

These β and α_0 will be used throughout this lecture. Remark that c and $h_{l,k}$ are invariant under $\beta \to \beta^{-1}$.

Detailed study of the Kac determinant shows that representations of the Virasoro algebra are classified into several classes. The most interesting class is the so-called minimal series [1]. In minimal series, β is a rational number (we take $\beta > 1$),

$$\beta = \frac{p''}{p'}, \quad p', p'' \in \mathbb{Z}_{>0},\ p'' > p',\ (p', p'') = 1, \tag{2.20}$$

and in this case eqs. (2.17), (2.18) become

$$c = 1 - 6\frac{(p''-p')^2}{p'p''}, \tag{2.21}$$

$$h_{l,k} = \frac{(p''l - p'k)^2 - (p''-p')^2}{4p'p''}. \tag{2.22}$$

An operator algebra of $\phi_h(z)$ closes for the following finite number of $h = h_{l,k}$,

$$1 \le l \le p'-1, \quad 1 \le k \le p''-1, \quad p''l - p'k > 0. \tag{2.23}$$

If one prefers $0 < \beta < 1$, it is achieved by $p' \leftrightarrow p''$ ($\beta \leftrightarrow \beta^{-1}$), and then $h_{l,k} \leftrightarrow h_{k,l}$.

Let $M_{l,k}$ denote the Verma module with $h = h_{l,k}$. From the Kac determinant and the property of $h_{l,k}$,

$$h_{l,k} = h_{-l,-k} = h_{l+p'n,k+p''n} \quad (n \in \mathbb{Z}), \tag{2.24}$$

$$h_{l,-k} = h_{l,k} + lk, \tag{2.25}$$

there are two basic singular vectors in $M_{l,k}$. One is at level lk ($h_{l,-k} = h_{l,k} + lk$, $h_{l,-k} = h_{-l,k}$) and the other is at level $(p'-l)(p''-k)$ ($h_{l,k} = h_{p'-l,p''-k}$, $h_{-p'+l,p''-k} = h_{l,k} + (p'-l)(p''-k)$, $h_{-p'+l,p''-k} = h_{-l+2p',k}$). Therefore we have

$$M_{l,k} \supset (M_{-l+2p',k} \cup M_{-l,k}). \tag{2.26}$$

To get an irreducible module we have to factor out invariant subspace, $M_{l,k}/(M_{-l+2p',k} \cup M_{-l,k})$. But the story has not ended yet because $M_{-l+2p',k} \cup M_{-l,k}$ does not coincide with $M_{-l+2p',k} \oplus M_{-l,k}$.

From the Kac determinant, $M_{-l,k}$ also has two singular vectors,

$$\begin{aligned} M_{-l,k} &= M_{p'-l,p''+k} \supset M_{-p'+l,p''+k} = M_{l-2p',k} \\ &= M_{p'+l,p''-k} \supset M_{-p'-l,p''-k} = M_{l+2p',k}. \end{aligned}$$

Here indices $(p'-l, p''+k)$ and $(p'+l, p''-k)$ are positive and minimal with respect to translation of (p', p'') (see (2.24)). Similarly

$$\begin{aligned} M_{-l+2p',k} &= M_{2p'-l,k} \supset M_{-2p'+l,k} = M_{l-2p',k} \\ &= M_{l,2p''-k} \supset M_{-l,2p''-k} = M_{l+2p',k}. \end{aligned}$$

These submodules of $M_{-l+2p',k}$ and $M_{-l,k}$ coincide, because if they do not coincide it implies more zeros of the Kac determinant. Therefore $M_{-l+2p',k}$ and $M_{-l,k}$ share two invariant submodules and we have

$$M_{-l,k} \cup M_{-l+2p',k} = (M_{-l,k} \oplus M_{-l+2p',k})/(M_{l-2p',k} \cup M_{l+2p',k}).$$

General embedding pattern is illustrated in the following figure

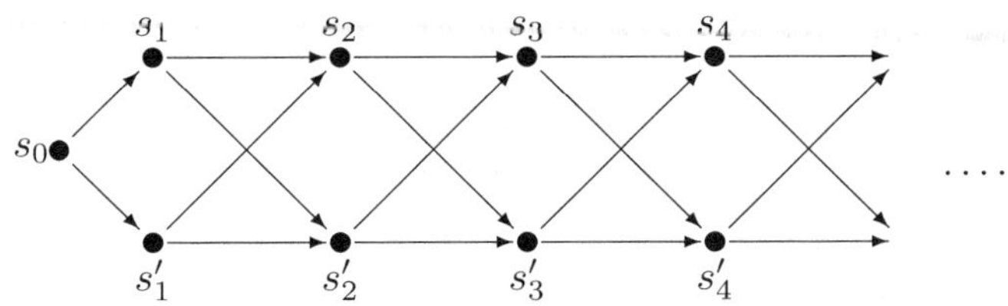

(2.27)

Conformal weights of singular vectors are given by

$$s_0: \quad L_0 = h_{l,k} = A(0) + \frac{c-1}{24}, \tag{2.28}$$

$$s'_{2m-1}: \quad L_0 = h_{-l+2mp',k} = B(-m) + \frac{c-1}{24} \quad (m \geq 1), \tag{2.29}$$

$$s_{2m+1}: \quad L_0 = h_{-l-2mp',k} = B(m) + \frac{c-1}{24} \quad (m \geq 0), \tag{2.30}$$

$$s'_{2m}: \quad L_0 = h_{l-2mp',k} = A(-m) + \frac{c-1}{24} \quad (m \geq 1), \tag{2.31}$$

$$s_{2m}: \quad L_0 = h_{l+2mp',k} = A(m) + \frac{c-1}{24} \quad (m \geq 1), \tag{2.32}$$

where $A(m)$ and $B(m)$ are given by

$$A(m) = \frac{(p''l - p'k + 2mp'p'')^2}{4p'p''}, \quad B(m) = \frac{(p''l + p'k + 2mp'p'')^2}{4p'p''}. \tag{2.33}$$

Irreducible module $\mathcal{L}_{l,k}$ is

$$\mathcal{L}_{l,k} = M_{l,k} - (M_{-l+2p',k} \oplus M_{-l,k}) + (M_{l-2p',k} \oplus M_{l+2p',k}) - \cdots. \tag{2.34}$$

This information is encoded into the character.

The character of a representation is a tool that counts the number of states at each level,

$$ch = \operatorname{tr} q^{L_0 - c/24} = \sum_{N=0}^{\infty} \dim\{\text{level } N \text{ states}\} \cdot q^{h+N-c/24}, \tag{2.35}$$

where $c/24$ is included for the modular transformation property. For the Verma module M with h we have

$$\mathrm{tr}_M\, q^{L_0-c/24} = q^{h-c/24}\prod_{n=1}^{\infty}\frac{1}{1-q^n}. \tag{2.36}$$

From this and (2.34), the character of the irreducible Virasoro module $\mathcal{L}_{l,k}$ is [28]

$$\chi_{l,k}(q) = \mathrm{tr}_{\mathcal{L}_{l,k}}\, q^{L_0-c/24} = \frac{1}{\eta(\tau)}\sum_{m\in\mathbb{Z}}(q^{A(m)}-q^{B(m)}), \tag{2.37}$$

where the Dedekind eta function $\eta(\tau)$ is

$$\eta(\tau) = q^{1/24}\prod_{n=1}^{\infty}(1-q^n), \quad q = e^{2\pi i\tau}. \tag{2.38}$$

The representation is called unitary if the inner product is positive definite. The unitary minimal series is [29]

$$\beta = \frac{m}{m-1} \text{ or } \frac{m-1}{m}, \quad c = 1-\frac{6}{m(m-1)} \quad (m=3,4,\dots). \tag{2.39}$$

This is shown by drawing the vanishing line of the Kac determinant in (c,h) plane. Use

$$\beta = \frac{1}{12}\left(13-c\pm\sqrt{(1-c)(25-c)}\right) \gtrless 1, \tag{2.40}$$

$$h_{l,k} = \frac{1-c}{24}(lk-1)+\frac{13-c}{48}\pm\frac{l^2-k^2}{48}\sqrt{(1-c)(25-c)}. \tag{2.41}$$

2.2.3 Primary Fields and Correlation Functions

Primary fields are fundamental fields in CFT. Under the conformal transformation, a primary field with the conformal weight h, $\phi_h(z)$, transforms as a rank h form ($\phi_h(z)\,dz^h$). We have the OPE,

$$L(z)\phi_h(w) = \frac{h\phi_h(w)}{(z-w)^2}+\frac{\partial\phi_h(w)}{z-w}+\text{reg.}, \tag{2.42}$$

or in mode

$$[L_n,\phi_{h,m}] = \big((h-1)n-m\big)\phi_{h,n+m}, \tag{2.43}$$

where $\phi_h(z) = \sum_{n\in\mathbb{Z}-h}\phi_{h,n}z^{-n-h}$. ($\phi_h(z)$ is called a quasiprimary field if (2.43) holds for $n=1,0,-1$.) We remark that the zero mode of a $h=1$ field commutes with the Virasoro algebra, $[L_n,\phi_{1,0}]=0$.

States are created by fields from the vacuum, $|\psi\rangle = \lim_{z\to 0}\psi(z)|\mathbf{0}\rangle$, and they are one-to-one correspondence. The highest weight state corresponds to the primary field and descendant states correspond to secondary fields. For example,

$$|h\rangle \longleftrightarrow \phi_h(z),$$
$$L_{-1}|h\rangle \longleftrightarrow \partial\phi_h(z),$$
$$L_{-3}L_{-2}|h\rangle \longleftrightarrow \hat{L}_{-3}\hat{L}_{-2}\phi_h(z),$$

where $\hat{L}_n$ is

$$\hat{L}_{-n}\psi(z) = \oint_z \frac{dy}{2\pi i}(y-z)^{1-n}L(y)\psi(z). \tag{2.44}$$

In quantum field theory physical information is obtained from the Green functions (correlation functions). In CFT conformal symmetry imposes that correlation functions obey the conformal Ward identity [1], which enables us to reduce calculation of correlation function of secondary fields to that of primary fields. Moreover if null states exist, the correlation functions satisfy differential equations [1]. We illustrate this by taking an example. For $h = h_{1,2}$ or $h_{2,1}$, from (2.15), we have a null field

$$\chi(z) = \left(\hat{L}_{-2} - \frac{3}{2(2h+1)}\hat{L}_{-1}^2\right)\phi_h(z). \tag{2.45}$$

Since a null field decouples from all the fields, correlation functions including it vanish,

$$\begin{aligned} 0 &= \langle\chi(z)X\rangle \\ &= \left(\mathcal{L}_{-2}(z) - \frac{3}{2(2h+1)}\mathcal{L}_{-1}(z)^2\right)\langle\phi_h(z)X\rangle \\ &= \left(\sum_{i=1}^{N}\left(\frac{h_i}{(z-z_i)^2} + \frac{1}{z-z_i}\partial_{z_i}\right) - \frac{3}{2(2h+1)}\partial_z^2\right)\langle\phi_h(z)X\rangle, \end{aligned} \tag{2.46}$$

where $X = \phi_{h_1}(z)\cdots\phi_{h_N}(z)$. Here we have used

$$\begin{aligned} \langle\hat{L}_{-n}\phi_h(z)X\rangle &= \oint_z \frac{dy}{2\pi i}(y-z)^{1-n}\langle L(y)\phi_h(z)X\rangle \quad (n \geq 1) \\ &= -\sum_{i=1}^{N}\oint_{z_i}\frac{dy}{2\pi i}(y-z)^{1-n}\left(\frac{h_i}{(y-z_i)^2}\langle\phi_h(z)X\rangle + \frac{1}{y-z_i}\partial_{z_i}\langle\phi_h(z)X\rangle\right) \\ &= \mathcal{L}_{-n}(z)\langle\phi_h(z)X\rangle, \end{aligned} \tag{2.47}$$

$$\mathcal{L}_{-n}(z) = \sum_{i=1}^{N}\left(\frac{(n-1)h_i}{(z_i-z)^n} - \frac{1}{(z_i-z)^{n-1}}\partial_{z_i}\right). \tag{2.48}$$

Due to translational invariance, we have $\mathcal{L}_{-1}(z) = \partial_z$. By solving this differential equation the correlation function can be calculated. Another method for calculation of correlation functions is free field realization, see [2, 30].

2.3 Free Field Realization

In the previous subsection the Virasoro algebra is treated in abstract way. In this subsection we treat it more explicitly by using our familiar free boson.

2.3.1 Free Field Realization

Let us introduce the free boson oscillators a_n ($n \in \mathbb{Z}_{\neq 0}$) and the zero modes a_0, Q,

$$[a_n, a_m] = 2n\delta_{n+m}, \quad [a_n, Q] = 2\delta_{n,0}. \tag{2.49}$$

Here 2 is the Cartan matrix of A_1 Lie algebra. The Fock space with momentum α, $\mathcal{F}_\alpha$, is defined by

$$\mathcal{F}_\alpha = \bigoplus_{l\geq 0} \bigoplus_{n_1\geq\cdots\geq n_l>0} \mathbb{C}a_{-n_1}\cdots a_{-n_l}|\alpha\rangle_B, \tag{2.50}$$

where $|\alpha\rangle_B$ is characterized by

$$a_n|\alpha\rangle_B = 0 \ (n>0), \quad a_0|\alpha\rangle_B = \alpha|\alpha\rangle_B. \tag{2.51}$$

$|\alpha\rangle_B$ is obtained from $|0\rangle_B$ ($a_n|0\rangle_B = 0$ for $n \geq 0$),

$$|\alpha\rangle_B = e^{\alpha Q/2}|0\rangle_B. \tag{2.52}$$

A boson field $\phi(z)$ is

$$\phi(z) = Q + a_0 \log z - \sum_{n\neq 0} \frac{1}{n} a_n z^{-n}, \quad \partial\phi(z) = \sum_{n\in\mathbb{Z}} a_n z^{-n-1}. \tag{2.53}$$

(the $\phi(z)$ here and the usual one used in string theory are related as $\phi(z) = i\sqrt{2}\phi^{\text{string}}(z)$. Q here is anti-Hermitian.) Normal ordering prescription $: \, :$ is defined by

$$\begin{cases} \text{move } a_n \ (n>0) \text{ and } a_0 \text{ to right,} \\ \text{move } a_n \ (n<0) \text{ and } Q \text{ to left.} \end{cases} \tag{2.54}$$

The Virasoro algebra is realized by a free boson in the following way,

$$L(z) = \frac{1}{4}{:}\partial\phi(z)\partial\phi(z){:} + \frac{1}{2}\alpha_0\partial^2\phi(z), \tag{2.55}$$

where the background charge α_0 is given in (2.19). In mode it is

$$L_n = \frac{1}{4}{:}\sum_m a_{n-m}a_m{:} - \frac{1}{2}\alpha_0(n+1)a_n, \tag{2.56}$$

or more explicitly,

$$L_n = \frac{1}{4} \sum_{m \neq 0, n} a_{n-m} a_m + \frac{1}{2} a_n a_0 - \frac{1}{2} \alpha_0 (n+1) a_n \quad (n \neq 0), \tag{2.57}$$

$$L_0 = \frac{1}{2} \sum_{m>0} a_{-m} a_m + \frac{1}{4} \big((a_0 - \alpha_0)^2 - \alpha_0^2 \big). \tag{2.58}$$

This Virasoro current realizes the Virasoro algebra with the central charge (2.17). $|\alpha\rangle_B$ is the highest weight state of the Virasoro algebra with the conformal weight $h = h(\alpha)$,

$$|\alpha\rangle_B = |h(\alpha)\rangle, \tag{2.59}$$

$$h(\alpha) = \tfrac{1}{4} \big((\alpha - \alpha_0)^2 - \alpha_0^2 \big), \tag{2.60}$$

and the corresponding primary field is

$$V_\alpha(z) = :e^{\frac{1}{2}\alpha\phi(z)}:. \tag{2.61}$$

Due to the background charge α_0, $\partial\phi(z)$ is not a primary field,

$$[L_n, a_m] = -m a_{n+m} - \alpha_0 n(n+1) \delta_{n+m,0}. \tag{2.62}$$

The dual Fock space $\mathcal{F}_\alpha^*$ also has the Virasoro module structure by the pairing $\langle\ ,\ \rangle\colon \mathcal{F}_\alpha^* \times \mathcal{F}_\alpha \to \mathbb{C}$,

$$\langle {}^t\! A f, v \rangle = \langle f, A v \rangle, \quad f \in \mathcal{F}_\alpha^*,\ v \in \mathcal{F}_\alpha,\ A\text{: operator}. \tag{2.63}$$

Here ${}^t L_n$ and ${}^t a_n$ are given by

$${}^t L_n = L_{-n}, \tag{2.64}$$

$${}^t a_n = -a_{-n} \qquad (n \neq 0), \tag{2.65}$$

$${}^t a_0 = -a_0 + 2\alpha_0, \tag{2.66}$$

and

$$\langle |\alpha\rangle_B^*, |\alpha\rangle_B \rangle = 1. \tag{2.67}$$

We write the following pairings as in the last lines,

$$\begin{aligned} &\langle {}^t L_{m_1} \cdots {}^t L_{m_k} |\alpha\rangle_B^*, L_{-n_1} \cdots L_{-n_l} |\alpha\rangle_B \rangle \\ &\qquad = \langle |\alpha\rangle_B, L_{m_k} \cdots L_{m_1} L_{-n_1} \cdots L_{-n_l} |\alpha\rangle_B \rangle \\ &\qquad =: {}_B^*\langle \alpha | L_{m_k} \cdots L_{m_1} L_{-n_1} \cdots L_{-n_l} |\alpha\rangle_B, \end{aligned} \tag{2.68}$$

$$\begin{aligned} &\langle {}^t a_{m_1} \ldots {}^t a_{m_k} |\alpha\rangle_B^*, a_{-n_1} \ldots a_{-n_l} |\alpha\rangle_B \rangle \\ &\qquad = \langle |\alpha\rangle_B, a_{m_k} \cdots a_{m_1} a_{-n_1} \ldots a_{-n_l} |\alpha\rangle_B \rangle \\ &\qquad =: {}_B^*\langle \alpha | a_{m_k} \cdots a_{m_1} a_{-n_1} \cdots a_{-n_l} |\alpha\rangle_B. \end{aligned} \tag{2.69}$$

By (2.65) and (2.66), $\mathcal{F}_\alpha^*$ is isomorphic to $\mathcal{F}_{2\alpha_0-\alpha}$ as a Virasoro module,

$$\mathcal{F}_\alpha^* \cong \mathcal{F}_{2\alpha_0-\alpha} \qquad \text{(Vir. module)} \tag{2.70}$$

$$|\alpha\rangle_B^* \leftrightarrow |2\alpha_0-\alpha\rangle_B. \tag{2.71}$$

Of course $h(\alpha)=h(2\alpha_0-\alpha)$ by (2.60).

For later use we define a_0' by,

$$a_0'=a_0-\alpha_0, \quad a_0'|\alpha\rangle_B=(\alpha-\alpha_0)|\alpha\rangle_B, \quad {}^t a_0'=-a_0'. \tag{2.72}$$

The Virasoro current (2.55) can be rewritten as

$$z^2L(z)=\tfrac{1}{4}{:}D\phi'(z)D\phi'(z){:}+\tfrac{1}{2}\alpha_0 D^2\phi'(z), \tag{2.73}$$

where $\phi'(z)$ and D are

$$\phi'(z)=\phi(z)|_{a_0\to a_0'}, \quad \partial\phi(z)=\partial\phi'(z)+\alpha_0 z^{-1}, \tag{2.74}$$

$$D=z\frac{\partial}{\partial z}. \tag{2.75}$$

2.3.2 Singular Vectors and Kac Determinant

In the case of $h=h_{l,k}$ there are singular vectors. In the free boson realization they can be expressed by using the screening currents. Screening currents $S_\pm(z)$ are primary fields with $h=1$,

$$S_\pm(z)=V_{2\alpha_\pm}={:}e^{\alpha_\pm\phi(z)}{:}, \quad \alpha_+=\sqrt{\beta}, \alpha_-=\frac{-1}{\sqrt{\beta}}. \tag{2.76}$$

Hence their zero modes (screening charge) commute with the Virasoro algebra,

$$\left[L_n, \oint_0 \frac{dz}{2\pi i}S_\pm(z)\right]=0. \tag{2.77}$$

We will use $S_+(z)$ in what follows. $S_-(z)$ can be treated similarly. We follow the method in [31].

The representation with $h=h_{l,k}$ is realized on $\mathcal{F}_{\alpha_{l,k}}$,

$$h_{l,k}=h(\alpha_{l,k}), \tag{2.78}$$

$$\alpha_{l,k}=\sqrt{\beta}(1-l)-\frac{1}{\sqrt{\beta}}(1-k). \tag{2.79}$$

In the following we write $\mathcal{F}_{\alpha_{l,k}}=\mathcal{F}_{l,k}$. To contact with our previous paper [32] ($\alpha_{r,s}$ there is $\propto \alpha_{-r,-s}$ here), we consider dual space $\mathcal{F}_{l,k}^*=\mathcal{F}_{-l,-k}$. In the Verma module with $h_{-l,-k}=h_{l,k}$, the singular vector at level lk is expressed as

$$
\begin{aligned}
|\chi_{-l,-k}\rangle &= \int \prod_{j=1}^{l} \frac{dz_j}{2\pi i} \cdot S_+(z_1)\dots S_+(z_l)|\alpha_{l,-k}\rangle_B \\
&= \int \prod_{j=1}^{l} \underline{dz}_j \cdot \prod_{1\le i<j\le l} (z_i - z_j)^{2\beta} \cdot \prod_{i=1}^{l} z_i^{\beta(1-l)-k} \\
&\qquad \cdot \prod_{j=1}^{l} \exp\Big(\sqrt{\beta}\sum_{n>0} \frac{1}{n} a_{-n} z_j^n\Big) \cdot |\alpha_{-l,-k}\rangle_B. \quad (2.80)
\end{aligned}
$$

Here $\underline{dz}_j$ is given in (A.14) and the integration contour is summarized in Appendix A.4 (for example take $I_{\rm BMP}$). Since this integral of S_+'s, which is a map from $\mathcal{F}_{l,-k}$ to $\mathcal{F}_{-l,-k}$, commutes with L_n, $|\chi_{-l,-k}\rangle$ is annihilated by L_n $(n>0)$. This state $|\chi_{-l,-k}\rangle$ is non zero because of (A.48).

To prove the Kac determinant formula, we first study the relation between states in the Verma module and Fock space. Let us introduce the following notation,

$$
\begin{aligned}
&I = \{n_1,\dots,n_l\},\ n_1 \ge \cdots \ge n_l > 0, && \ell(I) = l, && (2.81)\\
&L_I = L_{n_l}\cdots L_{n_1}, && L_{-I} = L_{-n_1}\cdots L_{-n_l}, && (2.82)\\
&a_I = a_{n_l}\cdots a_{n_1}, && a_{-I} = a_{-n_1}\cdots a_{-n_l}, && (2.83)
\end{aligned}
$$

and I is ordered by the reverse lexicographic ordering. States at level N in the Verma module are linear combinations of those of the Fock space,

$$
L_{-I}|\alpha\rangle_B = \sum_J C(N,\alpha)_{I,J} a_{-J}|\alpha\rangle_B. \quad (2.84)
$$

For example,

$$
\begin{aligned}
&L_{-1}|\alpha\rangle_B = \tfrac{1}{2}\alpha a_{-1}|\alpha\rangle_B,\\
&\begin{pmatrix} L_{-2}|\alpha\rangle_B \\ cL_{-1}^2|\alpha\rangle_B \end{pmatrix} = \begin{pmatrix} \tfrac12(\alpha+\alpha_0) & \tfrac14 a_{-1}^2 \\ \tfrac12\alpha & \tfrac14\alpha^2 \end{pmatrix} \begin{pmatrix} a_{-2}|\alpha\rangle_B \\ a_{-1}^2|\alpha\rangle_B \end{pmatrix},\\
&\begin{pmatrix} L_{-3}|\alpha\rangle_B \\ cL_{-2}L_{-1}|\alpha\rangle_B \\ cL_{-1}^3|\alpha\rangle_B \end{pmatrix} = \begin{pmatrix} \tfrac12\alpha+\alpha_0 & \tfrac12 & 0 \\ \tfrac12\alpha & \tfrac14\alpha(\alpha+\alpha_0) & \tfrac18\alpha \\ \alpha & \tfrac34\alpha^2 & \tfrac18\alpha^3 \end{pmatrix} \begin{pmatrix} a_{-3}|\alpha\rangle_B \\ a_{-2}a_{-1}|\alpha\rangle_B \\ a_{-1}^3|\alpha\rangle_B \end{pmatrix}
\end{aligned}
$$

The determinant of matrix $C(N,\alpha)$ is given by

$$
\det C(N,\alpha)_{I,J} = \prod_{\substack{l,k\ge1 \\ lk\le N}} \Big(\frac12(\alpha-\alpha_{l,k})\Big)^{p(N-lk)}. \quad (2.85)
$$

Proof. Since $L_{-n}=\frac{1}{2}a_{-n}a_0+\cdots$, the leading term of α in $\det C(N,\alpha)$ is

$$\prod_{\substack{\{k_i\}\\ \sum_i ik_i=N}}\prod_i\left(\frac{\alpha}{2}\right)^{k_i}=\prod_{\substack{l,k\ge 1\\ lk\le N}}\left(\frac{\alpha}{2}\right)^{p(N-lk)}.$$

Here we have used (A.30). In the last paragraph we constructed a number of singular vectors equal to this order of α. □

Next we consider the dual $\mathcal{F}_\alpha^*$. With the notation,

$$({}^tL)_I={}^t(L_I)={}^tL_{n_1}\cdots{}^tL_{n_l},\tag{2.86}$$

$$({}^ta)_I={}^t(a_I)={}^ta_{n_1}\cdots{}^ta_{n_l},\tag{2.87}$$

we define a matrix $C'(N,\alpha)$,

$$({}^tL)_I|\alpha\rangle_B^*=\sum_J C'(N,\alpha)_{I,J}({}^ta)_J|\alpha\rangle_B^*.\tag{2.88}$$

By (2.64) and (2.65), the left and right-hand sides of this equation are

$$\text{LHS}=L_{-I}|\alpha\rangle_B^*=\sum_K C(N,2\alpha_0-\alpha)_{I,K}a_{-K}|\alpha\rangle_B^*,\tag{2.89}$$

$$\text{RHS}=\sum_{J,K}C'(N,\alpha)_{I,J}D_{J,K}a_{-K}|\alpha\rangle_B^*,\tag{2.90}$$

where the matrix D is

$$D_{J,K}=\delta_{J,K}(-1)^{\ell(J)}.\tag{2.91}$$

Hence C' can be expressed in terms of C,

$$C'(N,\alpha)_{I,J}=\sum_K C(N,2\alpha_0-\alpha)_{I,K}D_{K,J},\tag{2.92}$$

and its determinant is

$$\begin{aligned}\det C'(N,\alpha)_{I,J}&=\det C(N,2\alpha_0-\alpha)\cdot\det D\\ &=\prod_{\substack{l,k\ge 1\\ lk\le N}}\left(\frac{1}{2}(\alpha-\alpha_{-l,-k})\right)^{p(N-lk)}.\end{aligned}\tag{2.93}$$

By gathering these results, the inner product of two states in the Verma module becomes

$$\begin{aligned}\langle h|L_IL_{-J}|h\rangle&=\langle({}^tL)_I|\alpha\rangle_B^*,L_{-J}|\alpha\rangle_B\rangle\\ &=\left\langle\sum_K C'(N,\alpha)_{I,K}({}^ta)_K|\alpha\rangle_B^*,\sum_L C(N,\alpha)_{J,L}a_{-L}|\alpha\rangle_B\right\rangle\\ &=\sum_{K,L}C'(N,\alpha)_{I,K}G_{K,L}C(N,\alpha)_{J,L}.\end{aligned}\tag{2.94}$$

Here $G_{K,L}$ is

$$\begin{aligned} G_{K,L} &= \big\langle ({}^t a)_K|\alpha\rangle_B^*, a_{-L}|\alpha\rangle_B\big\rangle \\ &= {}_B^*\langle\alpha|a_{m_k}\cdots a_{m_1}a_{-n_1}\ldots a_{-n_l}|\alpha\rangle_B \\ &= \delta_{K,L}{}_B^*\langle\alpha|\cdots a_2^{k_2}a_1^{k_1}a_{-1}^{k_1}a_{-2}^{k_2}\cdots|\alpha\rangle_B \\ &= \delta_{K,L}\prod_i (2i)^{k_i}k_i!, \end{aligned} \tag{2.95}$$

where $(m_1,\ldots,m_k) = 1^{k_1}2^{k_2}\cdots$. By using (A.28) and (A.30), its determinant becomes

$$\det G_{K,L} = \prod_{\substack{\{k_i\}\\ \sum_i ik_i=N}} \prod_i (2i)^{k_i}k_i! = \prod_{\substack{l,k\geq 1\\ lk\leq N}} (2lk)^{p(N-lk)}. \tag{2.96}$$

Therefore we obtain the Kac determinant (2.16),

$$\begin{aligned} \det\langle h|L_I L_{-J}|h\rangle &= \det C'(N,\alpha)\cdot\det G\cdot\det{}^tC(N,\alpha) \\ &= \prod_{\substack{l,k\geq 1\\ lk\leq N}} \big(2lk(h-h_{l,k})\big)^{p(N-lk)}. \end{aligned} \tag{2.97}$$

Here we have used $(\alpha-\alpha_{l,k})(\alpha-\alpha_{-l,-k}) = 4\big(h-h(\alpha_{l,k})\big)$.

2.3.3 Felder Complex

Irreducible Virasoro module $\mathcal{L}_{l,k}$ is realized on the Fock space $\mathcal{F}_{l,k}$, but this $\mathcal{F}_{l,k}$ is bigger than $\mathcal{L}_{l,k}$. So we have to discuss how $\mathcal{L}_{l,k}$ is obtained from $\mathcal{F}_{l,k}$. The Virasoro structure of $\mathcal{F}_{l,k}$ was investigated by Feigin–Fuchs [33] and Felder [34]. Here we review it following [34]. We consider minimal series eqs. (2.20)–(2.23).

The Becchi–Rouet–Stora–Tyupin (BRST) charge Q_m is defined by using the screening current $S_+(z)$,

$$\begin{aligned} Q_m &= \int\prod_{j=1}^m \frac{dz_j}{2\pi i}\cdot S_+(z_1)\ldots S_+(z_m), \\ &= \int\prod_{j=1}^m \frac{dz_j}{2\pi i}\cdot\prod_{\substack{i,j=1\\ i<j}}^m (z_i-z_j)^{2\beta}\cdot{:}S_+(z_1)\ldots S_+(z_m){:}, \end{aligned} \tag{2.98}$$

where the integration contour is taken as I_{BMP} in Appendix A.4. On $\mathcal{F}_{l',k'}$ conditions (A.32) and (A.33) become

$$\text{(i)}\quad m\equiv 0 \pmod{p'}, \qquad\qquad \text{(ii)}\quad m\equiv l' \pmod{p'}. \tag{2.99}$$

For example Q_l is well defined on $\mathcal{F}_{l',k}$ ($l' \equiv l \pmod{p'}$), and the contour can be deformed to I_F type by (A.47). An important property of the BRST charge is commutativity with the Virasoro algebra,

$$Q_l \colon \mathcal{F}_{l',k} \to \mathcal{F}_{l'-2l,k}, \qquad l' \equiv l \pmod{p'}, \qquad [L_n, Q_l] = 0, \tag{2.100}$$
$$Q_{p'-l} \colon \mathcal{F}_{l',k} \to \mathcal{F}_{l'-2(p'-l),k}, \quad l' \equiv -l \pmod{p'}, \quad [L_n, Q_{p'-l}] = 0, \tag{2.101}$$

and the nilpotency,

$$Q_l Q_{p'-l} = Q_{p'-l} Q_l = Q_{p'} = 0. \tag{2.102}$$

Here $Q_l Q_{p'-l} = Q_{p'}$ and $Q_{p'-l} Q_l = Q_{p'}$ are shown by I_{BMP} type contour, and $Q_{p'} = 0$ is shown by (A.47). The singular vector (2.80) can be written as $|\chi_{-l,-k}\rangle = Q_l |\alpha_{l,-k}\rangle_B$.

Let us consider the Felder complex $C_{l,k}$,

$$\cdots \xrightarrow{X_{-3}} C_{-2} \xrightarrow{X_{-2}} C_{-1} \xrightarrow{X_{-1}} C_0 \xrightarrow{X_0} C_1 \xrightarrow{X_1} C_2 \xrightarrow{X_2} \cdots, \tag{2.103}$$

where C_j and $X_j \colon C_j \to C_{j+1}$ ($j \in \mathbb{Z}$) are

$$C_{2j} = \mathcal{F}_{l-2p'j,k}, \qquad C_{2j+1} = \mathcal{F}_{-l-2p'j,k}, \tag{2.104}$$
$$X_{2j} = Q_l, \qquad X_{2j+1} = Q_{p'-l}. \tag{2.105}$$

X satisfies the BRST property,

$$X_j X_{j-1} = 0. \tag{2.106}$$

Structure of the Felder complex is illustrated in the following figure:

$$\cdots \longrightarrow C_{-2} \xrightarrow{X_{-2}} C_{-1} \xrightarrow{X_{-1}} C_0 \xrightarrow{X_0} C_1 \xrightarrow{X_1} C_2 \xrightarrow{X_2} \cdots$$

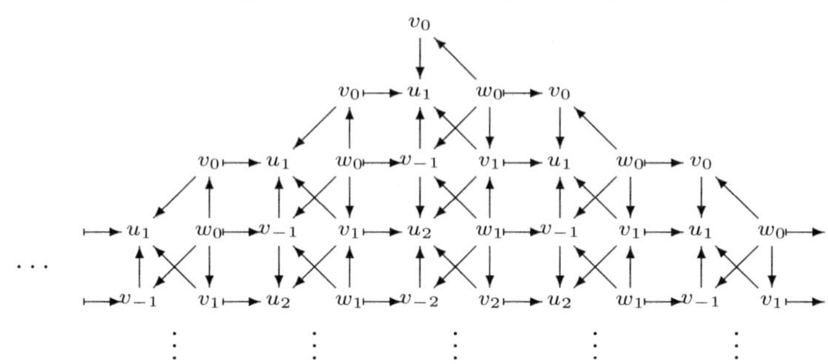

(2.107)

Each Fock space has the structure

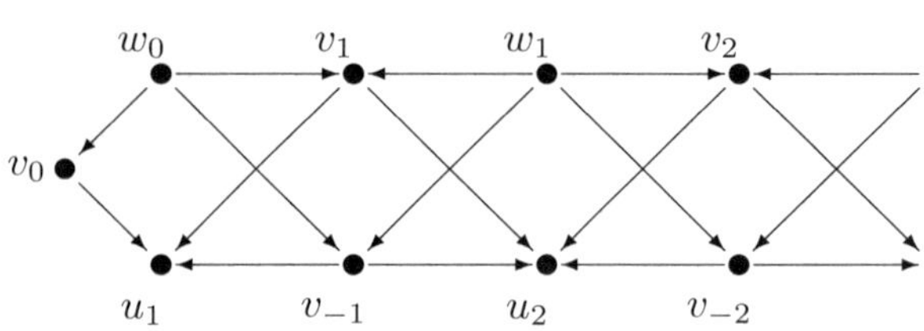

$\cdots,$ (2.108)

which is obtained by combining the information from $C(N,\alpha)$

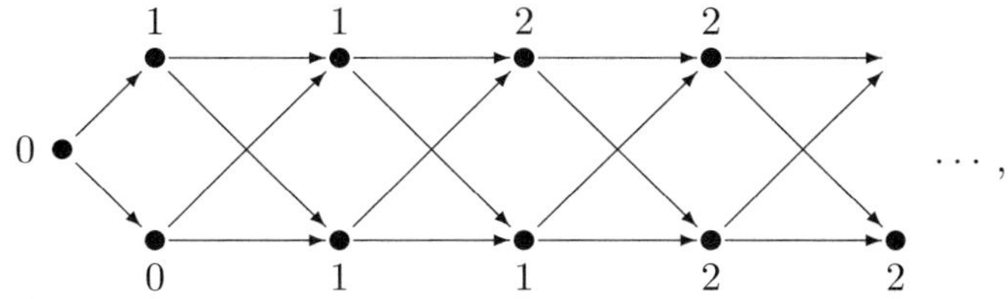

and the information from $C'(N,\alpha)$

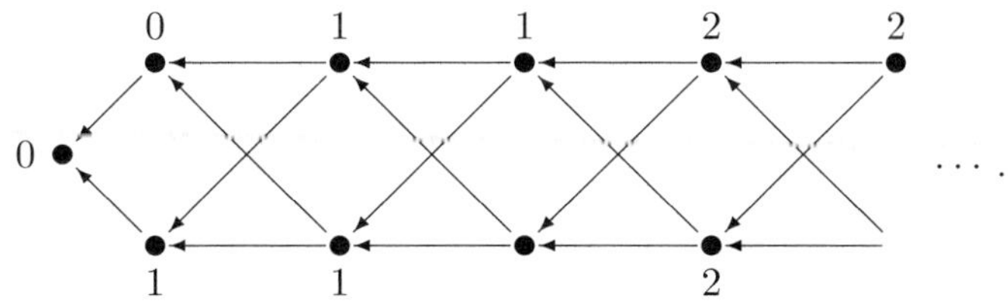

v_0, u_m, w_{m-1}, v_{-m}, v_m $(m \geq 1)$ have the same conformal weight as s_0, s'_{2m-1}, s_{2m-1}, s'_{2m}, s_{2m} in (2.27) respectively. s_{2m-1} is u_m but s'_{2m-1} vanishes on the Fock space. At the level of s'_{2m-1}, there is a state which does not belong to the Verma module. That state is w_{m-1}.

We write states in C_j with superfix (j). Conformal weights of v, u, w are the following:

$$v_0^{(2j)}: \quad L_0 = A(-j) + \frac{c-1}{24}, \tag{2.109}$$

$$v_0^{(2j+1)}: \quad L_0 = B(j) + \frac{c-1}{24}, \tag{2.110}$$

$$u_m^{(2j)}: \quad L_0 = B(-|j|-m) + \frac{c-1}{24} \qquad (m \geq 1), \tag{2.111}$$

$$u_m^{(2j+1)}: \quad L_0 = A\left(\frac{|2j+1|-1}{2} + m\right) + \frac{c-1}{24} \qquad (m \geq 1), \tag{2.112}$$

$$v_{-m}^{(2j)}: \quad L_0 = A(-|j|-m) + \frac{c-1}{24} \qquad (m \geq 1), \tag{2.113}$$

$$v_{-m}^{(2j+1)}: \quad L_0 = B\left(\frac{|2j+1|-1}{2} + m\right) + \frac{c-1}{24} \qquad (m \geq 1), \tag{2.114}$$

$$v_m^{(2j)}: \quad L_0 = A(|j|+m) + \frac{c-1}{24} \qquad (m \geq 1), \tag{2.115}$$

$$v_m^{(2j+1)}: \quad L_0 = B\left(-\frac{|2j+1|+1}{2} - m\right) + \frac{c-1}{24} \qquad (m \geq 1), \tag{2.116}$$

$$w_m^{(2j)}: \quad L_0 = B(|j|+m) + \frac{c-1}{24} \qquad (m \geq 0), \tag{2.117}$$

$$w_m^{(2j+1)}: \quad L_0 = A\left(-\frac{|2j+1|+1}{2} - m\right) + \frac{c-1}{24} \qquad (m \geq 0), \tag{2.118}$$

where A and B are given in (2.33). BRST charge X_j maps u, v, w in the following way:

$$
\begin{aligned}
j<0: \quad & X_j w_m^{(j)} = v_{-m-1}^{(j+1)} \ (m \geq 0), && X_j v_m^{(j)} = u_{m+1}^{(j+1)} \ (m \geq 0), \\
& X_j u_m^{(j)} = 0 \qquad\quad (m \geq 1), && X_j v_{-m}^{(j)} = 0 \qquad\quad (m \geq 1),
\end{aligned}
\tag{2.119}
$$

$$
\begin{aligned}
j \geq 0: \quad & X_j w_m^{(j)} = v_{-m}^{(j+1)} \quad (m \geq 0), && X_j v_m^{(j)} = u_m^{(j+1)} \ (m \geq 1), \\
& X_j u_m^{(j)} = 0 \qquad\quad (m \geq 1), && X_j v_{-m}^{(j)} = 0 \qquad\quad (m \geq 0).
\end{aligned}
\tag{2.120}
$$

The cohomology groups of the complex $C_{l,k}$ are [34]

$$
H^j(C_{l,k}) = \operatorname{Ker} X_j / \operatorname{Im} X_{j-1} = \begin{cases} 0 & j \neq 0, \\ \mathcal{L}_{l,k} & j = 0. \end{cases} \tag{2.121}
$$

Using this fact, the trace of operator $\mathcal{O}$ over the irreducible Virasoro module can be converted to the alternated sum of those over the Fock spaces. If we can draw the commutative diagram

$$
\begin{array}{ccccccccc}
 & \longrightarrow & C_{-1} & \xrightarrow{X_{-1}} & C_0 & \xrightarrow{X_0} & C_1 & \xrightarrow{X_1} & \\
\cdots & & \big\downarrow{\scriptstyle \mathcal{O}^{(-1)}} & & \big\downarrow{\scriptstyle \mathcal{O}^{(0)}} & & \big\downarrow{\scriptstyle \mathcal{O}^{(1)}} & & \cdots, \\
 & \longrightarrow & C_{-1} & \xrightarrow{X_{-1}} & C_0 & \xrightarrow{X_0} & C_1 & \xrightarrow{X_1} &
\end{array}
\tag{2.122}
$$

where $\mathcal{O}^{(j)}$ is an operator $\mathcal{O}$ realized on C_j, then by Euler–Poincaré principle, we have

$$
\operatorname{tr}_{\mathcal{L}_{l,k}} \mathcal{O} = \operatorname{tr}_{H^0(C_{l,k})} \mathcal{O}^{(0)} = \operatorname{tr}_{H^*(C_{l,k})} \mathcal{O} = \sum_{j \in \mathbb{Z}} (-1)^j \operatorname{tr}_{C_j} \mathcal{O}^{(j)}. \tag{2.123}
$$

For example let us calculate the character. Since the trace over the Fock space is

$$
\begin{aligned}
\operatorname{tr}_{C_{2m}} q^{L_0 - c/24} &= q^{-1/24 + A(-m)} \prod_{n>0} \frac{1}{1-q^n}, \\
\operatorname{tr}_{C_{2m+1}} q^{L_0 - c/24} &= q^{-1/24 + B(m)} \prod_{n>0} \frac{1}{1-q^n},
\end{aligned}
$$

we obtain the Virasoro character (2.37),

$$
\begin{aligned}
\operatorname{tr}_{\mathcal{L}_{l,k}} q^{L_0 - c/24} &= \sum_{j \in \mathbb{Z}} (-1)^j \operatorname{tr}_{C_j} q^{L_0 - c/24} \\
&= \frac{1}{\eta(\tau)} \sum_{m \in \mathbb{Z}} \left(q^{A(-m)} - q^{B(m)} \right).
\end{aligned}
\tag{2.124}
$$

If we use $S_-(z)$ instead of $S_+(z)$, another complex in which k' of $\mathcal{F}_{l,k'}$ changes is obtained.

Example: $c = \frac{1}{2}$ (Ising model)

Parameters and conformal weights are

$$c = \frac{1}{2},\ p' = 3,\ p'' = 4,\ \beta = \frac{4}{3},\ \alpha_0 = \frac{1}{2\sqrt{3}},\ \alpha_+ = \frac{2}{\sqrt{3}},\ \alpha_- = -\frac{\sqrt{3}}{2},$$
$$h_{1,1} = 0,\ h_{2,1} = \frac{1}{2},\ h_{2,2} = \frac{1}{16},\ \alpha_{1,1} = 0,\ \alpha_{2,1} = -\frac{2}{\sqrt{3}},\ \alpha_{2,2} = -\frac{1}{2\sqrt{3}}.$$

$h = 0, \frac{1}{2}, 1/16$ correspond to the identity operator, energy operator, spin operator respectively. Remark that

$$h_{2,3} = 0,\ h_{1,3} = \frac{1}{2},\ h_{1,2} = \frac{1}{16},\ \alpha_{2,3} = \frac{1}{\sqrt{3}},\ \alpha_{1,3} = \sqrt{3},\ \alpha_{1,2} = \frac{\sqrt{3}}{2}.$$

Basic two singular vectors are

$h = 0:$

$$|\chi\rangle = L_{-1}|0\rangle, \tag{2.125}$$
$$|\chi'\rangle = \left(L_{-6} + \frac{22}{9}L_{-4}L_{-2} - \frac{31}{36}L_{-3}^2 - \frac{16}{27}L_{-2}^3\right)|0\rangle$$
$$+ (b_1 L_{-5} + b_2 L_{-4}L_{-1} + b_3 L_{-3}L_{-2} + b_4 L_{-3}L_{-1}^2$$
$$+ b_5 L_{-2}^2 L_{-1} + b_6 L_{-2}L_{-1}^3 + b_7 L_{-1}^5)|\chi\rangle, \tag{2.126}$$

$h = \frac{1}{2}:$

$$|\chi\rangle = (L_{-2} - \tfrac{3}{4}L_{-1}^2)|\tfrac{1}{2}\rangle, \tag{2.127}$$
$$|\chi'\rangle = (L_{-3} - \tfrac{1}{3}L_{-1}^3)|\tfrac{1}{2}\rangle + aL_{-1}|\chi\rangle, \tag{2.128}$$

$h = 1/16:$

$$|\chi\rangle = \left(L_{-2} - \frac{4}{3}L_{-1}^2\right)\left|\frac{1}{16}\right\rangle, \tag{2.129}$$
$$|\chi'\rangle = \left(L_{-4} + \frac{40}{31}L_{-3}L_{-1} - \frac{256}{465}L_{-1}^4\right)\left|\frac{1}{16}\right\rangle + (bL_{-2} + b'L_{-1}^2)|\chi\rangle, \tag{2.130}$$

where

$$b_1 = -\frac{41}{225},\ b_2 = -\frac{78}{25},\ b_3 = \frac{14}{75},\ b_4 = \frac{34}{25},\ b_5 = \frac{172}{75},\ b_6 = -\frac{8}{5},\ b_7 = \frac{4}{25},$$
$$a = -\frac{4}{5},\quad b = \frac{147}{3100},\quad b' = -\frac{401}{775},$$

determined by the requirement that $|\chi'\rangle$ is orthogonal (with double zero) to descendants of $|\chi\rangle$.

In the free field realization these singular vectors become as follows. For $h = 0$,

$|0\rangle = |\alpha_{1,1}\rangle_B :$

$$|\chi\rangle = 0, \tag{2.131}$$

$$|\chi'\rangle=-\frac{1}{216\sqrt{3}}\Big(a_{-6}+6\sqrt{3}a_{-5}a_{-1}-32\sqrt{3}a_{-4}a_{-2}-51a_{-4}a^2_{-1}+\frac{51\sqrt{3}}{2}a^2_{-3}+75a_{-3}a_{-2}a_{-1}-18\sqrt{3}a_{-3}a^3_{-1}-\frac{97}{3}a^3_{-2}+\frac{31\sqrt{3}}{2}a^2_{-2}a^2_{-1}+6a_{-2}a^4_{-1}+2\sqrt{3}a^6_{-1}\Big)|\alpha_{1,1}\rangle_B, \tag{2.132}$$

$|0\rangle = |\alpha_{2,3}\rangle_B$:

$$|\chi\rangle = \frac{1}{2\sqrt{3}}a_{-1}|\alpha_{2,3}\rangle_B, \tag{2.133}$$

$$|\chi'\rangle = 0, \tag{2.134}$$

for $h = \frac{1}{2}$,

$|\frac{1}{2}\rangle = |\alpha_{2,1}\rangle_B$:

$$|\chi\rangle = 0, \tag{2.135}$$

$$|\chi'\rangle = \frac{1}{6\sqrt{3}}\Big(a_{-3}+\sqrt{3}a_{-2}a_{-1}+\frac{2}{3}a^3_{-1}\Big)|\alpha_{2,1}\rangle_B, \tag{2.136}$$

$|\frac{1}{2}\rangle = |\alpha_{1,3}\rangle_B$:

$$|\chi\rangle = \frac{5}{8\sqrt{3}}\Big(a_{-2}-\frac{\sqrt{3}}{2}a^2_{-1}\Big)|\alpha_{1,3}\rangle_B, \tag{2.137}$$

$$|\chi'\rangle = 0, \tag{2.138}$$

and for $h = 1/16$,

$|1/16\rangle = |\alpha_{2,2}\rangle_B$:

$$|\chi\rangle = \frac{1}{3\sqrt{3}}\Big(a_{-2}+\frac{2}{\sqrt{3}}a^2_{-1}\Big)|\alpha_{2,2}\rangle_B, \tag{2.139}$$

$$|\chi'\rangle = 0, \tag{2.140}$$

$|1/16\rangle = |\alpha_{1,2}\rangle_B$:

$$|\chi\rangle = 0, \tag{2.141}$$

$$|\chi'\rangle = -\frac{\sqrt{3}}{310}\Big(a_{-4}-8\sqrt{3}a_{-3}a_{-1}+\frac{37}{2\sqrt{3}}a^2_{-2}-2a_{-2}a^2_{-1}+2\sqrt{3}a^4_{-1}\Big)|\alpha_{1,3}\rangle_B. \tag{2.142}$$

On $\mathcal{F}_\alpha$, BRST charges are

$$Q_1 = \int_{C_{\rm KM}}\frac{dz_1}{2\pi i}S_+(z_1) = \oint_0 \frac{dz}{2\pi i}\exp\Big(\frac{2}{\sqrt{3}}\sum_{n>0}\frac{1}{n}a_{-n}z^n\Big) \times\exp\Big(-\frac{2}{\sqrt{3}}\sum_{n>0}\frac{1}{n}a_nz^{-n}\Big)e^{2Q/\sqrt{3}}z^{2a_0\sqrt{3}}, \tag{2.143}$$

$$\begin{aligned}
Q_2 &= \frac{1}{2}(e^{2\pi i\alpha_+^2}+1)(e^{2\pi i\alpha_+\alpha}-1)\int_{C_{\rm KM}}\frac{dz_1}{2\pi i}\frac{dz_2}{2\pi i}S_+(z_1)S_+(z_2)\\
&= \frac{1}{2}(e^{2\pi i/3}+1)(e^{2\pi i\cdot 2\alpha/\sqrt{3}}-1)\\
&\quad\times\frac{1}{2\pi i}\oint_0\frac{dz}{2\pi i z}\int_0^1 du\, z^{14/3}(1-u)^{8/3}e^{4Q/\sqrt{3}}z^{4a_0/\sqrt{3}}u^{2a_0/\sqrt{3}}\\
&\qquad\qquad\times\exp\Big(\frac{2}{\sqrt{3}}\sum_{n>0}\frac{1}{n}a_{-n}z^n(1+u^n)\Big)\\
&\qquad\qquad\times\exp\Big(-\frac{2}{\sqrt{3}}\sum_{n>0}\frac{1}{n}a_n z^{-n}(1+u^{-n})\Big),
\end{aligned}$$

where we have taken $I_{\rm KM}$ contour (see Appendix A.4) and changed integration variables $z_1=z$, $z_2=zu$. We give lower level examples. We perform $\oint_0 dz$ first, and next use $\int_0^1 du u^{a-1}(1-u)^{b-1}=B(a,b)$ for 'arbitrary' values a, b (analytic continuation). For $h=\frac{1}{2}$ and $|\frac{1}{2}\rangle=|\alpha_{2,1}\rangle_B$,

$$L_{-2}|\alpha_{2,1}\rangle_B=\frac{3}{4}L_{-1}^2|\alpha_{2,1}\rangle_B=-\frac{\sqrt{3}}{4}\Big(a_{-2}-\frac{1}{\sqrt{3}}a_{-1}^2\Big)|\alpha_{2,1}\rangle_B, \tag{2.144}$$

$$Q_1|\alpha_{4,1}\rangle_B=\frac{2}{3\sqrt{3}}\Big(a_{-3}+\sqrt{3}a_{-2}a_{-1}+\frac{2}{3}a_{-1}^3\Big)|\alpha_{2,1}\rangle_B, \tag{2.145}$$

$$Q_2(Aa_{-2}+Ba_{-1}^2)|\alpha_{2,1}\rangle_B=-\frac{120}{7\pi}\frac{\Gamma(\frac{2}{3})^2}{\Gamma(\frac{1}{3})}(A+\sqrt{3}B)|\alpha_{-2,1}\rangle_B, \tag{2.146}$$

$$\begin{aligned}
&Q_2(A'a_{-3}+B'a_{-2}a_{-1}+C'a_{-1}^3)|\alpha_{2,1}\rangle_B\\
&\qquad=-\frac{90\sqrt{3}}{7\pi}\frac{\Gamma(\frac{2}{3})^2}{\Gamma(\frac{1}{3})}\Big(A'+\frac{1}{\sqrt{3}}B'-3C'\Big)a_{-1}|\alpha_{-2,1}\rangle_B,
\end{aligned} \tag{2.147}$$

therefore we have

$$\begin{aligned}
Q_1|\alpha_{4,1}\rangle_B=4|\chi'\rangle,\quad Q_2L_{-2}|\alpha_{2,1}\rangle_B&=Q_2L_{-1}^2|\alpha_{2,1}\rangle_B=0,\\
Q_2Q_1|\alpha_{4,1}\rangle_B&=0.
\end{aligned} \tag{2.148}$$

For the complex $C_{2,1}$, we have

$$v_0^{(-1)}=|\alpha_{4,1}\rangle_B,\quad v_0^{(0)}=|\alpha_{2,1}\rangle_B,\quad v_0^{(1)}=|\alpha_{-2,1}\rangle_B, \tag{2.149}$$

$$u_1^{(0)}=Q_1v_0^{(-1)}=\frac{2}{3\sqrt{3}}\Big(a_{-3}+\sqrt{3}a_{-2}a_{-1}+\frac{2}{3}a_{-1}^3\Big)|\alpha_{2,1}\rangle_B, \tag{2.150}$$

$$Q_2w_0^{(0)}=v_0^{(1)},\quad w_0^{(0)}=-\frac{7\pi}{300}\frac{\Gamma(\frac{1}{3})}{\Gamma(\frac{2}{3})^2}\Big(a_{-2}+\frac{\sqrt{3}}{2}a_{-1}^2\Big)|\alpha_{2,1}\rangle_B. \tag{2.151}$$

Remark that the last equation is the dual of the following state,

$$Q_2|\alpha_{2,-1}\rangle_B=\frac{30}{7\pi}\frac{\Gamma(\frac{2}{3})^2}{\Gamma(\frac{1}{3})}\Big(a_{-2}-\frac{\sqrt{3}}{2}a_{-1}^2\Big)|\alpha_{-2,-1}\rangle_B. \tag{2.152}$$

2.3.4 Calogero–Sutherland Model and Jack Symmetric Polynomial

In this subsection we review the relation between the Virasoro singular vector and the Calogero–Sutherland model (CSM) [32, 35]. Calogero–Sutherland model is a many body quantum mechanical system on a circle with length L under the $1/r^2$ potential. Its Hamiltonian and momentum are

$$H_{\mathrm{CS}} = \sum_{j=1}^{N_0} \frac{1}{2m}\hat{p}_j^2 + \frac{\bar{\beta}(\bar{\beta}-\hbar)}{m}\left(\frac{\pi}{L}\right)^2 \sum_{\substack{i,j=1\\ i<j}}^{N_0} \frac{1}{\sin^2(\pi(q_i-q_j)/L)}, \tag{2.153}$$

$$P_{\mathrm{CS}} = \sum_{j=1}^{N_0} \hat{p}_j, \quad \hat{p}_j = \frac{\hbar}{i}\frac{\partial}{\partial q_j}, \tag{2.154}$$

where q_j is a coordinate of jth particle, p_j is its momentum, N_0 is a number of particles, m is the mass of the particles, $\bar{\beta}$ is a (dimensionful) coupling constant and $\hbar$ is the Planck constant. Since this Hamiltonian can be rewritten as

$$H_{\mathrm{CS}} = \sum_j \frac{1}{2m}\Pi_j^\dagger \Pi_j + \frac{\bar{\beta}^2}{m}\left(\frac{\pi}{L}\right)^2 \frac{N_0^3 - N_0}{6}, \tag{2.155}$$

$$\Pi_j = \hat{p}_j + i\bar{\beta}\frac{\pi}{L}\sum_{k\neq j} \cot\frac{\pi}{L}(q_j - q_k), \tag{2.156}$$

the ground state Ψ_0 is determined by $\Pi_j \Psi_0 = 0$,

$$\Psi_0 = \mathcal{N}\Bigl(\prod_{i<j}\sin\frac{\pi}{L}(q_i-q_j)\Bigr)^{\beta} \propto \Bigl(\prod_{i<j}\sqrt{\frac{x_i}{x_j}}\bigl(1-\frac{x_j}{x_i}\bigr)\Bigr)^{\beta}, \tag{2.157}$$

where $\mathcal{N}$ is a normalization constant, and β and x_j are dimensionless quantities,

$$\beta = \frac{\bar{\beta}}{\hbar}, \quad x_j = e^{2\pi i q_j/L}. \tag{2.158}$$

Under the interchange of particle, this ground state gives a phase $(-1)^\beta$ which means that this system obeys fractional statistics. Excited states that we seek for have the form $\Psi = \psi\Psi_0$, where ψ is a symmetric function of x_j because Ψ has same statistical property as Ψ_0. By removing the contribution from the ground state,

$$\Psi_0^{-1}\circ H_{\mathrm{CS}} \circ \Psi_0 = \frac{1}{2m}\left(\frac{2\pi\hbar}{L}\right)^2\Bigl(H_\beta + \frac{1}{12}\beta^2(N_0^3 - N_0)\Bigr), \tag{2.159}$$

$$\Psi_0^{-1}\circ P_{\mathrm{CS}} \circ \Psi_0 = \frac{2\pi\hbar}{L}P, \tag{2.160}$$

Hamiltonian and momentum that act directly ψ are

$$H_\beta = \sum_{i=1}^{N_0} D_i^2 + \beta \sum_{\substack{i,j=1\\ i<j}}^{N_0} \frac{x_i + x_j}{x_i - x_j}(D_i - D_j), \tag{2.161}$$

$$P = \sum_{i=1}^{N_0} D_i, \quad D_i = x_i \frac{\partial}{\partial x_i}. \tag{2.162}$$

Here we recall basic definitions of symmetric polynomials [36]:

partition:

$$\begin{aligned} &\lambda = (\lambda_1, \lambda_2, \dots) = 1^{m_1} 2^{m_2} \cdots, \\ &\lambda_1 \geq \lambda_2 \geq \cdots \geq 0, \quad m_i \geq 0, \\ &|\lambda| = \sum_i \lambda_i = \sum_i i m_i, \quad \ell(\lambda) = \max_i \{\lambda_i > 0\}, \end{aligned} \tag{2.163}$$

dominance (partial) ordering:

$$\lambda \geq \mu \iff |\lambda| = |\mu|, \; \lambda_1 + \cdots + \lambda_i = \mu_1 + \cdots + \mu_i \; (\forall i), \tag{2.164}$$

monomial symmetric function:

$$m_\lambda = \sum_\alpha x^\alpha, \quad x^\alpha = x_1^{\alpha_1} x_2^{\alpha_2} \cdots,$$
$$\alpha : \text{all distinct permutation of } \lambda = (\lambda_1, \lambda_2, \dots), \tag{2.165}$$

power sum symmetric function:

$$p_\lambda = p_{\lambda_1} p_{\lambda_2} \cdots p_{\lambda_{\ell(\lambda)}}, \quad p_n = \sum_i x_i^n, \tag{2.166}$$

inner product:

$$\langle p_\lambda, p_\mu \rangle_\beta = \delta_{\lambda,\mu} \prod_i i^{m_i} m_i! \cdot \beta^{-\sum_i m_i}. \tag{2.167}$$

The Jack symmetric polynomial $J_\lambda = J_\lambda(x;\beta)$ is uniquely determined by the following two conditions [36, 37],

$$\text{(i)} \qquad J_\lambda(x) = \sum_{\mu \leq \lambda} u_{\lambda,\mu} m_\mu(x), \quad u_{\lambda,\lambda} = 1, \tag{2.168}$$

$$\text{(ii)} \qquad \langle J_\lambda, J_\mu \rangle_\beta = 0 \quad \text{if } \lambda \neq \mu. \tag{2.169}$$

The condition (ii) can be replaced by (ii)′,

$$\text{(ii)'} \quad H_\beta J_\lambda - \varepsilon_{\beta,\lambda} J_\lambda, \quad \varepsilon_{\beta,\lambda} = \sum_{i=1}^{N_0} (\lambda_i^2 + \beta(N_0 + 1 - 2i)\lambda_i). \tag{2.170}$$

Therefore excited states of Calogero–Sutherland model are described by the Jack symmetric polynomials. Using properties of the Jack symmetric polynomial, some dynamical correlation functions were calculated [38].

There exists another inner product,

$$\langle f, g\rangle'_{l,\beta} = \frac{1}{l!} \oint \prod_{j=1}^{l} \underline{dx}_j \cdot \bar{\Delta}(x) f(\bar{x}) g(x), \tag{2.171}$$

$$\bar{\Delta}(x) = \prod_{i\neq j} \left(1 - \frac{x_j}{x_i}\right)^\beta, \quad f(\bar{x}) = f\left(\frac{1}{x_1}, \frac{1}{x_2}, \dots\right), \quad \underline{dx}_j = \frac{dx_j}{2\pi i x_j}. \tag{2.172}$$

This is a usual inner product in quantum mechanics. These two inner products are proportional,

$$\langle\ ,\ \rangle'_{l,\beta} \propto \langle\ ,\ \rangle_\beta. \tag{2.173}$$

Let us introduce two transformations for symmetric polynomials,

$$\mathcal{G}_k : \quad (\mathcal{G}_k f)(x_1, \dots, x_l) = \prod_{i=1}^{l} x^k \cdot f(x_1, \dots, x_l), \tag{2.174}$$

$$\begin{aligned} \mathcal{N}_{l',l} : \quad & (\mathcal{N}_{l',l} f)(x'_1, \dots, x'_{l'}) \\ & = \oint \prod_{j=1}^{l} \underline{dx}_j \cdot \overline{\Pi}(x', \bar{x}) \bar{\Delta}(x) f(x_1, \dots, x_l), \end{aligned} \tag{2.175}$$

$$\overline{\Pi}(x, y) = \prod_i \prod_j (1 - x_i y_j)^{-\beta}. \tag{2.176}$$

Then the Jack symmetric polynomial satisfies the following two properties,

$$J_{(k^l)+\lambda} = \mathcal{G}_k J_\lambda \qquad (l \text{ variables}), \tag{2.177}$$

$$J_\lambda = \frac{\langle J_\lambda, J_\lambda\rangle_\beta}{l!\,\langle J_\lambda, J_\lambda\rangle'_{l,\beta}} \mathcal{N}_{l',l} J_\lambda \qquad \begin{aligned}&(J_\lambda \text{ in LHS: } l' \text{ variables})\\ &(J_\lambda \text{ in RHS: } l \text{ variables}).\end{aligned} \tag{2.178}$$

By successive action of these two transformations to $J_\phi(x) = 1$, an integral representation of the Jack symmetric polynomial can be obtained [32, 35],

$$J_\lambda(x) \propto \mathcal{N}_{l,l_1} \mathcal{G}_{k_1} \mathcal{N}_{l_1,l_2} \mathcal{G}_{k_2} \cdots \mathcal{N}_{l_{N-2},l_{N-1}} \mathcal{G}_{k_{N-1}} \mathcal{N}_{l_{N-1},0} \cdot 1, \tag{2.179}$$

where the partition λ is $\lambda' = \left((l_1)^{k_1}, (l_2)^{k_2}, \ldots, (l_{N-1})^{k_{N-1}}\right)$, namely corresponds to the following Young diagram,

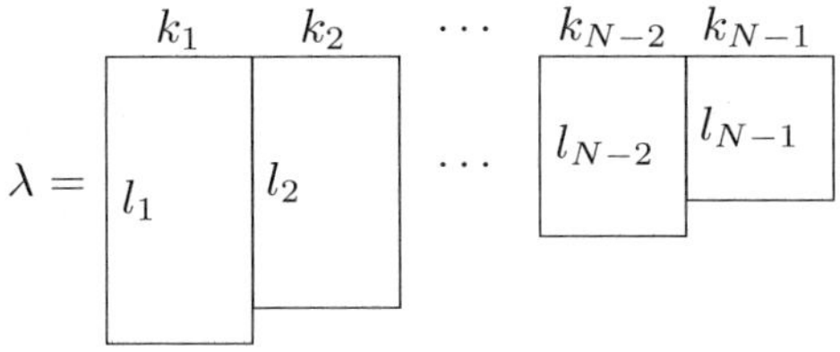

For example, the Jack symmetric polynomial with a rectangular Young diagram (k^l) is

$$J_{(k^l)}(x) \propto \oint \prod_{j=1}^{l} dz_j \cdot \prod_{j=1}^{l} \exp\left(\beta \sum_{n>0} \frac{1}{n} z_j^n \sum_i x_i^n\right) \\ \times \prod_{i \neq j} \left(1 - \frac{z_j}{z_i}\right)^{\beta} \cdot \prod_{j=1}^{l} z_j^{-k}. \quad (2.180)$$

We will show the relation between the Jack symmetric polynomials and the Virasoro singular vectors [32]. States in the Fock space and symmetric polynomials have the one-to-one correspondence,

$$\begin{aligned} \mathcal{F}_\alpha &\to \{\text{symmetric function}\} \\ |f\rangle &\mapsto f(x) = \langle \alpha | \exp\left(\frac{1}{2}\sqrt{\beta} \sum_{n>0} \frac{1}{n} a_n p_n\right) |f\rangle. \end{aligned} \quad (2.181)$$

Namely, the oscillator a_n and the power sum p_n correspond as,

$$\frac{a_{-n}}{\sqrt{\beta}} \leftrightarrow p_n, \qquad \frac{a_n}{\sqrt{\beta}} \leftrightarrow \frac{2n}{\beta} \frac{\partial}{\partial p_n}. \quad (2.182)$$

Using this correspondence, the Hamiltonian H_β (2.161), which is a differential operator with respect to x_j (or p_n), is bosonized,

$$H_\beta \langle \alpha | \exp\left(\frac{1}{2}\sqrt{\beta} \sum_{n>0} \frac{1}{n} a_n p_n\right) = \langle \alpha | \exp\left(\frac{1}{2}\sqrt{\beta} \sum_{n>0} \frac{1}{n} a_n p_n\right) \widehat{H}_\beta. \quad (2.183)$$

Bosonized Hamiltonian $\widehat{H}_\beta$ is an operator on $\mathcal{F}_\alpha$. It is cubic in a_n and can be rewritten using the Virasoro generator L_n (2.56),

$$\begin{aligned} \widehat{H}_\beta &= \frac{1}{4}\sqrt{\beta} \sum_{n,m>0} \left(a_{-n-m} a_n a_m + 2 a_{-n} a_{-m} a_{n+m}\right) \\ &\quad + \frac{1}{2} \sum_{n>0} a_{-n} a_n \left((1-\beta) n + N_0 \beta\right) \\ &= \sqrt{\beta} \sum_{n>0} a_{-n} L_n + \frac{1}{2} \sum_{n>0} a_{-n} a_n (N_0 \beta + \beta - 1 - \sqrt{\beta} a_0). \end{aligned} \quad (2.184)$$

When $\widehat{H}_\beta$ acts on the singular vector (2.80), the first term of $\widehat{H}_\beta$ vanishes because of the property of singular vector, and the second term is already diagonal,

$$\begin{aligned}\widehat{H}_\beta|\chi_{-l,-k}\rangle &= lk(N_0\beta+\beta-1-\sqrt{\beta}\alpha_{-l,-k})|\chi_{-l,-k}\rangle\\ &= \varepsilon_{\beta,(k^l)}|\chi_{-l,-k}\rangle.\end{aligned} \tag{2.185}$$

Therefore $|\chi_{-l,-k}\rangle$ is an eigenstate of $\widehat{H}_\beta$, i.e., it gives the Jack symmetric polynomial by the map (2.181). In fact the polynomial obtained from (2.80) by (2.181) agrees with the integral representation of the Jack symmetric polynomial with the partition (k^l), (2.180).

The Jack symmetric polynomial with general partition whose Young diagram is composed of $N-1$ rectangles is related to the singular vector of W_N algebra [32].

3 Deformed Virasoro Algebra ($A_1^{(1)}$ Type)

3.1 Definition and Consistency

Definition

Deformed Virasoro algebra (DVA) ($A_1^{(1)}$ type) is an associative algebra over $\mathbb{C}$ generated by T_n ($n\in\mathbb{Z}$) with two parameters x and r, and their relation is [17] (see (3.128) for correspondence of parameters)

$$\begin{aligned}[T_n,T_m] = -\sum_{\ell=1}^{\infty} f_\ell(T_{n-\ell}T_{m+\ell}-T_{m-\ell}T_{n+\ell})&\\ -(x-x^{-1})^2[r]_x[r-1]_x[2n]_x\delta_{n+m,0},&\end{aligned} \tag{3.1}$$

where the structure constants f_ℓ is given by

$$\begin{aligned}f(z) &= \sum_{\ell=0}^{\infty} f_\ell z^\ell\\ &= \exp\left(-\sum_{n>0}\frac{z^n}{n}\frac{(x^{rn}-x^{-rn})(x^{(r-1)n}-x^{-(r-1)n})}{x^n+x^{-n}}\right)\\ &= \frac{1}{1-z}\frac{(x^{2r}z,x^{-2(r-1)}z;x^4)_\infty}{(x^{2+2r}z,x^{2-2(r-1)}z;x^4)_\infty}.\end{aligned} \tag{3.2}$$

Here we have used the notation (A.1), (A.3). By introducing DVA current $T(z)=\sum_{n\in\mathbb{Z}}T_nz^{-n}$, the above relation can be written as a formal power series,

$$f\left(\frac{w}{z}\right)T(z)T(w) - T(w)T(z)f\left(\frac{z}{w}\right)$$
$$= -(x-x^{-1})[r]_x[r-1]_x\left(\delta\left(x^2\frac{w}{z}\right) - \delta\left(x^{-2}\frac{w}{z}\right)\right). \tag{3.3}$$

For later use we add a grading operator d,

$$[d, T_n] = -nT_n. \tag{3.4}$$

The above relation (3.1) is invariant under

$$T_n \mapsto -T_n. \tag{3.5}$$

It is also invariant under the following two transformations:

(i) $$\theta\colon x \mapsto x^{-1},\ r \mapsto r, \tag{3.6}$$

(ii) $$\omega\colon x \mapsto x,\ r \mapsto 1-r. \tag{3.7}$$

In the case of (i) $f(z)$ is understood as the first line of (3.2). Let us introduce β as

$$\beta = \frac{r}{r-1}, \tag{3.8}$$

then

$$\theta\cdot\beta = \beta, \quad \omega\cdot\beta = \beta^{-1}. \tag{3.9}$$

α_0 in (2.19) is

$$\alpha_0 = \frac{1}{\sqrt{r(r-1)}}. \tag{3.10}$$

Consistency

In Section 2.2.1 the central term of the Virasoro algebra is determined by the Jacobi identity. Here we will show that the structure function $f(z)$ is determined by associativity [39].

Let us consider the following relation,

$$f\left(\frac{w}{t}\right)T(z)T(w) - T(w)T(z)f\left(\frac{z}{w}\right) = c_0\left(\delta\left(x^2\frac{w}{z}\right) - \delta\left(x^{-2}\frac{w}{z}\right)\right), \tag{3.11}$$

where c_0 is a normalization constant and x is a parameter, and $f(z)$ is an unknown Taylor series $f(z) = \sum_{\ell=0}^{\infty} f_\ell z^\ell$. By using this relation, $f(z_2/z_1)f(z_3/z_1)f(z_3/z_2)T(z_1)T(z_2)T(z_3)$ is related to $f(z_1/z_2)f(z_1/z_3)\times f(z_2/z_3)T(z_3)T(z_2)T(z_1)$ in two ways,

$$\begin{array}{ccccc}
(123) & \longrightarrow & (132) & \longrightarrow & (312) \\
\downarrow & & & & \downarrow \\
(213) & \longrightarrow & (231) & \longrightarrow & (321).
\end{array}$$

These two results should agree. So we obtain an equation containing delta functions,

$$c_0T(z_1)\left(\delta\left(x^2\frac{z_3}{z_2}\right)-\delta\left(x^{-2}\frac{z_3}{z_2}\right)\right)\left(f\left(\frac{z_2}{z_1}\right)f\left(\frac{z_3}{z_1}\right)-f\left(\frac{z_1}{z_2}\right)f\left(\frac{z_1}{z_3}\right)\right)$$
$$+\text{ cyclic}=0. \quad (3.12)$$

This is equivalent to

$$c_0T(z_1)\left(\delta\left(x^2\frac{z_3}{z_2}\right)g\left(\frac{z_2}{z_1}\right)-\delta\left(x^{-2}\frac{z_3}{z_2}\right)g\left(\frac{z_3}{z_1}\right)\right)+\text{ cyclic}=0, \quad (3.13)$$

where $g(z)$ is

$$g(z)=f(z)f(x^{-2}z)-f(z^{-1})f(x^2z^{-1})=-g(x^2z^{-1}). \quad (3.14)$$

In mode expansion $g(z)=\sum_{n\in\mathbb{Z}}g_nz^n$, this equation becomes

$$(x^{2n}-x^{2m})g_{n+m}+(x^{2m}-x^{2l})g_{m+l}+(x^{2l}-x^{2n})g_{l+n}=0, \quad (3.15)$$
$$g_{-n}=-x^{2n}g_n. \quad (3.16)$$

From (3.16), we have $g_0=0$. From (3.15) with $(m,l)=(n-1,1-n)$, $(m,l)=(n-2,2-n)$, we have $g_{2n-1}=(1-x^{-2(2n-1)})g_1/(1-x^{-2})$ and $g_{2n-2}=(1-x^{-2(2n-2)})g_2/(1-x^{-4})$. And from (3.15) with $(m,l)=(n-1,n+1)$ we have $g_2=(1+x^{-2})g_1$. Combining these we get $g_n=(1-x^{-2n})g_1$ and this satisfies both (3.15) and (3.16). Therefore the solution is

$$g_n=c_0'(1-x^{-2n}),\quad g(z)=c_0'\big(\delta(z)-\delta(x^{-2}z)\big), \quad (3.17)$$

where c_0' is a constant. By setting $F(z)=f(z)f(x^2z)$, (3.14) and (3.17) become

$$F(z)-F(x^{-2}z^{-1})=-c_0'\sum_{n\in\mathbb{Z}}(1-x^{2n})z^n. \quad (3.18)$$

Since $F(z)$ is also a Taylor series, this equation implies

$$F(z)=-c_0'\left(\alpha+\sum_{n>0}(1-x^{2n})z^n\right), \quad (3.19)$$

where a new parameter α has appeared. If we express this α by a new parameter r as

$$\alpha=\frac{x-x^{-1}}{(x^r-x^{-r})(x^{r-1}-x^{-(r-1)})}, \quad (3.20)$$

then $F(z)$ becomes

$$F(z)=-c_0'\alpha\frac{(1-x^{2r}z)(1-x^{-2(r-1)})}{(1-z)(1-x^2z)}. \quad (3.21)$$

From this equation $f(z)$ is calculated as ($|x| < 1$)

$$f(z) = \frac{F(z)}{F(x^2 z)} f(x^4 z) = \frac{F(z)}{F(x^2 z)} \frac{F(x^4 z)}{F(x^6 z)} \times \cdots \times f(0)$$
$$= f(0) \exp\Bigg(\sum_{n>0} \frac{z^n}{n} \frac{(1 - x^{2rn})(1 - x^{-2(r-1)n})}{1 + x^{2n}} \Bigg). \tag{3.22}$$

We choose the normalization of $f(z)$ as $f(0) = 1$ ($c_0' = -\alpha^{-1}$), then this $f(z)$ is just (3.2). We fix the normalization of $T(z)$ by $c_0 = -\alpha^{-1}$.

3.2 Conformal Limit

Quantum group(algebra) $U_q(\mathfrak{g})$ is a deformation of $U(\mathfrak{g})$ and it reduces to undeformed one in the $q \to 1$ limit,

$$U_q(\mathfrak{g}) \xrightarrow{q \to 1} U(\mathfrak{g}).$$

The deformed Virasoro algebra defined in the previous subsection is a deformation of the Virasoro algebra as expected from its name. Then in what limit it reduces to the usual Virasoro algebra?

$$\text{DVA} \xrightarrow{??} \text{Vir}.$$

Since DVA contains two parameters x and r (or β (3.8)), it admits various limits [39]. In this subsection we will show that DVA reduces to the Virasoro algebra in the conformal limit.

The conformal limit is

$$x \to 1, \quad r \text{ (or } \beta\text{): fixed.} \tag{3.23}$$

To study this limit we write x as

$$x = e^{-\alpha_0 \hbar / 2}, \tag{3.24}$$

where α_0 is given in (2.19) and $\hbar$ is a fictitious Plank constant, and take $\hbar \to 0$. $f(z)$ has the expansion,

$$f(z) = 1 + \hbar^2 f^{[2]}(z) + \hbar^4 f^{[4]}(z) + \cdots . \tag{3.25}$$

Taking into account the invariance under (3.6), we assume the expansion

$$T(z) = 2 + \hbar^2 T^{[2]}(z) + \hbar^4 T^{[4]}(z) + \hbar^6 T^{[6]}(z) + \cdots , \tag{3.26}$$

(remark that $-T(z)$ is also a solution) and we set

$$T^{[2]}(z) = z^2 L(z) + \tfrac{1}{4}\alpha_0^2. \tag{3.27}$$

Then LHS of (3.3) is

$$\begin{aligned}
&\hbar^2 4\big(f^{[2]}(\zeta) - f^{[2]}(\zeta^{-1})\big) \\
&+ \hbar^4 \Big([T^{[2]}(z), T^{[2]}(w)] + 2\big(f^{[2]}(\zeta) - f^{[2]}(\zeta^{-1})\big)\big(T^{[2]}(z) + T^{[2]}(w)\big) \\
&\qquad\qquad + 4\big(f^{[4]}(\zeta) - f^{[4]}(\zeta^{-1})\big)\Big) + \cdots \\
&= \hbar^2\big(- 2\zeta\delta'(\zeta)\big) + \hbar^4 \bigg([z^2 L(z), w^2 L(w)] - \zeta\delta'(\zeta)\big(z^2 L(z) + w^2 L(w)\big) \\
&\qquad - \zeta\delta'(\zeta)\frac{1+2\alpha_0^2}{6} - \zeta^2\delta'''(\zeta)\frac{1-2\alpha_0^2}{12}\bigg) + \cdots , \qquad (3.28)
\end{aligned}$$

where $\zeta = w/z$. On the other hand RHS of (3.3) is

$$\hbar^2\big(- 2\zeta\delta'(\zeta)\big) + \hbar^4 \bigg(-\zeta\delta'(\zeta)\frac{1+2\alpha_0^2}{6} - \zeta^2\delta'''(\zeta)\frac{\alpha_0^2}{3}\bigg) + \cdots . \qquad (3.29)$$

Comparing these equations and (2.5) shows that $L(z)$ in (3.27) is the Virasoro current with the central charge (2.17).

Therefore the Virasoro current $L(z)$ is found in $\hbar^2$ term of the DVA current $T(z)$. And also (conformal) spin 4 current $T^{[4]}(z)$, spin 6 current $T^{[6]}(z)$, ...exist in the DVA current. In massless case only one Virasoro current $L(z)$ controls (chiral part of) CFT. In massive case, however, $L(z)$ and infinitely many higher spin currents $T^{[2n]}(z)$ are needed to control massive theory, and they gather and form the DVA current $T(z)$. DVA current is a 'dressed' Virasoro current.

3.3 Representation Theory

Let us consider the highest weight representation. The highest weight state $|\lambda\rangle$ $(\lambda \in \mathbb{C})$ is characterized by

$$T_n|\lambda\rangle = 0 \ (n > 0), \quad T_0|\lambda\rangle = \lambda|\lambda\rangle, \qquad (3.30)$$

and the Verma module is

$$M = \bigoplus_{l\geq 0} \bigoplus_{n_1\geq\cdots\geq n_l>0} \mathbb{C}T_{-n_1}\cdots T_{-n_l}|\lambda\rangle. \qquad (3.31)$$

M is a graded module with grading d,

$$\begin{aligned}
d|\lambda\rangle &= d_\lambda|\lambda\rangle \quad (d_\lambda \in \mathbb{C}), \\
dT_{-n_1}\cdots T_{-n_l}|\lambda\rangle &= \Big(d_\lambda + \sum_{i=1}^{l} n_l\Big) T_{-n_1}\cdots T_{-n_l}|\lambda\rangle,
\end{aligned}$$

and we call $\sum_{i=1}^{l} n_l$ as a level.

To obtain the irreducible module from the Verma module we have to quotient out invariant submodules, which are generated by singular vectors. Singular vector at level N, $|\chi\rangle$, is

$$T_n|\chi\rangle = 0 \quad (n > 0), \quad T_0|\chi\rangle = \lambda_\chi|\chi\rangle, \quad d|\chi\rangle = (d_\lambda + N)|\chi\rangle, \tag{3.32}$$

where λ_χ is some eigenvalue. Existence of singular vectors is detected by zeros of the Kac determinant. To define the Kac determinant let us introduce the dual module M^* on which the DVA act as $T_n^\dagger = T_{-n}$. M^* is generated by $\langle\lambda|$ which satisfies $\langle\lambda|T_n = 0$ $(n < 0)$, $\langle\lambda|T_0 = \lambda\langle\lambda|$ and $\langle\lambda|\lambda\rangle = 1$. At level N there are $p(N)$ states $|\lambda; N, 1\rangle = T_{-N}|\lambda\rangle$, $|\lambda; N, 2\rangle = T_{-N+1}T_{-1}|\lambda\rangle$, $\ldots$, $|\lambda; N, p(N)\rangle = T_{-1}^N|\lambda\rangle$ in M, and $\langle\lambda; N, 1| = \langle\lambda|T_N$, $\langle\lambda; N, 2| = \langle\lambda|T_1T_{N-1}$, $\ldots$, $\langle\lambda; N, p(N)| = \langle\lambda|T_1^N$ in M^*.

The Kac determinant at level N is given by [17, 40]

$$\det\langle\lambda; N, i|\lambda; N, j\rangle = \prod_{\substack{l,k\geq 1 \\ lk\leq N}} \left(\frac{(x^{rl} - x^{-rl})(x^{(r-1)l} - x^{-(r-1)l})}{x^l + x^{-l}}(\lambda^2 - \lambda_{l,k}^2)\right)^{p(N-lk)}, \tag{3.33}$$

where $\lambda_{l,k}$ is

$$\lambda_{l,k} = x^{rl-(r-1)k} + x^{-rl+(r-1)k}. \tag{3.34}$$

λ dependence appears through λ^2 because of the symmetry (3.5). Conformal limit (3.23) of this determinant is

$$\prod_{\substack{l,k\geq 1 \\ lk\leq N}} \left(\frac{\hbar^2 l^2}{2}\hbar^2(h - h_{l,k})\cdot 4\right)^{p(N-lk)} + \cdots = \hbar^A \prod_{\substack{l,k\geq 1 \\ lk\leq N}} \left(2lk(h - h_{l,k})\right)^{p(N-lk)} + \cdots,$$
$$A = 4\sum_{\substack{l,k\geq 1 \\ lk\leq N}} p(N - lk). \tag{3.35}$$

This is just the Kac determinant of the Virasoro algebra (2.16), and the order of $\hbar$ is consistent with (3.26) (use (A.30)).

For generic values of x and r, this Kac determinant (3.33) has essentially the same structure as the Virasoro one (2.16). Therefore embedding pattern is same as (2.27), and the character which counts the degeneracy at each level ($\mathrm{tr}\, q^d$) is also the same. For special value of x and r, for example x^r is a root of unity, the Kac determinant has more zeros. In this case we need special study, see [40].

3.4 Free Field Realization

3.4.1 Free Field Realization

Let us introduce free boson oscillator h_n ($n \in \mathbb{Z}_{\neq 0}$),

$$[h_n, h_m] = (x - x^{-1})^2 \frac{1}{n} \frac{[n]_x [rn]_x [(r-1)n]_x}{[2n]_x} \delta_{n+m,0}, \tag{3.36}$$

and use zero mode a_0 and Q defined in (2.49) (or a_0' in (2.72)). The Fock space $\mathcal{F}_\alpha$ is defined by

$$\mathcal{F}_\alpha = \bigoplus_{l \geq 0} \bigoplus_{n_1 \geq \cdots \geq n_l > 0} \mathbb{C} h_{-n_1} \ldots h_{-n_l} |\alpha\rangle_B, \tag{3.37}$$

where $|\alpha\rangle_B$ is given by (2.51) with replacing a_n by h_n.

The DVA current $T(z)$ is realized as follows:

$$\begin{aligned} T(z) &= \Lambda_+(z) + \Lambda_-(z), \\ \Lambda_\pm(z) &= :\!\exp\Big(\pm \sum_{n \neq 0} h_n (x^{\pm 1} z)^{-n} \Big)\!: \times\, x^{\pm \sqrt{r(r-1)} a_0'}. \end{aligned} \tag{3.38}$$

To prove this we need the OPE formula,

$$\begin{aligned} f\Big(\frac{w}{t}\Big) \Lambda_\pm(z) \Lambda_\pm(w) &= :\!\Lambda_\pm(z) \Lambda_\pm(w)\!:, \\ f\Big(\frac{w}{t}\Big) \Lambda_\pm(z) \Lambda_\mp(w) &= :\!\Lambda_\pm(z) \Lambda_\mp(w)\!: \gamma\Big(x^{\mp 1} \frac{w}{z} \Big), \end{aligned} \tag{3.39}$$

and the relation of $\Lambda_\pm$,

$$:\!\Lambda_+(x^{-1} z) \Lambda_-(xz)\!: = 1. \tag{3.40}$$

Here $\gamma(z)$ is

$$\gamma(z) = \frac{(1 - x^{2r-1} z)(1 - x^{-(2r-1)} z)}{(1 - xz)(1 - x^{-1} z)}, \tag{3.41}$$

and from (A.24) we have

$$\gamma(z) - \gamma(z^{-1}) = -(x - x^{-1}) [r]_x [r-1]_x \big(\delta(xz) - \delta(x^{-1} z) \big). \tag{3.42}$$

The grading operator d is realized by

$$d = \sum_{n > 0} \frac{n^2 [2n]_x}{(x - x^{-1})^2 [n]_x [rn]_x [(r-1)n]_x} h_{-n} h_n + \frac{1}{4} a_0'^2 - \frac{1}{24}, \tag{3.43}$$

which satisfies

$$[d, h_n] = -n h_n, \quad [d, Q] = a_0', \quad d |\alpha_{l,k}\rangle_B = \Big(h_{l,k} - \frac{c}{24} \Big) |\alpha_{l,k}\rangle_B, \tag{3.44}$$

where c and $h_{l,k}$ are given by (2.17) and (2.18) respectively.

$|\alpha\rangle_B$ is the highest weight state of DVA with $\lambda = \lambda(\alpha)$,

$$|\alpha\rangle_B = |\lambda(\alpha)\rangle, \tag{3.45}$$

$$\lambda(\alpha) = x^{\sqrt{r(r-1)}(\alpha-\alpha_0)} + x^{-\sqrt{r(r-1)}(\alpha-\alpha_0)}. \tag{3.46}$$

The dual space $\mathcal{F}_\alpha^*$ becomes a DVA module by (2.63) with

$${}^tT_n = T_{-n}, \tag{3.47}$$

$${}^th_n = -h_{-n}, \quad \text{eq. (2.66)}, \tag{3.48}$$

and (2.67). By (3.48), $\mathcal{F}_\alpha^*$ is isomorphic to $\mathcal{F}_{2\alpha_0-\alpha}$ as a DVA module,

$$\mathcal{F}_\alpha^* \cong \mathcal{F}_{2\alpha_0-\alpha} \qquad \text{(DVA module)}. \tag{3.49}$$

$$|\alpha\rangle_B^* \leftrightarrow |2\alpha_0 - \alpha\rangle_B. \tag{3.50}$$

Remark that $\lambda(\alpha) = \lambda(2\alpha_0 - \alpha)$ by (3.46).

In the conformal limit (3.23), oscillator h_n is expressed by a_n in (2.49) as follows:

$$h_n = \frac{\hbar}{2}\sqrt{\frac{2}{x^n + x^{-n}}\,\frac{x^{rn} - x^{-rn}}{rn\alpha_0\hbar}\,\frac{x^{(r-1)n} - x^{-(r-1)n}}{(r-1)n\alpha_0\hbar}}\,a_n. \tag{3.51}$$

Substituting this expression into (3.38) and expanding in $\hbar$, we get (3.26) with $L(z)$ in (2.55) (or (2.73)).

3.4.2 Singular Vectors and Kac Determinant

In the case of $\lambda = \lambda_{l,k}$ there are singular vectors. In the free boson realization they can be expressed by using the screening currents. For later convenience we denote screening currents as $x_\pm(z)$. $x_\pm(z)$ is defined by

$$x_+(z) = :\exp\Bigl(-\sum_{n\neq 0}\frac{\alpha_n}{[n]_x}z^{-n}\Bigr): \times e^{\sqrt{r/(r-1)}Q} z^{\sqrt{r/(r-1)}a_0' + r/(r-1)}, \tag{3.52}$$

$$x_-(z) = :\exp\Bigl(\sum_{n\neq 0}\frac{\alpha_n'}{[n]_x}z^{-n}\Bigr): \times e^{-\sqrt{(r-1)/r}Q} z^{-\sqrt{(r-1)/r}a_0' + (r-1)/r}, \tag{3.53}$$

where oscillators α_n, α_n' ($n \in \mathbb{Z}_{\neq 0}$) are related to h_n as

$$h_n = (x - x^{-1})(-1)^n \frac{[(r-1)n]_x}{[2n]_x}\alpha_n = (x - x^{-1})\frac{[rn]_x}{[2n]_x}\alpha_n'. \tag{3.54}$$

Conformal limit (3.23) of $x_-(z)$ is $zS_-(z)$ in (2.76) and that of $x_+(z)$ is $zS_+(-z)$ (up to phase) due to $(-1)^n$ factor in (3.54).

Commutation relation of DVA generator and screening currents is a total difference

$$[T_n, x_+(w)] = (x^r - x^{-r})\big((-x^{r-1}w)^n A_+(x^{r-1}w)$$
$$- (-x^{-(r-1)}w)^n A_+(x^{-(r-1)}w)\big), \quad (3.55)$$
$$[T_n, x_-(w)] = (x^{r-1} - x^{-(r-1)})\big((x^r w)^n A_-(x^r w)$$
$$- (x^{-r}w)^n A_-(x^{-r}w)\big), \quad (3.56)$$

where $A_\pm(w)$ is

$$A_+(w) = x^{\pm r}{:}\Lambda_\pm(-w)x_+(x^{\mp(r-1)}w){:}, \quad (3.57)$$
$$A_-(w) = x^{\pm(r-1)}{:}\Lambda_\mp(w)x_-(x^{\mp r}w){:}. \quad (3.58)$$

Hence their zero modes (screening charge) commute with DVA,

$$\left[T_n, \oint_0 \frac{dz}{2\pi i z} x_\pm(z)\right] = 0. \quad (3.59)$$

(Exactly speaking we have to specify the Fock space on which they act. See Section 3.4.3.) We will use $x_+(z)$ in what follows. $x_-(z)$ can be treated similarly.

The representation with $\lambda = \lambda_{l,k}$ is realized on $\mathcal{F}_{l,k} = \mathcal{F}_{\alpha_{l,k}}$,

$$\lambda_{l,k} = \lambda(\alpha_{l,k}), \quad (3.60)$$

where $\alpha_{l,k}$ is given in (2.79). To contact with our previous papers [17, 18] ($\alpha_{r,s}$ there is $\propto \alpha_{-r,-s}$ here), we consider dual space $\mathcal{F}^*_{l,k} = \mathcal{F}_{-l,-k}$. Like as the Virasoro case (2.80), singular vectors of DVA are expressed by product of screening charges. Moreover this product becomes more clear for the deformed case, if we include Lukyanov's zero mode factor (see Section 3.4.3). In the Verma module with $\lambda_{-l,-k}$, the singular vector at level lk is expressed as

$$|\chi_{-l,-k}\rangle = Q_l|\alpha_{l,-k}\rangle_B$$
$$= \int \prod_{j=1}^{l} \underline{dz}_j \cdot \prod_{\substack{i,j=1 \\ i\neq j}}^{l} \frac{(z_i/z_j; x^{2(r-1)})_\infty}{(x^{2r}z_i/z_j; x^{2(r-1)})_\infty}$$
$$\times C(z) \prod_{i=1}^{l} z_i^{-k} \cdot \prod_{j=1}^{l} e^{\sum_{n>0} \alpha_{-n} z_j^n/[n]_x} |\alpha_{-l,-k}\rangle_B, \quad (3.61)$$

where $\underline{dz}_j$ is given in (A.14), the BRST charge Q_l will be given in (3.77), and $C(z)$ is

$$C(z) = \prod_{\substack{i,j=1 \\ i<j}}^{l} \frac{(x^{2r} z_i/z_j; x^{2(r-1)})_\infty}{(z_i/z_j; x^{2(r-1)})_\infty} \frac{(x^{-2} z_j/z_i; x^{2(r-1)})_\infty}{(x^{2(r-1)} z_j/z_i; x^{2(r-1)})_\infty} \times \prod_{i=1}^{l} z_i^{r(l+1-2i)/(r-1)} \cdot \prod_{i=1}^{l} \frac{[-u_i + \frac{1}{2} - (2i-l)]^*}{[-u_i - \frac{1}{2}]^*}. \quad (3.62)$$

Here we have used the notation (A.6) with $r^* = r - 1$ and $z_i = x^{2u_i}$. Since the BRST charge Q_l, which is a map from $\mathcal{F}_{l,-k}$ to $\mathcal{F}_{-l,-k}$, commutes with T_n, $|\chi_{-l,-k}\rangle$ is annihilated by T_n $(n > 0)$. This state $|\chi_{-l,-k}\rangle$ is non zero because its conformal limit gives non zero state (2.80).

Like as in Section 2.3.2 let us introduce matrices $C(N, \alpha)$ and $C'(N, \alpha)$,

$$T_{-I}|\alpha\rangle_B = \sum_J C(N,\alpha)_{I,J} h_{-J}|\alpha\rangle_B, \quad (3.63)$$

$$({}^tT)_I|\alpha\rangle_B^* = \sum_J C'(N,\alpha)_{I,J} ({}^th)_J|\alpha\rangle_B^*. \quad (3.64)$$

C' can be expressed by C as (2.92) with (2.91). Their determinants are given by

$$\det C(N,\alpha)_{I,J} = \prod_{\substack{l,k\geq 1 \\ lk\leq N}} \left(x^{\sqrt{r(r-1)}(\alpha-\alpha_{l,k})} - x^{-\sqrt{r(r-1)}(\alpha-\alpha_{l,k})}\right)^{p(N-lk)}, \quad (3.65)$$

$$\begin{aligned} \det C'(N,\alpha)_{I,J} &= \det C(N, 2\alpha_0 - \alpha) \det D \\ &= \prod_{\substack{l,k\geq 1 \\ lk\leq N}} \left(x^{\sqrt{r(r-1)}(\alpha+\alpha_{l,k}-2\alpha_0)} - x^{-\sqrt{r(r-1)}(\alpha+\alpha_{l,k}-2\alpha_0)}\right)^{p(N-lk)}. \end{aligned} \quad (3.66)$$

The inner product of two states in the Verma module becomes (see (2.94))

$$\begin{aligned} \langle\lambda|T_I T_{-J}|\lambda\rangle &= \left\langle ({}^tT)_I|\alpha\rangle_B^*, T_{-J}|\alpha\rangle_B \right\rangle \\ &= \sum_{K,L} C'(N,\alpha)_{I,K} G_{K,L} C(N,\alpha)_{J,L}. \end{aligned} \quad (3.67)$$

Here $G_{K,L}$ is

$$\begin{aligned} G_{K,L} &= \left\langle ({}^th)_K|\alpha\rangle_B^*, h_{-L}|\alpha\rangle_B \right\rangle \\ &= \delta_{K,L} \prod_i \left(\frac{1}{i} \frac{(x^{ri} - x^{-ri})(x^{(r-1)i} - x^{-(r-1)i})}{x^i + x^{-i}} \right)^{k_i} k_i!, \end{aligned} \quad (3.68)$$

and its determinant is (use (A.28) and (A.30))

$$\det G_{K,L} = \prod_{\substack{\{k_i\} \\ \sum_i ik_i=N}} \prod_i \Big(\frac{1}{i} \frac{(x^{ri}-x^{-ri})(x^{(r-1)i}-x^{-(r-1)i})}{x^i+x^{-i}} \Big)^{k_i} k_i! \\ = \prod_{\substack{l,k\geq 1 \\ lk\leq N}} \Big(\frac{(x^{rl}-x^{-rl})(x^{(r-1)l}-x^{-(r-1)l})}{x^l+x^{-l}} \Big)^{p(N-lk)} . \quad (3.69)$$

Therefore we obtain the Kac determinant (3.33) [17, 40],

$$\det\langle\lambda|T_I T_{-J}|\lambda\rangle = \det C'(N,\alpha)\cdot\det G\cdot\det{}^t C(N,\alpha) \\ = \prod_{\substack{l,k\geq 1 \\ lk\leq N}} \Big(\frac{(x^{rl}-x^{-rl})(x^{(r-1)l}-x^{-(r-1)l})}{x^l+x^{-l}} (\lambda^2-\lambda_{l,k}^2) \Big)^{p(N-lk)} . \quad (3.70)$$

3.4.3 Felder Complex

We consider the representation of $\lambda=\lambda_{l,k}$ in (3.60) with (2.20) and (2.23), i.e.,

$$r=\frac{p''}{p''-p'} . \quad (3.71)$$

We set

$$r^*=r-1.$$

Let us define the screening operator $X(z)$ as the integral of $x_+(z')$ plus extra zero mode factor [41],

$$X(z)=\oint_C \underline{dz}' x_+(z') \frac{[u-u'+\frac{1}{2}-\hat{l}]^*}{[u-u'-\frac{1}{2}]^*}, \quad (3.72)$$

where $z=x^{2u}$, $z'=x^{2u'}$, $\underline{dz}'=dz'/(2\pi i z')$, and the integration contour C is a simple closed curve that encircles $z'=x^{-1+2(r-1)n}z$ but not $z'=x^{-1-2(r-1)(n+1)}z$ $(n=0,1,2,\dots)$. $\hat{l}$ is $\hat{l}=l\times\mathrm{id}$ on $\mathcal{F}_{l,k}$ (see Section 5.2). z is an arbitrary point, for example we take $z=1$.

This $X(z)$ is well defined on $\mathcal{F}_{l',k'}$ $(\forall l',k')$

$$X(z)\colon \mathcal{F}_{l',k'} \mapsto \mathcal{F}_{l'-2,k'} . \quad (3.73)$$

Product of $X(z)$'s is [41]

$$
\begin{aligned}
&X(z)^m\colon \mathcal{F}_{l',k'} \mapsto \mathcal{F}_{l'-2m,k'} \\
&= \oint \prod_{j=1}^m \underline{dz}_j \cdot x_+(z_1)\ldots x_+(z_m) \prod_{i=1}^m \frac{\left[u-u_i+\frac{1}{2}-(\hat{l}-2(m-i))\right]^*}{[u-u_i-\frac{1}{2}]^*} \\
&= \oint \prod_{j=1}^m \underline{dz}_j \cdot x_+(z_1)\cdots x_+(z_m) \\
&\quad \times \frac{1}{m!}\sum_{\sigma\in S_m}\prod_{i=1}^m \frac{\left[u-u_{\sigma(i)}+\frac{1}{2}-(\hat{l}-2(m-i))\right]^*}{[u-u_{\sigma(i)}-\frac{1}{2}]^*} \prod_{\substack{1\le i<j\le m\\ \sigma(i)>\sigma(j)}} h^*(u_{\sigma(i)}-u_{\sigma(j)}) \\
&= \oint \prod_{j=1}^m \underline{dz}_j \cdot x_+(z_1)\cdots x_+(z_m) \frac{1}{m!}\prod_{i=1}^m \frac{[i]^*}{[1]^*} \\
&\qquad \times \prod_{1\le i<j\le m} \frac{[u_i-u_j]^*}{[u_i-u_j+1]^*} \cdot \prod_{i=1}^m \frac{[u-u_i-\frac{1}{2}+m-\hat{l}]^*}{[u-u_i-\frac{1}{2}]^*},
\end{aligned}
\tag{3.74}
$$

where $z=x^{2u}$, $z_i=x^{2u_i}$ and $\underline{dz}_j$ is given in (A.14). Here we have used $x_+(z_1)x_+(z_2)=h^*(u_1-u_2)x_+(z_2)x_+(z_1)$ (as a meromorphic function) with $h^*(u)=\frac{[u+1]^*}{[u-1]^*}$ (see Section 5.2), and the formula (A.12) with $r\to r-1$. From the factor $[i]^*$ in the last line of (3.74), we have (use (A.10))

$$X(z)^{p'}=0. \tag{3.75}$$

Moreover $X(z)$ has the property,

$$[T_n, X(z)^l]=0 \text{ on } \mathcal{F}_{l',k} \quad l'\equiv l \pmod{p'}. \tag{3.76}$$

This can be proved by (3.55) and careful analysis of location of poles [41]. We define the BRST charge Q_m as

$$Q_m = X(1)^m. \tag{3.77}$$

Let us consider the Felder complex $C_{l,k}$,

$$\cdots \xrightarrow{X_{-3}} C_{-2} \xrightarrow{X_{-2}} C_{-1} \xrightarrow{X_{-1}} C_0 \xrightarrow{X_0} C_1 \xrightarrow{X_1} C_2 \xrightarrow{X_2} \cdots, \tag{3.78}$$

where C_j and $X_j\colon C_j\to C_{j+1}$ $(j\in\mathbb{Z})$ are

$$C_{2j}=\mathcal{F}_{l-2p'j,k}, \qquad C_{2j+1}=\mathcal{F}_{-l-2p'j,k}, \tag{3.79}$$

$$X_{2j}=Q_l, \qquad X_{2j+1}=Q_{p'-l}. \tag{3.80}$$

X satisfies the BRST property,

$$X_jX_{j-1}=0. \tag{3.81}$$

We assume that this Felder complex has the same structure as the Virasoro case because it formally tends to Virasoro one in the conformal limit ($x \to 1$ and r and $z = x^{2u}$ fixed kept). Then the cohomology groups of the complex $C_{l,k}$ are

$$H^j(C_{l,k}) = \operatorname{Ker} X_j / \operatorname{Im} X_{j-1} = \begin{cases} 0 & j \neq 0, \\ \mathcal{L}_{l,k} & j = 0, \end{cases} \tag{3.82}$$

where $\mathcal{L}_{l,k}$ is the irreducible DVA module of $\lambda = \lambda_{l,k}$. The trace of operator $\mathcal{O}$ over $\mathcal{L}_{l,k}$ can be written as (see (2.122))

$$\operatorname{tr}_{\mathcal{L}_{l,k}} \mathcal{O} = \operatorname{tr}_{H^0(C_{l,k})} \mathcal{O}^{(0)} = \operatorname{tr}_{H^*(C_{l,k})} \mathcal{O} = \sum_{j \in \mathbb{Z}} (-1)^j \operatorname{tr}_{C_j} \mathcal{O}^{(j)}, \tag{3.83}$$

where $\mathcal{O}^{(j)}$ is an operator $\mathcal{O}$ realized on C_j.

If we use $x_-(z)$ instead of $x_+(z)$, another complex $C'_{l,k}$ in which k' of $\mathcal{F}_{l,k'}$ changes is obtained. The corresponding screening operator is

$$X'(z) = \oint_{C'} \underline{dz'} x_-(z') \frac{[u - u' - \frac{1}{2} + \hat{k}]}{[u - u' + \frac{1}{2}]}, \tag{3.84}$$

where $[u]$ in (A.5), $z = x^{2u}$, $z' = x^{2u'}$, $\underline{dz'} = dz'/(2\pi i z')$, and the integration contour C' is a simple closed curve that encircles $z' = x^{1+2rn} z$ but not $z' = x^{1-2r(n+1)} z$ $(n = 0, 1, 2, \dots)$. $\hat{k}$ is $\hat{k} = k \times \mathrm{id}$ on $\mathcal{F}_{l,k}$ (see Section 5.2). $X'(z)$ is well defined on $\mathcal{F}_{l',k'}$ $(\forall l', k')$, $X'(z)\colon \mathcal{F}_{l',k'} \mapsto \mathcal{F}_{l',k'-2}$, and $X'(z)$ satisfies

$$X'(z)^{p''} = 0, \tag{3.85}$$

$$[T_n, X'(z)^k] = 0 \text{ on } \mathcal{F}_{l,k'} \quad k' \equiv k \pmod{p''}. \tag{3.86}$$

By setting the BRST charge as $Q'_m = X'(1)^m$, another Felder complex $C'_{l,k} = \{C'_j, X'_j\colon C'_j \to C'_{j+1}\}$ is $C'_{2j} = \mathcal{F}_{l,k-2p''j}$, $C'_{2j+1} = \mathcal{F}_{l,-k-2p''j}$, $X'_{2j} = Q'_k$, $X'_{2j+1} = Q'_{p''-k}$, and the cohomology is $H^j(C'_{l,k}) = \operatorname{Ker} X'_j / \operatorname{Im} X'_{j-1} = \delta_{j,0} \mathcal{L}_{l,k}$.

3.4.4 Trigonometric Ruijsenaars–Schneider Model and Macdonald Symmetric Polynomial

The trigonometric Ruijsenaars–Schneider model (tRSM) is a relativistic version of the Calogero–Sutherland model. The Ruijsenaars–Schneider model is a many body system with sufficiently enough conserved quantities $H_{\pm k}$ $(k = 1, \dots, N_0)$ [42] (see also [43]),

$$H_{\pm k} = \sum_{\substack{I \subset \{1,\dots,N_0\} \\ |I| = k}} \prod_{\substack{i \in I \\ j \notin I}} h\big(\pm(q_i - q_j)\big)^{1/2} \cdot e^{\mp \sum_{i \in I} \theta_i} \times \prod_{\substack{i \in I \\ j \notin I}} h\big(\mp(q_i - q_j)\big)^{1/2}, \tag{3.87}$$

where θ_i $(i = 1, \ldots, N_0)$ is a rapidity and $\bar{q}_i$ is its conjugate variable and $q_i = \bar{q}_i/(mc)$ is a coordinate. Their dimensions are $[\theta_i] = 1$, $[\bar{q}_i] = \hbar$ and $[q_i] = L$. (Exactly speaking θ_i and $\bar{q}_i$ are related to the coordinate q_i and momentum p_i by $\bar{q}_i = mcq_i \cosh\theta_i$ and $p_i = mc\sinh\theta_i$. But we skip this procedure. See [44].) θ_j is

$$\theta_j = \frac{\hbar}{i}\frac{\partial}{\partial \bar{q}_j} = \frac{1}{i}\frac{\hbar}{mc}\frac{\partial}{\partial q_j}, \tag{3.88}$$

and $h(q)$ is

$$h(q) = \frac{\sigma(\bar{q} + i\bar{\beta})}{\sigma(\bar{q})}, \tag{3.89}$$

where $\bar{\beta}$ is a coupling constant and $\sigma(z)$ is the Weierstrass σ function. These conserved quantities commute mutually

$$[H_k, H_l] = 0 \quad (k, l = -N_0, \ldots, N_0). \tag{3.90}$$

This model is a 'relativistic' model because it has Poincaré invariance,

$$\begin{aligned} H &= mc^2\tfrac{1}{2}(H_{-1} + H_1), & [H, P] &= 0, \\ P &= mc(H_{-1} - H_1), & [H, B] &= i\hbar P, \\ B &= -\frac{1}{c}\sum_{i=1}^{N_0} \bar{q}_i, & [P, B] &= i\hbar\frac{1}{c^2}H. \end{aligned} \tag{3.91}$$

In the "nonrelativistic" limit (the speed of light $c \to \infty$), the Hamiltonian H becomes

$$\lim_{c\to\infty}(H - N_0 mc^2) = \sum_{j=1}^{N_0}\frac{1}{2m}\left(\frac{\hbar}{i}\frac{\partial}{\partial q_j}\right)^2 + \frac{1}{m}\bar{\beta}(\bar{\beta} - \hbar)\sum_{\substack{i,j=1\\ i<j}}^{N_0}\wp(q_i - q_j), \tag{3.92}$$

where $\wp(z)$ is the Weierstrass $\wp$ function and we have rescaled the periods of $\sigma(z)$.

tRSM is a trigonometric case of RSM,

$$h(q) = \frac{\sin[\pi(\bar{q} + i\bar{\beta})/(mcL)]}{\sin[\pi\bar{q}/(mcL)]}. \tag{3.93}$$

Its "nonrelativistic" limit is

$$\lim_{c\to\infty}(H_{tRS} - N_0 mc^2) = H_{\rm CS}, \tag{3.94}$$

where $H_{\rm CS}$ is given in (2.153). Let us introducing dimensionless quantities x_j and β in (2.158) and

$$q = e^{-2\pi\hbar/(mcL)}, \quad t = q^\beta. \tag{3.95}$$

Removing the contribution from the ground state, we obtain

$$\Delta^{-1/2} H_{\pm k} \Delta^{1/2} = t^{\mp k(N_0-1)/2} D_k(q^{\pm 1}, t^{\pm 1}). \tag{3.96}$$

Here Δ is

$$\Delta = \Delta(x) = \Delta(x; q, t) = \prod_{\substack{i,j=1 \\ i \neq j}}^{N_0} \frac{(x_i/x_j; q)_\infty}{(tx_i/x_j; q)_\infty}, \tag{3.97}$$

and $D_k(q,t)$ ($k = 1, \ldots, N_0$) is the Macdonald operator

$$D_k(q,t) = t^{k(k-1)/2} \sum_{\substack{I \subset \{1,\ldots,N_0\} \\ |I|=k}} \prod_{\substack{i \in I \\ j \notin I}} \frac{tx_i - x_j}{x_i - x_j} \prod_{i \in I} q^{D_i}, \tag{3.98}$$

where $D_i = x_i \partial/\partial x_i$. q^{D_i} is a q-shift, $q^{D_i} f(x_1, \ldots, x_i, \ldots, x_{N_0}) = f(x_1, \ldots, qx_i, \ldots, x_{N_0})$.

The Macdonald symmetric polynomial $P_\lambda = P_\lambda(x; q, t)$ is a multivariable orthogonal polynomial with two parameters q and t, which is determined uniquely by the following conditions [36],

$$\text{(i)} \qquad P_\lambda = \sum_{\mu \leq \lambda} u_{\lambda,\mu} m_\mu(x), \qquad u_{\lambda,\lambda} = 1, \tag{3.99}$$

$$\text{(ii)} \qquad \langle P_\lambda, P_\mu \rangle_{q,t} = 0 \qquad \text{if } \lambda \neq \mu, \tag{3.100}$$

where the inner product is

$$\langle p_\lambda, p_\mu \rangle_{q,t} = \delta_{\lambda,\mu} \prod_i i^{m_i} m_i! \cdot \prod_{i=1}^{\ell(\lambda)} \frac{1 - q^{\lambda_i}}{1 - t^{\lambda_i}}. \tag{3.101}$$

The condition (ii) can be replaced by (ii)′,

$$\text{(ii)}' \qquad D_1(q,t) P_\lambda = \sum_{i=1}^{N_0} t^{N_0 - i} q^{\lambda_i} \cdot P_\lambda. \tag{3.102}$$

Moreover the Macdonald symmetric polynomial is the simultaneous eigenfunction of the Macdonald operators,

$$\sum_{k=0}^{N_0} (-u)^k D_k(q,t) P_\lambda(x; q, t) = \prod_{i=1}^{N_0} (1 - u t^{N_0 - i} q^{\lambda_i}) \cdot P_\lambda(x; q, t), \tag{3.103}$$

or

$$D_k(q^{\pm 1}, t^{\pm 1}) P_\lambda(x; q, t) = \sum_{1 \leq i_1 < \cdots < i_k \leq N_0} \prod_{l=1}^{k} t^{N_0 - i_l} q^{\lambda_{i_l}} \cdot P_\lambda(x; q, t). \tag{3.104}$$

Remark that $P_\lambda(x;q,t) = P_\lambda(x;q^{-1},t^{-1})$ and $\langle f,g\rangle_{q^{-1},t^{-1}} = (qt^{-1})^{N_0} \times \langle f,g\rangle_{q,t}$. In the 'conformal limit',

$$t = q^\beta, \quad q \to 1, \ \beta\text{: fixed}, \tag{3.105}$$

which corresponds to the nonrelativistic limit ($c \to \infty$) (see (3.95)), the Macdonald polynomial reduces to the Jack polynomial

$$\lim_{\substack{q\to 1\\ t=q^\beta}} P_\lambda(x;q,t) = J_\lambda(x;\beta). \tag{3.106}$$

$\beta \to 1$ limit of the Jack polynomial is the Schur polynomial $s_\lambda(x) = \lim_{\beta\to 1} J_\lambda(x;\beta)$. $t \to q$ limit of the Macdonald polynomial is also $s_\lambda(x) = \lim_{t\to q} P_\lambda(x;q,t)$.

There exists another inner product,

$$\langle f,g\rangle'_{l;q,t} = \frac{1}{l!}\oint \prod_{j=1}^{l} \underline{dx}_j \cdot \Delta(x;q,t) f(\bar{x}) g(x), \quad \underline{dx}_j = \frac{dx_j}{2\pi i x_j}, \tag{3.107}$$

which satisfies

$$\langle\ ,\ \rangle'_{l;q,t} \propto \langle\ ,\ \rangle_{q,t}. \tag{3.108}$$

Like the Jack symmetric polynomial, the Macdonald symmetric polynomial has an integral representation obtained by the following two transformations,

$$\mathcal{G}_k: \qquad (\mathcal{G}_k f)(x_1,\ldots,x_l) = \prod_{i=1}^{l} x^k \cdot f(x_1,\ldots,x_l), \tag{3.109}$$

$$\begin{aligned}\mathcal{N}_{l',l}: \qquad &(\mathcal{N}_{l',l} f)(x'_1,\ldots,x'_{l'})\\ &= \oint \prod_{j=1}^{l} \underline{dx}_j \cdot \Pi(x',\bar{x})\Delta(x) f(x_1,\ldots,x_l),\end{aligned} \tag{3.110}$$

$$\Pi(x,y) = \Pi(x,y;q,t) = \prod_i \prod_j \frac{(tx_iy_j;q)_\infty}{(x_iy_j;q)_\infty}. \tag{3.111}$$

The Macdonald symmetric polynomial with λ is expressed as [45]

$$P_\lambda(x) \propto \mathcal{N}_{l,l_1}\mathcal{G}_{k_1}\mathcal{N}_{l_1,l_2}\mathcal{G}_{k_2}\cdots\mathcal{N}_{l_{N-2},l_{N-1}}\mathcal{G}_{k_{N-1}}\mathcal{N}_{l_{N-1},0}\cdot 1 \tag{3.112}$$

($\lambda' = ((l_i)^{k_i},(l_2)^{k_2},\ldots,(l_{N-1})^{k_{N-1}})$), because the Macdonald symmetric polynomial has two properties

$$P_{(k^l)+\lambda} = \mathcal{G}_k P_\lambda \qquad (l \text{ variables}), \tag{3.113}$$

$$P_\lambda = \frac{\langle P_\lambda,P_\lambda\rangle_{q,t}}{l!\,\langle P_\lambda,P_\lambda\rangle'_{l;q,t}}\mathcal{N}_{l',l}P_\lambda \qquad \begin{array}{l}(P_\lambda \text{ in LHS: } l' \text{ variables})\\ (P_\lambda \text{ in RHS: } l \text{ variables}).\end{array} \tag{3.114}$$

This second property is valid if $\mathcal{N}_{l',l}$ is modified in the following way [18],

$$P_\lambda \propto \mathcal{N}'_{l',l} P_\lambda,$$
$$(\mathcal{N}'_{l',l} f)(x'_1, \ldots, x'_{l'}) = \oint \prod_{j=1}^{l} \underline{dx}_j \cdot \Pi(x', \bar{x}) \Delta(x) C(x) f(x_1, \ldots, x_l), \quad (3.115)$$

where $C(x) = C(x_1, \ldots, x_l)$ is an arbitrary pseudoconstant with respect to the q-shift, $q^{D_i} C(x) = C(x)$ $(\forall i)$. (Remark that $\langle D_1(q,t) f, g \rangle'_{l;q,t} = \langle f, D_1(q,t) g \rangle'_{l;q,t}$.)

For example, the Macdonald symmetric polynomial with a rectangular Young diagram (k^l) is

$$P_{(k^l)}(x) \propto \oint \prod_{j=1}^{l} \underline{dz}_j \cdot \prod_{j=1}^{l} \exp\biggl(\sum_{n>0} \frac{1}{n} \frac{1-t^n}{1-q^n} z_j^n \sum_i x_i^n \biggr)$$
$$\times \prod_{\substack{i,j=1 \\ i \neq j}}^{l} \frac{(z_i/z_j; q)_\infty}{(t z_i/z_j; q)_\infty} \cdot \prod_{j=1}^{l} z_j^{-k}. \quad (3.116)$$

Next let us consider bosonization of the Macdonald operator $\mathcal{D} = D_1(q,t)$ [17, 18]. $\mathcal{D}$ can be expressed by power sum polynomials,

$$\mathcal{D} = \sum_{i=1}^{N_0} \prod_{j \neq i} \frac{t x_i - x_j}{x_i - x_j} \cdot q^{D_i}$$
$$= \frac{1-t^{N_0}}{1-t} + \frac{t^{N_0}}{t-1} \biggl(\oint_0 \underline{dz} \exp\Bigl(\sum_{n>0} \frac{1}{n} (1-t^{-n}) p_n z^n \Bigr)$$
$$\times \exp\Bigl(\sum_{n>0} (q^n - 1) \frac{\partial}{\partial p_n} z^{-n} \Bigr) - 1 \biggr). \quad (3.117)$$

Proof. Since $q^{D_i} p_n = p_n + x_i^n (q^n - 1)$, q^{D_i} can be written as

$$q^{D_i} = \exp\Bigl(\sum_{n>0} x_i^n (q^n - 1) \frac{\partial}{\partial p_n} \Bigr) = \sum_{n \geq 0} x_i^n \tilde{q}_n \Bigl(-m \frac{\partial}{\partial p_n}; q \Bigr), \quad (3.118)$$

where $\tilde{q}_n(p_m; t) = q_n(x_m; t)$ is (see [36, p. 209])

$$q_n(x;t) = (1-t) \sum_{i=1}^{N_0} x_i^n \prod_{j \neq i} \frac{x_i - t x_j}{x_i - x_j},$$
$$\sum_{n \geq 0} q_n(x;t) z^n = \exp\Bigl(\sum_{n>0} \frac{1}{n} (1-t^n) p_n z^n \Bigr). \quad (3.119)$$

Then we have

$$\begin{aligned}\mathcal{D} &= \sum_{i=1}^{N_0} \prod_{j\neq i} \frac{tx_i - x_j}{x_i - x_j} \cdot \sum_{n\geq 0} x_i^n \tilde{q}_n\Big(-m\frac{\partial}{\partial p_n}; q\Big) \\ &= \frac{1-t^{N_0}}{1-t} + \frac{t^{N_0}}{t-1} \sum_{n>0} q_n(x_m; t^{-1}) \tilde{q}_n\Big(-m\frac{\partial}{\partial p_m}; q\Big) \end{aligned} \tag{3.120}$$

and get the result by (3.119). □

To bosonize this operator we introduce boson oscillator $\tilde{a}_n$ $(n \in \mathbb{Z}_{\neq 0})$,

$$[\tilde{a}_n, \tilde{a}_m] = n\frac{1-q^{|n|}}{1-t^{|n|}} \delta_{n+m,0}. \tag{3.121}$$

A state in the Fock space is mapped to a symmetric function by

$$|f\rangle \mapsto f(x) = \langle \alpha | \exp\Big(\sum_{n>0} \frac{1}{n} \frac{1-t^n}{1-q^n} \tilde{a}_n p_n\Big) |f\rangle, \tag{3.122}$$

namely

$$\tilde{a}_{-n} \leftrightarrow p_n, \quad \tilde{a}_n \leftrightarrow n\frac{1-q^n}{1-t^n}\frac{\partial}{\partial p_n}. \tag{3.123}$$

This normalization for $\tilde{a}_n$ is chosen so that it produces the inner product (3.101). Then $\mathcal{D}$ is realized on the Fock space:

$$\mathcal{D}\langle \alpha | \exp\Big(\sum_{n>0} \frac{1}{n} \frac{1-t^n}{1-q^n} \tilde{a}_n p_n\Big) = \langle \alpha | \exp\Big(\sum_{n>0} \frac{1}{n} \frac{1-t^n}{1-q^n} \tilde{a}_n p_n\Big) \widehat{\mathcal{D}}. \tag{3.124}$$

The bosonized operator $\widehat{\mathcal{D}}$ is

$$\begin{aligned}\widehat{\mathcal{D}} &= \frac{1-t^{N_0}}{1-t} + \frac{t^{N_0}}{t-1}\Big(\oint_0 \underline{dz}\, \exp\Big(\sum_{n>0} \frac{1}{n}(1-t^{-n})\tilde{a}_{-n} z^n\Big) \\ &\qquad \times \exp\Big(-\sum_{n>0}(1-t^n)\tilde{a}_n z^{-n}\Big) - 1\Big) \\ &= \frac{1-t^{N_0}}{1-t} + \frac{t^{N_0}}{t-1}\Big(\oint_0 \underline{dz}\, \Big(\psi(z)T(z) \\ &\qquad - \exp\Big(-\sum_{n>0} h_n x^n z^{-n}\Big) x^{-2\sqrt{r(r-1)}a_0'}\Big) - 1\Big) \\ &= \frac{1-t^{N_0}}{1-t} + \frac{t^{N_0}}{t-1}\Big(\sum_{n\geq 0} \psi_{-n} T_n - x^{-2\sqrt{r(r-1)}a_0'} - 1\Big), \end{aligned} \tag{3.125}$$

where $T(z)$ is the DVA current (3.38) and $\psi(z)$ is

$$\psi(z) = \sum_{n\geq 0} \psi_{-n} z^n = \exp\Big(\sum_{n>0} h_{-n} x^{-n} z^n\Big) x^{-\sqrt{r(r-1)}a_0'}. \tag{3.126}$$

Here $\tilde{a}_n$ is expressed in terms of h_n introduced in (3.36)

$$\tilde{a}_n = n\frac{1}{x^{rn} - x^{-rn}}h_n, \ \tilde{a}_{-n} = nx^{-n}\frac{x^n + x^{-n}}{x^{rn} - x^{-rn}}h_{-n}, \quad (n > 0), \tag{3.127}$$

and parameters are identified as

$$q = x^{2(r-1)}, \quad t = q^\beta = x^{2r}, \quad p = qt^{-1} = x^{-2}, \quad \beta = \frac{r}{r-1}. \tag{3.128}$$

When $\widehat{\mathcal{D}}$ acts on the singular vector (3.61), $\sum_{n>0}\psi_{-n}T_n$ term vanishes because of the property of singular vector and the remaining terms are already diagonal,

$$\widehat{\mathcal{D}}|\chi_{-l,-k}\rangle = \sum_{i=1}^{N_0} t^{N_0-i}q^{\lambda_i}|\chi_{-l,-k}\rangle, \tag{3.129}$$

where the partition λ is $\lambda = (k^l, 0^{N_0-l})$. Therefore $|\chi_{-l,-k}\rangle$ is an eigenstate of $\widehat{\mathcal{D}}$, i.e., it gives the Macdonald symmetric polynomial by the map (3.122). In fact the polynomial obtained from (3.61) by (3.122) agrees with the integral representation of the Macdonald symmetric polynomial with partition (k^l) in (3.116) up to $x_i \to -x_i$ and $C(z)$ in (3.62), which is irrelevant because $C(z)$ is a pseudoconstant with respect to the q-shift, $q^{D_{z_i}}C(z) = C(z)$ $(\forall i)$.

The Macdonald symmetric polynomial with general partition whose Young diagram is composed of $N-1$ rectangles is related to the singular vector of the deformed W_N algebra [18].

We have established the relation between the Macdonald symmetric polynomials and the singular vectors of DVA in free field realization. When the DVA was firstly formulated in [17], this relation was not the derived property but the guiding principle to find the DVA. At that time we knew the two facts:

(i) In the free field realization, the singular vectors of the Virasoro and W_N algebras realize the Jack symmetric polynomials [32, 35].

(ii) The Jack symmetric polynomials have the good q-deformation, the Macdonald symmetric polynomials [36].

Based on these, we set up the following 'natural' problem:

- Construct the algebras whose singular vectors in the free field realization realize the Macdonald symmetric polynomials.

The resultant algebra are worth being called quantum deformation (q-deformation) of the Virasoro and W_N algebras in this sense. This scenario is

illustrated in the following figure,

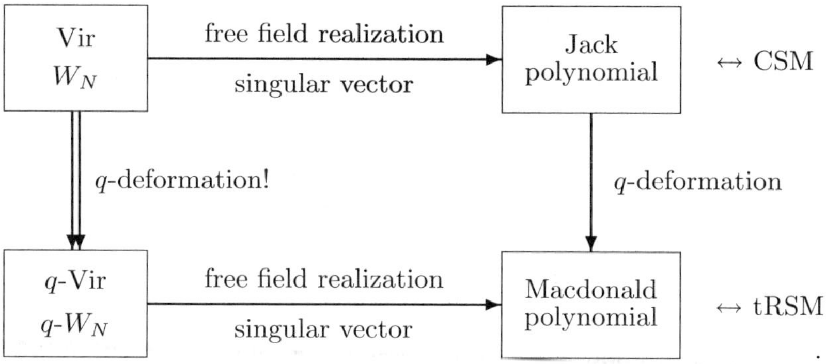

See [17] how the DVA was found.

3.5 Higher DVA Currents

In this subsection we present higher DVA currents. We define higher DVA currents $T_{(n)}(z)$ $(n = 1, 2, \dots)$ by 'fusion',

$$\begin{aligned} &T_{(1)}(z) = T(z), \quad T_{(0)}(z) = 1, \\ &T_{(n)}(z) = f_{(1),(n-1)}\left(x^{rn}\frac{z}{z'}\right)T_{(1)}(x^{-r(n-1)}z')T_{(n-1)}(x^{r}z)|_{z'\to z}, \end{aligned} \tag{3.130}$$

where $f_{(n),(m)}(z)$ is given by

$$f_{(n),(m)}(z) = \prod_{i=0}^{n-1}\prod_{j=0}^{m-1} f(x^{r(2j-m+1)-r(2i-n+1)}z) = f_{(m),(n)}(z), \tag{3.131}$$

with $f(z)$ in (3.2). In free boson realization (3.38), we have

$$T_{(n)}(z) = \sum_{i=0}^{n} {}_nB_i\Lambda_{n,i}(z), \tag{3.132}$$

where ${}_nB_i$ and $\Lambda_{n,i}(z)$ are

$${}_nB_i = \frac{b_i\, b_{n-i}}{b_n}, \quad b_i = \prod_{i'=0}^{i-1}\frac{[ri'+1]_x}{[r(i'+1)]_x}, \tag{3.133}$$

$$\Lambda_{n,i}(z) = :\prod_{i'=0}^{i-1}\Lambda_-(x^{r(2i'-n+1)}z)\cdot\prod_{i'=i}^{n-1}\Lambda_+(x^{r(2i'-n+1)}z):. \tag{3.134}$$

$T_{(n)}$ consists of $n+1$ terms, which corresponds to the spin $\frac{n}{2}$ representation of A_1. $T_{(n)}(z)$ satisfies the relation

$$
\begin{aligned}
&f_{(n),(m)}\Big(\frac{z_2}{z_1}\Big)T_{(n)}(z_1)T_{(m)}(z_2) - T_{(m)}(z_2)T_{(n)}(z_1)f_{(m),(n)}\Big(\frac{z_1}{z_2}\Big) \\
&= -(x-x^{-1})\sum_{a=1}^{n} {}_nB_{am}B_a[ra]_x[ra+1]_x\prod_{b=1}^{a}\frac{[rb-1]_x}{[rb+1]_x} \\
&\quad\times\Big(\delta\Big(x^{r(n+m-2a)+2}\frac{z_2}{z_1}\Big)\,{}^{\circ}_{\circ}T_{(n-a)}(x^{ra}z_1)T_{(m-a)}(x^{-ra}z_2){}^{\circ}_{\circ} \\
&\qquad-\delta\Big(x^{-r(n+m-2a)-2}\frac{z_2}{z_1}\Big)\,{}^{\circ}_{\circ}T_{(n-a)}(x^{-ra}z_1)T_{(m-a)}(x^{ra}z_2){}^{\circ}_{\circ}\Big). \\
&\hspace{20em}(n\le m) \qquad (3.135)
\end{aligned}
$$

Here normal ordering ${}^{\circ}_{\circ}T_{(a)}(\alpha z)T_{(b)}(z){}^{\circ}_{\circ}$ is

$$
\begin{aligned}
&{}^{\circ}_{\circ}T_{(a)}(\alpha z)T_{(b)}(z){}^{\circ}_{\circ} \\
&\stackrel{\text{def}}{=}\sum_{n=-\infty}^{\infty}\sum_{m=0}^{\infty}\sum_{\ell=0}^{m} f_{(a),(b),\ell}(\alpha^{m-\ell}T_{(a),-m}T_{(b),n+m} \\
&\qquad\qquad + \alpha^{\ell-m-1}T_{(b),n-m-1}T_{(a),m+1})z^{-n} \\
&= \oint\frac{dy}{2\pi iy}\Big(\sum_{j=0}^{\infty}\Big(\frac{\alpha z}{y}\Big)^j\cdot f_{(a),(b)}\Big(\frac{z}{y}\Big)T_{(a)}(y)T_{(b)}(z) \\
&\qquad+\sum_{j=0}^{\infty}\Big(\frac{y}{\alpha z}\Big)^{j+1}\cdot T_{(b)}(z)T_{(a)}(y)f_{(a),(b)}\Big(\frac{y}{z}\Big)\Big) \\
&= \oint_{|y|>|\alpha z|}\frac{dy}{2\pi iy}\frac{1}{1-\alpha z/y}\cdot f_{(a),(b)}\Big(\frac{z}{y}\Big)T_{(a)}(y)T_{(b)}(z) \\
&\quad+\oint_{|y|<|\alpha z|}\frac{dy}{2\pi iy}\frac{y/(\alpha z)}{1-y/(\alpha z)}\cdot T_{(b)}(z)T_{(a)}(y)f_{(a),(b)}\Big(\frac{y}{z}\Big) \\
&= \oint_{\alpha z}\frac{dy}{2\pi i}\frac{1}{y-\alpha z}\cdot f_{(a),(b)}\Big(\frac{z}{y}\Big)T_{(a)}(y)T_{(b)}(z), \qquad (3.136)
\end{aligned}
$$

where mode expansions are

$$
f_{(a),(b)}(z)=\sum_{\ell=0}^{\infty}f_{(a),(b),\ell}z^{\ell}, \quad T_{(a)}(z)=\sum_{n\in\mathbb{Z}}T_{(a),n}z^{-n}. \qquad (3.137)
$$

This is a generalization of the normal ordering for currents used in the CFT,

$$
(AB)(z)=\oint_z\frac{dy}{2\pi i}\frac{1}{y-z}\cdot A(y)B(z). \qquad (3.138)
$$

We remark that $T_{(n)}(z)$ is a composite field of $T(z)$,

$$
{}^{\circ}_{\circ}T_{(n)}(x^{-rm}z)T_{(m)}(x^{rn}z){}^{\circ}_{\circ}=T_{(n+m)}(z). \qquad (3.139)
$$

We remark also that other higher currents are obtained by replacing r by $1-r$.

Concerning the higher currents for the deformed W_N algebra, see Appendix C in [46] where delta function terms are neglected. (The arguments of $W_{(n)}(u)$ and $f_{(n),(m)}(u,v)$ in [46] should be shifted in order to compare them with $T_{(n)}(z)$ and $f_{(n),(m)}(z)$ here.)

4 Solvable Lattice Models and Elliptic Algebras

4.1 Solvable Lattice Models and Yang–Baxter Equation

A statistical lattice model is a statistical mechanical system on a lattice [6]. In this lecture we consider classical systems on two-dimensional space square lattice. There are two types of lattice models, vertex models and face models. For vertex models dynamical variables 'spin' are located on edges and the Boltzmann weights are assigned to each vertex. On the other hand, for face models, dynamical variables 'height' are located on vertices and the Boltzmann weights are assigned to each face.

Vertex model: 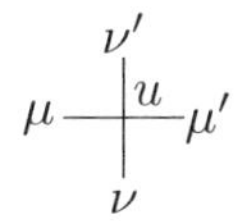

$R_{\mu'\nu',\mu\nu}(u)$, $\quad u=u_1-u_2$ 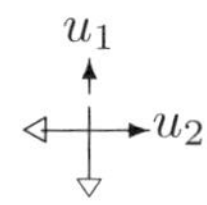

Face model: 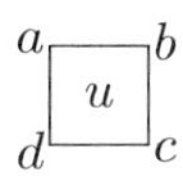

$W\left(\begin{matrix} a & b \\ c & d \end{matrix}\middle| u\right)$, $\quad u=u_1-u_2$ 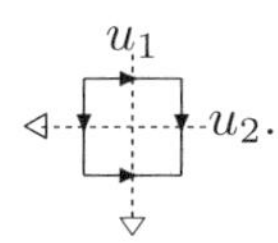

(4.1)

Here μ, ν, μ', ν' are spins, a, b, c, d are heights, and $R_{\mu'\nu',\mu\nu}(u)$ and $W\left(\begin{smallmatrix} a & b \\ c & d \end{smallmatrix}\middle| u\right)$ are Boltzmann weights. We have assumed that the spectral parameters enter in the Boltzmann weight only through their difference u. Roughly speaking, a vertex model on a lattice is a face model on its dual lattice. A face model is also called an interaction round a face (IRF) model or a solid on solid (SOS) model. If the height variable takes a finite number of states, an SOS model is called a restricted SOS (RSOS) model.

So-called solvable lattice model is a statistical lattice model whose Boltzmann weight satisfies the Yang–Baxter equation (YBE) [5–7],

$$R^{(12)}(u-v)R^{(13)}(u)R^{(23)}(v)=R^{(23)}(v)R^{(13)}(u)R^{(12)}(u-v), \tag{4.2}$$

for a vertex model, and

$$\begin{aligned}
&\sum_g W\left(\begin{matrix} b & c \\ g & d \end{matrix}\middle| u-v\right) W\left(\begin{matrix} a & b \\ f & g \end{matrix}\middle| u\right) W\left(\begin{matrix} f & g \\ e & d \end{matrix}\middle| v\right) \\
&\qquad = \sum_g W\left(\begin{matrix} a & b \\ g & c \end{matrix}\middle| v\right) W\left(\begin{matrix} g & c \\ e & d \end{matrix}\middle| u\right) W\left(\begin{matrix} a & g \\ f & e \end{matrix}\middle| u-v\right),
\end{aligned} \tag{4.3}$$

for a face model. Here $R(u)=\big(R_{\mu'\nu',\mu\nu}(u)\big)$ is a matrix on $V\otimes V$, (4.2) is an equation on $V\otimes V\otimes V$ and the superscripts (12), (13), etc., refer to the tensor components, e.g., $R^{(12)}(u)=R(u)\otimes\mathrm{id}$. YBE (4.3) is also satisfied by the new Boltzmann weight obtained by the following transformation (gauge transformation),

$$W'\left(\begin{matrix}a & b\\ c & d\end{matrix}\,\middle|\, u\right)=\frac{F(a,b)F(b,d)}{F(a,c)F(c,d)}W\left(\begin{matrix}a & b\\ c & d\end{matrix}\,\middle|\, u\right), \tag{4.4}$$

where $F(a,b)$ is an arbitrary u-independent function of a and b. The following graphical expression of the Yang–Baxter equation will help us.

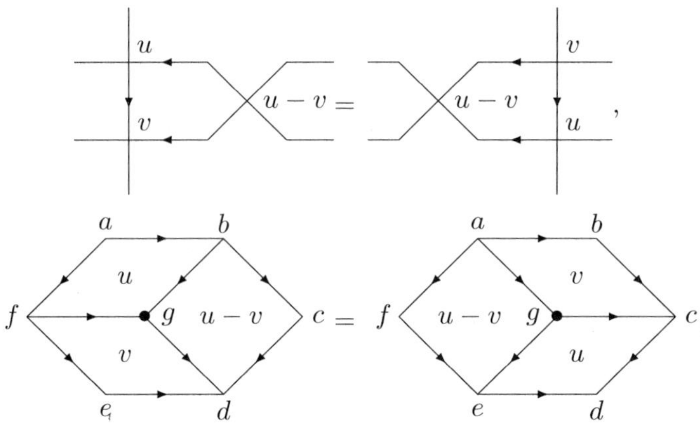

(4.5)

where summation is taken over the height g at site $\bullet$.

Solutions of the Yang–Baxter equation were studied by many people and the relation to the Lie algebras was clarified [5, 8–10]. Three types of solutions are known; rational, trigonometric and ellipic. Associated for each type of solution (R matrix), algebras are defined [11],

$$\begin{aligned}&\text{rational} &&\longrightarrow \text{Yangian},\\ &\text{trigonometric} &&\longrightarrow \text{quantum group (quantum algebra)},\\ &\text{elliptic} &&\longrightarrow \text{elliptic quantum group (elliptic algebra)}.\end{aligned} \tag{4.6}$$

Elliptic quantum groups will be introduced in Section 4.4.

Example: Vector representation of A_n algebra [10]

$$\begin{aligned}&W\left(\begin{matrix}a & a+\hat{\mu}\\ a+\hat{\mu} & a+2\hat{\mu}\end{matrix}\,\middle|\, u\right)=\frac{[1+u]}{[1]},\\ &W\left(\begin{matrix}a & a+\hat{\mu}\\ a+\hat{\mu} & a+\hat{\mu}+\hat{\nu}\end{matrix}\,\middle|\, u\right)=\frac{[a_\mu-a_\nu-u]}{[a_\mu-a_\nu]}\quad(\mu\neq\nu),\end{aligned} \tag{4.7}$$

$$W\left(\begin{matrix} a & a+\hat{\nu} \\ a+\hat{\mu} & a+\hat{\mu}+\hat{\nu} \end{matrix}\,\middle|\, u\right) = \frac{[u]}{[1]}\left(\frac{[a_\mu - a_\nu + 1][a_\mu - a_\nu - 1]}{[a_\mu - a_\nu]^2}\right)^{1/2} \quad (\mu \neq \nu).$$

Notation is as follows. a is an element of the weight space of $A_n^{(1)}$. $\hat{\mu}$ is $\hat{\mu} = \varepsilon_\mu - (\varepsilon_1 + \cdots + \varepsilon_{n+1})/(n+1)$ $(\mu = 1, 2, \ldots, n+1)$, where ε_μ is an orthonomal basis of $\mathbb{C}^{n+1}$. a_μ is $\langle a + \rho, \hat{\mu}\rangle$, where ρ is a sum of fundamental weight of $A_1^{(1)}$. So $\bar{a} + \bar{\rho} = \sum_{\mu=1}^{n+1} a_\mu \varepsilon_\mu$, $\sum_{\mu=1}^{n+1} a_\mu = 0$.

The partition function of a lattice model is

$$Z = \sum_{\text{config.}} \prod (\text{Boltzmann weight}), \tag{4.8}$$

where the summation is taken over all configurations and the product is taken over all vertices or faces. This partition function can be calculated by using the transfer matrix. The row-to-row transfer matrix $\mathcal{T}(u)$ is

$$\mathcal{T}(u) = \left(\mathcal{T}(u)_{\underline{\nu}',\underline{\nu}}\right) = \sum_{\mu_1,\ldots,\mu_N}$$ 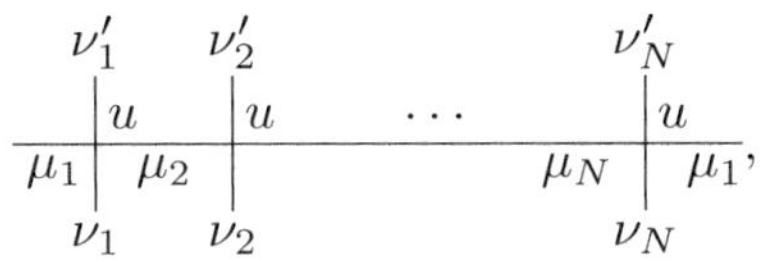

$$\mathcal{T}(u) = \left(\mathcal{T}(u)_{\underline{a},\underline{b}}\right) =$$ 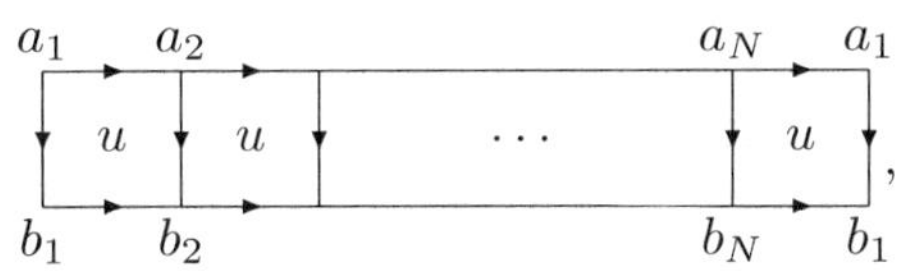 (4.9)

where $\underline{\nu} = (\nu_1, \ldots, \nu_N)$, $\underline{\nu}' = (\nu_1', \ldots, \nu_N')$, $\underline{a} = (a_1, \ldots, a_N)$ and $\underline{b} = (b_1, \ldots, b_N)$. Here lattice size is N (horizontal) by M (vertical), and we assume the periodic boundary condition. Then the partition function is

$$Z = \operatorname{tr}\left(\mathcal{T}(u)^M\right). \tag{4.10}$$

In the thermodynamic limit $(N, M \to \infty)$, only the maximal eigenvalue of the transfer matrix contributes to the partition function. Once we know information about the eigenvalues and eigenvectors of the transfer matrix, we can calculate various physical quantities.

If the Boltzmann weight satisfies the Yang–Baxter equation, the transfer matrix has good properties. For example the trasnfer matrices with different spectral parameters commute with each other,

$$[\mathcal{T}(u), \mathcal{T}(v)] = 0. \tag{4.11}$$

To show this let us introduce the monodromy matrix $\widehat{T}(u)$

$$\widehat{T}(u) = \sum_{\mu_2,\ldots,\mu_N}$$
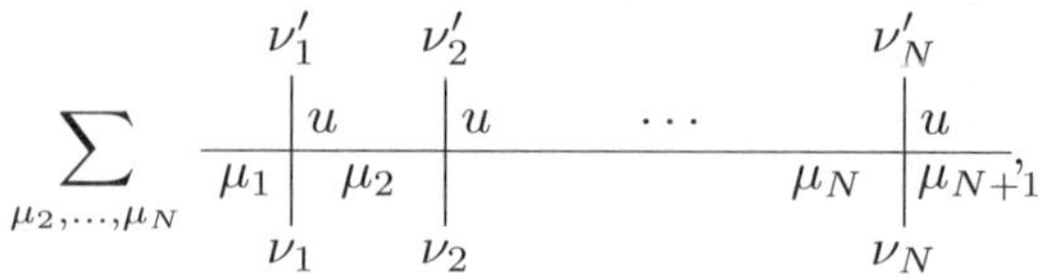

$$\widehat{T}(u) =$$
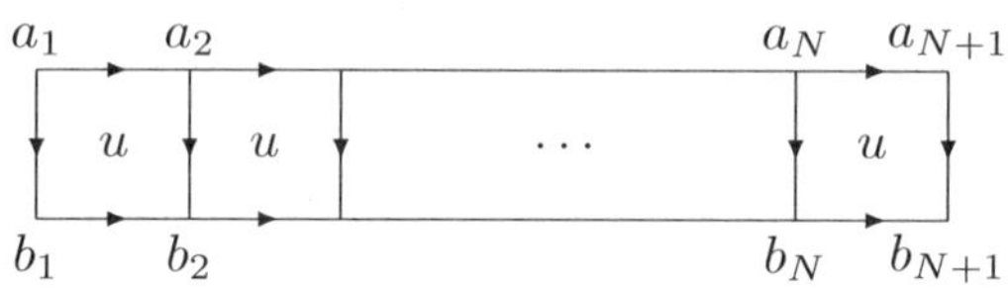

(4.12)

Then the transfer matrix is expressed as a trace of the monodromy matrix

$$T(u) = \operatorname{tr}\widehat{T}(u) \quad \Bigl(\operatorname{tr} = \sum_{\mu_1,\mu_{N+1}} \delta_{\mu_1,\mu_{N+1}} \text{ or } \sum_{a_{N+1}} \delta_{a_1,a_{N+1}}\Bigr). \tag{4.13}$$

The following figure (for face model see similar figure in (4.50))

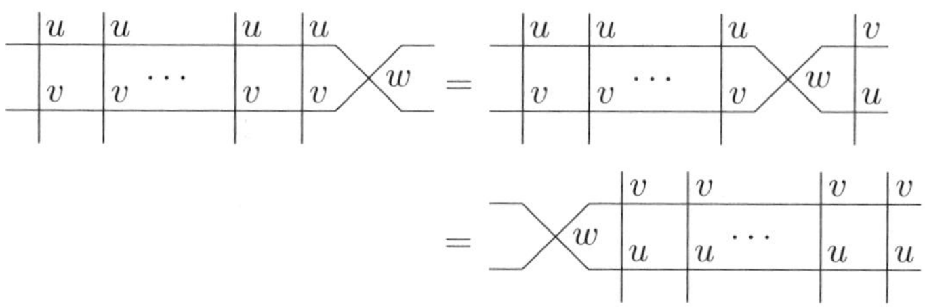

$(w = u - v)$ (4.14)

shows

$$R(u-v)\widehat{T}(u)\widehat{T}(v) = \widehat{T}(v)\widehat{T}(u)R(u-v) \tag{4.15}$$

or

$$W(u-v)\widehat{T}(u)\widehat{T}(v) = \widehat{T}(v)\widehat{T}(u)W(u-v). \tag{4.16}$$

(Here we do not write tensor product explicitly.) Therefore we obtain (4.11) by taking the trace of $R(u-v)\widehat{T}(u)\widehat{T}(v)R(u-v)^{-1} = \widehat{T}(v)\widehat{T}(u)$.

If we interpret an R matrix as an L matrix (L operator) whose entries are operators acting on vertical vector space,

$$L(u) = \bigl(L(u)_{\mu',\mu}\bigr), \quad \bigl(L(u)_{\mu',\mu}\bigr)_{\nu',\nu} =$$
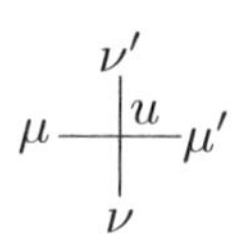

(4.17)

then (4.2) is rewritten as

$$R^{(12)}(u-v)L^{(1)}(u)L^{(2)}(v) = L^{(2)}(v)L^{(1)}(u)R^{(12)}(u-v). \tag{4.18}$$

The monodoromy matrix is

$$\widehat{\mathcal{T}}(u) = L_1(u)L_2(u)\cdots L_N(u), \tag{4.19}$$

where $L_i(u)$ acts nontrivially on ith component, $L_i(u) = 1 \otimes \cdots \otimes \underbrace{L(u)}_{i} \otimes \cdots \otimes 1$.

The R matrix obtained from the Boltzmann weight of the face model satisfies the dynamical Yang–Baxter equation [24],

$$\begin{aligned} &R^{(12)}(u-v;\lambda+h^{(3)})R^{(13)}(u;\lambda)R^{(23)}(v;\lambda+h^{(1)}) \\ &\qquad = R^{(23)}(v;\lambda)R^{(13)}(u;\lambda+h^{(2)})R^{(12)}(u-v;\lambda). \end{aligned} \tag{4.20}$$

Here h is an element of the Cartan subalgebra (see Sections 4.3, 4.4). For simplicity we consider the A_n vector representation case. Associated to the face Boltzmann weight (4.7), the R matrix is introduced by

$$R(u,\lambda) = \sum_{\substack{\mu,\nu,\mu',\nu' \\ \mu+\nu=\mu'+\nu'}} W\left(\begin{matrix} \lambda & \lambda+\hat{\nu}' \\ \lambda+\hat{\mu} & \lambda+\hat{\mu}+\hat{\nu} \end{matrix} \,\middle|\, u\right) E_{\mu',\mu} \otimes E_{\nu',\nu}, \tag{4.21}$$

where $E_{\mu',\mu}$ is a matrix unit, $(E_{\mu',\mu})_{\nu',\nu} = \delta_{\mu',\mu}\delta_{\nu',\nu}$. h acts on $E_{\mu',\mu}$ as $hE_{\mu',\mu} = \hat{\mu}' E_{\mu',\mu}$, $E_{\mu',\mu}h = \hat{\mu} E_{\mu',\mu}$. Using this and (4.3) we can show (4.20). Similarly the matrix $L(u,\lambda)$

$$\begin{aligned} L(u,\lambda) &= \big(L(u,\lambda)_{\mu',\mu}\big) = \sum_{\mu,\mu'} L(u,\lambda)_{\mu',\mu} E_{\mu',\mu}, \\ L(u,\lambda)_{\mu',\mu} &= \sum_{\substack{\nu,\nu' \\ \mu+\nu=\mu'+\nu'}} W\left(\begin{matrix} \lambda & \lambda+\hat{\nu}' \\ \lambda+\hat{\mu} & \lambda+\hat{\mu}+\hat{\nu} \end{matrix} \,\middle|\, u\right) E_{\nu',\nu}, \end{aligned} \tag{4.22}$$

satisfies the dynamical RLL-relation

$$\begin{aligned} &R^{(12)}(u-v;\lambda+h)L^{(1)}(u;\lambda)L^{(2)}(v;\lambda+h^{(1)}) \\ &\qquad = L^{(2)}(v;\lambda)L^{(1)}(u;\lambda+h^{(2)})R^{(12)}(u-v;\lambda). \end{aligned} \tag{4.23}$$

4.2 Corner Transfer Matrices and Vertex Operators

In the previous subsection the row-to-row transfer matrix is explained. There is another powerful method, corner transfer matrix (CTM) method, which was developed by Baxter [6]. In the following we consider face models in which heights take values in a set I. Vertex models are also treated similarly.

Let us consider the square lattice. We take the central site as the reference site O and devide a lattice into four quadrants.

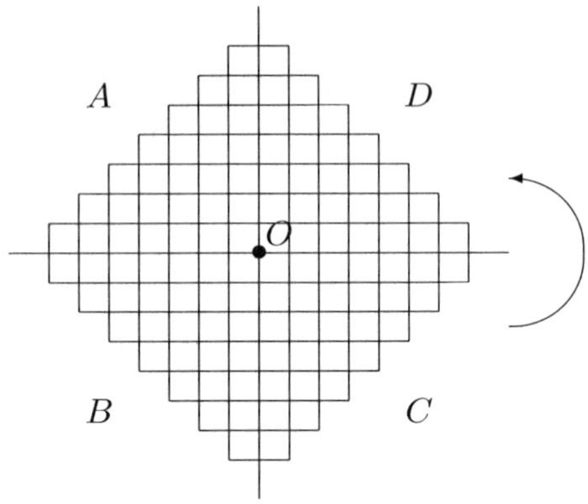

(4.24)

To each quadrant we assign a matrix A, B, C, D, which are called corner transfer matrices. We take a multiplication of the matrices counterclockwise. Sites on each column (row) are numbered increasingly from south to north (from east to west), $\ldots, -2, -1, 0, 1, 2, \ldots$, where the zeroth site is the reference site O. For example $A(u) = \big(A_{\underline{a},\underline{b}}(u)\big)$ is

$$A_{\underline{a},\underline{b}}(u) = \delta_{a_0,b_0}$$

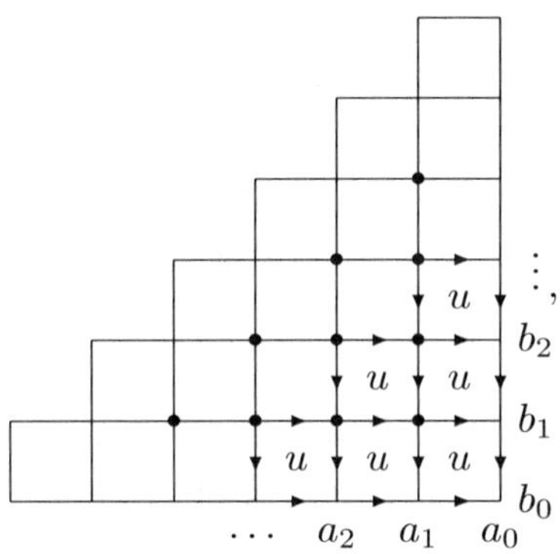

(4.25)

where $\underline{a} = (a_0, a_1, a_2, \ldots)$, $\underline{b} = (b_0, b_1, b_2, \ldots)$, and summation is taken over inner sites $\bullet$. Using these CTM's, the partition function is expressed as

$$Z = \mathrm{tr}(DCBA), \tag{4.26}$$

and one-point local height probability (LHP), which is the probability of finding local height at O to be k, is

$$P_k = Z^{-1}\,\mathrm{tr}_{a_0=k}(DCBA). \tag{4.27}$$

To discuss these quantities, however, we have to specify the ground state.

We restrict ourselves to consider the following ground state configuration

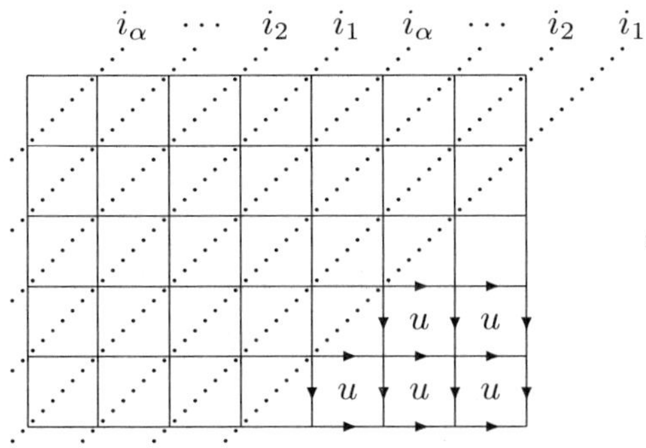

$$,\qquad (4.28)$$

which is invariant under the shift in the NE-SW direction. Many interesting models have this type of ground states. We label this ground state configuration (i.e., an ordered set $(i_1, i_2, \ldots, i_\alpha)$ whose cyclic permutations are identified) as an integer l. On each column or row the height sequence is $\ldots, i_1, i_2, \ldots, i_\alpha, i_1, i_2, \ldots, i_\alpha, \ldots$, but there are α configurations because the height on the zeroth site takes one of α values i_{m+1}. We distinguish these configurations by an integer $m \in \mathbb{Z}/\alpha\mathbb{Z}$. Namely the ground state configuration (l, m), which we denote by $\bar{a}(l, m)$, is

$$\begin{aligned} \bar{a}_{j-m}(l,m) &= i_{j+1} \qquad (0 \le j \le \alpha - 1), \\ \bar{a}_{j+\alpha}(l,m) &= \bar{a}_j(l,m) \quad (j \in \mathbb{Z}). \end{aligned} \tag{4.29}$$

When we discuss this face model, we impose the condition that the heights at the boundary (or sites far away from O) take one of the ground state configurations.

In the north and west directions the ground state configurations are $\ldots, i_1, i_2, \ldots, i_\alpha, \ldots$. Let $\mathcal{H}^{(k)}_{l,m}$ denote the space of states of the half-infinite lattice where the central height (i.e., the one on the zeroth site) is fixed to k and the boundary heights are in the ground state (l, m). Formally it is

$$\mathcal{H}^{(k)}_{l,m} = \mathrm{Span}\{(a_j)_{j=0}^{\infty} \mid a_j \in I, a_0 = k, a_j = \bar{a}_j(l,m)\,(j \gg 1)\}. \tag{4.30}$$

In the south and east directions we relabel the numbering of sites: $0, -1, -2, \ldots \mapsto 0, 1, 2, \ldots$. Then the ground state configurations in the south and east directions are $\ldots, i_\alpha, \ldots, i_2, i_1, \ldots$. So we introduce

$$\mathcal{H}'^{(k)}_{l,m} = \mathrm{Span}\{(a_j)_{j=0}^{\infty} \mid a_j \in I, a_0 = k, a_j = \bar{a}'_j(l,m)\,(j \gg 1)\}, \tag{4.31}$$

where $\bar{a}'_j(l,m) = \bar{a}_{-j}(l,m)$. We have

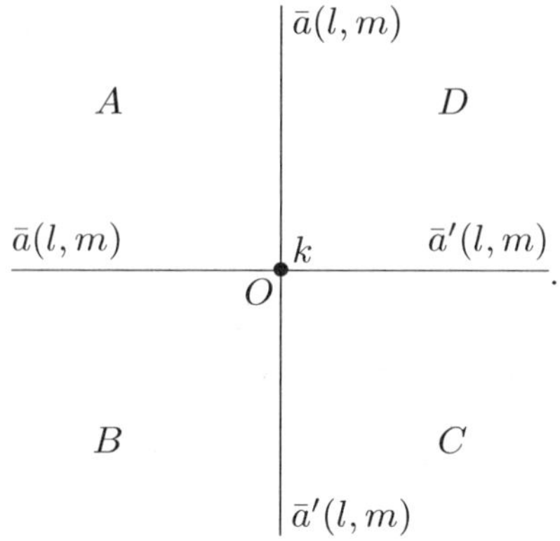

(4.32)

For later convenience we define

$$\mathcal{H}_l^{(k)} = \bigoplus_m \mathcal{H}_{l,m}^{(k)}, \qquad \mathcal{H}_l = \bigoplus_k \mathcal{H}_l^{(k)},$$
$$\mathcal{H}_l'^{(k)} = \bigoplus_m \mathcal{H}_{l,m}'^{(k)}, \qquad \mathcal{H}_l' = \bigoplus_k \mathcal{H}_l'^{(k)}. \tag{4.33}$$

Because of the factor δ_{a_0,b_0} in (4.25), $A(u)$ has a block structure

$$A = \bigoplus_k A^{(k)}, \tag{4.34}$$

where $A^{(k)}$ is a matrix in $a_0 = k$ sector. B, C and D have also this block structure. CTM's are linear maps:

$$A^{(k)} : \mathcal{H}_{l,m}^{(k)} \to \mathcal{H}_{l,m}^{(k)}, \qquad B^{(k)} : \mathcal{H}_{l,m}^{(k)} \to \mathcal{H}_{l,m}'^{(k)},$$
$$C^{(k)} : \mathcal{H}_{l,m}'^{(k)} \to \mathcal{H}_{l,m}'^{(k)}, \qquad D^{(k)} : \mathcal{H}_{l,m}'^{(k)} \to \mathcal{H}_{l,m}^{(k)}. \tag{4.35}$$

One-point LHP $P_k(l,m)$, which is the probability of finding local height at O to be k and the boundary heights in the north direction on the column through O to be in the ground state (l,m), is

$$P_k(l,m) = Z_{l,m}^{-1} \operatorname{tr}_{\mathcal{H}_{l,m}^{(k)}} (D^{(k)} C^{(k)} B^{(k)} A^{(k)}), \tag{4.36}$$

where $Z_{l,m}$ is the partition function for the ground state (l,m)

$$Z_{l,m} = \sum_k \operatorname{tr}_{\mathcal{H}_{l,m}^{(k)}} (D^{(k)} C^{(k)} B^{(k)} A^{(k)}). \tag{4.37}$$

Baxter's important observation is the following. Let us assume that the Boltzmann weight W satisfies the YBE (4.3),
the initial condition

$$W\left(\begin{matrix} a & b \\ c & d \end{matrix} \,\middle|\, 0\right) = \delta_{b,c}, \tag{4.38}$$

the unitarity condition

$$\sum_g W\left(\begin{matrix} a & b \\ g & d \end{matrix} \middle| u\right) W\left(\begin{matrix} a & g \\ c & d \end{matrix} \middle| -u\right) = \delta_{b,c}, \tag{4.39}$$

and the second inversion relation

$$\sum_g G_g W\left(\begin{matrix} a & b \\ c & g \end{matrix} \middle| \lambda - u\right) W\left(\begin{matrix} d & c \\ b & g \end{matrix} \middle| \lambda + u\right) = \frac{G_b G_c}{G_a}\delta_{a,d}, \tag{4.40}$$

where λ and G_a are some constants, and W is a double periodic function in u. Then, in the thermodynamic limit (infinite lattice size limit), the u-dependence of CTM becomes

$$\begin{aligned}
A^{(a)}(u) &= x^{-2uH_C^{(a)}} &&: \mathcal{H}_{l,m}^{(a)} \to \mathcal{H}_{l,m}^{(a)}, \\
B^{(a)}(u) &= \sqrt{G_a} P^{(a)^{-1}} A^{(a)}(\lambda - u) &&: \mathcal{H}_{l,m}^{(a)} \to \mathcal{H}_{l,m}^{\prime(a)}, \\
C^{(a)}(u) &= P^{(a)^{-1}} A^{(a)}(u) P^{(a)} &&: \mathcal{H}_{l,m}^{\prime(a)} \to \mathcal{H}_{l,m}^{\prime(a)}, \\
D^{(a)}(u) &= \sqrt{G_a} A^{(a)}(\lambda - u) P^{(a)} &&: \mathcal{H}_{l,m}^{\prime(a)} \to \mathcal{H}_{l,m}^{(a)},
\end{aligned} \tag{4.41}$$

where $H_C^{(a)}$ and $P^{(a)}$ are u-independent matrices and x is a parameter of the model. We have neglected multiplicative constant factors (normalization of CTM's) because they do not contribute to LHP. $P^{(a)}$ gives an isomorphism of vector spaces

$$P^{(a)}: \mathcal{H}_{l,m}^{\prime(a)} \to \mathcal{H}_{l,m}^{(a)}. \tag{4.42}$$

The spectrum of the corner Hamiltonian H_C has the following properties

bounded below, discrete, equidistance.

The multiplicity of the spectrum of H_C depends on the model and it is summarized in a character

$$\chi_{l,m,k}(q) = \mathrm{tr}_{\mathcal{H}_{l,m}^{(k)}} q^{H_C^{(k)}}. \tag{4.43}$$

By using these, one-point LHP (4.36) becomes

$$P_k(l,m) = Z_{l,m}^{-1} G_k \, \mathrm{tr}_{\mathcal{H}_{l,m}^{(k)}} x^{-4\lambda H_C^{(k)}} = Z_{l,m}^{-1} G_k \chi_{l,m,k}(x^{-4\lambda}), \tag{4.44}$$

where the partition function (4.37) is

$$Z_{l,m} = \sum_k G_k \chi_{l,m,k}(x^{-4\lambda}). \tag{4.45}$$

We remark that this $Z_{l,m}$ is independent on m because we are considering an infinitely large lattice.

To formulate multi-point LHP, let us introduce the vertex operator (VO) of type I [47, 48]. Foda et al. [47] presented a formulation of VO's in solvable lattice models. We explain a face model case following [47, 48]. Here we use intuitive graphical argument. For precise representation theoretical argument, see Section 4.4.

Graphically the Boltzmann weight is

$$W\left(\begin{matrix} a & b \\ c & d \end{matrix} \middle| u_1 - u_2\right) = $$

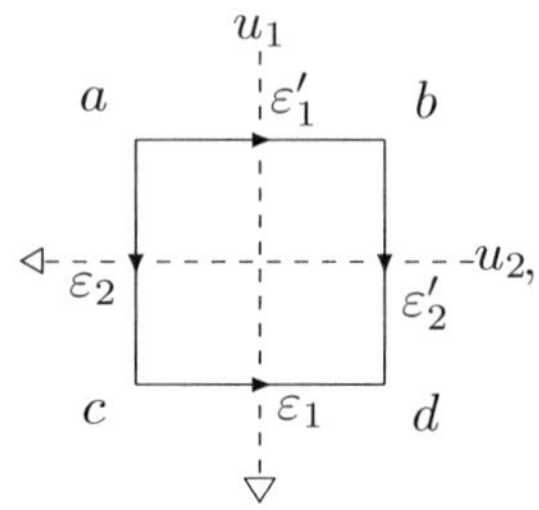

$$b = a + \varepsilon_1', \quad c = a + \varepsilon_2, \quad d = b + \varepsilon_2' = c + \varepsilon_1. \tag{4.46}$$

Let $\Phi_N^{(a,b)}(z)$, $\Phi_W^{(a,b)}(z^{-1})$, $\Phi_S^{(a,b)}(z)$ and $\Phi_E^{(a,b)}(z^{-1})$ be the half-infinite transfer matrices extending to infinity in the north, west, south and east direction respectively ($z = x^{2u}$):

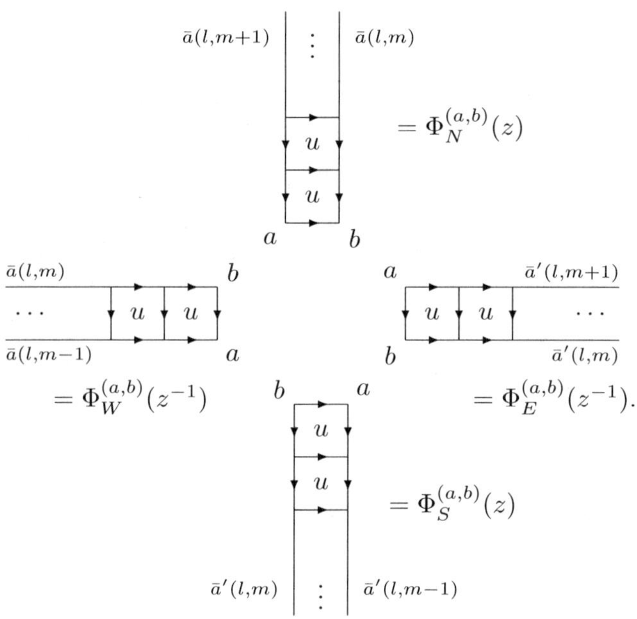

(4.47)

They are linear maps:

$$\begin{aligned} &\Phi_N^{(a,b)}(z) \colon \mathcal{H}_{l,m}^{(b)} \to \mathcal{H}_{l,m+1}^{(a)}, && \Phi_S^{(a,b)}(z) \colon \mathcal{H}_{l,m}^{\prime(b)} \to \mathcal{H}_{l,m-1}^{\prime(a)}, \\ &\Phi_W^{(a,b)}(z) \colon \mathcal{H}_{l,m}^{(b)} \to \mathcal{H}_{l,m-1}^{(a)} && \Phi_E^{(a,b)}(z) \colon \mathcal{H}_{l,m}^{\prime(b)} \to \mathcal{H}_{l,m+1}^{\prime(a)}. \end{aligned} \tag{4.48}$$

For CTM's and type I VO's we suppress the l dependence.

$\Phi_N^{(a,b)}(z)$ satisfies the commutation relation

$$\Phi_N^{(a,b)}(z_2)\Phi_N^{(b,c)}(z_1) = \sum_g W\left(\begin{matrix} a & g \\ b & c \end{matrix}\middle| u_1 - u_2\right)\Phi_N^{(a,g)}(z_1)\Phi_N^{(g,c)}(z_2) \qquad (z_j = x^{2u_j}), \quad (4.49)$$

which can be shown by repeated use of (4.3),

RHS =

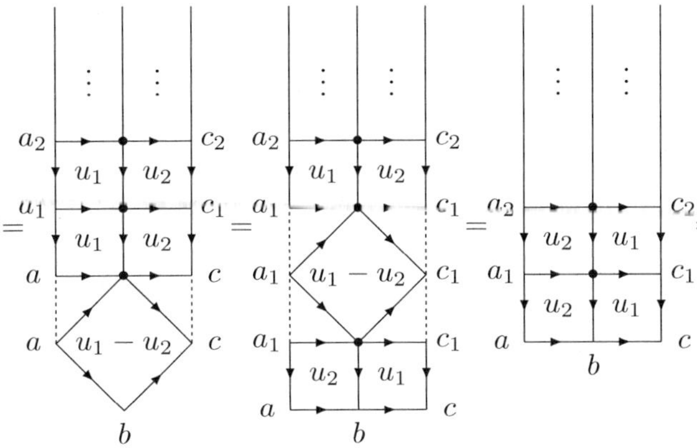

= LHS. (4.50)

Similarly we have $(z_j = x^{2u_j})$

$$\Phi_W^{(a,b)}(z_2)\Phi_W^{(b,c)}(z_1) = \sum_g W\left(\begin{matrix} c & b \\ g & a \end{matrix}\middle| u_1 - u_2\right)\Phi_W^{(a,g)}(z_1)\Phi_W^{(g,c)}(z_2), \quad (4.51)$$

$$\Phi_N^{(a,b)}(z_1)\Phi_W^{(b,c)}(z_2) = \sum_g W\left(\begin{matrix} g & c \\ a & b \end{matrix}\middle| u_1 - u_2\right)\Phi_W^{(a,g)}(z_2)\Phi_N^{(g,c)}(z_1), \quad (4.52)$$

and

$$\Phi_S^{(a,b)}(z_2)\Phi_S^{(b,c)}(z_1) = \sum_g W\left(\begin{matrix} c & b \\ g & a \end{matrix}\middle| u_1 - u_2\right)\Phi_S^{(a,g)}(z_1)\Phi_S^{(g,c)}(z_2), \quad (4.53)$$

etc.. The initial condition (4.38) implies

$$\begin{aligned}
A^{(a)}(u)\Phi_N^{(a,b)}(z) &= \Phi_N^{(a,b)}(1)A^{(b)}(u), \\
\Phi_W^{(a,b)}(z^{-1})A^{(b)}(u) &= A^{(a)}(u)\Phi_W^{(a,b)}(1), \\
B^{(a)}(u)\Phi_W^{(a,b)}(1) &= \Phi_S^{(a,b)}(z)B^{(b)}(u), \\
\Phi_S^{(a,b)}(1)B^{(b)}(u) &= B^{(a)}(u)\Phi_W^{(a,b)}(z^{-1}), \\
C^{(a)}(u)\Phi_S^{(a,b)}(z) &= \Phi_S^{(a,b)}(1)C^{(b)}(u), \\
\Phi_E^{(a,b)}(z^{-1})C^{(b)}(u) &= C^{(a)}(u)\Phi_E^{(a,b)}(1), \\
D^{(a)}(u)\Phi_E^{(a,b)}(1) &= \Phi_N^{(a,b)}(z)D^{(b)}(u), \\
\Phi_N^{(a,b)}(1)D^{(b)}(u) &= D^{(a)}(u)\Phi_E^{(a,b)}(z^{-1}),
\end{aligned} \qquad (4.54)$$

and therefore by combining with (4.41) we have

$$\Phi_N^{(a,b)}(z) = x^{2uH_C^{(a)}} \Phi_N^{(a,b)}(1) x^{-2uH_C^{(b)}}, \tag{4.55}$$

$$\Phi_W^{(a,b)}(z) = x^{2uH_C^{(a)}} \Phi_W^{(a,b)}(1) x^{-2uH_C^{(b)}}, \tag{4.56}$$

$$P^{(a)}\Phi_S^{(a,b)}(z){P^{(b)}}^{-1} = x^{2uH_C^{(a)}} P^{(a)}\Phi_S^{(a,b)}(1){P^{(b)}}^{-1} x^{-2uH_C^{(b)}}, \tag{4.57}$$

$$P^{(a)}\Phi_E^{(a,b)}(z){P^{(b)}}^{-1} = x^{2uH_C^{(a)}} P^{(a)}\Phi_E^{(a,b)}(1){P^{(b)}}^{-1} x^{-2uH_C^{(b)}}, \tag{4.58}$$

$$P^{(a)}\Phi_S^{(a,b)}(z){P^{(b)}}^{-1} = \sqrt{\frac{G_a}{G_b}}\,\Phi_W^{(a,b)}(x^{-2\lambda}z), \tag{4.59}$$

$$P^{(a)}\Phi_E^{(a,b)}(z){P^{(b)}}^{-1} = \sqrt{\frac{G_b}{G_a}}\,\Phi_N^{(a,b)}(x^{2\lambda}z). \tag{4.60}$$

The unitarity (4.39) gives

$$\sum_g \Phi_W^{(a,g)}(z)\Phi_N^{(g,a)}(z) = \mathrm{id}, \tag{4.61}$$

$$\sum_g \Phi_E^{(a,g)}(z)\Phi_S^{(g,a)}(z) = \mathrm{id}. \tag{4.62}$$

So (4.49), (4.38) and (4.61) imply

$$\Phi_N^{(a,b)}(z)\Phi_W^{(b,a)}(z) = \mathrm{id}. \tag{4.63}$$

The second inversion relation (4.40) entails

$$G_a^{-1}\sum_g G_g \Phi_N^{(a,g)}(x^{2\lambda}z)\Phi_W^{(g,a)}(x^{-2\lambda}z) = \mathrm{id}. \tag{4.64}$$

Consider $n+1$ neighboring sites in a row, whose most east site is the reference site O. We divide a lattice into $2n+4$ parts,

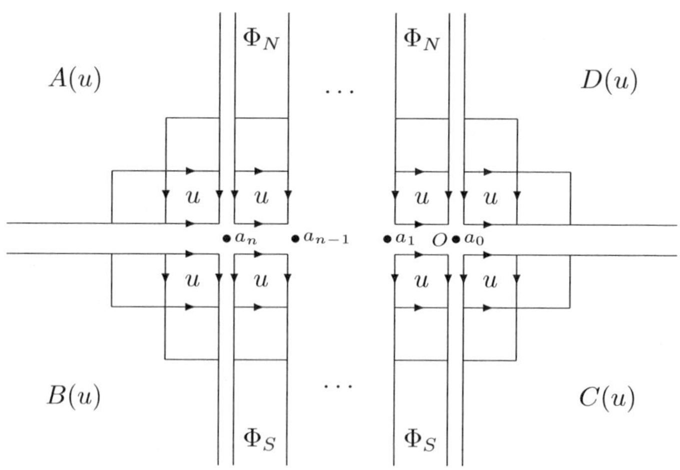

. (4.65)

Let $P_{a_n,\dots,a_1,a_0}(l,m)$ denote the probability of finding these local variables to be $(a_n,\dots,a_0)$ under the condition that the boundary heights in the north direction on the column through O are in the ground state (l,m). $P_{a_n,\dots,a_0}(l,m)$ satisfies obvious recursion relations and the normalization condition

$$\sum_{a_n} P_{a_n,\dots,a_0}(l,m) = P_{a_{n-1},\dots,a_0}(l,m),$$
$$\sum_{a_0} P_{a_n,\dots,a_0}(l,m) = P_{a_n,\dots,a_1}(l,m+1), \tag{4.66}$$

$$\sum_{a_0,\dots,a_n} P_{a_n,\dots,a_0}(l,m) = 1. \tag{4.67}$$

From (4.65) we have

$$\begin{aligned} P_{a_n,\dots,a_0}(l,m) &= Z_{l,m}^{-1}\mathrm{tr}_{\mathcal{H}_{l,m}^{(a_0)}}\big(D^{(a_0)}(u)C^{(a_0)}(u)\Phi_S^{(a_0,a_1)}(z)\cdots\Phi_S^{(a_{n-1},a_n)}(z) \\ &\qquad \times B^{(a_n)}(u)A^{(a_n)}(u)\Phi_N^{(a_n,a_{n-1})}(z)\cdots\Phi_N^{(a_1,a_0)}(z)\big). \end{aligned} \tag{4.68}$$

Using

$$\Phi_S^{(a,b)}(w)B^{(b)}(u)A^{(b)}(u) = B^{(a)}(u)A^{(a)}(u)\Phi_W^{(a,b)}(w), \tag{4.69}$$

which is derived from (4.41) and (4.59), one can write it as

$$\begin{aligned} P_{a_n,\dots,a_0}(l,m) &= Z_{l,m}^{-1}\mathrm{tr}_{\mathcal{H}_{l,m}^{(a_0)}}\big(D^{(a_0)}(u)C^{(a_0)}(u)B^{(a_0)}(u)A^{(a_0)}(u)\Phi_W^{(a_0,a_1)}(z)\cdots \\ &\qquad \times \Phi_W^{(a_{n-1},a_n)}(z)\Phi_N^{(a_n,a_{n-1})}(z)\cdots\Phi_N^{(a_1,a_0)}(z)\big) \\ &= Z_{l,m}^{-1}G_{a_0}\mathrm{tr}_{\mathcal{H}_{l,m}^{(a_0)}}\big(x^{-4\lambda H_C^{(a_0)}}\Phi_W^{(a_0,a_1)}(z)\cdots\Phi_W^{(a_{n-1},a_n)}(z) \\ &\qquad \times \Phi_N^{(a_n,a_{n-1})}(z)\cdots\Phi_N^{(a_1,a_0)}(z)\big), \end{aligned} \tag{4.70}$$

where the partition function $Z_{l,m}$ is given in (4.45). Using this expression, (4.61), the cyclic property of trace, (4.55) and (4.64), we can reproduce the recursion relations (4.66). By changing spectral parameters of VO's we can formulate more general quantities

$$\begin{aligned} P_{a_n,\dots,a_0}&\left(l,m;\begin{matrix} z_n,\dots,z_1 \\ z'_n,\dots,z'_1\end{matrix}\right) \\ &= Z_{l,m}^{-1}G_{a_0}\,\mathrm{tr}_{\mathcal{H}_{l,m}^{(a_0)}}\big(x^{-4\lambda H_C^{(a_0)}}\Phi_W^{(a_0,a_1)}(z'_1)\cdots\Phi_W^{(a_{n-1},a_n)}(z'_n) \\ &\qquad \times \Phi_N^{(a_n,a_{n-1})}(z_n)\dots\Phi_N^{(a_1,a_0)}(z_1)\big). \end{aligned} \tag{4.71}$$

We remark that these LHP's are independent of $z = x^{2u}$ because

$$\begin{aligned}\operatorname{tr}\bigl(x^{-4\lambda H_C}\mathcal{O}(w_i)\bigr) &= \operatorname{tr}\bigl(w^{H_C}w^{-H_C}x^{-4\lambda H_C}\mathcal{O}(w_i)\bigr)\\ &= \operatorname{tr}\bigl(x^{-4\lambda H_C}w^{-H_C}\mathcal{O}(w_i)w^{H_C}\bigr)\\ &\qquad\qquad\times \operatorname{tr}\left(x^{-4\lambda H_C}\mathcal{O}\left(\frac{w_i}{w}\right)\right). \end{aligned} \tag{4.72}$$

By averaging $P_{a_n,\dots,a_0}(l,m)$ over m we obtain

$$\begin{aligned}P_{a_n,\dots,a_0}(l) &= \sum_m \frac{Z_{l,m}}{Z_l}P_{a_n,\dots,a_0}(l,m)\\ &= Z_l^{-1}G_{a_0}\operatorname{tr}_{\mathcal{H}_l^{(a_0)}}\bigl(x^{-4\lambda H_C^{(a_0)}}\Phi_W^{(a_0,a_1)}(z)\cdots\Phi_W^{(a_{n-1},a_n)}(z)\\ &\qquad\qquad\times\Phi_N^{(a_n,a_{n-1})}(z)\dots\Phi_N^{(a_1,a_0)}(z)\bigr),\end{aligned} \tag{4.73}$$

where

$$Z_l = \sum_m Z_{l,m} = \sum_a G_a \operatorname{tr}_{\mathcal{H}_l^{(a)}} x^{-4\lambda H_C^{(a)}} = \sum_{a,m} G_a\chi_{l,m,a}(x^{-4\lambda}). \tag{4.74}$$

For later use we define the following quantities:

$$\begin{aligned}Q_{a_n,\dots,a_0}(l,m;z\mid\mathcal{O}) &= Z_{l,m}^{-1}G_{a_0}\operatorname{tr}_{\mathcal{H}_{l,m}^{(a_0)}}\bigl(x^{-4\lambda H_C^{(a_0)}}\mathcal{O}_\Phi(z)\mathcal{O}\bigr),\\ Q_{a_n,\dots,a_0}(l;z\mid\mathcal{O}) &= \sum_m\frac{Z_{l,m}}{Z_l}Q_{a_n,\dots,a_0}(l,m;z\mid\mathcal{O})\\ &= Z_l^{-1}G_{a_0}\operatorname{tr}_{\mathcal{H}_l^{(a_0)}}\bigl(x^{-4\lambda H_C^{(a_0)}}\mathcal{O}_\Phi(z)\mathcal{O}\bigr),\end{aligned} \tag{4.75}$$

where $\mathcal{O}_\Phi(z) = \Phi_W^{(a_0,a_1)}(z)\cdots\Phi_W^{(a_{n-1},a_n)}(z)\Phi_N^{(a_n,a_{n-1})}(z)\cdots\Phi_N^{(a_1,a_0)}(z)$ and $\mathcal{O}$ is a linear map $\mathcal{O}\colon \mathcal{H}_{l,m}^{(a)} \to \mathcal{H}_{l,m}^{(a)}$. Note that $Q_{a_n,\dots,a_0}(l,m;z\mid 1) = P_{a_n,\dots,a_0}(l,m)$. Like $P_{a_n,\dots,a_0}(l,m)$, $Q_{a_n,\dots,a_0}(l,m;z\mid\mathcal{O})$ satisfies recursion relations. From (4.61) we have

$$\sum_{a_n} Q_{a_n,\dots,a_0}(l,m;z\mid\mathcal{O}) = Q_{a_{n-1},\dots,a_0}(l,m;z\mid\mathcal{O}). \tag{4.76}$$

If $\mathcal{O}$ satisfies $\Phi_N^{(a,b)}(z)\mathcal{O} = f(z)\mathcal{O}\Phi_N^{(a,b)}(z)$ where $f(z)$ is a function independent on a and b, then we have another recursion relation from (4.64)

$$\sum_{a_0} Q_{a_n,\dots,a_0}(l,m;z\mid\mathcal{O}) = f(z)Q_{a_{n-1},\dots,a_1}(l,m+1;z\mid\mathcal{O}). \tag{4.77}$$

Type I VO's are half-infinite transfer matrices but a 'physical' transfer matrix is the row-to-row (or column-to-column) transfer matrix discussed in the previous subsection. The column-to-column transfer matrix $\mathcal{T}_{\rm col}$, which adds one column from west to east, is a linear map

$$\mathcal{T}_{\rm col}(u)\colon \widetilde{\mathcal{H}}_l \to \widetilde{\mathcal{H}}_l, \tag{4.78}$$

where $\widetilde{\mathcal{H}}_l$ is defined by

$$\widetilde{\mathcal{H}}_l = \bigoplus_k \widetilde{\mathcal{H}}_l^{(k)}, \quad \widetilde{\mathcal{H}}_l^{(k)} = \bigoplus_m \mathcal{H}_{l,m}^{\prime(k)} \otimes \mathcal{H}_{l,m}^{(k)}. \tag{4.79}$$

Here, the numbering of sites in the south direction is the original one (i.e., increasingly from south to north $\ldots,-2,-1,0,1,2,\ldots$) and the zeroth site is identified. Its matrix element is

$$\begin{aligned}\mathcal{T}_{\rm col}(u)&_{\cdots a_{-1}a_0a_1\cdots,\cdots b_{-1}b_0b_1\cdots}\\ &= \Phi_S^{(a_0,b_0)}(z)_{a_0a_{-1}\cdots,b_0b_{-1}\cdots}\Phi_N^{(b_0,a_0)}(z)_{b_0b_1\cdots,a_0a_1\cdots},\end{aligned} \tag{4.80}$$

and its matrix form is

$$\bigl(\mathcal{T}_{\rm col}(u)\bigr)^{(a,b)} = \Phi_S^{(a,b)}(z) \otimes {}^t\Phi_N^{(b,a)}(z), \tag{4.81}$$

where a, b are heights on the zeroth site. $\mathcal{T}_{\rm col}$ with different spectral parameters commute with each other

$$[\mathcal{T}_{\rm col}(u), \mathcal{T}_{\rm col}(v)] = 0, \tag{4.82}$$

because

$$\begin{aligned}\bigl(\mathcal{T}_{\rm col}(u_2)&\mathcal{T}_{\rm col}(u_1)\bigr)^{(a,c)}\\ &= \sum_b \bigl(\mathcal{T}_{\rm col}(u_2)\bigr)^{(a,b)}\bigl(\mathcal{T}_{\rm col}(u_1)\bigr)^{(b,c)}\\ &= \sum_b \Phi_S^{(a,b)}(z_2)\Phi_S^{(b,c)}(z_1) \otimes {}^t(\Phi_N^{(c,b)}(z_1)\Phi_N^{(b,a)}(z_2))\\ &= \sum_g \Phi_S^{(a,g)}(z_1)\Phi_S^{(g,c)}(z_2) \otimes {}^t\bigl(\Phi_N^{(c,g)}(z_2)\Phi_N^{(g,a)}(z_1)\bigr)\\ &= \bigl(\mathcal{T}_{\rm col}(u_1)\mathcal{T}_{\rm col}(u_2)\bigr)^{(a,c)},\end{aligned} \tag{4.83}$$

where we have used (4.49), (4.53) and (4.39).

When we consider the action of $\mathcal{T}_{\rm col}$ in this vertex operator approach, it is convenient to regard a state of $\widetilde{\mathcal{H}}_l \subset \mathcal{H}_l' \otimes \mathcal{H}_l$ as a linear map on $\mathcal{H}_l$. In general for two vector spaces V_i $(i=1,2)$ a vector of $V_1 \otimes V_2$ can be seen as a linear map from V_2 to V_1 by $f_1 \otimes f_2 \in V_1 \otimes V_2 \mapsto f_1{}^t f_2$ where f_1 and f_2 are understood as column vectors. When there is an isomorphism $P\colon V_1 \to V_2$, a vector of $V_1 \otimes V_2$ can be regarded as a linear map on V_2,

$$|f\rangle = f_1 \otimes f_2 \in V_1 \otimes V_2 \mapsto P f_1{}^t f_2 \in \mathrm{End}(V_2). \tag{4.84}$$

Similarly

$$\langle f| = {}^t f_1 \otimes {}^t f_2 \in (V_1 \otimes V_2)^* \mapsto f_2 {}^t f_1 P^{-1} \in \mathrm{End}(V_2). \tag{4.85}$$

The inner product (pairing) of two vectors $|f\rangle = f_1 \otimes f_2$, $|g\rangle = g_1 \otimes g_2 \in V_1 \otimes V_2$ can be expressed as a trace over V_2,

$$\langle f|g\rangle = {}^t f_1 g_1 \cdot {}^t f_2 g_2 = \mathrm{tr}_{V_2}(f_2 {}^t f_1 P^{-1} \cdot P g_1 {}^t g_2), \tag{4.86}$$

For $A = A_1 \otimes A_2$ ($A_i \in \mathrm{End}(V_i)$ $(i = 1, 2)$), $|f\rangle = f_1 \otimes f_2$ and $\langle f| = {}^t f_1 \otimes {}^t f_2$, we have the following correspondence

$$\begin{aligned} A|f\rangle &= A_1 f_1 \otimes A_2 f_2 \mapsto P(A_1 f_1)^t (A_2 f_2) \\ &= P A_1 P^{-1} \cdot P f_1 {}^t f_2 \cdot {}^t A_2, \\ \langle f|A &= {}^t f_1 A_1 \otimes {}^t f_2 A_2 \mapsto {}^t({}^t f_2 A_2)({}^t f_1 A_1) P^{-1} \\ &= {}^t A_2 \cdot f_2 {}^t f_1 P^{-1} \cdot P A_1 P^{-1}, \end{aligned} \tag{4.87}$$

and

$$\langle f|A|g\rangle = \langle f|Ag\rangle = \mathrm{tr}_{V_2}\big(f_2 {}^t f_1 P^{-1} \cdot (P A_1 P^{-1} \cdot P g_1 {}^t g_2 \cdot {}^t A_2)\big). \tag{4.88}$$

Using this correspondence $\mathcal{T}_{\mathrm{col}}$ acts on a vector $|f\rangle \in \widetilde{\mathcal{H}}_l$ as

$$\big(\mathcal{T}_{\mathrm{col}}(u)|f\rangle\big)^{(a)} = \sum_b P^{(a)} \Phi_S^{(a,b)}(z) {P^{(b)}}^{-1} \cdot f \cdot \Phi_N^{(b,a)}(z), \tag{4.89}$$

where in the RHS f is understood as a linear map on $\mathcal{H}_l$, which does not change the central height ($f \colon \mathcal{H}_l^{(a)} \to \mathcal{H}_l^{(a)}$). The translation invariant vacuum state $|\mathrm{vac}\rangle$ is given by

$$|\mathrm{vac}\rangle^{(a)} = \sqrt{G_a} x^{-2\lambda H_C^{(a)}}. \tag{4.90}$$

In fact it is translationally invariant

$$\begin{aligned} \big(\mathcal{T}_{\mathrm{col}}(u)|\mathrm{vac}\rangle\big)^{(a)} &= \sum_b P^{(a)} \Phi_S^{(a,b)}(z) {P^{(b)}}^{-1} \cdot \sqrt{G_b} x^{-2\lambda H_C^{(b)}} \cdot \Phi_N^{(b,a)}(z) \\ &= \sum_b \sqrt{G_b} x^{-2\lambda H_C^{(a)}} P^{(a)} \Phi_S^{(a,b)}(x^{2\lambda} z) {P^{(b)}}^{-1} \Phi_N^{(b,a)}(z) \\ &= \sqrt{G_a} x^{-2\lambda H_C^{(a)}} \sum_b \Phi_W^{(a,b)}(z) \Phi_N^{(b,a)}(z) \\ &= \sqrt{G_a} x^{-2\lambda H_C^{(a)}} = |\mathrm{vac}\rangle^{(a)}. \end{aligned} \tag{4.91}$$

For $\widetilde{\mathcal{O}} = 1 \otimes {}^t\mathcal{O}$ where $\mathcal{O}$ is a linear map $\mathcal{O} \colon \mathcal{H}_l^{(a)} \to \mathcal{H}_l^{(b)}$, we have

$$\begin{aligned} &\langle \mathrm{vac}|\widetilde{\mathcal{O}}|\mathrm{vac}\rangle \\ &= \sum_{a,b} \mathrm{tr}_{\mathcal{H}_l^a}\big(\sqrt{G_a} x^{-2\lambda H_C^{(a)}} \cdot P^{(a)} {P^{(b)}}^{-1} \cdot \sqrt{G_b} x^{-2\lambda H_C^{(b)}} \cdot \mathcal{O}^{(b,a)}\big). \end{aligned} \tag{4.92}$$

In particular when $\mathcal{O}$ is a linear map $\mathcal{O}\colon \mathcal{H}_l^{(a)} \to \mathcal{H}_l^{(a)}$, we have

$$\begin{aligned}\langle \mathrm{vac}|\widetilde{\mathcal{O}}|\mathrm{vac}\rangle &= \sum_a {}^{(a)}\langle \mathrm{vac}|\widetilde{\mathcal{O}}|\mathrm{vac}\rangle^{(a)}, \\ {}^{(a)}\langle \mathrm{vac}|\widetilde{\mathcal{O}}|\mathrm{vac}\rangle^{(a)} &= G_a \operatorname{tr}_{\mathcal{H}_l^{(a)}}(x^{-4\lambda H_C^{(a)}}\mathcal{O}),\end{aligned} \tag{4.93}$$

$$\langle \mathrm{vac}|\mathrm{vac}\rangle = \sum_a G_a \operatorname{tr}_{\mathcal{H}_l^{(a)}}(x^{-4\lambda H_C^{(a)}}) = Z_l. \tag{4.94}$$

Therefore LHP (4.73) can be written as a vacuum expectation value with the fixed central value

$$P_{a_n,\dots,a_0}(l) = \frac{{}^{(a_0)}\langle \mathrm{vac}|\widetilde{\mathcal{O}}|\mathrm{vac}\rangle^{(a_0)}}{\langle \mathrm{vac}|\mathrm{vac}\rangle}, \tag{4.95}$$

where $\widetilde{\mathcal{O}} = 1 \otimes {}^t\mathcal{O}$ with $\mathcal{O} = \Phi_W^{(a_0,a_1)}(z)\cdots\Phi_W^{(a_{n-1},a_n)}(z)\Phi_N^{(a_n,a_{n-1})}(z)\cdots\times \Phi_N^{(a_1,a_0)}(z)$.

Similarly the row-to-row transfer matrix, which adds one row from south to north, can be written as

$$\big(\mathcal{T}_{\mathrm{row}}(u)\big)^{(a,b)} = \Phi_E^{(a,b)}(z^{-1}) \otimes {}^t\Phi_W^{(b,a)}(z^{-1})\colon \widetilde{\mathcal{H}}_l \to \widetilde{\mathcal{H}}_l, \tag{4.96}$$

and it acts on a state $|f\rangle \in \widetilde{\mathcal{H}}_l$ as

$$\big(\mathcal{T}_{\mathrm{row}}(u)|f\rangle\big)^{(a)} = \sum_b P^{(a)}\Phi_E^{(a,b)}(z^{-1}){P^{(b)}}^{-1}\cdot f\cdot \Phi_W^{(b,a)}(z^{-1}). \tag{4.97}$$

The state (4.90) is translationally invariant

$$\begin{aligned}&\big(\mathcal{T}_{\mathrm{row}}(u)|\mathrm{vac}\rangle\big)^{(a)} \\ &\quad= \sum_b P^{(a)}\Phi_E^{(a,b)}(z^{-1}){P^{(b)}}^{-1}\cdot \sqrt{G_b}x^{-2\lambda H_C^{(b)}}\cdot \Phi_W^{(b,a)}(z^{-1}) \\ &\quad= \sum_b \sqrt{G_b}x^{-2\lambda H_C^{(a)}}P^{(a)}\Phi_E^{(a,b)}(x^{2\lambda}z^{-1}){P^{(b)}}^{-1}\Phi_W^{(b,a)}(z^{-1}) \\ &\quad= \sqrt{G_a}x^{-2\lambda H_C^{(a)}}\sum_b \frac{G_b}{G_a}\Phi_N^{(a,b)}(x^{4\lambda}z^{-1})\Phi_W^{(b,a)}(z^{-1}) \\ &\quad= \sqrt{G_a}x^{-2\lambda H_C^{(a)}} = |\mathrm{vac}\rangle^{(a)}.\end{aligned} \tag{4.98}$$

We remark that these $\mathcal{T}_{\mathrm{col}}$ and $\mathcal{T}_{\mathrm{row}}$ are defined on an infinitely large lattice from the beginning in contrast to those in Section 4.1. Their excited states are created by another type of vertex operators, type II VO's, from the vacuum state (note that we need also the translation noninvariant 'vacuum' states to obtain a complete set of excited states) [48]. We will give an example in Section 5.4. Commutation relations for type II VO will be given in Section 4.4, Section 5 and Section 6.

In the rest of this subsection we assume that W has the crossing symmetry,

$$W\left(\begin{matrix} b & d \\ a & c \end{matrix} \middle| \lambda - u\right) = \sqrt{\frac{G_a G_d}{G_b G_c}} W\left(\begin{matrix} a & b \\ c & d \end{matrix} \middle| u\right). \tag{4.99}$$

Remark that the unitarity (4.39) and the crossing symmetry (4.99) imply the second inversion relation (4.40). We also remark that there are models which enjoy the second inversion relation but do not have crossing symmetry. Here we assume the crossing symmetry. Then graphical argument shows that CTM's in the thermodynamic limit are ((4.41) with $P^{(a)} = 1$)

$$\begin{aligned} A^{(a)}(u) = C^{(a)}(u) &= x^{-2uH_C^{(a)}} \\ B^{(a)}(u) = D^{(a)}(u) &= \sqrt{G_a} x^{-2(\lambda-u)H_C^{(a)}} \end{aligned} : \mathcal{H}_{l,m}^{(a)} \to \mathcal{H}_{l,m}^{(a)}, \tag{4.100}$$

and type I VO's are related to each other,

$$\Phi_W^{(a,b)}(z) = \sqrt{\tfrac{G_b}{G_a}} \Phi_N^{(a,b)}(x^{2\lambda} z), \tag{4.101}$$

$$\Phi_S^{(a,b)}(z) = \Phi_N^{(a,b)}(z), \tag{4.102}$$

$$\Phi_E^{(a,b)}(z) = \sqrt{\tfrac{G_b}{G_a}} \Phi_N^{(a,b)}(x^{2\lambda} z) = \Phi_W^{(a,b)}(z). \tag{4.103}$$

We write

$$\Phi^{(a,b)}(z) = \Phi_N^{(a,b)}(z) \colon \mathcal{H}_{l,m}^{(b)} \to \mathcal{H}_{l,m+1}^{(a)}, \tag{4.104}$$

$$\Phi^{*(a,b)}(z) = \Phi_W^{(a,b)}(z) \colon \mathcal{H}_{l,m}^{(b)} \to \mathcal{H}_{l,m-1}^{(a)}. \tag{4.105}$$

Then, $\Phi^{(a,b)}(z)$ and $\Phi^{*(a,b)}(z)$ enjoy the following properties:

$$\Phi^{*(a,b)}(z) = \sqrt{\frac{G_b}{G_a}} \Phi^{(a,b)}(x^{2\lambda} z), \tag{4.106}$$

$$\begin{aligned} &\Phi^{(a,b)}(z_2)\Phi^{(b,c)}(z_1) \\ &\qquad = \sum_g W\left(\begin{matrix} a & g \\ b & c \end{matrix} \middle| u_1 - u_2\right) \Phi^{(a,g)}(z_1)\Phi^{(g,c)}(z_2), \end{aligned} \tag{4.107}$$

$$w^{H_C} \Phi^{(a,b)}(z) w^{-H_C} = \Phi^{(a,b)}(wz), \tag{4.108}$$

$$\sum_g \Phi^{*(a,g)}(z)\Phi^{(g,a)}(z) = 1, \tag{4.109}$$

$$\Phi^{(a,b)}(z)\Phi^{*(b,c)}(z) = \delta_{a,c}. \tag{4.110}$$

The first equation is (4.101), the second one is (4.49), the third one is obtained by (4.55), the fourth one is (4.61), and the fifth one is derived by using (4.106), (4.107), (4.99), (4.38) and (4.109). Multi-point LHP (4.70) becomes

$$\begin{aligned} P_{a_n,\ldots,a_0}(l,m) = Z_{l,m}^{-1} G_{a_0} \operatorname{tr}_{\mathcal{H}_{l,m}^{(a_0)}} \big(& x^{-4\lambda H_C^{(a_0)}} \Phi^{*(a_0,a_1)}(z) \cdots \Phi^{*(a_{n-1},a_n)}(z) \\ & \times \Phi^{(a_n,a_{n-1})}(z) \cdots \Phi^{(a_1,a_0)}(z)\big). \end{aligned} \tag{4.111}$$

Since the Boltzmann weight is invariant under $180°$ rotation $W\left(\begin{smallmatrix} a & b \\ c & d \end{smallmatrix} \,\middle|\, u\right) = W\left(\begin{smallmatrix} d & c \\ b & a \end{smallmatrix} \,\middle|\, u\right)$, $P_{a_n,\ldots,a_0}(l)$ satisfies

$$P_{a_n,\ldots,a_0}(l) = P_{a_0,\ldots,a_n}(l). \tag{4.112}$$

Equation (4.89) becomes

$$\big(\mathcal{T}_{\mathrm{col}}(u)|f\rangle\big)^{(a)} = \sum_b \Phi^{(a,b)}(z) \cdot f \cdot \Phi^{(b,a)}(z). \tag{4.113}$$

In Section 5 and Section 6.4 we will calculate LHP's by using bosonization technique for the VO's.

4.3 Introduction to Quasi-Hopf Algebra

In this subsection we illustrate an outline of a quasi-Hopf algebra. For more details we refer the readers to Refs. [11, 26].

In quantum mechanics, we know an addition of angular momentums very well. For two particles system the total angular momentum $\vec{J}$ is obtained simply by an addition of each angular momentum $\vec{J}^{(1)}$ and $\vec{J}^{(2)}$,

$$\vec{J} = \vec{J}^{(1)} + \vec{J}^{(2)}. \tag{4.114}$$

In mathematics, this formula is written in the following way; $\vec{J}^{(1)}$ acts on the representation space V_1, $\vec{J}^{(2)}$ acts on V_2, and the total angular momentum $\vec{J}$ acts on $V_1 \otimes V_2$ by,

$$\vec{J} = \vec{J} \otimes 1 + 1 \otimes \vec{J}. \tag{4.115}$$

This is called the tensor product representation of Lie algebra so(3). If a system has rotational symmetry, for example the Heisenberg spin chain (XXX spin chain), one can apply the representation theory of the rotational group SO(3) (or its Lie algebra so(3)) to it. But if the system is perturbed and loses the rotational symmetry, then one can not apply so(3) to it. Some models, however, have a good property. For example the XXZ spin chain has the same degeneracy of energy as the XXX spin chain. To treat such models we need some deformation of the Lie algebra or some deformation of the tensor product representation.

4.3.1 Algebra and Coalgebra

Let us begin with the definitions of an algebra and a coalgebra. For simplicity we take the complex field $\mathbb{C}$ as a base field. An *algebra* A is a vector space with two operations, product (multiplication) m and unit u, which satisfy

$$\begin{array}{ccc} A\otimes A\otimes A & \xrightarrow{m\otimes \mathrm{id}} & A\otimes A \\ {\scriptstyle \mathrm{id}\otimes m}\downarrow & & \downarrow{\scriptstyle m} \\ A\otimes A & \xrightarrow[m]{} & A \end{array} \qquad \begin{array}{ll} \text{product} & m\colon A\otimes A\to A \\ \text{unit} & u\colon \mathbb{C}\to A \\ \text{associativity} & m\circ(m\otimes \mathrm{id}) \\ & \quad = m\circ(\mathrm{id}\otimes m), \end{array} \tag{4.116}$$

and $m\circ(\mathrm{id}\otimes u)=\mathrm{id}=m\circ(u\otimes\mathrm{id})$ ($A\otimes\mathbb{C}$, A and $\mathbb{C}\otimes A$ are identified). If we write $m(a\otimes b)=ab$, the associativity becomes a usual form $(ab)c=a(bc)$.

A coalgebra is defined by reversing the arrows. A coalgebra A is a vector space with two operations, coproduct Δ and counit ε, which satisfy

$$\begin{array}{ccc} A\otimes A\otimes A & \xleftarrow{\Delta\otimes\mathrm{id}} & A\otimes A \\ {\scriptstyle \mathrm{id}\otimes\Delta}\uparrow & & \uparrow{\scriptstyle\Delta} \\ A\otimes A & \xleftarrow[\Delta]{} & A \end{array} \qquad \begin{array}{ll} \text{coproduct} & \Delta\colon A\to A\otimes A \\ \text{counit} & \varepsilon\colon A\to\mathbb{C} \\ \text{coassociativity} & (\Delta\otimes\mathrm{id})\circ\Delta \\ & \quad =(\mathrm{id}\otimes\Delta)\circ\Delta, \end{array} \tag{4.117}$$

and $(\mathrm{id}\otimes\varepsilon)\circ\Delta=\mathrm{id}=(\varepsilon\otimes\mathrm{id})\circ\Delta$ ($A\otimes\mathbb{C}$, A and $\mathbb{C}\otimes A$ are identified).

Let us introduce σ ($\sigma\colon A\otimes A\to A\otimes A$, $\sigma(a\otimes b)=b\otimes a$) and define $m'=m\circ\sigma$ and $\Delta'=\sigma\circ\Delta$. An algebra is called commutative if $m'=m$, and a coalgebra is called cocommutative if $\Delta'=\Delta$.

4.3.2 Hopf Algebra

A *Hopf algebra* is a set $(A,m,u,\Delta,\varepsilon,S)$ satisfying the following conditions: A is an algebra and a coalgebra; m, u, Δ, ε are homomorphism; antipode $S\colon A\to A$ satisfies $m\circ(S\otimes\mathrm{id})\circ\Delta=u\circ\varepsilon=m\circ(\mathrm{id}\otimes S)\circ\Delta$. S is an anti-homomorphism.

We give two examples of a Hopf algebra, a group G and a Lie algebra $\mathfrak{g}$, exactly speaking a function algebra of group $\mathrm{Fun}(G)$ and an enveloping algebra of Lie algebra $U(\mathfrak{g})$ respectively. Their Hopf algebra structures are

	$\mathrm{Fun}(G)=\mathrm{Map}(G,\mathbb{C})$	$U(\mathfrak{g})$
product	$(f_1f_2)(x)=f_1(x)f_2(x)$	XY
unit	$\big(u(a)\big)(x)=a$	$u(a)=a1$
coproduct	$\big(\Delta(f)\big)(x_1,x_2)=f(x_1x_2)$	$\Delta(X)=X\otimes 1+1\otimes X$
counit	$\big(\varepsilon(f)\big)(x)=f(e)$	$\varepsilon(X)=0$
antipode	$\big(S(f)\big)(x)=f(x^{-1})$	$S(X)=-X,$

(4.118)

and $\big(m(g)\big)(x)=g(x,x)$ for $g(x_1,x_2)\in\mathrm{Map}(G\times G,\mathbb{C})$, and $\Delta(1)=1\otimes 1$, $\varepsilon(1)=1$, $S(1)=1$ for $U(\mathfrak{g})$. Roughly speaking Δ, ε and S correspond to

$$\begin{array}{cclll} & & \mathrm{Fun}(G) & U(\mathfrak{g}) \\ \Delta & \longleftrightarrow & \text{product of } G & \text{tensor product rep. of } U(\mathfrak{g}) \\ \varepsilon & \longleftrightarrow & \text{unit element of } G & \text{trivial rep. of } U(\mathfrak{g}) \\ S & \longleftrightarrow & \text{inverse element of } G & \text{contragredient rep. of } U(\mathfrak{g}). \end{array} \tag{4.119}$$

$\mathrm{Fun}(G)$ is a commutative (and noncocommutative) Hopf algebra and $U(\mathfrak{g})$ is a cocommutative (and noncommutative) Hopf algebra. Noncommutative and noncocommutative Hopf algebra may be regarded as an extension (deformation) of group or Lie algebra in this sense. This is the idea of quantum group (quantum algebra).

The quantum group (quantum algebra) is a Hopf algebra. We give an example of the quantum group, $U_q(\mathfrak{sl}_2)$, which is a deformation of $U(\mathfrak{sl}_2)$. $U_q(\mathfrak{sl}_2)$ is generated by $t = q^h$, e and f, which satisfy

$$\begin{array}{lll} [h,e] = 2e & [h,f] = -2f & [e,f] = \dfrac{t-t^{-1}}{q-q^{-1}} \\ \Delta(h) = h\otimes 1 + 1\otimes h & \Delta(e) = e\otimes 1 + t\otimes e & \Delta(f) = f\otimes t^{-1} + 1\otimes f \\ \varepsilon(h) = 0 & \varepsilon(e) = 0 & \varepsilon(f) = 0 \\ S(h) = -h & S(e) = -t^{-1}e & S(f) = -ft. \end{array} \tag{4.120}$$

This quantum algebra appears in the XXZ spin chain as a symmetry,

$$\begin{aligned} &[H_{\mathrm{XXZ}}, U_q(\mathfrak{sl}_2)] = 0, \\ &H_{\mathrm{XXZ}} = J\sum_{i=1}^{N-1}\left(s_i^x s_{i+1}^x + s_i^y s_{i+1}^y + \frac{q+q^{-1}}{2} s_i^z s_{i+1}^z\right) \\ &\qquad\qquad + J\frac{q-q^{-1}}{4}(s_1^z - s_N^z), \\ &h = \sum_i 2s_i^z, \quad e = \sum_i q^{\sum_{j<i} 2s_j^z} s_i^+, \quad f = \sum_i s_i^- q^{-\sum_{j>i} 2s_j^z}, \end{aligned} \tag{4.121}$$

where $\vec{s} = \frac{1}{2}\vec{\sigma}$ and $s^{\pm} = s^1 \pm i s^2$. In the $q \to 1$ limit, $U_q(\mathfrak{sl}_2)$ reduces to $U(\mathfrak{sl}_2)$.

4.3.3 Quasi-Triangular Hopf Algebra

Using the coproduct, a tensor product representation of two representations (π_i, V_i) $(i = 1, 2)$ of the Hopf algebra can be defined in the following way,

$$\big((\pi_1 \otimes \pi_2)\circ\Delta, V_1 \otimes V_2\big). \tag{4.122}$$

Coassociativity implies the isomorphism,

$$(V_1 \otimes V_2) \otimes V_3 \cong V_1 \otimes (V_2 \otimes V_3) \quad (\text{as } A \text{ module}). \tag{4.123}$$

But cocommutativity does not hold in general, so the following isomorphism depends on the detail of the Hopf algebra:

$$V_1 \otimes V_2 \overset{?}{\cong} V_2 \otimes V_1 \quad (\text{as } A \text{ module}). \tag{4.124}$$

Of course we have $V_1 \otimes V_2 \cong V_2 \otimes V_1$ as vector space, by $P_{V_1V_2} : V_1 \otimes V_2 \overset{\cong}{\longrightarrow} V_2 \otimes V_1$, $P_{V_1V_2}(v_1 \otimes v_2) = v_2 \otimes v_1$. But the problem is the commutativity of the action of A and $P_{V_1V_2}$.

Drinfeld considered the situation where the isomorphism (4.124) does hold. A *quasi-triangular Hopf algebra* $(A, m, u, \Delta, \varepsilon, S, \mathcal{R})$ (we abbreviate it as $(A, \Delta, \mathcal{R})$) is a Hopf algebra with a universal R matrix $\mathcal{R}$, which satisfies

$$
\begin{aligned}
&\mathcal{R} \in A \otimes A : \textit{universal } R \textit{ matrix} \\
&\quad \Delta'(a) = \mathcal{R}\Delta(a)\mathcal{R}^{-1} \quad (\forall a \in A), \\
&\quad (\Delta \otimes \mathrm{id})\mathcal{R} = \mathcal{R}^{(13)}\mathcal{R}^{(23)}, \quad (\varepsilon \otimes \mathrm{id})\mathcal{R} = 1, \\
&\quad (\mathrm{id} \otimes \Delta)\mathcal{R} = \mathcal{R}^{(13)}\mathcal{R}^{(12)}, \quad (\mathrm{id} \otimes \varepsilon)\mathcal{R} = 1.
\end{aligned}
\tag{4.125}
$$

Then we have an intertwiner

$$
R_{V_1V_2} = P_{V_1V_2} \circ (\pi_1 \otimes \pi_2)(\mathcal{R}) : V_1 \otimes V_2 \overset{\cong}{\longrightarrow} V_2 \otimes V_1 \quad (\text{as } A \text{ module}), \tag{4.126}
$$

and $\mathcal{R}$ satisfies the Yang–Baxter equation,

$$
\mathcal{R}^{(12)}\mathcal{R}^{(13)}\mathcal{R}^{(23)} = \mathcal{R}^{(23)}\mathcal{R}^{(13)}\mathcal{R}^{(12)}. \tag{4.127}
$$

4.3.4 Quasi-Triangular Quasi-Hopf Algebra

As presented in 4.3.2, the quantum group is obtained from the Lie algebra by relaxing one condition, cocommutativity. Here we relax one more condition, coassociativity. Coassociativity (4.117) is modified by a coassociator Φ in the following way:

$$
\begin{aligned}
&\Phi \in A \otimes A \otimes A : \textit{coassociator} \\
&\quad (\mathrm{id} \otimes \Delta)\Delta(a) = \Phi(\Delta \otimes \mathrm{id})\Delta(a)\Phi^{-1} \quad (\forall a \in A), \\
&\quad (\mathrm{id} \otimes \mathrm{id} \otimes \Delta)\Phi \cdot (\Delta \otimes \mathrm{id} \otimes \mathrm{id})\Phi \\
&\qquad\qquad = (1 \otimes \Phi) \cdot (\mathrm{id} \otimes \Delta \otimes \mathrm{id})\Phi \cdot (\Phi \otimes 1), \\
&\quad (\mathrm{id} \otimes \varepsilon \otimes \mathrm{id})\Phi = 1.
\end{aligned}
\tag{4.128}
$$

A *quasi-triangular quasi-Hopf algebra* $(A, m, u, \Delta, \varepsilon, \Phi, S, \alpha, \beta, \mathcal{R})$ (we abbreviate it as $(A, \Delta, \Phi, \mathcal{R})$) satisfies (4.128) and

$$
\begin{aligned}
&\mathcal{R} \in A \otimes A \\
&\quad \Delta'(a) = \mathcal{R}\Delta(a)\mathcal{R}^{-1} \quad (\forall a \in A), \\
&\quad (\Delta \otimes \mathrm{id})\mathcal{R} = \Phi^{(312)}\mathcal{R}^{(13)}{\Phi^{(132)}}^{-1}\mathcal{R}^{(23)}\Phi^{(123)}, \\
&\quad (\mathrm{id} \otimes \Delta)\mathcal{R} = {\Phi^{(231)}}^{-1}\mathcal{R}^{(13)}\Phi^{(213)}\mathcal{R}^{(12)}{\Phi^{(123)}}^{-1},
\end{aligned}
\tag{4.129}
$$

where $\Phi^{(312)} = \sum_i Z_i \otimes X_i \otimes Y_i$ for $\Phi = \sum_i X_i \otimes Y_i \otimes Z_i$. $\alpha, \beta \in A$ satisfy

$$
\sum_i X_i \beta S(Y_i) \alpha Z_i = 1, \tag{4.130}
$$

$$
\sum_i S(b_i)\alpha c_i = \varepsilon(a)\alpha, \ \sum_i b_i \beta S(c_i) = \varepsilon(a)\beta \quad (\forall a \in A), \tag{4.131}
$$

where $\Delta(a) = \sum_i b_i \otimes c_i$. Then $\mathcal{R}$ enjoys the Yang–Baxter type equation,

$$\mathcal{R}^{(12)}\Phi^{(312)}\mathcal{R}^{(13)}\Phi^{(132)^{-1}}\mathcal{R}^{(23)}\Phi^{(123)} = \Phi^{(321)}\mathcal{R}^{(23)}\Phi^{(231)^{-1}}\mathcal{R}^{(13)}\Phi^{(213)}\mathcal{R}^{(12)}. \quad (4.132)$$

A quasi-Hopf algebra with $\Phi = 1$ is nothing but a Hopf algebra.

4.3.5 Twist

Quasi-Hopf algebras admit an important operation, *twist*. For any invertible element $F \in A \otimes A$ ($(\varepsilon \otimes \mathrm{id})F = (\mathrm{id} \otimes \varepsilon)F = 1$), which is called a twistor, there is a map from quasi-Hopf algebras to quasi-Hopf algebras:

$$\begin{array}{c}\text{quasi-Hopf algebra} \longrightarrow \text{quasi-Hopf algebra} \\ (A, \Delta, \Phi, \mathcal{R}) \xrightarrow{F} (A, \tilde{\Delta}, \widetilde{\Phi}, \widetilde{\mathcal{R}}).\end{array} \quad (4.133)$$

New coproduct, coassociator, R matrix, etc., are given by

$$\begin{aligned}
&F \in A \otimes A : \textit{twistor} \\
&\quad \widetilde{\Delta} = F\Delta(a)F^{-1} \quad (\forall a \in A), \\
&\quad \widetilde{\Phi} = \big(F^{(23)}(\mathrm{id} \otimes \Delta)F\big)\Phi\big(F^{(12)}(\Delta \otimes \mathrm{id})F\big)^{-1}, \\
&\quad \widetilde{\mathcal{R}} = F^{(21)}\mathcal{R}F^{-1}, \\
&\quad \widetilde{\varepsilon} = \varepsilon, \quad \widetilde{S} = S, \quad \widetilde{\alpha} = \sum_i S(d_i)\alpha e_i, \quad \tilde{\beta} = \sum_i f_i\beta S(g_i),
\end{aligned} \quad (4.134)$$

where $\sum_i d_i \otimes e_i = F^{-1}$ and $\sum_i f_i \otimes g_i = F$. We remark that an algebra A itself is unchanged. If a twistor F satisfies the cocycle condition, this twist operation maps a Hopf algebra to a Hopf algebra. However, for a general twistor F,, a Hopf algebra is mapped to a quasi-Hopf algebra:

$$\begin{array}{c}\text{Hopf algebra} \longrightarrow \text{quasi-Hopf algebra} \\ (A, \Delta, \mathcal{R}) \xrightarrow{F} (A, \widetilde{\Delta}, \widetilde{\Phi}, \widetilde{\mathcal{R}}).\end{array} \quad (4.135)$$

Let H be an Abelian subalgebra of A, with the product written additively. A twistor $F(\lambda) \in A \otimes A$ depending on $\lambda \in H$ is a shifted cocycle if it satisfies the relation (shifted cocycle condition),

$$\begin{aligned}
&F(\lambda) : \textit{shifted cocycle} \\
&\quad \Longleftrightarrow F^{(12)}(\lambda)(\Delta \otimes \mathrm{id})F(\lambda) = F^{(23)}(\lambda + h^{(1)})(\mathrm{id} \otimes \Delta)F(\lambda),
\end{aligned} \quad (4.136)$$

for some $h \in H$.

When a twistor $F(\lambda)$ satisfies the shifted cocycle condition, we obtain a quasi-triangular quasi-Hopf algebra from a quasi-triangular Hopf algebra by twisting,

$$\begin{aligned} &\text{Hopf algebra} \longrightarrow \text{quasi-Hopf algebra} \\ &(A,\Delta,\mathcal{R}) \xrightarrow{F(\lambda)} \big(A,\Delta_\lambda,\Phi(\lambda),\mathcal{R}(\lambda)\big), \end{aligned} \tag{4.137}$$

and we have

$$\begin{aligned} \Phi(\lambda) &= F^{(23)}(\lambda)F^{(23)}(\lambda+h^{(1)})^{-1}, \\ (\Delta_\lambda\otimes\mathrm{id})\mathcal{R}(\lambda) &= \Phi^{(312)}(\lambda)\mathcal{R}^{(13)}(\lambda)\mathcal{R}^{(23)}(\lambda+h^{(1)}), \\ (\mathrm{id}\otimes\Delta_\lambda)\mathcal{R}(\lambda) &= \mathcal{R}^{(13)}(\lambda+h^{(2)})\mathcal{R}^{(12)}(\lambda)\Phi^{(123)}(\lambda)^{-1}, \end{aligned} \tag{4.138}$$

and the R matrix satisfies the dynamical Yang–Baxter equation,

$$\begin{aligned} &\mathcal{R}^{(12)}(\lambda+h^{(3)})\mathcal{R}^{(13)}(\lambda)\mathcal{R}^{(23)}(\lambda+h^{(1)}) \\ &\qquad = \mathcal{R}^{(23)}(\lambda)\mathcal{R}^{(13)}(\lambda+h^{(2)})\mathcal{R}^{(12)}(\lambda). \end{aligned} \tag{4.139}$$

4.4 Elliptic Quantum Groups

The solutions of YBE (R-matrices) are classified into two types, vertex-type and face-type. Corresponding to these two there are two types of elliptic quantum groups (algebras).

The vertex-type elliptic algebras are associated with the R-matrix $R(u)$ of Baxter [49] and Belavin [50]. The first example of this sort is the Sklyanin algebra [51], designed as an elliptic deformation of the Lie algebra $\mathfrak{sl}_2$. It is presented by the 'RLL'-relation

$$R^{(12)}(u_1-u_2)L^{(1)}(u_1)L^{(2)}(u_2) = L^{(2)}(u_2)L^{(1)}(u_1)R^{(12)}(u_1-u_2), \tag{4.140}$$

together with a specific choice of the form for $L(u)$. R and L depend on an elliptic modulus r. Foda et al. [23] proposed its affine version, $\mathcal{A}_{q,p}(\widehat{\mathfrak{sl}}_2)$,

$$\begin{aligned} &R^{(12)}(u_1-u_2,r)L^{(1)}(u_1)L^{(2)}(u_2) \\ &\qquad = L^{(2)}(u_2)L^{(1)}(u_1)R^{(12)}(u_1-u_2,r-c), \end{aligned} \tag{4.141}$$

whose main point of is the shift of r by a central element c.

The face-type algebras are based on R-matrices of Andrews, Baxter, Forrester [52] and generalizations [8, 10, 53]. In this case, besides the elliptic modulus, R and L depend also on extra parameter(s) λ. As Felder has shown [24], the RLL relation undergoes a 'dynamical' shift by elements h of the Cartan subalgebra,

$$\begin{aligned} &R^{(12)}(u_1-u_2,\lambda+h)L^{(1)}(u_1,\lambda)L^{(2)}(u_2,\lambda+h^{(1)}) \\ &\qquad = L^{(2)}(u_2,\lambda)L^{(1)}(u_1,\lambda+h^{(2)})R^{(12)}(u_1-u_2,\lambda), \end{aligned} \tag{4.142}$$

and the YBE itself is modified to a dynamical one (4.139), (4.20). As we shall see, a central extension of this algebra is obtained by introducing further a shift of the elliptic modulus analogous to (4.141) (see (4.196)–(4.197)and the remark following them).

These two algebras, RLL relations (4.141) and (4.142), seemed to be different but Frønsdal [25] pointed out that they have a common structure; they are quasi-Hopf algebras obtained by twisting quantum affine algebras. Namely, there exist two types of twistors which give rise to different comultiplications on the quantum affine algebras $U_q(\mathfrak{g})$, and the resultant quasi-Hopf algebras are nothing but the two types of elliptic quantum groups. In [26] explicit formulas for the twistors satisfying the shifted cocycle condition were presented and two types of elliptic quantum groups, $\mathcal{A}_{q,p}(\widehat{\mathfrak{sl}}_n)$ and $\mathcal{B}_{q,\lambda}(\mathfrak{g})$, were defined:

$$\begin{array}{cccc} \text{type} & \text{twistor} & \text{Hopf algebra} & \text{quasi-Hopf algebra} \\ \text{face} & F(\lambda) & U_q(\mathfrak{g}) \xrightarrow{F(\lambda)} & \mathcal{B}_{q,\lambda}(\mathfrak{g}) \\ \text{vertex} & E(r) & U_q(\widehat{\mathfrak{sl}}_n) \xrightarrow{E(r)} & \mathcal{A}_{q,p}(\widehat{\mathfrak{sl}}_n). \end{array} \tag{4.143}$$

The face-type algebra has also been given an alternative formulation in terms of the Drinfeld currents. This is the approach adopted by Enriquez and Felder [54] and Konno [55]. The Drinfeld currents are suited to deal with infinite-dimensional representations.

4.4.1 Quantum Group

First let us fix the notation. Let $\mathfrak{g}$ be the Kac-Moody Lie algebra associated with a symmetrizable generalized Cartan matrix $A = (a_{ij})_{i,j\in I}$ [56]. We fix an invariant inner product $(\ ,\ )$ on the Cartan subalgebra $\mathfrak{h}$ and identify $\mathfrak{h}^*$ with $\mathfrak{h}$ via $(\ ,\ )$. If $\{\alpha_i\}_{i\in I}$ denotes the set of simple roots, then $(\alpha_i, \alpha_j) = d_i a_{ij}$, where $d_i = \frac{1}{2}(\alpha_i, \alpha_i)$.

Consider the corresponding quantum group $U = U_q(\mathfrak{g})$. Hereafter we fix a complex number $q \neq 0$, $|q| < 1$. The algebra U has generators e_i, f_i $(i \in I)$ and h $(h \in \mathfrak{h})$, satisfying the standard relations

$$[h, h'] = 0 \qquad (h, h' \in \mathfrak{h}), \tag{4.144}$$

$$[h, e_i] = (h, \alpha_i) e_i, \quad [h, f_i] = -(h, \alpha_i) f_i \qquad (i \in I, h \in \mathfrak{h}), \tag{4.145}$$

$$[e_i, f_j] = \delta_{ij} \frac{t_i - t_i^{-1}}{q_i - q_i^{-1}} \qquad (i, j \in I), \tag{4.146}$$

and the Serre relations which we omit. In (4.146) we have set $q_i = q^{d_i}$, $t_i = q^{\alpha_i}$. We adopt the Hopf algebra structure given as follows.

$$\Delta(h) = h \otimes 1 + 1 \otimes h, \tag{4.147}$$

$$\Delta(e_i) = e_i \otimes 1 + t_i \otimes e_i, \quad \Delta(f_i) = f_i \otimes t_i^{-1} + 1 \otimes f_i, \tag{4.148}$$

$$\varepsilon(e_i) = \varepsilon(f_i) = \varepsilon(h) = 0, \tag{4.149}$$

$$S(e_i) = -t_i^{-1} e_i, \quad S(f_i) = -f_i t_i, \quad S(h) = -h, \tag{4.150}$$

where $i \in I$ and $h \in \mathfrak{h}$.

Let $\mathcal{R} \in U^{\otimes 2}$ denote the universal R matrix of U. It has the form

$$\mathcal{R} = q^{-T}\mathcal{C}, \tag{4.151}$$

$$\begin{aligned} \mathcal{C} &= \sum_{\beta \in Q^+} q^{(\beta,\beta)} (q^{-\beta} \otimes q^{\beta}) \mathcal{C}_\beta \\ &= 1 - \sum_{i \in I} (q_i - q_i^{-1}) e_i t_i^{-1} \otimes t_i f_i + \cdots . \end{aligned} \tag{4.152}$$

Here the notation is as follows. Take a basis $\{h_l\}$ of $\mathfrak{h}$, and its dual basis $\{h^l\}$. Then

$$T = \sum_l h_l \otimes h^l \tag{4.153}$$

denotes the canonical element of $\mathfrak{h} \otimes \mathfrak{h}$. The element $\mathcal{C}_\beta = \sum_j u_{\beta,j} \otimes u^j_{-\beta}$ is the canonical element of $U^+_\beta \otimes U^-_{-\beta}$ with respect to a certain Hopf pairing, where U^+ (resp. U^-) denotes the subalgebra of U generated by the e_i (resp. f_i), and $U^\pm_{\pm\beta}$ ($\beta \in Q^+$) signifies the homogeneous components with respect to the natural gradation by $Q^+ = \sum_i \mathbb{Z}_{\geq 0}\alpha_i$. (For the details the reader is referred, e.g., to [11, 57].) Basic properties of the universal R matrix is given in (4.125).

4.4.2 Face Type Algebra

Let $\rho \in \mathfrak{h}$ be an element such that $(\rho, \alpha_i) = d_i$ for all $i \in I$. Let ϕ be an automorphism of U given by

$$\phi = \mathrm{Ad}\Big(q^{(\sum_l h_l h^l)/2 - \rho}\Big), \tag{4.154}$$

where $\{h_l\}$, $\{h^l\}$ are as in (4.153). In other words,

$$\phi(e_i) = e_i t_i, \quad \phi(f_i) = t_i^{-1} f_i, \quad \phi(q^h) = q^h. \tag{4.155}$$

Since

$$\mathrm{Ad}(q^T) \circ (\phi \otimes \phi) = \mathrm{Ad}\Big(q^{(\sum_l \Delta(h_l h^l))/2 - \Delta(\rho)}\Big), \tag{4.156}$$

we have

$$\mathrm{Ad}(q^T) \circ (\phi \otimes \phi) \circ \Delta = \Delta \circ \phi. \tag{4.157}$$

For $\lambda \in \mathfrak{h}$, we introduce an automorphism

$$\varphi_\lambda = \mathrm{Ad}\Big(q^{\sum_l h_l h^l + 2(\lambda - \rho)}\Big) = \phi^2 \circ \mathrm{Ad}(q^{2\lambda}). \tag{4.158}$$

Then the expression $(\varphi_\lambda \otimes \mathrm{id})(q^T \mathcal{R})$ is a formal power series in the variables $x_i = q^{2(\lambda,\alpha_i)}$ ($i \in I$) of the form $1 - \sum_i (q_i - q_i^{-1}) x_i e_i t_i \otimes t_i f_i + \cdots$. We define the twistor $F(\lambda)$ as follows:

Face type twistor.

$$F(\lambda) = \prod_{k\geq 1}^{\curvearrowleft} (\varphi_\lambda^k \otimes \mathrm{id})(q^T \mathcal{R})^{-1}. \tag{4.159}$$

Here and after, we use the ordered product symbol $\prod^{\curvearrowleft}_{k\geq 1} A_k = \cdots A_3 A_2 A_1$. Note that the kth factor in the product (4.159) is a formal power series in the x_i^k with leading term 1, and hence the infinite product makes sense. Then the face type twistor (4.159) satisfies the shifted cocycle condition

$$F^{(12)}(\lambda)(\Delta \otimes \mathrm{id})F(\lambda) = F^{(23)}(\lambda + h^{(1)})(\mathrm{id} \otimes \Delta)F(\lambda), \tag{4.160}$$

and

$$(\varepsilon \otimes \mathrm{id})F(\lambda) = (\mathrm{id} \otimes \varepsilon)F(\lambda) = 1. \tag{4.161}$$

A proof of this property and examples are given in [26]. In (4.160), if $\lambda = \sum_l \lambda_l h^l$, then $\lambda + h^{(1)}$ means $\sum_l (\lambda_l + h_l^{(1)})h^l$. Hence we have, for example,

$$\begin{aligned}
\mathrm{Ad}\big(q^{2lT^{(12)}}\big) F_k^{(23)}(\lambda) &= F_k^{(23)}\big(\lambda + l h^{(1)}\big), \\
\mathrm{Ad}\big(q^{2lT^{(13)}}\big) F_k^{(23)}(\lambda) &= F_k^{(23)}\big(\lambda - l h^{(1)}\big).
\end{aligned}$$

For convenience, let us give a name to the quasi-Hopf algebra associated with the twistor (4.159).

Face type algebra. We define the quasi-Hopf algebra $\mathcal{B}_{q,\lambda}(\mathfrak{g})$ of face type to be the set $\big(U_q(\mathfrak{g}), \Delta_\lambda, \Phi(\lambda), \mathcal{R}(\lambda)\big)$ together with $\alpha_\lambda = \sum_i S(d_i)e_i$, $\beta_\lambda = \sum_i f_i S(g_i)$, the antiautomorphism S defined by (4.150)) and ε defined by (4.149),

$$\Delta_\lambda(a) = F^{(12)}(\lambda)\,\Delta(a)\,F^{(12)}(\lambda)^{-1}, \tag{4.162}$$

$$\mathcal{R}(\lambda) = F^{(21)}(\lambda)\,\mathcal{R}\,F^{(12)}(\lambda)^{-1}, \tag{4.163}$$

$$\Phi(\lambda) = F^{(23)}(\lambda)F^{(23)}\big(\lambda + h^{(1)}\big)^{-1}, \tag{4.164}$$

and $\sum_i d_i \otimes e_i = F(\lambda)^{-1}$, $\sum_i f_i \otimes g_i = F(\lambda)$.

Let us consider the case where $\mathfrak{g}$ is of affine type, in which we are mainly interested. Let c be the canonical central element and d the scaling element. We set

$$\lambda - \rho = rd + s'c + \bar{\lambda} - \bar{\rho} \quad (r, s' \in \mathbb{C}), \tag{4.165}$$

where $\bar{\lambda}$ stands for the classical part of $\lambda \in \mathfrak{h}$. Denote by $\{\bar{h}_j\}$, $\{\bar{h}^j\}$ the classical part of the dual basis of $\mathfrak{h}$. Since c is central, φ_λ is independent of s'. Writing $p = q^{2r}$, we have

$$\varphi_\lambda = \mathrm{Ad}(p^d q^{2cd}) \circ \bar{\varphi}_\lambda, \quad \bar{\varphi}_\lambda = \mathrm{Ad}\big(q^{\sum \bar{h}_j \bar{h}^j + 2(\bar{\lambda} - \bar{\rho})}\big). \tag{4.166}$$

Set further

$$\mathcal{R}(z) = \mathrm{Ad}(z^d \otimes 1)(\mathcal{R}), \tag{4.167}$$

$$F(z,\lambda) = \mathrm{Ad}(z^d \otimes 1)\big(F(\lambda)\big), \tag{4.168}$$

$$\mathcal{R}(z,\lambda) = \mathrm{Ad}(z^d \otimes 1)\big(\mathcal{R}(\lambda)\big) = \sigma\big(F(z^{-1},\lambda)\big)\mathcal{R}(z)F(z,\lambda)^{-1}. \tag{4.169}$$

(4.167) and (4.168) are formal power series in z, whereas (4.169) contains both positive and negative powers of z. Note that $q^{c\otimes d+d\otimes c}\mathcal{R}(z)|_{z=0}$ reduces to the universal R matrix of $U_q(\bar{\mathfrak{g}})$ corresponding to the underlying finite dimensional Lie algebra $\bar{\mathfrak{g}}$. From the definition (4.159) of $F(\lambda)$ we have the difference equation

$$F(pq^{2c^{(1)}}z,\lambda) = (\bar{\varphi}_\lambda \otimes \mathrm{id})^{-1}\big(F(z,\lambda)\big)\cdot q^T\mathcal{R}(pq^{2c^{(1)}}z), \tag{4.170}$$

with the initial condition $F(0,\lambda) = F_{\bar{\mathfrak{g}}}(\bar{\lambda})$, where $F_{\bar{\mathfrak{g}}}(\bar{\lambda})$ signifies the twistor corresponding to $\bar{\mathfrak{g}}$.

4.4.3 Vertex Type Algebra

When $\mathfrak{g} = \widehat{\mathfrak{sl}}_n$, it is possible to construct a different type of twistor. We call it *vertex type*.

Let us write $h_i = \alpha_i$ $(i = 0,\ldots,n-1)$. A basis of $\mathfrak{h}$ is $\{h_0,\ldots,h_{n-1},d\}$. The element d gives the homogeneous grading,

$$[d,e_i] = \delta_{i0}e_i, \quad [d,f_i] = -\delta_{i0}f_i, \tag{4.171}$$

for all $i = 0,\ldots,n-1$. Let the dual basis be $\{\Lambda_0,\ldots,\Lambda_{n-1},c\}$. The Λ_i are the fundamental weights and c is the canonical central element. Let τ be the automorphism of $U_q(\widehat{\mathfrak{sl}}_n)$ such that

$$\tau(e_i) = e_{i+1 \bmod n}, \quad \tau(f_i) = f_{i+1 \bmod n}, \quad \tau(h_i) = h_{i+1 \bmod n} \tag{4.172}$$

and $\tau^n = \mathrm{id}$. Then we have

$$\tau(\Lambda_i) = \Lambda_{i+1 \bmod n} - \frac{n-1-2i}{2n}c. \tag{4.173}$$

The element $\rho = \sum_{i=0}^{n-1}\Lambda_i$ is invariant under τ. It gives the principal grading

$$[\rho,e_i] = e_i, \quad [\rho,f_i] = -f_i, \tag{4.174}$$

for all $i = 0,\ldots,n-1$. Note also that

$$(\tau\otimes\tau)\circ\Delta = \Delta\circ\tau, \quad (\tau\otimes\tau)(\mathcal{C}_\beta) = \mathcal{C}_{\tau(\beta)}. \tag{4.175}$$

For $r \in \mathbb{C}$, we introduce an automorphism

$$\widetilde{\varphi}_r = \tau\circ\mathrm{Ad}(q^2(r+c)\rho/n), \tag{4.176}$$

and set

$$\widetilde{T} = \frac{1}{n}\left(\rho \otimes c + c \otimes \rho - \frac{n^2-1}{12} c \otimes c\right). \tag{4.177}$$

Then $(\widetilde{\varphi}_r \otimes \mathrm{id})(q^{\widetilde{T}}\mathcal{R})^{-1}$ is a formal power series in $p^{1/n}$ where $p = q^{2r}$. Unlike the face case, this is a formal series with a nontrivial leading term $q^{T-\widetilde{T}}(1+\cdots)$. Nevertheless, the n-fold product

$$\prod_{n \geq k \geq 1}^{\curvearrowleft} (\widetilde{\varphi}_r^k \otimes \mathrm{id})(q^{\widetilde{T}}\mathcal{R})^{-1} \tag{4.178}$$

takes the form $1+\cdots$, because of the relation

$$\sum_{k=1}^{n} (\tau^k \otimes \mathrm{id})(T - \widetilde{T}) = 0. \tag{4.179}$$

We now define the vertex type twistor $E(r)$ as follows.

Vertex type twistor

$$E(r) = \prod_{k \geq 1}^{\curvearrowleft} (\widetilde{\varphi}_r^k \otimes \mathrm{id})(q^{\widetilde{T}}\mathcal{R})^{-1}. \tag{4.180}$$

The infinite product $\overset{\curvearrowleft}{\prod}_{k \geq 1}$ is to be understood as $\lim_{N\to\infty} \overset{\curvearrowleft}{\prod}_{nN \geq k \geq 1}$. In view of the remark made above, $E(r)$ is a well defined formal series in $p^{1/n}$. Then the vertex type twistor (4.180) satisfies the shifted cocycle condition

$$E^{(12)}(r)(\Delta \otimes \mathrm{id})E(r) = E^{(23)}(r + c^{(1)})(\mathrm{id} \otimes \Delta)E(r), \tag{4.181}$$

and

$$(\varepsilon \otimes \mathrm{id})E(r) = (\mathrm{id} \otimes \varepsilon)E(r) = 1. \tag{4.182}$$

A proof of this property and an example are given in [26].

Vertex type algebra We define the quasi-Hopf algebra $\mathcal{A}_{q,p}(\widehat{\mathfrak{sl}}_n)$ ($p = q^{2r}$) of vertex type to be the set $\big(U_q(\widehat{\mathfrak{sl}}_n), \Delta_r, \Phi(r), \mathcal{R}(r)\big)$ together with $\alpha_r = \sum_i S(d_i)e_i$, $\beta_r = \sum_i f_i S(g_i)$, the antiautomorphism S defined by (4.150) and ε defined by (4.149),

$$\Delta_r(a) = E^{(12)}(r)\,\Delta(a)\,E^{(12)}(r)^{-1}, \tag{4.183}$$

$$\mathcal{R}(r) = E^{(21)}(r)\,\mathcal{R}\,E^{(12)}(r)^{-1}, \tag{4.184}$$

$$\Phi(r) = E^{(23)}(r)E^{(23)}\big(r + c^{(1)}\big)^{-1}, \tag{4.185}$$

and $\sum_i d_i \otimes e_i = E(r)^{-1}$, $\sum_i f_i \otimes g_i = E(r)$.

Let us set

$$\widetilde{\mathcal{R}}'(\zeta) = \big(\mathrm{Ad}(\zeta^{\rho}) \otimes \mathrm{id}\big)(q^{\widetilde{T}}\mathcal{R}), \tag{4.186}$$

$$E(\zeta, r) = \big(\mathrm{Ad}(\zeta^{\rho}) \otimes \mathrm{id}\big)E(r). \tag{4.187}$$

In just the same way as in the face type case, the definition (4.180) can be alternatively described as the unique solution of the difference equation

$$E(p^{1/n}q^{2c^{(1)}/n}\zeta, r) = (\tau \otimes \mathrm{id})^{-1}\big(E(\zeta, r)\big) \cdot \widetilde{\mathcal{R}}'(p^{\frac{1}{n}}q^{\frac{2}{n}c^{(1)}}\zeta), \tag{4.188}$$

with the initial condition $E(0, r) = 1$, where $p = q^{2r}$.

4.4.4 Dynamical *RLL*-Relations and Vertex Operators

The L-operators and vertex operators for the elliptic algebras can be constructed from those of $U_q(\mathfrak{g})$ by 'dressing' the latter with the twistors. In this subsection, we examine various commutation relations among these operators. We shall mainly discuss the case of the face type algebra $\mathcal{B}_{q,\lambda}(\mathfrak{g})$ where $\mathfrak{g}$ is of affine type.

Hereafter we write $U = U_q(\mathfrak{g})$, $\mathcal{B} = \mathcal{B}_{q,\lambda}(\mathfrak{g})$. By a representation of the quasi-Hopf algebra $\mathcal{B}$ we mean that of the underlying associative algebra U. Let (π_V, V) be a finite dimensional module over U, and $(\pi_{V,z}, V_z)$ be the evaluation representation associated with it where $\pi_{V,z} = \pi_V \circ \mathrm{Ad}(z^d)$.

We define L-operators for $\mathcal{B}$ by

$$L_V^{\pm}(z, \lambda) = (\pi_{V,z} \otimes \mathrm{id})\mathcal{R}'^{\pm}(\lambda), \tag{4.189}$$

$$\mathcal{R}'^{+}(\lambda) = q^{c\otimes d + d\otimes c}\mathcal{R}(\lambda), \tag{4.190}$$

$$\mathcal{R}'^{-}(\lambda) = \mathcal{R}^{(21)}(\lambda)^{-1}q^{-c\otimes d - d\otimes c}. \tag{4.191}$$

Likewise we set

$$R_{VW}^{\pm}\left(\frac{z_1}{z_2}, \lambda\right) = (\pi_{V,z_1} \otimes \pi_{W,z_2})\mathcal{R}'^{\pm}(\lambda). \tag{4.192}$$

Setting further

$$\mathcal{R}'^{\pm}(z, \lambda) = \mathrm{Ad}(z^d \otimes 1)\mathcal{R}'^{\pm}(\lambda), \tag{4.193}$$

we find from the dynamical YBE (4.139) that

$$\begin{aligned}&\mathcal{R}'^{\pm(12)}\left(\frac{z_1}{z_2}, \lambda + h^{(3)}\right)\mathcal{R}'^{\pm(13)}\left(q^{\mp c^{(2)}}\frac{z_1}{z_3}, \lambda\right)\mathcal{R}'^{\pm(23)}\left(\frac{z_2}{z_3}, \lambda + h^{(1)}\right)\\ &= \mathcal{R}'^{\pm(23)}\left(\frac{z_2}{z_3}, \lambda\right)\mathcal{R}'^{\pm(13)}\left(q^{\pm c^{(2)}}\frac{z_1}{z_3}, \lambda + h^{(2)}\right)\mathcal{R}'^{\pm(12)}\left(\frac{z_1}{z_2}, \lambda\right),\end{aligned} \tag{4.194}$$

$$\begin{aligned}&\mathcal{R}'^{+(12)}\left(q^{c^{(3)}}\frac{z_1}{z_2}, \lambda + h^{(3)}\right)\mathcal{R}'^{+(13)}\left(\frac{z_1}{z_3}, \lambda\right)\mathcal{R}'^{-(23)}\left(\frac{z_2}{z_3}, \lambda + h^{(1)}\right)\\ &= \mathcal{R}'^{-(23)}\left(\frac{z_2}{z_3}, \lambda\right)\mathcal{R}'^{+(13)}\left(\frac{z_1}{z_3}, \lambda + h^{(2)}\right)\mathcal{R}'^{+(12)}\left(q^{-c^{(3)}}\frac{z_1}{z_2}, \lambda\right).\end{aligned} \tag{4.195}$$

Applying $\pi_V \otimes \pi_W \otimes \mathrm{id}$, we obtain the dynamical RLL relation,

$$R^{\pm(12)}_{VW}\Big(\frac{z_1}{z_2}, \lambda + h\Big) L^{\pm(1)}_V(z_1, \lambda) L^{\pm(2)}_W(z_2, \lambda + h^{(1)}) \\ = L^{\pm(2)}_W(z_2, \lambda) L^{\pm(1)}_V(z_1, \lambda + h^{(2)}) R^{\pm(12)}_{VW}\Big(\frac{z_1}{z_2}, \lambda\Big), \tag{4.196}$$

$$R^{+(12)}_{VW}\Big(q^c\frac{z_1}{z_2}, \lambda + h\Big) L^{+(1)}_V(z_1, \lambda) L^{-(2)}_W(z_2, \lambda + h^{(1)}) \\ = L^{-(2)}_W(z_2, \lambda) L^{+(1)}_V(z_1, \lambda + h^{(2)}) R^{+(12)}_{VW}\Big(q^{-c}\frac{z_1}{z_2}, \lambda\Big). \tag{4.197}$$

Here the index (1) (resp. (2)) refers to V (resp. W), and h, c (without superfix) are elements of $\mathfrak{h} \subset \mathcal{B}$. If we write

$$\lambda - \rho = rd + s'c + \bar{\lambda} - \bar{\rho} \quad (r, s' \in \mathbb{C}, \bar{\lambda} \in \bar{\mathfrak{h}}), \tag{4.198}$$

then

$$\lambda + h^{(1)} = \big(r + h^\vee + c^{(1)}\big)d + \big(s' + d^{(1)}\big)c + \big(\bar{\lambda} + \bar{h}^{(1)}\big), \tag{4.199}$$

where $h^\vee$ is the dual Coxeter number. The parameter r plays the role of the elliptic modulus. Note that, in (4.196)–(4.197), r also undergoes a shift depending on the central element c.

Actually the two L-operators (4.189) are not independent. We have

$$L^+_V(pq^c z, \lambda) = q^{-2\overline{T}_{V,\bullet}} \big(\mathrm{Ad}(\overline{X}_\lambda) \otimes \mathrm{id}\big)^{-1} L^-_V(z, \lambda), \tag{4.200}$$

where

$$\overline{T}_{V,\bullet} = \sum_j \pi(\bar{h}_j) \otimes \bar{h}^j, \quad \overline{X}_\lambda = \pi(q^{\sum \bar{h}_j \bar{h}^j + 2(\bar{\lambda} - \bar{\rho})}). \tag{4.201}$$

Next let us consider vertex operators. Let $(\pi_{V,z}, V_z)$ be as before, and let $V(\mu)$ be a highest weight module with highest weight μ. Consider intertwiners of U-modules of the form

$$\Phi^{(\nu,\mu)}_V(z)\colon V(\mu) \to V(\nu) \otimes V_z, \tag{4.202}$$

$$\Psi^{*(\nu,\mu)}_V(z)\colon V_z \otimes V(\mu) \to V(\nu), \tag{4.203}$$

which are called vertex operators of type I and type II respectively. Define the corresponding VO's for $\mathcal{B}$ as follows [58]:

$$\Phi^{(\nu,\mu)}_V(z, \lambda) = (\mathrm{id} \otimes \pi_z) F(\lambda) \circ \Phi^{(\nu,\mu)}_V(z), \tag{4.204}$$

$$\Psi^{*(\nu,\mu)}_V(z, \lambda) = \Psi^{*(\nu,\mu)}_V(z) \circ (\pi_z \otimes \mathrm{id}) F(\lambda)^{-1}. \tag{4.205}$$

When there is no fear of confusion, we often drop the sub(super)scripts V or (ν,μ). It is clear that (4.204) and (4.205) satisfy the intertwining relations relative to the coproduct Δ_λ (4.162),

$$\Delta_\lambda(a)\Phi(z,\lambda)=\Phi(z,\lambda)a \qquad (\forall a\in\mathcal{B}), \tag{4.206}$$

$$a\Psi^*(z,\lambda)=\Psi^*(z,\lambda)\Delta_\lambda(a) \quad (\forall a\in\mathcal{B}). \tag{4.207}$$

These intertwining relations can be encapsulated to commutation relations with the L-operators.

The "dressed" VO's (4.204), (4.205) satisfy the following dynamical intertwining relations (see the diagram below):

$$\begin{aligned}&\Phi_W(z_2,\lambda)L_V^+(z_1,\lambda)\\ &\quad=R_{VW}^+\left(q^c\frac{z_1}{z_2},\lambda+h\right)L_V^+(z_1,\lambda)\Phi_W(z_2,\lambda+h^{(1)}),\end{aligned} \tag{4.208}$$

$$\begin{aligned}&\Phi_W(z_2,\lambda)L_V^-(z_1,\lambda)\\ &\quad=R_{VW}^-\left(\frac{z_1}{z_2},\lambda+h\right)L_V^-(z_1,\lambda)\Phi_W(z_2,\lambda+h^{(1)}),\end{aligned} \tag{4.209}$$

$$\begin{aligned}&L_V^+(z_1,\lambda)\Psi_W^*(z_2,\lambda+h^{(1)})\\ &\quad=\Psi_W^*(z_2,\lambda)L_V^+(z_1,\lambda+h^{(2)})R_{VW}^+\left(\frac{z_1}{z_2},\lambda\right),\end{aligned} \tag{4.210}$$

$$\begin{aligned}&L_V^-(z_1,\lambda)\Psi_W^*(z_2,\lambda+h^{(1)})\\ &\quad=\Psi_W^*(z_2,\lambda)L_V^-(z_1,\lambda+h^{(2)})R_{VW}^-\left(q^c\frac{z_1}{z_2},\lambda\right).\end{aligned} \tag{4.211}$$

$$\begin{CD}V_{z_1}\otimes V(\mu) @>{\Phi_W}>> V_{z_1}\otimes V(\nu)\otimes W_{z_2} @>{L_V^\pm}>> V_{z_1}\otimes V(\nu)\otimes W_{z_2}\\ @V{L_V^\pm}VV @. @VV{R_{VW}^\pm}V\\ V_{z_1}\otimes V(\mu) @>>{\Phi_W}> @. V_{z_1}\otimes V(\nu)\otimes W_{z_2}\end{CD} \tag{4.212}$$

$$\begin{CD}V_{z_1}\otimes W_{z_2}\otimes V(\mu) @>{R_{VW}^\pm}>> V_{z_1}\otimes W_{z_2}\otimes V(\mu) @>{L_V^\pm}>> V_{z_1}\otimes W_{z_2}\otimes V(\mu)\\ @V{\Psi_W^*}VV @. @VV{\Psi_W^*}V\\ V_{z_1}\otimes V(\nu) @>>{L_V^\pm}> @. V_{z_1}\otimes V(\nu)\end{CD} \tag{4.213}$$

From the theory of q-KZ-equation [59], we know the VO's for U satisfy the commutation relations of the form

$$\begin{aligned}&\check{R}_{VV}\left(\frac{z_1}{z_2}\right)\Phi_V^{(\nu,\mu)}(z_1)\Phi_V^{(\mu,\kappa)}(z_2)\\ &\quad=\sum_{\mu'}\Phi_V^{(\nu,\mu')}(z_2)\Phi_V^{(\mu',\kappa)}(z_1)W_{\mathrm{I}}\left(\begin{matrix}\kappa & \mu\\ \mu' & \nu\end{matrix}\middle|\,\frac{z_1}{z_2}\right),\end{aligned} \tag{4.214}$$

$$\Psi_V^{*(\nu,\mu)}(z_1)\Psi_V^{*(\mu,\kappa)}(z_2)\check{R}_{VV}(\tfrac{z_1}{z_2})^{-1}$$
$$= \sum_{\mu'} W_{\mathrm{II}}\left(\begin{matrix}\kappa & \mu \\ \mu' & \nu\end{matrix}\,\middle|\,\frac{z_1}{z_2}\right)\Psi_V^{*(\nu,\mu')}(z_2)\Psi_V^{*(\mu',\kappa)}(z_1), \tag{4.215}$$
$$\Phi_V^{(\nu,\mu)}(z_1)\Psi_V^{*(\mu,\kappa)}(z_2)$$
$$= \sum_{\mu'} W_{\mathrm{I,II}}\left(\begin{matrix}\kappa & \mu \\ \mu' & \nu\end{matrix}\,\middle|\,\frac{z_1}{z_2}\right)\Psi_V^{*(\nu,\mu')}(z_2)\Phi_V^{(\mu',\kappa)}(z_1). \tag{4.216}$$

Here

$$\check{R}_{VV}(z) = PR_{VV}(z), \quad P(v\otimes v') = v'\otimes v, \tag{4.217}$$
$$R_{VV}\left(\frac{z_1}{z_2}\right) = (\pi_{V,z_1}\otimes\pi_{V,z_2})\mathcal{R} \tag{4.218}$$

is the "trigonometric" R matrix. In (4.214)–(4.216) we used a slightly abbreviated notation. For example, the left hand side of (4.214) means the composition

$$V(\kappa) \xrightarrow{\Phi(z_2)} V(\mu)\otimes V_{z_2} \xrightarrow{\Phi(z_1)\otimes\mathrm{id}} V(\nu)\otimes V_{z_1}\otimes V_{z_2} \xrightarrow{\mathrm{id}\otimes\check{R}(z_1/z_2)} V(\nu)\otimes V_{z_2}\otimes V_{z_1}.$$

Similarly (4.215), (4.216) are the maps

$$V_{z_2}\otimes V_{z_1}\otimes V(\kappa) \to V(\nu), \quad V_{z_2}\otimes V(\kappa) \to V(\nu)\otimes V_{z_1}, \tag{4.219}$$

respectively. For $U_q(\widehat{\mathfrak{sl}}_2)$, the formulas for the W-factors in the simplest case can be found, e.g., in [60].

The "dressed" VO's satisfy similar relations with appropriate dynamical shift. Setting $\check{R}_{VV}(z,\lambda) = PR_{VV}(z,\lambda)$, we have

$$\check{R}_{VV}\left(\frac{z_1}{z_2},\lambda+h^{(1)}\right)\Phi_V^{(\nu,\mu)}(z_1,\lambda)\Phi_V^{(\mu,\kappa)}(z_2,\lambda)$$
$$= \sum_{\mu'} \Phi_V^{(\nu,\mu')}(z_2,\lambda)\Phi_V^{(\mu',\kappa)}(z_1,\lambda)W_I\left(\begin{matrix}\kappa & \mu \\ \mu' & \nu\end{matrix}\,\middle|\,\frac{z_1}{z_2}\right), \tag{4.220}$$
$$\Psi_V^{*(\nu,\mu)}(z_1,\lambda)\Psi_V^{*(\mu,\kappa)}(z_2,\lambda+h^{(1)})\check{R}_{VV}(\tfrac{z_1}{z_2},\lambda)^{-1}$$
$$= \sum_{\mu'} W_{II}\left(\begin{matrix}\kappa & \mu \\ \mu' & \nu\end{matrix}\,\middle|\,\frac{z_1}{z_2}\right)\Psi_V^{*(\nu,\mu')}(z_2,\lambda)\Psi_V^{*(\mu',\kappa)}(z_1,\lambda+h^{(1)}), \tag{4.221}$$
$$\Phi_V^{(\nu,\mu)}(z_1,\lambda)\Psi_V^{*(\mu,\kappa)}(z_2,\lambda)$$
$$= \sum_{\mu'} W_{I,II}\left(\begin{matrix}\kappa & \mu \\ \mu' & \nu\end{matrix}\,\middle|\,\tfrac{z_1}{z_2}\right)\Psi_V^{*(\nu,\mu')}(z_2,\lambda)\Phi_V^{(\mu',\kappa)}(z_1,\lambda+h^{(1)}). \tag{4.222}$$

Notice that the W-factors stay the same with the trigonometric case, and are not affected by a dynamical shift.

The case of vertex type algebras can be treated in a parallel way. See [26].

4.4.5 Drinfeld Currents

The Drinfeld currents are suited to deal with infinite-dimensional representations. We explain it by taking $U_q(\widehat{\mathfrak{sl}}_2)$ as an example. (See [61] for general case.)

First let us recall the Drinfeld currents of $U_q(\widehat{\mathfrak{sl}}_2)$ [62]. Let $x_n^\pm$ $(n \in \mathbb{Z})$, a_n $(n \in \mathbb{Z}_{\neq 0})$, h, c, d denote the Drinfeld generators of $U_q(\widehat{\mathfrak{sl}}_2)$. In terms of the generating functions

$$x^\pm(z) = \sum_{n\in\mathbb{Z}} x_n^\pm z^{-n}, \tag{4.223}$$

$$\psi(q^{c/2}z) = q^h \exp\Big((q-q^{-1})\sum_{n>0} a_n z^{-n}\Big), \tag{4.224}$$

$$\varphi(q^{-c/2}z) = q^{-h} \exp\Big(-(q-q^{-1})\sum_{n>0} a_{-n} z^{n}\Big), \tag{4.225}$$

the defining relations read as follows:

$$c : \text{central}, \tag{4.226}$$

$$[h,d]=0, \quad [d,a_n]=na_n, \quad [d,x_n^\pm]=nx_n^\pm, \tag{4.227}$$

$$[h,a_n]=0, \quad [h,x^\pm(z)] = \pm 2x^\pm(z), \tag{4.228}$$

$$[a_n,a_m] = \frac{[2n]_q[cn]_q}{n} q^{-c|n|}\delta_{n+m,0}, \tag{4.229}$$

$$[a_n,x^+(z)] = \frac{[2n]_q}{n} q^{-c|n|} z^n x^+(z), \ [a_n,x^-(z)] = -\frac{[2n]_q}{n} z^n x^-(z), \tag{4.230}$$

$$(z-q^{\pm 2}w)x^\pm(z)x^\pm(w) = (q^{\pm 2}z-w)x^\pm(w)x^\pm(z), \tag{4.231}$$

$$[x^+(z),x^-(w)] = \frac{1}{q-q^{-1}}\Big(\delta\Big(q^c\frac{w}{z}\Big)\psi(q^{c/2}w) - \delta\Big(q^{-c}\frac{w}{z}\Big)\varphi(q^{-c/2}w)\Big), \tag{4.232}$$

and the Serre relation which we omit.

For the level 1 case $(c=1)$, $x^\pm(z)$ has the following free boson realization [63],

$$x^+(z) = :\exp\Big(-\sum_{n\neq 0}\frac{a_n}{[n]_q}z^{-n}\Big): e^Q z^{a_0+1}, \tag{4.233}$$

$$x^-(z) = :\exp\Big(\sum_{n\neq 0}\frac{a_n}{[n]_q}z^{-n}q^{|n|}\Big): e^{-Q} z^{-a_0+1}, \tag{4.234}$$

where a_n $(n \in \mathbb{Z}_{\neq 0})$ is given in (4.229), a_0 and Q are given in (2.49), and $h = a_0$. For general level c, $U_q(\widehat{\mathfrak{sl}}_2)$ has the Wakimoto realization (see, e.g., [64] and references therein).

We now introduce a new parameter p and modify (4.223)–(4.225) to define another set of currents. For notational convenience,we will frequently write

$$p = q^{2r}, \quad p^* = pq^{-2c} = q^{2r^*}, \tag{4.235}$$

where r^* is

$$r^* = r - c. \tag{4.236}$$

Let us introduce two currents $u^{\pm}(z,p) \in U_q(\widehat{\mathfrak{sl}}_2)$ depending on p by

$$u^+(z,p) = \exp\Big(\sum_{n>0} \frac{1}{[r^*n]_q} a_{-n}(q^r z)^n\Big), \tag{4.237}$$

$$u^-(z,p) = \exp\Big(-\sum_{n>0} \frac{1}{[rn]_q} a_n(q^{-r} z)^{-n}\Big). \tag{4.238}$$

We define the "dressed" currents $x_+(z,p)$, $x_-(z,p)$, $\psi^{\pm}(z,p)$ by

$$x_+(z,p) = u^+(z,p)x^+(z), \tag{4.239}$$

$$x_-(z,p) = x^-(z)u^-(z,p), \tag{4.240}$$

$$\psi^+(z,p) = u^+(q^{c/2}z,p)\psi(z)u^-(q^{-c/2}z,p), \tag{4.241}$$

$$\psi^-(z,p) = u^+(q^{-\frac{c}{2}}z,p)\varphi(z)u^-(q^{c/2}z,p). \tag{4.242}$$

We call these as elliptic currents. They are 'Drinfeld currents' of $U_q(\widehat{\mathfrak{sl}}_2)$ or $\mathcal{B}_{q,\lambda}(\widehat{\mathfrak{sl}}_2)$ which is nothing but $U_q(\widehat{\mathfrak{sl}}_2)$ equipped with a different coproduct. We will often drop p, and write $x_-(z,p)$ as $x_-(z)$ and so forth.

The merit of these currents is that they obey the following 'elliptic' commutation relations:

$$\psi^{\pm}(z)\psi^{\pm}(w) = \frac{\Theta_p(q^{-2}z/w}{\Theta_p(q^2z/w)}\frac{\Theta_{p^*}(q^2z/w)}{\Theta_{p^*}(q^{-2}z/w)}\psi^{\pm}(w)\psi^{\pm}(z), \tag{4.243}$$

$$\psi^+(z)\psi^-(w) = \frac{\Theta_p(pq^{-c-2}z/w)}{\Theta_p(pq^{-c+2}z/w)}\frac{\Theta_{p^*}(p^*q^{c+2}z/w)}{\Theta_{p^*}(p^*q^{c-2}z/w)}\psi^-(w)\psi^+(z), \tag{4.244}$$

$$\psi^{\pm}(z)x_+(w)\psi^{\pm}(z)^{-1} = q^{-2}\frac{\Theta_{p^*}(q^{\pm c/2+2}z/w)}{\Theta_{p^*}(q^{\pm c/2-2}z/w)}x_+(w), \tag{4.245}$$

$$\psi^{\pm}(z)x_-(w)\psi^{\pm}(z)^{-1} = q^{2}\frac{\Theta_{p}(q^{\mp c/2-2}z/w)}{\Theta_{p}(q^{\mp c/2+2}z/w)}x_-(w), \tag{4.246}$$

$$\begin{aligned}&[x_+(z), x_-(w)]\\ &\quad = \frac{1}{q-q^{-1}}\left(\delta(q^c w/z)\psi^+(q^{c/2}w) - \delta(q^{-c}w/z)\psi^-(q^{-c/2}w)\right),\end{aligned} \tag{4.247}$$

where $\Theta_p(z)$ is given in (A.4). It is convenient to consider also the current

$$k(z) = \exp\Bigg(\sum_{n>0} \frac{[n]_q}{[2n]_q[r^*n]_q} a_{-n}(q^c z)^n\Bigg) \times \exp\Bigg(-\sum_{n>0} \frac{[n]_q}{[2n]_q[rn]_q} a_n z^{-n}\Bigg). \tag{4.248}$$

The $\psi^{\pm}(z)$ are related to $k(z)$ by the formula

$$\psi^{\pm}(p^{\mp(r-c/2)}z) = \kappa q^{\pm h} k(qz)k(q^{-1}z), \tag{4.249}$$

$$\kappa = \frac{\xi(z;p^*,q)}{\xi(z;p,q)}\bigg|_{z=q^{-2}}, \tag{4.250}$$

where the function

$$\xi(z;p,q) = \frac{(q^2z, pq^2z; p, q^4)_\infty}{(q^4z, pz; p, q^4)_\infty} \tag{4.251}$$

is a solution of the difference equation

$$\xi(z;p,q)\xi(q^2z;p,q) = \frac{(q^2z;p)_\infty}{(pz;p)_\infty}. \tag{4.252}$$

We have the commutation relations supplementing (4.243)–(4.247),

$$k(z)k(w) = \frac{\xi(w/z;p,q)}{\xi(w/z;p^*,q)} \frac{\xi(z/w;p^*,q)}{\xi(z/w;p,q)} k(w)k(z), \tag{4.253}$$

$$k(z)x_+(w)k(z)^{-1} = \frac{\Theta_{p^*}(p^{*1/2}qz/w)}{\Theta_{p^*}(p^{*1/2}q^{-1}z/w)} x_+(w), \tag{4.254}$$

$$k(z)x_-(w)k(z)^{-1} = \frac{\Theta_p(p^{1/2}q^{-1}z/w)}{\Theta_p(p^{1/2}qz/w)} x_-(w). \tag{4.255}$$

Elliptic algebra $U_{q,p}(\widehat{\mathfrak{sl}}_2)$ is obtained as a tensor product of $U_q(\widehat{\mathfrak{sl}}_2)$ ($\mathcal{B}_{q,\lambda}(\widehat{\mathfrak{sl}}_2)$) and a Heisenberg algebra $[Q,P]=1$. For details see [55, 61]. (See also [65].)

5 Free Field Approach to ABF Model

Lukyanov and Pugai studied the Andrews–Baxter–Forrester (ABF) model and calculated LHS's by bosonizing VO's in Section 4.2 [21]. Here we explain their work.

r, β, $x_\pm(z)$, $T(z)$, α_n, α'_n, h_n, $C_{l,k}$, etc., in this section are the same as those in Section 3. In this section r is a positive integer,

$$r \in \mathbb{Z}, \quad r \geq 4, \tag{5.1}$$

and r^* is

$$r^* = r - 1. \tag{5.2}$$

So we have

$$\beta = \frac{r}{r-1}, \quad \alpha_0 = \frac{1}{\sqrt{r(r-1)}}, \quad p' = r-1, \quad p'' = r. \tag{5.3}$$

5.1 ABF Model

The ABF model [52] is a RSOS model associated to the vector representation of $A_1^{(1)}$ algebra [10]. The height variables $a, b, \dots$ take one of the $r-1$ states $1, 2, \dots, r-1$ and those on neighboring sites are subject to the condition $|a-b| = 1$. We write the Boltzmann weight in the form

$$W\left(\begin{matrix} a & b \\ c & d \end{matrix} \middle| u\right) = \rho(u)\overline{W}\left(\begin{matrix} a & b \\ c & d \end{matrix} \middle| u\right), \tag{5.4}$$

where an overall scalar factor $\rho(u)$ is given in (5.134) and chosen so that the partition function per site equals to 1. Nonzero components of $\overline{W}$ are given by

$$\begin{aligned}
\overline{W}\left(\begin{matrix} a & a\pm 1 \\ a\pm 1 & a\pm 2 \end{matrix} \middle| u\right) &= 1, \\
\overline{W}\left(\begin{matrix} a & a\pm 1 \\ a\pm 1 & a \end{matrix} \middle| u\right) &= \frac{[a\mp u]}{[a]}\frac{[1]}{[1+u]}, \\
\overline{W}\left(\begin{matrix} a & a\pm 1 \\ a\mp 1 & a \end{matrix} \middle| u\right) &= \frac{\sqrt{[a+1][a-1]}}{[a]}\frac{[-u]}{[1+u]}.
\end{aligned} \tag{5.5}$$

Here $[u]$ is given in (A.5) and we use a parameter x which is obtained from the original nome p by the modular transformation ($p = e^{2\pi i\tau} \mapsto x^{2r} = e^{-2\pi i/\tau}$). Low temperature limit corresponds to $x \sim 0$ ($p \sim 1$).

This Boltzmann weight W enjoys YBE (4.3), initial condition (4.38), unitarity (4.39) and crossing symmetry (4.99) ($\lambda = -1$, $G_a = [a]$),

$$W\left(\begin{matrix} b & d \\ a & c \end{matrix} \middle| -1-u\right) = \sqrt{\frac{[a][d]}{[b][c]}}\, W\left(\begin{matrix} a & b \\ c & d \end{matrix} \middle| u\right). \tag{5.6}$$

In this gauge W enjoys also a reflection symmetry

$$W\left(\begin{matrix} a & c \\ b & d \end{matrix} \middle| u\right) = W\left(\begin{matrix} a & b \\ c & d \end{matrix} \middle| u\right). \tag{5.7}$$

Along the additive variable u, we often use the multiplicative variable $z = x^{2u}$.

We will restrict ourselves to the "regime III" region defined by

$$0<x<1, \quad -1<u<0. \tag{5.8}$$

In this region W is positive. On the critical point ($x \to 1$ $(p \to 0)$), a correlation length becomes infinite and this model becomes scale invariant, i.e., it has conformal symmetry. So it is described by CFT, i.e., Virasoro algebra. In this case it corresponds to the minimal unitary series with $\beta = r/(r-1)$. On the off-critical point ($x<1$), which corresponds to the $(1,3)$-perturbation of the minimal unitary CFT, the Virasoro symmetry is lost but the DVA symmetry remains. In the low temperature limit ($x \to 0$ with $z = x^{2u}$ (≥ 1) kept fixed), W behaves as

$$\begin{aligned}
&\rho(u) \to z^{-rs/(2r)}, \\
&\overline{W}\left(\begin{matrix} a & a\pm 1 \\ a\pm 1 & a\pm 2 \end{matrix}\,\middle|\, u\right) \to 1, \quad \overline{W}\left(\begin{matrix} a & a\pm 1 \\ a\mp 1 & a \end{matrix}\,\middle|\, u\right) \to 0, \\
&\overline{W}\left(\begin{matrix} a & a+1 \\ a+1 & a \end{matrix}\,\middle|\, u\right) \to z^{(r-a-1)/r} \quad (\geq 1), \\
&\overline{W}\left(\begin{matrix} a & a-1 \\ a-1 & a \end{matrix}\,\middle|\, u\right) \to z^{(a-1)/r} \quad (\geq 1).
\end{aligned} \tag{5.9}$$

Therefore the ground state configuration is

$$\begin{array}{|c|c|c|c|}
l & l+1 & l & l+1 \\ \hline
l+1 & l & l+1 & l \\ \hline
l & l+1 & l & l+1 \\ \hline
l+1 & l & l+1 & l \\ \hline
\end{array} \;. \tag{5.10}$$

In the notation of Section 4.2 this corresponds to $(i_1, i_2) = (l, l+1)$. We label this ground state by an integer l ($l = 1, 2, \ldots, r-2$) and m takes two values $0, 1 \in \mathbb{Z}/2\mathbb{Z}$. Since heights on neighboring sites differ by ± 1, we have

$$\mathcal{H}^{(k)}_{l,m} = 0 \quad \bigl(m \not\equiv l-k \pmod 2\bigr). \tag{5.11}$$

Therefore we can identify

$$\mathcal{H}^{(k)}_{l} = \bigoplus_{m=0}^{1} \mathcal{H}^{(k)}_{l,m} \cong \mathcal{H}^{(k)}_{l,l-k}. \tag{5.12}$$

This space of states $\mathcal{H}^{(k)}_l$, on which the corner Hamiltonian $H^{(k)}_C$ acts, has two labels l and k with range $1 \leq l \leq r-2$, $1 \leq k \leq r-1$. This range is the same as (2.23).

In the thermodynamic limit CTM's (4.100) become

$$\begin{aligned} A^{(k)}(u) &= C^{(k)}(u) = x^{-2uH_C^{(k)}}, \\ B^{(k)}(u) &= D^{(k)}(u) = \sqrt{[k]}x^{2(u+1)H_C^{(k)}}. \end{aligned} \tag{5.13}$$

Careful study of the corner Hamiltonian shows that the character (4.43) agrees with the Virasoro minimal unitary character [9, 52],

$$\chi_{l,l-k,k}(q) = \chi_{l,k}^{\mathrm{Vir}}(q), \tag{5.14}$$

where $\chi_{l,k}^{\mathrm{Vir}}(q)$ is given in (2.37). This character $\chi_{l,k}^{\mathrm{Vir}}(q)$, which agrees with DVA one, should be understood as a character of the representation of DVA. Comparing the free field realization given in Section 3.4, we make the identification [21]

$$\mathcal{H}_l^{(k)} = \mathcal{L}_{l,k}, \tag{5.15}$$

$$H_C^{(k)} = d|_{\mathcal{L}_{l,k}}, \tag{5.16}$$

where $\mathcal{L}_{l,k}$ is an irreducible DVA module given in (3.82) as a cohomology of the Felder complex and d is realized in (3.43). We remark that $Z_{l,m}$ (4.45) is in fact m-independent, $Z_{l,0} = Z_{l,1}$.

For later use we define W^*,

$$W^*\left(\begin{matrix} a & b \\ c & d \end{matrix}\middle| u\right) = \overline{W}\left(\begin{matrix} a & b \\ c & d \end{matrix}\middle| u\right)\Bigg|_{r\to r^*} \times \rho^*(u), \tag{5.17}$$

where $\rho^*(u)$ is given in (5.135).

5.2 Vertex Operators

Although bosons are already introduced in Section 3.4, we present their definitions again.

Let us introduce free boson oscillator α_n $(n \in \mathbb{Z}_{\neq 0})$,

$$[\alpha_n, \alpha_m] = \frac{[n]_x[2n]_x}{n}\frac{[rn]_x}{[r^*n]_x}\delta_{n+m,0}, \tag{5.18}$$

and use zero mode a_0 and Q defined in (2.49) (or a_0' in (2.72)). The Fock space $\mathcal{F}_{l,k}$ is defined by

$$\mathcal{F}_{l,k} = \bigoplus_{m\geq 0}\bigoplus_{n_1\geq\cdots\geq n_m>0} \mathbb{C}\alpha_{-n_1}\cdots\alpha_{-n_m}|\alpha_{l,k}\rangle_B, \tag{5.19}$$

where $|\alpha\rangle_B$ is given by (2.51) with replacing a_n by h_n,

$$a_0'|\alpha_{l,k}\rangle_B = (\alpha_{l,k}-\alpha_0)|\alpha_{l,k}\rangle_B, \quad \alpha_{l,k}-\alpha_0 = -\sqrt{\frac{r}{r-1}}l + \sqrt{\frac{r-1}{r}}k. \tag{5.20}$$

We use also free boson oscillator α'_n ($n \in \mathbb{Z}_{\neq 0}$),

$$\alpha'_n = (-1)^n \frac{[r^*n]_x}{[rn]_x}\alpha_n, \quad [\alpha'_n, \alpha'_m] = \frac{[n]_x[2n]_x}{n}\frac{[r^*n]_x}{[rn]_x}\delta_{n+m,0}. \tag{5.21}$$

Operators $\hat{l}, \hat{k}\colon \mathcal{F}_{l,k} \to \mathcal{F}_{l,k}$ are defined by

$$\hat{l}|_{\mathcal{F}_{l,k}} = l \times \mathrm{id}_{\mathcal{F}_{l,k}}, \quad \hat{k}|_{\mathcal{F}_{l,k}} = k \times \mathrm{id}_{\mathcal{F}_{l,k}}. \tag{5.22}$$

If $\hat{l}$ and $\hat{k}$ appear in arguments of $[u]^*$ and $[u]$ respectively, they can be realized by a'_0,

$$[u + \hat{l}]^* = (-1)^{l-k}[u - \sqrt{rr^*}a'_0]^* \quad \text{on } \mathcal{F}_{l,k}, \tag{5.23}$$

$$[u + \hat{k}] = (-1)^{l-k}[u - \sqrt{rr^*}a'_0] \quad \text{on } \mathcal{F}_{l,k}. \tag{5.24}$$

Elliptic currents $x_\pm(z)$ for $U_x(\widehat{\mathfrak{sl}}_2)$ (or $\mathcal{B}_{x,\lambda}(\widehat{\mathfrak{sl}}_2)$) of level 1 ($c = 1$) are obtained by a "dressing" procedure described in 4.4.5

$x_+(z)\colon \mathcal{F}_{l,k} \to \mathcal{F}_{l-2,k}$

$$x_+(z) = :\exp\Bigl(-\sum_{n\neq 0}\frac{\alpha_n}{[n]_x}z^{-n}\Bigr): \times e^{\sqrt{r/r^*}Q} z^{\sqrt{r/r^*}a'_0 + r/r^*}, \tag{5.25}$$

$x_-(z)\colon \mathcal{F}_{l,k} \to \mathcal{F}_{l,k-2}$

$$x_-(z) = :\exp\Bigl(\sum_{n\neq 0}\frac{\alpha'_n}{[n]_x}z^{-n}\Bigr): \times e^{-\sqrt{r^*/r}Q} z^{-\sqrt{r^*/r}a'_0 + r^*/r}, \tag{5.26}$$

and they are interpreted as screening currents in Section 3.4.

The elliptic versions of VO's (of type I and type II) are defined in terms of their trigonometric ones and a "twistor" given by an infinite product of the universal R matrix described in Section 4.4.4. They satisfy the commutation relations of the type (5.33)–(5.35) below. Since we do not know how to evaluate the twistor in the bosonic realization, we have solved the relations (5.33)–(5.35) directly for $\Phi_\varepsilon(z)$, $\Psi^*_\varepsilon(z)$ ($\varepsilon = \pm 1$). We write $\Phi_\pm(z) = \Phi_{\pm 1}(z)$ and $\Psi^*_\pm(z) = \Psi^*_{\pm 1}(z)$. We obtain the following: [21, 66]

type I

$\Phi_\varepsilon(z)\colon \mathcal{F}_{l,k} \to \mathcal{F}_{l,k-\varepsilon}$

$$\Phi_-(z) = \sqrt{g}{:}\exp\Bigl(-\sum_{n\neq 0}\frac{\alpha'_n}{[2n]_x}z^{-n}\Bigr){:} \times e^{\sqrt{r^*/r}\,Q/2} z^{\sqrt{r^*/r}\,a'_0/2 + r^*/(4r)}, \tag{5.27}$$

$$\Phi_+(z) = \oint_{C_\Phi(z)} \underline{dz'}\, \Phi_-(z)x_-(z')\frac{[u - u' - \frac{1}{2} + \hat{k}]}{[u - u' + \frac{1}{2}]}\frac{1}{\sqrt{[\hat{k}][\hat{k}-1]}}, \tag{5.28}$$

type II

$\Psi^*_\varepsilon(z)\colon \mathcal{F}_{l,k} \to \mathcal{F}_{l-\varepsilon,k}$

$$\Psi^*_-(z)=\frac{1}{\sqrt{g^*}}:\exp\Big(\sum_{n\neq 0}\frac{\alpha_n}{[2n]_x}z^{-n}\Big):\times e^{\sqrt{r/r^*}Q/2}z^{-\sqrt{r/r^*}a'_0/2+r/(4r^*)}, \quad (5.29)$$

$$\Psi^*_+(z)=\oint_{C_{\Psi^*}(z)}\underline{dz'}\,\Psi^*_-(z)x_+(z')\frac{[u-u'+\frac{1}{2}-\hat{l}]^*}{[u-u'-\frac{1}{2}]^*}\frac{1}{\sqrt{[\hat{l}]^*[\hat{l}-1]^*}}. \quad (5.30)$$

Here $z=x^{2u}$, $z'=x^{2u'}$, $\underline{dz'}=dz'/(2\pi i z')$, and normalization constants g and g^* will be given in (5.48) and (5.49). For $k=1$ $(l=1)$, we adopt the following prescription; set $k=1+\epsilon$ $(l=1+\epsilon)$ and take a limit $\epsilon\to 0$ after all the calculation. (Another method of getting rid of this factor is a gauge transformation, see Section 5.4.) The poles of the integrand of (5.27)–(5.30) and the integration contours are listed in the following table. For example, $C_\Phi(z)$ is a simple closed contour that encircles $x^{1+2rn}z$ $(n\geq 0)$ but not $x^{-1-2rn}z$ $(n\geq 0)$.

	inside	outside
$C_\Phi(z)$	$z'=x^{1+2rn}z$	$z'=x^{-1-2rn}z$
$C_{\Psi^*}(z)$	$z'=x^{-1+2r^*n}z$	$z'=x^{1-2r^*n}z$

$(n=0,1,2,\ldots)$. (5.31)

OPE formulas are given in Section 5.A. In the conformal limit (3.23), $g^{-1/2}\Phi_-(z)$ reduces to a primary field $V_{\alpha_{1,2}}(z)$ and $g^{*\frac{1}{2}}\Psi^*_-(z)$ reduces to $V_{\alpha_{2,1}}(-z)$ (up to phase factor) where $V_\alpha(z)$ is given in (2.61). $g^{-1/2}\Phi_+(z)$ and $g^{*\frac{1}{2}}\Psi^*_+(z)$ reduce to primary fields with screening current (screened vertex operator [34]). We can show that type I VO $\Phi_\varepsilon(z)$ commutes with BRST operator (3.80)

$$[X_j,\Phi_\varepsilon(z)]=0. \quad (5.32)$$

Type II VO $\Psi^*_\varepsilon(z)$ commutes with another BRST operator obtained by using(3.84), see Section 5.B.

The VO's given above satisfy the following commutation relations (ε_1, $\varepsilon_2=\pm 1$),

$$\Phi_{\varepsilon_2}(z_2)\Phi_{\varepsilon_1}(z_1)=\sum_{\substack{\varepsilon'_1,\varepsilon'_2=\pm1\\ \varepsilon'_1+\varepsilon'_2=\varepsilon_1+\varepsilon_2}}W\left(\begin{matrix}\hat{k} & \hat{k}+\varepsilon'_1\\ \hat{k}+\varepsilon_2 & \hat{k}+\varepsilon_1+\varepsilon_2\end{matrix}\,\middle|\, u_1-u_2\right)\Phi_{\varepsilon'_1}(z_1)\Phi_{\varepsilon'_2}(z_2), \quad (5.33)$$

$$\Psi^*_{\varepsilon_1}(z_1)\Psi^*_{\varepsilon_2}(z_2)=\sum_{\substack{\varepsilon'_1,\varepsilon'_2=\pm1\\ \varepsilon'_1+\varepsilon'_2=\varepsilon_1+\varepsilon_2}}W^*\left(\begin{matrix}\hat{l} & \hat{l}+\varepsilon_1\\ \hat{l}+\varepsilon'_2 & \hat{l}+\varepsilon_1+\varepsilon_2\end{matrix}\,\middle|\, u_1-u_2\right)\Psi^*_{\varepsilon'_2}(z_2)\Psi^*_{\varepsilon'_1}(z_1), \quad (5.34)$$

$$\Phi_{\varepsilon_2}(z_2)\Psi^*_{\varepsilon_1}(z_1)=\tau(u_1-u_2)\Psi^*_{\varepsilon_1}(z_1)\Phi_{\varepsilon_2}(z_2), \quad (5.35)$$

where $z_i=x^{2u_i}$ and $\tau(u)$ is given in (5.136).

Proof. First let us show (5.33). For $(\varepsilon_1,\varepsilon_2)=(-1,-1)$, (5.33) is

$$\Phi_-(z_2)\Phi_-(z_1)=\rho(u_1-u_2)\Phi_-(z_1)\Phi_-(z_2).$$

This is an OPE rule (5.131) itself. For $(\varepsilon_1,\varepsilon_2)=(-1,1)$, (5.33) is

$$\Phi_+(z_2)\Phi_-(z_1)=W\left(\begin{matrix}\hat{k} & \hat{k}+1\\ \hat{k}+1 & \hat{k}\end{matrix}\,\middle|\, u_1-u_2\right)\Phi_+(z_1)\Phi_-(z_2) + W\left(\begin{matrix}\hat{k} & \hat{k}-1\\ \hat{k}+1 & \hat{k}\end{matrix}\,\middle|\, u_1-u_2\right)\Phi_-(z_1)\Phi_+(z_2).$$

We set (see Section 5.A)

$$\Phi_+(z)=\oint \underline{dz'}\,\Phi_-(z)x_-(z')\varphi_+(u,u';\hat{k}),\qquad \varphi_+(u,u';\hat{k})=\frac{[u-u'-\frac{1}{2}+\hat{k}]}{[u-u'+\frac{1}{2}]}\frac{1}{\sqrt{[\hat{k}]\,[\hat{k}-1]}},\tag{5.36}$$

$$x_-(z_1)\Phi_-(z_2)=f(u_1-u_2)\Phi_-(z_2)x_-(z_1),\quad f(u)=\frac{[u+\frac{1}{2}]}{[-u+\frac{1}{2}]},\tag{5.37}$$

$$x_-(z_1)x_-(z_2)=h(u_1-u_2)x_-(z_2)x_-(z_1),\quad h(u)=\frac{[u-1]}{[u+1]}.\tag{5.38}$$

By using OPE rules in Section 5.A, the above equation becomes

$$\begin{aligned}&\oint \underline{dz'}\,\Phi_-(z_2)\Phi_-(z_1)x_-(z')\varphi_+(u_2,u';\hat{k}+1)f(u'-u_1)\\ &=\oint \underline{dz'}\,\Phi_-(z_2)\Phi_-(z_1)x_-(z')\overline{W}\left(\begin{matrix}\hat{k} & \hat{k}+1\\ \hat{k}+1 & \hat{k}\end{matrix}\,\middle|\, u_1-u_2\right)\times\varphi_+(u_1,u';\hat{k}+1)f(u'-u_2)\\ &\quad+\oint \underline{dz'}\Phi_-(z_2)\Phi_-(z_1)x_-(z')\overline{W}\left(\begin{matrix}\hat{k} & \hat{k}-1\\ \hat{k}+1 & \hat{k}\end{matrix}\,\middle|\, u_1-u_2\right)\varphi_+(u_2,u';\hat{k}).\end{aligned}$$

Careful analysis of the location of poles shows that we can take a common integration contour. Therefore it is sufficient to compare the integrands,

$$\begin{aligned}&\varphi_+(u_2,u';\hat{k}+1)f(u'-u_1)\\ &=\overline{W}\left(\begin{matrix}\hat{k} & \hat{k}+1\\ \hat{k}+1 & \hat{k}\end{matrix}\,\middle|\, u_1-u_2\right)\varphi_+(u_1,u';\hat{k}+1)f(u'-u_2)\\ &\quad+\overline{W}\left(\begin{matrix}\hat{k} & \hat{k}-1\\ \hat{k}+1 & \hat{k}\end{matrix}\,\middle|\, u_1-u_2\right)\varphi_+(u_2,u';\hat{k}).\end{aligned}$$

This equation does hold by the Riemann identity (A.11). $(\varepsilon_1,\varepsilon_2)=(1,-1)$ case is similar. For $(\varepsilon_1,\varepsilon_2)=(1,1)$, (5.33) is

$$\Phi_+(z_2)\Phi_+(z_1)=\rho(u_1-u_2)\Phi_+(z_1)\Phi_+(z_2).$$

By using OPE rules, this becomes

$$\oint \underline{dz'_2} \oint \underline{dz'_1}\, \Phi_-(z_2)\Phi_-(z_1)x_-(z'_2)x_-(z'_1)\varphi_+(u_2,u'_2;\hat{k}-1) \times \varphi_+(u_1,u'_1;\hat{k})f(u'_2-u_1)$$
$$= \oint \underline{dz'_1} \oint \underline{dz'_2}\, \Phi_-(z_2)\Phi_-(z_1)x_-(z'_2)x_-(z'_1) \times \varphi_+(u_1,u'_1;\hat{k}-1)\varphi_+(u_2,u'_2;\hat{k})f(u'_1-u_2)h(u'_1-u'_2).$$

Detailed analysis of the location of poles shows that we can take a common integration contour which is symmetric in z'_1 and z'_2. Consequently it is sufficient to compare the integrands after symmetrization in z'_1 and z'_2. Here symmetrization means

$$\oint \underline{dz'_1} \oint \underline{dz'_2}\, x_-(z'_1)x_-(z'_2)F(u'_1,u'_2)$$
$$= \oint \underline{dz'_1} \oint \underline{dz'_2}\, x_-(z'_1)x_-(z'_2)\frac{1}{2}\big(F(u'_1,u'_2)+F(u'_2,u'_1)h(u'_1-u'_2)\big). \tag{5.39}$$

We remark that $h(-u)=h(u)^{-1}$. Therefore we are enough to show

$$\varphi_+(u_2,u'_2;\hat{k}-1)\varphi_+(u_1,u'_1;\hat{k})f(u'_2-u_1)+(u'_1\leftrightarrow u'_2)\times h(u'_1-u'_2)$$
$$=\varphi_+(u_1,u'_1;\hat{k}-1)\varphi_+(u_2,u'_2;\hat{k})f(u'_1-u_2)h(u'_1-u'_2) + (u'_1\leftrightarrow u'_2)\times h(u'_1-u'_2),$$

and this is correct due to the Riemann identity (A.11).

Next let us show (5.34). We set (see Section 5.A)

$$\Psi^*_+(z)=\oint \underline{dz'}\,\Psi^*_-(z)x_+(z')\varphi^*_+(u,u';\hat{l}),$$
$$\varphi^*_+(u,u';\hat{l})=\frac{[u-u'+\frac{1}{2}-\hat{l}]^*}{[u-u'-\frac{1}{2}]^*}\frac{1}{\sqrt{[\hat{l}]^*[\hat{l}-1]^*}}, \tag{5.40}$$

$$x_+(z_1)\Psi^*_-(z_2)=f^*(u_1-u_2)\Psi^*_-(z_2)x_+(z_1),\quad f^*(u)=\frac{[u-\frac{1}{2}]^*}{[-u-\frac{1}{2}]^*}, \tag{5.41}$$

$$x_+(z_1)x_+(z_2)=h^*(u_1-u_2)x_+(z_2)x_+(z_1),\quad h^*(u)=\frac{[u+1]^*}{[u-1]^*}. \tag{5.42}$$

We remark that

$$\varphi^*_+(u,u';a)=\varphi_+(-u,-u';a)|_{r\to r^*}, \tag{5.43}$$
$$f^*(u)=f(-u)|_{r\to r^*}, \tag{5.44}$$
$$h^*(u)=h(-u)|_{r\to r^*}, \tag{5.45}$$

and (5.17) and (5.7). Integrands of (5.34) are obtained from those of (5.33) by replacement $r\to r^*$, $u\to -u$ and $\hat{k}\to\hat{l}$. Careful analysis of the location of poles shows that we can take a common (symmetric) integration contour for each case. Therefore (5.34) holds.

(5.35) is easily shown, although we have to take care of integration contours. □

The dual VO's are realized in the following way,

$$\Phi^*_\varepsilon(z) = \sqrt{[\hat{k}]}^{-1} \Phi_{-\varepsilon}(x^{-2}z)\sqrt{[\hat{k}]}, \tag{5.46}$$

$$\Psi_\varepsilon(z) = \sqrt{[\hat{l}]^*}\Psi^*_{-\varepsilon}(x^{-2}z)\sqrt{[\hat{l}]^*}^{-1}, \tag{5.47}$$

and normalization constants g and g^* in (5.27) and (5.29) are

$$g^{-1} = x^{r^*/(2r)} \frac{1}{(x^2, x^{2r}, x^{2r}; x^{2r})_\infty} \frac{(x^4, x^{2r}; x^4, x^{2r})_\infty}{(x^2, x^{2r+2}; x^4, x^{2r})_\infty}, \tag{5.48}$$

$$g^* = x^{-r/(2r^*)} \frac{1}{(x^{-2}, x^{2r^*}, x^{2r^*}; x^{2r^*})_\infty} \frac{(x^2, x^{2r^*+2}; x^4, x^{2r^*})_\infty}{(x^4, x^{2r^*}; x^4, x^{2r^*})_\infty}. \tag{5.49}$$

Then we have

$$\sum_{\varepsilon=\pm 1} \Phi^*_\varepsilon(z)\Phi_\varepsilon(z) = \mathrm{id}, \tag{5.50}$$

$$\Phi_{\varepsilon_2}(z)\Phi^*_{\varepsilon_1}(z) = \delta_{\varepsilon_1,\varepsilon_2} \times \mathrm{id}, \tag{5.51}$$

$$\sum_{\varepsilon=\pm 1} \Psi^*_\varepsilon(z_2)\Psi_\varepsilon(z_1) = \frac{1}{1 - z_1/z_2} \times \mathrm{id} + \cdots \qquad (z_1 \to z_2), \tag{5.52}$$

$$\Psi_{\varepsilon_1}(z_1)\Psi^*_{\varepsilon_2}(z_2) = \frac{\delta_{\varepsilon_1,\varepsilon_2}}{1 - z_1/z_2} \times \mathrm{id} + \cdots \qquad (z_1 \to z_2), \tag{5.53}$$

and for d in (3.43)

$$w^d \mathcal{O}(z) w^{-d} = \mathcal{O}(wz), \quad \text{for } \mathcal{O} = \Phi_\varepsilon, \Phi^*_\varepsilon, \Psi^*_\varepsilon, \Psi_\varepsilon, x_\pm. \tag{5.54}$$

We identify type I VO's in Section 4.2 and those here in the following way:

$$\Phi^{(a-\varepsilon,a)}(z) = \Phi_\varepsilon(z)|_{\mathcal{L}_{l,a}}, \tag{5.55}$$

$$\Phi^{*(a+\varepsilon,a)}(z) = \Phi^*_\varepsilon(z)|_{\mathcal{L}_{l,a}}. \tag{5.56}$$

Then eqs. (4.106)–(4.110) correspond to (5.46), (5.33), (5.54), (5.50) and (5.51) respectively.

Next let us see how the DVA current is obtained from VO's. Let introduce free boson oscillator h_n ($n \in \mathbb{Z}_{\neq 0}$),

$$h_n = (x - x^{-1})(-1)^n \frac{[r^*n]_x}{[2n]_x}\alpha_n = (x - x^{-1})\frac{[rn]_x}{[2n]_x}\alpha'_n, \tag{5.57}$$

$$[h_n, h_m] = (x - x^{-1})^2 \frac{1}{n}\frac{[n]_x}{[2n]_x}[rn]_x[r^*n]_x\delta_{n+m,0}. \tag{5.58}$$

As explained in Section 3.4, the DVA current $T(z)$ is realized as

$$T(z) = \Lambda_+(z) + \Lambda_-(z),$$
$$\Lambda_\pm(z) = :\exp\left(\pm \sum_{n\neq 0} h_n (x^{\pm 1} z)^{-n}\right): \times x^{\pm\sqrt{rr^*}a_0'}. \tag{5.59}$$

This $T(z)$ is obtained from type I VO's by fusing them [67],

$$\Phi_{\varepsilon_2}(x^{1+r}z')\Phi^*_{\varepsilon_1}(x^{1-r}z) = \left(1-\frac{z}{z'}\right)\delta_{\varepsilon_1,\varepsilon_2}T(z)\cdot x^{-r^*/2}\frac{(x^4,x^{4-2r};x^4)_\infty}{(x^2,x^{2-2r};x^4)_\infty} + \cdots \quad (z'\to z), \tag{5.60}$$

or from type II VO's

$$\Psi_{\varepsilon_1}(x^{1+r^*}z')\Psi^*_{\varepsilon_2}(x^{1-r^*}z) = \frac{1}{1-z'/z}\delta_{\varepsilon_1,\varepsilon_2}\bigl(-T(-z)\bigr)\cdot(-x^{-r/2})\frac{(x^2,x^{2-2r^*};x^4)_\infty}{(x^4,x^{-2r^*};x^4)_\infty} + \cdots \quad (z'\to z). \tag{5.61}$$

Higher DVA currents (3.132) in Section 3.5 are also obtained by fusion,

$$\Phi_{\varepsilon_2}(x^{1+rj}z')\Phi^*_{\varepsilon_1}(x^{1-rj}z) = \delta_{\varepsilon_1,\varepsilon_2}\begin{cases}\mathrm{id} & (j=0)\\ (1-z/z')T_{(j)}(z)A_j & (j\geq 1)\end{cases} + \cdots \quad (z'\to z), \tag{5.62}$$

where A_j is

$$A_j = x^{-r^*j/2}\frac{(x^4,x^{4-2rj};x^4)_\infty}{(x^2,x^{2-2rj};x^4)_\infty}\prod_{i=1}^{j-1}\frac{(x^{4-2ri},x^{-2ri};x^4)_\infty}{(x^{2-2ri},x^{2-2ri};x^4)_\infty}, \tag{5.63}$$

and we have assumed that r is generic.

5.3 Local Height Probability

Let us consider local height probabilities $P_{a_n,\ldots,a_0}(l)$. As remarked in Section 5.1, we have

$$P_{a_n,\ldots,a_0}(l) = \sum_{m=0}^{1}\frac{Z_{l,m}}{Z_l}P_{a_n,\ldots,a_0}(l,m) = \frac{1}{2}P_{a_n,\ldots,a_0}(l,l-a_0). \tag{5.64}$$

One-point LHP is already obtained in (4.44),

$$P_k(l) = Z_l^{-1}[k]\chi_{l,k}(x^4). \tag{5.65}$$

Here the partition function Z_l (4.74) is

$$Z_l = \sum_{k=1}^{r-1} [k] \chi_{l,k}(x^4), \tag{5.66}$$

where $\chi_{l,k}(q)$ is given in (2.37). Two-point LHP's satisfy recursion relations (4.66)

$$\sum_a P_{a,b}(l) = P_b(l), \quad \sum_b P_{a,b}(l) = P_a(l). \tag{5.67}$$

Since $P_{a,b}(l)$ vanishes unless $|a-b| = 1$ and $1 \le a, b \le r-1$, two-point LHP can be determined uniquely by these recursion relations and expressed in terms of one-point LHP. Starting from $P_{0,1}(l) = P_{1,0}(l) = 0$, we obtain

$$\begin{aligned} P_{k+1,k}(l) = P_{k,k+1}(l) &= \sum_{a=1}^{k} (-1)^{k-a} P_a(l) \\ &= Z_l^{-1} \sum_{a=1}^{k} (-1)^{k-a} [a] \chi_{l,a}(x^4), \end{aligned} \tag{5.68}$$

and this satisfies $P_{r,r-1}(l) = P_{r-1,r}(l) = 0$ (which is equivalent to $Z_{l,0} = Z_{l,1}$). $P_{a,b}(l) = P_{b,a}(l)$ agrees with physical requirement (4.112). For higher-point LHP's, however, the recursion relations can not determine them. So we will use vertex operator approach.

As explained in Section 4.2, local height probabilities can be expressed in terms of corner transfer matrices and type I VO's. In this case (4.73) with (4.111) becomes

$$\begin{aligned} P_{a_n,\dots,a_0}(l) = Z_l^{-1} [a_0] \operatorname{tr}_{\mathcal{H}_l^{(a_0)}} \big(x^{4H_C^{(a_0)}} \Phi^{*(a_0,a_1)}(z) \cdots \Phi^{*(a_{n-1},a_n)}(z) \\ \times \Phi^{(a_n,a_{n-1})}(z) \cdots \Phi^{(a_1,a_0)}(z) \big). \end{aligned} \tag{5.69}$$

We evaluate this LHP using a free field realization of VO. Our identification of the space of state and operators are (5.15), (5.16), (5.55) and (5.56). Then we have

$$\begin{aligned} &P_{a_n,\dots,a_0}(l) \\ &\quad = Z_l^{-1} [a_0] \operatorname{tr}_{\mathcal{L}_{l,a_0}} \big(x^{4d} \Phi^*_{\varepsilon_1}(z) \cdots \Phi^*_{\varepsilon_n}(z) \Phi_{\varepsilon_n}(z) \cdots \Phi_{\varepsilon_1}(z) \big), \end{aligned} \tag{5.70}$$

where $\varepsilon_i = a_{i-1} - a_i$. Since the type I VO has the property (5.32) (see also Section 5.B), we can apply the formula (3.83) to the trace in the above LHP.

We illustrate this free field calculation by taking two-point LHP as an example. Two-point LHP is ($\varepsilon = \pm 1$)

$$\begin{aligned} P_{k-\varepsilon,k}(l) &= Z_l^{-1} [k] \operatorname{tr}_{\mathcal{L}_{l,k}} \big(x^{4d} \Phi^*_\varepsilon(z) \Phi_\varepsilon(z) \big) \\ &= Z_l^{-1} \sqrt{[k][k-\varepsilon]} \operatorname{tr}_{\mathcal{L}_{l,k}} \big(x^{4d} \Phi_{-\varepsilon}(x^{-2} z) \Phi_\varepsilon(z) \big). \end{aligned} \tag{5.71}$$

From (5.71), (5.48) and OPE in Section 5.A, we have

$$P_{k-1,k}(l) = Z_l^{-1}(x^2,x^2,x^{2r};x^{2r})_\infty \oint_C \underline{dz'}\, \frac{(x^{2r-1}\zeta, x^{2r+1}\zeta; x^{2r})_\infty}{(x\zeta, x^3\zeta; x^{2r})_\infty}$$
$$\times \operatorname{tr}_{\mathcal{L}_{l,k}}\Bigg(x^{4d} g^{-1} {:}\Phi_-(x^{-2}z)\Phi_-(z)x_-(z'){:} z^{-3r^*/(2r)}$$
$$\times\, x^{r^*/(2r)} \sqrt{\frac{[k][k-1]}{[\hat{k}][\hat{k}-1]}}\, \frac{[u-u'-\frac{1}{2}+\hat{k}]}{[u-u'+\frac{1}{2}]}\Bigg), \tag{5.72}$$

where $z = x^{2u}$, $z' = x^{2u'}$ and $\zeta = z'/z$. Calculation of trace is separated into two parts, oscillator part and zero mode part. The oscillator parts are common for all Fock spaces

$$\operatorname{tr}^{\rm osc}_{\mathcal{F}_{l',k}}\left(x^{4d} {:}\exp\Big(\sum_{n\neq 0} \frac{\alpha'_n}{[n]_x}(z'^{-n} - x^n z^{-n})\Big){:}\right)$$
$$= \frac{1}{(x^4;x^4)_\infty} \frac{(x^4,x^4,x^{2r+3}\zeta, x^{2r+1}\zeta^{-1}; x^2, x^{2r})_\infty}{(x^{2r+2},x^{2r+2},x^5\zeta, x^3\zeta^{-1}; x^2, x^{2r})_\infty}, \tag{5.73}$$

where $\zeta = z'/z$ and we have used (A.53) (or use the formula in Section 5.A). On the other hand, the zero mode part,

$$\operatorname{tr}^{\rm zero}_{\mathcal{L}_{l,k}}\Bigg(x^{4d} \left(\frac{z'}{z}\right)^{-\sqrt{r^*/r}\,a'_0 + r^*/r} x^{-\sqrt{r^*/r}\,a'_0}$$
$$\times \sqrt{\frac{[k][k-1]}{[\hat{k}][\hat{k}-1]}}\, \frac{[u-u'-\frac{1}{2}+\hat{k}]}{[u-u'+\frac{1}{2}]}\Bigg), \tag{5.74}$$

can be calculated by (3.83) with the Felder complex (3.79)

$$C_{2j} = \mathcal{F}_{l-2r^*j,k}, \quad C_{2j+1} = \mathcal{F}_{-l-2r^*j,k}. \tag{5.75}$$

(5.74) becomes

$$\frac{[u-u'-\frac{1}{2}+k]}{[u-u'+\frac{1}{2}]} \mathcal{O}_{l,k}\left(\frac{z'}{z}\right), \tag{5.76}$$

where $\mathcal{O}_{l,k}(\zeta)$ is (this definition is different from $\mathcal{O}^{\rm LP}_{l,k}(\zeta^{-2}) = \zeta^{-l+r^*(k-1)/r} \times x^{r^*/r}\mathcal{O}_{l,k}(x^{-1}\zeta)/(x^4;x^4)_\infty$ [21])

$$\begin{aligned}
\mathcal{O}_{l,k}(\zeta) &= \sum_{j\in\mathbb{Z}} (-1)^j \operatorname{tr}^{\rm zero}_{C_j}\big(x^{4d}\zeta^{-\sqrt{r^*/r}\,a'_0 + r^*/r} x^{-\sqrt{r^*/r}\,a'_0}\big) \\
&= \sum_{j\in\mathbb{Z}} \Big(x^{4(h_{l-2r^*j,k} - c/24)} (x\zeta)^{l-2r^*j - kr^*/r} \\
&\qquad - x^{4(h_{-l-2r^*j,k} - c/24)} (x\zeta)^{-l-2r^*j-kr^*/r}\Big)\zeta^{r^*/r} \\
&= x^{4(h_{l,k}-c/24)} (x\zeta)^{l-(k-1)r^*/r} x^{-r^*/r} \\
&\quad \times \Big(\Theta_{8rr^*}(-x^{4(r^*k-rl+rr^*)}(x\zeta)^{-2r^*}) \\
&\qquad - \Theta_{8rr^*}(-x^{4(r^*k+rl+rr^*)}(x\zeta)^{-2r^*}) x^{4lk}(x\zeta)^{-2l}\Big).
\end{aligned} \tag{5.77}$$

So we have obtained an integral representation of the two-point LHP,

$$P_{k-1,k}(l) = Z_l^{-1}\frac{(x^2;x^2)_\infty^3}{(x^4;x^4)_\infty} I_{l,k}, \tag{5.78}$$

where $I_{l,k}$ is

$$I_{l,k} = \oint_{|z'|=|z|} \underline{dz'}\, F_{l,k}(\zeta) = \oint_{|\zeta|=1} \underline{d\zeta}\, F_{l,k}(\zeta), \tag{5.79}$$

$$F_{l,k}(\zeta) = \frac{\Theta_{x^{2r}}(x^{-1+2k}\zeta^{-1})}{\Theta_{x^{2r}}(x\zeta)} \mathcal{O}_{l,k}(\zeta)\zeta^{-(k-1)/r} x^{k(k-1)/r+1-k}. \tag{5.80}$$

Next let us evaluate this integral. $F_{l,k}(\zeta)$ has simple poles at $\zeta = x^{1+2n}$ $(n \in \mathbb{Z})$. Since $\mathcal{O}_{l,k}(\zeta)$ satisfies

$$\mathcal{O}_{l,k}(x^{4rm}\zeta) = \mathcal{O}_{l,k}(\zeta)\zeta^{-2r^*m} x^{-4rr^*m^2+2r^*m} \quad (m \in \mathbb{Z}), \tag{5.81}$$

$F_{l,k}(\zeta)$ has the property

$$F_{l,k}(x^{4r}\zeta) = F_{l,k}(\zeta). \tag{5.82}$$

Consequently $I_{l,k}$ requires some regularization. We regularize $I_{l,k}$ in the following way,

$$I_{l,k} = \lim_{\epsilon\to 0} I_{l,k}^{\epsilon}, \quad I_{l,k}^{\epsilon} = \oint_{|\zeta|=1} \underline{d\zeta}\, \zeta^{\epsilon} F_{l,k}(\zeta). \tag{5.83}$$

Deforming the contour

$$I_{l,k}^{\epsilon} = \oint_{\zeta=x,x^3,\ldots,x^{4r-1}} \underline{d\zeta}\, \zeta^{\epsilon} F_{l,k}(\zeta) + \oint_{|\zeta|=x^{4r}} \underline{d\zeta}\, \zeta^{\epsilon} F_{l,k}(\zeta), \tag{5.84}$$

and using (5.82), we obtain

$$I_{l,k}^{\epsilon} = \frac{1}{1-x^{4r\epsilon}} \oint_{\zeta=x,x^3,\ldots,x^{4r-1}} \underline{d\zeta}\, \zeta^{\epsilon} F_{l,k}(\zeta). \tag{5.85}$$

We remark that

$$\oint_{\zeta=x,x^3,\ldots,x^{4r-1}} \underline{d\zeta}\, F_{l,k}(\zeta) = 0, \tag{5.86}$$

namely

$$\sum_{a=0}^{2r-1} (-1)^a [k-1-a] x^{a^2 r^*/r} \mathcal{O}_{l,k}(x^{1+2a}) = 0. \tag{5.87}$$

Using these and picking up residues at $\zeta = x, x^3, \ldots, x^{4r-1}$, we obtain

$$P_{k-1,k}(l) = \frac{1}{Z_l (x^4;x^4)_\infty} \sum_{a=0}^{2r-1} \frac{-a}{2r} (-1)^a [k-1-a] x^{a^2 r^*/r} \mathcal{O}_{l,k}(x^{1+2a}). \tag{5.88}$$

We can check that this answer (and $P_{k,k-1}(l) = P_{k-1,k}(l)$) satisfies (5.67) and $P_l(0,1) = 0$ (use (5.87)). Therefore this answer obtained by free field approach agrees with (5.68).

For multi-point LHP (5.70), we can write down its integral representation in a similar way. (Evaluation of the integral is another problem.) See [21].

5.4 Form Factor

Let us recall the translation invariant vacuum state (4.90)

$$|\mathrm{vac}\rangle = \sqrt{[\hat{k}]}x^{2d}, \tag{5.89}$$

and the action of the column-to-column transfermatrix (4.113)

$$\big(\mathcal{T}_{\mathrm{col}}(u)|f\rangle\big)^{(k)} = \sum_{\varepsilon=\pm1} \Phi_\varepsilon(z)\cdot f\cdot \Phi_{-\varepsilon}(z)\big|_{\mathcal{L}_{l,k}}, \tag{5.90}$$

where f is a linear map $f\colon \mathcal{L}_{l,a} \to \mathcal{L}_{l,a}$.

For the time being, we consider operators on the Fock spaces $\mathcal{F}_{l,k}$ not on the cohomology $\mathcal{L}_{l,k}$. Action of $\mathcal{T}_{\mathrm{col}}$ is $\mathcal{T}_{\mathrm{col}}(u)|f\rangle = \sum_{\varepsilon=\pm1}\Phi_\varepsilon(z)\cdot f\cdot\Phi_{-\varepsilon}(z)$. Let us define an 'excited' state

$$\begin{aligned}&|w_m,\ldots,w_1\rangle_{\varepsilon_m,\ldots,\varepsilon_1}\\ &\qquad = \Psi^*_{\varepsilon_m}(w_m)\cdots\Psi^*_{\varepsilon_1}(w_1)|\mathrm{vac}\rangle\colon \mathcal{F}_{l,k}\to\mathcal{F}_{l-\sum_i\varepsilon_i,k}.\end{aligned} \tag{5.91}$$

(We remark that the original description of this Ψ^* is $\Psi^*\otimes 1$, see (4.87).) This state is an eigenstate of $\mathcal{T}_{\mathrm{col}}$

$$\begin{aligned}&\mathcal{T}_{\mathrm{col}}(u)|w_m,\ldots,w_1\rangle_{\varepsilon_m,\ldots,\varepsilon_1}\\ &\quad = \sum_{\varepsilon=\pm1}\Phi_\varepsilon(z)\cdot\Psi^*_{\varepsilon_m}(w_m)\cdots\Psi^*_{\varepsilon_1}(w_1)\sqrt{[\hat{k}]}x^{2d}\cdot\Phi_{-\varepsilon}(z)\\ &\quad = \prod_{j=1}^m \tau(v_j-u)\cdot|w_m,\ldots,w_1\rangle_{\varepsilon_m,\ldots,\varepsilon_1},\end{aligned} \tag{5.92}$$

where $z = x^{2u}$, $w_j = x^{2v_j}$ and we have used (5.35). Therefore type II VO creates (or annihilates) 'particles'. But we remark that this state is a 'true' eigenstate of $\mathcal{T}_{\mathrm{col}}$ because it is not a linear map on $\mathcal{L}_{l,k}$ in general. We will show when this state becomes a true eigenstate.

To avoid $1/\sqrt{[\hat{k}][\hat{k}-1]}$ and $1/\sqrt{[\hat{l}]^*[\hat{l}-1]^*}$ factors we perform a gauge transformation (see (4.4))

$$\begin{aligned} &\widetilde{W}\begin{pmatrix} a & b \\ c & d \end{pmatrix} u \Big) = \frac{F(a,b)F(b,d)}{F(a,c)F(c,d)} W\begin{pmatrix} a & b \\ c & d \end{pmatrix} u \Big), \\ &F(a,b) = ([a][b])^{-1/4}, \end{aligned} \tag{5.93}$$

$$\begin{aligned} &\widetilde{W}^*\begin{pmatrix} a & b \\ c & d \end{pmatrix} u \Big) = \frac{F^*(a,b)F^*(b,d)}{F^*(a,c)F^*(c,d)} W^*\begin{pmatrix} a & b \\ c & d \end{pmatrix} u \Big), \\ &F^*(a,b) = ([a]^*[b]^*)^{1/4}, \end{aligned} \tag{5.94}$$

$$\widetilde{\Phi}_-(z) = \Phi_-(z), \quad \widetilde{\Phi}_+(z) = \Phi_+(z)\sqrt{[\hat{k}][\hat{k}-1]}, \tag{5.95}$$

$$\widetilde{\Psi}^*_-(z) = \Psi^*_-(z), \quad \widetilde{\Psi}^*_+(z) = \Psi^*_+(z)\sqrt{[\hat{l}]^*[\hat{l}-1]^*}. \tag{5.96}$$

$\widetilde{W}$ differs from W only in the following component

$$\widetilde{W}\begin{pmatrix} a & a\pm 1 \\ a\mp 1 & a \end{pmatrix} u \Big) = \frac{[a\mp 1]}{[a]}\frac{[-u]}{[1+u]}\rho(u). \tag{5.97}$$

In this gauge the reflection symmetry (5.7) is lost. $\widetilde{W}^*$ is

$$\widetilde{W}^*\begin{pmatrix} a & b \\ c & d \end{pmatrix} u \Big) = \Big(\rho(u)^{-1}\widetilde{W}\begin{pmatrix} a & c \\ b & d \end{pmatrix} u \Big)\Big)\Big|_{r\to r^*} \times \rho^*(u). \tag{5.98}$$

$\widetilde{\Phi}_\varepsilon(z)$ and $\widetilde{\Psi}^*_\varepsilon(z)$ satisfy (5.33)–(5.35) with (Φ, Ψ^*, W, W^*) replaced by $(\widetilde{\Phi}, \widetilde{\Psi}^*, \widetilde{W}, \widetilde{W}^*)$ respectively. (5.35) also holds by replacing Ψ^* with $\widetilde{\Psi}^*$ and Φ unchanged. Therefore the state

$$\begin{aligned} &|w_m, \dots, w_1\rangle^{\sim}_{\varepsilon_m,\dots,\varepsilon_1} \\ &\qquad = \widetilde{\Psi}^*_{\varepsilon_m}(w_m)\dots\widetilde{\Psi}^*_{\varepsilon_1}(w_1)|\mathrm{vac}\rangle : \mathcal{F}_{l,k} \to \mathcal{F}_{l-\sum_i \varepsilon_i, k}, \end{aligned} \tag{5.99}$$

which differs from (5.91) by a multiplicative constant factor dependent on $\hat{l}$, is also an eigenstate of $\mathcal{T}_{\mathrm{col}}$.

How to realize the type II VO on the cohomology is discussed in Section 5.B. In order to interpret the state (5.99) as a linear map on $\mathcal{L}_{l,k}$, the first requirement is $\sum_{i=1}^m \varepsilon_i = 0$. Hence m is an even integer, $m = 2n$. Then $\overset{\curvearrowleft}{\prod}{}_{i=1}^{2n}\widetilde{\Psi}^*_{\varepsilon_i}(w_i)$ and the BRST charge satisfy the intertwining property

$$Q_m \cdot \overset{\curvearrowleft}{\prod_{i=1}^{2n}} \widetilde{\Psi}^*_{\varepsilon_i}(w_i) = (-1)^n \overset{\curvearrowleft}{\prod_{i=1}^{2n}} \widetilde{\Psi}^*_{-\varepsilon_i}(w_i)\cdot Q_m \quad (m = l, r^*-l). \tag{5.100}$$

Therefore this product of type II VO's is a well-defined operator on the cohomology $\mathcal{L}_{l,k}$. Explicitly it is

$$\overset{\curvearrowleft}{\prod_{i=1}^{2n}} \widetilde{\Psi}^*_{\varepsilon_i}(w_i) \text{ on } C_{2j}, \quad (-1)^n \overset{\curvearrowleft}{\prod_{i=1}^{2n}} \widetilde{\Psi}^*_{-\varepsilon_i}(w_i) \text{ on } C_{2j+1}. \tag{5.101}$$

Consequently the state $|w_{2n},\ldots,w_1\rangle^{\sim}_{\varepsilon_{2n},\ldots,\varepsilon_1}$ with $\sum_{i=1}^{2n}\varepsilon_i = 0$ is a true eigenstate of $\mathcal{T}_{\rm col}$. (If m is an odd integer or $\sum_i \varepsilon_i$ does not vanish, then $|w_m,\ldots,w_1\rangle_{\varepsilon_m,\ldots,\varepsilon_1}$ is a map from $\mathcal{F}_{l,k}$ to $\mathcal{F}_{l',k}$ ($l' = l - \sum_i \varepsilon_i \neq l$), namely in the original description it corresponds to a state whose half is in the ground state l and the other half in l'. As remarked in Section 4.2, to obtain a complete set of excited states, we need not only the translation invariant vacuum state (5.89) but also the translation noninvariant vacuum states.) The form factors of local operator $\widetilde{\mathcal{O}}$ are defined by

$$\langle \mathrm{vac}|\widetilde{\mathcal{O}}|w_m,\ldots,w_1\rangle_{\varepsilon_m,\ldots,\varepsilon_1}. \tag{5.102}$$

Next let us study (4.75) with the following $\mathcal{O} = \mathcal{O}_{\Psi^*}$

$$\mathcal{O}_{\Psi^*} = \prod_{i=1}^{\overset{\frown}{2n}} \widetilde{\Psi}^*_{\varepsilon_i}(w_i) \quad \left(w_j = x^{2v_j}, \sum_{i=1}^{2n}\varepsilon_i = 0\right). \tag{5.103}$$

In this model we have

$$\begin{aligned} Q_{a_n,\ldots,a_0}(l;z|\mathcal{O}_{\Psi^*}) &= \sum_{m=0}^{1} \frac{Z_{l,m}}{Z_l} Q_{a_n,\ldots,a_0}(l,m;z|\mathcal{O}_{\Psi^*}) \\ &= \tfrac{1}{2} Q_{a_n,\ldots,a_0}(l,l-a_0;z|\mathcal{O}_{\Psi^*}). \end{aligned} \tag{5.104}$$

This is the form factor of $\Phi^*_{a_0-a_1}(z)\cdots\Phi^*_{a_{n-1}-a_n}(z)\Phi_{a_{n-1}-a_n}(z)\cdots\Phi_{a_0-a_1}(z)$. (4.76) and (4.77) with $n=1$ become

$$\sum_{\varepsilon=\pm 1} Q_{a+\varepsilon,a}(l;z \mid \mathcal{O}_{\Psi^*}) = Q_a(l \mid \mathcal{O}_{\Psi^*}), \tag{5.105}$$

$$\sum_{\varepsilon=\pm 1} Q_{a,a+\varepsilon}(l;z \mid \mathcal{O}_{\Psi^*}) = G(z) Q_a(l \mid \mathcal{O}_{\Psi^*}), \tag{5.106}$$

where $Q_a(l \mid \mathcal{O}_{\Psi^*}) = Q_a(l;z \mid \mathcal{O}_{\Psi^*})$ and $G(z)$ is

$$G(z) = \prod_{j=1}^{2n} \tau(w_j - u). \tag{5.107}$$

This $Q_{a,b}(l;z \mid \mathcal{O}_{\Psi^*})$ is uniquely determined by these recursion relations and it is expressed in terms of $Q_a(l \mid \mathcal{O}_{\Psi^*})$. Starting from $Q_{0,1}(l;z \mid \mathcal{O}_{\Psi^*}) = Q_{1,0}(l;z \mid \mathcal{O}_{\Psi^*}) = 0$, we obtain

$$\begin{aligned} Q_{a+1,a}(l;z \mid \mathcal{O}_{\Psi^*}) = &\sum_{\substack{b=1 \\ b\equiv a \ (\mathrm{mod}\, 2)}}^{a} Q_b(l \mid \mathcal{O}_{\Psi^*}) \\ &- G(z) \sum_{\substack{b=1 \\ b\not\equiv a \ (\mathrm{mod}\, 2)}}^{a} Q_b(l \mid \mathcal{O}_{\Psi^*}), \end{aligned} \tag{5.108}$$

$$Q_{a,a+1}(l;z\mid\mathcal{O}_{\Psi^*})=\sum_{\substack{b=1\\ b\not\equiv a\ (\mathrm{mod}\,2)}}^{a}Q_b(l\mid\mathcal{O}_{\Psi^*})+G(z)\sum_{\substack{b=1\\ b\equiv a\ (\mathrm{mod}\,2)}}^{a}Q_b(l\mid\mathcal{O}_{\Psi^*}).\qquad(5.109)$$

This answer should satisfy $Q_{r,r-1}(l;z|\mathcal{O}_{\Psi^*})=Q_{r-1,r}(l;z|\mathcal{O}_{\Psi^*})=0$.

As an example let us calculate $Q_k\big(l|\widetilde{\Psi}^*_+(w_2)\widetilde{\Psi}^*_-(w_1)\big)$ by free field realization. Using (3.83) it becomes

$$\begin{aligned}Q_k\big(l\mid\widetilde{\Psi}^*_+(w_2)\widetilde{\Psi}^*_-(w_1)\big)&=Z_l^{-1}[k]\,\mathrm{tr}_{\mathcal{L}_{l,k}}\big(x^{4d}\widetilde{\Psi}^*_+(w_2)\widetilde{\Psi}^*_-(w_1)\big)\\&=Z_l^{-1}[k]\sum_{j\in\mathbb{Z}}\Big(\mathrm{tr}_{C_{2j}}\big(x^{4d}\widetilde{\Psi}^*_+(w_2)\widetilde{\Psi}^*_-(w_1)\big)\\&\qquad+\mathrm{tr}_{C_{2j+1}}\big(x^{4d}\widetilde{\Psi}^*_-(w_2)\widetilde{\Psi}^*_+(w_1)\big)\Big).\end{aligned}\qquad(5.110)$$

By using OPE and trace rule listed in Section 5.A, we obtain

$$\begin{aligned}&Q_k\big(l\mid\widetilde{\Psi}^*_+(w_2)\widetilde{\Psi}^*_-(w_1)\big)=Z_l^{-1}[k]g^{*-1}\\&\times\Bigg(\oint_{C_{\Psi^*}(w_2)}\underline{dw'}\,(w_1w_2)^{rl/(2r^*)}w'^{-rl/r^*}\langle\!\langle x_+(w')\Psi^*_-(w_1)\rangle\!\rangle\\&\qquad\times\frac{[v_2-v'-\frac{1}{2}-l]^*}{[v_2-v'-\frac{1}{2}]^*}\Theta_{x^{8rr^*}}\bigg(-\Big(\frac{w'^2}{w_1w_2}\Big)^r x^{4(-rl+r^*k+rr^*)}\bigg)\\&\quad+\oint_{C_{\Psi^*}(w_1)}\underline{dw'}\,(w_1w_2)^{-rl/(2r^*)}w'^{rl/r^*}\langle\!\langle\Psi^*_-(w_1)x_+(w')\rangle\!\rangle\\&\qquad\times\frac{[v_1-v'+\frac{1}{2}+l]^*}{[v_1-v'-\frac{1}{2}]^*}\Theta_{x^{8rr^*}}\bigg(-\Big(\frac{w'^2}{w_1w_2}\Big)^r x^{4(rl+r^*k+rr^*)}\bigg)x^{4lk}\Bigg)\\&\times(w_1w_2)^{-k/2+r/(4r^*)}w'^{k+r/r^*}x^{4h_{l,k}-c/24}\langle\!\langle\Psi^*_-(w_2)\Psi^*_-(w_1)\rangle\!\rangle\langle\!\langle\Psi^*_-(w_2)x_+(w')\rangle\!\rangle\\&\times(x^4;x^4)^{-1}_\infty\prod_{i,j=1}^{2}F_{\Psi^*_-,\Psi^*_-}\Big(\frac{w_i}{w_j}\Big)\cdot\prod_{i=1}^{2}F_{\Psi^*_-,x_+}\Big(\frac{w_i}{w'}\Big)F_{x_+,\Psi^*_-}\Big(\frac{w'}{w_i}\Big),\end{aligned}\qquad(5.111)$$

where $F_{A,B}(z)$ is given in Section 5.A. $Q_{k-1,k}\big(l;z\mid\widetilde{\Psi}^*_+(w_2)\widetilde{\Psi}^*_-(w_1)\big)$ can be obtained by using this result and (5.109). If you want to calculate it directly, you will follow

$$\begin{aligned}&Q_{k-1,k}\big(l;z\mid\widetilde{\Psi}^*_+(w_2)\widetilde{\Psi}^*_-(w_1)\big)\\&=Z_l^{-1}[k]\,\mathrm{tr}_{\mathcal{L}_{l,k}}\big(x^{4d}\Phi^*_+(z)\Phi_+(z)\widetilde{\Psi}^*_+(w_2)\widetilde{\Psi}^*_-(w_1)\big)\\&=Z_l^{-1}[k]\sum_{j\in\mathbb{Z}}\Big(\mathrm{tr}_{C_{2j}}\big(x^{4d}\Phi^*_+(z)\Phi_+(z)\widetilde{\Psi}^*_+(w_2)\widetilde{\Psi}^*_-(w_1)\big)\\&\qquad+\mathrm{tr}_{C_{2j+1}}\big(x^{4d}\Phi^*_+(z)\Phi_+(z)\widetilde{\Psi}^*_-(w_2)\widetilde{\Psi}^*_+(w_1)\big)\Big)\\&=\cdots.\end{aligned}\qquad(5.112)$$

5.A OPE and Trace

OPE

We list the normal ordering relations used in Section 5. Recall

$$r^* = r - 1.$$

For operators $A(z)$, $B(w)$ that have the form :exp(linear in boson):, we use the notation

$$A(z)B(w) = \langle\!\langle A(z)B(w)\rangle\!\rangle :A(z)B(w):, \tag{5.113}$$

and write down only the part $\langle\!\langle A(z)B(w)\rangle\!\rangle$:

$$\langle\!\langle x_+(z_1)x_+(z_2)\rangle\!\rangle = z_1^{2r/r^*}(1-\zeta)\frac{(x^{-2}\zeta; x^{2r^*})_\infty}{(x^{2r^*+2}\zeta; x^{2r^*})_\infty}, \tag{5.114}$$

$$\langle\!\langle x_-(z_1)x_-(z_2)\rangle\!\rangle = z_1^{2r^*/r}(1-\zeta)\frac{(x^{2}\zeta; x^{2r})_\infty}{(x^{2r-2}\zeta; x^{2r})_\infty}, \tag{5.115}$$

$$\langle\!\langle x_\pm(z_1)x_\mp(z_2)\rangle\!\rangle = z_1^{-2}\frac{1}{(1+x\zeta)(1+x^{-1}\zeta)}, \tag{5.116}$$

$$\langle\!\langle \Phi_-(z_1)x_+(z_2)\rangle\!\rangle = \langle\!\langle x_+(z_1)\Phi_-(z_2)\rangle\!\rangle = z_1 + z_2, \tag{5.117}$$

$$\langle\!\langle \Phi_-(z_1)x_-(z_2)\rangle\!\rangle = \langle\!\langle x_-(z_1)\Phi_-(z_2)\rangle\!\rangle = z_1^{-r^*/r}\frac{(x^{2r-1}\zeta; x^{2r})_\infty}{(x\zeta; x^{2r})_\infty}, \tag{5.118}$$

$$\langle\!\langle \Psi_-^*(z_1)x_+(z_2)\rangle\!\rangle = \langle\!\langle x_+(z_1)\Psi_-^*(z_2)\rangle\!\rangle = z_1^{-r/r^*}\frac{(x^{2r^*+1}\zeta; x^{2r^*})_\infty}{(x^{-1}\zeta; x^{2r^*})_\infty}, \tag{5.119}$$

$$\langle\!\langle \Psi_-^*(z_1)x_-(z_2)\rangle\!\rangle = \langle\!\langle x_-(z_1)\Psi_-^*(z_2)\rangle\!\rangle = z_1 + z_2, \tag{5.120}$$

$$\langle\!\langle \Phi_-(z_1)\Phi_-(z_2)\rangle\!\rangle = z_1^{r^*/(2r)}\frac{(x^2\zeta, x^{2r+2}\zeta; x^4, x^{2r})_\infty}{(x^4\zeta, x^{2r}\zeta; x^4, x^{2r})_\infty}, \tag{5.121}$$

$$\langle\!\langle \Psi_-^*(z_1)\Psi_-^*(z_2)\rangle\!\rangle = z_1^{r/2r^*}\frac{(\zeta, x^{2r^*+4}\zeta; x^4, x^{2r^*})_\infty}{(x^2\zeta, x^{2r^*+2}\zeta; x^4, x^{2r^*})_\infty}, \tag{5.122}$$

$$\langle\!\langle \Phi_-(z_1)\Psi_-^*(z_2)\rangle\!\rangle = \langle\!\langle \Psi_-^*(z_1)\Phi_-(z_2)\rangle\!\rangle = z_1^{-1/2}\frac{(-x^3\zeta; x^4)_\infty}{(-x\zeta; x^4)_\infty}, \tag{5.123}$$

where $\zeta = z_2/z_1$ and we have used (A.50).

As meromorphic functions we have ($z_i = x^{2u_i}$)

$$x_+(z_1)x_+(z_2) = x_+(z_2)x_+(z_1)\frac{[u_1-u_2+1]^*}{[u_1-u_2-1]^*}, \tag{5.124}$$

$$x_-(z_1)x_-(z_2) = x_-(z_2)x_-(z_1)\frac{[u_1-u_2-1]}{[u_1-u_2+1]}, \tag{5.125}$$

$$x_\pm(z_1)x_\mp(z_2) = x_\mp(z_2)x_\pm(z_1), \tag{5.126}$$

$$\Phi_-(z_1)x_+(z_2) = x_+(z_2)\Phi_-(z_1), \tag{5.127}$$

$$\Phi_-(z_1)x_-(z_2) = x_-(z_2)\Phi_-(z_1)\frac{[u_1-u_2+\frac{1}{2}]}{[-u_1+u_2+\frac{1}{2}]}, \tag{5.128}$$

$$\Psi^*_-(z_1)x_+(z_2)=x_+(z_2)\Psi^*_-(z_1)\frac{[u_1-u_2-\frac{1}{2}]^*}{[-u_1+u_2-\frac{1}{2}]^*}, \tag{5.129}$$

$$\Psi^*_-(z_1)x_-(z_2)=x_-(z_2)\Psi^*_-(z_1), \tag{5.130}$$

$$\Phi_-(z_1)\Phi_-(z_2)=\Phi_-(z_2)\Phi_-(z_1)\rho(u_2-u_1), \tag{5.131}$$

$$\Psi^*_-(z_1)\Psi^*_-(z_2)=\Psi^*_-(z_2)\Psi^*_-(z_1)\rho^*(u_1-u_2), \tag{5.132}$$

$$\Phi_-(z_1)\Psi^*_-(z_2)=\Psi^*_-(z_2)\Phi_-(z_1)\tau(u_2-u_1). \tag{5.133}$$

Here $\rho(u)$, $\rho^*(u)$ and $\tau(u)$ are given by

$$z^{r^*/(2r)}\rho(u)=\frac{\rho_+(u)}{\rho_+(-u)}, \qquad \rho_+(u)=\frac{(x^2z,x^{2r+2}z;x^4,x^{2r})_\infty}{(x^4z,x^{2r}z;x^4,x^{2r})_\infty}, \tag{5.134}$$

$$z^{-r/(2r^*)}\rho^*(u)=\frac{\rho^*_+(u)}{\rho^*_+(-u)}, \quad \rho^*_+(u)=\frac{(x^2z,x^{2r^*+2}z;x^4,x^{2r^*})_\infty}{(z,x^{2r^*+4}z;x^4,x^{2r^*})_\infty}, \tag{5.135}$$

$$\tau(u)=z^{1/2}\frac{\Theta_{x^4}(-xz^{-1})}{\Theta_{x^4}(-xz)}. \tag{5.136}$$

Note that

$$\rho^*(u)=-\rho(u)|_{r\to r^*}. \tag{5.137}$$

Trace

For an operator $A(z)$ that have the form :exp(linear in boson):, we write

$$\begin{aligned} A(z)&=A^{\rm osc}(z)A^{\rm zero}(z),\\ A^{\rm osc}(z)&=\exp\Big(\sum_{n>0}f^A_{-n}\alpha_{-n}z^n\Big)\exp\Big(\sum_{n>0}f^A_n\alpha_n z^{-n}\Big), \end{aligned} \tag{5.138}$$

where f^A_n is a coefficient. In this model A is $x_\pm$, Φ_-, Ψ^*_-. For example,

$$\begin{aligned} \Phi^{\rm osc}_-(z)&=\exp\Big(\sum_{n>0}\frac{\alpha'_n}{[2n]_x}z^n\Big)\exp\Big(-\sum_{n>0}\frac{\alpha'_n}{[2n]_x}z^{-n}\Big),\\ \Phi^{\rm zero}_-(z)&=\sqrt{g}e^{\sqrt{r^*/r}Q/2}z^{\sqrt{r^*/r}a'_0/2+r^*/(4r)}. \end{aligned}$$

For such operators Wick's theorem tells us

$$A_1(z_1)\cdots A_n(z_n)=\prod_{i<j}\langle\langle A_i(z_i)A_j(z_j)\rangle\rangle\times:\prod_i A_i(z_i):. \tag{5.139}$$

The trace of oscillator parts over the Fock space $\mathcal{F}=\mathcal{F}_{l,k}$ can be calculated by using the trace technique (A.53). We have

$$\mathrm{tr}_{\mathcal{F}}\Big(x^{4d}:\prod_i A^{\rm osc}_i(z_i):\Big)=\frac{1}{(x^4;x^4)_\infty}\prod_{i,j}F_{A_i,A_j}\Big(\frac{z_i}{z_j}\Big), \tag{5.140}$$

where $F_{A,B}(z)$ is given by

$$F_{A,B}(z) = \exp\left(\sum_{n>0} \frac{1}{n}[n]_x[2n]_x \frac{[rn]_x}{[r^*n]_x} \frac{x^{4n}}{1-x^{4n}} f^A_{-n} f^B_n z^n \right). \tag{5.141}$$

We write down $F_{A,B}(z)$ (Remark $F_{A,B}(z) = F_{B,A}(z)$):

$$F_{x_+,x_+}(z) = \frac{(x^2 z; x^2, x^{2r^*})_\infty}{(x^{2r^*+4} z; x^2, x^{2r^*})_\infty}, \tag{5.142}$$

$$F_{x_-,x_-}(z) = \frac{(x^4 z; x^2, x^{2r})_\infty}{(x^{2r+2} z; x^2, x^{2r})_\infty}, \tag{5.143}$$

$$F_{x_+,x_-}(z) = \frac{1}{(-x^3 z, -x^5 z; x^4)_\infty}, \tag{5.144}$$

$$F_{\Phi_-,x_+}(z) = (-x^4 z; x^4)_\infty, \tag{5.145}$$

$$F_{\Phi_-,x_-}(z) = \frac{(x^{2r+3} z; x^4, x^{2r})_\infty}{(x^5 z; x^4, x^{2r})_\infty}, \tag{5.146}$$

$$F_{\Psi^*_-,x_+}(z) = \frac{(x^{2r^*+5} z; x^4, x^{2r^*})_\infty}{(x^3 z; x^4, x^{2r^*})_\infty}, \tag{5.147}$$

$$F_{\Psi^*_-,x_-}(z) = (-x^4 z; x^4)_\infty, \tag{5.148}$$

$$F_{\Phi_-,\Phi_-}(z) = \frac{(x^6 z, x^{2r+6} z; x^4, x^4, x^{2r})_\infty}{(x^8 z, x^{2r+4} z; x^4, x^4, x^{2r})_\infty}, \tag{5.149}$$

$$F_{\Psi^*_-,\Psi^*_-}(z) = \frac{(x^4 z, x^{2r^*+8} z; x^4, x^4, x^{2r^*})_\infty}{(x^6 z, x^{2r^*+6} z; x^4, x^4, x^{2r^*})_\infty}, \tag{5.150}$$

$$F_{\Phi_-,\Psi^*_-}(z) = \frac{(-x^7 z; x^4, x^4)_\infty}{(-x^5 z; x^4, x^4)_\infty}. \tag{5.151}$$

5.B Screening Operators and Vertex Operators

All the formulas in this subsection are understood as meromorphic functions.

Let us recall the screening operators (3.72) and (3.84)

$$X(z) = \oint_{C_X(z)} \underline{dz'}\, x_+(z') \frac{[u-u'+\frac{1}{2}-\hat{l}]^*}{[u-u'-\frac{1}{2}]^*} : \mathcal{F}_{l,k} \to \mathcal{F}_{l-2,k}, \tag{5.152}$$

$$X'(z) = \oint_{C_{X'}(z)} \underline{dz'}\, x_-(z') \frac{[u-u'-\frac{1}{2}+\hat{k}]}{[u-u'+\frac{1}{2}]} : \mathcal{F}_{l,k} \to \mathcal{F}_{l,k-2}, \tag{5.153}$$

where integration contours are

	inside	outside
$C_X(z)$	$z' = x^{-1+2r^*n} z$	$z' = x^{-1-2r^*(n+1)} z$
$C_{X'}(z)$	$z' = x^{1+2rn} z$	$z' = x^{1-2r(n+1)} z$

$(n = 0, 1, 2, \dots)$. (5.154)

These screening operators satisfy the following exchange relations [61]

$$X(z_1)X(z_2)\frac{[u_1-u_2-1]^*}{[u_1-u_2]^*}-X(z_2)^2\frac{[u_1-u_2-\hat{l}+2]^*}{[u_1-u_2]^*}\frac{[1]^*}{[\hat{l}-2]^*} = (u_1\leftrightarrow u_2), \quad (5.155)$$

$$X'(z_1)X'(z_2)\frac{[u_1-u_2+1]}{[u_1-u_2]}-X'(z_2)^2\frac{[u_1-u_2+\hat{k}-2]}{[u_1-u_2]}\frac{[1]}{[\hat{k}-2]} = (u_1\leftrightarrow u_2), \quad (5.156)$$

$$X(z_1)X'(z_2)=X'(z_2)X(z_1). \quad (5.157)$$

$\widetilde{\Phi}_+$ and $\widetilde{\Psi}^*_+$ can be expressed as (see (5.28), (5.30), (5.95), (5.96))

$$\widetilde{\Phi}_+(z)=\widetilde{\Phi}_-(z)X'(z), \quad \widetilde{\Psi}^*_+(z)=\widetilde{\Psi}^*_-(z)X(z). \quad (5.158)$$

Here RHS's are understood as the analytic continuation of $A(z)B(z')$ from the region $|z|\gg|z'|$ and hence integration contours become (5.31).

Since screening operators and VO's satisfy

$$X(z)\widetilde{\Phi}_\varepsilon(w)=\widetilde{\Phi}_\varepsilon(w)X(z), \quad (5.159)$$

$$X'(z)\widetilde{\Psi}^*_\varepsilon(w)=\widetilde{\Psi}^*_\varepsilon(w)X'(z), \quad (5.160)$$

$\prod_i\widetilde{\Phi}_{\varepsilon_i}(w_i)$ and $\prod_i\widetilde{\Psi}^*_{\varepsilon_i}(w_i)$ ($\sum_i\varepsilon_i=0$) are well-defined operators on the cohomology $H^0(C_{l,k})$ and $H^0(C'_{l,k})$ respectively where the BRST charges are $Q_m=X(1)^m$ and $Q'_m=X'(1)^m$ (see Section 3.4.3). On the other hand $\widetilde{\Phi}_\varepsilon(w)$ and $\widetilde{\Psi}^*_\varepsilon(w)$ do not commute with $X'(z)$ and $X(z)$ respectively, so they are not well-defined operators on the cohomology $H^0(C'_{l,k})$ and $H^0(C_{l,k})$ in general. However we can find the following intertwining properties.

We can show that

$$X'(z)\widetilde{\Phi}_-(w)=-\widetilde{\Phi}_-(w)X'(z)\frac{[u-v+1]}{[u-v]}\frac{[\hat{k}]}{[\hat{k}-1]} + \widetilde{\Phi}_+(w)\frac{[u-v+\hat{k}]}{[u-v]}\frac{[1]}{[\hat{k}-1]}, \quad (5.161)$$

$$X(z)\widetilde{\Psi}^*_-(w)=-\widetilde{\Psi}^*_-(w)X(z)\frac{[u-v-1]^*}{[u-v]^*}\frac{[\hat{l}]^*}{[\hat{l}-1]^*} + \widetilde{\Psi}^*_+(w)\frac{[u-v-\hat{l}]^*}{[u-v]^*}\frac{[1]^*}{[\hat{l}-1]^*}, \quad (5.162)$$

where $z=x^{2u}$ and $w=x^{2v}$. By induction we obtain

$$X'(z)^n\widetilde{\Phi}_-(w)=\widetilde{\Phi}_-(w)X'(z)^nA_n(u-v,\hat{k}) + \widetilde{\Phi}_+(w)X'(z)^{n-1}B_n(u-v,\hat{k}), \quad (5.163)$$

$$X'(z)^n\widetilde{\Phi}_+(w) = \widetilde{\Phi}_-(w)X'(z)^{n+1}B_n(v-u,\hat{k}-2) + \widetilde{\Phi}_+(w)X'(z)^n A_n(v-u,\hat{k}-2), \tag{5.164}$$

$$X(z)^n\widetilde{\Psi}^*_-(w) = \widetilde{\Psi}^*_-(w)X(z)^n A^*_n(u-v,\hat{l}) + \widetilde{\Psi}^*_+(w)X(z)^{n-1}B^*_n(u-v,\hat{l}), \tag{5.165}$$

$$X(z)^n\widetilde{\Psi}^*_+(w) = \widetilde{\Psi}^*_-(w)X(z)^{n+1}B^*_n(v-u,\hat{l}-2) + \widetilde{\Psi}^*_+(w)X(z)^n A^*_n(v-u,\hat{l}-2), \tag{5.166}$$

where coefficients are

$$A_n(u,k) = (-1)^n\frac{[u+n]}{[u]}\frac{[k-n+1]}{[k-2n+1]}, \tag{5.167}$$

$$B_n(u,k) = (-1)^{n-1}\frac{[u+k-n+1]}{[u]}\frac{[n]}{[k-2n+1]}, \tag{5.168}$$

$$A^*_n(u,l) = (-1)^n\frac{[u-n]^*}{[u]^*}\frac{[l-n+1]^*}{[l-2n+1]^*}, \tag{5.169}$$

$$B^*_n(u,l) = (-1)^{n-1}\frac{[u-k+n-1]^*}{[u]^*}\frac{[n]^*}{[l-2n+1]^*}. \tag{5.170}$$

Noticing the following properties of these coefficients

$$A_{k+1}(u,k') = 0, \quad B_{k+1}(u,k') = (-1)^{k+1} \quad \text{for } k' \equiv k \pmod r, \tag{5.171}$$

$$A^*_{l+1}(u,l') = 0, \quad B^*_{l+1}(u,l') = (-1)^{l+1} \quad \text{for } l' \equiv l \pmod{r^*}, \tag{5.172}$$

we obtain the intertwining properties ($1 \le l \le r^*-1$, $1 \le k \le r-1$)

$$X'(z)^{k-\varepsilon}\widetilde{\Phi}_\varepsilon(w) = (-1)^{k-\varepsilon}\widetilde{\Phi}_{-\varepsilon}(w)X'(z)^k \quad \text{on } \mathcal{F}_{l',k'}\ k' \equiv k \pmod r, \tag{5.173}$$

$$X'(z)^{r-k-\varepsilon}\widetilde{\Phi}_\varepsilon(w) = (-1)^{r-k-\varepsilon}\widetilde{\Phi}_{-\varepsilon}(w)X'(z)^{r-k} \quad \text{on } \mathcal{F}_{l',k'}\ k' \equiv -k \pmod r, \tag{5.174}$$

$$X(z)^{l-\varepsilon}\widetilde{\Psi}^*_\varepsilon(w) = (-1)^{l-\varepsilon}\widetilde{\Psi}^*_{-\varepsilon}(w)X(z)^l \quad \text{on } \mathcal{F}_{l',k'}\ l' \equiv l \pmod{r^*}, \tag{5.175}$$

$$X(z)^{r^*-l-\varepsilon}\widetilde{\Psi}^*_\varepsilon(w) = (-1)^{r^*-l-\varepsilon}\widetilde{\Psi}^*_{-\varepsilon}(w)X(z)^l \quad \text{on } \mathcal{F}_{l',k'}\ l' \equiv -l \pmod{r^*}. \tag{5.176}$$

Therefore $\prod_i \widetilde{\Phi}_{\varepsilon_i}(w_i)$ and $\prod_i \widetilde{\Psi}^*_{\varepsilon_i}(w_i)$ ($\sum_i \varepsilon_i = 0$) are well-defined operators on the cohomology $H^0(C'_{l,k})$ and $H^0(C_{l,k})$ respectively. However we remark that on C'_{2j+1} and C_{2j+1} those operators should be replaced by $\pm\prod_i \widetilde{\Phi}_{-\varepsilon_i}(w_i)$ and $\pm\prod_i \widetilde{\Psi}^*_{-\varepsilon_i}(w_i)$ respectively.

6 DVA ($A_2^{(2)}$ Type) and Dilute A_L Models

In this section we introduce another deformed Virasoro algebra and study dilute A_L models by free field approach [46]. Both of them are associated to $A_2^{(2)}$ algebra.

r, β, $x_\pm(z)$, $T(z)$, α_n, α'_n, h_n, $C_{l,k}$, etc., in this section are different form those in Section 5.

6.1 DVA($A_2^{(2)}$)

Brazhnikov and Lukyanov [22] pointed out that one can associate to the algebra $A_2^{(2)}$ a deformed Virasoro algebra which is different from the one discussed in Section 3. We denote it DVA($A_2^{(2)}$).

DVA($A_2^{(2)}$) is an associative algebra over $\mathbb{C}$ generated by T_n ($n\in\mathbb{Z}$) with two parameters x and r, and their relation is [22]

$$\begin{aligned}
[T_n,T_m] = &-\sum_{\ell=1}^{\infty} f_\ell (T_{n-\ell}T_{m+\ell}-T_{m-\ell}T_{n+\ell}) \\
&+(x-x^{-1})^2\frac{[r+\frac{1}{2}]_x[r]_x[r-1]_x[r-\frac{3}{2}]_x}{[\frac{1}{2}]_x[\frac{3}{2}]_x}[3n]_x\delta_{n+m,0} \\
&\qquad+(x-x^{-1})^2\frac{[r]_x[r-\frac{1}{2}]_x[r-1]_x}{[\frac{1}{2}]_x}[n-m]_xT_{n+m}, \qquad (6.1)
\end{aligned}$$

where the structure constants f_ℓ is given by

$$\begin{aligned}
f(z) = \sum_{\ell=0}^{\infty} f_\ell z^\ell &= \exp\Bigl(-\sum_{n>0}(x-x^{-1})^2\frac{1}{n}\frac{[n]_x[rn]_x[(r-1)n]_x}{[2n]_x-[n]_x}z^n\Bigr) \\
&= \frac{1}{1-z}\frac{(x^{2-2r}z,x^{3-2r}z,x^4z,x^5z,x^{2r}z,x^{2r+1}z;x^6)_\infty}{(x^{5-2r}z,x^{6-2r}z,xz,x^2z,x^{2r+3}z,x^{2r+4}z;x^6)_\infty}. \qquad (6.2)
\end{aligned}$$

By introducing DVA current $T(z)=\sum_{n\in\mathbb{Z}}T_nz^{-n}$, the above relation can be written as a formal power series,

$$\begin{aligned}
&f\Bigl(\frac{w}{t}\Bigr)T(z)T(w)-T(w)T(z)f\Bigl(\frac{z}{w}\Bigr) \\
&=(x-x^{-1})\frac{[r+\frac{1}{2}]_x[r]_x[r-1]_x[r-\frac{3}{2}]_x}{[\frac{1}{2}]_x[\frac{3}{2}]_x}\Bigl(\delta\Bigl(x^3\frac{w}{z}\Bigr)-\delta\Bigl(x^{-3}\frac{w}{z}\Bigr)\Bigr) \\
&\quad+(x-x^{-1})\frac{[r]_x[r-\frac{1}{2}]_x[r-1]_x}{[\frac{1}{2}]_x} \\
&\qquad\times\Bigl(\delta\Bigl(x^2\frac{w}{z}\Bigr)T(xw)-\delta\Bigl(x^{-2}\frac{w}{z}\Bigr)T(x^{-1}w)\Bigr). \qquad (6.3)
\end{aligned}$$

The notation of [22] is related to ours by $x_{\rm BL} = x^{3/2}$, $b/Q = r$, $1/(Qb) = 1-r$, $g(z) = f(z)$, $\mathbf{V}(z) = T(z)$. For later use we add a grading operator d,

$$[d, T_n] = -nT_n. \tag{6.4}$$

The above relation (6.1) is invariant under

$$\text{(i)} \qquad x \mapsto x^{-1}, \quad r \mapsto r, \quad T(z) \mapsto T(z), \tag{6.5}$$

$$\text{(ii)} \qquad x \mapsto x, \quad r \mapsto 1-r, \quad T(z) \mapsto -T(z). \tag{6.6}$$

In the case of (i) $f(z)$ is understood as the first line of (6.2). Let us introduce β as

$$\beta = \frac{r}{2(r-1)}, \tag{6.7}$$

then (2.19) becomes

$$\alpha_0 = \frac{2-r}{\sqrt{2r(r-1)}}. \tag{6.8}$$

In the conformal limit ($x = e^{\hbar} \to 1$, r: fixed), (6.1) admits two limits [22] related by (6.6),

$$T(z) = 3 - 2r + 8r(r-1)\hbar^2\left(z^2L(z) + \frac{1}{48}(1-2r) + \frac{(2-r)^2}{8r(r-1)}\right) + O(\hbar^4), \tag{6.9}$$

$$T(z) = -1 - 2r - 8r(r-1)\hbar^2\left(z^2\widetilde{L}(z) - \frac{1}{48}(1-2r) + \frac{(1+r)^2}{8r(r-1)}\right) + O(\hbar^4), \tag{6.10}$$

where $L(z)$, $\widetilde{L}(z)$ are the Virasoro currents with the central charges c, $\tilde{c}$ respectively,

$$c = 1 - \frac{3(2-r)^2}{r(r-1)} = 1 - 6\alpha_0^2, \quad \tilde{c} = 1 - \frac{3(1+r)^2}{r(r-1)}. \tag{6.11}$$

The highest weight representation is defined by the same manner presented in Section 3.3. The Kac determinant at level N is [46]

$$\det(\langle\lambda; N, i \mid \lambda; N, j\rangle)_{1\le i,j\le p(N)} = \prod_{\substack{l,k\ge 1\\ lk\le N}} \left(\frac{(x^{rl}-x^{-rl})(x^{(r-1)l}-x^{-(r-1)l})}{x^l - 1 + x^{-l}}(\lambda-\lambda_{l,k})(\lambda-\tilde{\lambda}_{l,k})\right)^{p(N-lk)}, \tag{6.12}$$

where $\lambda_{l,k}$ and $\tilde{\lambda}_{l,k}$ are given by

$$\lambda_{l,k} = x^{-lr+2k(r-1)} + x^{lr-2k(r-1)} - \frac{[r-\frac{1}{2}]_x}{[\frac{1}{2}]_x}, \tag{6.13}$$

$$\tilde{\lambda}_{l,k} = -x^{l(r-1)-2kr} - x^{-l(r-1)+2kr} - \frac{[r-\frac{1}{2}]_x}{[\frac{1}{2}]_x}, \tag{6.14}$$

and they are related by (6.6). In the conformal limit this Kac determinant reduces to the Virasoro one with the proportional constant expected from (6.9) or (6.10).

6.2 Free Field Realization

6.2.1 Free Field Realization

Let us introduce free boson oscillator h_n ($n \in \mathbb{Z}_{\neq 0}$),

$$[h_n, h_m] = (x - x^{-1})^2 \frac{1}{n} \frac{[n]_x [rn]_x [(r-1)n]_x}{[2n]_x - [n]_x} \delta_{n+m,0}, \tag{6.15}$$

and use zero mode a_0 and Q defined in (2.49) (or a'_0 in (2.72)). Notice that $[2n]_x - [n]_x = [3n]_x [\frac{n}{2}]_x / [\frac{3}{2}n]_x$. The Fock space $\mathcal{F}_\alpha$ is defined as before.

The DVA($A_2^{(2)}$) current $T(z)$ is realized as follows:

$$\begin{aligned}
T(z) &= \Lambda_+(z) + \Lambda_0(z) + \Lambda_-(z), \\
\Lambda_\pm(z) &= :\!\exp\Bigl(\pm \sum_{n \neq 0} h_n (x^{\pm 3/2} z)^{-n}\Bigr)\!: \times x^{\pm\sqrt{2r(r-1)}a'_0}, \\
\Lambda_0(z) &= -\frac{[r-\frac{1}{2}]_x}{[\frac{1}{2}]_x} :\!\exp\Bigl(\sum_{n \neq 0} h_n (x^{-n/2} - x^{n/2}) z^{-n}\Bigr)\!:
\end{aligned} \tag{6.16}$$

To prove this we need (A.23) and the OPE formula,

$$\begin{aligned}
f\Bigl(\frac{w}{t}\Bigr)\Lambda_\pm(z)\Lambda_\pm(w) &= :\!\Lambda_\pm(z)\Lambda_\mp(w)\!:, \\
f\Bigl(\frac{w}{t}\Bigr)\Lambda_\pm(z)\Lambda_\mp(w) &= :\!\Lambda_\pm(z)\Lambda_\mp(w)\!:\gamma\Bigl(x^{\mp 1}\frac{w}{z}\Bigr)\gamma\Bigl(x^{\mp 2}\frac{w}{z}\Bigr), \\
f\Bigl(\frac{w}{t}\Bigr)\Lambda_0(z)\Lambda_0(w) &= :\!\Lambda_0(z)\Lambda_0(w)\!:\gamma\Bigl(\frac{w}{t}\Bigr), \\
f\Bigl(\frac{w}{t}\Bigr)\Lambda_\pm(z)\Lambda_0(w) &= :\!\Lambda_\pm(z)\Lambda_0(w)\!:\gamma\Bigl(x^{\mp 1}\frac{w}{z}\Bigr), \\
f\Bigl(\frac{w}{t}\Bigr)\Lambda_0(z)\Lambda_\pm(w) &= :\!\Lambda_0(z)\Lambda_\pm(w)\!:\gamma\Bigl(x^{\pm 1}\frac{w}{z}\Bigr),
\end{aligned} \tag{6.17}$$

where $\gamma(z)$ is given in (3.41). The grading operator d is realized by

$$d = \sum_{n>0} \frac{n^2([2n]_x - [n]_x)}{(x - x^{-1})^2 [n]_x [rn]_x [(r-1)n]_x} h_{-n} h_n + \frac{1}{4} a_0'^2 - \frac{1}{24}, \tag{6.18}$$

which satisfies

$$[d, h_n] = -n h_n, \quad [d, Q] = a'_0, \quad d|\alpha_{l,k}\rangle_B = \Bigl(h_{l,k} - \frac{c}{24}\Bigr)|\alpha_{l,k}\rangle_B, \tag{6.19}$$

where c and $h_{l,k}$ are given by (2.17) and (2.18) respectively. We remark that $\widetilde{T}(z) = -\Lambda_+(z) + \Lambda_0(z) - \Lambda_-(z)$ also satisfies (6.1)).

$|\alpha\rangle_B$ is the highest weight state of DVA with $\lambda = \lambda(\alpha)$,

$$|\alpha\rangle_B = |\lambda(\alpha)\rangle, \tag{6.20}$$

$$\lambda(\alpha) = x^{\sqrt{2r(r-1)}(\alpha-\alpha_0)} + x^{-\sqrt{2r(r-1)}(\alpha-\alpha_0)} - \frac{[r-\frac{1}{2}]_x}{[\frac{1}{2}]_x}. \tag{6.21}$$

The dual space $\mathcal{F}_\alpha^*$ becomes a DVA module by (2.63) with

$$^tT_n = T_{-n}, \tag{6.22}$$

$$^th_n = -h_{-n},, \tag{6.23}$$

eq. (2.66) and (2.67). By (6.23), $\mathcal{F}_\alpha^*$ is isomorphic to $\mathcal{F}_{2\alpha_0-\alpha}$ as a DVA module,

$$\mathcal{F}_\alpha^* \cong \mathcal{F}_{2\alpha_0-\alpha} \qquad \text{(DVA module)}. \tag{6.24}$$

$$|\alpha\rangle_B^* \leftrightarrow |2\alpha_0 - \alpha\rangle_B. \tag{6.25}$$

Note that $\lambda(\alpha) = \lambda(2\alpha_0 - \alpha)$ by (6.21).

In the conformal limit ($x = e^\hbar \to 1$, r: fixed), oscillator h_n is expressed by a_n in (2.49) as follows:

$$h_n = \hbar\sqrt{2r(r-1)} \times \sqrt{\frac{1}{x^n - 1 + x^{-n}} \frac{x^{rn} - x^{-rn}}{2rn\hbar} \frac{x^{(r-1)n} - x^{-(r-1)n}}{2(r-1)n\hbar}}\, a_n. \tag{6.26}$$

Substituting this expression into (6.16) and expanding in $\hbar$, we get (6.9) with $L(z)$ in (2.55) (or (2.73)).

The representation with $\lambda = \lambda_{l,k}$ is realized on $\mathcal{F}_{l,k} = \mathcal{F}_{\alpha_{l,k}}$,

$$\lambda_{l,k} = \lambda(\alpha_{l,k}), \tag{6.27}$$

where $\alpha_{l,k}$ is given in (2.79).

6.2.2 Kac Determinant

In the free boson realization the singular vectors can be expressed by screening currents. A screening current $x_+(z)$ is defined by

$$x_+(z) = :\exp\Big(-\sum_{n\neq 0} \frac{\alpha_n}{[n]_x} z^{-n}\Big): \times e^{\sqrt{r/[2(r-1)]}Q} z^{\sqrt{r/[2(r-1)]}a_0' + r/[2(r-1)]}, \tag{6.28}$$

where oscillators α_n ($n \in \mathbb{Z}_{\neq 0}$) are related to h_n as

$$h_n = (x - x^{-1})(-1)^n \frac{[(r-1)n]_x}{[2n]_x - [n]_x} \alpha_n. \tag{6.29}$$

Like as (3.61) singular vector is obtained by the BRST operator. Screening charges and BRST charges will be discussed in the next subsection.

Like as in Section 3.4.2 let us introduce matrices $C(N,\alpha)$ and $C'(N,\alpha)$ by (3.63) and (3.64). Then their determinants are given by

$$\begin{aligned}
&\det C(N,\alpha)_{I,J} \\
&\quad = \prod_{\substack{l,k\geq 1\\ lk\leq N}} \Big((z^{1/2} x^{lr/2-k(r-1)} - z^{-1/2} x^{-lr/2+k(r-1)}) \\
&\qquad\qquad \times (z^{1/2} x^{-l(r-1)/2+kr} + z^{-1/2} x^{l(r-1)/2-kr}) \Big)^{p(N-lk)}, \qquad (6.30)\\
&\det C'(N,\alpha)_{I,J} \\
&\quad = \det C(N, 2\alpha_0 - \alpha) \cdot \det D \\
&\quad = \prod_{\substack{l,k\geq 1\\ lk\leq N}} \Big((z^{1/2} x^{-lr/2+k(r-1)} - z^{-1/2} x^{lr/2-k(r-1)}) \\
&\qquad\qquad \times (z^{1/2} x^{l(r-1)/2-kr} + z^{-1/2} x^{-l(r-1)/2+kr}) \Big)^{p(N-lk)}, \qquad (6.31)
\end{aligned}$$

where $z = x^{\sqrt{2r(r-1)}(\alpha-\alpha_0)}$. The inner product of two states in the Verma module is given by (3.67) where $G_{K,L}$ is

$$\begin{aligned}
G_{K,L} &= \big\langle ({}^t h)_K |\alpha\rangle_B^*, h_{-L}|\alpha\rangle_B \big\rangle \\
&= \delta_{K,L} \prod_i \left(\frac{1}{i} \frac{(x^{ri} - x^{-ri})(x^{(r-1)i} - x^{-(r-1)i})}{x^i - 1 + x^{-i}} \right)^{k_i} k_i!,
\end{aligned} \tag{6.32}$$

and its determinant is (use (A.28) and (A.30))

$$\begin{aligned}
\det G_{K,L} &= \prod_{\substack{\{k_i\}\\ \sum_i ik_i = N}} \prod_i \left(\frac{1}{i} \frac{(x^{ri} - x^{-ri})(x^{(r-1)i} - x^{-(r-1)i})}{x^i - 1 + x^{-i}} \right)^{k_i} k_i! \\
&= \prod_{\substack{l,k\geq 1\\ lk\leq N}} \left(\frac{(x^{rl} - x^{-rl})(x^{(r-1)l} - x^{-(r-1)l})}{x^l - 1 + x^{-l}} \right)^{p(N-lk)}.
\end{aligned} \tag{6.33}$$

Therefore we obtain the Kac determinant (6.12),

$$\begin{aligned}
&\det \langle h | T_I T_J | h \rangle \\
&\quad = \det C'(N,\alpha) \cdot \det G \cdot \det {}^t C(N,\alpha) \\
&\quad = \prod_{\substack{l,k\geq 1\\ lk\leq N}} \left(\frac{(x^{rl} - x^{-rl})(x^{(r-1)l} - x^{-(r-1)l})}{x^l - 1 + x^{-l}} (\lambda - \lambda_{l,k})(\lambda - \tilde{\lambda}_{l,k}) \right)^{p(N-lk)}.
\end{aligned} \tag{6.34}$$

6.2.3 Felder Complex

We consider the representation of $\lambda = \lambda_{l,k}$ in (6.27) with (2.20) and (2.23), i.e.,

$$r = \frac{2p''}{2p'' - p'}. \tag{6.35}$$

Let us consider the Felder complex $C_{l,k}$,

$$\cdots \xrightarrow{X_{-3}} C_{-2} \xrightarrow{X_{-2}} C_{-1} \xrightarrow{X_{-1}} C_0 \xrightarrow{X_0} C_1 \xrightarrow{X_1} C_2 \xrightarrow{X_2} \cdots, \tag{6.36}$$

where C_j and $X_j \colon C_j \to C_{j+1}$ ($j \in \mathbb{Z}$) are

$$C_{2j} = \mathcal{F}_{l-2p'j,k}, \qquad C_{2j+1} = \mathcal{F}_{-l-2p'j,k}, \tag{6.37}$$

$$X_{2j} = Q_l, \qquad X_{2j+1} = Q_{p'-l}. \tag{6.38}$$

We assume that p' is odd (we have not obtained the result for general even p'). BRST charge Q_m ($1 \le m \le p'-1$) is defined by

$$Q_{2m+1} = Q_1 Q_2^{(1)} Q_2^{(2)} \cdots Q_2^{(m)}, \tag{6.39}$$

$$Q_{2m} = Q_2^{([p'+1]/2-m)} \cdots Q_2^{([p'-3]/2)} Q_2^{([p'-1]/2)}. \tag{6.40}$$

Here Q_1 and $Q_2^{(a)}$ are

$$Q_1 = \oint_{|z|=1} \underline{dz}\, x_+(z) \frac{[u + \frac{1}{2}\hat{l}]^*}{[u + \frac{1}{2}]^*}, \tag{6.41}$$

$$\begin{aligned} Q_2^{(a)} &= \oint\oint_{|z_1|=|z_2|=1} \underline{dz}_1\, \underline{dz}_2\, x_+(z_1) x_+(z_2) \frac{1}{[u_1 + \frac{1}{2}]^* [u_2 + \frac{1}{2}]^*} \\ &\quad \times \frac{[u_1 - u_2]^*}{[u_1 - u_2 + 1]^* [u_1 - u_2 - \frac{1}{2}]^*} f_2^{(a)}(u_1 + \tfrac{1}{2}\hat{l}, u_2 + \tfrac{1}{2}\hat{l}), \end{aligned} \tag{6.42}$$

where $[u]^*$ is given in (A.6) with $r^* = r - 1$ and $z = x^{2u}$, $z_i = x^{2u_i}$ and

$$\begin{aligned} f_2^{(a)}(u_1, u_2) &= [2a+1]^* [a - \tfrac{1}{2}]^* [u_1 - a]^* [u_2 + a - 1]^* [u_1 - u_2 + a - \tfrac{1}{2}]^* \\ &\quad - [2a-1]^* [a + \tfrac{1}{2}]^* [u_1 + a]^* [u_2 - a - 1]^* [u_1 - u_2 - a - \tfrac{1}{2}]^*. \end{aligned} \tag{6.43}$$

X satisfies the BRST property,

$$X_j X_{j-1} = 0. \tag{6.44}$$

We assume that this Felder complex has the same structure as the Virasoro case because it formally tends to Virasoro one in the conformal limit ($x \to 1$, while r and $z = x^{2u}$ are kept fixed). Then the cohomology groups of the complex $C_{l,k}$ are

$$H^j(C_{l,k}) = \operatorname{Ker} X_j / \operatorname{Im} X_{j-1} = \begin{cases} 0 & j \neq 0, \\ \mathcal{L}_{l,k} & j = 0, \end{cases} \tag{6.45}$$

where $\mathcal{L}_{l,k}$ is the irreducible DVA module of $\lambda = \lambda_{l,k}$. The trace of operator $\mathcal{O}$ over $\mathcal{L}_{l,k}$ can be written as (3.83).

The BRST charge commutes with DVA [46]

$$[T_n, Q_l] = 0 \quad \text{on } \mathcal{F}_{l',k} \; l' \equiv l \pmod{p'}. \tag{6.46}$$

Singular vector can be obtained by Q_l like as (3.61).

6.3 Dilute A_L Models

In the following, we fix a positive integer $L \geq 3$. The dilute A_L model [68, 69] is an integrable RSOS model obtained by restricting the face model of type $A_2^{(2)}$ [70]. In the dilute A_L model, the local fluctuation variables $a, b, \ldots$ take one of the L states $1, 2, \ldots, L$, and those on neighboring lattice sites are subject to the condition $a - b = 0, \pm 1$. The Boltzmann weights can be found in [69, eq. (3.1)]. For our purpose it is convenient to use the parametrization given in Appendix A of [69], which is suitable in the 'low-temperature' regime. With some change of notation we recall the formula below [46].

Let $x = e^{-2\pi\lambda/\varepsilon}$, $r = \pi/(2\lambda)$ and $u = -u_{\text{orig}}/(2\lambda)$, where λ, ε are the variables used in [69] and u_{orig} stands for 'u' there. We shall restrict ourselves to the 'regime 2^+' defined by

$$0 < x < 1, \quad r = 2\frac{L+1}{L+2}, \quad -\frac{3}{2} < u < 0. \tag{6.47}$$

We have taken $p' = L$, $p'' = L+1$ in (6.35), which corresponds to minimal unitary series. (6.7) and (6.8) become

$$\beta = \frac{L+1}{L}, \quad \alpha_0 = \frac{1}{\sqrt{L(L+1)}}, \quad p' = L, \quad p'' = L+1, \tag{6.48}$$

and we set

$$r^* = r - 1. \tag{6.49}$$

On the critical point ($x \to 1$), this model is described by CFT, i.e., the minimal unitary series with $\beta = (L+1)/L$. On the off-critical point ($x < 1$), which corresponds to the $(1,2)$-perturbation of the minimal unitary CFT, the Virasoro symmetry is lost but the DVA$(A_2^{(2)})$ symmetry remains. Changing an overall scalar factor we put the Boltzmann weights in the form

$$W\left(\begin{matrix} a & b \\ c & d \end{matrix}\middle| u\right) = \rho(u)\overline{W}\left(\begin{matrix} a & b \\ c & d \end{matrix}\middle| u\right), \tag{6.50}$$

where $\rho(u)$ is given in (6.127) and chosen so that the partition function per site of the model equals to 1. $\overline{W}$ is

$$\begin{aligned}
&\overline{W}\left(\begin{matrix} a\pm 1 & a \\ a & a\mp 1 \end{matrix}\middle| u\right) = 1,\\
&\overline{W}\left(\begin{matrix} a & a\pm 1 \\ a & a\pm 1 \end{matrix}\middle| u\right) = \overline{W}\left(\begin{matrix} a\pm 1 & a\pm 1 \\ a & a \end{matrix}\middle| u\right)\\
&\qquad = -\left(\frac{[\pm a+\frac{3}{2}]_+[\pm a-\frac{1}{2}]_+}{[\pm a+\frac{1}{2}]_+^2}\right)^{1/2}\frac{[u]}{[1+u]},\\
&\overline{W}\left(\begin{matrix} a\pm 1 & a \\ a & a \end{matrix}\middle| u\right) = \overline{W}\left(\begin{matrix} a & a \\ a & a\pm 1 \end{matrix}\middle| u\right) = \frac{[\pm a+\frac{1}{2}+u]_+}{[\pm a+\frac{1}{2}]_+}\frac{[1]}{[1+u]},\\
&\overline{W}\left(\begin{matrix} a & a\mp 1 \\ a\pm 1 & a \end{matrix}\middle| u\right) = (G_a^+G_a^-)^{1/2}\frac{[\frac{1}{2}+u]}{[\frac{3}{2}+u]}\frac{[u]}{[1+u]},\\
&\overline{W}\left(\begin{matrix} a & a \\ a\pm 1 & a \end{matrix}\middle| u\right) = \overline{W}\left(\begin{matrix} a & a\pm 1 \\ a & a \end{matrix}\middle| u\right)\\
&\qquad = -(G_a^\pm)^{1/2}\frac{[\pm a-1-u]_+}{[\pm a+\frac{1}{2}]_+}\frac{[1]}{[1+u]}\frac{[u]}{[\frac{3}{2}+u]},\\
&\overline{W}\left(\begin{matrix} a & a\pm 1 \\ a\pm 1 & a \end{matrix}\middle| u\right) = \frac{[\pm 2a+1-u]}{[\pm 2a+1]}\frac{[1]}{[1+u]}\\
&\qquad\qquad - G_a^\pm\frac{[\pm 2a-\frac{1}{2}-u]}{[\pm 2a+1]}\frac{[u]}{[\frac{3}{2}+u]}\frac{[1]}{[1+u]},\\
&\overline{W}\left(\begin{matrix} a & a \\ a & a \end{matrix}\middle| u\right) = \frac{[3+u]}{[3]}\frac{[1]}{[1+u]}\frac{[\frac{3}{2}-u]}{[\frac{3}{2}+u]} + H_a\frac{[1]}{[3]}\frac{[u]}{[1+u]}.
\end{aligned} \tag{6.51}$$

Here

$$\begin{aligned}
G_a^\pm &= \frac{S(a\pm 1)}{S(a)}, \quad S(a) = (-1)^a\frac{[2a]}{[a]_+},\\
H_a &= G_a^+\frac{[a-\frac{5}{2}]_+}{[a+\frac{1}{2}]_+} + G_a^-\frac{[a+\frac{5}{2}]_+}{[a-\frac{1}{2}]_+},
\end{aligned} \tag{6.52}$$

where $[u]$ and $[u]_+$ are given in (A.5) and (A.7).

This Boltzmann weight W enjoys YBE (4.3), initial condition (4.38), unitarity (4.39) and crossing symmetry (4.99) ($\lambda = -\frac{3}{2}$, $G_a = S(a)$),

$$W\left(\begin{matrix} b & d \\ a & c \end{matrix}\middle| -\frac{3}{2}-u\right) = \sqrt{\frac{S(a)S(d)}{S(b)S(c)}}\,W\left(\begin{matrix} a & b \\ c & d \end{matrix}\middle| u\right). \tag{6.53}$$

In this gauge, W enjoys also a reflection symmetry

$$W\left(\begin{matrix} a & c \\ b & d \end{matrix}\middle| u\right) = W\left(\begin{matrix} a & b \\ c & d \end{matrix}\middle| u\right). \tag{6.54}$$

Along with the additive variable u, we often use the multiplicative variable $z = x^{2u}$. For later use we define W^*,

$$W^*\left(\begin{matrix} a & b \\ c & d \end{matrix}\middle| u\right) = \overline{W}\left(\begin{matrix} a & b \\ c & d \end{matrix}\middle| u\right)\Big|_{r\to r^*} \times \rho^*(u), \tag{6.55}$$

where $\rho^*(u)$ is given in (6.128).

Hereafter we assume that L is odd. The model has ground states labeled by odd integers $l = 1, 3, \ldots, L-2$ [69]. They are characterized as configurations in which all heights take the same value b. If $L = 4n \pm 1$, then the possible values are $b = l$ $(1 \le l \le 2n-1,\ l\text{: odd})$ or $b = l+1$ $(2n+1 \le l \le L-2,\ l\text{: odd})$. Therefore in the notation of Section 4.2 we have $(i_1) = (b)$, $m = 0 \in \mathbb{Z}/\mathbb{Z}$ and $\mathcal{H}_l^{(k)} = \mathcal{H}_{l,0}^{(k)}$. In the thermodynamic limit CTM's (4.100) become

$$\begin{aligned} A^{(k)}(u) &= C^{(k)}(u) = x^{-2uH_C^{(k)}}, \\ B^{(k)}(u) &= D^{(k)}(u) = \sqrt{S(k)}x^{2(u+3/2)H_C^{(k)}}. \end{aligned} \tag{6.56}$$

Careful study of the corner Hamiltonian shows that the character (4.43) coincides with the Virasoro minimal unitary character [69],

$$\chi_{l,0,k}(q) = \chi_{l,k}^{\mathrm{Vir}}(q), \tag{6.57}$$

where $\chi_{l,k}^{\mathrm{Vir}}(q)$ is given in (2.37). Comparing the free field realization given in Section 6.2, we make an identification

$$\mathcal{H}_l^{(k)} = \mathcal{L}_{l,k}, \tag{6.58}$$

$$H_C^{(k)} = d|_{\mathcal{L}_{l,k}}, \tag{6.59}$$

where $\mathcal{L}_{l,k}$ is given in (6.45) as a cohomology of the Felder complex and d is realized in (6.18).

6.4 Free Field Approach

Although bosons are already introduced in Section 6.2, we present their definitions again. Recall that

$$r = 2\frac{L+1}{L+2}, \quad r^* = r - 1.$$

Let us introduce free boson oscillator α_n $(n \in \mathbb{Z}_{\neq 0})$

$$[\alpha_n, \alpha_m] = \frac{[n]_x([2n]_x - [n]_x)}{n}\frac{[rn]_x}{[r^*n]_x}\delta_{n+m,0}, \tag{6.60}$$

and use zero mode a_0 and Q defined in (2.49) (or a'_0 in (2.72)). The Fock space $\mathcal{F}_{l,k}$ is defined by

$$\mathcal{F}_{l,k} = \bigoplus_{m\geq 0} \bigoplus_{n_1\geq\cdots\geq n_m>0} \mathbb{C}\alpha_{-n_1}\ldots\alpha_{-n_m}|\alpha_{l,k}\rangle_B, \tag{6.61}$$

where $|\alpha\rangle_B$ is given by (2.51) with replacing a_n by h_n,

$$\begin{aligned} a'_0|\alpha_{l,k}\rangle_B &= (\alpha_{l,k}-\alpha_0)|\alpha_{l,k}\rangle_B, \\ \alpha_{l,k}-\alpha_0 &= -\sqrt{\frac{L+1}{L}}\,l + \sqrt{\frac{L}{L+1}}\,k. \end{aligned} \tag{6.62}$$

We use also free boson oscillator α'_n ($n \in \mathbb{Z}_{\neq 0}$)

$$\begin{aligned} \alpha'_n &= (-1)^n \frac{[r^*n]_x}{[rn]_x}\alpha_n, \\ [\alpha'_n, \alpha'_m] &= \frac{[n]_x([2n]_x-[n]_x)}{n}\frac{[r^*n]_x}{[rn]_x}\delta_{n+m,0}. \end{aligned} \tag{6.63}$$

Operators $\hat{l}, \hat{k} \colon \mathcal{F}_{l,k} \to \mathcal{F}_{l,k}$ are defined by (5.22).

Elliptic currents $x_\pm(z)$ for $U_x(A_2^{(2)})$ (or $\mathcal{B}_{x,\lambda}(A_2^{(2)})$) of level 1 ($c=1$) are obtained by a 'dressing' procedure described in Section 4.4 **5**

$$x_+(z)\colon \mathcal{F}_{l,k} \to \mathcal{F}_{l-2,k}$$

$$x_+(z) = :\!\exp\Big(-\sum_{n\neq 0}\frac{\alpha_n}{[n]_x}z^{-n}\Big)\!: \times\, e^{\sqrt{r/(2r^*)}Q} z^{\sqrt{r/(2r^*)}a'_0 + r/2r^*}, \tag{6.64}$$

$$x_-(z)\colon \mathcal{F}_{l,k} \to \mathcal{F}_{l,k-1}$$

$$x_-(z) = :\!\exp\Big(\sum_{n\neq 0}\frac{\alpha'_n}{[n]_x}z^{-n}\Big)\!:\times e^{-\sqrt{r^*/(2r)}Q} z^{-\sqrt{r^*/(2r)}a'_0 + r^*/(2r)}. \tag{6.65}$$

$x_+(z)$ is interpreted as a screening current in Section 6.2. Note that $x_-(z)$ is not a screening current.

VO's are obtained by solving the relations (6.74)–(6.76) below directly for $\Phi_\varepsilon(z)$, $\Psi^*_\varepsilon(z)$ ($\varepsilon = 0, \pm 1$). We write $\Phi_\pm(z) = \Phi_{\pm 1}(z)$ and $\Psi^*_\pm(z) = \Psi^*_{\pm 1}(z)$. Results are

type I $\quad \Phi_\varepsilon(z)\colon \mathcal{F}_{l,k} \to \mathcal{F}_{l,k-\varepsilon}$

$$\begin{aligned} \Phi_-(z) = \sqrt{g}:\!\exp\Big(-\sum_{n\neq 0}\frac{\alpha'_n}{[2n]_x-[n]_x}z^{-n}\Big)\!: \\ \times\, e^{\sqrt{r^*/(2r)}Q} z^{\sqrt{r^*/(2r)}a'_0 + r^*/(2r)}, \end{aligned} \tag{6.66}$$

$$\begin{aligned} \Phi_0(z) = x^{-r^*/(2r)} \oint_{C_0(z)} \underline{dz}_1\, \Phi_-(z)x_-(z_1) \\ \times \frac{1}{\sqrt{[\hat{k}+\frac{1}{2}]_+[\hat{k}-\frac{1}{2}]_+}}\frac{[u-u_1+\hat{k}]_+}{[u-u_1+\frac{1}{2}]}, \end{aligned} \tag{6.67}$$

$$\Phi_+(z) = x^{-r^*/r} \oint\oint_{C_+(z)} \underline{dz}_1\, \underline{dz}_2\, \Phi_-(z) x_-(z_1) x_-(z_2)$$
$$\times \sqrt{\frac{S(\hat{k}\quad 1)}{S(\hat{k})}\frac{1}{[\hat{k}-\frac{1}{2}]_+[2\hat{k}-2]}}$$
$$\times \frac{[u-u_1+2\hat{k}-\frac{3}{2}]}{[u-u_1+\frac{1}{2}]}\frac{[u_1-u_2+\hat{k}]_+}{[u_1-u_2+\frac{1}{2}]}, \qquad (6.68)$$

type II $\Psi^*_\varepsilon(z)\colon \mathcal{F}_{l,k} \to \mathcal{F}_{l-2\varepsilon,k}$

$$\Psi^*_-(z) = \frac{1}{\sqrt{g^*}} {:}\exp\Bigl(\sum_{n\neq 0} \frac{\alpha_n}{[2n]_x - [n]_x} z^{-n}\Bigr){:}$$
$$\times\, e^{-\sqrt{r/(2r^*)}Q} z^{-\sqrt{r/(2r^*)}a'_0 + r/(2r^*)}, \qquad (6.69)$$

$$\Psi^*_0(z) = i x^{r/(2r^*)} \oint_{C^*_0(z)} \underline{dz}_1\, \Psi^*_-(z) x_+(z_1)$$
$$\times \frac{1}{\sqrt{[\frac{1}{2}(\hat{l}+1)]^*_+[\frac{1}{2}(\hat{l}-1)]^*_+}} \frac{[u-u_1-\frac{1}{2}\hat{l}]^*_+}{[u-u_1-\frac{1}{2}]^*}, \qquad (6.70)$$

$$\Psi^*_+(z) = x^{r/r^*} \oint\oint_{C^*_+(z)} \underline{dz}_1\, \underline{dz}_2\, \Psi^*_-(z) x_+(z_1) x_+(z_2)$$
$$\times \sqrt{\frac{S^*(\hat{l}/2-1)}{S^*(\hat{l}/2)}\frac{1}{[\frac{1}{2}(\hat{l}-1)]^*_+[\hat{l}-2]^*}}$$
$$\times \frac{[u-u_1-\hat{l}+\frac{3}{2}]^*}{[u-u_1-\frac{1}{2}]^*}\frac{[u_1-u_2-\frac{1}{2}\hat{l}]^*_+}{[u_1-u_2-\frac{1}{2}]^*}. \qquad (6.71)$$

Here $z = x^{2u}$, $z_j = x^{2u_j}$, $\underline{dz}_j = \frac{dz_j}{2\pi i z_j}$, and normalization constants g and g^* will be given in (6.79) and (6.80). The poles of the integrand of (6.67)–(6.71) and the integration contours are listed in the following table. For example, $C_0(z)$ is a simple closed contour that encircles $x^{1+2rn}z$ ($n \geq 0$) but not $x^{-1-2rn}z$ ($n \geq 0$).

	inside	outside
$C_0(z)$	$z_1 = x^{1+2rn}z$	$z_1 = x^{-1-2rn}z$
$C_+(z)$	$z_1 = x^{1+2rn}z$ $z_2 = x^{1+2rn}z_1$	$z_1 = x^{-1-2rn}z$ $z_2 = x^{-1-2rn}z, x^{-1-2rn}z_1, x^{2-2r(n+1)}z_1$
$C^*_0(z)$	$z_1 = x^{-1+2r^*n}z$	$z_1 = x^{1-2r^*n}z$
$C^*_+(z)$	$z_1 = x^{-1+2r^*n}z$ $z_2 = x^{-1+2r^*n}z_1$	$z_1 = x^{1-2r^*n}z$ $z_2 = x^{1-2r^*n}z, x^{1-2r^*n}z_1, x^{-2-2r^*(n+1)}z_1$

$$(n = 0, 1, 2, \ldots). \qquad (6.72)$$

OPE formulas are given in Section 6.A. In the conformal limit ($x = e^\hbar \to 1$, r: fixed), $x_+(z) \to zV_{\alpha_{-1,1}}(-z)$ (up to phase), $x_-(z) \to V_{\alpha_{1,0}}(z)$ (up to power of z), $g^{-1/2}\Phi_-(z) \to V_{\alpha_{1,2}}(z)$ (up to power of z), and $g^{*\frac{1}{2}}\Psi^*_-(z) \to$

$V_{\alpha_{3,1}}(-z)$ (up to phase and power of z), where $V_\alpha(z)$ is given in (2.61). Type I VO $\Phi_\varepsilon(z)$ commutes with BRST operator (6.38)

$$[X_j, \Phi_\varepsilon(z)] = 0. \tag{6.73}$$

The VO's given above satisfy the following commutation relations (ε_1, $\varepsilon_2 = 0, \pm 1$),

$$\Phi_{\varepsilon_2}(z_2)\Phi_{\varepsilon_1}(z_1) = \sum_{\substack{\varepsilon_1',\varepsilon_2'=0,\pm1 \\ \varepsilon_1'+\varepsilon_2'=\varepsilon_1+\varepsilon_2}} W\left(\begin{matrix}\hat{k} & \hat{k}+\varepsilon_1' \\ \hat{k}+\varepsilon_2 & \hat{k}+\varepsilon_1+\varepsilon_2\end{matrix}\middle| u_1-u_2\right)\Phi_{\varepsilon_1'}(z_1)\Phi_{\varepsilon_2'}(z_2), \tag{6.74}$$

$$\Psi^*_{\varepsilon_1}(z_1)\Psi^*_{\varepsilon_2}(z_2) = \sum_{\substack{\varepsilon_1',\varepsilon_2'=0,\pm1 \\ \varepsilon_1'+\varepsilon_2'=\varepsilon_1+\varepsilon_2}} W^*\left(\begin{matrix}\frac{1}{2}\hat{l} & \frac{1}{2}\hat{l}+\varepsilon_1 \\ \frac{1}{2}\hat{l}+\varepsilon_2' & \frac{1}{2}\hat{l}+\varepsilon_1+\varepsilon_2\end{matrix}\middle| u_1-u_2\right)\Psi^*_{\varepsilon_2'}(z_2)\Psi^*_{\varepsilon_1'}(z_1), \tag{6.75}$$

$$\Phi_{\varepsilon_2}(z_2)\Psi^*_{\varepsilon_1}(z_1) = \tau(u_1-u_2)\Psi^*_{\varepsilon_1}(z_1)\Phi_{\varepsilon_2}(z_2), \tag{6.76}$$

where $z_i = x^{2u_i}$ and $\tau(u)$ is given in (6.129). We do not present the tedious but straightforward verification of (6.74)–(6.76).

For the description of correlation functions we need also the 'dual' VO's. Define

$$\Phi^*_\varepsilon(z) = \sqrt{S(\hat{k})}^{\,-1}\Phi_{-\varepsilon}(x^{-3}z)\sqrt{S(\hat{k})}, \tag{6.77}$$

$$\Psi_\varepsilon(z) = \sqrt{S^*(\tfrac{1}{2}\hat{l})}\Psi^*_{-\varepsilon}(x^{-3}z)\sqrt{S^*(\tfrac{1}{2}\hat{l})}^{\,-1}, \tag{6.78}$$

and normalization constants g and g^* in (6.66) and (6.69) are

$$g^{-1} = \frac{(x;x^{2r})_\infty}{(x^2;x^{2r})^2_\infty(x^{2r-1};x^{2r})_\infty(x^{2r};x^{2r})^4_\infty} \times \frac{(x^5,x^6,x^{2r},x^{2r+1};x^6,x^{2r})_\infty}{(x^2,x^3,x^{2r+3},x^{2r+4};x^6,x^{2r})_\infty}, \tag{6.79}$$

$$g^* = \frac{(x^{-1};x^{2r^*})_\infty}{(x^{-2};x^{2r^*})^2_\infty(x^{2r^*+1};x^{2r^*})_\infty(x^{2r^*};x^{2r^*})^5_\infty} \times \frac{(x^3,x^4,x^{2r^*+2},x^{2r^*+3};x^6,x^{2r^*})_\infty}{(x,x^6,x^{2r^*+5},x^{2r^*+6};x^6,x^{2r^*})_\infty}. \tag{6.80}$$

Then we have

$$\sum_{\varepsilon=0,\pm1}\Phi^*_\varepsilon(z)\Phi_\varepsilon(z) = \mathrm{id}, \tag{6.81}$$

$$\Phi_{\varepsilon_2}(z)\Phi^*_{\varepsilon_1}(z) = \delta_{\varepsilon_1,\varepsilon_2}\times\mathrm{id}, \tag{6.82}$$

$$\sum_{\varepsilon=0,\pm1} \Psi^*_\varepsilon(z_2)\Psi_\varepsilon(z_1) = \frac{1}{1-z_1/z_2} + \cdots \quad (z_1 \to z_2), \tag{6.83}$$

$$\Psi_{\varepsilon_1}(z_1)\Psi^*_{\varepsilon_2}(z_2) = \frac{\delta_{\varepsilon_1,\varepsilon_2}}{1-z_1/z_2} + \cdots \quad (z_1 \to z_2), \tag{6.84}$$

and for d in (6.18)

$$w^d \mathcal{O}(z) w^{-d} = \mathcal{O}(wz), \quad \text{for } \mathcal{O} = \Phi_\varepsilon, \Phi^*_\varepsilon, \Psi^*_\varepsilon, \Psi_\varepsilon, x_\pm. \tag{6.85}$$

We identify type I VO's in Section 4.2 and those here in the following way:

$$\Phi^{(a-\varepsilon,a)}(z) = \Phi_\varepsilon(z)|_{\mathcal{L}_{l,a}}, \tag{6.86}$$

$$\Phi^{*(a+\varepsilon,a)}(z) = \Phi^*_\varepsilon(z)|_{\mathcal{L}_{l,a}}. \tag{6.87}$$

Then eqs. (4.106)–(4.110) correspond to (6.77), (6.74), (6.85), (6.81) and (6.82) respectively.

Next let us see how the DVA($A_2^{(2)}$) current is obtained from VO's. Let introduce free boson oscillator h_n ($n \in \mathbb{Z}_{\neq 0}$),

$$\lambda_n = (-1)^n (x - x^{-1}) \frac{[r^* n]_x}{[2n]_x - [n]_x} \alpha_n \tag{6.88}$$

$$= (x - x^{-1}) \frac{[rn]_x}{[2n]_x - [n]_x} \alpha'_n, \tag{6.89}$$

$$[\lambda_n, \lambda_m] = (x - x^{-1})^2 \frac{1}{n} \frac{[n]_x [rn]_x [r^* n]_x}{[2n]_x - [n]_x} \delta_{m+n,0}. \tag{6.90}$$

As explained in Section 6.2, the DVA($A_2^{(2)}$) current $T(z)$ is realized as

$$\begin{aligned} T(z) &= \Lambda_+(z) + \Lambda_0(z) + \Lambda_-(z), \\ \Lambda_\pm(z) &= :\!\exp\Bigl(\pm \sum_{n\neq 0} h_n (x^{\pm 3/2} z)^{-n}\Bigr)\!: \times x^{\pm\sqrt{2rr^*}\alpha'_0}, \\ \Lambda_0(z) &= -\frac{[r-\frac{1}{2}]_x}{[\frac{1}{2}]_x} :\!\exp\Bigl(\sum_{n\neq 0} h_n (x^{-n/2} - x^{n/2}) z^{-n}\Bigr)\!:. \end{aligned} \tag{6.91}$$

This $T(z)$ is obtained from type I VO's by fusing them,

$$\begin{aligned} &\Phi_{\varepsilon_2}(x^{3/2+r} z')\Phi^*_{\varepsilon_1}(x^{3/2-r} z) \\ &= \Bigl(1 - \frac{z}{z'}\Bigr)(-1)^{\varepsilon_1+1}\delta_{\varepsilon_1,\varepsilon_2} T(z) \cdot x^{1-r} \frac{(x, x^6, x^{5-2r}, x^{6-2r}; x^6)_\infty}{(x^3, x^4, x^{2-2r}, x^{3-2r}; x^6)_\infty} + \cdots \\ &\qquad (z' \to z). \end{aligned} \tag{6.92}$$

$\widetilde{T}(z) = -\Lambda_+(z) + \Lambda_0(z) - \Lambda_-(z)$ also satisfies (6.1), and it is obtained from type II VO's,

$$\begin{aligned}
&\Psi_{\varepsilon_1}(x^{3/2+r^*}z')\Psi^*_{\varepsilon_2}(x^{3/2-r^*}z) \\
&= \frac{1}{1-z'/z}(-1)^{\varepsilon_1+1}\delta_{\varepsilon_1,\varepsilon_2}\widetilde{T}(-z)\cdot(-x^{-r})\frac{(x^2,x^3,x^{3-2r^*},x^{4-2r^*};x^6)_\infty}{(x^5,x^6,x^{-2r^*},x^{1-2r^*};x^6)_\infty}+\cdots \\
&\qquad (z' \to z). \qquad (6.93)
\end{aligned}$$

Using these bosonized VO's let us calculate LHP (4.73) with (4.111),

$$\begin{aligned}
&P_{a_n,\ldots,a_0}(l) \\
&= Z_l^{-1}S(a_0)\,\mathrm{tr}_{\mathcal{H}_l^{(a_0)}}\big(x^{6H_C^{(a_0)}}\Phi^{*(a_0,a_1)}(z)\cdots\Phi^{*(a_{n-1},a_n)}(z) \\
&\qquad\qquad \times\Phi^{(a_n,a_{n-1})}(z)\cdots\Phi^{(a_1,a_0)}(z)\big) \\
&= Z_l^{-1}S(a_0)\,\mathrm{tr}_{\mathcal{L}_{l,a_0}}\big(x^{6d}\Phi^*_{\varepsilon_1}(z)\cdots\Phi^*_{\varepsilon_n}(z)\Phi_{\varepsilon_n}(z)\cdots\Phi_{\varepsilon_1}(z)\big), \qquad (6.94)
\end{aligned}$$

where $\varepsilon_i = a_{i-1} - a_i$. Here we have identified the space of state and operators as (6.58), (6.59), (6.86) and (6.87).

One-point LHP is already obtained in (4.44),

$$P_k(l) = Z_l^{-1}S(k)\chi_{l,k}(x^6), \qquad (6.95)$$

where $\chi_{l,k}(q)$ is given in (2.37) and Z_l (4.74) is

$$Z_l = \sum_{k=1}^{L} S(k)\chi_{l,k}(x^6), \qquad (6.96)$$

which can be expressed in product of theta functions with conjugate modulus [69]. Two-point LHP satisfies (4.112) and (4.66)

$$P_{a,b}(l) = P_{b,a}(l), \quad \sum_b P_{a,b}(l) = P_a(l). \qquad (6.97)$$

In contrast with Section 5.3, however, this recursion relation does not determine $P_{a,b}(l)$ uniquely. So we will use free field realization of vertex operator approach. Two-point LHP $P_{k-\varepsilon,k}(l)$ is $(\varepsilon = 0, \pm 1)$

$$\begin{aligned}
P_{k-\varepsilon,k}(l) &= Z_l^{-1}S(k)\,\mathrm{tr}_{\mathcal{L}_{l,k}}\big(x^{6d}\Phi^*_\varepsilon(z)\Phi_\varepsilon(z)\big) \\
&= Z_l^{-1}\sqrt{S(k)S(k-\varepsilon)}\,\mathrm{tr}_{\mathcal{L}_{l,k}}\big(x^{6d}\Phi_{-\varepsilon}(x^{-3}z)\Phi_\varepsilon(z)\big). \qquad (6.98)
\end{aligned}$$

Since this is independent on z, we take $z=1$. The evaluation of the trace yields the following expressions:

$$P_{k-1,k}(l) = -\frac{g}{Z_l}\frac{S(k-1)x^{-r^*/r}}{[k-\frac{1}{2}]_+[2k-2]}\oint\oint_{C_+(1)}\underline{dw_1}\,\underline{dw_2}\,\mathcal{I}(w_1,w_2)$$
$$\times\frac{[v_1-2k+\frac{3}{2}][v_1-v_2+k]_+}{[v_1+\frac{1}{2}][v_1-v_2+\frac{1}{2}]}, \tag{6.99}$$

$$P_{k,k}(l) = \frac{g}{Z_l}\frac{S(k)x^{-r^*/r}}{[k+\frac{1}{2}]_+[k-\frac{1}{2}]_+}\oint_{C_0(x^{-3})}\underline{dw_1}\oint_{C_0(1)}\underline{dw_2}\,\mathcal{I}(w_1,w_2)$$
$$\times\frac{[v_1-k+\frac{3}{2}]_+[v_2-k]_+}{[v_1+1][v_2-\frac{1}{2}]}, \tag{6.100}$$

$$P_{k+1,k}(l) = -\frac{g}{Z_l}\frac{S(k)x^{-r^*/r}}{[k+\frac{1}{2}]_+[2k]}\oint\oint_{C_+(x^{-3})}\underline{dw_1}\,\underline{dw_2}\,\mathcal{I}(w_1,w_2)$$
$$\times\frac{[v_1-2k+1][v_1-v_2+k+1]_+[v_2+\frac{1}{2}]}{[v_1+1][v_1-v_2+\frac{1}{2}][v_2-\frac{1}{2}]}, \tag{6.101}$$

where $w_i = x^{2v_i}$ $(i=1,2)$ and $\mathcal{I}(w_1,w_2)$ is

$$\begin{aligned}\mathcal{I}(w_1,w_2) &= g^{-1}\,\mathrm{tr}_{\mathcal{L}_{l,k}}\big(x^{6d}\Phi_-(x^{-3})x_-(w_1)\Phi_-(1)x_-(w_2)\big)\\ &= \mathcal{I}^{\mathrm{OPE}}(w_1,w_2)\mathcal{I}^{\mathrm{osc}}(w_1,w_2)\mathcal{I}^{\mathrm{zero}}(w_1,w_2).\end{aligned} \tag{6.102}$$

Here $\mathcal{I}^{\mathrm{OPE}}(w_1,w_2)$ is the OPE contribution

$$\begin{aligned}&\mathcal{I}^{\mathrm{OPE}}(w_1,w_2)\\ &\quad= \langle\!\langle\Phi_-(x^{-3})x_-(w_1)\rangle\!\rangle\langle\!\langle\Phi_-(x^{-3})\Phi_-(1)\rangle\!\rangle\langle\!\langle\Phi_-(x^{-3})x_-(w_2)\rangle\!\rangle\\ &\qquad\times\langle\!\langle x_-(w_1)\Phi_-(1)\rangle\!\rangle\langle\!\langle x_-(w_1)x_-(w_2)\rangle\!\rangle\langle\!\langle\Phi_-(1)x_-(w_2)\rangle\!\rangle,\end{aligned} \tag{6.103}$$

and $\mathcal{I}^{\mathrm{osc}}(w_1,w_2)$ is the oscillator contribution

$$\begin{aligned}&\mathcal{I}^{\mathrm{osc}}(w_1,w_2)\\ &\quad= \mathrm{tr}_{\mathcal{F}}\big(x^{6d} : \Phi_-^{\mathrm{osc}}(x^{-3})x_-^{\mathrm{osc}}(w_1)\Phi_-^{\mathrm{osc}}(1)x_-^{\mathrm{osc}}(w_2) :\big)\\ &\quad= \frac{1}{(x^6;x^6)_\infty}F_{\Phi_-,\Phi_-}(1)^2F_{\Phi_-,\Phi_-}(x^3)F_{\Phi_-,\Phi_-}(x^{-3})F_{x_-,x_-}(1)^2\\ &\qquad\times\prod_{i=1,2}F_{\Phi_-,x_-}(w_i)F_{\Phi_-,x_-}(w_i^{-1})F_{\Phi_-,x_-}(x^3w_i)F_{\Phi_-,x_-}(x^{-3}w_i^{-1})\\ &\qquad\qquad\times F_{x_-,x_-}\left(\frac{w_2}{w_1}\right))F_{x_-,x_-}\left(\frac{w_1}{w_2}\right),\end{aligned} \tag{6.104}$$

and $\mathcal{I}^{\rm zero}(w_1, w_2)$ is the zero mode contribution

$$\begin{aligned}
&\mathcal{I}^{\rm zero}(w_1, w_2)\\
&\quad = g^{-1}\, {\rm tr}^{\rm zero}_{\mathcal{L}_{l,k}}\left(x^{6d}{:}\Phi^{\rm zero}_{-}(x^{-3})x^{\rm zero}_{-}(w_1)\Phi^{\rm zero}_{-}(1)x^{\rm zero}_{-}(w_2){:}\right)\\
&\quad = {\rm tr}^{\rm zero}_{\mathcal{L}_{l,k}}\left(x^{6d}(x^3w_1w_2)^{-\sqrt{r^*/(2r)}a_0'}(x^{-3}w_1w_2)^{r^*/(2r)}\right)\\
&\quad = \sum_{j\in\mathbb{Z}}\Big(x^{6(h_{l-2Lj,k}-c/24)}(x^3w_1w_2)^{(l-2Lj)/2-kr^*/r}\\
&\qquad\qquad - x^{6(h_{-l-2Lj,k}-c/24)}(x^3w_1w_2)^{(-l-2Lj)/2-kr^*/r}\Big)(x^{-3}w_1w_2)^{r^*/(2r)}\\
&\quad = x^{6(h_{l,k}-c/24)}x^{-3rs/r}(x^3w_1w_2)^{l/2-kr^*/r+r^*/(2r)}\\
&\qquad \times\Big(\Theta_{x^{12L(L+1)}}\left(-x^{6(-(L+1)l+Lk+L(L+1))}(x^3w_1w_2)^{-L}\right)\\
&\qquad\quad - \Theta_{x^{12L(L+1)}}\left(-x^{6((L+1)l+Lk+L(L+1))}(x^3w_1w_2)^{-L}\right)\\
&\qquad\qquad\qquad\qquad \times x^{6lk}(x^3w_1w_2)^{-l}\Big). \qquad (6.105)
\end{aligned}$$

Using formulas in Section 6.A we have

$$\begin{aligned}
&\mathcal{I}^{\rm OPE}(w_1, w_2)\mathcal{I}^{\rm osc}(w_1, w_2)\\
&\quad = x^{3r^*/r}\frac{(x^5, x^5, x^6, x^6, x^8, x^8, x^{2r+5}, x^{2r+5}; x^6, x^{2r})_\infty}{(x^7, x^7, x^{2r+3}, x^{2r+3}, x^{2r+4}, x^{2r+4}, x^{2r+4}, x^{2r+4}; x^6, x^{2r})_\infty}\\
&\quad \times \frac{(x^{2r-1}w_1^{-1}, x^{2r+2}w_1; x^3, x^{2r})_\infty}{(xw_1^{-1}, x^4w_1; x^3, x^{2r})_\infty}\frac{(x^{2r-1}w_2, x^{2r+2}w_2^{-1}; x^3, x^{2r})_\infty}{(xw_2, x^4w_2^{-1}; x^3, x^{2r})_\infty}\\
&\quad \times \Theta_{x^6}(z)\frac{(x^2z, x^8z^{-1}, x^{2r-1}z, x^{2r+5}z^{-1}; x^6, x^{2r})_\infty}{(xz, x^7z^{-1}, x^{2r-2}z, x^{2r+4}z^{-1}; x^6, x^{2r})_\infty}\Bigg|_{z=w_2/w_1}. \qquad (6.106)
\end{aligned}$$

The contours $C_+(1)$, $C_0(x^{-3}) \cup C_0(1)$, $C_+(x^{-3})$ are chosen as follows ($n,\ m \geq 0$): For all the contours, the poles $w_1 = x^{4+3m+2rn}$, $w_2 = x^{4+3m+2rn}$, $x^{4+6m+2r(n+1)}w_1$, $x^{1+6(m+1)+2rn}w_1$ are inside and the poles $w_1 = x^{-4-3m-2rn}$, $w_2 = x^{-1-3m-2rn}$, $x^{2-6m-2r(n+1)}w_1$, $x^{-1-6m-2rn}w_1$ are outside. In addition,

	inside	outside
$C_+(1)$	$w_1 = x^{-1+2r(n+1)}$ $w_2 = x^{1+2rn}w_1$	$w_1 = x^{-1-2rn}$ $w_2 = \begin{cases} x^{-1-2rn}, \\ x^{-1-2rn}w_1, \\ x^{2-2r(n+1)}w_1 \end{cases}$
$C_0(x^{-3}) \cup C_0(1)$	$w_1 = x^{-2+2rn}$ $w_2 = x^{1+2rn}$	$w_1 = x^{-4-2rn}$ $w_2 = x^{-1-2rn}$
$C_+(x^{-3})$	$w_1 = x^{-2+2rn}$ $w_2 = \begin{cases} x^{1+2rn}, \\ x^{1+2rn}w_1 \end{cases}$	$w_1 = x^{-4-2rn}$ $w_2 = \begin{cases} x^{1-2r(n+1)}, \\ x^{-1-2rn}w_1, \\ x^{2-2r(n+1)}w_1 \end{cases}$

.

For integral representations of general LHP, see [46].

Excited states are obtained by using type II VO and traces of type I and type II VO's are calculated similarly like in Section 5.4.

6.A OPE and trace

OPE

We list the normal ordering relations used in Section 6. r^* is

$$r^* = r - 1.$$

Notation $\langle\!\langle A(z)B(w)\rangle\!\rangle$ is given in (5.113):

$$\langle\!\langle x_+(z_1)x_+(z_2)\rangle\!\rangle = z_1^{r/r^*}(1-\zeta)\frac{(x^{-2}\zeta, x^{2r^*+1}\zeta; x^{2r^*})_\infty}{(x^{-1}\zeta, x^{2r^*+2}\zeta; x^{2r^*})_\infty}, \tag{6.107}$$

$$\langle\!\langle x_-(z_1)x_-(z_2)\rangle\!\rangle = z_1^{r^*/r}(1-\zeta)\frac{(x^{2}\zeta, x^{2r-1}\zeta; x^{2r})_\infty}{(x\zeta, x^{2r-2}\zeta; x^{2r})_\infty}, \tag{6.108}$$

$$\langle\!\langle x_\pm(z_1)x_\mp(z_2)\rangle\!\rangle = z_1^{-1}\frac{1+\zeta}{(1+x\zeta)(1+x^{-1}\zeta)}, \tag{6.109}$$

$$\langle\!\langle \Phi_-(z_1)x_+(z_2)\rangle\!\rangle = \langle\!\langle x_+(z_1)\Phi_-(z_2)\rangle\!\rangle = z_1 + z_2, \tag{6.110}$$

$$\langle\!\langle \Phi_-(z_1)x_-(z_2)\rangle\!\rangle = \langle\!\langle x_-(z_1)\Phi_-(z_2)\rangle\!\rangle = z_1^{-r^*/r}\frac{(x^{2r-1}\zeta; x^{2r})_\infty}{(x\zeta; x^{2r})_\infty}, \tag{6.111}$$

$$\langle\!\langle \Psi^*_-(z_1)x_+(z_2)\rangle\!\rangle = \langle\!\langle x_+(z_1)\Psi^*_-(z_2)\rangle\!\rangle = z_1^{-r/r^*}\frac{(x^{2r^*+1}\zeta; x^{2r^*})_\infty}{(x^{-1}\zeta; x^{2r^*})_\infty}, \tag{6.112}$$

$$\langle\!\langle \Psi^*_-(z_1)x_-(z_2)\rangle\!\rangle = \langle\!\langle x_-(z_1)\Psi^*_-(z_2)\rangle\!\rangle = z_1 + z_2, \tag{6.113}$$

$$\langle\!\langle \Phi_-(z_1)\Phi_-(z_2)\rangle\!\rangle = z_1^{r^*/r}\frac{(x^2\zeta, x^3\zeta, x^{2r+3}\zeta, x^{2r+4}\zeta; x^6, x^{2r})_\infty}{(x^5\zeta, x^6\zeta, x^{2r}\zeta, x^{2r+1}\zeta; x^6, x^{2r})_\infty}, \tag{6.114}$$

$$\langle\!\langle \Psi^*_-(z_1)\Psi^*_-(z_2)\rangle\!\rangle = z_1^{r/r^*}\frac{(\zeta, x\zeta, x^{2r^*+5}\zeta, x^{2r^*+6}\zeta; x^6, x^{2r^*})_\infty}{(x^3\zeta, x^4\zeta, x^{2r^*+2}\zeta, x^{2r^*+3}\zeta; x^6, x^{2r^*})_\infty}, \tag{6.115}$$

$$\langle\!\langle \Phi_-(z_1)\Psi^*_-(z_2)\rangle\!\rangle = \langle\!\langle \Psi^*_-(z_1)\Phi_-(z_2)\rangle\!\rangle = z_1^{-1}\frac{(-x^4\zeta, -x^5\zeta; x^6)_\infty}{(-x\zeta, -x^2\zeta; x^6)_\infty}, \tag{6.116}$$

where $\zeta = z_2/z_1$ and we have used (A.50).

As meromorphic functions we have ($z_i = x^{2u_i}$)

$$x_+(z_1)x_+(z_2) = x_+(z_2)x_+(z_1)\frac{[u_1-u_2+1]^*}{[u_1-u_2-1]^*}\frac{[u_1-u_2-\frac{1}{2}]^*}{[-u_1+u_2-\frac{1}{2}]^*}, \tag{6.117}$$

$$x_-(z_1)x_-(z_2) = x_-(z_2)x_-(z_1)\frac{[u_1-u_2-1]}{[u_1-u_2+1]}\frac{[u_1-u_2+\frac{1}{2}]}{[-u_1+u_2+\frac{1}{2}]}, \tag{6.118}$$

$$x_\pm(z_1)x_\mp(z_2) = x_\mp(z_2)x_\pm(z_1), \tag{6.119}$$

$$\Phi_-(z_1)x_+(z_2) - x_+(z_2)\Phi_-(z_1), \tag{6.120}$$

$$\Phi_-(z_1)x_-(z_2) = x_-(z_2)\Phi_-(z_1)\frac{[u_1 - u_2 + \frac{1}{2}]}{[-u_1 + u_2 + \frac{1}{2}]}, \tag{6.121}$$

$$\Psi^*_-(z_1)x_+(z_2) = x_+(z_2)\Psi^*_-(z_1)\frac{[u_1 - u_2 - \frac{1}{2}]^*}{[-u_1 + u_2 - \frac{1}{2}]^*}, \tag{6.122}$$

$$\Psi^*_-(z_1)x_-(z_2) = x_-(z_2)\Psi^*_-(z_1), \tag{6.123}$$

$$\Phi_-(z_1)\Phi_-(z_2) = \Phi_-(z_2)\Phi_-(z_1)\rho(u_2 - u_1), \tag{6.124}$$

$$\Psi^*_-(z_1)\Psi^*_-(z_2) = \Psi^*_-(z_2)\Psi^*_-(z_1)\rho^*(u_1 - u_2), \tag{6.125}$$

$$\Phi_-(z_1)\Psi^*_-(z_2) = \Psi^*_-(z_2)\Phi_-(z_1)\tau(u_2 - u_1). \tag{6.126}$$

Here $\rho(u)$, $\rho^*(u)$ and $\tau(u)$ are given by

$$\begin{aligned} z^{r^*/r}\rho(u) &= \frac{\rho_+(u)}{\rho_+(-u)}, \\ \rho_+(u) &= \frac{(x^2z, x^3z, x^{2r+3}z, x^{2r+4}z; x^6, x^{2r})_\infty}{(x^5z, x^6z, x^{2r}z, x^{2r+1}z; x^6, x^{2r})_\infty}, \end{aligned} \tag{6.127}$$

$$\begin{aligned} z^{-r/r^*}\rho^*(u) &= \frac{\rho^*_+(u)}{\rho^*_+(-u)}, \\ \rho^*_+(u) &= \frac{(x^3z, x^4z, x^{2r^*+2}z, x^{2r^*+3}z; x^6, x^{2r^*})_\infty}{(z, xz, x^{2r^*+5}z, x^{2r^*+6}z; x^6, x^{2r^*})_\infty}, \end{aligned} \tag{6.128}$$

$$\tau(u) = z\frac{\Theta_{x^6}(-xz^{-1})\Theta_{x^6}(-x^2z^{-1})}{\Theta_{x^6}(-xz)\Theta_{x^6}(-x^2z)}. \tag{6.129}$$

Note that

$$\rho^*(u) = -\rho(u)|_{r\to r^*} \times z\frac{\Theta_{x^6}(xz^{-1})\Theta_{x^6}(x^2z^{-1})}{\Theta_{x^6}(xz)\Theta_{x^6}(x^2z)}. \tag{6.130}$$

Trace

We use the same notation as the second part of Section 5.A. The trace of oscillator parts over the Fock space $\mathcal{F} = \mathcal{F}_{l,k}$ is

$$\mathrm{tr}_{\mathcal{F}}\left(x^{6d}:\prod_i A_i^{\mathrm{osc}}(z_i):\right) = \frac{1}{(x^6;x^6)_\infty}\prod_{i,j} F_{A_i,A_j}\left(\frac{z_i}{z_j}\right), \tag{6.131}$$

where $F_{A,B}(z)$ is given by

$$F_{A,B}(z) = \exp\left(\sum_{n>0}\frac{1}{n}[n]_x([2n]_x - [n]_x)\frac{[rn]_x}{[r^*n]_x}\frac{x^{6n}}{1-x^{6n}}f^A_{-n}f^B_n z^n\right). \tag{6.132}$$

We write down $F_{A,B}(z)$ (Remark $F_{A,B}(z) = F_{B,A}(z)$):

$$F_{x_+,x_+}(z) = (x^6z;x^6)_\infty\frac{(x^4z, x^{2r^*+7}z; x^6, x^{2r^*})_\infty}{(x^5z, , x^{2r^*+8}; x^6, x^{2r^*})_\infty}, \tag{6.133}$$

$$F_{x_-,x_-}(z) = (x^6z;x^6)_\infty \frac{(x^8z, x^{2r+5}z; x^6, x^{2r})_\infty}{(x^7z, x^{2r+4}z; x^6, x^{2r})_\infty}, \tag{6.134}$$

$$F_{x_+,x_-}(z) = \frac{(-x^6z;x^0)_\infty}{(-x^5z, -x^7z; x^6)_\infty}, \tag{6.135}$$

$$F_{\Phi_-,x_+}(z) = (-x^6z;x^6)_\infty, \tag{6.136}$$

$$F_{\Phi_-,x_-}(z) = \frac{(x^{2r+5}z; x^6, x^{2r})_\infty}{(x^7z; x^6, x^{2r})_\infty}, \tag{6.137}$$

$$F_{\Psi^*_-,x_+}(z) = \frac{(x^{2r^*+7}z; x^6, x^{2r^*})_\infty}{(x^5z; x^6, x^{2r^*})_\infty}, \tag{6.138}$$

$$F_{\Psi^*_-,x_-}(z) = (-x^6z;x^6)_\infty, \tag{6.139}$$

$$F_{\Phi_-,\Phi_-}(z) = \frac{(x^8z, x^9z, x^{2r+9}z, x^{2r+10}z; x^6, x^6, x^{2r})_\infty}{(x^{11}z, x^{12}z, x^{2r+6}z, x^{2r+7}z; x^6, x^6, x^{2r})_\infty}, \tag{6.140}$$

$$F_{\Psi^*_-,\Psi^*_-}(z) = \frac{(x^6z, x^7z, x^{2r^*+11}z, x^{2r^*+12}z; x^6, x^6, x^{2r^*})_\infty}{(x^9z, x^{10}z, x^{2r^*+8}z, x^{2r^*+9}z; x^6, x^6, x^{2r^*})_\infty}, \tag{6.141}$$

$$F_{\Phi_-,\Psi^*_-}(z) = \frac{(-x^{10}z, -x^{11}z; x^6, x^6)_\infty}{(-x^7z, -x^8z; x^6, x^6)_\infty}. \tag{6.142}$$

7 Conclusion

In this lecture we have explained deformed Virasoro algebras ($A_1^{(1)}$ type and $A_2^{(2)}$ type) and elliptic quantum groups (face type algebra $\mathcal{B}_{q,\lambda}(\mathfrak{g})$ and vertex type algebra $\mathcal{A}_{q,p}(\widehat{\mathfrak{sl}}_n)$) and studied solvable lattice models (ABF model in regime III and dilute A_L model in regime 2^+) by using vertex operators and free field realizations.

We close this lecture by mentioning some related topics.

Deformed W algebras (DWA's)

In CFT there are several extensions of the Virasoro algebra, e.g., superconformal algebras, current algebras (affine Lie algebras), W algebras, parafermions, which contain the Virasoro algebra as a subalgebra. For W algebras, see review [71]. W_N algebra is a W algebra associated to A_{N-1} algebra. Deformation of W_N algebra, which we denote DWA(A_{N-1}), was obtained in [18] (see also [72]) and [20] by using correspondence between singular vectors and Macdonald symmetric polynomials or quantization of the deformed W_N Poisson algebra respectively. The deformed W_n Poisson algebra was obtained from the Wakimoto realization of $U_q(\widehat{\mathfrak{sl}}_N)$ at the critical level by E. Frenkel and Reshetikhin [19] and they pointed out that deformed

W currents in a free field realization have the same forms of transfer matrices in analytic Bethe ansatz, dressed vacuum form. (Bethe ansatz is also a powerful method to study solvable models [6, 73].) Based on this observation, DWA's for arbitrary simple Lie algebras were constructed [74]. See [75] for screening currents, [76] for relation to $\mathcal{A}_{q,p}(\widehat{\mathfrak{sl}}_N)$, [77] for connection to q-difference version of the Drinfeld–Sokorov reduction, and see [46, 78, 79] for higher currents. DWA(A_{N-1}) appears in the $A^{(1)}_{N-1}$ face model [75, 80] and also in the ABF model in regime II [81]. Since $A^{(1)}_1$ and $A^{(2)}_2$ are the only affine Lie algebras of rank 1, those DWA closes for one current $T(z)$, i.e., DVA. In CFT, W_N algebra does not contain W_n $(n \leq N)$ algebra as a subalgebra explicitly except for $n = 2$ case which corresponds to the Virasoro algebra. In deformed case, DWA(A_{N-1}) does not contain even DVA$(A^{(1)}_1)$ explicitly. As shortly explained in Section 2.2.3, singular vectors of Virasoro and W algebras imply that the correlation functions containing corresponding primary fields satisfy the differential equations. For singular vectors of DVA and DWA, do the correlation functions satisfy some difference equations?

KZ and q-KZ Equations

In CFT Wess–Zumino–Novikov–Witten (WZNW) model has gauge symmetries, i.e., affine Lie algebra symmetries [2]. Virasoro current is realized as a quadratic form of affine Lie algebra currents (Sugawara construction). Consequently conformal Ward identity has rich structure, which is known as the Knizhnik–Zamolodchikov (KZ) equation [82, 83]. Since the Virasoro algebra is a Lie algebra, we know a rule for its tensor product representation. Moreover we obtain new realizations and character formulas of the Virasoro algebra by coset construction [84]. On the other hand DVA is not a Lie algebra and its tensor product representation is unknown. (We remark that tensor product representations of W_N algebras in CFT are also unknown.) If there exists some deformation of the Sugawara construction, it could give tensor product representations of DVA but we do not know it at present. Rather we derive DVA or DWA from (elliptic) quantum groups by fusion of VO's [67]. q-deformation of KZ equation were presented by I. Frenkel and Reshetikhin [59]. q-KZ equations are holonomic q-difference equations for the matrix coefficients of the products of intertwining operators for representations of quantum affine algebra. Connection matrix of their solutions gives a face type elliptic solution of YBE. See [16, 59, 60, 85, 86].

Massive Integrable Models

Integrable perturbations of CFT were studied in [12–15] and $(1,3)$-perturbed CFT is described by the sine-Gordon model. Sine-Gordon model is a typical massive integrable model. S-matrix was obtained by Zamlodochikov's bootstrap approach [4]. For form factors see Smirnov's bootstrap approach [87] (see also [88]). Lukyanov pointed out that DVA($A_1^{(1)}$) current $T(z)$ in certain scaling limit gives the Zamolodchikov–Faddeev (ZF) algebra of sine-Gordon model (before taking a scaling limit, $T(z)$ is interpreted as the ZF algebra for basic scalar excitation of XYZ spin chain) [89]. In Section 3, z of $T(z)$ is introduced as a formal parameter, but here z of $T(z)$ is related to the spectral parameter of the particle, like as lattice models in Sections 4, 5 ($z = x^{2u}$). This is contrasted with the CFT case, where z of $L(z)$ is interpreted as a complex coordinate of the Riemann surface. We can obtain integrable massive field theory models from solvable lattice models by taking appropriate scaling limit. Particles in a field theory are created by type II VO's. For XXZ model, sine-Gordon model, Bullough-Dodd model and affine Toda model see [22, 66, 90–92]. Field theory analog of transfer matrix and Baxter's Q-operator is studied in [93].

Eight Vertex Model

ABF model was studied by Lukyanov and Pugai by bosonization of type I VO's. An algebraic approach to the fusion ABF models was presented in [55, 61] on the basis of the quasi-Hopf algebra $\mathcal{B}_{q,\lambda}(\widehat{\mathfrak{sl}}_2)$ and the elliptic algebra $U_{q,p}(\widehat{\mathfrak{sl}}_2)$. Bosonization of VO's for the $A_{N-1}^{(1)}$ face model was given in [80]. Another interesting direction is to study Baxter's eight vertex model and Belavin's generalization. Lashkevich and Pugai proposed a remarkable bosonization formula of the type I VO for the eight vertex model [94]. They succeeded in reducing the problem to the already known bosonization for the ABF model through the use of intertwining vectors and Lukyanov's screening operators. To understand their bosonization scheme, it seems necessary to clarify the relationship between the intertwining vectors and the two twistors $F(\lambda)$ and $E(r)$, which define $\mathcal{B}_{q,\lambda}(\widehat{\mathfrak{sl}}_n)$ and $\mathcal{A}_{q,p}(\widehat{\mathfrak{sl}}_n)$ respectively. It is also interesting to seek a more direct bosonization, which is intrinsically connected with the quasi-Hopf structure of $\mathcal{A}_{q,p}(\widehat{\mathfrak{sl}}_2)$ and does not rely on the bosonization of the ABF model.

Supersymmetry

In string theory supersymmetries are essential to cancelation of divergence and consistency of theory, and $N = 2$ superconformal algebra is related to many interesting topics, e.g., chiral ring, mirror symmetry, topological field theory. Are there "good" deformations of superconformal algebras? Super version ($\mathbb{Z}_2$ graded algebra) of elliptic quantum group was formulated in [95] along the line of [26] (see also [96]). Can we obtain deformed superconformal currents by fusion of VO's of this elliptic superalgebra or higher level VO's of $U_q(\widehat{\mathfrak{sl}}_2)$ [97]? See also [98].

As explained in the introduction our motivation is to find the symmetry of massive integrable models, but present status is far from satisfactory. We hope that this lecture can help the study in this (and also other) field.

Acknowledgments: This lecture is based on our collaboration and I would like to thank Hidetoshi Awata, Yuji Hara, Michio Jimbo, Harunobu Kubo, Hitoshi Konno, Yaroslav Pugai and Jun'ichi Shiraishi for valuable discussions and comments. I thank also the organizers and participants of this summer school for their kind hospitality, and CRM for financial support.

8 References

1. A.A. Belavin, A.M. Polyakov, and A.B. Zamolodchikov, "Infinite conformal symmetry in two-dimensional quantum field theory," *Nucl. Phys.* **B241** (1984), 333–380.

2. C. Itzykson, H. Saleur, and J.-B. Zuber (eds.), *Conformal Invariance and Applications to Statistical Mechanics*, World Scientific, Singapore, 1988; P. Ginsparg, "Applied conformal field theory," 1–168, in *Les Houches 1988, fields, strings and critical phenomena* edited by E. Brézin and J. Zinn-Justin, North-Holland, 1990; C. Itzykson and J.-M. Drouffe, *Statistical Field Theory* vols. 1, 2, Cambridge Univ. Press, 1989; S.V. Ketov, *Conformal Field Theory*, World Scientific, 1995; P. Di Francesco, P. Mathieu, and D. Sénéchal, *Conformal Field Theory*, Springer, 1997; V.G. Kac and A.K. Raina, *Bombay Lectures on Highest Weight Representations of Infinite Dimensional Lie Algebras*, World Scientific, 1987.

3. M.B. Green, J.H. Schwarz and E. Witten, *Superstring Theory*, vols. 1, 2, Cambridge Univ. Press, 1987; J. Polchinski, *String Theory*, vol. I, II, Cambridge Univ. Press, 1998.

4. A.B. Zamolodchikov and Al.B. Zamolodchikov, "Factorized S-matrices in two dimensions as the exact solutions of certain relativistic quantum field theory models," *Ann. Phys.* **120** (1979), 253–291.

5. M. Jimbo (ed.), *Yang–Baxter Equation in Integrable Systems*, World Scientific, 1989.

6. R.J. Baxter, *Exactly Solved Models in Statistical Mechanics*, Academic Press, 1982.

7. M. Jimbo, "Introduction to the Yang–Baxter equation," *Internat. J. Modern Phys.* **A4** (1989), 3759–3777.

8. M. Jimbo, T. Miwa, and M. Okado, "Solvable lattice models whose states are dominant integral weights of $A^{(1)}_{n-1}$," *Lett. Math. Phys.* **14** (1987), 123–131.

9. E. Date, M. Jimbo, A. Kuniba, T. Miwa ,and M. Okado, "Exactly solvable SOS models. II. Proof of the star-triangle relation and combinatorial identities," *Adv. Stud. Pure Math.* **16** (1988), 17–122.

10. M. Jimbo, T. Miwa, and M. Okado, "Solvable lattice models related to the vector representation of classical simple Lie algebra," *Comm. Math. Phys.* **116** (1988), 507–525.

11. V.G. Drinfeld, "Quasi-Hopf algebras," *Leningrad Math. J.* **1** (1990), 1419–1457; "On quasitriangular quasi-Hopf algebras and a group closely connected with Gal($\overline{\mathbb{Q}}/\mathbb{Q}$)," *Leningrad Math. J.* **2** (1991), 829–860; "Quantum groups," *Proc. ICM*, Berkeley, CA, Amer. Math. Soc.,1986, 798–820.

12. A.B. Zamolodchikov, "Higher order integrals of motion in two-dimensional models of the field theory with a broken conformal symmetry," *JETP Lett.* **46** (1987), 160–164; "Integrable field theory from conformal field theory," *Adv. Stud. Pure Math.* **19** (1989), 641–674.

13. T. Eguchi and S.-K. Yang, "Deformations of conformal field theories and soliton equations," *Phys. Lett.* **B224** (1989), 373–378; "Sine-Gordon theory at rational values of the coupling constant and minimal conformal models," *Phys. Lett.* **B235** (1990), 282–286.

14. A.B. Zamolodchikov, "Integrals of motion and S-matrix of the (scaled) $T = T_c$ Ising model with magnetic field," *Internat. J. Modern Phys.* **A4** (1989), 4235–4248.

15. F.A. Smirnov, "Exact S-matrices for $\phi_{1,2}$-perturbed minimal models of conformal field theory," *Internat. J. Modern Phys.* **A6** (1991), 1407–1428.

16. B. Davies, O. Foda, M. Jimbo, T. Miwa, and A. Nakayashiki, "Diagonalization of the XXZ Hamiltonian by vertex operators," *Comm. Math. Phys.* **151** (1993), 89–153 (hep-th/9204064).

17. J. Shiraishi, H. Kubo, H. Awata, and S. Odake, "A quantum deformation of the Virasoro algebra and the Macdonald symmetric functions", *Lett. Math. Phys.* **38** (1996), 33–51 (q-alg/9507034).

18. H. Awata, H. Kubo, S. Odake, and J. Shiraishi, "Quantum $\mathcal{W}_N$ algebras and Macdonald polynomials", *Comm. Math. Phys.* **179** (1996) 401–416 (q-alg/9508011).

19. E. Frenkel and N. Reshetikhin, "Quantum affine algebras and deformations of the Virasoro and $\mathcal{W}$-algebra," *Comm. Math. Phys.* **178** (1996), 237–264 (q-alg/9505025).

20. B. Feigin and E. Frenkel, "Quantum $\mathcal{W}$-algebras and elliptic algebras," *Comm. Math. Phys.* **178** (1996), 653–678 (q-alg/9508009).

21. S. Lukyanov and Y. Pugai, "Multipoint local height probabilities in the integrable RSOS model," *Nuclear Phys.* **B473** [FS] (1996), 631–658 (hep-th/9602074).

22. V. Brazhnikov and S. Lukyanov, "Angular quantization and form factors in massive integrable models," *Nuclear Phys.* **B512** (1998), 616–636 (hep-th/9707091).

23. O. Foda, K. Iohara, M. Jimbo, R. Kedem, T. Miwa, and H. Yan, "An elliptic quantum algebra for $\widehat{\mathfrak{sl}}_2$," *Lett. Math. Phys.* **32** (1994), 259–268 (hep-th/9403094); "Notes on highest weight modules of the elliptic algebra $\mathcal{A}_{q,p}(\widehat{\mathfrak{sl}}_2)$," *Prog. Theor. Phys. Suppl.* **118** (1995), 1–34 (hep-th/9405058).

24. G. Felder, "Elliptic quantum groups," *Proc. ICMP*, Paris 1994, Internat. Press, 1995, 211–218 (hep-th/9412207).

25. C. Frønsdal, "Generalization and exact deformations of quantum groups," *Publ. RIMS* **33** (1997), 91–149 (q-alg/9606020); "Quasi-Hopf deformation of quantum groups," *Lett. Math. Phys.* **40** (1997), 117–134 (q-alg/9611028).

26. M. Jimbo, H. Konno, S. Odake, and J. Shiraishi, "Quasi-Hopf twistors for elliptic quantum groups," *Transformation Groups* **4** (1999), 303–327 (q-alg/9712029).

27. V.G. Kac, "Contravariant form for infinite-dimensional Lie Algebras and superalgebras," in *Group Theoretical Methods in Physics*, Lecture Notes in Phys., vol. 94, Springer, 1979, 441–445.

28. A. Rocha-Caridi, "Vacuum vector representations of the Virasoro algebra," in *Vertex Operators in Mathematics and Physics*, MSRI Publications #3, Springer, 1984, 451–473.

29. D. Friedan, Z. Qiu, and S. Shenker, "Conformal invariance, unitarity, and critical exponents in two dimensions," *Phys. Rev. Lett.* **52** (1984), 1575–1578.

30. Vl.S. Dotsenko and V.A. Fateev, "Four-point correlation functions and the operator algebra in $2d$ conformal invariant theories with central charge $c \leq 1$," *Nuclear Phys.* **B251** [FS13] (1985), 691–734.

31. M. Kato and S. Matsuda, "Null field construction in conformal and superconformal algebras," *Adv. Stud. Pure Math.* **16** (1988), 205–254.

32. H. Awata, Y. Matsuo, S. Odake, and J. Shiraishi, "Excited states of Calogero–Sutherland model and singular vectors of the W_N algebra", *Nuclear Phys.* **B449** (1995), 347–374 (hep-th/9503043).

33. B.L. Feigin and D.B. Fuchs, Moscow preprint (1983); "Differential operators on the line and Verma modules over the Virasoro algebra," *Funct. Anal. Appl.* **16** (1982), 114–126; "Verma modules over the Virasoro algebra," *Funct. Anal. Appl.* **17** (1983), 241–242.

34. G. Felder, "BRST approach to minimal models," *Nuclear Phys.* **B317** (1989), 215–236; Erratum *ibid.* **B324** (1989), 548.

35. K. Mimachi and Y. Yamada, "Singular vectors of the Virasoro algebra in terms of Jack symmetric polynomials", *Comm. Math. Phys.* **174** (1995), 447–455; "Singular vectors of Virasoro algebra in terms of Jack symmetric polynomials," *Kokyuroku* **919** (1995), 68–78 (in Japanese).

36. I.G. Macdonald, *Symmetric Functions and Hall Polynomials*, 2nd ed., Oxford Univ. Press, 1995.

37. R.P. Stanley, "Some combinatorial properties of Jack symmetric functions," *Adv. Math.* **77** (1989), 76–115.

38. Z.N.C. Ha, "Fractional statistics in one-dimension: View from exactly solvable model," *Nuclear Phys.* **B435** [FS] (1995), 604–636 (cond-mat/9410101).

39. H. Awata, H. Kubo, S. Odake, and J. Shiraishi, "Virasoro-type symmetries in solvable models," preprint hep-th/9612233, to appear in CRM Series in Mathematical Physics, Springer.

40. P. Bouwknegt and K. Pilch, "The deformed Virasoro algebra at roots of unity," *Comm. Math. Phys.* **196** (1998), 249–288 (q-alg/9710026).

41. M. Jimbo, M. Lashkevich, T. Miwa, and Y. Pugai, "Lukyanov's screening operators for the deformed Virasoro algebra," *Phys. Lett.* **A229** (1997), 285–292 (hep-th/9607177).

42. S.N.M. Ruijsenaars and H. Schneider, "A new class of integrable systems and its relation to solitons," *Ann. Physics* **170** (1986), 370–405; S.N.M. Ruijsenaars, "Complete integrability of relativistic Calogero–Moser systems and elliptic function identities," *Comm. Math. Phys.* **110** (1987), 191–213.

43. H. Konno, "Relativistic Calogero–Sutherland model: Spin generalization, quantum affine symmetry and dynamical correlation functions," *J. Phys.* **A29** (1996), L191–L198 (hep-th/9508016); "Dynamical correlation functions and finite size scaling in Ruijsenaars–Schneider model," *Nuclear Phys.* **B473** (1996), 579–600 (hep-th/9606012).

44. H.W. Braden and R. Sasaki, "The Ruijsenaars–Schneider model," *Prog. Theor. Phys.* **97** (1997), 1003–1018 (hep-th/9702182).

45. H. Awata, S. Odake, and J. Shiraishi, "Integral representations of the Macdonald symmetric polynomials", *Comm. Math. Phys.* **179** (1996), 647–666 (q-alg/9506006).

46. Y. Hara, M. Jimbo, H. Konno, S. Odake, and J. Shiraishi, "Free field approach to the dilute A_L models," *J. Math. Phys.* **40** (1999), 3791–3826 (math.QA/9902150).

47. O. Foda, M. Jimbo, T. Miwa, K. Miki, and A. Nakayashiki, "Vertex operators in solvable lattice models," *J. Math. Phys.* **35** (1994) 13–46 (hep-th/9305100).

48. M. Jimbo and T. Miwa, *Algebraic Analysis of Solvable Lattice Models.* CBMS Regional Conference Series in Mathematics, vol. 85, Amer. Math. Soc., 1994.

49. R.J. Baxter, "Partition function of the eight-vertex lattice model," *Ann. Physics* **70** (1972), 193–228.

50. A. Belavin, "Dynamical symmetry of integrable quantum systems," *Nuclear Phys.* **B180** [FS2] (1981), 189–200.

51. E. K. Sklyanin, "Some algebraic structure connected with the Yang–Baxter equation," *Funct. Anal. Appl.* **16** (1982), 263–270.

52. G. E. Andrews, R. J. Baxter and P. J. Forrester, "Eight-vertex SOS model and generalized Rogers–Ramanujan-type identities," *J. Stat. Phys.* **35** (1984), 193–266.

53. E. Date, M. Jimbo, T. Miwa and M. Okado, "Fusion of the eight vertex SOS model," *Lett. Math. Phys.* **12** (1986), 209–215.

54. B. Enriquez and G. Felder, "Elliptic quantum groups $E_{\tau,\eta}(\widehat{\mathfrak{sl}}_2)$ and quasi-Hopf algebras," preprint q-alg/9703018.

55. H. Konno, "An elliptic algebra $U_{q,p}(\widehat{\mathfrak{sl}}_2)$ and the fusion RSOS model," *Comm. Math. Phys.* **195** (1998), 373–403 (q-alg/9709013).

56. V. G. Kac, *Infinite-dimensional Lie algebra*, 3rd ed., Cambridge Univ. Press, 1990.

57. T. Tanisaki, "Killing forms, Harish-Chandra isomorphisms, and universal R-matrices for quantum algebras," *Internat. J. Modern Phys.* **A7** (1992), Suppl. 1B, 941–961.

58. C. Frønsdal and A. Galindo, "8-vertex correlation functions and twist covariance of q-KZ equation," preprint q-alg/9709028.

59. I.B. Frenkel and N.Yu. Reshetikhin, "Quantum affine algebras and holonomic difference equations," *Comm. Math. Phys.* **146** (1992), 1–60.

60. M. Idzumi, K. Iohara, M. Jimbo, T. Miwa, T. Nakashima, and T. Tokihiro, "Quantum affine symmetry in vertex models," *Internat. J. Modern Phys.* **A8** (1993) 1479–1511 (hep-th/9208066).

61. M. Jimbo, H. Konno, S. Odake, and J. Shiraishi, "Elliptic algebra $U_{q,p}(\widehat{\mathfrak{sl}}_2)$: Drinfeld currents and vertex operators," *Comm. Math. Phys.* **199** (1999), 605–647 (math.QA/9802002).

62. V. G. Drinfeld, "A new realization of Yangians and quantized affine algebras," *Soviet Math. Dokl.* **36** (1988), 212–216.

63. I.B. Frenkel and N. Jing, "Vertex representations of quantum affine algebra," *Proc. Natl. Acad. Sci. USA* **85** (1988), 9373–9377.

64. H. Awata, S. Odake, and J. Shiraishi, "Free boson realization of $U_q(\widehat{\mathfrak{sl}}_N)$", *Comm. Math. Phys.* **162** (1994), 61–83 (hep-th/9305146).

65. B. Hou and W. Yang, "Dynamically twisted algebra $A_{q,p;\hat{\pi}}(\widehat{gl}_2)$ as current algebra generalizing screening currents of q-deformed Virasoro algebra," *J. Phys. A: Math. Gen.* **31** (1998), 5349–5369 (q-alg/9709024).

66. T. Miwa and R. Weston, "Boundary ABF models," *Nuclear Phys.* **B486** [PM] (1997), 517–545 (hep-th/9610094).

67. M. Jimbo and J. Shiraishi, "A coset-type construction for the deformed Virasoro algebra," *Lett. Math. Phys.* **43** (1998), 173–185; Errata, ibid. **44** (1998), 349–352 (q-alg/9709037).

68. S.O. Warnaar, B. Nienhuis, and K.A. Seaton, "New construction of solvable lattice models including an Ising model in a field," *Phys. Rev. Lett.* **69** (1992), 710–712.

69. S. O. Warnaar, P. A. Pearce, K. A. Seaton and B. Nienhuis, "Order parameters of the dilute A models," *J. Stat. Phys.* **74** (1994), 469–531 (hep-th/9305134).

70. A. Kuniba, "Exact solution of solid-on-solid models for twisted affine Lie algebras $A^{(2)}_{2n}$ and $A^{(2)}_{2n-1}$," *Nuclear Phys.* **B355** (1991), 801–821.

71. P. Bouwknegt and K. Schoutens, "$\mathcal{W}$-symmetry in conformal field theory", *Phys. Rep.* **223** (1993), 183–276 (hep-th/9210010).

72. H. Awata, H. Kubo, S. Odake, and J. Shiraishi, "Quantum deformation of the W_N Algebra," preprint q-alg/9612001, to appear in CRM Series in Mathematical Physics, Springer.

73. V. E. Korepin, N. M. Bogoliubov and A. G. Izergin, *Quantum Inverse Scattering Method and Correlation Functions*, Cambridge Univ. Press, 1993.

74. E. Frenkel and N. Reshetikhin, "Deformations of W-algebras associated to simple Lie algebras," preprint q-alg/9708006.

75. B. Feigin, M. Jimbo, T. Miwa, A. Odesskii, and Y. Pugai, "Algebra of screening operators for the deformed W_n algebra," *Comm. Math. Phys.* **191** (1998), 501–541 (q-alg/9702029).

76. J. Avan, L. Frappat, M. Rossi, and P. Sorba, "Poisson structures on the center of the elliptic algebra $\mathcal{A}_{q,p}(\widehat{\mathfrak{sl}}(2)_c)$," *Phys. Lett.* **A235** (1997), 323–334 (q-alg/9705012); "New $\mathcal{W}_{q,p}(\mathfrak{sl}(2))$ algebras from the elliptic algebra $\mathcal{A}_{q,p}(\widehat{\mathfrak{sl}}(2)_c)$," *Phys. Lett.* **A239** (1998), 27–35 (q-alg/9706013); "Deformed $\mathcal{W}_N$ algebras from elliptic $\mathfrak{sl}(N)$ algebras," *Comm. Math. Phys.* **199** (1999), 697–728 (math.QA/9801105); "Universal construction of $W_{p,q}$ algebras," preprint math.QA/9807048.

77. E. Frenkel, N. Reshetikhin, and M.A. Semenov-Tian-Shansky, "Drinfeld–Sokolov reduction for difference operators and deformations of $\mathcal{W}$-Algebras I. The case of Virasoro algebra," preprint q-alg/9704011; M.A. Semenov-Tian-Shansky and A.V. Sevostyanov, "Drinfeld–Sokolov reduction for difference operators and deformations of $\mathcal{W}$-algebras. II. General semisimple case," preprint q-alg/9702016.

78. P. Bouwknegt and K. Pilch, "On deformed $\mathcal{W}$-algebras and quantum affine algebras," *Adv. Theor. Math. Phys.* **2** (1998), 125–165 (math.QA/9801112).

79. E. Frenkel and N. Reshetikhin, "The q-characters of representations of quantum affine algebras and deformations of W-algebras," preprint math.QA/9810055.

80. Y. Asai, M. Jimbo, T. Miwa, and Y. Pugai, "Bosonization of vertex operators for the $A^{(1)}_{N-1}$ face model," *J. Phys.* **A29** (1996), 6595–6616 (hep-th/9606095).

81. M. Jimbo, H. Konno, S. Odake, Y. Pugai and J. Shiraishi, "Remarks on the free field construction for the ABF model in regime II," in preparation.

82. V.G. Knizhnik and A.B. Zamolodchikov, "Current algebra and Wess–Zumino model in two dimensions," *Nuclear Phys.* **B247** (1984), 83–103.

83. A. Tsuchiya and Y. Kanie, "Vertex operators in conformal field theory on P^1 and monodromy representations of braid group," *Adv. Stud. Pure Math.* **16** (1988), 297–372; Errata, ibid. **19** (1989), 675–682.

84. P. Goddard, A. Kent, and D. Olive, "Unitary representations of the Virasoro and super-Virasoro algebras," *Comm. Math. Phys.* **103** (1986), 105–119.

85. K. Aomoto, Y. Kato and K. Mimachi, "A solution of Yang–Baxter equation as connection coefficients of a holonomic q-difference system," *Duke Math. J.* **65** (1992), 7–15.

86. F.A. Smirnov, "Dynamical symmetries of massive integrable models. I," *Internat. J. Modern Phys.* **A7** Suppl. 1B (1992), 813–837; II, 839–858.

87. F.A. Smirnov, *Form Factors in Completely Integrable Models of Quantum Field Theory*, World Scientific, 1992.

88. S. Lukyanov, "Free field representation for massive integrable models," *Comm. Math. Phys.* **167** (1995), 183–226 (hep-th/9509037).

89. S. Lukyanov, "A note on the deformed Virasoro algebra," *Phys. Lett.* **B367** (1996), 121–125 (hep-th/9509037).

90. M. Jimbo, H. Konno, and T. Miwa, "Massless XXZ model and degeneration of the elliptic algebra $\mathcal{A}_{q,p}(\widehat{\mathfrak{sl}}_2)$," *Deformation theory and symplectic geometry*, Eds. D. Sternheimer, J. Rawnsley and S. Gutt, Math. Phys. Studies, vol. 20, Kluwer, 1997, 117–138 (hep-th/9610079); H. Konno, "Degeneration of the elliptic algebra $\mathcal{A}_{q,p}(\widehat{\mathfrak{sl}}_2)$ and form factors in the sine-Gordon theory," preprint hep-th/9701034, to appear in CRM Series in Mathematical Physics, Springer.

91. S. Lukyanov, "Form factors of exponential fields in the sine-Gordon model," *Mod. Phys. Lett.* **A12** (1997) 2543–2550 (hep-th/9703190); V. Fateev, D. Fradkin, S. Lukyanov, A. Zamolodchikov, and Al. Zamolodchikov, "Expectation values of descendant fields in the sine-Gordon model," *Nuclear Phys.* **B540** [FS] (1999), 587–609 (hep-th/9807236); V. Fateev, S. Lukyanov, A. Zamolodchikov, and Al. Zamolodchikov, "Expectation values of local fields in the Bullough–Dodd model and integrable conformal field theories," *Nuclear Phys.* **B516** [FS] (1998), 652–674 (hep-th/9709034);

S. Lukyanov, "Form-factors of exponential fields in the affine $A_{N-1}^{(1)}$ Toda model," *Phys. Lett.* **B408** (1997), 192–200 (hep-th/9704213).

92. T. Oota, "Functional equations of form factors for diagonal scattering theories," *Nuclear Phys.* **B466** [FS] (1996), 361–382.

93. V.V. Bazhanov, S.L. Lukyanov and A.B. Zamolodchikov, "Integrable structure of conformal field theory, quantum KdV theory and thermodynamic Bethe ansatz," *Comm. Math. Phys.* **177** (1996), 381–398 (hep-th/9412229); "Integrable structure of conformal field theory. II. Q-operator and DDV equation," *Comm. Math. Phys.* **190** (1997), 247–278 (hep-th/9604044); "Integrable structure of conformal field theory. III. The Yang Baxter relation," *Comm. Math. Phys.* **200** (1999), 297–324 (hep-th/9805008).

94. M. Lashkevich and Y. Pugai, "Free field construction for correlation functions of the eight-vertex model," *Nuclear Phys.* **B516** (1998), 623–651 (hep-th/9710099).

95. Y.-Z. Zhang and M.D. Gould, "Quasi-Hopf superalgebras and elliptic quantum supergroups," preprint math.QA/9809156.

96. D. Arnaudon, E. Buffenoir, E. Ragoucy, and Ph. Roche, "Universal solutions of quantum dynamical Yang–Baxter equations," *Lett. Math. Phys.* **44** (1998), 201–214 (q-alg/9712037).

97. Y. Hara, "Free field realization of vertex operators for level two modules of $U_q(\widehat{\mathfrak{sl}}_2)$," *J. Phys.* **A31** (1998), 8483–8494 (math.QA/9809078).

98. J. Ding and B. Feigin, "Quantized W-algebra of $\mathfrak{sl}(2,1)$: A construction from the quantization of screening operators," preprint math.QA/9801084.

99. P. Bouwknegt, J. McCarthy, and K. Pilch, "Quantum group structure in the Fock space resolutions of $\widehat{\mathfrak{sl}}(n)$ representations," *Comm. Math. Phys.* **131** (1990), 125–155; "Free field approach to 2-dimensional conformal field theories," *Prog. Theor. Phys. Suppl.* **102** (1990), 67–135.

100. L. Clavelli and J.A. Shapiro, "Pomeron factorization in general dual models," *Nuclear Phys.* **B57** (1973), 490–535. (see Appendix C.1)

A Some Formulas

In this appendix we give a summary of notations and formulas used throughout this lecture.

A.1 Some Functions

Let us fix x, r, r^*. The following functions are used in this lecture:

$$[n]_x = \frac{x^n - x^{-n}}{x - x^{-1}}, \tag{A.1}$$

$$(z; p_1, \ldots, p_k)_\infty = \prod_{n_1,\ldots,n_k=0}^{\infty} (1 - p_1^{n_1} \cdots p_k^{n_k} z), \tag{A.2}$$

$$(z_1, \ldots, z_n; p_1, \ldots, p_k)_\infty = \prod_{j=1}^{n} (z_j; p_1, \ldots, p_k)_\infty, \tag{A.3}$$

$$\Theta_p(z) = (p, z, pz^{-1}; p)_\infty = \sum_{n\in\mathbb{Z}} (-1)^n z^n p^{n(n-1)/2}, \tag{A.4}$$

$$[u] = x^{u^2/r-u}\Theta_{x^{2r}}(x^{2u}), \tag{A.5}$$

$$[u]^* = x^{u^2/r^*-u}\Theta_{x^{2r^*}}(x^{2u}). \tag{A.6}$$

$$[u]_+ = x^{u^2/r-u}\Theta_{x^{2r}}(-x^{2u}), \tag{A.7}$$

$$[u]_+^* = x^{u^2/r^*-u}\Theta_{x^{2r^*}}(-x^{2u}). \tag{A.8}$$

$[u]$ and $[u]^*$, $[u]_+$ and $[u]_+^*$, are related by

$$[u]^* = [u]\big|_{r\to r^*}, \quad [u]_+^* = [u]_+\big|_{r\to r^*}. \tag{A.9}$$

$[u]$ satisfies

$$[-u] = -[u], \quad [u+r] = -[u], \quad [u+\tau] = -[u]e^{2\pi i(u+\tau/2)/r} \qquad (x = e^{\pi i/\tau}), \tag{A.10}$$

and the Riemann identity

$$\begin{aligned}
&[2u_1][2u_2][2u_3][2u_4] \\
&\quad = [u_1+u_2+u_3+u_4][u_1-u_2-u_3+u_4] \\
&\qquad\quad [u_1+u_2-u_3-u_4][u_1-u_2+u_3-u_4] \\
&\quad + [-u_1+u_2+u_3+u_4][u_1-u_2+u_3+u_4] \\
&\qquad\quad [u_1+u_2-u_3+u_4][u_1+u_2+u_3-u_4].
\end{aligned} \tag{A.11}$$

Lemma 4 in [41] is

$$\begin{aligned}
&\frac{1}{m!}\sum_{\sigma\in S_m}\prod_{i=1}^{m}[v_{\sigma(i)} - 2i + 2] \cdot \prod_{\substack{1\le i<j\le m \\ \sigma(i)>\sigma(j)}} \frac{[v_{\sigma(i)} - v_{\sigma(j)} - 1]}{[v_{\sigma(i)} - v_{\sigma(j)} + 1]} \\
&\quad = \frac{1}{m!}\prod_{i=1}^{m}\frac{[i]}{[1]} \cdot \prod_{\substack{i,j=1 \\ i<j}}^{m} \frac{[v_i - v_j]}{[v_i - v_j - 1]} \cdot \prod_{i=1}^{m}[v_i - m + 1].
\end{aligned} \tag{A.12}$$

Along with the additive variable u, we often use the multiplicative variable

$$z = x^{2u}, \quad z_j = x^{2u_j}, \tag{A.13}$$

and the following abbreviation for an integration measure

$$\underline{dz} = \frac{dz}{2\pi i z}, \quad \underline{dz}_j = \frac{dz_j}{2\pi i z_j}. \tag{A.14}$$

$(z;p)_\infty$, $\Theta_p(z)$ and $[u]$ have no poles and have simple zeros,

$$(z;p)_\infty : z = p^{-n} \quad (n \in \mathbb{Z}_{\geq 0}), \tag{A.15}$$

$$\Theta_p(z) : z = p^m \quad (m \in \mathbb{Z}), \tag{A.16}$$

$$[u] : u = rm \quad (m \in \mathbb{Z}) \quad (\text{i.e., } z = x^{2rm}). \tag{A.17}$$

For $m \in \mathbb{Z}$ we have

$$\oint_{u=rm+a} \underline{dz} \frac{1}{[u-a]} f(u) = \frac{(-1)^{m-1}}{(x^{2r};x^{2r})^3_\infty} f(rm+a), \tag{A.18}$$

where $f(u)$ is regular at $u = rm + a$, and $1/[u-a]f(u)$ does not contain a fractional power of z.

A.2 Delta Function

Delta function $\delta(z)$ is a formal power series

$$\delta(z) = \sum_{n \in \mathbb{Z}} z^n, \tag{A.19}$$

and has the property

$$\oint_0 \frac{dz}{2\pi i z} f(z) \delta\left(\frac{z}{a}\right) = f(a). \tag{A.20}$$

Let $F(z;q)$ be the following Taylor series in z,

$$F(z;q) = \prod_{i=1}^{n} \frac{1 - q^{\beta_i} z}{1 - q^{\alpha_i} z}. \tag{A.21}$$

Here $\frac{1}{1-z}$ means $\sum_{m \geq 0} z^m$ and parameters α_i, β_i satisfy

$$\sum_{i=1}^{n} \alpha_i = \sum_{i=1}^{n} \beta_i, \quad \text{where the } \alpha_i \text{ are all distinct.} \tag{A.22}$$

Then we have a formula,

$$F(z;q) - F(z^{-1};q^{-1}) = \sum_{i=1}^{n} a_i \delta(q^{\alpha_i} z), \quad a_i = \frac{\prod_{j=1}^{n}(1 - q^{\beta_j - \alpha_i})}{\prod_{\substack{j=1 \\ j \neq i}}^{n}(1 - q^{\alpha_j - \alpha_i})}. \tag{A.23}$$

A.2.1 Example

For $a \neq a'$, we have

$$\frac{(1-q^{2b}z)(1-q^{2(a+a'-b)}z)}{(1-q^{2a}z)(1-q^{2a'}z)} - \frac{(1-q^{-2b}z^{-1})(1-q^{-2(a+a'-b)}z^{-1})}{(1-q^{-2a}z^{-1})(1-q^{-2a'}z^{-1})} \\ -(q-q^{-1})\frac{[b-a]_q[b-a']_q}{[a-a']_q}\big(\delta(q^{2a}z)-\delta(q^{2a'}z)\big). \quad \text{(A.24)}$$

By taking a limit $a' \to a$, we get

$$\frac{(1-q^{2b}z)(1-q^{2(2a-b)}z)}{(1-q^{2a}z)^2} - \frac{(1-q^{-2b}z^{-1})(1-q^{-2(2a-b)}z^{-1})}{(1-q^{-2a}z^{-1})^2} \\ = -(q-q^{-1})^2[b-a]_q^2 q^{2a}z\delta'(q^{2a}z). \quad \text{(A.25)}$$

A.3 Some Summations

As it is well known in statistical mechanics, in order to calculate the following summation

$$\sum_{\substack{\{k_i \geq 0\} \\ sum_{i=1}^N ik_i = N}} f(k_1, k_2, \dots), \quad \text{(A.26)}$$

it is convenient to introduce its generating function

$$\sum_{N=0}^{\infty} \sum_{\substack{\{k_i\} \\ \sum_i ik_i = N}} f(k_1, k_2, \dots) y^N = \sum_{k_1=0}^{\infty} \sum_{k_2=0}^{\infty} \cdots f(k_1, k_2, \dots) y^{k_1+2k_2+\cdots}. \quad \text{(A.27)}$$

We have two formulas:

$$\prod_{\substack{\{k_i\} \\ \sum_i ik_i = N}} \prod_i k_i! = \prod_{\substack{\{k_i\} \\ \sum_i ik_i = N}} \prod_i i^{k_i}, \quad \text{(A.28)}$$

$$\sum_{\substack{\{k_i\} \\ \sum_i ik_i = N}} k_l = \sum_{\substack{k \geq 1 \\ lk \leq N}} p(N - lk) \qquad (1 \leq l \leq N), \quad \text{(A.29)}$$

where $p(N)$ is the number of partition (see (2.13)). Generating function of log(A.28) is $\prod_{n>0} 1/(1-y^n) \times \sum_{i>0} y^i/(1-y^i) \log i$, and that of (A.29) is $\prod_{n>0} 1/(1-y^n) \times y^l/(1-y^l)$. From (A.29) we obtain

$$\prod_{\substack{\{k_i\} \\ \sum_i ik_i = N}} \prod_i F(i)^{k_i} = \prod_{\substack{l,k \geq 1 \\ lk \leq N}} F(l)^{p(N-lk)}, \quad \text{(A.30)}$$

where $F(z)$ is any function.

A.4 Some Integrals

We summarize the relations among the following integral with various contours [30, 31, 34, 99],

$$I = \int dz_1 \dots dz_m\, F,$$
$$F = \prod_{\substack{i,j=1\\ i<j}}^{m} (z_i - z_j)^{2\alpha} \cdot \prod_{i=1}^{m} z_i^{\alpha'} \cdot f(z_1, \dots, z_m). \tag{A.31}$$

Here $f(z_1, \dots, z_m)$ is a symmetric function and has no pole at $z_i = z_j$. Parameters α and α' are assumed to satisfy the condition (i) (and (ii) for (A.47)),

$$\text{(i)} \qquad m(m-1)\alpha + m\alpha' \in \mathbb{Z}, \tag{A.32}$$
$$\text{(ii)} \qquad m(m-1)\alpha + m\alpha' \in m\mathbb{Z}. \tag{A.33}$$

Let us consider the contours illustrated in Figure 5.1:

$$I_{\mathrm{KM}} = \oint dz_1 \int_0^{z_1} dz_2 \int_0^{z_2} dz_3 \cdots \int_0^{z_{m-1}} dz_m\, F, \tag{A.34}$$
$$I_{\mathrm{F}} = \oint dz_1 \int dz_2 \dots dz_m\, F, \tag{A.35}$$
$$I_{\mathrm{BMP}} = \int dz_1 \dots dz_m\, F, \tag{A.36}$$
$$I'_{\mathrm{BMP}} = \int dz_1 \dots dz_m\, F, \quad (z_j = r_j e^{i\theta_j},\ 0 < \theta_1 < \theta_2 < \cdots < \theta_m < 2\pi), \tag{A.37}$$
$$I_{\mathrm{DF}} = \int dz_1 \dots dz_m\, F, \tag{A.38}$$
$$I'_{\mathrm{DF}} = \int_0^1 dz_1 \int_0^{z_1} dz_2 \int_0^{z_2} dz_3 \cdots \int_0^{z_{m-1}} dz_m\, F. \tag{A.39}$$

For I_{KM} and I_{F} the contour for z_1 closes.

By analytic continuation these integrals are related as follows:

$$\begin{aligned} I_{\mathrm{F}} &= (-1)^{m-1} \prod_{j=0}^{m-2} (1 - a'a^j) \cdot \prod_{j=1}^{m-1} (1 + a + a^2 + \cdots + a^{j-1}) \cdot I_{\mathrm{KM}}, \\ &= \prod_{j=0}^{m-2} (a'a^j - 1) \cdot \prod_{j=1}^{m-1} \frac{a^j - 1}{a - 1} \cdot I_{\mathrm{KM}}, \end{aligned} \tag{A.40}$$

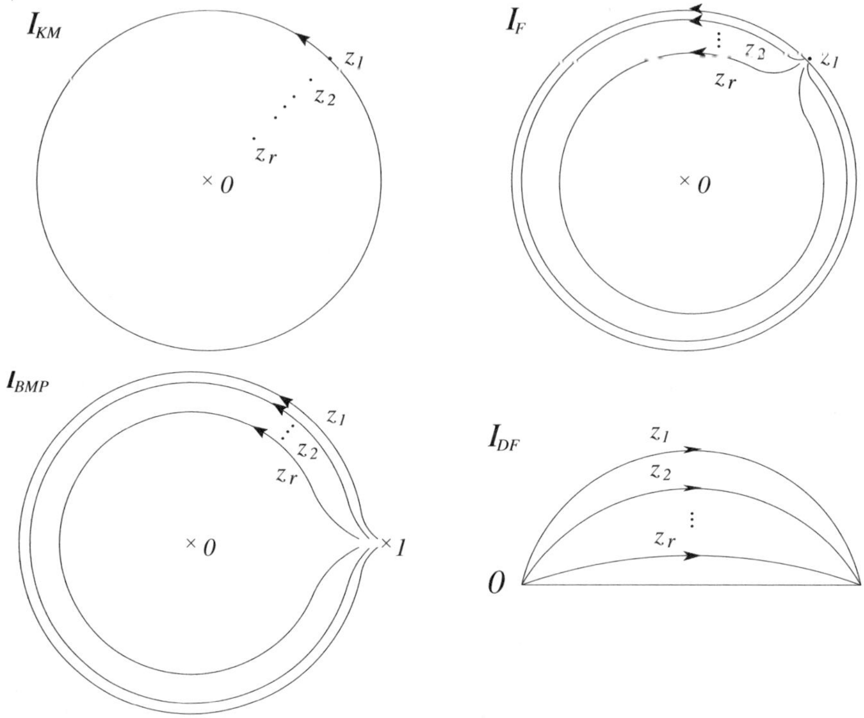

FIGURE 5.1. Some contour examples

$$I_{\mathrm{BMP}} = \prod_{j=1}^{m}(1+a+a^2+\cdots+a^{j-1})\cdot I'_{\mathrm{BMP}}$$
$$= \prod_{j=1}^{m}\frac{a^j-1}{a-1}\cdot I'_{\mathrm{BMP}}, \tag{A.41}$$

$$I_{\mathrm{BMP}} = (-1)^m\prod_{j=0}^{m-1}(1-a'a^j)\cdot I_{DF} = \prod_{j=0}^{m-1}(a'a^j-1)\cdot I_{DF}, \tag{A.42}$$

$$I_{\mathrm{DF}} = \prod_{j=1}^{m}(1+a+a^2+\cdots+a^{j-1})\cdot I'_{DF}$$
$$= \prod_{j=1}^{m}\frac{a^j-1}{a-1}\cdot I'_{DF}, \tag{A.43}$$

$$I_{\mathrm{F}} = \prod_{j=1}^{m-1}(1+a+a^2+\cdots+a^{j-1})$$
$$\times\left(1+a'a^{m-1}+(a'a^{m-1})^2+\cdots+(a'a^{m-1})^{m-1}\right)\cdot I'_{\mathrm{BMP}}, \tag{A.44}$$

where a and a' are

$$a = e^{2\pi i\alpha}, \quad a' = e^{2\pi i\alpha'}, \tag{A.45}$$

and we remark that

$$a'a^{m-1} = 1 \text{ for (ii)}, \quad a'a^{m-1} \neq 1 \text{ for (i) but not (ii)}. \tag{A.46}$$

Under the condition (ii), from (A.41) and (A.44), we have

$$I_{\mathrm{BMP}} = \frac{1}{m}\frac{a^m - 1}{a - 1} I_F. \tag{A.47}$$

We give an example of this kind of integral:

$$\begin{aligned} f(z_1, \dots, z_m) &= \prod_{i=1}^{m} (1 - z_i)^{\alpha''}, \\ I'_{\mathrm{DF}} &= \prod_{j=1}^{m} \frac{\Gamma(j\alpha)}{\Gamma(\alpha)} \frac{\Gamma((j-1)\alpha + \alpha' + 1)\Gamma((j-1)\alpha + \alpha'' + 1)}{\Gamma((m-2+j)\alpha + \alpha' + \alpha'' + 2)}. \end{aligned} \tag{A.48}$$

A.5 Hausdorff Formula

For two operators A and B, we have a formula

$$e^A B e^{-A} = e^{\mathrm{ad}\, A} B, \quad (\mathrm{ad}\, A) B = [A, B], \tag{A.49}$$

because two functions of t, $e^{tA} B e^{-tA}$ and $e^{t\, \mathrm{ad}\, A} B$, satisfy the same differential equation $f'(t) = Af(t) - f(t)A$ and the same initial condition $f(0) = B$. When A and $[A, B]$ commute each other, we have $e^A B e^{-A} = B + [A, B]$ and so $e^A e^B e^{-A} = e^{B + [A,B]}$. Therefore we have a formula

$$\big[A, [A, B]\big] = \big[B, [A, B]\big] = 0 \iff e^A e^B = e^{[A,B]} e^B e^A, \tag{A.50}$$

which is a special case of the Campbell–Baker–Hausdorff formula $e^A e^B = e^{A + B + [A,B]/2 + \cdots}$. A harmonic oscillator $[a, a^\dagger] = 1$ and this formula are basic tools of free field realization.

A.6 Trace Technique

Let us consider one free boson oscillator, $[a, a^\dagger] = 1$. The Fock space $\mathcal{F}$ is generated by $|0\rangle$ ($a|0\rangle = 0$), and its orthonormal basis is $|n\rangle = (n!)^{-1/2} \times a^{\dagger\, n}|0\rangle$ ($n \geq 0$). A trace over the Fock space ($\mathrm{tr}_{\mathcal{F}}\, \mathcal{O} = \sum_{n \geq 0} \langle n|\mathcal{O}|n\rangle$) can be expressed as a vacuum to vacuum amplitude. Let us introduce another oscillator b, $[b, b^\dagger] = 1$, which commute with a and satisfy $b|0\rangle = 0$. Then we have the following Clavelli-Shapiro's trace formula [100],

$$\mathrm{tr}_{\mathcal{F}}\big(y^{a^\dagger a} \mathcal{O}(a, a^\dagger)\big) = \frac{1}{1 - y} \langle 0|\mathcal{O}(d, \bar{d})|0\rangle. \tag{A.51}$$

Here d, $\bar{d}$ are

$$d = \frac{1}{1 - y} a + b^\dagger, \quad \bar{d} = a^\dagger + \frac{y}{1 - y} b, \quad [d, \bar{d}] = 1. \tag{A.52}$$

Especially we have

$$\mathrm{tr}_{\mathcal{F}}(y^{a^\dagger a}e^{Aa^\dagger}e^{Ba}) = \frac{1}{1-y}e^{ABy/(1-y)}, \tag{A.53}$$

where A, B are constants.

For reader's convenience, we give three direct proofs of (A.53).

(i) The first method uses a cyclic property of trace. We have

$$\begin{aligned} f(A,B) &= \mathrm{tr}_{\mathcal{F}}(y^{a^\dagger a}e^{Aa^\dagger}e^{Ba}) \qquad\qquad (|y|<1) \\ &= \mathrm{tr}_{\mathcal{F}}(e^{Aya^\dagger}y^{a^\dagger a}e^{Ba}) = \mathrm{tr}_{\mathcal{F}}(y^{a^\dagger a}e^{Ba}e^{Aya^\dagger}) \\ &= e^{ABy}\,\mathrm{tr}_{\mathcal{F}}(y^{a^\dagger a}e^{Aya^\dagger}e^{Ba}) = e^{ABy}f(Ay,B), \end{aligned} \tag{A.54}$$

and

$$\begin{aligned} f(A,B) &= e^{-AB}\,\mathrm{tr}_{\mathcal{F}}(y^{a^\dagger a}e^{Ba}e^{Aa^\dagger}) = e^{-AB}\,\mathrm{tr}_{\mathcal{F}}(e^{By^{-1}a}y^{a^\dagger a}e^{Aa^\dagger}) \\ &= e^{-AB}\,\mathrm{tr}_{\mathcal{F}}(y^{a^\dagger a}e^{Aa^\dagger}e^{By^{-1}a}) = e^{-AB}f(A,By^{-1}). \end{aligned} \tag{A.55}$$

(A.54) implies

$$f(A,B) = e^{AB(y+y^2+\cdots)}f(Ay^\infty,B) = e^{ABy/(1-y)}f(0,B),$$

and (A.55) implies

$$f(0,B) = f(0,By) = f(0,By^\infty) = f(0,0).$$

Since $f(0,0) = 1/(1-y)$, we obtain (A.53).

(ii) The second method uses a coherent state $|\alpha) = e^{\alpha a^\dagger}|0\rangle$ ($\alpha \in \mathbb{C}$), which satisfies $a|\alpha) = \alpha|\alpha)$, $(\alpha|\alpha') = e^{\bar{\alpha}\alpha'}$ and the completeness condition $1 = \sum_{n\geq 0}|n\rangle\langle n| = \int \frac{d^2\alpha}{\pi}e^{-|\alpha|^2}|\alpha)(\alpha|$, where $d^2\alpha = d\alpha_1\,d\alpha_2$ with $\alpha = \alpha_1 + i\alpha_2$. The trace becomes

$$\mathrm{tr}_{\mathcal{F}}(y^{a^\dagger a}e^{Aa^\dagger}e^{Ba}) = \int \frac{d^2\alpha}{\pi}e^{-|\alpha|^2}(\alpha|y^{a^\dagger a}e^{Aa^\dagger}e^{Ba}|\alpha).$$

Since $(\alpha|y^{a^\dagger a}e^{Aa^\dagger}e^{Ba}|\alpha) = e^{B\alpha}(\alpha|y^{a^\dagger a}e^{Aa^\dagger}|\alpha) = e^{B\alpha}(\alpha|y^{a^\dagger a}|\alpha+A) = e^{B\alpha}\big(\alpha|y(\alpha+A)\big) = e^{B\alpha}e^{\bar{\alpha}y(\alpha+A)}$, by completing the square and performing the Gauss integral, we obtain the result.

(iii) The third method is a direct calculation.

$$\begin{aligned} \mathrm{tr}_{\mathcal{F}}(y^{a^\dagger a}e^{Aa^\dagger}e^{Ba}) &= \sum_{n=0}^{\infty}\frac{1}{n!}\langle 0|a^n y^{a^\dagger a}e^{Aa^\dagger}e^{Ba}a^{\dagger n}|0\rangle \\ &= \sum_{n=0}^{\infty}\frac{1}{n!}y^n\langle 0|a^n e^{Aa^\dagger}(a^\dagger+B)^n|0\rangle \end{aligned}$$

$$
\begin{aligned}
&= \sum_{n=0}^{\infty} \frac{1}{n!} y^n \langle 0 | a^n \sum_{l=0}^{n} \frac{A^{n-l}}{(n-l)!} \binom{n}{l} B^{n-l} a^{\dagger\, n} | 0 \rangle \\
&= \sum_{n=0}^{\infty} y^n \sum_{l=0}^{n} \frac{1}{(n-l)!} \binom{n}{l} (AB)^{n-l}.
\end{aligned}
$$

By interchanging the order of summations $\sum_{n=0}^{\infty} \sum_{l=0}^{n} = \sum_{l=0}^{\infty} \sum_{n=l}^{\infty}$ and shifting $n = m + l$, it becomes

$$
\sum_{m=0}^{\infty} \frac{1}{m!} (ABy)^m \sum_{l=0}^{\infty} \binom{m+l}{l} y^l.
$$

Since $\sum_{l=0}^{\infty} \binom{m+l}{l} y^l$ equals to $(y\partial_y + m) \cdots (y\partial_y + 1)/m! \sum_{l=0}^{\infty} y^l = 1/(1-y)^{m+1}$, we obtain the result.

6

Mesoscopic Physics

B.D. Simons
A. Altland

ABSTRACT The following is a course of lectures on the manifestations of quantum phase coherence phenomena and interaction in mesoscopic structures. In discussing applications to a variety of modern topics in condensed matter physics emphasis is given to methods of statistical field theory.

1 Introduction to Mesoscopic Physics

As students we are usually introduced to the phenomena of solid state physics within the framework of the "nearly-free electron" theory of metals in which electrons are thought to interact only weakly with a *regular* (i.e., ordered) crystalline lattice potential [1]. The interaction between electrons is limited to the quantum mechanical exclusion principle. Yet, in the majority of metals, the Coulomb interaction between electrons is by no means small—the energy associated with Coulomb interaction is typically comparable to the kinetic energy. However, the validity of the noninteracting theory is, in many cases, assured by the *Fermi liquid theory.*

The fundamental principle underlying Fermi liquid theory is one of "adiabatic continuity" [2]: In the absence of an electronic phase transition (such as "Wigner crystallisation" of the electrons, or a "Mott–Hubbard transition" to a magnetic insulating state), a noninteracting ground state evolves smoothly or adiabatically into the interacting ground state as the strength of interaction is increased. An elementary excitation of the noninteracting system represents an "approximate excitation" of the interacting system (i.e., the "lifetime" of an elementary excitation is long).

However, the integrity of a fully ordered lattice potential is more questionable. Electrons in metals and semi-conductors typically experience an irregular lattice potential arising from defects or lattice imperfections, grain boundaries, vacancies, and doped impurities. At length scales in excess of the (temperature dependent) phase coherence length L_φ (i.e., the scale over which the electron dynamics is phase coherent), the transport properties in such a disordered environment can be simply understood within the framework of quasi-classical techniques such as the Kinetic or Boltzmann theory [3]. However, at length scales smaller than L_φ, the quantum phase coherence of the electron degrees of freedom dramatically influence the nature of the dynamics. Here at the "mesoscopic" scale, new physical principles emerge.

What is Mesoscopic Physics? As a subject, Mesoscopic physics involves the domain of length scales in between the microscopic and macroscopic where the influence of quantum phase coherence effects find manifestations in observed physical properties. Yet, as we will see in these lectures, the subject of mesoscopic physics is not confined to the study of electrons in disordered conductors.

Where can one find Mesoscopic structures? The development of modern fabrication techniques combined with the routine availability of millikelvin temperatures in the laboratory have exposed a novel class of materials macroscopic in their constitution, yet smaller than the typical scales over which phase coherence is established. The market forces that drive the semiconductor industry have driven this "nanostructure" technology to a level of unprecedented sophistication. These days, artificial semiconductor devices can be manufactured with a resolution of 100 nm or less (see below). Even "nanotubes" of carbon have been successfully incorporated into small conducting bridges.

Why Study Mesoscopic Physics? Leaving aside the obvious benefits brought about by the continual miniaturisation of quantum devices, such structures present a unique opportunity to study new and fundamental physical phenomena. Where phase coherence is established, the manifestations of quantum mechanics on the observed physical properties is often substantial. Therefore, the fabrication of mesoscopic structures such as low-dimensional, nanostructure or "quantum dot" devices provides a laboratory in which one can explore the fundamental properties of many-particle systems from the influence of disorder to strong electron interaction phenomena.

Aim of These Lectures

Mesoscopic physics is an old subject which, since the late '70's has engaged a vast number of researchers. To embark on a comprehensive review of the entire field would be foolhardy. An attempt to achieve such a goal in just a few lectures would be futile. Instead, the aim of these lectures will be to introduce some of the guiding principles, to highlight some of the more recent developments and generalisations, and, perhaps most importantly, expose many unresolved questions in the field. In doing so, we will exploit novel methods of quantum statistical field theory which will establish useful and, sometimes, surprising connections to other branches of physics from atomic physics to QCD. Finally, a disclaimer: In preparing this course, it is inevitable that the subjective choice of material reflects to a large extent our own polarised interests and for which we should apologise.

The outline of the course is as follows: In the remainder of this chapter we will introduce the central ideas and concepts that shape the subject of mesoscopic physics. In doing so, we aim to review the main questions presented by experiment as well as providing interpretations in terms of simple quasi-classical phenomenology. These qualitative ideas are put on a firm footing in Section 2 where a statistical field theory of weakly disordered metallic conductors is developed within the framework of a supersymmetric nonlinear σ-model. In applying the field theory, particular emphasis is placed on establishing connections which tie together the phenomenology of "weak localization" and "universality" (the random matrix theory).

To broaden the class of applications of the statistical approach, Section 3 is concerned with the generalisation of the quantum field theory to the properties of nonstochastic chaotic quantum structures—"quantum chaos." Here we will stress the importance of the intuition afforded by the characteristic mesoscopic phenomena observed in weakly disordered metallic systems, and point out some of the outstanding problems in the field.

Sections 1, 2 and 3 focus largely on the properties of *noninteracting* quantum structures. Yet, the influence of Coulomb interaction, particularly in the environment of disorder, is strong. In Section 4 we will establish a generalisation of the statistical field theory of disordered conductors to encompass the important effects of Coulomb interaction. Applications of this theory will be limited as much by the current state of the technology, as by the constraints placed on the scope of these lectures.

Phase coherence effects in weakly disordered structures can be usually classified into three universality classes which reflect the fundamental symmetries of the bare Hamiltonian. Yet, such a classification is not exhaustive: Recent attempts to extend theories of disordered systems to encompass superconducting structures and sublattice models have identified novel universality classes which exhibit qualitatively new phenomena. Surprisingly, it is mostly through the study of such systems that unusual connections to other branches of physics have been identified. In Section 5 we introduce and explore the unusual spectral and localization properties associated with these novel symmetry classes pointing out their close connection to the study of lattice QCD, and the dynamical properties of random *classical* operators.

The program is broad: It is therefore not the intention of these lectures to provide a complete or comprehensive review of the field. Rather, we hope to convey the central concepts of this diverse subject, as well as developing some of the general technology which has proved to be so successful in the exploration of this field. Finally, we have attempted to keep the prerequisites for this course to a minimum: We will, however, assume a familiarity with the fundamental principles of elementary solid state physics, and an exposure to some basic concepts in advanced statistical mechanics and quantum field theory.

1.1 Manifestations of Phase Coherence in Mesoscopics

To begin our investigation of mesoscopic physics we start with a survey of several of the classic experiments which explore mechanisms of quantum phase coherence. In doing so, where possible, we will separate those experiments which find their origin in Coulomb interaction effects from those which can be understood within the framework of a purely noninteracting theory.

The majority of the experiments which we discuss below are performed on low-dimensional semi-conducting GaAs or Si heterostructures. Without going into extensive detail, we remark that the latter involve "sandwich" structures where abrupt changes of doping concentration of donor and acceptor impurities trap electrons in a narrow regions. The inversion layer potential can be used to confine electrons (or holes) to two-dimensional regions, one-dimensional channels (wires), and zero-dimensional wells (known as "quantum dots" [4]). Here, by confinement, we mean that the Fermi energy lies in an interval of energy well below the energy required to excite the lowest "transverse" mode. (For a review of device technology see, e.g., Ref. 5.)

For simplicity, we will not dwell on specific fabrication techniques. Nor will we assess the integrity of device characteristics. However, to orient our discussion, we note the typical range of relevant physical parameters below:

Parameter	GaAs	Si	Units
Density n	4.0	4.0	10^{11} cm^{-2}
Mobility μ	10^5	10^4	cm^2/Vs
Scattering Time τ	3.8	1.1	10^{-12} s
Fermi Wavevector λ_F	1.6	1.6	10^6 cm^{-1}
Fermi Velocity v_F	2.76	0.97	10^7 cm/s
Elastic m.f.p. ℓ	1.05	0.107	10^{-4} cm

Before we turn to the experiments, as preparation for our discussion, let us first recall some elementary concepts in the theory of electron transport. Firstly, in a degenerate electron gas, the conductivity σ and the diffusion constant D are fundamentally connected by the Einstein relation [1]

$$\sigma = e^2 \nu D \tag{1.1}$$

where ν denotes the average density of states at the Fermi level. The diffusion constant can, in turn, be obtained from the time-integrated velocity correlation function

$$D = \frac{1}{d} \int_0^\infty dt \, \langle \mathbf{v}(t) \cdot \mathbf{v}(0) \rangle$$

where the average is taken over all initial velocities $\mathbf{v}(0)$.

Now, within the framework of classical transport or kinetic theory, the velocity of an electron is randomised on a time scale set by the mean free scattering time between collisions, τ. In this approximation, the diffusion constant is specified by $D = v_F^2\tau/d$, where $v_F = p_F/m$ denotes the Fermi velocity and d the dimensionality. The latter, in turn, leads to the classical Drude formula for the conductivity,

$$\sigma = \frac{ne^2\tau}{m}$$

where $n \sim \nu E_F$ represents the electron density, and E_F denotes the Fermi energy.

Finally, from the conductivity, we can define the conductance $G = \sigma L^{2-d}$ which, making use of the Einstein relation, (1.1), can be expressed as

$$G = \frac{e^2}{\hbar}\nu L^d \frac{\hbar D}{L^2} = \frac{e^2}{\hbar} g, \quad g \equiv \frac{E_c}{\Delta} \tag{1.2}$$

where $\Delta = 1/\nu L^d$ denotes the average energy level spacing, and $E_c = \hbar D/L^2$ represents the typical inverse diffusion time for an electron to cross a sample of dimension L^d. This result shows that the conductance of a metallic sample can be expressed as the product of the quantum unit of conductance $e^2/\hbar = (4.1k\Omega)^{-1}$, and a *dimensionless conductance* g equal to the number of levels inside an interval E_c. In a good metallic sample, the dimensionless conductance is large, $g \gg 1$.

Magneto-Fingerprints

With this background, we turn our attention to experiment: Fig. 6.1 shows a series of magneto-conductance measurements [8] (i.e., the dependence of G as a function of magnetic field) performed on a SiGaAs wire after repeated heating and cooling.

Although each individual trace appears to fluctuate randomly around some uniform value, the measurements are completely reproduceable (i.e., when held at a constant low temperature, the fluctuations of the conductance do not change with time). However, by thermal cycling, the accompanying rearrangement of the impurities completely changes the pattern of magneto-conductance fluctuations. These results, which are typical of those observed in phase coherent devices, suggest that sample-sample fluctuations provide a characteristic signature (a "magneto-fingerprint") of an individual system [9, 10].

Yet characteristic fluctuations of this kind are not limited to wires. Figs. 6.2 and 6.3 show the variation of two-terminal magneto-conductance measurements together with an applied gate voltage potential performed on a mesoscopic GaAs quantum dot. Again, the qualitative behavior is similar: The system shows a reproduceable magneto-fingerprint with the same characteristic scale of fluctuations, e^2/h.

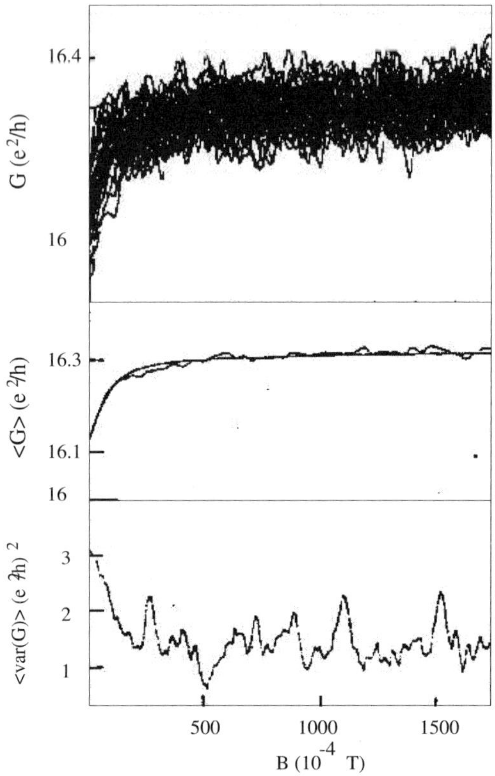

FIGURE 6.1. Magneto-conductance measurements of a SiGaAs wire taken from Ref. 8. (a) Conductance as a function of magnetic field for samples which differ by thermal cycling. (b) The conductance averaged over different realizations. (c) The variation of the variance of the conductance fluctuations measured as a function of magnetic field. (Courtesy of Ref. 8.)

• Firstly, why is the magnitude of the conductance fluctuations *universal*? That is, for any size or geometry of the sample, the conductance is found to fluctuate on a scale comparable to e^2/h. More precisely, defining $\langle \cdots \rangle$ as the ensemble average over realizations of the impurity potential (generated in the present case by thermal cycling), $\langle (\delta G)^2 \rangle \simeq (e^2/h)^2$, where $\delta G \equiv G - \langle G \rangle$—i.e., the characteristic scale of fluctuations is independent of the system size.

• Secondly, why are these fluctuations so large? Naively, in a metallic d-dimensional sample, it is reasonable to suppose that each block of dimension ℓ_c^d, where ℓ_c represents a macroscopic length scale (e.g., mean distance between impurities), contributes independently to the total conductance (see Fig. 6.4). With this assumption, one can expect the magnitude of conductance fluctuations to vanish in the thermodynamic limit as the power law $\langle (\delta G)^2 \rangle / \langle G \rangle^2 \sim (\ell_c/L)^d$. In this sense, the conductance is expected to be

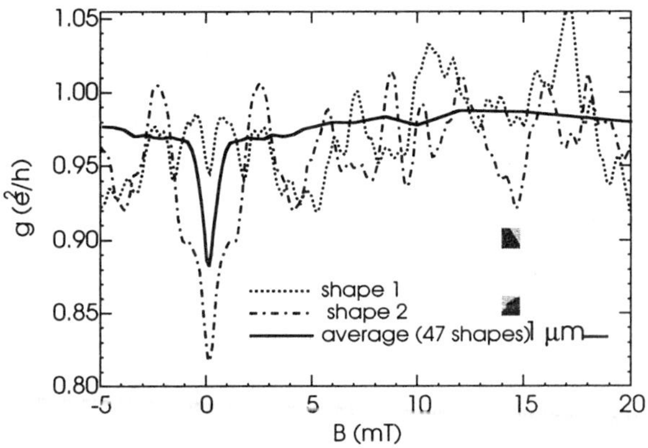

FIGURE 6.2. Magneto-conductance measurements of a GaAs quantum dot taken from Ref. 12. The device, shown inset, has a transport mean free path and phase coherence length greatly in excess of the dimensions of the device (ca. 1μm). The "shape" of the quantum dot can be changed by tuning a gate or "plunger". Separate traces are shown for two particular shapes together with the average over an ensemble of different realizations. Note the universal scale of the fluctuations, and the conductance minimum at zero field. (Courtesy of Ref. 12.)

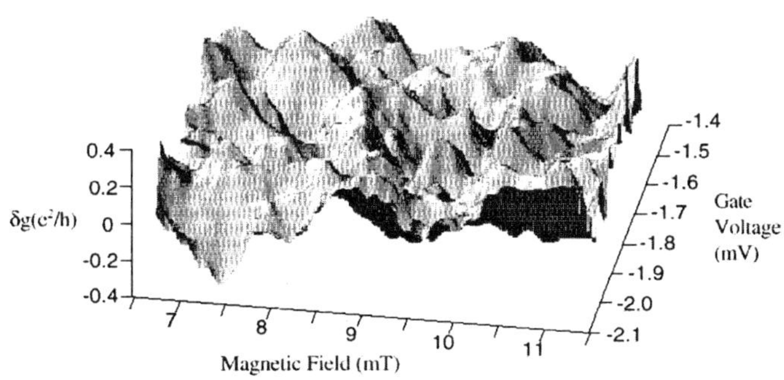

FIGURE 6.3. Conductance measurements as a function of magnetic field and external gate voltage for a quantum dot qualitatively similar to that shown in Fig. 6.2. (Courtesy of Ref. 12.)

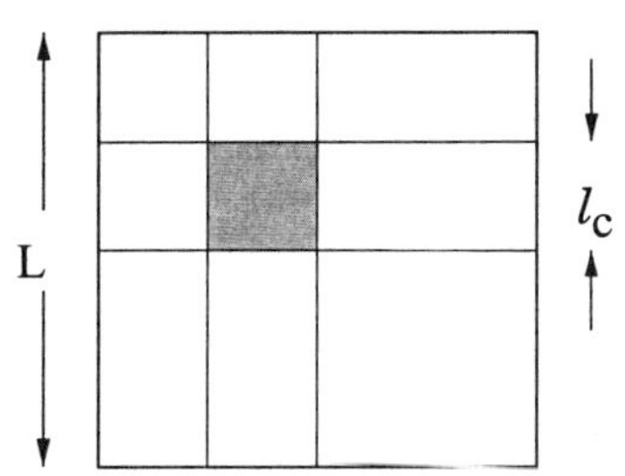

FIGURE 6.4. Schematic diagram showing small region over which fluctuations are correlated.

self-averaging. However, taking $\langle G\rangle = \sigma L^{2-d}$ (i.e., Ohm's law), the experimental measurements above are compatible with fluctuations of magnitude $\langle(\delta G)^2\rangle/\langle G\rangle^2 \sim L^{4-2d}$ which, at least in dimensions $d < 4$, are substantially larger! Evidently, there exist substantial *nonlocal* correlations which, in turn, lead to the absence of *self-averaging*.

• Finally, why is the strength of both the ensemble averaged conductance, as well as its fluctuation, magnetic field dependent? More precisely, the measurements above show two clear effects: Firstly, the conductivity increases with the application of a weak magnetic field (i.e., a field small enough that the influence of orbital effects can be ruled out). Secondly, on the same field scale, the magnitude of the fluctuations is substantially diminished.

Aharonov–Bohm Effects

A second classic experiment involves the Aharonov–Bohm effect. Fig. 6.5 shows the magneto-resistance measurements of a single gold metallic ring. As with the quantum wire, the results show a characteristic pattern of

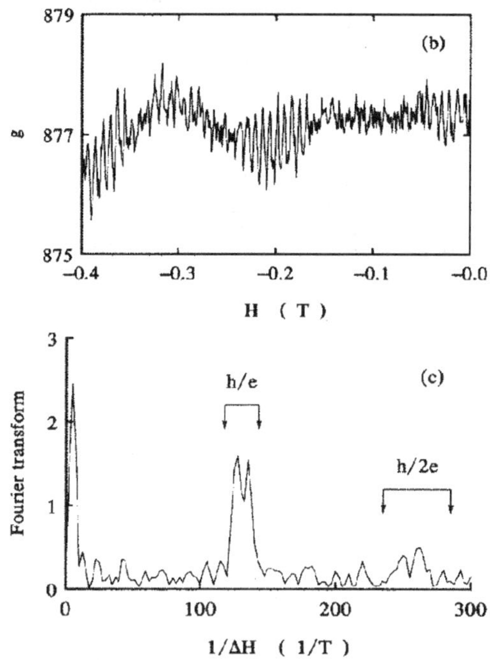

FIGURE 6.5. Magneto-conductance measurements of a single Au loop together with the corresponding power spectrum showing periodic oscillations at h/e and $h/2e$ (Taken from S. Washburn and R.A. Webb, *Anaronov–Bohm effect in normal metal quantum coherence and transport*, Adv. Phys. **35** (1986) 375–422, cf. `http://www.tandf.co.uk` (Ref. 13)).

magneto-fingerprint fluctuations. However, these fluctuations are modulated by periodic oscillations. These oscillations are manifest most clearly in the power spectrum which indicate Fourier components at a frequencies of $1/\Delta H$ corresponding to values of h/e and $h/2e$. Furthermore, separate measurements of a statistical ensemble of mesoscopic rings shows the h/e oscillations to vanish on averaging, while the $h/2e$ oscillations remain. To our list above, we can, therefore, add the questions:

- In the Aharonov–Bohm geometry, what is the physical origin of the h/e and $h/2e$ oscillations?
- And how can we understand the action of ensemble averaging on these two types of oscillations?

1.2 Qualitative Considerations: Feynman Paths

To find qualitative answers to the questions raised above, it is useful to consider a simplified model of electron dynamics in the background of a random scattering potential. In doing so, we should first recognise the crucial distinction between the two types of scattering an electron can undergo. *Elastic scattering* (e.g., by lattice defects or impurities) imparts on the electron a well-defined phase shift. By contrast, *inelastic scattering*, which involves the exchange of energy with other degrees of freedom (e.g., phonons, other electrons, and spins), introduces uncertainty in the phase. It is this uncertainty which is responsible for the destruction of quantum interference. In the following, we will assume that the time scale over which phase coherence is maintained, the inelastic scattering time τ_φ, is greatly in excess of the time scale on which an electron is scattered across the Fermi surface, the elastic scattering time τ.

At its simplest level, the dynamics of an electron in a phase coherent environment can be described by a noninteracting or single-particle Hamiltonian

$$\widehat{H} = \frac{\widehat{\mathbf{p}}^2}{2m} + V(\mathbf{r}), \tag{1.3}$$

where $\widehat{\mathbf{p}} = -i\hbar\partial$ denotes the momentum operator, and $V(\mathbf{r})$ represents a random impurity potential. Now, even for this simplified model, a given realization of the random impurity potential is associated with a complex spectrum. Yet, if the potential fluctuations are weak as compared with the typical energy scale of the particle, the influence of quantum mechanical interference on the particle dynamics can be determined both qualitatively and, as we shall see later, quantitatively.

To stay within the domain in which a quasi-classical analysis is valid, we will focus on the hierarchy of length scales shown in Fig. 6.6. Specifically, we will focus on the *semi-classical* regime in which the mean free path of the particle ℓ is greatly in excess of the corresponding wavelength of the

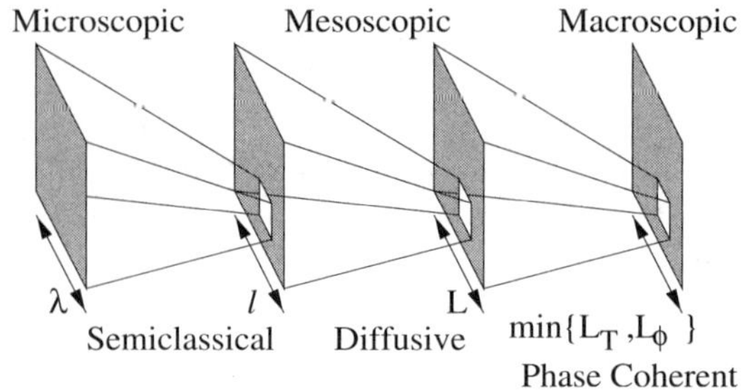

FIGURE 6.6. Hierarchy of length scales which electron dynamics is phase coherent and diffusive.

particle λ_F (or, equivalently $E_F\tau \gg \hbar$). Moreover, to stay within the regime in which the particle dynamics is diffusive, we will require the system size L to be greatly in excess of ℓ. Finally, we will take the system size to be much smaller than the typical length scale L_φ over which the particle dynamics is phase coherent.

With this definition, the quantum transfer *probability amplitude* for a particle to propagate from a point $\mathbf{r}_\mathrm{I}$ to a point $\mathbf{r}_F$ in a time t is specified by the quantum mechanical Feynman propagator

$$G(\mathbf{r}_F, \mathbf{r}_\mathrm{I}; t) = \int_{\mathbf{r}(0)=\mathbf{r}_\mathrm{I}}^{\mathbf{r}(t)=\mathbf{r}_F} D\mathbf{r}(t) e^{i\int_0^t dt\,(m\dot{\mathbf{r}}^2/2 - V(\mathbf{r}(t)))/\hbar}.$$

Now, for an arbitrary configuration of the impurity potential, an exact evaluation of the path integral is, of course, impossible. However, we can gain a qualitative understanding of the problem by visualising the path integral as an infinite sum of separate paths i each with its own amplitude A_i and phase φ_i,

$$G(\mathbf{r}_F, \mathbf{r}_\mathrm{I}; t) = \sum_i A_i e^{i\varphi_i},$$

i.e., the superposition of randomly scattered waves.

Typically, different paths have lengths that differ substantially implying a statistical independence of the phases φ_i (i.e., for two paths i and j, $|\varphi_i - \varphi_j| \gg 2\pi$). Therefore, applied to the propagator, an ensemble average over different realizations of the random impurity potential leads to a random phase cancellation which is manifest as an exponential decay of the average on the time scale of the mean free scattering time τ. Equivalently, the ensemble average of the propagator decays on a length scale comparable to the mean free path ℓ.[1]

---

[1]More formally, this can be seen by expanding the propagator in powers of the impurity potential V, and subjecting the resulting series expansion to an ensemble average

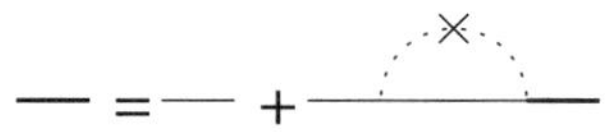

FIGURE 6.7. Diagrammatic representation of the Dyson equation for the ensemble average of the single-particle Green function in the *self-consistent Born approximation* (SCBA). Here the bold line represents the average Green function while the dotted line indicates a connection to an impurity scattering center represented by a cross. (Higher order contributions to the ensemble averaged Green function involve higher order corrections to the self-energy small in the quasi-classical parameter $\hbar/k_F\ell$, where $k_F = 2\pi/\lambda_F$ represents the magnitude of the Fermi wavevector of the incoming electron.)

By contrast, the corresponding transfer *probability density* of a particle from a position $\mathbf{r}_\mathrm{I}$ to $\mathbf{r}_F$

$$P(\mathbf{r}_F, \mathbf{r}_\mathrm{I}; t) \equiv |G(\mathbf{r}_F, \mathbf{r}_\mathrm{I}; t)|^2 = \sum_i A_i^2 + 2\sum_{i\neq j} A_i A_j \cos(\varphi_i - \varphi_j),$$

involves long-ranged correlations. More precisely, when subjected to an ensemble average, while the second "*interference*" contribution averages to zero, a long-ranged *diagonal* contribution survives:

$$\langle P(\mathbf{r}_F, \mathbf{r}_\mathrm{I}; t)\rangle = \sum_i \langle A_i^2\rangle.$$

Here the random phase accumulated by a particle propagating from a point $\mathbf{r}_\mathrm{I}$ to $\mathbf{r}_F$ is cancelled by the phase acquired on the return providing the two paths coincide on the scale of Fermi wavelength, λ_F of the particle (see Fig. 6.8, paths 2 and 3).

Again, a formal expression for the average transfer probability density can be obtained by subjected the series expansion to an ensemble average (for details, see Section 1.4). In the leading approximation, the diagonal contributions lead to a *diffusive* relaxation of the probability density,

$$\langle P(\mathbf{r}_F, \mathbf{r}_\mathrm{I}; t)\rangle = \frac{1}{(2\pi D t)^{d/2}} \exp\left[-\frac{(\mathbf{r}_F - \mathbf{r}_\mathrm{I})^2}{2Dt}\right], \tag{1.6}$$

---

over, say, a δ-correlated white-noise impurity potential (with zero mean), and correlator

$$\langle V(\mathbf{r})V(\mathbf{r}')\rangle = \frac{\hbar}{2\pi\nu\tau}\delta^d(\mathbf{r} - \mathbf{r}'). \tag{1.4}$$

The different contributions to the average can be organized into a diagrammatic series (see Fig. 6.7) within which one finds that, to within a good approximation,

$$\langle G(\mathbf{r}_\mathrm{f}, \mathbf{r}_\mathrm{i}; t)\rangle = G_0(\mathbf{r}_\mathrm{f} - \mathbf{r}_\mathrm{i}; t)e^{-t/2\tau},$$

where G_0 denotes the Green function for a free particle. (For completeness, a derivation of this result is presented at the end of this section.) Equivalently, Fourier transforming,

$$\langle G(\mathbf{p}; E_F - i0)\rangle = \frac{1}{E_F + \mathbf{p}^2/2m - i/2\tau}. \tag{1.5}$$

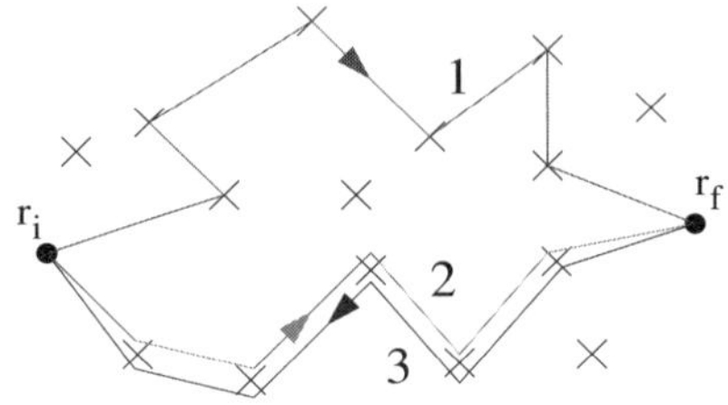

FIGURE 6.8. Typical Feynman paths connecting lattice sites $\mathbf{r}_\mathrm{I}$ and $\mathbf{r}_F$. When summed over all configurations of the random impurity potential, the contribution of paths such as 1 and 2 add incoherently to the average probability amplitude or Green function. However, the diagonal contribution from paths 2 and 3 (which follow the same path) contribute coherently to the average probability density.

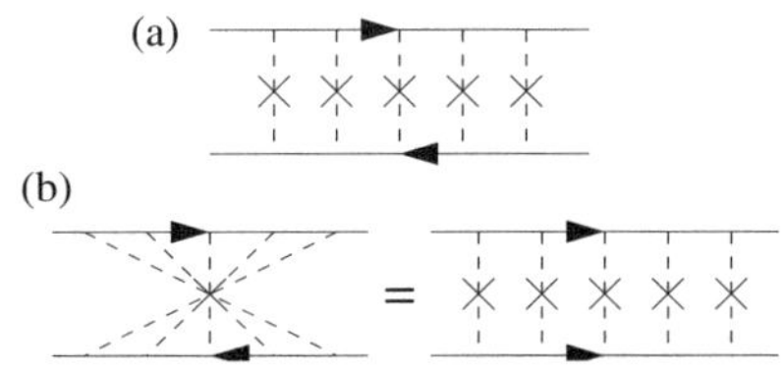

FIGURE 6.9. Diagrammatic representation of the diffusion modes: (a) Diffuson and (b) Cooperon. Note that the maximally crossed Cooperon diagram can be re-expressed as a ladder series by time-reversing the advanced Green function.

where $D = v^2\tau/d$ denotes the bare classical diffusion constant. At the lowest order, the long-time (i.e., $t \gg \tau$) dynamics of a particle in a random impurity potential is described by simple classical diffusion.[2]

A second phase coherent contribution which survives impurity averaging, and one that has no classical counterpart, can be identified by studying the return probability. In the absence of an external magnetic field, the coherent superposition of a path i with its time-reversed counterpart $j \equiv \bar{i}$ leads to an enhancement of the average return probability,

$$\langle P(\mathbf{r}_F = \mathbf{r}_\mathrm{I}, \mathbf{r}_\mathrm{I}; t)\rangle = 2 \times \sum_i \langle A_i^2 \rangle.$$

As with the diffuson mode, this long-range phase coherent contribution can also be identified with a diagrammatic series (see Fig. 6.9b) involving maximally crossed "impurity lines". Motivated by the analogy with the Cooper

[2]The phase cancellation of the "diagonal" paths shown in Fig. 6.8 can be interpreted as a cartoon of the dominant contributions to the perturbative expansion of the two-particle Green function. In particular, diagrammatically, these contributions can be identified with a summation over the "ladder series" shown in Fig. 6.9a. The latter, which is associated with the long-range mode of density relaxation is known as a *Diffuson*, $\Pi_\mathrm{D}(\mathbf{q}, \Omega) = (-i\Omega + \hbar D\mathbf{q}^2)^{-1}$.

channel in the BCS theory of superconductivity, the latter are known as *Cooperon modes*, $\Pi_{\mathrm{C}}(\mathbf{Q},\Omega)=(-i\Omega+\hbar D\mathbf{Q}^2)^{-1}$.

Together, the diffusion modes represents the fundamental "building blocks" out of which quantum coherence phenomena can be described. As a first example, let us identity the main quantum interference corrections to the classical conductivity.

Weak Localization

Applied to the transfer probability density, the phase coherent superposition of time-reversed paths (the Cooperon mode) provides the leading quantum correction to the transfer probability density or classical conductivity (see Fig. 6.10). The enhanced return probability has a tendency to "delay", or "weakly localise" the particle, and has important implications for localization and transport properties. An estimate of the quantum correction to the diffusion constant can be obtained by evaluating the reduction in transfer probability along a diffusing path due to redundant excursions (see Fig. 6.11). Requiring that a returning path stays within a wavelength of the outgoing path, an estimate for the relative change in the diffusion constant can be obtained by accounting for the return probability $P(0,0;t)=1/(Dt)^{d/2}$ at any point along the trajectory [14]

$$\begin{aligned}\frac{\delta D}{D} &= \frac{\delta P}{P}\simeq -v_F\lambda_F^{d-1}\int_\tau^{t_D}\frac{dt}{(Dt)^{d/2}}\\ &= -\frac{1}{\hbar\nu D}\frac{2}{2-d}(L^{2-d}-\ell^{2-d}).\end{aligned}\tag{1.7}$$

Here $t_D=L^2/D$ denotes the typical diffusion time across a sample of size L^d, and ν is the d-dimensional density of states (DoS).

Applying the Einstein relation for the conductivity, $\sigma=e^2\nu D$, this estimate of the weak localization correction (1.7) is in accord with a formal

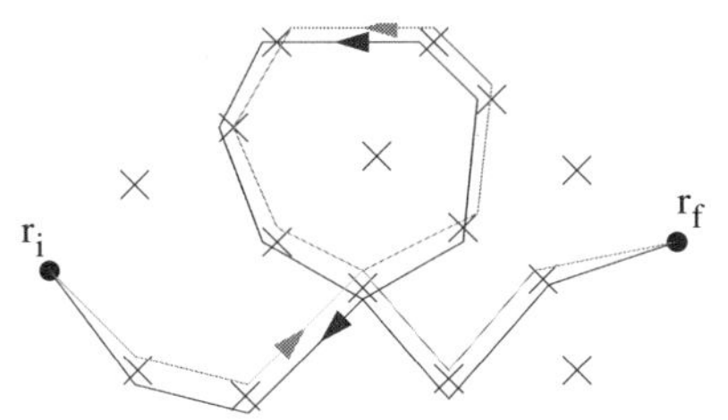

FIGURE 6.10. Quantum phase coherent paths contributing to the transfer probability. Note that the constructive interference of a path with its time-reversed counterpart leads to an enhancement of the return probability which renormalises down the diffusion constant.

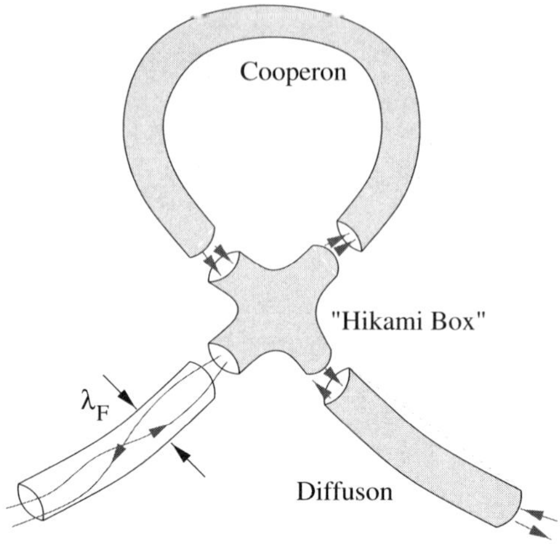

FIGURE 6.11. The weak localization correction shown in Fig. 6.10 can be separated into a three stage process: Two Diffuson ladders connected to a Cooperon ladder by a junction region known, for historical reasons, colloquially as an "Hikami box". (Note that the magnitude of the correction is constrained by the condition to return to the same point in real space with a resolution of the Fermi wavelength.)

diagrammatic calculation showing that, in two dimensions [15],

$$\delta\sigma = -\frac{e^2}{\pi^2\hbar}\ln\left(\frac{L}{\ell}\right). \tag{1.8}$$

This result resolves one of the puzzles presented by experiment: When subjected to a weak magnetic field, a given path no longer interferes constructively with its time-reversed counterpart. As a result, to leading order, the conductivity is predicted to increase in agreement with the observed negative magneto-resistance measurements.

The long-ranged Diffuson and Cooperon modes provide the key elements from which a consistent diagrammatic theory of quantum coherence phenomena in disordered conductors can be assembled. Later, in Section 2, their existence will be used to motivate the existence of a low-energy quantum field theory of disordered conductors.

Fluctuations

While the random phase cancellation of different Feynman paths restricts the long-range contributions to the transfer probability density to classical diagonal contributions, their influence on *fluctuations* is substantial. In particular, phase coherent contributions to the quadratic fluctuations can arise from paths that are nondiagonal (see Fig. 6.12). More precisely, since $\langle\cos^2(\varphi_i - \varphi_j)\rangle = 1/2$, fluctuations of the probability density differ from

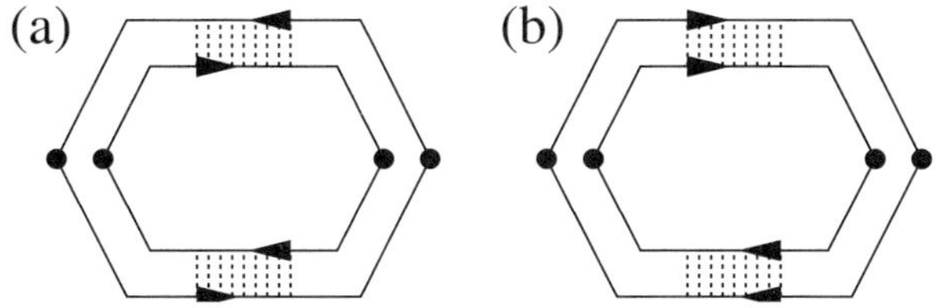

FIGURE 6.12. Diagrammatic representation of the main phase coherent contributions to the fluctuations of the transfer probability density from (a) the diffuson channel, and (b) the Cooperon channel.

the classical transfer probability by an amount

$$\left\langle \left(P(\mathbf{r}_F, \mathbf{r}_\mathrm{I}; t) - \langle P(\mathbf{r}_F, \mathbf{r}_\mathrm{I}; t)\rangle\right)^2 \right\rangle = 2\sum_{i\neq j} \langle A_i^2 A_j^2 \rangle.$$

Mechanisms of quantum interference between different Feynman trajectories induce fluctuations in the probability density which depend *nonlocally* on the particular realization of the impurity potential. These fluctuations resolve a second puzzle concerning the existence of the characteristic magneto-fingerprint pattern observed in the conductance measurements: The phase accumulated by a closed Feynman path depends sensitively on the external magnetic field according to the relation

$$\varphi = \frac{e}{\hbar c} \oint \mathbf{A} \cdot d\mathbf{l} \equiv \frac{e}{\hbar c} \int \mathbf{H} \cdot d\mathbf{a}. \tag{1.9}$$

Thus, a change of magnetic field by an amount ϕ_0/L^2, where $\phi_0 = hc/e$ denotes the *magnetic flux quantum*, effectively changes the configuration space of the system which in turn induce reproduceable, characteristic fluctuations of the conductance. Equivalently, defining the magnetic length $L_\mathrm{H} = \sqrt{\phi_0/H}$ as the typical length scale over which phase coherence of Cooperon modes is maintained, we can estimate

$$\langle \delta G(0) \delta G(H) \rangle \sim \left(\frac{L_\mathrm{H}}{L}\right)^d \left(\frac{e^2}{h}\right)^2.$$

Later we will find an explanation for the universality of the magnitude of the conductance fluctuations. However, before doing so, let us see how these ideas can be used to explain the nature of the Aharonov–Bohm oscillations.

Aharonov–Bohm Oscillations

Both mechanisms of quantum interference shown in Figs. 6.11 and 6.12 are manifest in the transport properties of a metallic ring threaded by a magnetic flux, ϕ. In the Aharonov–Bohm geometry (see Fig. 6.13) Feynman trajectories which pass anti-clockwise around the ring experience an additional phase shift relative to those that move clockwise. A contribution to

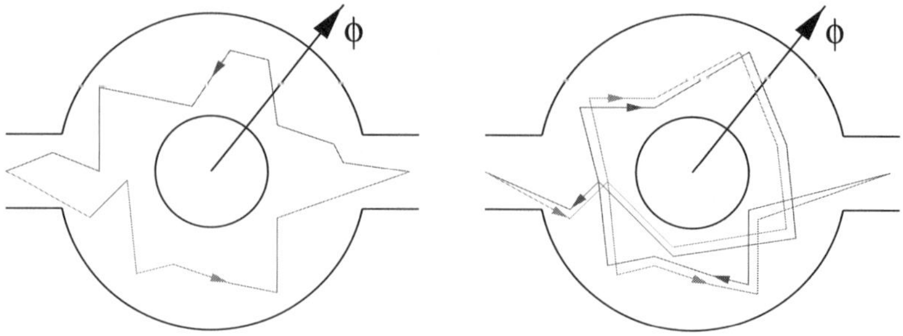

FIGURE 6.13. Schematic diagram showing the interference correction to a two-terminal conductance measurement of a device with the geometry of a ring. The second diagram shows a Cooperon correction from trajectories which circulate around the ring. The latter is responsible for the $h/2e$ oscillations.

the transfer probability or conductance which involves a pair of Feynman paths which traverse the ring in opposite directions will therefore experience an Aharonov–Bohm oscillation of period hc/e. Yet, such contributions add incoherently and do not appear in the ensemble average conductance. However, following on from our discussion above, phase coherent contributions to the fluctuations do survive. Altogether these results suggest a periodic modulation of the magneto-conductance fluctuations with a period set by the flux quantum $\langle \delta G(0)\delta G(\varphi)\rangle \sim \cos(2\pi\phi/\phi_0)$. These fluctuations account for the hc/e oscillations observed in the experiment.

The $hc/2e$ oscillations of the average conductance itself finds its origin in the mechanism of weak localization. In the Aharonov-Bohm geometry the application of a magnetic flux imparts a relative phase into the time-reversed paths encircling the ring (see Fig. 6.13). Here, each particle has to completely circumnavigate the ring once. As a result the average conductance itself is expected to exhibit oscillations periodic in $hc/2e$. This effect, predicted by Altshuler, Aronov and Spivak [16] was first observed in a classic experiment by Sharvin and Sharvin [17].

Energy Level Repulsion and Universality

Our considerations above were primarily concerned with understanding the influence of quantum phase coherence on the transport properties of mesoscopic systems. However, the same phase coherence phenomena are responsible for characteristic fluctuations in spectral properties. Although clearly there are no low-energy scale correlations between the energy levels belonging to completely different impurity configurations, within any given realization the eigenvalues are strongly correlated. An estimate of the degree of correlation can be established from the two-point correlator of DoS fluctuations,

$$R_2(\Omega) \equiv \Delta^2\langle \nu(E+\Omega/2)\nu(E-\Omega/2)\rangle - 1, \tag{1.10}$$

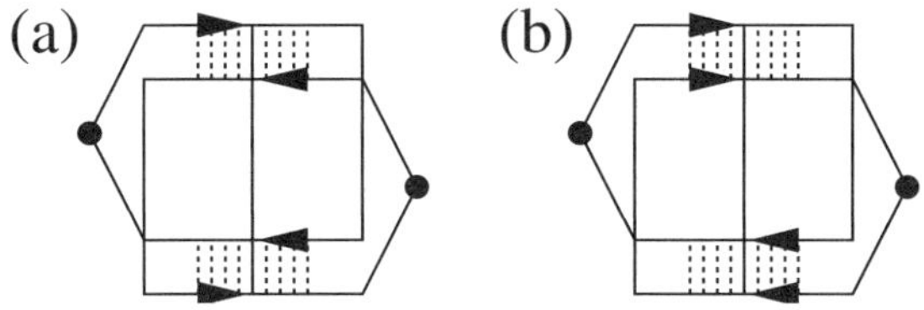

FIGURE 6.14. Diagrams showing the leading contribution to the two-point auto-correlator of DoS fluctuations. The first diagram (a) represents the exchange of two diffuson ladders between two closed loops while the second denotes the exchange of two Cooperon ladders.

where $\nu(E) = \operatorname{tr}\delta(E - \widehat{H})$ denotes the DoS operator, and $\Delta = 1/L^d\nu$ denotes the average level spacing. (For convenience, here and throughout, the smooth average DoS is expressed as the shorthand, $\nu \equiv \langle\nu(E)\rangle$.)

To interpret the origin of the statistical correlations within the framework of interfering trajectories, we can exploit the identity connecting the Feynman propagator with the DoS. Defining the advanced Green function $\widehat{G}^- = [E - \widehat{H} - i0]^{-1} \equiv (\widehat{G}^+)^\dagger$, the DoS is defined by

$$\nu(E) = \frac{1}{\pi}\operatorname{tr}\operatorname{Im}\widehat{G}^-(E). \tag{1.11}$$

In this representation, the DoS is seen to be associated with the return probability of Feynman trajectories. As before, when subjected to an ensemble average, the long-time contributions to the DoS vanish due to the random phase cancellation of Feynman paths. Only the very short-time $t \ll \tau$ trajectories, which give rise to a smooth part of the DoS survive.

However, as with the transfer probability, phase coherent long-time contributions to the fluctuations can be identified (see Fig. 6.14). The complete phase cancellation of Feynman trajectories is achieved by matching paths with two diffusion modes (cf. the diagrammatic representation of the fluctuation correction). Formally, the existence of two diffusion ladders arises from the exchange of two diffusion ladders between two closed loops, and gives the result [18]

$$R_2(\Omega) = \frac{1}{2\pi^2}\operatorname{Re}\sum_{\mathbf{q}}\frac{\Delta^2}{(-i\Omega + \hbar D\mathbf{q}^2)^2}. \tag{1.12}$$

In systems invariant under time-reversal, a second mechanism arising from the exchange of Cooperon ladders is accounted by a further factor of two (see Fig. 6.14b).

While this result shows that on energy scales Ω in excess of the "Thouless energy" (i.e., the inverse transport time),

$$E_c \equiv \frac{\hbar}{t_D} = \frac{\hbar D}{L^2}, \tag{1.13}$$

level statistics depend explicitly on material properties such as geometry and morphology, on scales $\Omega \ll E_c$, the zero spatial diffusion mode $\mathbf{q} = 0$ dominates, and

$$R_2(\Omega) \simeq -\frac{1}{2\pi^2}\left(\frac{\Delta}{\Omega}\right)^2 . \tag{1.14}$$

In this limit fluctuations of the DoS are *universal*, independent of detailed properties of the system. In fact Eq. (1.14) is a manifestation of "rigidity" in the spectrum: Taking $N(\Omega)$ to be the number of levels inside a band of width Ω, Eq. (1.14) shows fluctuations in N to increase only logarithmically with Ω,

$$\langle N^2(\Omega)\rangle_c \simeq \frac{1}{\pi^2}\ln\langle N\rangle . \tag{1.15}$$

This compares with a hypothetical random or Poissonian distribution of energy levels in which fluctuations of N grow as $\sqrt{\langle N\rangle}$.

Universal Conductance Fluctuations

Spectral rigidity (1.15) provides a simple physical interpretation of the characteristic scale of conductance fluctuations [19]: As we have seen, the bare conductance of a metallic sample of dimension L can be expressed through the Einstein relation (1.1) as a ratio of characteristic energy scales, $G = ge^2/\hbar$, where $g = E_c/\Delta$; i.e., the dimensionless conductance g is equivalent to the number of levels inside a window of energy E_c. Thus, while the average conductance $\langle G\rangle = ge^2/\hbar$, is large in a good metal ($g \gg 1$), Eq. (1.15) implies that the characteristic scale of fluctuations,

$$\sqrt{\langle(\delta G)^2\rangle_c} \sim O(e^2/\hbar),$$

is universal [9, 10], independent of the average conductance, system size, etc. Spectral rigidity of the spectrum of the disordered quantum Hamiltonian therefore provides a natural way to understand the characteristic universal conductance fluctuations observed in experiment.

1.2.1 Breakdown of Perturbation Theory: RMT

The estimate for the spectral correlation function (1.14) also signals a warning concerning the domain of validity of the diagrammatic perturbation theory which will be of relevance in future. At energy scales $\Omega \ll \Delta$, or equivalently at time scales in excess of the inverse level spacing (known as the "Heisenberg time"), $t \gg t_H = \hbar/\Delta$, response functions obtained with diagrammatic perturbation theory show unphysical IR divergences. Taking into account contributions higher order in perturbation theory doesn't help. Indeed, the strength of the divergence just becomes enhanced.

In fact, the validity of diagrammatic perturbation theory relies on two independent parameters: Firstly, the disorder potential should be sufficiently weak that a quasi-classical approximation is valid (i.e., $\lambda_F/\ell \ll 1$). Furthermore, the influence of quantum interference corrections should be small. More precisely, the validity of Eq. (1.12) relies on the further condition,

$$\left(\frac{\Delta}{E_c}\right)^{d/2}\left(\frac{\Delta}{\Omega}\right)^{(2-d)/2} \ll 1.$$

In the *quantum regime* (i.e., where $\Omega \lesssim \Delta$), this condition is not met. Here the intuition afforded by the diagrammatic picture breaks down.

In fact, the universality of the low-energy sector of the correlation function gives a clue to the true behavior in the quantum regime. On time scales in excess if $t_D = L^2/D$, the entire phase space of the system has been explored. In this "ergodic regime" (and with the benefit of a good deal of hindsight), one would expect the dynamics of the particle in a disordered environment to be indistinguishable from that of a "random matrix" Hamiltonian (i.e., a Hamiltonian in which *all* matrix elements between different basis states are drawn from some smooth unspecified distribution).

In fact, Random Matrix Theory (RMT), as it has come to be known, provides a convenient way of classifying disordered Hamiltonians according to their fundamental symmetries. Focusing on the ensemble of Gaussian distributed random matrices,

$$P(H)\, dH \propto \exp\left[-\frac{N\beta}{\lambda^2}\,\mathrm{tr}\, H^2\right] dH,$$

three principle universality classes can be identified [11] according to whether the $N \times N$ matrix H is constrained to be real symmetric ($\beta = 1$, *Orthogonal*), complex Hermitian ($\beta = 2$, *Unitary*), or real quaternion ($\beta = 4$, *Symplectic*). Hamiltonians invariant under time-reversal belong to the orthogonal ensemble, while those which are not belong to the unitary ensemble. Time-reversal invariant systems with half-integer spin and broken rotational symmetry belong to the third symplectic ensemble.

Now, expressed in the basis of eigenstates $H = U^\dagger \Lambda U$, where Λ denotes the matrix of eigenvalues, the probability distribution can be recast in the form

$$P(H)\, dH = J(\{E_i\}) \exp\left[-\frac{\beta N}{\lambda^2}\sum_i^N E_i^2\right]\prod_l^N dE_l\, dU$$

where the invariant measure $J(\{E_i\}) = \prod_{i<j}|E_i - E_j|^\beta$ reveals the characteristic repulsion of the energy levels. As with the disordered Hamiltonian, the rigidity of the complex eigenvalue spectrum is revealed in the two-point correlator of DoS fluctuations. In particular, applied to the unitary class of complex Hermitian matrices, to leading order in N random matrix analysis [20] reveals a universal (i.e., independent of the distribution) correlation function

$$R_2(\Omega) = 1 - \frac{\sin^2(\pi\Omega/\Delta)}{(\pi\Omega/\Delta)^2}. \tag{1.16}$$

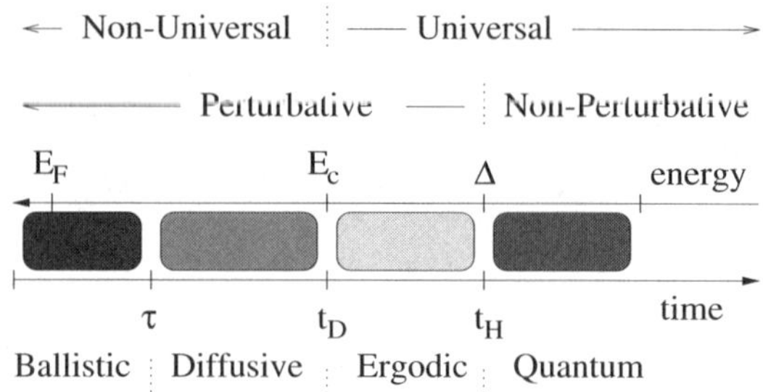

FIGURE 6.15. Schematic diagram showing the relevant time and energy scales in disordered transport. Also shown is the regime over which properties can be treated within a perturbative framework.

This result, nonperturbative in Δ/Ω, interpolates smoothly onto the result of Eq. (1.12) if we set $\sin^2(\pi\Omega/\Delta) \to \frac{1}{2}$.

In fact, the connection between RMT and the long-time or low-energy properties of disordered conductors is even more robust extending to statistical properties of the wavefunctions themselves.

Summary

This concludes our preliminary survey of the quantum interference phenomena which characterize mesoscopic structures. To summarise our findings, let us consider the influence of quantum interference processes on the dynamics of a spreading wavepacket in a background of weakly scattering random impurities. (Here we imagine a wavepacket whose spectral resolution is confined to a narrow interval (i.e., $\Delta \ll E_{\rm width} \ll \hbar/\tau$) around the Fermi level.) The different regimes encountered by the diffusing particle are summarised in Fig. 6.15.

Ballistic Regime Here, on time scales $t < \tau$ a wavepacket spreads ballistically at the Fermi velocity. Within this regime, particle dynamics is largely insensitive to the impurity background.

Diffusive Regime One time scales $\tau < t < t_D \equiv L^2/D$, the quantum interference due to scattering off the background impurity potential leads to diffusive dynamics of the wavepacket. Within the Diffusive regime, quantum weak localization corrections renormalise the diffusion constant below the bare value. In fact, soon we will see that in dimensions $d \leq 2$, these quantum corrections accumulate, and ultimately lead to the complete localization of the wavepacket.

Ergodic Regime If the particle does not localise, or if the localization length is greatly in excess of the system size $\xi \gg L$, on time scales $t > t_D \equiv L^2/D$ the wavepacket extends over the whole system. In this sense, the particle dynamics is said to be "ergodic". Here the dynamics becomes universal, independent of the geometrical and material

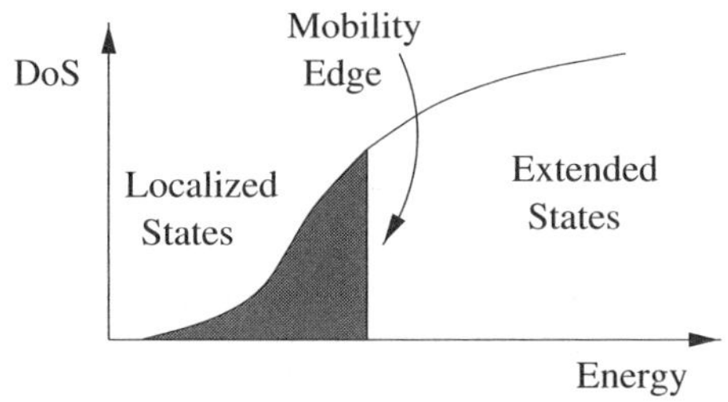

FIGURE 6.16. Schematic diagram showing the typical variation of the DoS with energy. A mobility edge separates regions of extended and localised states.

properties of the system. Spectral properties of the system become indistinguishable from those of random matrix ensembles.

Quantum Regime Finally, on time scales $t > t_H \equiv \hbar/\Delta$, the particle enters a regime in which its behaviour is dictated by the universal properties associated with the resolution of individual levels. In particular, the rigidity of the spectrum cooperates to produce a "dynamical echo" of the wavepacket [21] (cf. the dynamics of a wavepacket confined to an harmonic oscillator potential).

Scaling Theory of Localization

To close this section on coherence effects in noninteracting disordered systems, let us return to the phenomena of weak localization. Previously, we saw that the constructive interference of time-reversed paths gave rise to a quantum correction to the conductivity which was interpreted as the renormalization of the classical diffusion constant to a lower value. This invites the question, how does the accumulation of quantum interference corrections finally affect the transport in a disordered environment.

This question finds its origin in the pioneering work by Anderson in a seminal paper entitled "The absence of diffusion in certain random lattices" [22]. Based on the consideration of a simple random "tight-binding" Hamiltonian, Anderson proposed that, above a certain critical strength of the disorder potential, there is a qualitative change in the nature of the electronic wavefunction in which its envelope becomes exponentially localised.

Here a distinction should be drawn between qualitatively different mechanisms of localization. Evidently, the lowest energy states in a disordered system will be dictated by "optimal fluctuations" of the random impurity potential. These "tail states" can be interpreted as "bound states" localised in regions where the potential is unusually low and uniform [23, 24]. Instead, here we are interested in the localization properties of states with energies greatly in excess of the typical magnitude of the impurity potential, i.e., $E_F\tau \gg 1$. Here, the mechanism of localization can derive only from the accumulation of subtle interference corrections.

According to its energy E, Anderson proposed that a given state is either extended or localised depending on the degree of disorder. Following an argument due to Mott, it is also plausible that extended and localised states at the same energy do not coexist—in such circumstances, the admixture of the former due to arbitrarily small perturbations would lead to the delocalization of the latter. Moreover, it is natural that states at the band edge are more susceptible to the formation of bound or localised states. These considerations suggest a profile for the DoS in which the low-lying localised states are separated from higher-energy extended states by a *Mobility edge*. The transition signalled by the Mobility edge is known as the *Anderson transition*.

Since the pioneering work of Anderson, almost two decades passed before a consistent theory of localization emerged. Its basis was a simple yet power scaling hypothesis which followed naturally from a conjecture made by Thouless [25, 26]. Suppose that the conductance of a disordered metal is related to the sensitivity of its energy levels to changes in boundary condition. Typically, under a change of boundary condition, an energy level of a d-dimensional fragment of metal of size L^d ($L \gg \ell \gg \lambda_F$), is shifted by an amount

$$\delta E(L) \sim E_c(L) = \Delta(L) g(L).$$

If we assume that, under a change of the system size from L to $2L$, the change in the typical energy shift δE is a function of the sensitivity to such a change, and proportional to the fractional change, one expects

$$\frac{\delta E(2L)}{\Delta(2L)} - \frac{\delta E(L)}{\Delta(L)} = g(2L) - g(L) = 2f\big(g(L)\big).$$

These early ideas were assembled by Abrahams, Anderson, Licciardello, and Ramakrishnan [27] into a *one-parameter scaling theory of localization* according to which the dimensionless conductance obeys the Gell–Mann Low equation

$$\frac{d\ln g}{d\ln L} = \beta(g).$$

The scaling function $\beta(g) \equiv f(g)/g$ is universal, independent of the microscopic properties of the sample, such as the bare microscopic conductance and ℓ/λ_F. Reassuringly, this scaling picture fits neatly into the perturbative scheme of weak localization. In fact the weak localization correction (1.7) represents just the first term in a perturbative series which accumulate into a renormalization or scaling of the diffusion constant D.

The one-parameter scaling theory has profound consequences on the localization properties of disordered conductors. For an ohmic conductor, the scaling function takes a constant value $\beta = d - 2$, while deep in the insulating regime, where states are exponentially localised, $\beta(g) \sim \ln g$ (i.e., $g \sim e^{-L/\xi}$). A smooth interpolation between these limits (see Fig. 6.17) suggests that below two dimensions all states are localised, while above

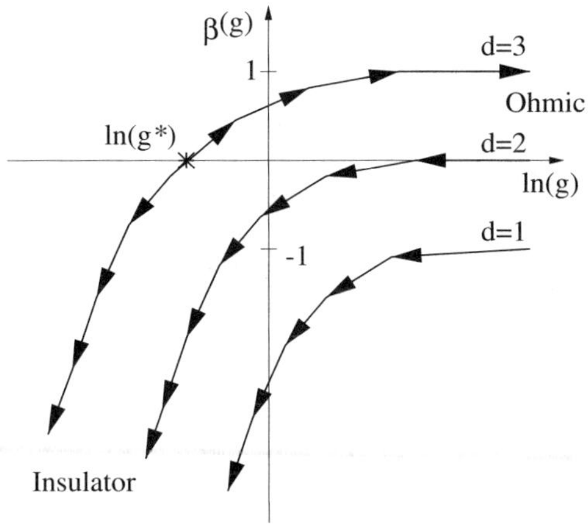

FIGURE 6.17. Scaling function of the dimensionless conductance, $\beta(g)$, in dimensions $d = 1$, 2 and 3. At large values of the conductance, the scaling function approaches the asymptotic $\beta(g) \to d - 2$ corresponding to Ohmic behavior. For very small values of conductance, the scaling function approaches $\beta(g) \to \ln(g)$ characteristic of insulating behavior. According to the weak localization expansion, a localization transition is predicted in dimensions *greater* than two.

there is a critical conductance, g_*, above which states are extended. The unstable fixed point is associated with the Anderson localization transition.

The situation in two dimensions is more delicate. Localization properties depend sensitively on the asymptotic approach of the β function to the metallic limit. However, taking the first quantum correction from Eq. (1.8), one obtains $\beta(g) = -1/\pi^2 g$ which is consistent with localization of all states in two dimensions. While the one-parameter scaling theory of localization has yet to find a truly rigorous mathematical proof, it nevertheless represents a important milestone in phenomenology. At the same time, trust in the phenomenology places important constraints on the theories of disordered materials.

1.3 Interaction Phenomena

Although, as we have seen, many of the characteristic mesoscopic phenomena can be satisfactorily understood within the framework of a noninteracting theory, the effects of Coulomb interaction give rise to a number of important manifestations in experiment including charging effects and dephasing. Indeed the slow dynamics of electrons in a disordered environment can lead to an enhancement of interaction effects. In this section, we will discuss several of the classic experiments in which interaction effects are important.

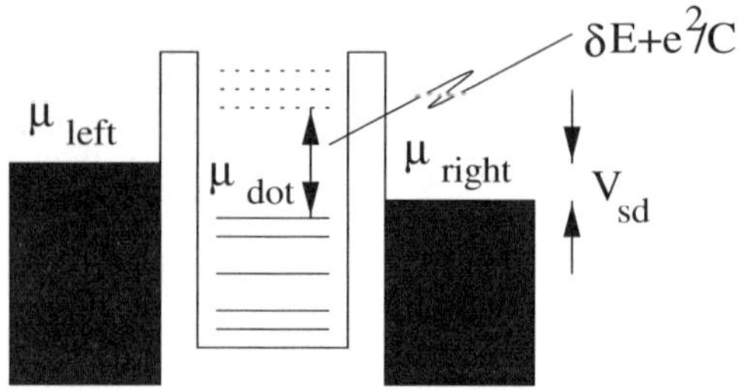

FIGURE 6.18. Schematic diagram showing the condition for charge transfer through a quantum dot. The continuum of states in the reservoirs are filled up to the chemical potential μ_{left} and μ_{right} which are related by the external Source-Drain voltage V_{sd}. Discrete states of the zero-dimensional quantum dot are filled up to a chemical potential μ_{dot}. Charge transfer is possible when the chemical potential μ_{dot} lies between μ_{right} and μ_{left}.

Coulomb Blockade

Perhaps the most direct and simplest manifestation of Coulomb interaction in charge transport arises in measurements of the resonant conductance through almost closed quantum structures. Previously, we saw that in open dots, two-terminal conductance measurements showed characteristic mesoscopic fluctuations—magneto-fingerprints—superposed on top of an average background (see, e.g., Fig. 6.1). Suppose that a changing gate potential narrows the constriction between the dot and the leads so that the number of open channels is substantially diminished. When the number of open channels falls below unity, an incoming electron experiences a tunnel barrier. In this regime, one would expect charge transport to be limited to resonant transmission through the discrete energy levels of the device when the chemical potential of the leads matches the dot.

Experimentally, resonant conductance peaks are observed as a function of the chemical potential (i.e., external gate voltage), but the separation of the peak spacing is typically controlled not by the single electron-energy level spacing of the dot Δ, but rather by a more substantial charging energy incurred when an electron is transferred onto the dot. In the simplest approximation, the Coulomb interaction of the electrons on the device can be taken into account phenomenologically by including in the Hamiltonian a "classical" charging energy $E_{\text{C}}(n) = n^2e^2/2C$, where C denotes the capacitance. According to this "orthodox model", charge transfer is forbidden unless the excess charging energy vanishes,

$$\big(E_c(n+1) - (n+1)eV_{\text{G}}\big) - \big(E_c(n) - eV_{\text{G}}n\big) = 0,$$

where V_{G} is a gate potential controlling the chemical potential of the dot. This condition under which charge transfer is allowed is met when $V_{\text{G}} = (n+\frac{1}{2})e^2/C$. Away from this value charge transfer is strongly suppressed—the "Coulomb Blockade".

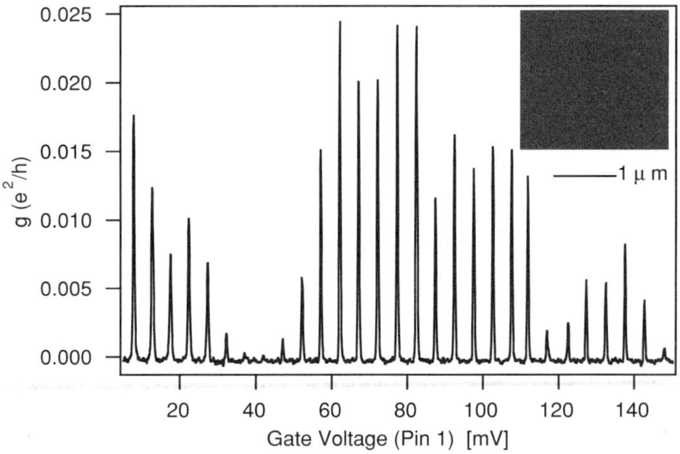

FIGURE 6.19. Resonant tunnelling conductance of a ballistic quantum dot (with the geometry as shown inset) (courtesy of Ref. 12).

Despite its simplicity, the orthodox model of the Coulomb Blockade gives a good qualitative explanation of experiment. Fig. 6.19 shows the resonant conductance measurements of a ballistic GaAs quantum dot. These results show resonant conductance peaks separated by a uniform energy scale reflecting the charging energy $e^2/2C$ of device. Fluctuations in the height of the peaks can be ascribed to mesoscopic fluctuations of the wavefunction amplitude at the contacts [28] (although their correlation from one peak to another is harder to explain).

Coulomb blockade oscillations are usually periodic in the charging energy. However, in a dot containing a very small number of electrons, both the electron-electron interaction and the effects of confinement become sufficiently strong that the spacings between the Coulomb blockade peaks becomes irregular. In such cases, the quantum dot behaves as an artificial atom showing "magic numbers" at which angular momentum states form a closed shell and electronic configurations are highly stable. Fig. 6.20 shows a grey-scale plot of the conductance as a function of gate and source-drain voltage of a device known as a vertical quantum dot [29]. At zero source-drain bias, the Coulomb blockade oscillations reveal stable configurations in which the total number of electrons on the dot are $N = 2$, 6 and 12 (cf. the "noble gas" configurations). At nonzero bias, features appear corresponding to the many-body electronic excitations in the dot.

Despite the apparent success of the orthodox theory, the agreement between theory and experiment reported in the early literature has, in recent years, been called into question. Correlation between the parametric dependences of the resonant peaks both as a function of changing gate voltage and magnetic field point at the existence of interaction effects which lie

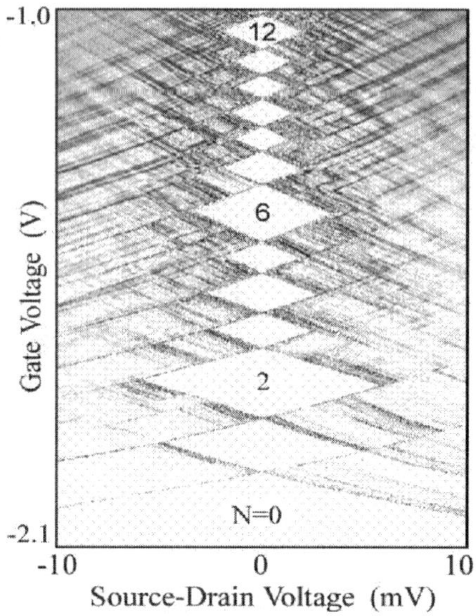

FIGURE 6.20. Measurements of the conductance of a vertical quantum dot as a function of gate and source-drain voltage (Reprinted with permission from Ref. 29. Copyright 1997 American Association for the Advancement of Science).

outside the orthodox scheme [30, 31]. The continuing attempts to reconcile these effects with many-body theories of electron interactions is the subject of on-going research.

Zero-Bias Anomaly

Surprisingly, an effect analogous to the Coulomb Blockade described above occurs in the tunnelling of a particle onto an *open* system (i.e., where the net charging energy vanishes). The transfer of an electron onto a metal involves two steps: Traversing the barrier, followed by spreading within the metal. Typically the traversal time is much shorter than the relaxation time in the electron liquid. Accordingly, it is legitimate to separate tunnelling into a single-electron, and a many-electron process. At small bias voltages, the first contribution, which is described by the transmission coefficient of the barrier, can typically be taken as constant. The second process by which the electron is accommodated on the device, involves the collective motion of a large number of electrons. At low bias voltages, the second collective effect may dominate the tunnelling rate.

The effect of charge relaxation on the tunnelling rate has a simple qualitative interpretation. Following the discussion of Ref. 32, let us consider a situation when at small bias voltages, one electron is transferred across a tunnel barrier into a disordered conductor. Since the traversal time is typically much shorter than the relaxation time in the metal, while the

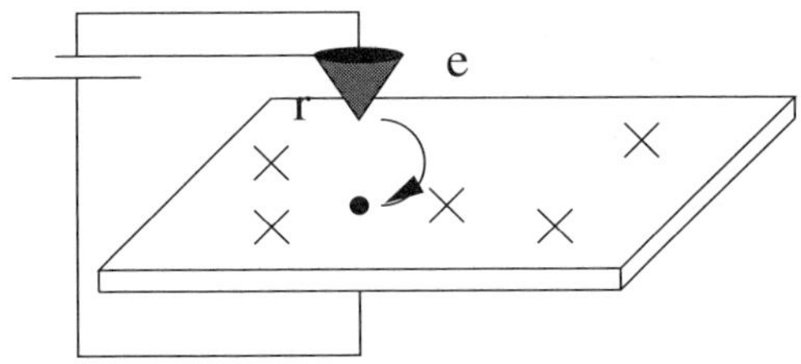

FIGURE 6.21. Schematic diagram showing the geometry for the measurement of the tunnelling conductance of a disordered metallic conductor.

electron traverses the barrier, other electrons practically do not move. Thus instantly, a large electrostatic potential is established, both due to the tunnelling electron itself, and due to the screening hole left behind. The jump in electrostatic energy by an amount much bigger than the bias eV means that immediately after the one electron transfer, the system is found to be in a classically forbidden state under the Coulomb barrier. In order to accomplish tunnelling, the charge yet has to spread over a large area, so that the potential of the charge fluctuation is reduced below eV. If the conductivity is finite, the spreading over large distances takes a long time, and thus the cooperative under-barrier action is much bigger than $\hbar$.

Taking the effective Coulomb interaction to be short-ranged, a crude estimate of the excess charging energy suggests $E_{\text{Coulomb}}(t) \sim \lambda \int d\mathbf{r}\, \rho(\mathbf{r},t)^2$ where λ denotes the interaction constant. If the charge relaxation is diffusive, the charge is distributed approximately uniformly over a radius $r(t) \sim (Dt)^{1/2}$ (i.e., $\rho(\mathbf{r},t) = 1/r(t)^d$). With this assumption, we obtain the following estimate for the excess action,

$$S_{\text{int}} \sim \int_\tau^{1/eV} dt\, E_{\text{Coulomb}} \sim \lambda \int_\tau^{1/eV} \frac{v_F dt}{k_F^{d-1}} \frac{1}{(Dt)^{d/2}}. \qquad (1.17)$$

Applied to the tunnelling conductance, $\delta G(V)/G \sim \delta\nu(V)/\nu \sim \exp[-S_{\text{int}}/\hbar]$ implying, for example, a logarithmic suppression of the tunnelling DoS in two dimensions, and a square root singularity in three [34].

Although we will not dwell upon it here, we remark that transport properties such as the conductivity and magneto-resistance are also strongly affected by interaction (for a review see Ref. 35). For example, if $k_B T > E_c$, the interaction correction to the conductivity is given by

$$\delta\sigma(T) \sim \begin{cases} \sqrt{T}, & d = 3 \\ \ln T, & d = 2, \end{cases}$$

and competes directly with the localization corrections.

Dephasing

A second important manifestation of Coulomb interaction is through the phenomenon of dephasing. In a clean metallic system, an estimate for the dephasing time associated with the electron-electron interaction is obtained from the "Golden rule". Taking into account the occupancy of the initial and final states, an estimate of the dephasing rate due to electron-electron interactions is given by

$$\frac{\hbar}{\tau_{\mathrm{ee}}(E)} \sim \int_0^E d\Omega \int_{-\Omega}^0 dE' \langle \mathbf{p}|\hat{U}|\mathbf{p}-\mathbf{q}\rangle\langle \mathbf{p}'+\mathbf{q}|\hat{U}|\mathbf{p}'\rangle \\ \times \delta(E-\zeta_{\mathbf{p}})\delta(E'-\zeta_{\mathbf{p}'})\delta(E-\Omega-\zeta_{\mathbf{p}-\mathbf{q}})\delta(E'+\Omega-\zeta_{\mathbf{p}'+\mathbf{q}}),$$

where $\zeta_{\mathbf{p}} \equiv \mathbf{p}^2/2m - E_F$. Defining the Fourier components of the screened Coulomb interaction $U_{\mathbf{q}} = 4\pi e^2/(\mathbf{q}^2+\kappa^2)$ where κ^{-1} represents the screening length, the scattering rate takes the form

$$\frac{\hbar}{\tau_{\mathrm{ee}}(E)} \sim \int_0^E \Omega\, d\Omega \int q^{d-1}\, dq \frac{|U_{\mathbf{q}}|^2}{(qv_F)^2} \sim \begin{cases} E^2/E_F, & d=3 \\ E^2\ln(E)/E_F, & d=2. \end{cases}$$

Here the parameter $1/qv_F$ can be identified with the typical time scale on which two electrons stay within a radius $1/q$ and interact. In the disordered environment, the typical interaction time is set, not by the ballistic scale, but instead by the diffusive scale $\mathrm{Re}[(1/(i\Omega+\hbar D\mathbf{q}^2)]$. Taking into account the screening of the Coulomb interaction, the dephasing rate takes the form

$$\frac{\hbar}{\tau_{\mathrm{ee}}(E)} = \frac{1}{\nu L_{\mathrm{E}}^d},$$

where $L_{\mathrm{E}} = (\hbar D/E)^{1/2}$ denotes the diffusion length.

Metal-Insulator Transition

Finally, perhaps the most striking manifestation of Coulomb interaction in the physics of disordered conductors concerns localization properties in two dimensions. As we have argued, according to the noninteracting theory, the mechanism of weak localization conspires to localise all states in two dimensions. However, for physical electrons, weak localization effects compete with Coulomb interaction corrections. The slow density relaxation in disordered conductors lead to an enhancement of Coulomb interaction effects which, in two dimensions, are manifest in logarithmic corrections to the conductivity.

Before reviewing the status of the theory of interaction effects and the metal-insulator transition in disordered media, we first turn to the experimental situation. Fig. 6.22a shows the resistivity of a two-dimensional electron doped Si MosFet device as a function of the charge density for various temperatures. For charge densities in excess of a critical density, the resistivity diminishes with increasing temperature characteristic of metallic behavior. While, below the critical density, the resistivity increases with

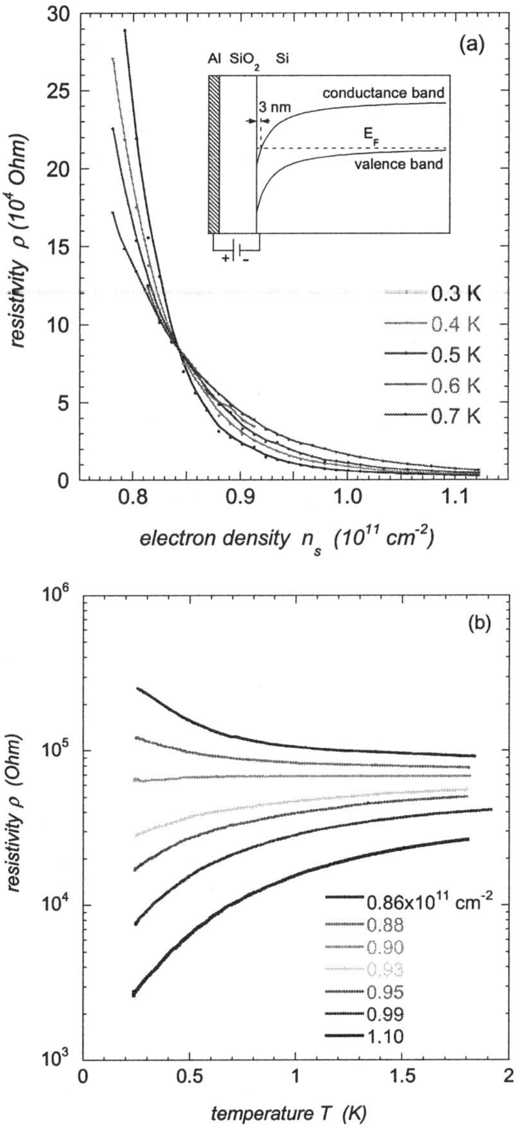

FIGURE 6.22. Resistivity measurements of Si MosFet device as a function of (a) changing density, and (b) temperature. (Courtesy of Ref. 36. Copyright 1999 National Academy of Sciences, U.S.A.)

temperature characteristic of insulating behavior. The critical density has been associated with a mobility edge signalling the existence of metal-insulator transition. The range over which scaling behavior of the resistivity is observed is indicated in Fig. 6.22b.

Since the identification of a metal-insulator transition in electron doped Si MosFet devices, apparent transitions have now been reported in both electron and hole doped GaAs structures. Nevertheless, a consensus on the existence of a metal-insulator transition in two dimensions has yet to be established.

The experimental detection a metal-insulator transition have motivated a reinvestigation of the influence of Coulomb interactions in a disordered environment. However, on the theoretical side, the situation is more poorly developed. Existing field theories of interaction and disorder are applicable only in weak coupling. These results indicate that interaction effects can compete with weak localization and lead to an enhancement of the conductivity in two dimensions [37]. However, the scaling theory flows into a regime of strong coupling outside the regime in which the field theory is controlled. The nature of the strong coupling fixed point is a matter for heated debate. The field is still very much open.

This completes our superficial survey of just some of the manifestations of interaction in disordered metallic systems. It is largely included here as a salutary reminder that the experimental environment is often more complex than the models which will address in the following lectures. Indeed, in the majority of the remainder of the course we will neglect the influence of interaction altogether. We remark that there exists a number of reviews on the influence of interaction and disorder including Refs. 38–42, and 43.

1.4 *Impurity Diagram Technique*

For completeness, as a footnote to this section, we include a more formal discussion of the "impurity diagram technique" [23, 44]. Our aim will be to find formal expressions for the ensemble average of the single-particle and two-particle Green function.

Single-Particle Green Function

The starting point of the diagrammatic perturbation theory is the expansion of the single-particle Green function as an infinite series in V,

$$\widehat{G}^- \equiv (\widehat{G}_0^{-1} - V)^{-1} = \widehat{G}_0^{-1} + \widehat{G}_0^{-1} \sum_{n=1}^{\infty} (V\widehat{G}_0^{-1})^n .$$

The latter can be arranged as a diagrammatic series in which the bare or noninteracting Green function is represented by a line and the potential V by a cross (see Fig. 6.23a). According to the properties of a Gaussian δ-

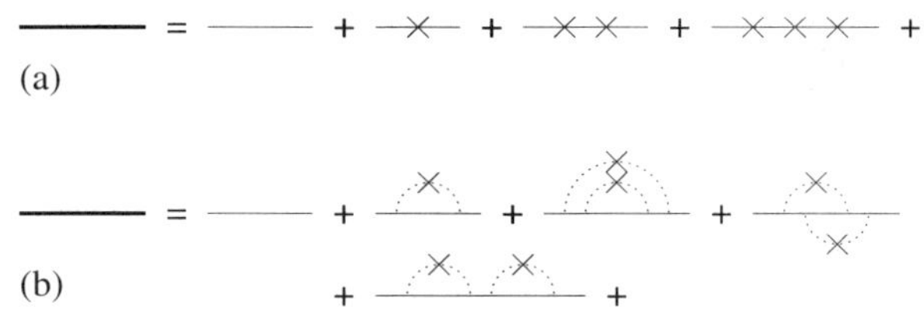

FIGURE 6.23. (a) Diagrammatic representation of the perturbative expansion of the single-particle Green function. (b) Representation of the ensemble average of the Green function. Impurities are connected pairwise by the Gaussian δ-correlated potential.

correlated impurity distribution (1.4), an ensemble average over the random potential generates all possible contractions of V (see Fig. 6.23b).

Separate contributions to the ensemble average can be arranged into reducible and irreducible parts. Diagrams that can be split into two by cutting a single G_0 line are called reducible (see, for example, the last diagram shown in Fig. 6.23b). The latter can be built up out of irreducible diagrams. Collecting together all the irreducible diagrams, the impurity averaged Green function can be written in the form of a *Dyson equation*

$$\langle\widehat{G}\rangle = \widehat{G}_0 + \widehat{G}_0\widehat{\Sigma}\langle\widehat{G}\rangle = \frac{1}{\widehat{G}_0^{-1} - \widehat{\Sigma}}, \tag{1.18}$$

where $\widehat{\Sigma}$ denotes the self-energy operator.

An exact evaluation of all diagrams contributing to the self-energy is impossible. However, in the semi-classical limit $1/k_F\ell \ll 1$, where $\ell = v_F\tau$ represents the mean-free path, the self-energy is, to a good approximation, given by the *self-consistent Born approximation* (SCBA) (see Fig. 6.7),

$$\Sigma^{\pm}(\mathbf{p}) \simeq \int (d\mathbf{q})\langle V(\mathbf{q})G_0^{\pm}(\mathbf{p}-\mathbf{q}; E \pm i0)V(-\mathbf{q})\rangle,$$

where we have adopted the shorthand, $(d\mathbf{q}) = d^dq/(2\pi)^d$. The real part of the self-energy simply leads to an irrelevant shift in energy, while the imaginary part is given by

$$\operatorname{Im}\Sigma^{\pm} = \frac{1}{2\pi\nu\tau}\int (d\mathbf{p}) \operatorname{Im} G_0(\mathbf{p}; E \pm i0) = \mp\frac{1}{2\tau}.$$

Diagrams excluded in the approximation to the self-energy are smaller by a factor $1/k_F\ell$. Substituting the self-energy into the expression for the impurity average Green function (1.18), we obtain the result of shown in Eq. (1.5).

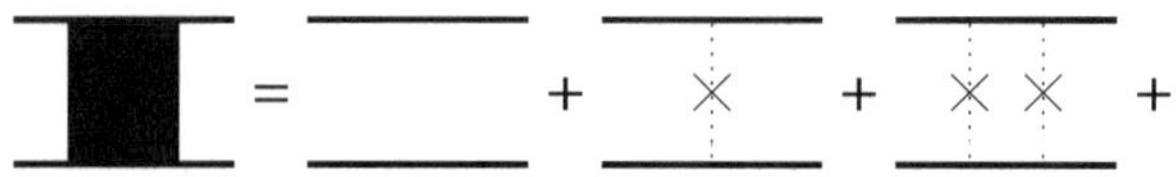

FIGURE 6.24. Diagrammatic representation of the ladder series associated with the average two-particle Green function.

Two-Particle Green Function

The transfer probability density involves the two-particle Green function $P(\mathbf{r},0;t) = |G(\mathbf{r},0;t)|^2$. In Fourier space, the latter can be represented as

$$P(\mathbf{q},\Omega) = \int (d\mathbf{p}_1) \int (d\mathbf{p}_2) \int (d\mathbf{p}) \int dE\, G(\mathbf{p}_1, \mathbf{p}+\mathbf{q}/2; E+\Omega_+) \\ \times G(\mathbf{p}-\mathbf{q}/2, \mathbf{p}_2; E-\Omega_+),$$

where $\Omega_+ = \Omega + i0$. Expanding the Green functions as a diagrammatic series in V and subjecting the product to an ensemble average, one can identify the dominant contribution as that arising from the connection of the Green function lines by a ladder series (see Fig. 6.24). Contributions from diagrams in which the impurity lines are crossed are small by a factor of $1/k_F\ell$. More precisely, one finds that

$$\begin{aligned}\langle G(\mathbf{p}_1, \mathbf{p}+\mathbf{q}/2; E+\Omega_+/2) G(\mathbf{p}_2, \mathbf{p}-\mathbf{q}/2; E-\Omega_+/2)\rangle_c \\ = \bar{G}(\mathbf{p}_1; E+\Omega_+/2)\bar{G}(\mathbf{p}-\mathbf{q}/2; E-\Omega_+/2) \\ \times \Gamma(\mathbf{q},\Omega)\bar{G}(\mathbf{p}+\mathbf{q}/2; E+\Omega_+/2)\bar{G}(\mathbf{p}_2; E-\Omega_+/2),\end{aligned}$$

where $\bar{G}(\mathbf{p}; E \pm i0) = \langle G(\mathbf{p}-\mathbf{q}, \mathbf{q}_1; E \pm i0)\rangle = (E - p^2/2m \pm i/2\tau)^{-1}$ represents the impurity average Green function, and

$$\begin{aligned}\Gamma(\mathbf{q},\Omega) &= \frac{1}{2\pi\nu\tau}[1 + \Pi(\mathbf{q},\Omega)\Gamma(\mathbf{q},\Omega)], \\ \Pi(\mathbf{q},\Omega) &= \int (d\mathbf{p})\, \bar{G}(\mathbf{p}+\mathbf{q}/2, E+\Omega_+/2)\bar{G}(\mathbf{p}-\mathbf{q}/2, E-\Omega_+/2).\end{aligned}$$

From this result we obtain $\Gamma(\mathbf{q},\Omega) = \big(2\pi\nu\tau - \Pi(\mathbf{q},\Omega)\big)^{-1}$. For small $\mathbf{q}$ and Ω,

$$\Pi(\mathbf{q},\Omega) \simeq 2\pi\nu\tau L^d(1 + i\Omega\tau - D\mathbf{q}^2\tau),$$

where $D = v_F^2\tau/d$ denotes the classical diffusion constant. Altogether, we obtain the diffusion propagator,

$$\Gamma(\mathbf{q},\Omega) = \frac{1}{L^d}\frac{1}{2\pi\nu\tau^2}\frac{1}{D\mathbf{q}^2 - i\Omega}.$$

2 Field Theory of Disordered Conductors

Having laid the conceptual foundation for mesoscopic physics it is now time to put flesh on these ideas by developing a formal theory. In the previous section, our terminology implies the existence of a diagrammatic perturbation theory whose validity depended on the quasi-classical parameter $1/k_F\ell \ll 1$. Indeed, within such an approach, a consistent theory of weak localization and coherence phenomena can be developed. However, the accumulation of quantum interference corrections within the framework of the one-parameter scaling theory, together with the appearance of IR divergences of the diagrammatic perturbation theory hints at the existence of a higher unifying scheme in which the diffusion modes of density relaxation are associated with soft modes of a quantum field theory.

In fact, the coherence properties of weakly disordered conductors can be described within the framework of a statistical field theory in which all the mesoscopic phenomena described in the previous section can be understood. Focusing on the properties of an "n-orbital" Hamiltonian, Wegner [45] established that the diffusion modes of the diagrammatic perturbation theory were associated with the soft modes of a functional nonlinear σ-model. The renormalization properties of this statistical field theory are compatible with diagrammatic perturbation theory and the one-parameter scaling. Within this theory, the breakdown of perturbation theory is understood as a consequence of the proliferation of massless Goldstone modes associated with the mechanism of quantum interference.

With the benefit of hindsight, we can draw a formal analogy between the disordered conductor and the classical O(3) ferromagnet. The classical statistical mechanics of the ferromagnet is also represented by a nonlinear σ-model, of the form

$$\mathcal{Z} = \int_{\mathbf{S}^2=1} \exp\left[-\int K_{\text{sw}}(\partial \mathbf{S})^2 - \mathbf{H}\cdot\mathbf{S} \right], \tag{2.1}$$

where K_{sw} denotes the spin wave stiffness, and H represents an external magnetic field. As with the disordered metal, the ferromagnet exhibits a branch of low-lying excitations describing spin wave modes, $(K_{\text{sw}}\mathbf{q}^2+H)^{-1}$. If the magnetic field is strong, a controlled diagrammatic perturbation theory of the ferromagnet can be formally established. At the same time, the renormalization properties of the spin wave stiffness can be developed by studying the influence of spin-wave interactions. However, by diminishing the magnetic field, the spin wave modes become massless and perturbation theory breaks down. Associating the spin wave stiffness K_{sw} with the diffusion constant D, and the magnetic field H with the frequency scale of the two-particle response $i\Omega$, the analogy is complete.

The development of the quantum field theory of disordered conductors has a history dating back to the pioneering work of Wegner [45]. In the following, for reasons that will become clear, we will develop a more recent version of the field theory introduced by Efetov [46]. Our analysis will concentrate on establishing the fundamental principles behind the theory (often at the expense of technical detail). For a more comprehensive discussion the reader is referred to the existing reviews on the subject, particularly those by Efetov [47], and Weidenmüller, Verbaarschot and Zirnbauer [48]. The material developed in this section will form the backbone for the remainder of the course.

2.1 Derivation of the σ-Model Action

As preparation for the formal analysis, we begin by presenting a rough outline of the derivation of the effective field theory. Previously we saw that spectral and transport properties can be expressed through the correlators of the Green function $\widehat{G}(E) = (E - \widehat{H})^{-1}$. The starting point of the field theoretic approach is the representation of the Green function (or rather their product) as a Gaussian functional field integral. When subjected to an ensemble average over configurations of the impurity potential, an interaction amongst the fields is generated. At this stage, a power series expansion of the action in the interaction can be assembled into the diagrammatic perturbation theory described qualitatively in Section 1. Instead, anticipating the existence of long-range diffusion modes, we will introduce a Hubbard–Stratonovich decoupling of the interaction where the auxiliary fields Q are associated with the soft modes of the theory.

Again, it is useful to draw a formal analogy, this time with the BCS theory of superconductivity. The quantum partition function of the bare BCS Hamiltonian, is also expressed in terms of an interacting quantum field theory. Anticipating the slow fields of the theory to be associated with the formation of the Cooper pair condensate, the interaction can be decoupled by a Hubbard–Stratonovich transformation involving the BCS order parameter, Δ. When subjected to a saddle-point, or mean-field approximation, the action recovers the "Gap equation" for Δ. Taking into account fluctuations around the saddle-point, a gradient expansion obtains the familiar Ginzburg–Landau theory. Neglecting the massive fluctuations of the amplitude of the order parameter, the massless phase fluctuations are described by an action of nonlinear σ-model type.

In the present case, the saddle-point field configuration of Q is associated with the self-consistent Born approximation for the average single-particle Green function. As with the superconductor, subjecting the action to a gradient expansion in the vicinity of the saddle-point we will in turn obtain a nonlinear σ-model in which the low-lying modes represent the diffusion modes of the theory.

Finally, before we embark on this program, we should keep in mind that the Ginzburg–Landau theory of superconductivity found its origin in phenomenological considerations. The structure of the theory is fully constrained by the fundamental symmetries of the microscopic Hamiltonian. In principle, identifying the fundamental symmetries, one could legitimately apply the same line reasoning to develop the "Ginzburg–Landau theory" of the disordered conductor. However, for the sake of clarity, we will stick to the more pedestrian route of presenting a formal derivation of the σ-model action.

Field Integral

Following the outline our first task is to represent the Green function of the stochastic Hamiltonian (1.3) as a Gaussian functional field integral. Applied to the advanced function,

$$\begin{aligned} G^-(\mathbf{r},\mathbf{r}';E_F) &\equiv \left\langle \mathbf{r}' \left| \frac{1}{E_F^- - \widehat{H}} \right| \mathbf{r} \right\rangle \equiv \sum_\nu \frac{\psi_\nu(\mathbf{r}')^* \psi_\nu(\mathbf{r})}{E_F^- - E_\nu} \\ &= \frac{i}{\mathcal{Z}} \int DS S(\mathbf{r}) S^*(\mathbf{r}') e^{i \int S^* (\widehat{H} - E_F^-) S} \end{aligned}$$

where $\mathcal{Z}[V]$ denotes the constant of normalisation, $E_F^- \equiv E_F - i0$, and $\widehat{H}$ represents the Hamiltonian (1.3) of a single particle subject to a weak impurity potential. Here we have deliberately introduced the notation E_F for the energy to remind us that the *noninteracting* theory developed below is tailored to the consideration of large Fermi energy scales in the disordered electron gas. Applied to a single-particle Hamiltonian, one might wonder why it is necessary to invoke a functional field integral as opposed to a Feynman path integral—indeed the latter formed the basis of our conceptual discussions in Section 1. In fact, while it is possible to reproduce the diagrammatic perturbation theory from ensemble averages of the Feynman path integral, for reasons that will become clear, a trajectory representation seems to be inappropriate as a vehicle to describe the soft modes.

Expressed in this form, the presence of the normalisation factor $\mathcal{Z}[V]$, which depends explicitly on the impurity potential V, makes ensemble averaging difficult. The neglect of this dependence leads to unphysical vacuum loops when G^- is subjected to an ensemble average. To circumvent these difficulties, various "tricks" have been introduced in the literature. Originally conceived to study the classical spin glass, the "Replica Trick" [49, 50] exploits the formal identity

$$\ln \mathcal{Z} = \lim_{n \to 0} \frac{\mathcal{Z}^n - 1}{n}, \tag{2.2}$$

to express ensemble averages of the generating function. However, experience has shown considerable difficulty in correctly implementing the analytic continuation $n \to 0$ in a way that respects the nonperturbative sector of the theory [51, 52]. While very recent advances hint at a resolution of this problem within the framework of "replica symmetry breaking" [53, 54], fortunately, at least for systems which are noninteracting, a second and more reliable approach can be developed.

The "supersymmetry method" exploits properties of the Grassmann algebra to implement a scheme of "book-keeping". In particular, expressed as a field integral over two-component superfields $\psi^T = (S, \chi)$, the Green function can be expressed as

$$G^-(E_F) = i \int D\psi S S^* e^{i\int \psi^\dagger(\widehat{H}-E_F^-)\psi}.$$

Here, exploiting the properties of Gaussian integrals over complex commuting or bosonic (B) variables, $\int DSe^{-S^*MS/2} = \det M^{-1}$, and anti-commuting or fermionic (F) variables, $\int D\chi e^{-\bar{\chi}M\chi/2} = \det M$, the normalisation $\mathcal{Z} = 1$ is assured. (A comprehensive introduction to superalgebra can be found in Ref. 55.)

A doubling of the field space, allows the representation of *two-point* correlators or response functions to be expressed. More precisely, the product of advanced (A) and retarded (R) Green functions can be expressed as

$$G^-(E_F - \Omega/2)G^+(E_F + \Omega/2) \\ = -\int D\psi S_- S_-^* S_+ S_+^* e^{i\int \bar{\psi}(\widehat{H}-E_F+\Omega^+\sigma_3^{\rm AR}/2)\psi}$$

where $\Omega^+ \equiv \Omega + i0$ denotes the frequency source, and the field integral involves four-component superfields

$$\psi = \begin{pmatrix} \psi_- \\ \psi_+ \end{pmatrix}_{\rm AR}, \quad \psi_\mp = \begin{pmatrix} S_\mp \\ \chi_\mp \end{pmatrix}_{\rm BF}.$$

Setting $L = \sigma_3^{\rm AR} \otimes E_{11}^{\rm BF} + \mathbf{1}^{\rm AR} \otimes E_{22}^{\rm BF}$, where $E_{11}^{\rm BF}$ and $E_{22}^{\rm BF}$ project onto the bosonic and fermionic sector respectively, the superfields have the property $\bar{\psi} = \psi^\dagger L$.

Pauli matrices $\sigma_3^{\rm BF}$ and $\sigma_3^{\rm AR}$ respectively break the symmetry between boson/fermion and advanced/retarded degrees of freedom. Convergence of the field integral determines the form of the metric L in the BB sector. While the choice in the fermionic sector seems arbitrary, a consideration of the saddle-point manifold of the σ-model determines the form above [48].

Finally, our preliminary discussion in the previous section classified two modes of density relaxation in the disordered metal, Diffusons and Cooperons. Both will be identified in the analysis that follows. Accordingly, anticipating the structure of the saddle-point action, it is convenient to further double the field space to include complex or time-reversed components,

$$\Psi = \frac{1}{\sqrt{2}} \begin{pmatrix} \psi \\ \psi^* \end{pmatrix}.$$

The elements of the newly defined supervector Ψ are not independent, but fulfill the "time-reversal" symmetry relation

$$\Psi^\dagger = (C\Psi)^T,$$

where $C = \sigma_1^{\rm TR} \otimes E_{11}^{\rm BF} + i\sigma_2^{\rm TR} \otimes E_{22}^{\rm BF}$. Under the transformation $\psi \mapsto \psi^*$, the Hamiltonian maps to the time-reversed counterpart $\widehat{H} \mapsto \widehat{H}^T$. In fact, any discrete symmetry of the microscopic Hamiltonian doubles the number of low-lying modes of density relaxation. In each case, it is convenient to double the field space [56].

Finally, instead of representing each vertex separately, it is often more convenient to introduce a "source term" into the effective action from which arbitrary correlators can be constructed,

$$\mathcal{Z}[J] = \int D\Psi e^{\int [i\bar{\Psi}(\widehat{H} - E_F + \Omega^+ \sigma_3^{\mathrm{AR}}/2)\Psi + \bar{J}\Psi + \bar{\Psi}J]}. \tag{2.3}$$

Here the sources $\bar{J}$ and J obey the same symmetry properties as $\bar{\Psi}$ and Ψ. Hereafter, for the sake of clarity, we will turn off the source (i.e., $\mathcal{Z}[J = 0]$), reinstating it only when essential.

Impurity Averaging

Expressed as a functional field integral involving superfields, the ensemble average over a Gaussian δ-correlated impurity distribution

$$P(V)DV = \exp\left[-\pi\nu\tau \int V^2\right] DV$$

can be applied straightforwardly. (Here, and henceforth, we set $\hbar = 1$.) Applied to the generating function,

$$\langle \mathcal{Z}[J = 0] \rangle = \int D\Psi e^{\int [i\bar{\Psi}(\widehat{\mathbf{p}}/(2m) - E_F + \Omega^+/2\sigma_3^{\mathrm{AR}})\Psi - \frac{1}{(}\bar{\Psi}\Psi)^2/(4\pi\nu\tau)]}$$

where $\langle \cdots \rangle = \int (\cdots) P(V) DV$ represents the ensemble average. In the absence of the symmetry breaking sources, Ω and J, the total action is seen to be invariant under pseudo-unitary rotations of the fields in superspace,

$$\Psi \to U\Psi, \quad U^\dagger L U = L, \quad U \in \mathrm{SU}(2, 2 \,|\, 4).$$

At this stage, an expansion of the total action in powers of the field interaction can be assembled into a diagrammatic series expansion from which the *perturbative* results of Section 1 can be established. However, anticipating the dominance of contributions arising from long-range fluctuations of the density, we will instead seek a mean-field decomposition of the action.

Hubbard–Stratonovich Decoupling

To establish a useful mean-field decoupling of the interaction requires the identification of the soft modes of the theory, i.e., those modes that arise from two-particle channels undergoing multiple scattering with the exchange of momenta smaller than the inverse of the elastic mean-free path, $\ell = v\tau$. The isolation of these modes is conveniently performed in the

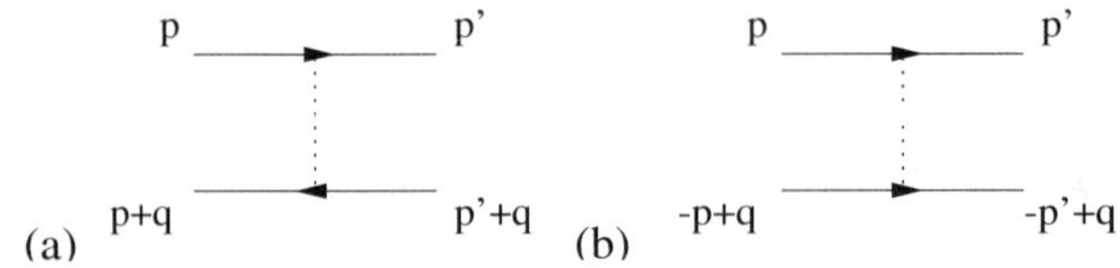

FIGURE 6.25. Diagrammatic representation of the terms in the Hubbard–Stratonovich decoupling that lead to (a) the diffuson mode, and (b) the Cooperon mode.

Fourier space. Applied to the interaction of the fields, the relevant contributions arise from the channels

$$\int(\bar\psi\psi)^2 \approx \sum_{\mathbf{p},\mathbf{p}'}\sum_{|\mathbf{q}|\ll\ell^{-1}} [(\bar\psi_{-\mathbf{p}'}\psi_{\mathbf{p}})(\bar\psi_{-\mathbf{p}-\mathbf{q}}\psi_{\mathbf{p}'+\mathbf{q}}) + (\bar\psi_{-\mathbf{p}'}\psi_{\mathbf{p}})(\bar\psi_{\mathbf{p}'-\mathbf{q}}\psi_{-\mathbf{p}+\mathbf{q}})]$$

where the first and second terms generate the contributions from the Diffuson and Cooperon respectively, and the summation over $\mathbf{q}$ is limited to low momentum transfer. Both processes are represented diagrammatically in Fig. 6.25.

Introducing the Hubbard–Stratonovich decoupling

$$\exp\left[-\int\frac{1}{4\pi\nu\tau}\,\mathrm{str}(\Psi\otimes\bar\Psi)^2\right] = \int DQ\exp\left[-\frac{1}{2\tau}\int\left(\bar\Psi Q\Psi - \frac{\pi\nu}{4}\,\mathrm{str}\,Q^2\right)\right], \quad (2.4)$$

the contributions to the Diffuson and Cooperon degrees of freedom are accounted for by *slow* fluctuations of the 8×8 supermatrix fields, Q. Here the supertrace operation is defined by $\mathrm{str}\,M = \mathrm{tr}\,M_{\mathrm{BB}} - \mathrm{tr}\,M_{\mathrm{FF}}$. (Note that, as written, the Hubbard–Stratonovich transformation is not exact. If *all* degrees of freedom of Q (fast and slow) are taken into account, the decoupling (2.4) involves an over-counting by a factor of 2. This is because the saddle-point corresponding to the Cooperonic sector (i.e., those degrees of freedom which anti-commute with the symmetry breaking matrix σ_3^{TR}) can be found in fast fluctuations of the diffuson sector, and *vice versa*.)

Finally, the symmetry properties of the dyadic product $\Psi\otimes\bar\Psi$ are reflected in the symmetry properties of Q by the constraint

$$Q = CLQ^T LC^T, \quad (2.5)$$

where the transpose of a supermatrix is defined by

$$M^T = \begin{pmatrix} M_{\mathrm{BB}}^T & M_{\mathrm{FB}}^T \\ -M_{\mathrm{BF}}^T & M_{\mathrm{FF}}^T \end{pmatrix}.$$

Gaussian in the fields Ψ, the functional integral obtains

$$\langle\mathcal{Z}[0]\rangle = \int DQ\exp\left[\int\mathrm{str}\left(\frac{\pi\nu}{8\tau}Q^2 - \frac{1}{2}\ln\widehat{\mathcal{G}}^{-1}\right)\right],$$

where

$$\widehat{\mathcal{G}}^{-1} = \frac{\widehat{\mathbf{p}}^2}{2m} - E_F + \frac{\Omega^+}{2}\sigma_3^{\mathrm{AR}} + \frac{i}{2\tau}Q, \tag{2.6}$$

represents the supermatrix Green function. In summary, the problem of computing impurity averaged Green functions has been reduced to the consideration of the Euclidean field theory which depends on an 8×8 supermatrix field Q.

Saddle-Point Equation

Following the procedure outlined at the beginning of this section, we now subject the field integral to a saddle-point analysis. The legitimacy of such a mean-field decomposition is ensured by the symmetry properties of the Hubbard–Stratonovich fields which were chosen to reflect the soft modes of the disordered Hamiltonian. A variation of the action with respect to Q generates the saddle-point equation

$$Q_{\mathrm{sp}}(\mathbf{r}) = \frac{i}{\pi\nu}\mathcal{G}(\mathbf{r},\mathbf{r}). \tag{2.7}$$

Taking the symmetry breaking sources Ω and J to be vanishingly small, and applying the ansatz that the saddle-point solution Q_{sp} is spatially constant, and diagonal in the internal indices, the saddle-point equation (2.7) takes the form

$$\begin{aligned} Q_{\mathrm{sp}} &= \frac{i}{\pi\nu}\int d\mathbf{p}\left[\frac{\mathbf{p}^2}{2m} - E_F + i0 + \frac{i}{2\tau}Q_{\mathrm{sp}}\right]^{-1} \\ &= \frac{i}{\pi}\int d\xi\left[\xi - E_F + \frac{i}{2\tau}Q_{\mathrm{sp}}\right]^{-1}. \end{aligned}$$

From this equation, one can attach to Q_{sp} the interpretation of the self-energy of the impurity averaged single-particle Green function. Performing the integral over ζ, we deduce that the elements of Q_{sp} take values of ± 1. To choose the signs correctly, we note that the expression on the right-hand side of the saddle-point equation relates to the Green function for the disordered system in the self-consistent Born approximation. The disorder preserves the causal (i.e., advanced versus retarded) character of the Green function, therefore the signs of the diagonal elements of Q_{sp} must coincide with the sign of the imaginary part of the energy. This singles out the solution $Q_{\mathrm{sp}} = \sigma_3^{\mathrm{AR}}$.

In the limit $\Omega \to 0$, the action is invariant under the global transformations $Q(\mathbf{r}) \mapsto TQ(\mathbf{r})T^{-1}$, where T represent pseudo-unitary rotations that are constant in space and compatible with the symmetry properties of Q (2.5). This implies that the saddle-point solution spans the nonlinear manifold $Q_{\mathrm{sp}}^2 = \mathbf{1}$ (i.e., $Q_{\mathrm{sp}} = T\sigma_3^{\mathrm{AR}}T^{-1}$). Dividing out rotations that leave σ_3^{AR} invariant, the degeneracy of the manifold is specified by the factor space $\mathrm{SU}(2,2\,|\,4)/\,\mathrm{SU}(2\,|\,2) \otimes \mathrm{SU}(2\,|\,2)$.

Substituting $Q_{\rm sp}$ into Eq. (2.6), we obtain the following expression for the supermatrix Green function at the saddle-point:

$$\mathcal{G}_{\rm sp}(\mathbf{r},\mathbf{r}') = -i\pi\nu f_d(|\mathbf{r}-\mathbf{r}'|)Q_{\rm sp},$$

where defining $G_0^-(\mathbf{r}) \equiv \langle G^-(\mathbf{r},0)\rangle$, f_d denotes the "Friedel function"

$$f_d(r) \equiv \frac{\operatorname{Im} G_0^-(\mathbf{r})}{\operatorname{Im} G_0^-(0)} = \Gamma(d/2)\left(\frac{2}{kr}\right)^{d/2-1} J_{d/2-1}(kr)e^{-r/(2\ell)},$$

which decays on a scale comparable to the mean free path.

Gradient Expansion

In the semi-classical limit (i.e., $1/E_F\tau \ll 1$ or, equivalently, $1/k_F\ell \ll 1$) spatial fluctuations in the vicinity of the saddle-point manifold can be separated into a massive and massless sector (see Fig. 6.26). To leading order in the parameter $1/E_F\tau$, the massive fluctuations do not contribute to the low-energy effective action. Instead, we will focus on the low-energy spatial fluctuations which preserve the nonlinear constraint $Q(\mathbf{r})^2 = \mathbf{1}$. Anticipating that the relevant field configurations $T(\mathbf{r})$ fluctuate only slowly as a function of $\mathbf{r}$, it makes sense to separate the momentum operator into two parts, $\widehat{\mathbf{p}} = \mathbf{p} + \widehat{\mathbf{q}}$, where the operator $\widehat{\mathbf{q}} = -i\partial$ acts exclusively on the fields $Q(\mathbf{r})$, while the momenta $\mathbf{p}$ are associated with the trace. Taking $\widehat{\mathbf{p}}^2/2m \simeq p^2/2m + \mathbf{p}\cdot\widehat{\mathbf{q}}/m$, an expansion of the action yields

$$S[Q] = \frac{1}{2}\,\mathrm{str}_{\mathbf{r},\mathbf{p}} \ln\left[G_0^{-1} - T^{-1}\left(i\frac{\Omega^+}{2}\sigma_3^{\rm AR} + \frac{1}{m}\mathbf{p}\cdot\widehat{\mathbf{q}}\right)T\right]$$

where $G_0^{-1}(\mathbf{p}) = E - \mathbf{p}^2/2m - i\sigma_3^{\rm AR}/2\tau$, and $\mathrm{str}_{\mathbf{r},\mathbf{p}}$ denotes a trace over internal indices, and the phase space coordinates.

Taking $1/E_F\tau$, $1/\Delta\tau$ and $L/\ell \ll 1$, where L denotes the typical scale at which the supermatrices fluctuate, an expansion in the 'slow' operators $\widehat{\mathbf{q}}$

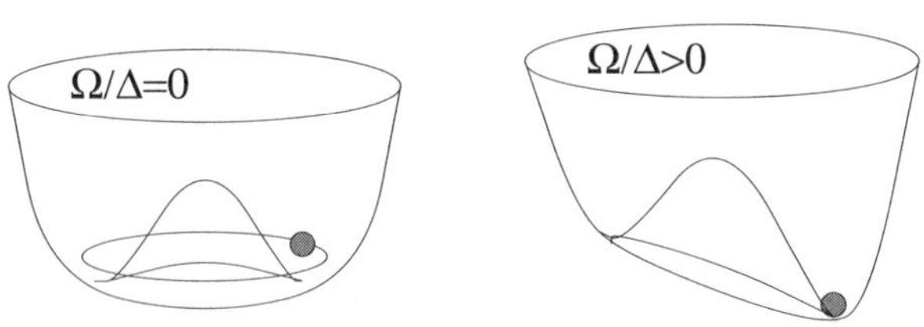

FIGURE 6.26. Schematic diagram showing the structure of the saddle-point manifold. For $\Omega/\Delta \gg 1$, there is a unique saddle-point. As $\Omega/\Delta \to 0$, the saddle-point opens to span an entire manifold. Spatial fluctuations around the saddle-point can be classified into a massive and massless sector. The massive sector is controlled by the semi-classical parameter $1/E_F\tau$.

to lowest nonvanishing order obtains

$$S[Q] \simeq \frac{1}{2}\,\mathrm{str}_{\mathbf{r},\mathbf{p}}\left[G_0T^{-1}i\frac{\Omega^+}{2}\sigma_3^{\mathrm{AR}}T + \frac{1}{2}\left(G_0T^{-1}\frac{\mathbf{p}\cdot\widehat{\mathbf{q}}}{m}T\right)^2\right].$$

(Note that the term linear in the fast momenta $\mathbf{p}$ vanishes under the trace.)

To prepare for the tracing out of the fast momenta, we first formulate some useful identities involving the bare average Green function G_0. All the relations below can be proved straightforwardly by explicitly performing the momentum integrations and using some Pauli-matrix identities.

- Firstly, in the momentum representation, the Green function can be presented as

$$G_0(\mathbf{p}) = \frac{1}{2}\sum_{s=\pm 1}\frac{1+s\sigma_3^{\mathrm{AR}}}{E_F - p^2/2m + is/2\tau}.$$

- In this form, the trace over internal momentum variables obtains:

$$\int (d\mathbf{p})G_0(\mathbf{p}) = \text{const.} \times \mathbf{1}^{\mathrm{AR}} - i\pi\nu\sigma_3^{\mathrm{AR}}.$$

- Finally, if operators $\hat{\mathbf{A}}$ and $\widehat{\mathbf{B}}$ vary slowly in space, then

$$\int (d\mathbf{p})\,\mathrm{str}[G_0(\mathbf{p})\mathbf{p}\cdot\hat{\mathbf{A}}G_0(\mathbf{p})\mathbf{p}\cdot\widehat{\mathbf{B}}] = \frac{m^2}{4}2\pi D\nu\sum_s \mathrm{str}[(1+s\sigma_3^{\mathrm{AR}})\hat{\mathbf{A}}\cdot(1-s\sigma_3^{\mathrm{AR}})\widehat{\mathbf{B}}], \quad (2.8)$$

  where $D = v_F^2\tau/d$ denotes the classical diffusion constant, with $v_F = p_F/m$.

A straightforward application of these identities to the effective action above leads to the nonlinear σ-model action,

$$S[Q] = -\frac{\pi\nu}{8}\int \mathrm{str}[D(\partial Q)^2 + 2i\Omega^+\sigma_3^{\mathrm{AR}}Q]. \quad (2.9)$$

We have thus succeeded in expressing the average two-particle properties of weakly disordered metallic grains in the form of a functional supersymmetric nonlinear σ-model. (Note that the generalisation of this formalism to account for higher-point response functions follows straightforwardly.) The derivation of the effective action relies only on the integrity of the semi-classical parameter $1/E_F\tau$ (cf. the discussion of Section 1.2.1).

Magnetic Field

An extension of the present theory to include an external magnetic field introduces a magnetic vector potential into the action,

$$\mathcal{Z}[J=0] = e^{i\int \bar{\Psi}[(\widehat{\mathbf{p}}-e\sigma_3^{\mathrm{TR}}\mathbf{A}/c)^2/(2m)+V-E_F+\Omega^+\sigma_3^{\mathrm{AR}}/2]\Psi}.$$

The magnetic field breaks time-reversal symmetry and couples to the symmetry breaking matrix σ_3^{TR}. Treating the magnetic field as a *weak* perturbation (see below), and proceeding as before (i.e., setting $\widehat{\mathbf{q}} \equiv -i\partial - (e/c)\sigma_3^{\mathrm{TR}}\mathbf{A}$), it is straightforward to show that the effective action takes the same form as Eq. (2.9), but where the derivative is replaced by a covariant derivative involving the magnetic vector potential, $\partial \mapsto \widetilde{\partial} \equiv \partial - i(e/c)\mathbf{A}[\sigma_3^{\mathrm{TR}},\ ]$.

According to this result, the degrees of freedom of Q which commute with σ_3^{TR} remain unperturbed by a *weak* magnetic field, while those degrees of freedom that do not commute acquire a mass. The former denote the diffuson modes while the latter represent the field sensitive Cooperon modes (recall the phenomenological discussion of Section 1). For strong enough magnetic fields, the influence of the Cooperon modes become negligible and can be neglected. Note that the Cooperon degrees of freedom are coupled to an effective charge of $2e$ reflecting the two-particle nature of the mode.

The influence of magnetic fields on the phase coherence properties of disordered conductors appears at minute fields (ca. one flux quantum through the system). However, if the magnetic length $L_{\mathrm{B}} = (\phi_0/B)^{1/2}$, equal to the cyclotron radius, becomes comparable to the mean-free path ℓ, the influence of *orbital* effects becomes significant. In this case we enter the *quantum Hall regime*, a limit to which we will return in Section 3. For simplicity we will postpone the consideration of this regime and focus on the limit of weak or zero field.

2.2 Applications of the σ-Model

Having established the basis of the statistical field theory, how can the phenomenology of disordered conductors be recovered? In the twenty years since its construction, the σ-model action has been subject of numerous investigations. In the following we will briefly review just a few of the principle applications from which connections to the conceptual discussions of Section 1 can be drawn.

Perturbation Theory

To help digest the form of the effective action it is useful to establish a contact between the field integral and the phenomenology of the diagrammatic perturbation theory. To do so, we focus on the short-time dynamics limiting our considerations to frequency scales Ω large as compared to the

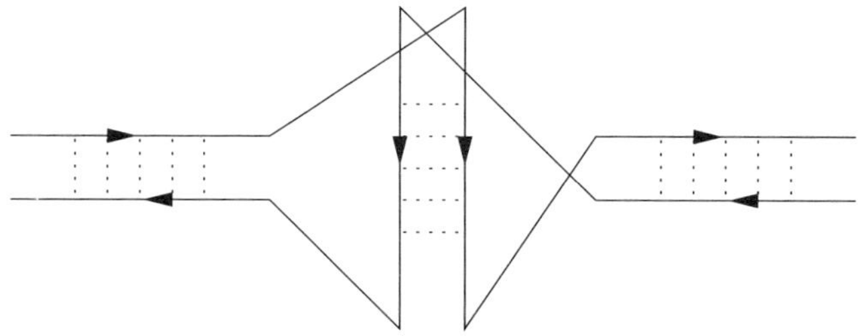

FIGURE 6.27. Diagrams contributing to the weak localization correction to the diffusion constant at one loop.

inverse transport or diffusion time $E_c = D/L^2$—i.e., the diffusive regime. In this case, the functional integral is dominated by field configurations of $Q(\mathbf{r})$ which deviate little from the global saddle-point, σ_3^{AR} (cf. the spin wave expansion of the ferromagnet).

With the parameterisation $T = e^{-W/2}$, where the generators obey the constraint $[W, \sigma_3^{\mathrm{AR}}]_+ = 0$, an expansion of the effective action obtains

$$S_0[W] = -\frac{\pi\nu}{8}\int \mathrm{str}\, W\hat{\Pi}^{-1}W + O(W^4),$$

where $\hat{\Pi}^{-1} = -D\partial^2 + i\Omega^+$. As anticipated, the slow modes of density relaxation are associated with diffusion modes. More precisely, the components of W which commute with σ_3^{TR} represent the diffuson modes while those which do not acquire a mass in the presence of a weak magnetic field, and represent the Cooperon modes. At zero magnetic field, both are controlled by the same propagator. Corrections to the action higher order in W induced by the nonlinear constraint $Q^2 = \mathbf{1}$ describe interactions of the diffusion modes, and reflect different mechanisms of quantum interference.

Weak Localization, Scaling and the RG

As with the O(3) σ-model for the ferromagnet, the action (2.9) can be subjected to a $2+\epsilon$ perturbative renormalization group analysis. Separating slow from fast degrees of freedom, an application of the momentum shell RG to one loop order obtains the renormalization,

$$D \mapsto D' = D\left[1 - \frac{(2-\beta)}{\pi\nu}\Pi(0,0)\right],$$

where $\beta = 1$ or 2 according to whether or not the Hamiltonian is invariant under time-reversal. A diagrammatic representation of this correction is shown in Fig. 6.27. This contribution represents just the first order correction to an infinite series which forms the one parameter scaling of the diffusion constant, or equivalently, of the conductivity.

This result, which predicts a quantum correction to the conductivity of $\delta\sigma/\sigma = -(2-\beta)\Pi(0,0)/\pi\nu$, is consistent with the estimate of the weak localization made in Section 1. (Note that, while the leading order weak localization correction is cancelled by a weak magnetic field, higher order contributions from diffusons alone, lead to a renormalization of the diffusion constant even in the unitary ensembles.)

Level Statistics

As a second application, let us consider the two-point correlator of DoS fluctuations $R_2(\Omega)$. In the σ-model representation, the latter is expressed in terms of the correlator

$$R_2(\Omega) = \frac{1}{64}\left\langle \left[\int \frac{d\mathbf{r}}{L^d} \,\mathrm{str}\big(\sigma_3^{\mathrm{AR}} \otimes \sigma_3^{\mathrm{BF}} Q(\mathbf{r})\big)\right]^2 \right\rangle_Q - 1.$$

To leading order, an expansion in terms of the generators obtains $R_2(\Omega) = \langle [\int \mathrm{str}(\sigma_3^{\mathrm{BF}} W^2)]^2 \rangle_W / 256$, where $\langle \cdots \rangle_W = \int DW \cdots e^{-S_0[W]}$. Focusing on the unitary symmetry class, $[Q, \sigma_3^{\mathrm{TR}}] = 0$, and making use of the contraction rules,

$$\langle \mathrm{str}[W(\mathbf{r}) P W(\mathbf{r}') R] \rangle_W = \frac{1}{\pi\nu} \Pi(\mathbf{r}, \mathbf{r}') [\mathrm{str}\, P \,\mathrm{str}\, R - \mathrm{str}\, P\sigma_3^{\mathrm{AR}} \,\mathrm{str}\, R\sigma_3^{\mathrm{AR}}],$$
$$\langle \mathrm{str}[W(\mathbf{r}) P] \,\mathrm{str}[W(\mathbf{r}') R] \rangle_W = \frac{1}{\pi\nu} \Pi(\mathbf{r}, \mathbf{r}') \,\mathrm{str}[PR - P\sigma_3^{\mathrm{AR}} R \sigma_3^{\mathrm{AR}}],$$

we recover the perturbative result of Eq. (1.12)

$$R_2(\Omega) = \frac{\Delta^2}{2\pi^2} \,\mathrm{tr}\, \hat{\Pi}^2. \tag{2.10}$$

Formally, in the language of diagrammatics, this contribution is associated with the exchange of two diffuson ladders between two closed loops (see Fig. 6.14). Performing the trace we find that, for $\Omega \gg E_c = D/L^2$, $R_2(\Omega) \simeq g^{-d/2} (\Delta/\Omega)^{2-d/2}$. In the zero-dimensional limit this result recovers the perturbative random matrix limit. Interestingly, in two dimensions, the constant of proportionality identically vanishes, and a nontrivial dependence of R_2 on Ω only emerges from higher order weak localization corrections.

Zero-Mode

Having examined the short-time perturbative dynamics, we now switch attention to consider the low energy quantum regime where diagrammatic analysis fails. Taking $\Omega \ll E_c$, contributions to the effective action involving spatial fluctuations of Q are strongly suppressed. In this limit, the contribution at leading order in $1/g$ is dominated by the spatially uniform zero-mode configuration $Q(\mathbf{r}) = Q_0$. In this case, the effective action takes the universal form

$$S[Q_0] = -i \frac{\pi \Omega^+}{4\Delta} \,\mathrm{str}(\sigma_3^{\mathrm{ar}} Q_0), \tag{2.11}$$

independent of dimensionality, the value of the diffusion constant, or geometry of the system. In fact, as will be shown below, Eq. (2.11) is an exact representation of statistical correlators of the corresponding random matrix ensembles.

Although the zero mode integration is definite, its evaluation relies on finding an explicit parameterisation for the Q matrix. While straightforward in principle, the practice is somewhat involved. As such, the detailed parameterisation of the zero-dimensional Q matrix has been included only as a footnote.[3]

$$Q_0 = UH\bar{U},$$

where matrices H and U have the block structure

$$H = \begin{pmatrix} \cos\hat{\theta} & i\sin\hat{\theta} \\ -i\sin\hat{\theta} & -\cos\hat{\theta} \end{pmatrix}_{\mathrm{AR}}, \qquad U = \begin{pmatrix} u_1 u_2 & 0 \\ 0 & v \end{pmatrix}_{\mathrm{AR}}.$$

For the unitary ensemble, these matrices are further separated into components

$$\hat{\theta} = \begin{pmatrix} i\theta_1 & 0 \\ 0 & \theta \end{pmatrix}_{\mathrm{BF}} \otimes \mathbf{1}^{\mathrm{TR}}, \quad \theta_1 > 0, 0 < \theta < \pi,$$

$$u_1 = \exp\begin{bmatrix} 0 & -2\bar{\eta} \\ 2\eta & 0 \end{bmatrix}_{\mathrm{BF}}, \quad \eta = \begin{pmatrix} \eta_1 & 0 \\ 0 & -\eta_1^* \end{pmatrix}_{\mathrm{TR}},$$

$$u_2 = \begin{pmatrix} e^{i\chi\sigma_3^{\mathrm{TR}}} & 0 \\ 0 & e^{i\phi\sigma_3^{\mathrm{TR}}} \end{pmatrix}_{\mathrm{BF}},$$

$$v = \exp\begin{bmatrix} 0 & -2i\bar{\kappa} \\ 2i\kappa & 0 \end{bmatrix}_{\mathrm{BF}}, \quad \kappa = \begin{pmatrix} \kappa_1 & 0 \\ 0 & -\kappa_1^* \end{pmatrix}_{\mathrm{TR}},$$

where $\bar{u}_i \equiv u_i^\dagger$, $\bar{v} \equiv \sigma_3^{\mathrm{BF}} v^\dagger \sigma_3^{\mathrm{BF}}$. Here the coordinates θ_1 and θ represent the eigenvalues of the noncompact bosonic and compact fermionic sector respectively.

Employing this parameterisation we find

$$\mathrm{str}[\sigma_3^{\mathrm{AR}} Q_0] = 4(\lambda_1 - \lambda),$$
$$\mathrm{str}[\sigma_3^{\mathrm{BF}} \otimes \sigma_3^{\mathrm{AR}} Q] = 4(\lambda_1 + \lambda) - 16(\lambda_1 - \lambda)[\eta_1^*\eta_1 - \kappa_1^*\kappa_1],$$

where $\lambda = \cos\theta$, $\lambda_1 = \cosh\theta_1$, $\mu = (1-\lambda^2)^{1/2}$, and $\mu_1 = (\lambda_1^2 - 1)^{1/2}$. Finally, taking into account the invariant measure,

$$dQ_0 = \frac{1}{128\pi^2}\frac{\mu\mu_1}{(\lambda_1 - \lambda)^2}\, d\theta\, d\theta_1\, d\phi\, d\chi\, d\eta_1^*\, d\eta_1\, d\kappa_1^*\, d\kappa_1,$$

---

[3]**Zero-Mode Integration**: The evaluation of the zero mode integrations is most readily performed by finding a convenient parameterisation of the Q_0 matrix. In constructing an explicit parameterisation, it is necessary to take into account the symmetry of the Q matrix. To keep our discussion concise, we shall not describe this construction in detail. Instead, we simply remark that the degrees of freedom of the zero dimensional Q matrix can be expressed in the form [47]

we obtain (2.12). Qualitatively, parameterising Q_0 in terms of the pseudo-unitary transformations which bring it to diagonal form, an expression for R_2 can be given in terms of eigenvalue integrations,

$$R_2(\Omega) = \frac{1}{2} \operatorname{Re} \int_1^\infty d\lambda_1 \int_{-1}^1 d\lambda \, e^{i\pi\Omega^+(\lambda_1-\lambda)/\Delta}. \tag{2.12}$$

Explicit integration over the eigenvalues λ, λ_1 recovers the random matrix result (1.16).

In conclusion, we have demonstrated that, by exploiting the ensemble average, an effective field theory can be developed which is capable of describing the response of weakly disordered conductors from time scales short as compared to the typical transport or diffusion time, to times in excess of the Heisenberg time $t_H = 1/\Delta$ where properties become universal. As an appendix to this section, we will employ the same supersymmetry approach to show that the zero-dimensional σ-model is equivalent to the supermatrix representation of correlators of random matrix ensembles.

2.3 Random Matrix Theory

To make explicit the connection between the long-time dynamics of disordered conductors and the properties of random matrix ensembles we can quickly repeat the field theoretic construction of the σ-model action. In particular, following the classification scheme of Section 1.2.1, the application of the field integral technique to a Gaussian distribution of $N \times N$ random matrices,

$$P(H)dH \propto \exp\left[-\frac{N\beta}{\lambda^2} \operatorname{tr} H^2\right] dH,$$

where β indicates the corresponding symmetry class, an ensemble average of the generating function leads to the expression

$$\langle \mathcal{Z}[0] \rangle = \int d[\Psi] \, e^{i\bar{\Psi}(\Omega^+\sigma_3^{\mathrm{AR}}/2-E)\Psi - \frac{\lambda^2}{N}(\bar{\Psi}\Psi)^2},$$

where Ψ represents $8 \times N$ component supervectors.

The interaction of the supervectors generated by the ensemble average can be decoupled with the introduction of *zero-dimensional* 8×8 supermatrices with the symmetry properties shown in Eq. (2.5). Integrating out the supervectors obtains

$$\langle \mathcal{Z}[0] \rangle = \int d[Q] \, \exp\left[\operatorname{str}\left(\frac{N}{4}Q^2 - \frac{1}{2}\ln \mathcal{G}^{-1}\right)\right]$$

where $\mathcal{G}^{-1} = -E + \Omega^+\sigma_3^{\mathrm{AR}}/2 + i\lambda Q$.

A variation of the action with respect to fluctuations of Q obtains the saddle-point equation $Q_{\rm sp} = i\lambda\mathcal{G}$ from which we obtain the solution

$$Q_{\rm sp} = -\frac{i}{2\lambda}E + \left(1 - \frac{E^2}{4\lambda^2}\right)^{1/2} \sigma_3^{\rm AR}.$$

Applied to the average DoS, the saddle-point solution recovers the semi-circular Wigner distribution $\langle \nu(E) \rangle = (N/\pi\lambda)\big(1 - (E/2\lambda)^2\big)^{1/2}$. Finally, taking the energy at the center of the band, $E = 0$, and expanding to leading order in N, we obtain the zero-dimensional nonlinear σ-model (2.11).

To summarise, the correspondence between the zero-dimensional field theories of the disordered metal and random matrix ensembles establishes universality of low-energy spectral correlations. I.e. when rescaled by the average energy level spacing $\Delta = 1/\nu L^d$, statistical correlations of the energy levels, $\epsilon_i = E_i/\Delta$, become independent of the detailed properties of the system. In fact, universality can be extended to include the response of energy levels to an arbitrary external perturbation such as a magnetic field. Characterising the field strength by some parameter X, when subjected to the rescaling

$$C(0) = \left\langle \left(\frac{\partial \epsilon_i(X)}{\partial X} \right)^2 \right\rangle, \quad x = \sqrt{C(0)}X, \tag{2.13}$$

the statistical properties of the entire random functions $\epsilon_i(x)$ becomes universal [57].[4]

The intuition afforded by the random matrix theory has proved to be of considerable use both qualitatively and quantitatively. As well as providing a complete description of the universal properties associated with energy level, the statistics of wavefunctions is also provided by the random matrix theory.

3 Quantum Chaos

In the previous section a statistical field theory of weakly disordered metallic systems was established. The nonlinear σ-model (2.9) provided a formal

[4]As a footnote, we remark on the existence of a surprising connection between the parametric eigenvalue correlations of random matrix ensembles, and the quantum dynamical properties of an interacting one-dimensional electron gas. The correspondence is drawn by interpreting the energy levels with the position of the particles, and their parametric dependence on some aribitrary external perturbation with time—i.e., the spaghetti of energy levels of a disordered quantum Hamiltonian can be viewed as the "world lines" of a gas of interacting quantum particles. In the simplest case, Eq. (2.12) can then be interpreted as the equal-time density-density correlation function of a one-dimensional free electron gas—the eigenvalue integrations correspond to momentum integrations for the particle/hole excitations. A review of these connections and a discussion of the implications on interacting theories can be found in Ref. 41.

framework from which the phenomenology of Section 1 could be developed: Weak Localization, Scaling, Universality, and the Random Matrix Theory. In this section, we will explore generalizations of the statistical approach which account firstly for the quantum properties of systems whose classical counterparts are *chaotic*. Later, in the following section, we will develop a consistent field theory of Coulomb interaction in the background of disorder.

The derivation of the low-energy effective action for the disordered conductor specifically involved the consideration of a single-particle Hamiltonian in which the random impurity potential was drawn from a Gaussian δ-correlated white-noise distribution. However, in the spirit of the Ginzburg–Landau phenomenology, we would expect the domain of applicability of the low-energy effective action to be much wider. In particular, for any model in which the long-time classical dynamics is fundamentally diffusive, one would expect the nonlinear σ-model to apply. For example, a particle subject to a random "classical" potential (i.e., one on which the scale of variation is greatly in excess of the Fermi wavelength) is expected to be described by the same diffusive σ-model on time scales in excess of the corresponding transport mean free time (i.e., the typical time scale on which the momentum of a particle becomes randomized). Yet, such a model, which has a well defined classical limit, belongs to the wider class of systems which are *chaotic*. Does the field theory tell us something about chaotic quantum structures?

According to the laws of classical Hamiltonian dynamics, a system in which the number of degrees of freedom exceed the number of constants of motion is defined as chaotic. In the simplest cases, the only constant of motion is the energy. For strongly chaotic systems, the fundamental nonintegrability of the Hamiltonian is reflected in an exponential sensitivity of the classical dynamics to the initial boundary condition (i.e., a classical path in phase space separates exponentially fast from an initially neighboring classical path). Such systems are known as "Axiom A" or "uniformly hyperbolic". More generally, chaotic structures usually exhibit a mixed phase space in which chaotic regions are intersected by islands of integrability.

The quantum description of systems which are chaotic in their classical limit is the subject of "Quantum Chaos" [58]. A wide variety of physical systems fall into this category. Amongst those most commonly studied are models of particles confined to irregular potentials, the so-called "Quantum Billiards" (see Fig. 6.28), resonances of microwave cavities, and Rydberg atoms in strong magnetic fields.

As a typical example, one can consider a hydrogen atom subject to a strong magnetic field (for a review see, e.g., Ref. 59). The corresponding nonrelativistic Hamiltonian is specified by (in atomic units)

$$\widehat{H} = \frac{p^2}{2} - \frac{1}{r} + \frac{B}{2}m + \frac{B^2}{8}(x^2 + y^2).$$

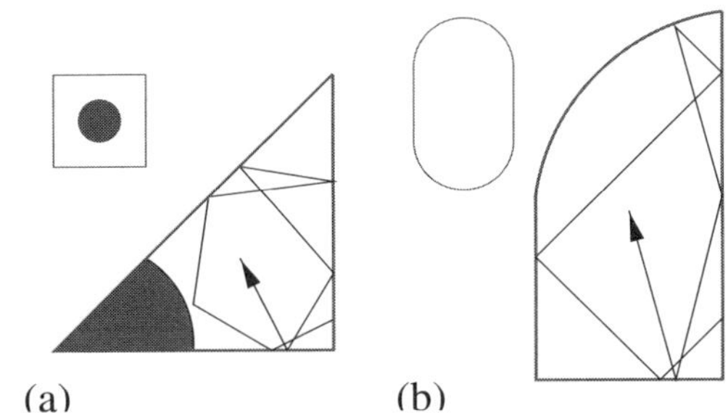

FIGURE 6.28. (a) Sketch of the chaotic "stadium" billiard, and (b) the Sinai billiard.

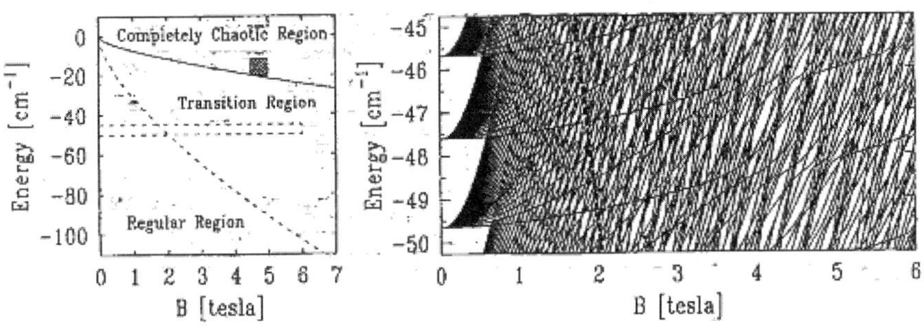

FIGURE 6.29. Typical section of the spectrum of the diamagnetic hydrogen atom spanning a region in which the phase space is mixed (shown dotted). The shaded region lies in a domain in which the classical dynamics is fully chaotic. Here the spectral correlations are universal (courtesy of Ref. 60).

The third term describes the Zeeman splitting separating levels according to the conserved z-component of angular momentum, m. This, together with the parity represent the only good quantum numbers. The final diamagnetic term prevents the usual separation of variables and is responsible for nonintegrability. For highly excited states, the diamagnetic energy can become comparable to the Coulomb energy at which point the classical dynamics becomes fully chaotic (see Fig. 6.29).

The second class of experimental systems which have been the subject of investigation are microwave cavities. Exploiting the equivalence of the Maxwell equation for the stationary TM modes of a two-dimensional cavity and a Schrödinger equation, $(\nabla^2 + k^2)\psi = 0$, the eigenfunctions and eigenspectra of chaotic "billiards" have been investigated. Fig. 6.30 shows typical eigenfunctions for modes of a "Sinai stadium" billiard, together with a device involving a disordered array of scatterers.

Over the years, the general properties of chaotic quantum structures have come under intense scrutiny from both mathematicians and physicists alike. All have been motivated in part by the question: What are the manifestations of chaotic dynamics in the quantum properties of classically

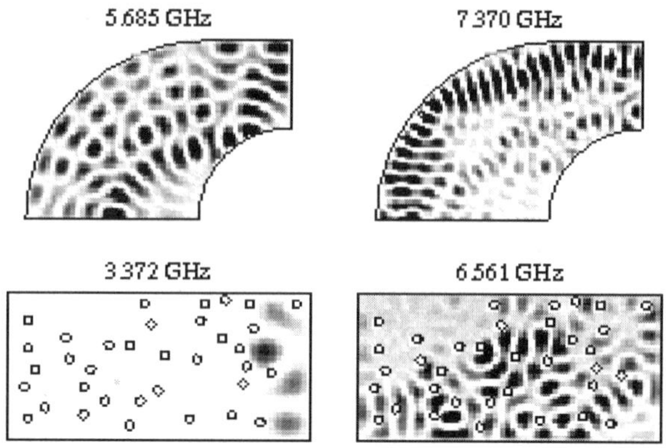

FIGURE 6.30. Measurements of the typical wavefunction amplitude of a "Sinai stadium" billiard and an array of scatterers taken from Ref. 61.

nonintegrable structures? We begin our investigation with a brief review of the important milestones that have helped to shape the general field of quantum chaos.

3.1 Spectral Statistics: A Brief History

Wigner Surmise

Perhaps the earliest investigation of the statistical properties of complex quantum systems began with the pioneering work of Wigner in the late '40's and early '50's. Guided by the very complexity of the system, Wigner introduced a statistical ansatz to explain phenomenologically the strong level correlations observed in the resonance spectra of complex nuclei [62, 63]. Taking matrix elements to be statistically uncorrelated Wigner obtained, within a two-level approximation, an estimate of the distribution of neighboring level spacings known as the Wigner surmise,

$$P(S) = \big\langle \delta\big(S - (E_{i+1} - E_i)\big) \big\rangle \sim \frac{\pi}{2}\frac{S}{\Delta}\exp\left[-\frac{\pi S^2}{4\Delta^2}\right],$$

where, as usual, $\Delta = \langle E_{i+1} - E_i \rangle$ denotes the average level spacing. More accurately, according to the statistical Ansatz, the spectral properties of complex nuclei coincide with those of (time-reversal invariant) random matrix ensembles.

Thus, by surrendering information specific to each individual system, Wigner was able to quantitatively account for the characteristic level repulsion observed in experiment. Forty years on, similar considerations are now frequently given to explain the ground state properties of artificial atoms, or quantum dots [4] (with so far only limited success!).

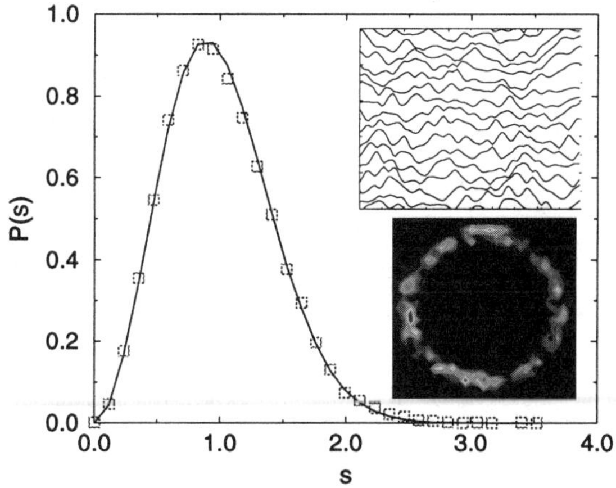

FIGURE 6.31. The distribution of level spacings for a tight-binding impurity Hamiltonian subject to a quasi-periodic boundary conditions. The estimate from the Wigner surmise ($P(s) = 32s^2 \exp[-4s^2/\pi]/\pi^2$ for unitary ensembles) is shown for comparison. Inset, a typical set of levels is shown as a function of the changing quasi-periodic boundary condition. Also shown are the Fourier components of a typical high energy wavefunction. The bulk of the weight lies around the energy shell $E = \mathbf{p}^2/2m$.

Gor′kov–Eliashberg Theory

While successfully studied and applied to the physics of complex nuclei, it was not until almost twenty years later that similar ideas were applied to the study of the quantum properties of weakly disordered metallic grains. By identifying spectral correlations of the grains with those of random Hermitian matrix ensembles, Gor′kov and Eliashberg investigated the dielectric response of metallic grains embedded in an insulating matrix [64]. Although often neglected in the more recent literature, this early work anticipated applications of random matrix theory to the field of mesoscopic physics by two decades.

As confirmation of the random matrix hypothesis for disordered metallic grains, Fig. 6.31 shows the distribution of neighboring level spacings $P(s)$ for an ensemble of tight-binding models with on-site disorder belonging to the unitary ensemble. The results compare favorably with the universal formula for the Wigner surmise.

Supersymmetry: Efetov's Nonlinear σ-Model

Belonging to a statistical ensemble, average properties of disordered quantum systems are more susceptible to analytical investigation than individual systems. As we have seen, providing the statistical average of an individual system (for example, over a range of energy levels) is statistically equivalent to an average over an ensemble of similar systems, analytical results can be obtained systematically and rigorously.

Relying on such an *ergodicity hypothesis*, Efetov developed a statistical field theory of level correlations in weakly disordered metallic grains [47]. As we have seen in Section 2, building upon the seminal work on Anderson localization by Wegner [45] and others [65–70], Efetov provided a rigorous mathematical foundation for the random matrix hypothesis introduced by Gor′kov and Eliashberg. However, perhaps more importantly, Efetov's approach provided a systematic way in which nonuniversal properties reflecting the intrinsic diffusive dynamics of the system could be studied.

Bohigas–Giannoni–Schmidt Conjecture

Until the early '80's, the application of random matrix theory had been largely limited to the study of complex many-body or disordered quantum systems. However, an extension of these ideas to a wider class of structures was suggested by Bohigas, Giannoni and Schmidt [71]. Based on an extensive numerical investigations the following conjecture was proposed:

- The spectral statistical properties of quantum systems which are chaotic in their classical limit (even those with few degrees of freedom) are UNIVERSAL, independent of material properties of the system, and coincide with those of random matrix ensembles.

(Being, in general, unable to define an ensemble, the statistics are assumed to involve averaging over a suitably wide range of states.) This conjecture is now acknowledged as a milestone in the phenomenology of quantum chaos. It suggested a common thread which united a whole class of qualitatively different systems, and lay down a gauntlet to mathematicians to establish a proof. So far, very few exceptions to this rule have been identified. Importantly, for those that have, it is usually simple to understand why they lie outside the present scheme. However, despite more than fifteen years of intense study, and the accumulation of a wealth of supporting data, this conjecture still awaits a rigorous mathematical proof.

Although compelling in its simplicity, the conjecture of Bohigas, Giannoni and Schmidt begs the question: Which properties are universal and, perhaps more importantly, *which are not*? Empirically, the study of spectral statistics typically shows level correlations to be universal on low energy scales comparable with the average energy level spacing Δ (see Fig. 6.32). However, universal correlations are typically observed to break down at some nonuniversal large energy scale, or equivalently, some short-time scale $E_c = 1/t_c$. With this observation, one can ask:

- What, in general, sets the range of universal correlations?

- How is chaos reflected in the statistics of the nonuniversal interval?

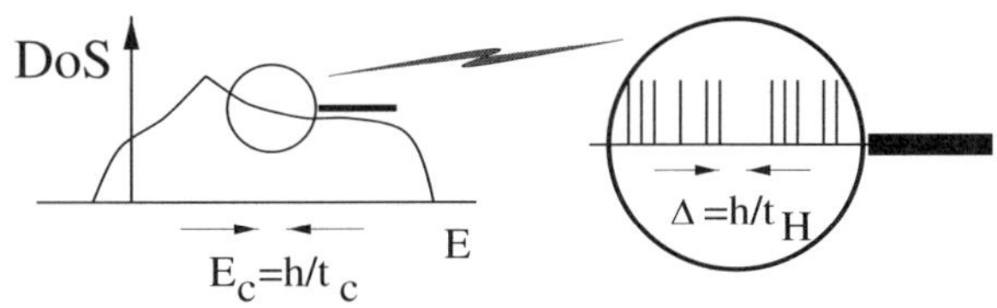

FIGURE 6.32. Time scales over which response functions such as $R_2(\Omega)$ are universal.

To address these questions and identify others we can gain useful insight by drawing on the statistical properties of weakly disordered conductors. There we saw that the low-energy response of spectral correlators were governed by the low-lying modes of density relaxation as specified by the nonlinear σ-model action (2.9). On energy scales Ω in excess of the Thouless energy E_C, the inverse diffusion time across the system $t_D = L^2/D$, the spectral properties depend sensitively on the spectrum of the diffusion modes. While on energy scales $\Omega < E_C$ spectral correlations become universal, independent of the particular geometry or morphology of the system. Within this ergodic regime, the statistical properties of the spectrum and wavefunctions become indistinguishable from those of random matrix ensembles.

Does an analogous scenario describe the spectral response of, say, an irregular or chaotic cavity (a quantum billiard) without impurities? In such systems there too exists some ergodic energy scale below which properties of the system become universal. How is the unstable nature of the classical dynamics reflected in the large energy scales properties? More precisely,

- What plays the role of diffusion in describing the low-lying relaxational degrees of freedom in general chaotic quantum systems?
- What is the analogue of *weak localization* and how are such quantum coherence effects manifest in experiment?
- What determines the domain of universality?

which provides a solid framework in which these questions can be answered (at least in principle). However, to motivate the form of the effective action (and, indeed, to answer the first of the questions above), we will begin with a semi-classical analysis based on the Feynman path integral. Here we will exploit ideas similar to the phenomenology discussed in Section 1.

3.2 Semi-Classics and the Trace Formula

As with the disordered conductors, expressed as the trace of the propagator, we can gain some intuition into spectral properties of chaotic structures by representing the DoS as a Feynman path integral. In the semi-classical limit

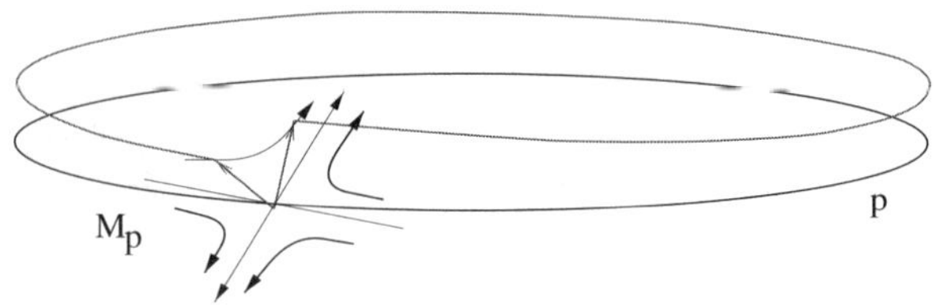

FIGURE 6.33. Schematic diagram showing the meaning of monodromy matrix.

($\hbar \to 0$), this representation identifies the special role played by closed (i.e., periodic) classical trajectories. (Recall $\nu(\epsilon) = \mathrm{tr}\,\mathrm{Im}\,G(\epsilon - i0)/\pi$.) Taking into account the dominant contribution to the Feynman propagator, the DoS assumes the form [58, 72–74]

$$\frac{\nu(E)}{\nu} = 1 + \mathrm{Re}\,\frac{\Delta}{\pi\hbar}\sum_p T_p \sum_{r=1}^{\infty} \frac{e^{iS_p(E)r/\hbar - i\nu_p r}}{|\det(M_p^r - \mathbf{1})|^{1/2}}, \tag{3.1}$$

where p labels a primitive orbit with a period $T_p \equiv \partial S_p(E)/\partial E$, action $S_p(E)$, and Maslov phase ν_p. (Here the first uniform contribution (known as the Weyl term) depends only on the phase space volume and derives from very short (nonclassical) Feynman trajectories.) The sum over r accounts for repetitions of the classical periodic trajectory. Finally the prefactor takes into account the leading fluctuations around the classical paths and involves the stability or *monodromy matrix* M_p associated with the linearized dynamics on the Poincaré section perpendicular to the orbit p (see Fig. 6.33). Crucially, this expression identifies the DoS with a sum over classical periodic orbits weighted by a phase which depends sensitively on the orbit.

Now, as with the disordered conductor, when subjected to an average over a range of energies greatly in excess of the average level spacing, the incoherent superposition of the phase associated with different classical paths removes the second contribution. However, applied to the two-point function, the average preserves a long-range phase coherent contribution. More precisely, substituting Eq. (3.1) into Eq. (1.10), the two-point correlator of DoS fluctuations is expressed as a double sum over periodic orbits. A random phase cancelation of long trajectories identifies the "diagonal contribution" as dominant. Taking this contribution alone, and expanding $S_p(E+\Omega) \simeq S_p(E) + T_p\Omega$, we obtain

$$R_2(\Omega) = \mathrm{Re}\,\frac{\Delta^2}{2\pi^2\hbar^2}\sum_p T_p^2 \sum_{r=1}^{\infty} \frac{e^{i\Omega T_p r/\hbar}}{|\det(M_p^r - \mathbf{1})|}. \tag{3.2}$$

In the disordered metal, the diagonal contribution to the two-point function was identified with the exchange of diffusion modes. To identify the physical relaxation mode to which the diagonal sum (3.2) corresponds we first consider the propagator of classical phase space density, the Liouville operator. Formally, the latter is defined by

$$\rho(\mathbf{x},t) = e^{\{H,\ \}t}\rho(\mathbf{x},0) = \int d\mathbf{x}'\, \delta^{2d}\big(\mathbf{x} - \mathbf{u}(\mathbf{x}';t)\big)\rho(\mathbf{x}',0)$$

where

$$\{H,\ \} = \partial_{\mathbf{r}} H \cdot \partial_{\mathbf{p}} - \partial_{\mathbf{p}} H \cdot \partial_{\mathbf{r}}$$

denotes the classical Poisson bracket, $\mathbf{x} = (\mathbf{r},\mathbf{p})$ represent phase space coordinates, and $\mathbf{u}(\mathbf{x};t)$ denotes the solution of the classical equations of motion of a particle starting at a position $\mathbf{x}$ after a time t.

From this definition, it is straightforward to obtain the following expression for the trace [75],

$$\operatorname{tr} e^{\{H,\ \}t} = \sum_p T_p \sum_r \frac{\delta(t - rT_p)}{|\det(M_p^r - \mathbf{1})|}.$$

The δ-function is associated with the constraint on the period of the orbit, while the prefactor involving the Monodromy matrix is associated with the Jacobian of the δ-function in the phase space directions transverse to the periodic orbit. Comparing this expression to Eq. (3.2), and *neglecting repetitions*, i.e., the contribution of short periodic orbits that are traversed repeatedly, we obtain [76]

$$\begin{aligned} R_2(\Omega) &= \frac{\Delta^2}{2\pi^2\hbar} \operatorname{Re} \frac{\partial}{\partial(i\Omega)} \int_0^\infty dt\, e^{i\Omega^+ t/\hbar} \operatorname{tr} e^{\{H,\ \}t} \\ &= \frac{1}{2\pi^2} \operatorname{Re} \sum_\mu \frac{\Delta^2}{(-i\Omega^+ + \hbar\gamma_\mu)^2}, \end{aligned}$$

where the second equality is based on $\operatorname{tr} e^{\{H,\ \}t} = \sum_\mu \exp(-t\gamma_\mu)$, $\{\gamma_\mu\}$ being the eigenvalues of the classical evolution operator. This result compares to the Altshuler-Shklovskii formula of Eq. (1.12) (with $\gamma_\mu \to D\mathbf{q}^2$), and suggests that the soft modes of density relaxation in a general chaotic structure are associated with the modes of the classical evolution operator.

But, if this identification is correct, do the modes of the Liouville operator relax? And how can we account for mechanisms of weak localization, and corrections nonperturbative in Δ/Ω? These questions can not be answered within the framework of the diagonal approximation. Recent attempts to go beyond this approximation within the framework of periodic theory have met with some limited success [77]. However, these studies have, as yet, failed to identify weak localization corrections. Instead, motivated by this finding, we will employ a different approach based on a nonperturbative field theory.

3.3 Ballistic σ-Model

The analysis of the trace formula above suggests that the analogy between weakly disordered conductors and the general class of chaotic quantum systems is not superficial. In fact, a very close correspondence between the phase coherence phenomena which characterize mesoscopic structures and those observed in quantum chaos can be (at least phenomenologically) established through a generalization of the statistical field theory.

The first suggestion that such a description was possible came in an insightful work by Muzykantskii and Khmel′nitskii [78]. Guided by the quasi-classical Boltzmann description of density relaxation in ballistic transport, they proposed a natural generalization of the diffusive σ-model in which the diffusive character of the action is replaced by a kinetic dependence,

$$S = -i\frac{\pi}{4\Delta}\int \mathrm{str}[\Omega^+\sigma_3^{\mathrm{AR}}Q - 2i\hbar T\sigma_3^{\mathrm{AR}}\{H, T^{-1}\}], \tag{3.3}$$

where, in this case, the supermatrix fields $Q = T\sigma_3^{\mathrm{AR}}T^{-1}$ depend on the $2d-1$ phase space coordinates parameterizing the constant energy shell, $\mathbf{x}_\| = (\mathbf{r}, \mathbf{p})_{2d-1}$. (With this definition the integration measure is normalized such that $\int \equiv \int d\mathbf{x}_\| = 1$.) Although still phenomenological, as we shall see below, their "ballistic action" establishes a crucial bridge between the semi-classical description of level statistics based on the Trace formula above, and the universal random matrix theory.

After its introduction, the phenomenology of Muzykantskii and Khmel′nitskii found further support in the work of Andreev, Simons, Agam, and Altshuler [79]. Specifically, recognizing that the spectrum of an individual system provides a statistical ensemble over which an average can be performed, an explicit derivation of the ballistic action was performed. Yet, it should be noted that the formal derivation of the action is flawed. In particular, the formal derivation of Ref. 79 assumes that fluctuations of the matrix degrees of freedom perpendicular to the constant energy shell are suppressed. However, the mechanism by which the matrix degrees of freedom become "locked" remains obscure, and has become the subject of some controversy in the literature (see, e.g., [80, 81]).

3.4 Perturbation Theory

Leaving aside the difficulties in deriving the ballistic action, there are separate problems in understanding its behavior. Indeed, to make sense of the functional integral (3.3) we must identify the low-lying modes of the action. To do so, we adopt a perturbative expansion of the effective action around the high-energy saddle-point σ_3^{AR}. Employing the parameterization $T = e^{-W/2}$, an expansion of the action leads to

$$S[W] = -\frac{\pi}{8\Delta}\int \mathrm{str}[W\hat{\Pi}^{-1}W] + \mathrm{O}(W^4) \tag{3.4}$$

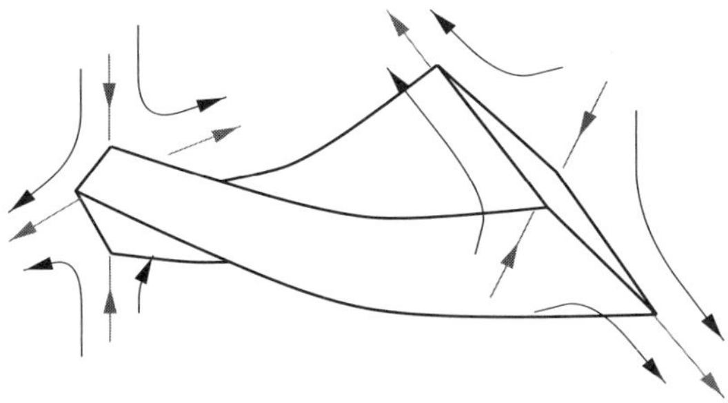

FIGURE 6.34. Schematic diagram showing the tendency for a hyperbolic system to stretch exponentially rapidly along the unstable direction and contract along the stable direction.

where $\hat{\Pi}^{-1} = \hbar\sigma_3^{\rm AR}\{H,\ \} + i\Omega^+$ shows the low-lying modes of density relaxation are described by modes of the classical evolution operator—a result compatible with that obtained in the analysis of the Trace formula. However this identification deserves some qualification.

As with any functional integral there is a need to define an appropriate regularization. For example the functional integral (3.3) may be understood as the limit $a \to 0$ of a product of definite integrations over a discretized space, where a denotes the discretization cell size. This admits to functions $T(\mathbf{x}_\parallel)$ which are *smooth* and square integrable. In seeking such a basis, the eigenfunctions of the classical evolution operator seem to be the natural choice. However, the intricate nature of chaotic classical evolution cause these eigenfunctions to lie outside the Hilbert space: The chaotic dynamics of probability densities involves contraction along stable manifolds, together with stretching along unstable ones (see Fig. 6.34). Thus, in the course of time, an initially nonuniform distribution turns into a singular function on the unstable manifold, which in turn covers the whole energy shell densely. Therefore the eigenfunctions of $\hat{\Pi}$ are not square integrable and their contribution to the functional integral cannot be directly recovered by the discretization procedure involved in evaluating the functional integral.

Thus, in choosing a convenient basis one has to take account of the regularization. Its primary effect is to truncate the contraction along the stable manifold and thereby render the classical evolution *irreversible*. It is the eigenfunctions of this regularized classical evolution operator that serve as a suitable basis for the quantum mechanical correlator. Remarkably, as the strength of the regulator is taken to zero, the spectrum, $\{\gamma_n\}$, of the resulting operator, known in the literature as the Perron–Frobenius operator, reflects intrinsic irreversible properties of the *purely classical dynamics* [82–86].

In ergodic systems, the leading eigenvalue $\gamma_0 = 0$ is nondegenerate, and manifests the conservation of probability density. Thus any initial density

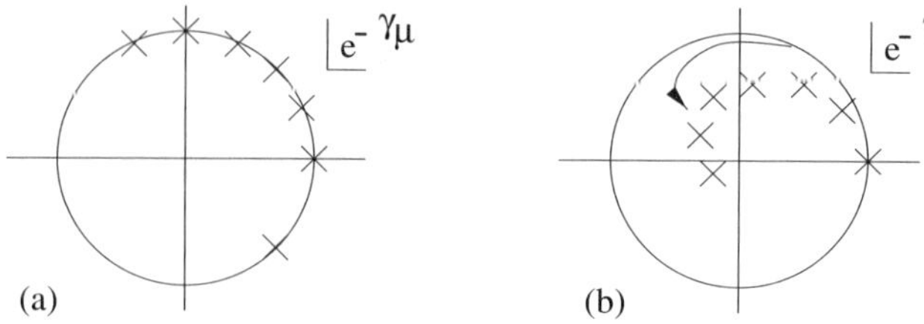

FIGURE 6.35. (a) Formal representation of the eigenvalues of the reversible evolution operator which lie on the unit circle. In nonintegrable systems, the eigenvalues form an infinitely degenerate and dense set. (b) Poles or resonances of the coarse-grained, regularized Perron–Frobenius operator. In systems which are uniformly hyperbolic, all resonances lie within the unit circle (apart from the zero mode), and correspond to modes of density relaxation.

relaxes, in the course of time, to the state associated with γ_0. If, in addition, this relaxation is exponential in time, then the Perron–Frobenius spectrum has a gap associated with the slowest decay rate. Thus, the first nonzero eigenvalue has the property $\gamma_1' \equiv \mathrm{Re}(\gamma_1) > 0$. This gap sets the ergodic time scale, $t_c = 1/\gamma_1'$ over which the classical dynamics relaxes to equilibrium. In the case of disordered metallic grains it coincides with the transport time, t_D, while in ballistic systems or quantum billiards it is of order of the flight time across the system.

In the limit $\Omega \gg \hbar\gamma_1'$ the perturbative expansion (3.4) is justified, and R_2 takes the form of Eq. (2.10) but where the diffusion operator is replaced by the Perron–Frobenius operator. This result coincides with that obtained within the "diagonal approximation" of periodic orbit theory [76] thereby establishing a direct correspondence of the field theoretic and semi-classical approaches. In the limit $\Omega \ll \hbar\gamma_1'$ the dominant contribution to the semiclassical action (3.3) arises from the ergodic classical distribution, the zero-mode $\{H, T\} = 0$. Taking this contribution alone, the action coincides with the *universal* action (2.11) from which one can deduce that correlations coincide with those of random matrix ensembles.

Although much attention has been paid to the spectral properties of the Perron–Frobenius operator, still little is known about the resonance spectra in dynamical systems. Uniformly hyperbolic (the so-called axiom A) systems, such as billiards with constant negative curvature, are characterized by exponential decay of classical correlation functions—i.e., the resonance spectrum of the Perron–Frobenius operator has a gap. Irregular boundary scattering in two-dimensional billiards on flat surfaces is, instead, characterized by isolated resonances together with gapless modes (of low spectral weight) associated with weakly unstable periodic orbits.

Finally, although the resonance spectra of the Perron–Frobenius operator are well defined, the eigenstates associated with the resonances are themselves singular. This has important implications on the role of quantum "weak localization" corrections which are beyond the scope of these lectures. On this point we refer the interested reader to the insightful work of Aleiner and Larkin [87]

In conclusion, we have argued that the quantum statistical properties of chaotic structures are expressed in terms of a ballistic nonlinear σ-model. Yet, although this model is a triumph of phenomenology, its formal derivation, as well as its properties are by no means fully understood. As such, it represents one of the many important areas of research discussed in these lectures which are still very much in a stage of early development.

To close this section we will leave behind the subject of quantum chaos, and explore the connection between the ballistic and diffusive action. In doing so, we will return the consideration of the influence of a strong magnetic field on the behavior of a diffusive particle; we aim to complete our discussion of the influence of disorder on the dynamics of noninteracting quantum particles.

3.5 Quantum Hall Effect

A particle moving in a random impurity potential and subject to a homogeneous weak magnetic field is specified by the Hamiltonian

$$\widehat{H} = \frac{1}{2m}\left(\widehat{\mathbf{p}} - \frac{e}{c}\mathbf{A}\right)^2 + V(\mathbf{r})$$

where $V(\mathbf{r})$ is drawn from some short-ranged impurity distribution, and $\mathbf{B} = \partial \times \mathbf{A}$. Previously, we have studied the influence of a *weak* magnetic field on coherence phenomena in the disordered conductor. In the following, we wish to study the influence of a *strong* magnetic field where the orbital effects become significant. To account for the latter we will make use of the ballistic formalism developed above.

Before turning to the formal analysis, we begin with some qualitative considerations concerning the clean system. Firstly, according to the classical theory of transport, an electron subject to a uniform magnetic field obeys the equation of motion

$$m\dot{\mathbf{v}} = -e\mathbf{E} - e\mathbf{v} \times \mathbf{B} - \frac{m}{\tau}\mathbf{v}.$$

Evaluating the current density $\mathbf{j} = -en\mathbf{v}$ for a stationary solution (i.e., $\dot{v} = 0$), one obtains the following formulae for the longitudinal and transverse conductivity

$$\sigma_{\mathrm{xx}} = \frac{\sigma}{1 + (\omega_c\tau)^2}, \qquad \sigma_{\mathrm{xy}} = \omega_c\tau\sigma_{\mathrm{xx}},$$

where $\omega_c = eB/mc$ denotes the cyclotron frequency, and $\sigma = ne^2\tau/m$ represents the zero-field Drude conductivity. Finally, an inversion of the conductivity tensor reveals that

$$\rho_{\mathrm{xx}} = \frac{1}{\sigma}, \qquad \rho_{\mathrm{xy}} = \frac{1}{\nu_F}\frac{h}{e^2},$$

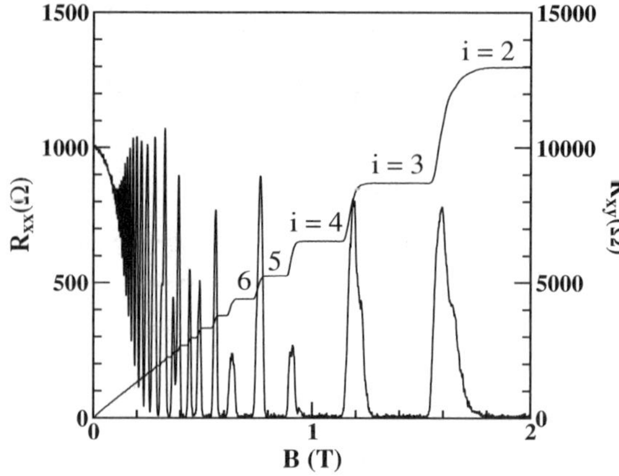

FIGURE 6.36. The change of longitudinal $R_{\rm xx}$ and Hall resistance $R_{\rm xy}$ as a function of magnetic field. The Hall resistance varies stepwise with changes in magnetic field B. The step height is given by the constant of resistance h/e^2 (ca. 25kΩ) divided by an integer i. The figure shows plateaus for $i = 2$, 3, 4, 5, 6, 8 and 10 (J.T. Nicholls, unublished).

where the "filling factor", which measures the ratio of the number of electrons to the total flux through the system, is connected to the electron density by the relation $\nu_F = 2\pi L_B^2 n$, with $L_B = \sqrt{\hbar c/eB}$ defining the magnetic length. Classically, the Hall resistivity scales linearly with the magnetic field.

Quantum mechanically, the situation is found to be quite different: Instead the Hall resistivity is found to take constant values

$$\rho_{xy} = \frac{1}{i}\frac{h}{e^2}, \qquad i = 1, 2, \ldots$$

within a certain range around each integer value of the filling factor ν_F (see Fig. 6.36). Simultaneously, the longitudinal resistivity is found to be vanishing in these ranges so that

$$\sigma_{\rm xx} = 0, \quad \sigma_{\rm xy} = i\frac{e^2}{h}, \qquad i = 1, 2, \ldots .$$

The discovery of this remarkable an unexpected phenomenon [88] earned von Klitzing a Nobel prize in early 1985.

To discover the origin of the quantum Hall effect and to understand how it fits into the framework of the statistical field theory of disordered conductors, we must recall some elementary facts about the influence of a strong magnetic field on the dynamics of electrons. In the absence of disorder, the spectrum of a (spinless) electron confined to two dimensions and subject to a strong uniform perpendicular magnetic field collapses into highly degenerate Landau levels separated in energy by $\hbar\omega_c$. Each Landau level is

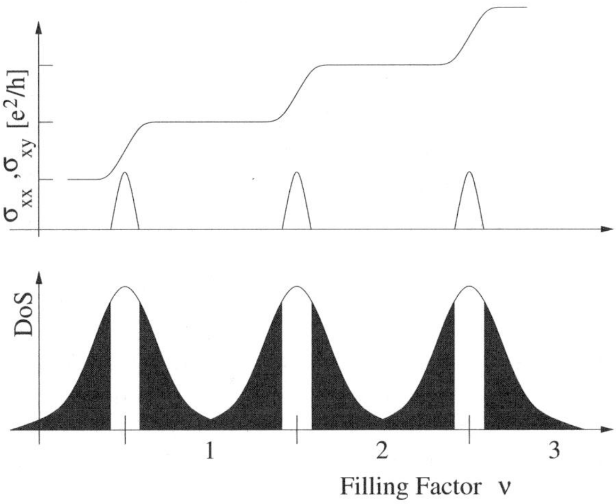

FIGURE 6.37. Schematic diagram showing the broadening of the lowest three Landau levels due to disorder. A mobility edge is indicated separating a region of extended states at the band center, from a region of localized states (shaded) at the band edges.

associated with an electron density of $n = 1/2\pi L_B^2$ (i.e., a filling factor $\nu_F = 1$ corresponds to the lowest Landau level being completely filled). When subjected to a weak random impurity potential, the degeneracy of each Landau level is lifted.

Now, according to the phenomenology of the one-parameter scaling theory of localization, the eigenstates of a quantum particle confined to two dimensions and subject to a weak impurity potential are localized. However, in the presence of a strong magnetic field, the localization length is seen to diverge on approaching the center of the Landau band. In a finite system, a region of extended states around the band center is separated from a region of localized states at the band edge by a sharp mobility edge (see Fig. 6.37). In the thermodynamic limit, this region shrinks leaving behind a single delocalized state at the band center.

Dissipative electric currents are due to transitions of charge carriers from occupied to empty states. At zero temperature, only states at the Fermi energy contribute to the conductivity. Therefore, when the Fermi energy lies in a region of localized states, the longitudinal conductivity $\sigma_{\rm xx}$ vanishes, while in the region of extended states, the conductivity assumes some nonzero value. By contrast, *every* occupied delocalized state contributes to the nondissipative Hall current. Therefore, once in the region of localized states, the Hall conductivity $\sigma_{\rm xy}$ assumes a *constant* value. The value of the constant can be determined by invoking arguments of adiabatic continuity. In the clean system, the Hall conductivity associated with a filled Landau level is quantized in units of e^2/h. Adiabatically increasing disorder, the value of this Hall conductivity at each plateau remains at the same quantized value.

With this preparation, we now ask how this qualitative picture fits within the framework of the statistical field theory of the disordered conductor. To account for the orbital effects associated with the strong magnetic field, we will turn to the ballistic action (3.3). However, before we can import the field theory proposed above, we must first learn how to deal with the quantum nature of the short-ranged scattering potential $V(\mathbf{r})$ due to the impurities. As with a Boltzmann equation, a quantum scattering potential can not be treated entirely within the framework of a kinetic action (3.3). Instead, one must explicitly take into account *irreversible* scattering processes. In such cases, a careful derivation of the effective action obtains an additional term which takes the form of a "collision integral" [78, 80]. Taking the energy shell to be specified by the condition $E_F = p_F^2/2m$, the latter takes the form

$$S_{\text{qu.}} = \frac{\pi^2}{\Delta^2}\int \frac{d\mathbf{r}}{L^d}\int \frac{d\mathbf{n}d\mathbf{n}'}{S_d^2}|V(\mathbf{n}-\mathbf{n}')|^2 \,\text{str}[Q(\mathbf{n})Q(\mathbf{n}')],$$

where $\mathbf{n} = \mathbf{p}/p_F$ denotes the unit momentum vector, S_d represents the total d-dimensional solid angle, and $V(\mathbf{n}-\mathbf{n}')$ represents the Fourier elements of the scattering potential.

In the presence of a strong magnetic field, the Cooperon degrees of freedom acquire a large mass and can be safely neglected. In this case we may focus on the reduced action for the Diffusons modes—the unitary model. Applied to a Gaussian δ-correlated white-noise impurity potential (1.4), taking into account the Lorentz force due to the strong magnetic field, the total ballistic action takes the form [78, 79]

$$S = -i\frac{\pi}{2\Delta}\int \frac{d\mathbf{r}}{L^d}\int \frac{d\mathbf{n}}{S_d}\,\text{str}\left[\Omega_+\sigma_3^{\text{AR}}Q + 2i\hbar\sigma_3^{\text{AR}}T^{-1}\left(\mathbf{v}\cdot\partial_{\mathbf{r}} - \frac{e}{c}\mathbf{v}\times\mathbf{B}\cdot\partial_{\mathbf{p}}\right)T\right] \\ - \frac{\pi\hbar}{4\tau\Delta}\int\frac{d\mathbf{r}}{L^d}\int\frac{d\mathbf{n}\,d\mathbf{n}'}{S_d^2}\,\text{str}[Q(\mathbf{n})Q(\mathbf{n}')],$$

where $\mathbf{v} = \mathbf{n}p_F/m$.

By analogy with the Boltzmann equation, for $\Delta\tau \ll h$ the ballistic action is easily converted to a diffusive form. To obtain the diffusive action we follow an approach similar to that employed by Muzykantskii and Khmel'nitskii [78]. Anticipating a rapid relaxation of the momentum dependent degrees of freedom of Q on the energy shell, and a slow variation of the spatial modes, we introduce a parameterization which involves the moment expansion $T(\mathbf{r},\mathbf{n}) = [\mathbf{1} + i\mathbf{n}\cdot\mathbf{W}(\mathbf{r})]T_0(\mathbf{r})$, where $[\sigma_3^{\text{AR}}, W]_+ = 0$.

Expanding the action to second order in $\mathbf{W}$ and performing integrals over $\mathbf{n}$ we obtain

$$S[Q] = -\pi\nu \int d\mathbf{r}\,\mathrm{str}\left[i\frac{\Omega^+}{2}\sigma_3^{\mathrm{AR}}Q - \frac{i\hbar|\mathbf{v}|}{3}\mathbf{W}\cdot T_0(\partial_{\mathbf{r}}Q)T_0^{-1} - \frac{\omega_c}{3}\sigma_3^{\mathrm{AR}}\mathbf{W}\times\mathbf{W} - \frac{\hbar}{3\tau}\mathbf{W}^2\right],$$

where $Q(\mathbf{r}) = T_0(\mathbf{r})\sigma_3^{\mathrm{AR}}T_0^{-1}(\mathbf{r})$, and $\omega_c = eB/mc$ denotes the cyclotron frequency. Finally, performing the Gaussian integration over $\mathbf{W}$, and treating the term arising from the Lorentz force perturbatively, we obtain the effective unitary action

$$S = -\frac{\pi\nu}{2}\int d\mathbf{r}\,\mathrm{str}[2i\Omega^+\sigma_3^{\mathrm{AR}}Q + D_{xx}(\partial Q)^2 + D_{xy}\epsilon_{\mu\nu}Q(\partial_{r_\mu}Q)(\partial_{r_\nu}Q)] \quad (3.5)$$

where $D_{xx} = v_F\ell/d$ and $D_{xy} = D_{xx}\omega_c\tau$ represent the classical longitudinal and transverse diffusion constant.

From this result, we find that, in the presence of a strong magnetic field, the conventional diffusive action is modified by a topological or boundary contribution which strongly modifies the renormalization properties of the model. In this context, the renormalization properties have been first considered by Pruisken and collaborators [89] within the framework of a replica field theory. These investigations were in qualitative agreement with the phenomenology of a two-parameter scaling theory proposed by Khmel'nitskii [90]. The latter, which can be viewed as a natural generalization of the one-parameter scaling theory, implies that both the longitudinal and transverse components of the conductivity obey renormalization group equations

$$\frac{d\ln\sigma_{\mathrm{xx}}}{d\ln L} = \beta_1(\sigma_{\mathrm{xx}}, \sigma_{\mathrm{xy}}), \qquad \frac{d\ln\sigma_{\mathrm{xy}}}{d\ln L} = \beta_2(\sigma_{\mathrm{xx}}, \sigma_{\mathrm{xy}}).$$

In two dimensions, these equations exhibit a nontrivial fixed point at $\sigma_{\mathrm{xx}} = e^2/2h$ and $\sigma_{\mathrm{xy}} = (n+1/2)e^2/h$ (see Fig. 6.38). (Note that the nature of the topological term places the constraint that the RG flow diagram must be periodic in σ_{xy}.)

Focusing on the lowest Landau level, the two-parameter scaling picture is in accord with the qualitative discussion above. For an electron density less than $\nu_F = \frac{1}{2}$, an increase of the system size leads to a renormalization of the longitudinal and Hall conductivities to zero. At a filling factor of $\frac{1}{2} < \nu_F < 1$, the longitudinal conductivity flows to zero, while the transverse part flows to e^2/h. Finally, at $\nu_F = \frac{1}{2}$, the system flows towards an unstable fixed point at a critical value of the longitudinal and Hall conductivity.

The success of the two-parameter scaling theory has been recorded in a number of papers (see, for example, Ref. 91), while recent discrepancies reported in the literature are likely to be associated with the nature of

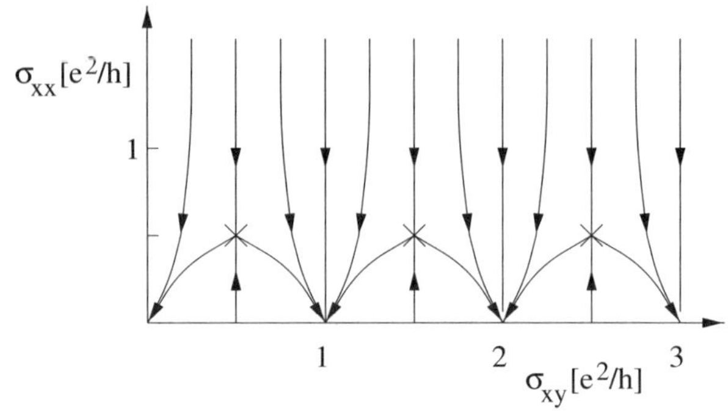

FIGURE 6.38. The two-parameter scaling flow diagram of the integer quantum Hall effect.

the crossover in very clean samples. On the theoretical side, very recent investigations of the supersymmetric version of the action by Zirnbauer [92] have claimed to identify the conformal field theory governing the transition between quantized Hall plateaus of a disordered noninteracting 2d electron gas. For a review of the general field, we refer to Refs. 94 and 93.

4 Coulomb Interaction Phenomena

In Section 1 we saw that many of the coherence phenomena that characterize mesoscopic structures can be understood within the framework of a noninteracting theory. However, experiments both on quantum dots as well as extended semiconducting electron heterostructures show that Coulomb interaction effects are by no means small. Indeed, dramatic manifestations of Coulomb interaction are seen in both confined and extended structures. Examples include Coulomb Blockade phenomena, and the fractional Quantum Hall effect. Moreover, it seems likely that interaction effects play a crucial role in supporting the recently observed two-dimensional metal-insulator transition.

Can the effects of Coulomb interaction be assimilated into the existing field theory? The full statistical field theory describing the interplay of disorder and interaction contains many elements. We have already seen that, in a noninteracting time-reversal invariant system, at least two soft relaxation modes operate, the Diffuson and Cooperon. When combined with the spin singlet and triplet channels associated with the Coulomb interaction, a complete theory is likely to be complicated. To keep our discussion of the influence of Coulomb interaction pedagogical, we will therefore focus on a limited scenario restricting attention to a fictitious gas of spinless electrons interacting via a Coulomb potential and subject to a weak (time-reversal symmetry breaking) magnetic field, i.e., we shall not consider orbital effects associated with strong magnetic fields). In focusing on spinless electrons, exchange effects associated with the electronic degrees of freedom do not arise.

Although the framework developed below is quite general, as a particular goal we will explore the phenomena of the zero-bias anomaly seen in the tunnelling conductance measurements in a disordered metal (and described in Section 1). However, our aim is more ambitious. In studying this particular problem, we hope to address the wider phenomenology of electron dynamics in a disordered environment. We will see that the influence of electron interaction effects are strongly enhanced by the slow diffusive dynamics due to the impurity potential.

The starting point for our analysis is the microscopic many-body Hamiltonian $\widehat{H} = \widehat{H}_0 + \widehat{H}_{\rm C}$, where

$$\widehat{H}_0 = \int d\mathbf{r}\, c^\dagger(\mathbf{r}) \left[\frac{1}{2m} \left(\widehat{\mathbf{p}} - \frac{e}{c}\mathbf{A} \right)^2 + V(\mathbf{r}) \right] c(\mathbf{r})$$

$$\widehat{H}_{\rm C} = \frac{1}{2} \int d\mathbf{r} \int d\mathbf{r}'\, c^\dagger(\mathbf{r}) c^\dagger(\mathbf{r}') \frac{e^2}{|\mathbf{r} - \mathbf{r}'|} c(\mathbf{r}') c(\mathbf{r}),$$

$c^\dagger(\mathbf{r})$ and $c(\mathbf{r})$ represent the particle creation and annihilation operator, and, as usual, $V(\mathbf{r})$ represents a Gaussian δ-correlated impurity potential.

Our aim is to establish a statistical field theory describing the thermodynamic properties of the electron system: i.e., to determine the ensemble average free energy $\langle F \rangle = -kT \langle \ln \mathcal{Z} \rangle$ of the electron gas. For this purpose we invoke the Replica Field Theory approach representing the free energy as the analytic continuation $\ln \mathcal{Z} = \lim_{n \to 0} (\mathcal{Z}^n - 1)/n$. Note that the Fermionic nature of the particles prohibits a supersymmetric formulation of the problem. Therefore, it is pertinent to ask whether we are equipped to deal properly with the analytic continuation $n \to 0$. Indeed this issue has become the subject of some controversy in the literature. Recent attempts to apply the analytic continuation in the *noninteracting* problem have met with some success [53, 54] (but see Ref. 52). Fortunately, it has become clear that the replica field theory approach can be safely applied to the study of perturbative properties of the system. And it is to those properties that we limit our consideration.

(For completeness, we remark that a separate field theoretic approach has been developed which circumvents the explicit need for replicas. The latter, which is based on a technique due to Keldysh [95], has been introduced by Horbach and Schön [96] and developed further by Kamenev and Andreev [97].)

4.1 Matsubara Field Integral

To gain some confidence in the replica approach applied to a many-body Hamiltonian, we begin by focusing attention on the purely *noninteracting* model $\widehat{H}_0$. Starting with the coherent state path integral, the replicated quantum partition function for a weakly disordered metallic conductor is given by

$$\mathcal{Z}_0^n = \int D[\bar{\psi},\psi] e^{-\int_0^\beta d\tau \int \bar{\psi}_\alpha [\partial_\tau + \frac{1}{2m}(\widehat{\mathbf{p}} - \frac{e}{c}\mathbf{A})^2 + V - E_F]\psi_\alpha}$$

where $\psi_\alpha(\mathbf{r},\tau)$ represent a set ($\alpha = 1,\ldots,n$) of Grassmann fields. Here the sum over repeated replica indices is assumed. (In the following, for simplicity, we will avoid explicitly writing out the replica indices.) Note that, were the Hamiltonian invariant under time-reversal, a further doubling of the field space would be necessary.

Switching to the Matsubara representation $\psi(\tau) = (1/\sqrt{\beta})\sum_{\epsilon_n} e^{i\epsilon_n\tau}\psi_{\epsilon_n}$, where $\epsilon_n = \pi(2n+1)/\beta$ denote Fermionic frequencies, the ensemble average over the Gaussian δ-correlated impurity potential (1.4) generates a quartic interaction of the fields. Following our experience with the noninteracting theory, the latter are decoupled with the introduction of matrix fields nonlocal in time,

$$\exp\left[\int \frac{1}{4\pi\nu\tau_{\rm tr}} \operatorname{tr}(\psi\otimes\bar{\psi})^2\right] = \int DQ \exp\left[\frac{1}{2\tau_{\rm tr}}\int\left(\sum_{mn} i\bar{\psi}_m Q_{mn}\psi_n - \frac{\pi\nu}{2}\operatorname{tr} Q^2\right)\right] \quad (4.1)$$

where the normalisation factor has been absorbed into the measure. Here note that the matrices $Q = Q_{nm}^{\alpha\beta}$ carry both replica α and Matsubara ϵ_n indices. As usual, the symmetry properties of the matrix fields reflect those of the dyadic product $\psi\otimes\bar{\psi}$ from which we deduce the hermiticity constraint,

$$Q^\dagger = Q. \quad (4.2)$$

Integrating over the Grassmann fields, the average replicated partition function takes the form

$$\langle\mathcal{Z}_0^n\rangle = \int DQ \exp\left[-\int \operatorname{tr}\left(\frac{\pi\nu}{4\tau_{\rm tr}} Q^2 - \ln\widehat{\mathcal{G}}^{-1}\right)\right]$$

where, defining the matrix of Matsubara frequencies $[\hat{\epsilon}]_{nm} = \epsilon_n\delta_{nm}\otimes\mathbf{1}$,

$$\widehat{\mathcal{G}} = i\hat{\epsilon} - \frac{1}{2m}\left(\widehat{\mathbf{p}} - \frac{e}{c}\mathbf{A}\right)^2 + E_F + \frac{i}{2\tau_{\rm tr}}Q.$$

From this action, the derivation of the nonlinear σ-model is straightforward. As with the single-particle theory, a variation of the action with respect to the matrix field Q obtains the usual saddle-point equation $Q_{\rm SP}(\mathbf{r}) = (i/\pi\nu)\mathcal{G}(\mathbf{r},\mathbf{r})$. Applying the ansatz, $Q_{nm} = Q_{nn}\delta_{nm}\otimes\mathbf{1}$, the diagonal elements are identified with the solutions of the self-consistent Born approximation for the self-energy. Taking into account the analyticity properties of the average Green function one obtains the solution $Q = \Lambda$ where

$\Lambda = \text{sgn}(\hat{\epsilon})$. As with the supersymmetric formulation, in the limit of low energy (i.e., $\epsilon_n \to 0$), the saddle-point solution expands to the degenerate manifold of homogeneous solutions, $Q = T\Lambda T^{-1}$.

Finally, subjecting the effective action to a gradient expansion stabilised by the condition $\epsilon_F \tau_{\text{tr}} \gg 1$, one obtains the nonlinear σ-model functional

$$\langle \mathcal{Z}_0^n \rangle = \int DQ e^{-\pi\nu(\int \text{tr}[D(\partial Q)^2 - 4\hat{\epsilon}Q])/4}$$

where $Q(\mathbf{r}) = T(\mathbf{r})\Lambda T^{-1}(\mathbf{r})$. As with the noninteracting theory, to identify the low-lying modes of the effective action with the diffusion modes, we employ the parameterisation

$$T = e^{-W/2}, \quad W_{nm} = \begin{matrix} \\ \epsilon_n > 0 \\ \epsilon_n < 0 \end{matrix} \begin{matrix} \epsilon_m > 0 & \epsilon_m < 0 \\ \left(\begin{matrix} & B \\ -B^\dagger & \end{matrix} \right. & \left. \right) \end{matrix},$$

i.e., $[W, \Lambda]_+ = 0$ and $W^\dagger = -W$. Then, expanding the action to quadratic order in the generators, we obtain the Gaussian functional

$$\langle \mathcal{Z}_0^n \rangle = \int DB e^{-\pi\nu(\int \sum_{\substack{n>0 \\ m<0}} [D\nabla B_{nm}^{\alpha\beta} \nabla B_{mn}^{\beta\alpha\,\dagger} + (\epsilon_n - \epsilon_m) B_{nm}^{\alpha\beta} B_{mn}^{\beta\alpha\,\dagger}])/2}$$

showing the diffusive nature of the propagator.

With the background of Section 2, one would be inclined to believe that the random matrix properties of the low-energy zero-dimensional action would be easy to recover. Yet, a straightforward analytic continuation of the replica to zero limit misses the true nonperturbative content of the theory. Here we remark on speculations in the literature [53, 54] that the correct nonperturbative behavior is given by stationary configurations of the action corresponding to replica symmetry broken solutions.

With these preliminaries, we turn now to the consideration of the influence of the Coulomb interaction. In this case, a separate Hubbard–Stratonovich field, local in time, must be introduced to decouple the Coulomb interaction of the fields. Applying a Hubbard–Stratonovich transformation to the replicated Coulomb interaction, one finds

$$\exp\left[-\int_0^\beta d\tau \frac{1}{2} \iint' \bar{\psi}(\mathbf{r})\bar{\psi}(\mathbf{r}') \frac{e^2}{|\mathbf{r} - \mathbf{r}'|} \psi(\mathbf{r}')\psi(\mathbf{r})\right]$$
$$= \int D\Phi \exp\left[-\int d\tau \int \left(\Phi \frac{\widehat{\Gamma}_0^{-1}}{2} \Phi + i\bar{\psi} e \Phi \psi\right)\right],$$

where the Fourier transform of the form factor for the Coulomb interaction takes the form $\Gamma_0^{-1}(\mathbf{q}) = \mathbf{q}^2/4\pi$ in three dimensions, and $\Gamma_0^{-1}(\mathbf{q}) = |\mathbf{q}|/2\pi$ in two. Here, as with ψ, the fields Φ carry replica indices.

4.2 Plasma Theory of the Free Electron Gas

Again, to keep our discussion simple, we begin by studying the conventional Gell-Mann Brückner theory [98] of a clean Hamiltonian free of disorder. After the Hubbard–Stratonovich decoupling, the quantum partition function for the interacting electron gas formally takes the form

$$\mathcal{Z}^n = \int D[\bar{\psi}, \psi] D\Phi \, e^{-\int d\tau \int [\Phi \widehat{\Gamma}_0^{-1} \Phi/2 - \bar{\psi} \widehat{G}^{-1} \psi]}$$

where

$$\widehat{G}^{-1} = E_F - \partial_\tau - \frac{\widehat{\mathbf{p}}^2}{2m} - ie\Phi.$$

Gaussian in the fermionic fields, the functional integral can be performed exactly and yields the formal expression

$$\mathcal{Z}^n = \int D\Phi e^{-\int d\tau \int \Phi \widehat{\Gamma}_0 \Phi + \ln\det \widehat{G}^{-1}}.$$

Setting $e = 0$, the auxiliary photon field decouples from the determinant and we recover the formal expression for the partition function of the non-interacting electron gas. To deal with the functional integral over Φ, further progress is possible within an approximation. In this case, we will take the strength of the Coulomb interaction to be weak and look for a perturbative expansion in Φ. (In fact, the constraints set by the Fermi liquid theory extend the validity of the phenomenology of this expansion beyond the regime in which it can be formally justified.)

Defining the bare free particle Green function (i.e., of the unperturbed system) as $\widehat{G}_0^{-1} = E_F - \partial_\tau - \widehat{\mathbf{p}}^2/2m$, and expanding in the strength of the interaction e, we obtain

$$\begin{aligned}\ln\det \widehat{G}^{-1} &\equiv \operatorname{tr}\ln(\widehat{G}_0^{-1} - ie\Phi) = \operatorname{tr}\ln \widehat{G}_0^{-1} + \operatorname{tr}\ln[1 - ie\widehat{G}_0\Phi] \\ &= \operatorname{tr}\ln \widehat{G}_0^{-1} - \operatorname{tr}\big[ie\widehat{G}_0\Phi + \tfrac{1}{2}(ie\widehat{G}_0\Phi)^2 + \cdots\big].\end{aligned}$$

Taking each term in order, we note that, since $\Phi(0) = 0$ (due to an implicit neutralising background),

$$\operatorname{tr}[\widehat{G}_0\Phi] = \frac{1}{\sqrt{\beta L^d}} \sum_{\omega_n, \mathbf{k}} \frac{1}{i\omega_n + \epsilon_\mathbf{k} - E_F} \Phi(0) = 0.$$

Similarly, defining the *density-density response function*.

$$\chi(\omega_n, \mathbf{q}) = -\frac{1}{\beta L^d} \sum_{\omega_n, \mathbf{k}} \frac{1}{i\omega_n + \epsilon_\mathbf{k} - E_F} \frac{1}{i\omega_n + i\omega_m + \epsilon_{\mathbf{k}+\mathbf{q}} - E_F} \tag{4.3}$$

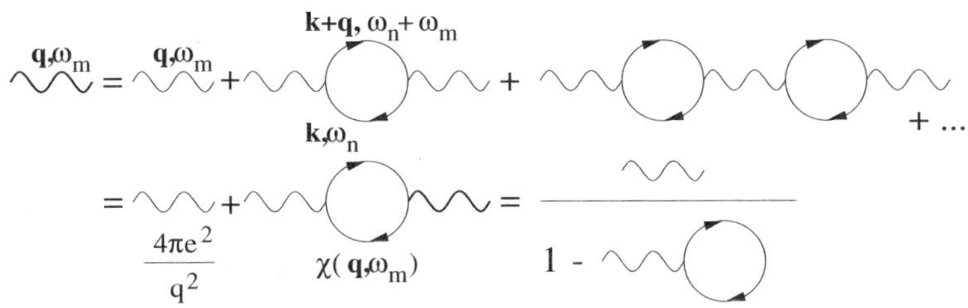

FIGURE 6.39. The modified screened Coulomb interaction, $D(\omega_n, \mathbf{q})$ can be viewed as the summation of an infinite 'diagrammatic' series expansion in the interaction: The bare Coulomb interaction vertex is 'dressed' by repeated particle-hole excitations of the electron gas. The corresponding summation of the geometric series (known as a *Dyson equation*) is shown schematically.

where $\epsilon_{\mathbf{k}} = \mathbf{k}^2/2m$, the second term in the expansion gives

$$\frac{e^2}{2}\,\mathrm{tr}[\widehat{G}_0\Phi]^2 = \sum_{\substack{\omega_n,\mathbf{k}\\ \omega_m,\mathbf{q}}} G_0(\omega_n,\mathbf{k})\frac{\Phi_{\omega_m,\mathbf{q}}}{\sqrt{\beta L^d}}G_0(\omega_n+\omega_m,\mathbf{k}+\mathbf{q})\frac{\Phi_{-\omega_m,-\mathbf{q}}}{\sqrt{\beta L^d}}$$
$$= \sum_{\omega_m,\mathbf{q}} \frac{e^2}{2}\chi(\omega_m,\mathbf{q})\Phi_{-\omega_m,-\mathbf{q}}\Phi_{\omega_m\mathbf{q}}.$$

Collecting together the bare photon action with this expansion, to leading order in e, the quantum partition function takes the form

$$\mathcal{Z} = \mathcal{Z}_0 \int D\Phi e^{-S[\Phi]},$$

where $\mathcal{Z}_0$ denotes the partition function of the noninteracting gas, and the effective action for the photon field is given by

$$S[\Phi] = \frac{1}{2}\sum_{\omega_n,\mathbf{q}}{}' \frac{e^2}{D(\omega_n,\mathbf{q})}|\Phi_{\omega_n,\mathbf{q}}|^2 + O(e^4).$$

This result has a clear physical interpretation: The interaction of the electron gas with the electromagnetic field induces a modified *screened Coulomb interaction* (see Fig. 6.39),

$$D(\omega_n,\mathbf{q}) = \frac{1}{\epsilon(\omega_n,\mathbf{q})}\frac{4\pi e^2}{\mathbf{q}^2}, \quad \epsilon(\omega_n,\mathbf{q}) = 1 - \frac{4\pi e^2}{\mathbf{q}^2}\chi(\omega_n,\mathbf{q}),$$

where $\epsilon(\omega_n, \mathbf{q})$ is the energy and momentum dependent effective *dielectric function*. This result, which is known in the literature as the *Random Phase Approximation (RPA)*, amounts to treating the long-range part of the Coulomb interaction as an "external" polarisation field, and $(4\pi e^2/\mathbf{q}^2)\chi(\omega_n, \mathbf{q})$ is known as the *screened polarisability*.

To explore the dielectric properties of the interacting electron gas it is necessary to understand the frequency and momentum dependence of the density-density response function (4.3). Making use of the analytical properties of the Green function, a summation over the Matsubara frequencies obtains the response function

$$\chi(\omega_m, \mathbf{q}) = -\frac{1}{L^d} \sum_{\mathbf{k}} \frac{n_{\mathbf{k}+\mathbf{q}} - n_{\mathbf{k}}}{i\omega_m + \epsilon_{\mathbf{k}} - \epsilon_{\mathbf{k}+\mathbf{q}}} \tag{4.4}$$

where

$$n_{\mathbf{k}} = \frac{1}{e^{\beta(E_F - \epsilon_{\mathbf{k}})} + 1}$$

denotes the Fermi-Dirac distribution function.

To analyse the nature of the screened Coulomb interaction $D(\omega_n, \mathbf{q})$, we divide our consideration into two limits:

Static Limit ($|\omega_n| \ll k_F|\mathbf{q}|/m$)

For frequencies small as compared to the momentum transfer (and temperatures $T \ll E_F$), the response function converges to the static limit

$$\chi(0, \mathbf{q}) \simeq \frac{1}{L^d} \sum_{\mathbf{k}} \frac{\mathbf{q} \cdot \partial_{\mathbf{k}} n_{\mathbf{k}}}{\mathbf{q} \cdot \partial_{\mathbf{k}} \epsilon_{\mathbf{k}}} \simeq \int \frac{d^3k}{(2\pi)^3} \partial_{\epsilon_{\mathbf{k}}} n_{\mathbf{k}} = \nu(E_F),$$

where $\nu(E_F)$ represents the DoS at the Fermi level. Substituting this result into the expression for $D(\omega_n, \mathbf{q})$, and applying a Fourier transform, one obtains the static screened Coulomb interaction

$$V_{\text{screen}}(r) = \frac{e^2}{r} e^{-r/\lambda_{\text{TF}}},$$

where $\lambda_{\text{TF}} = 4\pi e^2 \nu(E_F)$ defines the *Thomas-Fermi screening length*. Physically, on length scales in excess of λ_{TF}, the bare Coulomb interaction is screened by the collective fluctuations of the electron gas.

Taking into account higher order contributions in the interaction does not change the conclusions qualitatively. The screening behavior of the electron gas is preserved while the formula for the screen length λ_{TF} is modified by Fermi liquid corrections.

High Frequency Limit ($|\omega_n| \gg k_F|\mathbf{q}|/m$)

By contrast, in the high frequency limit, the density-density response function takes the form

$$\begin{aligned}
\chi(\omega_m, \mathbf{q}) &\simeq -\frac{1}{L^d}\sum_{\mathbf{k}} \frac{\mathbf{q}\cdot\partial_{\mathbf{k}} n_{\mathbf{k}}}{i\omega_m - \mathbf{q}\cdot\partial_{\mathbf{k}}\epsilon_{\mathbf{k}}} \\
&\simeq -\int \frac{d^dk}{(2\pi)^d}\frac{1}{i\omega_m}\left(1 - \frac{\mathbf{q}\cdot\mathbf{k}}{im\omega_m}\right)\mathbf{q}\cdot\partial_{\mathbf{k}} n_{\mathbf{k}} \\
&\simeq -\int \frac{d^dk}{(2\pi)^d}\frac{\mathbf{q}^2}{m\omega_m^2} n_{\mathbf{k}} = -\frac{\mathbf{q}^2}{m\omega_m^2} n,
\end{aligned}$$

where $n = N/L^d$ denotes the number density. Applying the analytic continuation to real frequencies, $i\omega_n \to \omega + i0$, obtains

$$\lim_{m\omega/k_F|\mathbf{q}|\to\infty} D^R(\omega, \mathbf{q}) = \frac{4\pi e^2}{\mathbf{q}^2}\left[1 - \frac{\omega_p^2}{\omega^2}\right]^{-1},$$

where $\omega_p = 4\pi e^2 n/m$ represents the *Plasma frequency*. This result has a natural physical interpretation: The response of the electron gas is not instantaneous but is retarded. At frequencies greatly in excess of the plasma frequency, the electrons are unable to screen an electromagnetic field and the bare Coulomb interaction is recovered.

How are these results modified when the particles are subject to a weak impurity potential? Qualitatively, we can expect the slow diffusive dynamics to have important manifestations on the screening and relaxation properties of the electron gas.

4.3 Plasma Theory of the Disordered Electron Gas

With this background, we turn to the formal consideration of the influence of disorder. Applied to the impurity averaged, replicated partition function, the quantum partition of the electron gas takes the form

$$\langle \mathcal{Z}^n \rangle = \int DQ \int D\Phi e^{-\int [\pi\nu \,\mathrm{tr}\, Q^2/(4\tau_{\mathrm{tr}}) - \mathrm{tr}\ln \widehat{\mathcal{G}}^{-1} + \int d\tau\, \Phi \widehat{\Gamma}_0^{-1}\Phi/2]}$$

where the full matrix Green function is defined by

$$\widehat{\mathcal{G}}^{-1} = i\hat{\epsilon} - \frac{1}{2m}\left(\widehat{\mathbf{p}} - \frac{e}{c}\mathbf{A}\right)^2 - ie\Phi + E_F + \frac{iQ}{2\tau_{\mathrm{tr}}}.$$

Previously, with the noninteracting theory, we saw that fluctuations of the matrix fields separate into massive and massless contributions. On energy scales $\epsilon \ll 1/\tau_{\mathrm{tr}}$, the latter, which are as associated with the long-range

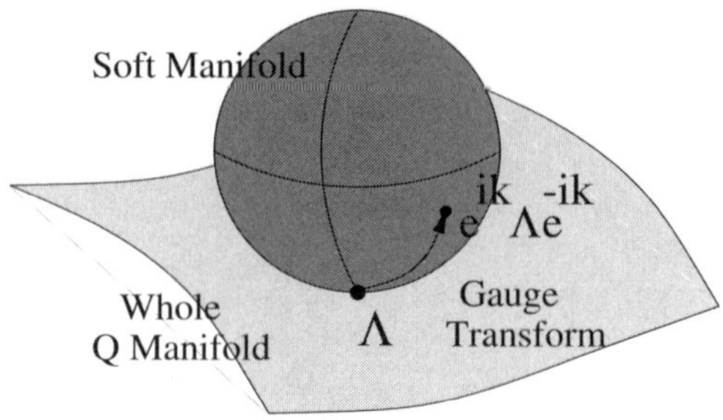

FIGURE 6.40. Schematic diagram showing the action of the gauge transformation described in the text.

diffusion mode contributions, dominate the low-energy effective action. By contrast, in the absence of the disorder potential, an expansion of the action in the fluctuations of Φ obtains the Plasma theory of the screened Coulomb potential. A consistent theory of the interacting disordered metal must necessarily combine these two limits. To achieve a correct decoupling of the fast and slow degrees of freedom it proves convenient to take into account the gauge invariance of the action [97]. Here, loosely speaking, we follow the analysis of Kamenev and Andreev [97].

In its present form, the functional is invariant under the time and space local gauge transformation

$$Q(\tau,\tau';\mathbf{r}) \mapsto \widetilde{Q}(\tau,\tau';\mathbf{r}) = e^{ik(\mathbf{r},\tau)}Q(\tau,\tau';\mathbf{r})e^{-ik(\mathbf{r},\tau')}.$$

Applied to the functional, taking k to be diagonal in the replica space, such a gauge transformation induces a vector potential,

$$\mathbf{A} \mapsto \tilde{\mathbf{A}} = \mathbf{A} - \frac{c}{e}\partial k, \quad e\Phi \mapsto e\widetilde{\Phi} = e\Phi - \partial_\tau k.$$

In the following, we will find it convenient to leave the specific choice of gauge unspecified. Later, we will use (an approximation for) the saddle-point of the low-energy functional to specify a particularly convenient gauge.

Considering the high energy sector of the theory, $\epsilon \gg 1/\tau_{\rm tr}$, an expansion of the action obtains

$$\int \mathrm{tr}_{\epsilon\gg 1/\tau_{\rm tr}} \ln\left[i\hat{\epsilon} - \frac{\widehat{\mathbf{p}}^2}{2m} + E_F - ie\widetilde{\Phi}\right] \simeq \frac{\nu}{2}\iint d\tau\, e^2\widetilde{\Phi}^2$$

where high frequency and momentum corrections to the screened Coulomb interaction are neglected against the low energy content of the remainder of the action. In the high-frequency limit, the disorder has no influence on the dynamics of the electron degrees of freedom.

Turning to the low energy sector, $\epsilon \lesssim 1/\tau_{\rm tr}$, setting $Q = T\Lambda T^{-1}$, and applying the gradient expansion

$$T^{-1}\widehat{\mathcal{G}}^{-1}T = E_F - \frac{\mathbf{p}^2}{2m} + \frac{i\Lambda}{2\tau_{\rm tr}} + T^{-1}\left[i\hat{\epsilon} - \frac{\mathbf{p}}{m}\cdot\widehat{\mathbf{q}} - ie\widetilde{\Phi}\right]T$$

where $\widehat{\mathbf{q}} = -i\partial - e\tilde{\mathbf{A}}/c$, we obtain the nonlinear σ-model action [37]

$$S[Q] = \frac{\pi\nu}{4}\int \mathrm{tr}[D(\widetilde{\partial}Q)^2 - 4(\hat{\epsilon} - e\widetilde{\Phi})Q]$$

where $\widetilde{\partial} = \partial - i(e/c)[\tilde{\mathbf{A}},\ ]$. Combined with the high energy part of the action, and the bare Coulomb interaction, we obtain

$$\langle \mathcal{Z}^n \rangle = \int DQ \int D\Phi e^{-S[Q]-(\iint d\tau\,[\Phi\widehat{\Gamma}_0^{-1}\Phi - e^2\nu\widetilde{\Phi}^2])/2}.$$

At this stage, it is tempting to immediately implement an expansion perturbative in the interaction, and in the diffusion modes. Indeed such an expansion recovers the results of diagrammatic perturbation theory contained in the literature [35]. However, the interaction field Φ couples linearly to Q, which implies that the saddle-point configuration of the Q field is rotated from Λ (see Fig. 6.40). Put differently, the ground state of the interacting electron gas is modified by the impurity potential. A consistent expansion of the interaction (which avoids an infinite re-summation of diagrams) requires the identification of the true saddle-point of the theory.

4.4 Gauge Fixing: Low-Energy Saddle-Point

A variation of the low energy action with respect to Q subject to the nonlinear constraint obtains the saddle-point equation,

$$D\widetilde{\partial}(Q_{\rm SP}\widetilde{\partial}Q_{\rm SP}) + [\hat{\epsilon} - e\widetilde{\Phi}, Q_{\rm SP}] = 0.$$

Ideally, one would like to find the solution $Q_{\rm SP}$ which solves the equation exactly. However, for an arbitrary (in time and space) configuration of the field $\widetilde{\Phi}$, such a program is unfeasible. Here, instead, we will fix the local gauge field $k(\mathbf{r}, \tau)$ such that the saddle-point solution for Q approaches the noninteracting solution Λ. Thus, setting $Q_{\rm SP} = \Lambda$, we obtain the nonlinear equation for k,

$$[iD\Lambda\partial^2 k - D\partial k\Lambda\partial k - e\Phi + \partial_\tau k, \Lambda] = 0.$$

Neglecting the term second order in k, the resulting linear equation has the (replica diagonal) solution

$$k(\omega, \mathbf{q}) = \frac{i\Phi(\omega, \mathbf{q})}{D\mathbf{q}^2\,\mathrm{sgn}(\omega) + \omega},$$

FIGURE 6.41. Diagrammatic representation of the RPA correction to the Coulomb interaction. The shading within the loop represents a diffusion ladder.

where ω denotes a bosonic Matsubara frequency. Substituting this solution back into the action, after some algebra, one obtains the modified Coulomb interaction,

$$S = \frac{1}{2} \sum_{\omega} \int \Phi \widehat{\Gamma}_{\mathrm{RPA}}^{-1} \Phi, \quad \Gamma_{\mathrm{RPA}}^{-1}(\omega, \mathbf{q}) = \Gamma_0^{-1}(\mathbf{q}) + \frac{\sigma \mathbf{q}^2}{D\mathbf{q}^2 + |\omega|}$$

where as usual $\sigma = e^2 \nu D$. This estimate, which in the present context is known as a *random phase approximation*, has a simple physical interpretation. The screening of the Coulomb interaction in the disordered environment is modified by the diffusive dynamics of the electrons. The modification can be interpreted diagrammatically as an infinite summation of polarisation loops dressed by diffuson modes (see Fig. 6.41).

Now, in principle, an expansion of the action in the fluctuations of Q around the saddle-point leads to higher order quantum interference corrections. However, already at this order, we can obtain the dominant contribution to the tunnelling anomaly since the latter can be interpreted physically as a consequence of diffusively slowed screening.

4.5 Zero-Bias Anomaly

To illustrate the utility of the field theoretic approach, we turn to an application of the functional to examine the voltage dependence of the tunnelling conductance $G(V)$ of a particle from a point contact (or STM tip) onto a disordered metallic background. In particular, if the charge accommodation time is much longer than the transfer time from contact onto the metal, the tunnelling conductance can be identified with the *single-particle* DoS, $G(V) \sim \nu(\epsilon = eV)$. The latter is defined in terms of the single-particle Green function as

$$\nu(\epsilon) = -\frac{1}{\pi} \operatorname{tr} \operatorname{Im} \widehat{G}(\mathbf{r}, \mathbf{r}; \epsilon_n) \Big|_{i\epsilon_n \to \epsilon + i0}.$$

The single-particle DoS can be expressed in terms of the functional integral above. Specifically, taking into account the gauge transformation, and performing the (trivial) replica to zero limit, the saddle-point approximation for the ensemble averaged local single-particle Green function is given by

$$\langle \widehat{G}^{>}(\mathbf{r}, \mathbf{r}; 0, \tau) \rangle_V = -i\pi\nu \langle e^{ik(0)} \Lambda_{-\tau} e^{-ik(\tau)} \rangle_{\Phi} = -i\pi\nu \Lambda_{-\tau} e^{-S(\tau)}$$

where $\Lambda_{-\tau}$ represents the single-particle propagator for a clean system, and

$$S(\tau) = \frac{e^2}{2} \int \sum_{\omega_n} 4\sin^2(\omega\tau) \frac{\Gamma_{\rm RPA}(\mathbf{q}, \omega)}{(D\mathbf{q}^2 + |\omega|)^2}.$$

Here the self-consistency of the approach demands that the Matsubara summation and momentum integration are limited to the range $|\omega|, v_F|\mathbf{q}| < 1/\tau_{\rm tr}$. Specifically, for a two-dimensional metal,

$$S(\tau) = \frac{e^2}{8\pi^2\sigma} \ln\left(\frac{\tau}{\tau_{\rm tr}}\right) \ln\left(\tau\tau_{\rm tr}\sigma^2(\nu e^2)^2\right).$$

Physically, the single-particle Green function $\langle \widehat{G}^{>}(\mathbf{r}, \mathbf{r}; 0, \tau)\rangle_V$ represents the probability for a particle injected into a disordered metal at a position $\mathbf{r}$ at time $\tau = 0$ to return to the point $\mathbf{r}$ after a time τ. The suppression in the return probability $S(\tau)$ represents an additional under barrier contribution to the action arising from the Coulomb interaction of the transferred charge with itself—the "self-energy".

With this definition, the single-particle DoS is given by

$$\langle \nu(\epsilon)\rangle = \nu \int d\tau e^{\epsilon\tau - S(\tau)} \Lambda_{-\tau}.$$

The latter is dominated by the stationary configuration of the total effective action, which (to a good approximation) is obtained from the condition

$$\left.\frac{\partial S}{\partial \tau}\right|_{\tau_*(\epsilon)} = \epsilon.$$

Here τ_* can be interpreted as the charge accommodation time. In two dimensions, the latter is given by

$$\tau_* = \frac{1}{4\pi^2\epsilon g} \ln\left(\frac{\sigma\nu e^2}{\epsilon}\right),$$

a result which is self-consistent in the hydrodynamic limit if $\tau_* \gg \tau_{\rm tr}$, i.e., at $\epsilon < e^2/\sigma\tau_{\rm tr}$. Substituting this expression back into the action, we obtain the tunnelling conductance for the two-dimensional system

$$\langle G(V)\rangle = G_0 \exp\left[-\frac{e^2}{8\pi^2\sigma} \ln\left(\frac{e^2}{4\pi^2\sigma eV\tau_{\rm tr}}\right) \ln\left(\frac{e^2\tau_{\rm tr}\sigma(\nu e^2)^2}{4\pi^2 eV}\right)\right]$$

where G_0 denotes a voltage independent constant conductance. Diagrammatically, this results can be interpreted as the summation of an infinite series of diagrams for which the first is shown in Fig. 6.42.

The manifestation of Coulomb interaction in the tunnelling anomaly is not unique. In particular, further analysis of the effective action, which goes

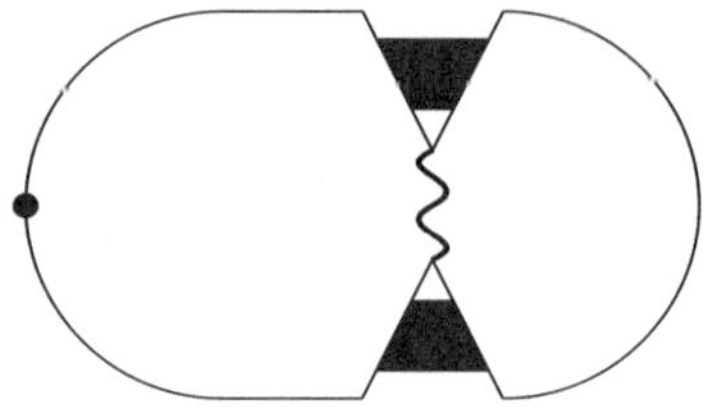

FIGURE 6.42. Diagrammatic representation of the leading contribution to the Zero-Bias Anomaly. The exponential formula given in the text represents an infinite re-summation of a set of diagrams of which this contribution is the first.

beyond the leading approximation above, leads to a logarithmic enhancement of the conductivity. This quantum correction competes with the weak localization correction. Whether this competition explains the recently observed two-dimensional metal-insulator transition remains the subject of intense investigation.

This completes our very brief survey of the influence of disorder on Coulomb interaction phenomena is metallic structures. Needless to say, we have touched upon just a fragment of the literature in this field. Nevertheless, the physical concepts exhibited by screening and the zero-bias anomaly are representative of a wider class of interaction phenomena in disordered conductors. For a more comprehensive review of the field, we refer to Refs. 35, 99, and 100.

5 Novel Symmetry Classes

Following on from the pioneering work of Wigner [62] on the properties of complex atomic nuclei, in a classic series of papers [11] Dyson introduced a complete classification of complex many-body systems according to their fundamental symmetries—"the three-fold way". Using arguments based largely on mathematical grounds, Dyson proposed the existence of three universality classes which are distinguished by their invariance properties under time-reversal and spin rotation. In Section 1.2.1 we saw that the universal statistical properties of these classes are characterized by three random matrix ensembles, Unitary, Orthogonal and Symplectic. At the same time, the importance of the Wigner-Dyson scheme in the classification of the low-energy statistical properties of random and complex chaotic phase coherent quantum structures was stressed.

But is this classification exhaustive? In the course of the last few years it has become clear that the answer is negative, i.e., that various classes of stochastic quantum systems of current interest do not fall into the threefold classification scheme of Wigner and Dyson. For a clear mathematical exposition of this fact, see the review by Zirnbauer [101]. Curiously, perhaps the most transparent ways of anticipating the existence these novel and *fundamental* symmetry classes is by purely formal mathematical reasoning:

Consider the N-dimensional Hamiltonian H of some (generally spinful) quantum system. Besides Hermiticity, one commonly considers two additional discrete symmetries, invariance under time reversal $\mathcal{T}$ and invariance under spin rotations $\mathcal{S}$:

$$\mathcal{T}: \quad H = H^T,$$

$$\mathcal{S}: \quad H = \sigma^2 H^T \sigma_2,$$

where σ_i are Pauli matrices acting on spin. Depending on whether or not the symmetries $\mathcal{T}$ and $\mathcal{S}$ are realized, one has to distinguish four different cases. How can one embed these symmetry structures into the quantitative description of stochastic quantum systems?

The original method, put forward by Wigner and Dyson, did not concentrate on the Hamiltonian itself but rather on its transformation behavior. The idea was to demand that under unitary transformations $H \mapsto UHU^\dagger$, $U \in \mathrm{U}(N)$ the fundamental symmetry of the Hamiltonian be unchanged. E.g. for the case of a spin invariant $(\mathcal{S} : +)$ and time reversal invariant $(\mathcal{T} : +)$ Hamiltonian, this requirement entails $U^* = U$, i.e., $U \in \mathrm{O}(N) \subset \mathrm{U}(N)$, the orthogonal group in N dimensions. A maximally entropic set of Hamiltonians, closing under all possible orthogonal transformations, forms the so-called Gaussian Orthogonal Ensemble, GOE, discussed above. (Here the term "Gaussian" refers to the nonuniversal probability distribution usually chosen. The notation is slightly unfortunate since other distributions such as the "circular ensembles" belong to the same fundamental symmetry class.) Similar considerations for the other symmetry cases led to the introduction of the Gaussian Unitary and Symplectic Ensemble, respectively. (It is straightforward to show that both cases $(\mathcal{T} : -, \mathcal{S} : +)$ and $(\mathcal{T} : -, \mathcal{S} : -)$ fall into the unitary class.)

Quite recently an alternative scheme of classifying symmetries has been introduced. The new method, which focuses on the Hamiltonian itself rather than on its transformation behavior, has the advantage of hinting at a possible generalisation of the three-fold Wigner Dyson classification. Considering the same Hamiltonian as before, we define $X \equiv iH$. The motivation for introducing X is that this object belongs to an *algebra*, viz. the algebra of antihermitian matrices, $\mathbf{u}(N)$ generating the unitary group $\mathrm{U}(N)$. (We will denote the algebra of any group A by the corresponding small Latin character 'a' throughout. $\mathrm{u}(N)$ generates $\mathrm{U}(N)$ in the sense that for X running through all of $\mathrm{U}(N)$, $\exp(X)$ runs through all of $\mathrm{U}(N)$.) As with the considerations above, we next observe that the objects X corresponding to Hamiltonians of higher symmetry can not be drawn from the full algebra but rather symmetry restricted subsets thereof. Using $(\mathcal{T} : +, \mathcal{S} : +)$ again as an example, we have $X^T = X$, implying that $X \in \mathrm{u}(N)/\,\mathrm{o}(N)$.

(Here we have used the fact that the quotient vector space $\mathrm{u}(N)/\,\mathrm{o}(N)$ forms the algebra of antihermitian symmetric matrices.) Exponentiating this sub-algebra we generate the *coset space* $\mathrm{U}(N)/\,\mathrm{O}(N) \subset \mathrm{U}(N)$. Similar considerations for the case $(\mathcal{T} : +, \mathcal{S} : +)$ lead to the identification of the coset space $\mathrm{U}(N)/\,\mathrm{Sp}(N)$. Why do we care to go through these, admittedly formal, manipulations? The motivation is that the sets $\mathrm{U}(N), \mathrm{U}(N)/\,\mathrm{O}(N)$ and $\mathrm{U}(N)/\,\mathrm{Sp}(N)$ identified for the case of unitary, orthogonal and symplectic symmetry, respectively, possess a very high degree of mathematical structure: They are representatives of so-called compact symmetric spaces. Of such structures only ten different types (indexed by an integer N) exist. For simplicity, we restrict ourselves to the consideration of even N. For N odd, an eleventh symmetric space exists. Rather than phrasing the comparatively complex definition of a symmetric space (cf., for example, Ref. 102) we merely display at the top of the next page their ten-fold classification, as derived and labelled by Cartan:

space	denotation
$\mathrm{U}(N)$	A
$\mathrm{U}(N)/\,\mathrm{O}(N)$	AI
$\mathrm{U}(2N)/\,\mathrm{Sp}(2N)$	AII
$\mathrm{U}(N+M)/\,\mathrm{U}(N)\times\mathrm{U}(M)$	AIII
$\mathrm{SO}(N+M)/\,\mathrm{SO}(N)\times\mathrm{SO}(M)$	BDI
$\mathrm{Sp}(N+M)/\,\mathrm{Sp}(N)\times\mathrm{Sp}(M)$	CII
$\mathrm{SO}(N)$	BD
$\mathrm{Sp}(N)$	C
$\mathrm{Sp}(N)/\,\mathrm{U}(N)$	CI
$\mathrm{SO}(2N)/\,\mathrm{U}(N)$	DIII

Above we have shown that the first three of these objects appear in the context of the established symmetry classes of Wigner and Dyson. One may now wonder whether the remaining seven spaces, AIII, ..., DIII, might have any physical significance. Of course we would not have gone through the formal discussion above if the answer were not positive. Research done in the last few years has shown that not only may these objects appear in principle, they even do so in rather interesting physical contexts. The principal connection between symmetric spaces on the one hand and their realization as symmetry classes of physical significance was first revealed in the seminal paper [101]. Focusing on those aspects motivated by real application, the rest of this section will be devoted to a discussion of the realizations of the novel symmetry classes, AIII, ..., in physics.

5.1 Realizations of Novel Symmetries

As a prototype system let us consider the example of a $2N \times 2N$ matrix Hamiltonian

$$H = \begin{pmatrix} & Z \\ Z^\dagger & \end{pmatrix} \tag{5.1}$$

where the matrix elements of the complex matrix Z are drawn at random from some unspecified distribution. As well as being Hermitian $H^\dagger = H$, the matrix Hamiltonian exhibits a *chiral symmetry* in the 2×2 block space,

$$H = -\sigma_3 H \sigma_3. \tag{5.2}$$

In consequence, the eigenvalues are either zero, or come in pairs of opposite sign. In other words, the energy $\epsilon = 0$ plays a special role implying that the average uniformity of the spectrum—a hallmark of the standard ensembles—is broken in the vicinity of $\epsilon = 0$. In particular, the average spectral density of H close to zero energy is nonstationary, yet nevertheless *universal* (i.e., for sufficiently low energies, the average density is independent of the nature of the distribution of the elements Z). Applying symmetry considerations like the ones outlined above to the chirally symmetric operator (5.1) one finds that H belongs to the symmetry class AIII.

In order to more explicitly assess the implications of the chiral structure of the Hamiltonian, we consider a random matrix implementation of the Hamiltonian H, i.e., a model where the blocks Z are drawn from some maximally entropic random distribution. (For the appearance of such models in physical applications, see below.) All information about the level statistics of this system is contained within the joint probability distribution function of the eigenvalues, $P(\{\epsilon\})$. Computing the Jacobian of the unitary transformations which bring the members of the ensemble of Gaussian distributed Hamiltonians to diagonal form,

$$H = U \begin{pmatrix} \hat{\epsilon} & \\ & -\hat{\epsilon} \end{pmatrix} U^{-1}, \quad \hat{\epsilon} = \mathrm{diag}(\epsilon_1, \epsilon_2, \ldots),$$

one obtains the distribution function [103]

$$P(\{\epsilon\}) d[\{\epsilon\}] = \prod_{i<j} |\epsilon_i^2 - \epsilon_j^2|^\beta \prod_k |\epsilon_k|^\alpha e^{-\epsilon_k^2/v^2} d\epsilon_k \tag{5.3}$$

where $\beta = 2$, $\alpha = 2$, and v represents the variance of the distribution. (We note that, by interpreting the Hamiltonian (5.1) as being tangent to the symmetric space AIII, the Jacobian can be read off immediately from the tabulated root system.) The structure of the distribution function makes both the breaking of spectral translational invariance and the special role played by $\epsilon = 0$ manifest. (The function does not only depend on energy differences but also on the energies themselves; for levels approaching $\epsilon = 0$, the function vanishes.) The vanishing of spectral weight in the vicinity of $\epsilon = 0$ has a simple physical interpretation: An energy level ϵ attempting to approach zero encounters its partner $-\epsilon$ approaching from the "other side." The randomness implemented in the Hamiltonian implies that the two levels repel each other and that the close vicinity of 0 becomes a "no go zone."

In addition to the fundamental chiral symmetry (5.2), the further symmetry under time-reversal (i.e., $H^T = H$ or, equivalently, $Z^* = Z$) can be imposed. According to the Cartan classification, a Hamiltonian of this type belongs to the class BDI. In this case, the corresponding distribution function takes the same general form as Eq. (5.3) but with $\beta = 1$ and $\alpha = 1$. (An extension of (5.1) to a symplectic structure identifies two further symmetry classes [101] which, for simplicity, we do not consider here.)

How does the chiral symmetry appear in the context of stochastic quantum physics? To mention one example, in the chiral (or massless) limit, a Dirac operator $\hat{D}$ anti-commutes with γ_5 (i.e., in the basis in which γ_5 is diagonal, the matrix elements of $\hat{D}$ are block off-diagonal). When subjected to a random gauge field, the low-energy properties of the Dirac operator are specified by one of the chiral random matrix ensembles. Here the different ensembles correspond to different choices of the gauge group and the number of flavours [104]. Later, in Section 5.5, we will see that operators of chiral symmetry appear in other physical contexts including disordered single-electron systems.

In addition to the four chiral symmetry classes described above, there exist a further three classes associated with "particle/hole"-like structures. Again, beginning with the lowest symmetry, the Hamiltonian

$$H = \begin{pmatrix} a & b \\ b^\dagger & -a^T \end{pmatrix} \tag{5.4}$$

where the block diagonal elements are Hermitian, $a^\dagger = a$, and the off-diagonal blocks are symmetric, $b^T = b$, exhibits the PH-symmetry

$$H = -\sigma_2 H^T \sigma_2. \tag{5.5}$$

In this case, according to the Cartan classification scheme, the Hamiltonian (5.4) belongs to symmetry class C. Again, the distribution function takes the general form of Eq. (5.3), but now with $\beta = 2$ and $\alpha = 2$ [105]. By imposing the further symmetry of time-reversal (i.e., $a^* = a$ and $b^* = b$), the symmetry is raised to class CI; $\beta = 1$ and $\alpha = 1$. Once again, an extension to a symplectic structure identifies a third universality class [101].

Hamiltonian's of symmetry class C and CI also find applications in various branches of physics. Of these, perhaps the one most directly motivated from the form of the Hamiltonian is superconductivity. However, as we will see below, various stochastic classical operators can often be classified as members of the symmetry class CI. A summary of the classification scheme for random matrix ensembles (without a symplectic structure), together with some common applications is tabulated below.

Class	RM Ensemble	Application: E.g.,
A	Dyson (GUE)	Disordered Conductor in B-Field
AI	Dyson (GOE)	T-invariant Conductor
AIII	Chiral GUE	Random Flux Model
BDI	Chiral GOE	Imaginary Vector Potential
C	NS	Superconductor in B-Field
CI	NS	T-invariant Gapless Superconductor

In the following, we will explore some of the unusual spectral and localization properties which accompany these different classes. Our first application is to the study of the quantum coherence properties of dirty superconductors.

5.2 *Dirty Superconductivity*

According to the phenomenology of the *BCS theory*, in the mean-field approximation, the quasi-particle Hamiltonian of a superconductor is specified by

$$\widehat{H}_{\rm BCS} = \int c_\sigma^\dagger \widehat{H} c_\sigma + \int [\Delta c_\uparrow^\dagger c_\downarrow^\dagger + \text{h.c.}] \tag{5.6}$$

where $\widehat{H} = \widehat{H}_0 + V$, ($\widehat{H}_0 = \left(\widehat{\mathbf{p}} - (e/c)\mathbf{A}\right)^2/2m - E_F$) denotes the bare Hamiltonian of the particle, and Δ denotes the (complex) order parameter. In the Nambu representation, the corresponding quasi-particle Hamiltonian takes the 2×2 matrix form $\widehat{H}_{\rm BCS} = \int \psi^\dagger \widehat{H}_{\rm Gorkov} \psi$ where $\psi^\dagger = (c_\uparrow^\dagger, c_\downarrow)$, and

$$\widehat{H}_{\rm Gorkov} = \begin{pmatrix} \widehat{H} & \Delta \\ \Delta^\dagger & -\widehat{H}^T \end{pmatrix} \tag{5.7}$$

defines the Gor′kov Hamiltonian. A comparison of $\widehat{H}_{\rm Gorkov}$ with Eq. (5.5) identifies the Gor′kov Hamiltonian as a member of symmetry class C. In the absence of an external magnetic field $\mathbf{A}$, a gauge can be specified in which the order parameter is real. In this case the Hamiltonian is symmetric under time-reversal $\widehat{H}^T_{\rm Gorkov} = \widehat{H}_{\rm Gorkov}$, and belongs to the higher symmetry class *CI*. Introducing Pauli matrices $\boldsymbol{\sigma}^{\rm PH}$ which operate in the matrix or PH-sector of $\widehat{H}_{\rm Gorkov}$, the PH-symmetry (5.5) translates to the condition, $\widehat{H}_{\rm Gorkov} = -\sigma_2^{\rm PH} \widehat{H}^T_{\rm Gorkov} \sigma_2^{\rm PH}$.

In the absence of an impurity potential V, the clean T-invariant Hamiltonian can be brought to the diagonal form

$$\epsilon_{\rm BCS}(\mathbf{p}) = \pm(\epsilon_{\mathbf{p}}^2 + \Delta^2)^{1/2}$$

where $\epsilon_{\mathbf{p}}$ represents the spectrum of the normal Hamiltonian $\widehat{H}_0$. The corresponding single-particle DoS is then given by

$$\nu_{\rm BCS}(\epsilon) = \frac{|\epsilon|}{|\epsilon^2 - \Delta^2|^{1/2}} \theta(|\epsilon| - |\Delta|). \tag{5.8}$$

How does disorder influence the quasi-particle spectrum? In the normal metallic conductor, disorder was shown to have no influence on single-particle properties such as the DoS. Recalling that nontrivial phenomena in mesoscopic physics typically originate in the interference of advanced and retarded Green functions, one might speculate that the average DoS of the superconductor, a quantity represented by the single Green function, $\widehat{G}^{\pm}(\epsilon) = (\epsilon_{\pm} - \widehat{H}_{\rm Gorkov})^{-1}$ does not exhibit interesting behavior. The fact is, however, that the average superconductor DoS is a highly nontrivial object, influenced by various mechanisms of quantum interference. The reason for the failure of the naive argument above is that it does not account for the high degree of symmetry of the Hamiltonian or, equivalently, for the structural information encoded in (5.7): Applied to the Gor′kov Green function, the PH-symmetry (5.5) translates to the property

$$G^{\pm}(\epsilon) = -\sigma_2^{\rm PH}[G^{\mp}(-\epsilon)]^T \sigma_2^{\rm PH} \tag{5.9}$$

which brings with it two implications: Firstly it shows that the PH-symmetry entails a pole ambiguity; retarded and advanced Green functions are equivalent up to a PH-transformation. This pole structure implies that a *single* Gor′kov Green function displays a degree of complexity comparable to a *pair* of normal Green functions, a phenomenon whose physical consequences we are going to discuss below. Secondly, (5.9) shows that finite energies ϵ break the PH-symmetry which is just another way of expressing the special role played by the band center. In the presence of disorder, these structures manifest themselves in the existence of soft modes which become massive for finite ϵ. To better understand the formation of these modes and the mesoscopic phenomena caused by them, we turn to the consideration of an effective low-energy field theory.

Field Theory of the Dirty Superconductor

The general construction of the low energy effective action for a bulk superconductor was first discussed by Oppermann [106], and later developed by Kravtsov and Oppermann [107] within the framework of a fermion replica field theory. Their approach was subsequently generalized to the supersymmetric scheme by Altland, Simons and Taras-Semchuk [108] in a study of the coherence properties of hybrid superconductor-normal structures. As usual, the field theoretic approach begins with the representation of the generating function of the Gor′kov Green function and its products as a functional field integral. As with the T-invariant normal metallic systems, to accommodate the PH-symmetry of the Hamiltonian, it is convenient to make use of the "charge conjugation" operation (5.9) to affect a doubling of the field space.

Following the notation of Ref. 109 (and the general scheme as outlined in Section 2), the generating function of the Gor′kov Green function can be represented as a functional field integral involving four component superfields ψ which carry a PH-index and BF-grading. As with T-invariant normal conductors, to take into account the particle/hole symmetry, $\int \bar{\psi}(\widehat{H}_{\text{Gorkov}}-\epsilon)\psi = -\int \psi^T\sigma_2^{\text{PH}}(\widehat{H}_{\text{Gorkov}}+\epsilon)\sigma_2^{\text{PH}}\bar{\psi}^T$, it is convenient to double the field space after which the generating function takes the form

$$\mathcal{Z}[J=0] = e^{-i\int \bar{\Psi}(\epsilon^-\sigma_3^{\text{CC}}-\widehat{H}_{\text{Gorkov}})\Psi},$$

where

$$\Psi = \frac{1}{\sqrt{2}}\begin{pmatrix}\psi \\ \sigma_2^{\text{PH}}\bar{\psi}^T\end{pmatrix}, \quad \bar{\Psi} = \frac{1}{\sqrt{2}}\begin{pmatrix}\bar{\psi} & -\psi^T\sigma_2^{\text{PH}}\end{pmatrix}.$$

It is in this additional sector, hereafter known as the "charge conjugation" (CC) sector, that the Pauli matrices $\boldsymbol{\sigma}^{\text{CC}}$ operate. With this definition, the superfields Ψ and $\bar{\Psi}$ obey the symmetry relations

$$\Psi = \sigma_2^{\text{PH}}\gamma\bar{\Psi}^T, \quad \bar{\Psi} = -\Psi^T\sigma_2^{\text{PH}}\gamma^{-1}$$

where $\gamma = \mathbf{1}^{\text{PH}}\otimes(E_{\text{BB}}\otimes\sigma_1^{\text{CC}} - E_{\text{FF}}\otimes i\sigma_2^{\text{CC}})$. Note that, in principle, one could further extend the field space to accommodate more than a single Green function in parallel. Moreover, according to Eq. (5.9), an extension of the field space to accommodate advanced and retarded degrees of freedom is unnecessary as one can be presented as a transform of the other (5.9).

Proceeding as with the normal conductor, taking the impurity potential to be Gaussian δ-correlated, the interaction of the fields generated by the ensemble averaging over the impurity configuration can be decoupled with the introduction of 8×8 component supermatrix fields Q. As a result, the generating functional takes the form

$$\langle\mathcal{Z}[J=0]\rangle = \int DQ e^{\int \text{str}(\pi\nu_{\text{N}}Q^2/(8\tau)-[\ln\widehat{\mathcal{G}}^{-1}]/2)} \tag{5.10}$$

where ν_{N} denotes the normal average DoS, and

$$\widehat{\mathcal{G}}^{-1} = \widehat{H}_{\text{Gorkov}}(V=0) - \epsilon^-\sigma_3^{\text{CC}} + \frac{i\sigma_3^{\text{PH}}}{2\tau}Q,$$

represents the supermatrix Green function. Here the symmetry properties of the superfields reflect those of the dyadic product $\psi\otimes\bar{\psi}$ and induce the symmetry constraint $Q = \sigma_1^{\text{PH}}\gamma Q^T\gamma^{-1}\sigma_1^{\text{PH}}$ [109].

Previously, with the normal conductor, progress was made by subjecting the action to a mean-field analysis and studying influence of slow fluctuations around the saddle-point by means of a gradient expansion.

Here we can attempt to formulate a similar approach. Taking the typical strength of the disorder potential $1/\tau$ to be greatly in excess of the strength of the order parameter $|\Delta|$, both the superconducting component as well as ϵ can be treated as a weak perturbation. In this case, a variation of the action with respect to Q obtains the saddle-point equation $Q_{\rm SP}(\mathbf{r}) = i\mathcal{G}(\mathbf{r},\mathbf{r})\sigma_3^{\rm PH}/\pi\nu$. As usual, identifying the saddle-point solution with the self-consistent Born approximation for the average Green function, we obtain the solution $Q_{\rm SP} = \sigma_3^{\rm PH} \otimes \sigma_3^{\rm CC}$.

However, as with the normal conductor, the invariance of the action as $\mathbf{A} \to 0$, $|\Delta| \to 0$ and $\epsilon \to 0$ under the spatially homogeneous transformation $Q(\mathbf{r}) \mapsto TQ(\mathbf{r})T^{-1}$ where T is compatible with the symmetry properties of Q, identifies an entire saddle-point manifold. A gradient expansion of the action which takes into account the soft modes of the theory associated with the saddle-point manifold obtains the nonlinear σ-model action,

$$S[Q] = -\frac{\pi\nu_{\rm N}}{8}\int [D(\widetilde{\partial}Q)^2 - 4i(\epsilon^- \sigma_3^{\rm CC} + \hat{\Delta})\sigma_3^{\rm PH} Q], \qquad (5.11)$$

where $\hat{\Delta} = |\Delta|\sigma_1^{\rm PH} e^{-i\phi\sigma_3^{\rm PH}}$, $\widetilde{\partial} = \partial - i(e/c)\mathbf{A}[\sigma_3^{\rm PH},\ ]$ and, as usual, the supermatrix fields are subject to the nonlinear constraint $Q(\mathbf{r})^2 = \mathbf{1}$.

As anticipated from our preliminary considerations of the Gor′kov Green function, the form of the effective action closely resembles that of the normal conductor. Indeed, removing the order parameter, and identifying the base point $\sigma_3^{\rm PH} \otimes \sigma_3^{\rm CC}$ of the manifold with $\sigma_3^{\rm AR} \otimes \mathbf{1}^{\rm TR}$ the action of the normal conductor (2.9) and (5.11) coincide. By analogy with our previous discussions, the action shows that the soft modes associated with density relaxation in the PH-channel are made massive by the frequency source, ϵ. However, in contrast to the normal metal, the order parameter Δ further reduces the degeneracy of the saddle-point manifold even at $\epsilon = 0$.

Mean-Field Theory

To determine the influence of the order parameter on the low-lying structure of $S[Q]$ it is necessary to subject the nonlinear σ-model action to a further saddle-point analysis. To keep our discussion simple, we will hereafter limit our consideration to a T-invariant Hamiltonian $\mathbf{A} = 0$, wherein we adopt a gauge in which the phase of the order parameter ϕ is set to zero. In this case, a variation of the action with respect to Q subject to the nonlinear constraint $Q^2 = \mathbf{1}$, obtains the low-energy saddle-point equation

$$D\partial(Q_{\rm SP}\partial Q_{\rm SP}) + i[Q_{\rm SP}, \mathbf{\Omega}\cdot\mathbf{\Sigma}] = 0 \qquad (5.12)$$

where we have defined the vectors $\mathbf{\Omega} = (0, -i|\Delta|, \epsilon)$ and $\mathbf{\Sigma} = (0, \sigma_2^{\rm PH}, \sigma_3^{\rm PH} \otimes \sigma_3^{\rm CC})$. Homogeneity of the gap function Δ implies homogeneity of the saddle-point solution, $Q_{\rm SP}(\mathbf{r}) = Q_{\rm SP}$. Applying the ansatz $Q_{\rm SP} = \widehat{\mathbf{q}}\cdot\mathbf{\Sigma}$, where $\widehat{\mathbf{q}}^2 = 1$, the saddle-point equation translates to the condition $\widehat{\mathbf{q}}\times\mathbf{\Omega} =$

0. From this equation we obtain the homogeneous solution $\widehat{\mathbf{q}} \propto \mathbf{\Omega}$. A consideration of the analytic properties of the Green function requires that the branch cut in the definition of $\mathbf{\Omega}$ lies along the negative real axis. With this choice

$$\widehat{\mathbf{q}} = \frac{\mathbf{\Omega}}{|\mathbf{\Omega}|} \times \begin{cases} 1, & |\epsilon| > |\Delta|, \\ -i\,\mathrm{sgn}(\epsilon), & |\epsilon| < |\Delta|. \end{cases} \tag{5.13}$$

Applied to the DoS, the saddle-point solution reproduces the DoS corresponding to that of a *clean* bulk superconductor,

$$\langle \nu(\epsilon) \rangle = \frac{\nu_{\mathrm{N}}}{8} \,\mathrm{Re}\langle \mathrm{str}(Q\sigma_3^{\mathrm{CC}} \otimes \sigma_3^{\mathrm{PH}} \otimes \sigma_3^{\mathrm{BF}}) \rangle_Q \overset{Q \to Q_{\mathrm{SP}}}{\simeq} \nu_{\mathrm{BCS}}(\epsilon).$$

On the level of the saddle-point or mean-field, disorder has no influence on the average DoS of the superconductor.

Fluctuations

In taking into account fluctuation corrections to the mean-field solution, it is first necessary to identify the different classes of soft modes that operate in the vicinity of the saddle-point. Based on our earlier considerations of the fundamental symmetries, we anticipate the existence of two separate universality classes, CI and C. With $\epsilon = 0$, $\mathbf{A} = 0$, and Δ (real and) arbitrary, the action is invariant under global transformations

$$Q(\mathbf{r}) \mapsto TQ(\mathbf{r})T^{-1}, \tag{5.14}$$

where T is constant in space and obeys the conditions $\gamma = T\gamma T^T$ and $\sigma_2^{\mathrm{PH}} = T\sigma_2^{\mathrm{PH}}T^T$. Formally, T belongs to the manifold $\mathrm{OSp}(2|2) \times \mathrm{OSp}(2|2)$. Dividing off those transformations $H\colon T \mapsto TH$ which leave $\sigma_3^{\mathrm{PH}} \otimes \sigma_3^{\mathrm{CC}}$ invariant (i.e., $H\sigma_3^{\mathrm{PH}} \otimes \sigma_3^{\mathrm{CC}} H^{-1} = \sigma_3^{\mathrm{PH}} \otimes \sigma_3^{\mathrm{CC}}$), the symmetry is reduced to the group manifold $\mathrm{OSp}(2\,|\,2) \times \mathrm{OSp}(2\,|\,2)/\,\mathrm{OSp}(2\,|\,2) = \mathrm{OSp}(2\,|\,2)$ which defines the symmetry class CI.[5]

Having specified the symmetry properties associated with the soft modes of the action, we are, at least in principle, in a position to investigate their influence on the DoS. Here, instead of carrying out that program in detail, to keep our discussion concise we merely state the outcome of such an analysis. Referring to Ref. 108 for technical details, an investigation of the influence of fluctuations on the average DoS reveals the existence of only small corrections to the mean-field solution, perturbative in the parameter

[5] Here, for completeness, we remark that under the influence of a nonvanishing external magnetic field $\mathbf{A}$, time-reversal symmetry is broken, the superconducting gap collapses, and the symmetry of the invariant manifold is reduced. Yet, for $\epsilon = 0$, the action remains invariant under global transformations $T = \mathbf{1}^{\mathrm{PH}} \otimes t$, where the elements $t \in \mathrm{OSp}(2\,|\,2)$ obey the condition $\gamma = t\gamma t^T$. Dividing off those transformations $h\colon t \mapsto th$ which leave σ_3^{CC} invariant (i.e., $h\sigma_3^{\mathrm{CC}}h^{-1} = \sigma_3^{\mathrm{CC}}$), the symmetry is reduced to the quotient or coset space $\mathrm{OSp}(2\,|\,2)/\,\mathrm{Gl}(1\,|\,1)$ which defines the symmetry class C.

E_c/Δ, *above* the gap edge. At least in the noninteracting model considered here, the integrity of the superconducting energy gap itself is *maintained* in the presence of an impurity potential!

In fact, this result is consistent with the celebrated *Anderson theorem* [110], and could have been anticipated from the outset: Going back to the quasi-particle Hamiltonian (5.6), even in the presence of disorder, a canonical transformation which diagonalizes the bare Hamiltonian $\widehat{H}_{\text{Gorkov}}$ can be used to bring the matrix Hamiltonian to a block diagonal form. Diagonalized, the quasi-particle Hamiltonian has the spectrum $E_i = \pm(\epsilon_i^2 + \Delta^2)^{1/2}$ where ϵ_i denotes the eigenvalues of the disordered Hamiltonian $\widehat{H}$. Evidently, the gap Δ is preserved by the impurity potential. Put differently, Cooper pairs can be formed out of the exact time-reversed eigenstates of the random single-particle Hamiltonian, the DoS of which is largely unaffected by static disorder.

This conclusion also has a natural interpretation at the level of the diffus on modes. The soft modes associated with the PH-symmetry rely on the constructive phase interference of a particle with energy ϵ and a hole with energy $-\epsilon$. Since the clean Hamiltonian exhibits no quasi-particle states below the gap, the "soft modes" are guaranteed to have a mass of at least Δ.

At first sight, the conclusion of this investigation might seem disappointing: Low-energy fluctuations associated with novel modes of density relaxation (in this case in the PH-sector) seem to have little role to play in the environment of a time-reversal invariant superconductor with a large uniform energy gap. However, as we will see shortly, such modes can have a dramatic effect on the phase coherence properties of gapless superconductors (e.g., those of d-wave symmetry), as well on the properties of a normal metal brought into contact with a superconductor.

5.3 Hybrid Superconductor-Normal Systems

With this preparation, let us now consider the influence of disorder on the spectral properties of a hybrid superconductor-normal system. A superconductor brought into contact with a normal metal tends to impart aspects of its superconducting character onto the normal region. Underlying this effect is the mechanism of *Andreev scattering* by which an electron from the normal region is reflected at the SN-interface as a hole, and a Cooper pair is added to the superconducting condensate. This phenomenon has a striking effect on spectral and transport properties, known collectively as the *Proximity Effect.*

SN Phenomenology

The essential features that distinguish Andreev scattering from normal scattering are summarised below:

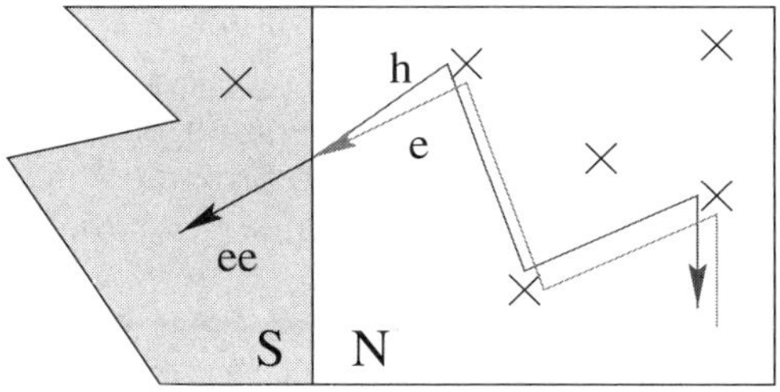

FIGURE 6.43. Schematic diagram showing a typical pair of Feynman paths which, by the mechanism of Andreev scattering, induce a nonzero expectation value of the anomalous average $\langle c_\uparrow^\dagger(\mathbf{r})c_\downarrow^\dagger(\mathbf{r})\rangle$.

• As opposed to ordinary specular reflection, Andreev reflection represents a process of *retro-reflection*. Apart from a slight angular mismatch proportional to the excitation energy ϵ of a quasi-particle above the Fermi energy E_F, the hole is reflected back along the trajectory of the incoming electron (see Fig. 6.43).

• An electron with excitation energy ϵ is scattered into a hole with energy $-\epsilon$.

• The hole acquires a scattering phase shift $\pi/2-\phi$ where ϕ is the phase of the superconducting order parameter at the interface.

An important consequence of Andreev processes is the formation of a Cooper pair amplitude $\langle c_\uparrow^\dagger(\mathbf{r})c_\downarrow^\dagger(\mathbf{r})\rangle$ in the normal metal region which, survives even in the disordered environment. The creation of a local pairing field expectation value can be heuristically understood from simple semiclassical considerations: Consider the creation of an electron somewhere at a point $\mathbf{r}$ inside a disordered metal adjacent to a superconductor (see Fig. 6.43). Due to the presence of disorder, the electron propagates diffusively. In doing so it may eventually strike the SN-interface and be Andreev reflected. In general the newly created hole may now diffuse along its own path. However, a particularly interesting situation arises if the hole happens to propagate along the path of the incoming electron back to the point of creation. As a result we not only obtain a nonvanishing pairing field expectation value but also a quantity that is insensitive to disorder averaging: During their propagation through the disordered background, both the incoming electron and the outgoing hole accumulate a quantum mechanical scattering phase which depends sensitively on microscopic details of the disorder. However, owing to the fact that the two particles propagate along the same path these phases cancel each other to a large extent. (For an excitation energy $\epsilon=0$ the cancellation is, in fact, perfect. For nonvanishing ϵ one obtains a phase mismatch $\sim \epsilon L^2/D$, where L denotes the separation of $\mathbf{r}$ from the interface.) Of course, more generic paths, where the electron and hole follow different trajectories, also contribute to the pairing field amplitude. However such contributions vanish upon disorder averaging due to their strong phase dependence.

The nonvanishing of the anomalous average $\langle c_\uparrow^\dagger(\mathbf{r})c_\downarrow^\dagger(\mathbf{r})\rangle$ even within the normal region is the basic content of the proximity effect. Besides being robust against disorder, the pairing field amplitude possesses a number of important features (all of which are related to the phase argument above):

- $\langle c_\uparrow^\dagger(\mathbf{r})c_\downarrow^\dagger(\mathbf{r})\rangle$ varies weakly as a function of $\mathbf{r}$. Specifically, it does not fluctuate on the scale of the wavelength, but rather on length scales set by the diffusion length $L_\epsilon \equiv (D/\epsilon)^{1/2}$.
- While phase coherent (i.e., $\epsilon \gg T, \tau_\varphi^{-1}$), the anomalous average $\langle c_\uparrow^\dagger(\mathbf{r})c_\downarrow^\dagger(\mathbf{r})\rangle$ decays exponentially as a function of $\epsilon L^2/D$ as one moves further into the normal region away from the SN interface.
- Quantitative expressions for the diffusive pairs of quantum paths entering the physics of the proximity effect are provided by diffuson modes in the PH-channel.
- Finally, the pairing field amplitude depends on the phases of the order parameters of the adjacent superconductors. If only a single superconducting terminal with constant phase ϕ is present, the phase is inessential and can be eliminated by means of a global gauge transformation. More interesting situations arise when more than one superconductor is present. In such cases, the phase sensitivity of the pairing amplitude provides the basic mechanism for the stationary *Josephson effect.*

The nonvanishing of the pairing field amplitude heavily influences the properties of the normal metal compounds of SN-systems. Widely known examples of proximity effect induced phenomena are the DC and AC Josephson effect, which allow for the accommodation of supercurrent across an SNS-junction. Another important phenomenon is the dependence of the normal conductance on the phases of adjacent superconductors—again triggered by the phase sensitivity of the proximity amplitude. However, for brevity, we again limit our consideration to the influence of the proximity effect on the quasi-particle spectrum inside the normal region.

Why should the local DoS be influenced by the Proximity Effect? An answer to this question, at least qualitatively, is provided by the identification of relevant phase coherent corrections. Referring to Fig. 6.44, a particle injected at a position $\mathbf{r}$ can propagate diffusively through the normal region. In the simplest case, a phase coherent contribution to the local DoS at $\mathbf{r}$ is provided by a closed path in which a particle is Andreev scattered into a hole and then, later, back to a particle. Taking into account the total π phase shift due to the Andreev scattering processes, the local DoS is diminished from the normal value by such a process.

At higher order in the number of Andreev scatterings, one can identify "weak localization" type processes (now in the PH-channel) which induce further corrections to the local DoS (see Fig. 6.44). Now, although such

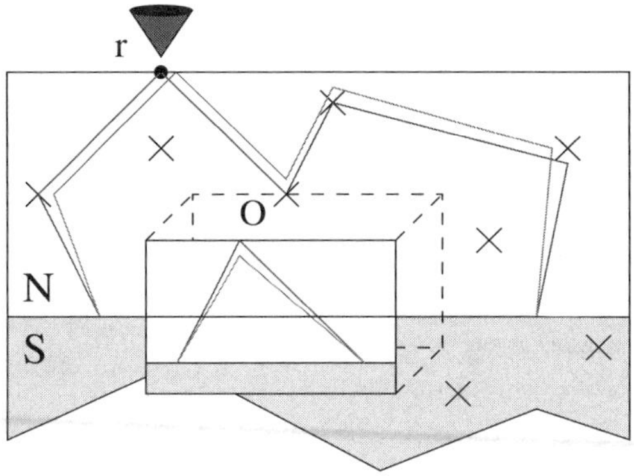

FIGURE 6.44. Schematic diagram showing a phase coherent path contributing to the renormalization of the DoS at the position **r**. A quantum correction to this path is shown schematically as an overlay. The trajectory shown in the overlay gives an enhanced return probability to the point O, and can be viewed as a kind of "weak localization" correction operating in the PH-channel. However, unlike the weak localization corrections in normal conductors, such contributions are here not small but accumulate. The infinite hierarchy of corrections is contained within the nonlinear Usadel equation (5.12).

contributions are higher order in the number of Andreev scatterings, they nevertheless contribute at the same order. As a result, to account quantitatively for the influence of the proximity effect on the local DoS, it is necessary to take into account the accumulation of the infinite hierarchy of such contributions. Fortunately, this is achieved more efficiently within the framework of an effective low-energy action.

Field Theory of SN-Structures

To investigate the influence of Andreev processes on the spectral properties of phase coherent SN devices we can again construct an effective low-energy field theory. For this purpose, we can immediately import the formalism developed for the bulk superconductor above. In particular, if the contact between the normal metal and superconductor is metallic, it suffices to consider the quasi-particle Hamiltonian (5.6) with an order parameter which takes a nonzero value only within the superconducting regions. (Strictly speaking, a rigorous approach would demand a self-consistent treatment of the order parameter particularly in the vicinity of the interface. However, at least for s-wave symmetry, the self-consistent solution is only slightly modified and the phenomenology of our discussion below applies.) By taking, for simplicity, the normal average DoS ν_{N}, and scattering time τ to be uniform throughout the S and N regions, the effective low-energy theory is represented by the same nonlinear σ-model action as (5.11), with the inhomogeneity of Δ understood.

Similarly, varying the action with respect to Q, the low-energy mean-field equation is given by Eq. (5.12). Historically, within this context, the saddle-point equation was first derived within the framework of a quasi-classical theory of the average Green function [111], and is known in the literature as the *Usadel equation* [112]. (For a review, see Ref. 113.)

The qualitative form of the solution can be easily deduced from the asymptotic dependences: Deep within the superconducting region, the solution to the saddle-point equation converges to the bulk value (5.13). While, deep within the normal region, $Q_{\rm SP} = \widehat{\mathbf{q}} \cdot \mathbf{\Sigma}$ is dominated by the quasi-particle energy scale ϵ, and converges on the bulk solution for the normal region, $Q_{\rm SP} = \sigma_3^{\rm PH} \otimes \sigma_3^{\rm CC}$. To accommodate the spatial inhomogeneity of the gap function across the SN-interface, the matrix $Q_{\rm SP}$ "rotates" on a length scale set by the diffusion length L_ϵ. Finally, current conservation at the SN-interface implies the boundary condition [114], $\sigma_{\rm S} Q\partial_\perp Q|_{\rm SN} = \sigma_{\rm N} Q\partial_\perp Q|_{\rm SN}$, where $\sigma_{\rm S}$ and $\sigma_{\rm N}$ denote the normal conductivity of the S and N regions.

As with the bulk superconductor, an explicit solution of the saddle-point equation can be obtained by applying the ansatz, $Q_{\rm SP} = \widehat{\mathbf{q}} \cdot \mathbf{\Sigma}$, after which Eq. (5.12) takes the form

$$\widehat{\mathbf{q}} \times (D\partial^2 \widehat{\mathbf{q}} - 2i\mathbf{\Omega}) = 0.$$

An angular parameterisation of $\mathbf{q}$ brings the nonlinear Usadel equation to the form of a sine-Gordon equation. A solution for the latter can be found in closed form. Again, leaving the details to Ref. 108, we focus on the qualitative form of the solution. Applied to an SNS-junction (with the geometry shown in Fig. 6.45) with a zero phase difference between the superconducting terminals, the solution to the Usadel equation is shown in Fig. 6.45 for a typical range of parameters. The corresponding local DoS, obtained as the projection of $\widehat{\mathbf{q}}$ onto $\sigma_3^{\rm PH}$, $\nu(\mathbf{r}) = \nu_{\rm N}\,{\rm Re}[\hat{q}(\mathbf{r})]_3$ is shown in Fig. 6.46.

The most striking feature of the solution is the appearance of a spatially constant minigap in the local quasi-particle DoS of the normal metal region. While, within the superconductor, in the vicinity of the SN-interface, a nonzero local DoS develops below the gap. Away from the interface, the

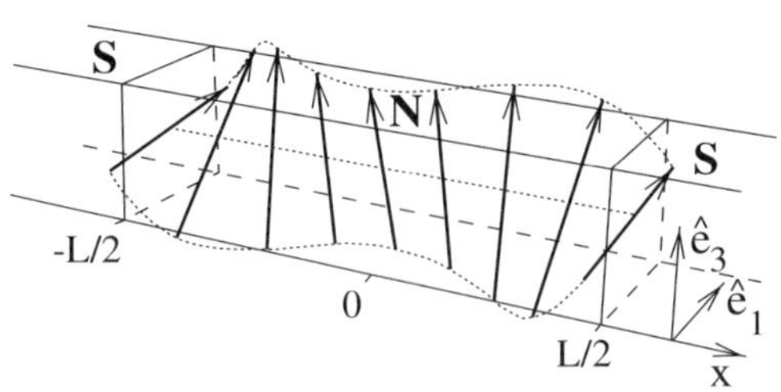

FIGURE 6.45. Arrows showing the real part of the solution $\widehat{\mathbf{q}}$ applied to an SNS junction for a typical range of parameters (courtesy of Ref. 108).

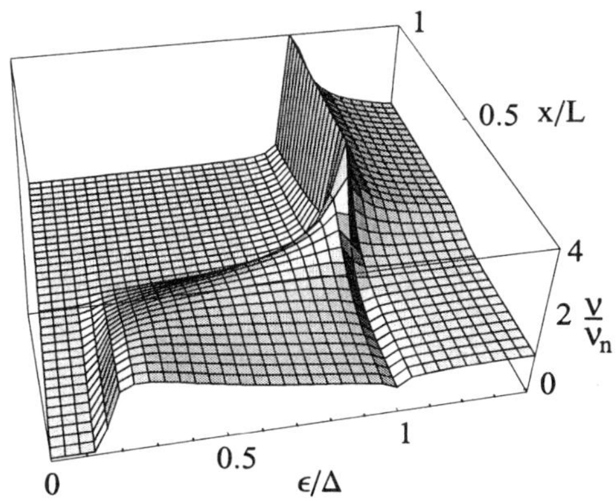

FIGURE 6.46. Local DoS ν of the SNS structure of Fig. 6.45 shown as a function of energy and position for a typical range of parameters (courtesy of Ref. 108).

latter decays exponentially on a scale set by the bulk coherence length $\xi = (D/2\Delta)^{1/2}$.

This concludes our preliminary survey of the influence of quantum coherence phenomena in phase coherent disordered hybrid SN-structures. Our discussion has focussed specifically on the influence of the proximity effect on the single-particle properties of the system. A discussion of the two-particle properties and the interplay of fluctuations in different channels can be found in Ref. 108. Following from our discussion of the coherence properties of chaotic quantum systems, it is interesting to note that a theory of the proximity effect in nonstochastic structures can be cast in the form of the ballistic σ-model. The corresponding "kinetic equation" associated with the saddle-point also finds a counterpart in the early literature and is known as the *Eilenberger equation* [111].

Finally, we should note that Coulomb interaction effects can play an important role in the shaping of the properties of dirty superconductors. Again, a generalisation of the statistical field to account for interaction effects is achieved straightforwardly, while its analysis remains the subject of on-going research.

The investigation of quantum coherence phenomena in hybrid SN–structures emphasises the importance of discrete symmetries on the properties of weakly disordered systems. While the low-lying diffusion modes in the PH-channel were rendered massive by the gap function in the bulk superconductor, in the SN-environment the influence of these modes in the normal region can dramatically influence the quasi-particle spectrum. These results also beg the question as to whether the disorder can influence the properties of bulk superconductors with a nontrivial symmetry.

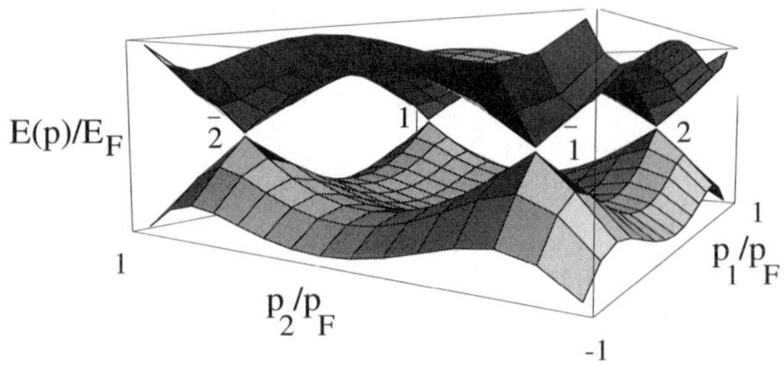

FIGURE 6.47. The clean quasi-particle spectrum of a d-wave superconductor showing the existence of four Dirac nodes at the Fermi level.

5.4 Dirty d-Wave Superconductivity

We have seen that the integrity of the gap in a conventional bulk superconductor make the soft low-energy diffuson modes redundant. Their influence becomes visible only when there exist quasi-particle states at zero energy, as in the hybrid SN systems. However, a second mechanism to induce quasi-particle states at zero energy arises in bulk superconductors of unconventional symmetry. To close our survey of coherence properties of disordered superconductors, we therefore turn to the investigation of the long-range properties of superconductors of unconventional d-wave symmetry. In recent years, the latter have come under intense scrutiny [115–120] in relation to the hole-doped high temperature cuprate superconductors. Indeed, these attempts to resolve the impact of impurity scattering on the low temperature spectral and transport properties of a disordered d-wave superconductor have ignited great controversy in the literature.

The key feature which distinguishes d-wave from conventional s-wave superconductors, and makes the influence of disorder harder to resolve, is the existence of low-lying gapless Dirac-like quasi-particle excitations at the Fermi energy (see Fig. 6.47). Beginning with the work of Gor′kov and Kalugin [115], and later by Lee [116], early considerations based on approximate self-consistent treatments suggested that an arbitrarily weak impurity potential induces a finite DoS at the Fermi surface, and leads to weak localization of all quasi-particle states in the two-dimensional system. These results were finally put on a firm footing by Senthil et al. [120]. As expected from our qualitative considerations above, their results showed that the disordered d-wave superconductor is described by a diffusive nonlinear σ-model of symmetry class CI.

In principle, such an effective field theory can be derived from first principles. However, based on fundamental symmetries of the Hamiltonian, we can propose an effective low-energy action simply on phenomenological grounds. Previously, we saw that, for the bulk s-wave superconductor, the low-energy field configurations belonged to the group manifold $\mathrm{OSp}(2\,|\,2)$. Focusing on these modes alone, and parameterising $Q =$

$e^{(\sigma_2^{\rm PH}/2)\ln T}\sigma_3^{\rm PH}\otimes\sigma^{\rm CC}e^{-(\sigma_2^{\rm PH}/2)\ln T}$, a comparison with Eq. (5.11) suggests the low energy action

$$S=-\int \mathrm{str}\left[\frac{1}{4\pi g}\partial T\partial T^{-1}+i\kappa\epsilon(T+T^{-1})\right],$$

where g and κ represent nonuniversal coupling constants.

According to this phenomenology, at the lowest energy scales, the spectral properties of the d-wave superconductor are described by a nonlinear σ-model of symmetry class CI, whose fields belong to the *group* manifold OSp(2|2). At very low energies $\epsilon \ll 1/gL^2$, (i.e., within the ergodic regime) the action is dominated by the zero spatial mode. In this regime, the average DoS coincides with the respective random matrix ensemble. At intermediate energy scales, diffusive corrections lead to a renormalization of the DoS. More importantly, the renormalization properties of the action imply that influence of disorder leads to a weak localisation of all quasi-particle states.

As a footnote, we should add that, the consideration of a soft scattering potential (i.e., one that includes only forward intranode scattering) leads to a low energy action of a different symmetry class [121]. However, a discussion of these results lies beyond the scope of this review.

The analysis above emphasises the importance of discrete PH-symmetries in providing novel mechanisms of quantum coherence. Yet, as we have discussed earlier, PH-like symmetries represent just one class of models in which discrete symmetries play a crucial role. In the following we will examine how sublattice or chiral symmetries lead to unusual spectral and localization properties in general stochastic sublattice models.

5.5 *Chiral Symmetry and Sublattice Models*

The majority of disordered quantum systems are described by microscopic Hamiltonians of the general form $\widehat{H}=\widehat{H}_0+\hat{V}$, where $\widehat{H}_0$ models the underlying "clean" system (e.g., $\widehat{H}_0=\widehat{\mathbf{p}}^2/2m$), and disorder is introduced via the randomly distributed *Hermitian* operator $\hat{V}$ (e.g., $\hat{V}=V(\mathbf{r})$ random). However, in lattice models, disorder is often implemented in terms of stochastic operators which operate in purely off-diagonal elements,

$$\widehat{H}=-\sum_{\langle ij\rangle}[c^\dagger_{i\alpha}U^{\alpha\beta}_{ij}c_{j\beta}+\mathrm{h.c.}],\tag{5.15}$$

where $\langle ij\rangle$ denote neighbouring sites of a d-dimensional hypercubic lattice, the N-component operators $c^\dagger_{i\alpha}$, $\alpha=1,\ldots,N$, create lattice Fermions, and $U^{\alpha\beta}_{ij}$ represent N-dimensional matrices residing on the links of the lattice. Stochasticity is introduced by drawing the (generally complex) matrix elements U from some random distribution (subject to the Hermiticity requirement $U_{ij}=U^\dagger_{ji}$). For $N=1$, Hamiltonians of the type (5.15) are

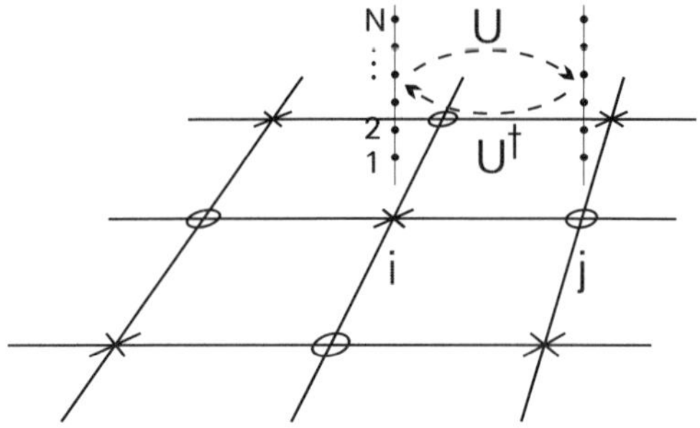

FIGURE 6.48. The generalized random flux model.

commonly referred to as *random flux* (RF) models, a denotation we will hereafter adopt for the cases $N \neq 1$.

In fact RF-models appear in a variety of different physical contexts: The two-dimensional $N = 1$ version describes the dynamics of lattice fermions subject to a random magnetic field or, more accurately, a random vector potential [122–126]. This model has also been discussed in connection with the physics of the half-filled fractional quantum Hall phase [127], the physics of the spin-split Landau level [124, 128], and the gauge theory of high temperature superconductivity [129]. Identifying the two fermion components of the $N = 2$ RF-model with a spin degree of freedom, (5.15) describes the propagation of lattice electrons on a spin-disordered background, a situation that occurs, for example, in connection with the physics of manganese oxides [130].

At first sight it would seem that the random flux Hamiltonian falls into the general category of Anderson Hamiltonians for disordered systems. Being Hermitian but not real, the Hamiltonian (5.15) is therefore expected to belong to the general class of models with unitary symmetry. As such, according to the one-parameter scaling theory, all states are expected to be Anderson localised in two dimensions. However, as will be discussed below, this picture is correct only as long as one stays away from zero energy.

Being of pure nearest neighbour type (i.e., having no matrix elements connecting orbitals on the same site), the RF-Hamiltonian (5.15) possesses an implicit discrete symmetry which, as will be explored below, heavily influences the physical properties of the model. The nature of the symmetry can easily be understood by inspection of Fig. 6.48. When subdivided into two sublattices (the ∘'s and the ×'s in Fig. 6.48), the lattice Hamiltonian becomes purely off-diagonal,

$$\widehat{H} = \begin{pmatrix} & \hat{H}_{\circ,\times} \\ \widehat{H}_{\times,\circ} & \end{pmatrix},$$

exposing an implicit chiral symmetry (5.2) of the Hamiltonian, and showing the spectrum to be symmetric, $\epsilon \to -\epsilon$. Therefore, according to the Cartan classification introduced in Section 5.1, a generic member of the random flux Hamiltonian belongs to the chiral symmetry class AIII.

As with the *d*-wave superconductor, the existence of a chiral symmetry has far reaching consequences for the physics of the model at the band-center. Although the Hamiltonian is *always* chiral, i.e., not just in the band center, for finite energies the chiral symmetry is ineffective. The reason is that the information about both thermodynamic and transport properties of the system is encoded in the single-particle Green function $G^{\pm}(\epsilon) \equiv (\epsilon^{\pm} - \widehat{H})^{-1}$ rather than in the Hamiltonian itself. Now, as with the superconductor, since $G^{\pm}(\epsilon) = -\sigma_3 G^{\mp}(-\epsilon)\sigma_3$, where the Pauli matrices operate in the space of the sublattice, the inversion property under adjungation with σ_3 is broken by finite ϵ.

In the vicinity of the band center, the chiral symmetry of the Hamiltonian gives rise to the existence of soft modes. To explore the influence of these modes on the spectral and delocalisation properties of the states near the band center we again look for an effective low-energy field theory. To keep our discussion concrete, we focus on a particular model in which the off-diagonal matrix elements U are drawn at random from the unitary distribution $\mathrm{U}(N)$.

Field Theory of the RF-Model

Previously, the construction of an effective field theory has relied on a separation of energy scales in which the influence of disorder is assumed to be weak (i.e., $1/E_F\tau \ll 1$). However, in the present case, there is no separation of energy scales—the disorder potential is manifestly strong. In principle, to implement the usual scheme, we could deliberately limit our considerations to sublattice models of the general form of Eq. (5.15) but where the disordered components of the matrix elements U_{ij} are small as compared to some uniform part. Indeed, beginning with the seminal work of Gade [131], sublattice models of this kind have been studied in the literature within the framework of conventional (replica) field theory approaches [132].

However, instead, we will introduce a second kind of approach tailored to the study of *unitary* stochastic operators. This approach, known in the literature as the 'colour-flavour transformation', provides a formal method for mapping model systems of this kind onto functional integrals containing a low-energy sector describing the large distance physics [133]. One could legitimately argue that such a generalisation is gratuitous—if we believe that the form of the low-energy action is universal, dictated solely by fundamental symmetries, we could work with a weak coupling theory, and ultimately, in the spirit of Ginzburg–Landau theory, extend the effective action into the strong coupling regime. However, apart from an obvious aesthetic appeal, our motivation for developing this novel approach is to emphasize the bridge that exists between mesoscopic physics, and other branches of theoretical physics. At the same time, one can view the coincidence of the strong and weak coupling theories as a microscopic or "*ab initio*" confirmation of universality.

As a first step towards establishing the transformed theory, we consider the ensemble average of the usual generating functional $\langle \mathcal{Z}[J] \rangle$ where

$$\mathcal{Z}[J=0] = \int D\psi e^{i\sum_i \bar{\psi}_i \epsilon^+ \psi_i + i \sum_{\langle i\in A, j \in B\rangle} (\bar{\psi}_i U_{ij} \psi_j + \bar{\psi}_j U^\dagger_{ij} \psi_i)} .$$

Here $\langle \cdots \rangle \equiv \prod_{\langle i,j\rangle} \int dU_{ij}(\cdots)$ represents the non-Abelian generalisation of a phase average, and dU_{ij} denotes the Haar measure on $\mathrm{U}(N)$. As usual, the fields $\psi = \{\psi^\alpha_{i,a}\}$ carry three types of indices: A lattice index i, a 'colour' index α (coupling to the internal group indices of the U's), and a 'flavour' index $a = \mathrm{B}, \mathrm{F}$ referring to the boson/fermion grading. (An extension of the field space to accommodate n-point functions is easily achieved by increasing the number of flavour components. Here, for simplicity, we focus on the one-point function.)

Colour-Flavour Transformation

The ensemble average over the bond matrices U_{ij} induces a complicated interaction of the fields. The colour-flavour transformation provides a method of decoupling the interaction effectively trading the group integrals over the U_{ij} for integrations over a set of auxiliary field variables Z_{ij}. Remarkably, despite the complicated nature of the interaction, the transformation is exact! (And straightforwardly generalized to other groups.) Following Ref. 133, an application of the transformation to link ij of the lattice, one obtains

$$\langle e^{i(\bar{\psi}_i U_{ij} \psi_j + \bar{\psi}_j U^\dagger_{ij} \psi_i)} \rangle = \int D(Z, \widetilde{Z}) M_{ij} e^{i(\bar{\psi}_i Z_{ij} \psi_i + \bar{\psi}_j \widetilde{Z}_{ij} \psi_j)},$$

where $M_{ij} = \mathrm{sdet}(1 - Z_{ij} \widetilde{Z}_{ij})^N$ represents the invariant measure, and the fields

$$Z_{ij} \in \mathrm{GL}(1\,|\,1), \quad \widetilde{Z}_{ij} \in \mathrm{GL}(1\,|\,1) \tag{5.16}$$

are 2×2-dimensional supermatrices living on the directed link from the A site i to the nearest neighbour B-site j. (By 'directed' we mean that there is no Z_{ji}.) The integration $\int D(Z, \widetilde{Z})$ over all pairs $(Z_{ij}, \widetilde{Z}_{ij})$ is subject to the further set of constraints:

$$Z_{ij,\mathrm{FF}} = -\widetilde{Z}^*_{ij,\mathrm{FF}}, \quad \widetilde{Z}_{ij,\mathrm{BB}} = Z_{ij,\mathrm{BB}}, \tag{5.17}$$

while the Grassmann valued components, $Z_{ij,\mathrm{BF}}$, $Z_{ij,\mathrm{FB}}$, $\widetilde{Z}_{ij,\mathrm{BF}}$, and $\widetilde{Z}_{ij,\mathrm{FB}}$ are independent.

Conceptually, the colour-flavour transformation has a status similar to that of the Hubbard–Stratonovich transformation employed in the study

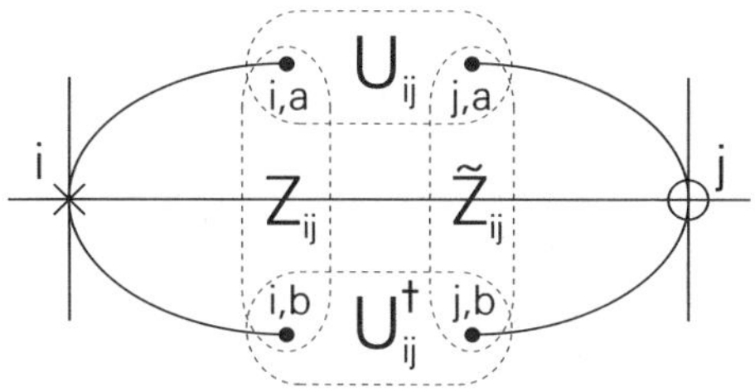

FIGURE 6.49. Visualisation of the different coupling of the U's and the Z's, respectively, to the field variables ψ (represented by the solid dots).

of continuum Hamiltonians. The motivation for introducing the Z-fields is that they rearrange the coupling between the fields (see Fig. 6.49). Qualitatively, the Z's connect pairs of fields that, by construction, represent segments of 'paired paths'. Segments of this structure are stabilised by mechanisms of quantum interference, implying that the Z's connect to a sector of the theory which contains information about the long-range behavior. In addition, the Z's couple nontrivially to the 'flavour' space. The remainder of the analysis will essentially amount to extracting the low-energy sector of the Z-functional stabilised by the interference.

Applying the colour-flavour transformation to each link of the lattice, and integrating over the ψ-fields, we obtain the functional

$$\langle \mathcal{Z}[J=0]\rangle \int D(Z,\widetilde{Z}) \prod_{\langle i\in A, j\in B\rangle} M_{ij} e^{-N\sum_{i\in A}\operatorname{str}\ln(\epsilon^+ + \sum_{j\in N_i} Z_{ij})} \times e^{-N\sum_{j\in B}\operatorname{str}\ln(\epsilon^+ + \sum_{i\in N_j}\widetilde{Z}_{ij})}, \quad (5.18)$$

where the symbol $\sum_{j\in N_i}$ denotes a summation over all sites j which neighbour i.

So far all manipulations have been exact. Further progress is made possible by subjecting the functional (5.18) to a saddle-point analysis. At $\epsilon = 0$, the variation of the action with respect to Z_{ij} and $\widetilde{Z}_{ij}$ obtains a saddle-point equation with the homogeneous solution $\widetilde{Z}_{\text{SP}} = Z_{\text{SP}} \simeq ix\mathbf{1}^{\text{BF}}$ where $x \equiv (2d-1)^{-1/2}$. However, as usual, this solution is not unique. The isotropy of the $\epsilon = 0$ action in the BF-sector, which follows directly from the chiral structure of the Hamiltonian, implies a manifold of degeneracy spanned by the transformations

$$Z_{\text{SP}} = ix\overline{T}T, \quad \widetilde{Z}_{\text{SP}} = ix(\overline{T}T)^{-1}, \quad (5.19)$$

where T, $\hat{T} \in \text{Gl}(1 \mid 1)$. Taking into account the fact that $\text{GL}(1 \mid 1) \times \text{GL}(1 \mid 1) = \text{Gl}(1 \mid 1)$, we conclude that the full extent of the saddle-point manifold is given by $\text{GL}(1 \mid 1)$ [134].

Equation (5.19) defines the maximum manifold of solutions of the saddle-point equations. In order to complete the determination of the global structure of the theory, we next need to explore how the manifold (5.19) intersects with the domain of the field integration (as specified by (5.16), and (5.17)). To this end, we represent a general pair of matrices as

$$(Z, \widetilde{Z}) \equiv ix(PT, T^{-1}P), \tag{5.20}$$

where $P, T \in \mathrm{GL}(1|1)$ and we have omitted the site index (ij) for notational transparency. The restriction of Eq. (5.17) leads to the relations,

$$\begin{aligned} &P_{\mathrm{FF}} \in \mathrm{GL}(1)/\,\mathrm{U}(1), \qquad &&T_{\mathrm{FF}} \in \mathrm{U}(1) \\ &P_{\mathrm{BB}} \in \mathrm{U}(1), &&T_{\mathrm{BB}} \in \mathrm{GL}(1)/\,\mathrm{U}(1), \end{aligned} \tag{5.21}$$

while the off-diagonal elements P_{BF}, P_{FB}, T_{BF} and T_{FB} are independent. Since Eqs. (5.19) and (5.20) imply that the saddle-point configurations of the theory are specified by $P = \mathbf{1}$, the measure of the P-integration is flat Euclidean, whilst the T-integration is over the invariant measures on the manifolds (5.21).

The saddle-point decomposition (5.21) can now be used to reduce the exact functional (5.18) to a simpler effective functional describing the long-range behavior of the model. Technically, the procedure is based on a continuum approximation in combination with a gradient expansion around the spatially constant saddle-point manifold. It is important to realize that this evaluation scheme is, in fact, problematic: As compared to analogous treatments of *weakly* disordered models, the gradient expansion is not stabilised by the presence of a small parameter (e.g., $1/E_F\tau$). Apart from the external parameter ϵ, all energy scales are rather of the same order. With the non-Abelian versions of the RF-model, $N > 1$, the situation is less problematic, the reason being that the mean field analysis is stabilised by the parameter $1/N$.

Keeping these words of caution in mind, we now turn to the discussion of the continuum expansion of the theory. On general grounds, it may be anticipated that after integration over massive modes P, the effective action takes the form $S[P,T] \longrightarrow S_{\mathrm{eff}}[T] = S_\epsilon[T] + S_{\mathrm{fl}}[T] + S_{\mathrm{m}}[T]$, where $S_\epsilon[T]$ is the contribution due to finite energies ϵ, $S_{\mathrm{fl}}[T]$ is the action associated with fluctuations of the Goldstone modes (the T's), and $S_{\mathrm{m}}[T]$ represents a residual action induced by the interaction between massive and Goldstone modes, respectively. (Note that, in conventional disordered models, to leading order in $1/k_F\ell$, no coupling between the massless and massive modes survives.) Assuming that the field configurations relevant to the long-range physics are smooth, the different contributions to the action can then be computed by a gradient expansion followed by a continuum limit. Albeit conceptually straightforward, the explicit formulation of this program is somewhat involved and we therefore choose to omit it here. Referring instead to the original literature [135] for technical details, we simply note that the outcome of the gradient expansion is the effective *group* action

$$S[T] = -\int \left[\operatorname{str}\left(c_1 \partial T^{-1} \partial T + i c_2 \epsilon (T + T^{-1})\right) + c_3 \left(\operatorname{str}(T^{-1}\partial T)\right)^2\right] + S_{\mathrm{b}}[T], \quad (5.22)$$

where $T \in \mathrm{GL}(1|1)$ (i.e., T belongs to the group of invertible supermatrices of dimension 2). Defining N_i, $i = 1, \ldots, d$ as the number of sites in the $\widehat{\mathbf{e}}_i$-direction,

$$S_{\mathrm{b}}[T] = \frac{N}{2^d} \sum_{s_i=0,1} (-)^{\sum_i (N_i+1)s_i} \operatorname{str} \ln\left(T(s_1 L_1, \ldots, s_d L_d)\right),$$

represents a boundary action that depends on the values of the fields T at the corner points of the lattice. The coupling constants are defined by $c_1 = Na^{2-d}/8d$, $c_2 = N(2d-1)^{1/2}a^{-d}/4d$, $c_3 = a^{2-d}C/16d$ where a represents the lattice spacing, and C denotes a geometry-dependent numerical constant $O(1)$.

Thus, for the ensemble of random flux lattice models whose matrix elements belong to the unitary group $\mathrm{U}(N)$, we have succeeded in formulating a low energy effective field theory. Based on the fundamental symmetries of the action, we can expect (5.22) to describe the long-range properties of any stochastic Hamiltonian which belongs to the symmetry class AIII, a conjecture which finds support in the analysis of the weak coupling version of the random flux model [131]. As a caveat, we remark that, as with the class CI action for the disordered d-wave superconductor, the group structure of the field manifold, in principle, admits the existence of a WZWN-term. However, at least for the random flux model considered here, such a term does not appear.

Properties of the Field Theory

What information can be drawn from the field theory? To answer this question, it is convenient to distinguish the very low energy, or "ergodic regime" from the diffusive. As usual, on energy scales $\epsilon \ll E_c \equiv c_1/(c_2 L^2)$, the functional is dominated by the spatially constant zero mode configuration $T_0 = \text{const.}$,

$$S_0[T_0] = -\frac{\pi\nu}{2} \operatorname{str}\left(\hat{\epsilon}(T_0 + T_0^{-1})\right) + S_{\mathrm{b}}[T_0], \quad (5.23)$$

where ν is the bulk mean DoS of the system. As expected, correlation functions computed with respect to the zero mode action coincide with those otherwise obtained from an analysis of the chiral unitary random matrix ensemble ChGUE [103, 104, 136, 137], (the matrix ensemble of symmetry class AIII in the classification scheme of Ref. 101). In particular, an explicit calculation of the mean DoS shows that [104]

$$\nu(\epsilon) = \frac{\pi^2 \epsilon \nu^2}{2} \left(\mathrm{J}_0^2(\pi\epsilon\nu) + \mathrm{J}_1^2(\pi\epsilon\nu)\right),$$

if the total number of lattice sites is even (and therefore $S_b = 0$), while, for N orbitals,

$$\nu(\epsilon) = \frac{\pi^2 \epsilon \nu^2}{2} \left(J_N^2(\pi\epsilon\nu) - J_{N-1}(\pi\epsilon\nu) J_{N+1}(\pi\epsilon\nu) \right),$$

if the number of sites is odd. These results show that, as $\epsilon \to 0$, the DoS vanishes on a scale set by the mean level spacing. (Moreover, in contrast to conventional disordered systems, the limit $\epsilon \to 0$ does not behave *absolutely* universally: The fine structure of the DoS close to $\epsilon = 0$ depends sensitively on the 'parity' of the lattice, i.e., on whether the number of sites is even or odd [104, 138–140]. Without going into detail, we remark that the information about this effect is encoded in the boundary term S_b.)

Leaving the random matrix regime and turning to the more subtle diffusive regime, various conclusions regarding the physical behavior of the RF-model, most notably about its localization behavior, can be directly inferred from the RG analysis of Ref. 131. There it was shown that the conductance of the weakly disordered $2d$ model at the band center (which is essentially determined by the coupling constant c_1) did not change under one-loop perturbative renormalization. This observation suggests that a nonlocalised state might exist in the middle of the band. Since the stability of the perturbative RG merely relies on the smallness of the parameters $a^{2-d}c_1^{-1}$, and $a^{4-d}c_2/c_1^2 \ll 1$, its results can be straightforwardly carried over to the $N \gg 1$ non-Abelian RF model: The one-loop renormalization indicates that for $N \gg 1$ the strongly disordered RF model exhibits metallic behavior at the band center.

As for the validity the RG results in the Abelian case ($N = 1$), from the results presented here, there appears to be only one useful test of consistency: The existence of a metallic state for $N = 1$ with strong disorder would at least be compatible with the behavior in the limiting cases: $N = 1$ and weak disorder, and $N \gg 1$ with arbitrary disorder. Moreover, this conclusion does also seem to be consistent with the most recent numerical investigations [122]. More concrete analytical evidence in favour of band center delocalisation is not currently available.

Besides the conductance, the DoS of the extended model also behaves in an unusual manner: In particular, the analysis of Ref. 131 predicts divergent behavior upon approaching the band center, where the detailed functional form of the divergence depends on the dimensionality of the system. In particular, at energy scales $E_c \ll |\epsilon| \ll 1$, in two dimensions the RG analysis predicts

$$\nu(\epsilon) \sim \frac{e^{-\kappa\sqrt{\ln(-|\epsilon|)}}}{|\epsilon|}.$$

The scale below which the $|\epsilon|^{-1}$ divergence begins to dominate over the exponential factor is determined by the constant κ. Unfortunately the dependence of that constant on the relevant parameters of the model can not

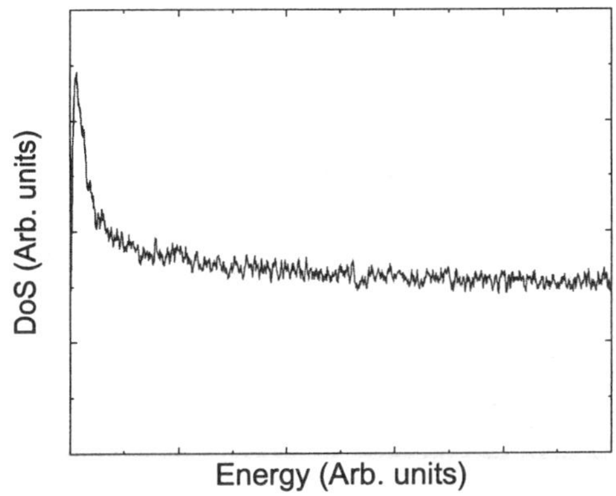

FIGURE 6.50. Measurement of the DoS of an $N = 1$ random flux Hamiltonian for a 20×20 tight-binding lattice Hamiltonian. The data clearly shows a divergence in the vicinity of zero energy together with a "microgap" corresponding to the ergodic regime of the model.

be reliably extracted from the perturbative RG analysis. The only statement that can safely be made is that a divergence forms at some point upon approaching the band center. Ultimately, the divergent behavior is cut-off by the onset of the nonperturbative ergodic regime discussed above. A quantitative understanding of both the formation of the divergence and its truncation in the close vicinity of the band center would necessitate a combination of an RG analysis with a nonperturbative treatment of the zero mode contribution. Nevertheless, these conclusions bear at least qualitative comparison with numerics (see Fig. 6.50).

A third piece of information that can be extracted from Ref. 131 is how finite energy arguments ϵ affect the localization properties of the model: In the spirit of the RG, finite energies have the significance of relevant perturbations. Thus, the band center represents a critical manifold whilst for finite ϵ, the model flows into a different universality class. More precisely, the band center field theory—belonging to the universality class AIII—develops into an ordinary GUE type model. (Note that this conclusion is in accord with the analysis in Ref. 125 where a continuum random field model far away from the band center was shown to be described by a supersymmetric unitary model.)

In that context, one may ask how the degrees of freedom of the model of reduced, unitary symmetry, denoted by Q [47], emerge from the fields T of the present model upon crossing over to the unitary regime. In fact the full RF model could have been formulated in terms of Q-matrices (albeit Q's of a symmetry different from the standard unitary symmetry) from the outset: Defining $Q = \widetilde{T}\sigma_3\widetilde{T}^{-1}$, where $\widetilde{T} = \exp[W\sigma_2]$ and the σ_i operate in sublattice space, the matrices $\widetilde{T}$ are then related to our T's by the identity

$T = \exp[W]$. The price to pay for this more common Q-formulation of the theory would be a certain amount of redundancy. A description in terms of structures $Q \sim \widetilde{T}\sigma_3\widetilde{T}^{-1}$ is the natural one for problems with *coset space* symmetry. However, the present problem has the *group* symmetry $\mathrm{Gl}(1\,|\,1)$ implying that a modelling directly in terms of group degrees of freedom is more natural.

Besides the weakly disordered sublattice model explored by Gade, the field theory (5.22) has at least two other close relatives: In QCD, (5.22) has been suggested on phenomenological grounds as relevant for the determination of the low energy spectrum of the Dirac operator [141]. (In QCD, 'randomness' is represented by gauge field fluctuations in the Yang-Mills Hamiltonian.) The simplest lattice action of a matter field with N 'colour' indices subject to a non-Abelian local gauge symmetry is given by

$$S = -\sum_{\langle i,j\rangle} c_i^{a\dagger} U_{ij} c_j^a - \frac{1}{2g}\sum_{\text{plaq.}}\left[\left(1 - \frac{1}{N}\,\mathrm{tr}_{\text{plaq.}}\, U^4\right) + \text{h.c.}\right].$$

where $a = 1,\ldots,n$ has the physical significance of a fermionic flavour index, $\sum_{\text{plaq.}}$ is a sum over all plaquettes of the lattice, and $\mathrm{tr}_{\text{plaq.}}(U^4) \equiv \mathrm{tr}(U_{ij}U_{jk}U_{kl}U_{li})$ represents the trace over all U's residing on the links of the plaquette. Physically, the plaquette action represents the discrete version of a continuum field strength tensor. In the strong coupling limit, $g \to \infty$ (i.e., the long distance, confinement phase of the theory) the gauge fields become free and the Hamiltonian is governed by a structure like (5.15). In this context, the base manifold is $4+1$-dimensional whilst the fields $T \in \mathrm{U}(n_f+1|1)$, where n_f is the number of quark flavours. For a comprehensive discussion of the QCD-analogue of (5.22), its connection with ChRMT and its relevance for lattice QCD-analyses, we refer the reader to Ref. 142. The similarity between the theories is again a manifestation of the universality of chiral σ-models or, more physically, the universal consequences chiral symmetries have for the long-range properties of random systems.

This concludes our investigation of the unusual low-energy properties of the ensembles that make up the novel symmetry classes. Our survey has been incomplete missing the ensembles exhibiting a symplectic structure. However, the generic features of these models and the unusual localization properties have been explored. As a final application of the scheme outlined above for the chiral ensembles, we consider the influence of coherence phenomena in the unusual setting of random classical dynamics.

5.6 *Random Classical Dynamics*

As we have seen, phase coherence phenomena play a crucial role in controlling the spectral and transport properties of disordered or complex chaotic quantum structures. Such effects are manifest in characteristic mesoscopic fluctuation phenomena (weak localization), as well as universal spectral correlations. With this background, it is pertinent to ask, are there analogous mechanisms of coherence operating in random *classical* structures? Here, in contrast to the behavior of closed quantum systems, the time evolution of classical distributions is typically specified by a dissipative dynamics (e.g., the Navier-Stokes equation). Although the equations of motion usually depend nonlinearly on the classical distributions, often the dynamics is specified by a linear equation (e.g., Fokker-Planck dynamics). In such circumstances, a parallel between classical and quantum dynamics can be developed.

The spectral properties of linear non-Hermitian operators have come under scrutiny in connection with a variety of physical applications including problems as diverse as anomalous diffusion and advection in classical random media [143], scattering in open quantum systems [144], neural networks [145], and the statistical mechanics of flux lines in dirty superconductors [146]. In each case, a correspondence can be drawn between the classical dynamics of the non-Hermitian operators and the quantum properties of *Hermitian* Schrödinger operators with a chiral symmetry (see, for example, Ref. 147).

To illustrate the general connection, we will focus on two particular applications (chosen largely for their simplicity). The first concerns the classical diffusion of a particle subject to a time-independent *imaginary* vector potential [146, 147],

$$\hat{\mathcal{L}}_{\mathrm{IV}} = \frac{1}{2m}\left(\widehat{\mathbf{p}} + i\frac{e}{c}\mathbf{A}\right)^2 + V(\mathbf{r}), \tag{5.24}$$

where $\widehat{\mathbf{p}} \equiv -i\partial$ defines the gradient operator, $1/2m$ denotes the effective "diffusion" constant, and $V(\mathbf{r})$ represents a random potential drawn, as usual, from a Gaussian δ-correlated impurity distribution (1.4). (Here our notation is deliberately chosen to emphasize the connection between the classical and the quantum.) The second example concerns the classical diffusion of a particle subject to a *complex* random scalar potential

$$\hat{\mathcal{L}}_{\mathrm{IS}} = \frac{\widehat{\mathbf{p}}^2}{2m} + i\Gamma(\mathbf{r}), \tag{5.25}$$

where Γ represents a real random potential drawn from some unspecified distribution. Both of these models have been studied in connection with specific physical applications.

In the first case, the motion of a quantum particle subject to a random imaginary vector potential has been discussed in connection with the statistical mechanics of flux lines in a type II superconductor pinned by a background of quenched random impurities [146]. More precisely, when expressed as a Feynman path integral, a matrix element of the Euclidean time evolution operator $\mathcal{Z} = \langle\psi_F|\exp[-T\hat{\mathcal{L}}_{\mathrm{IV}}/\hbar]|\psi_{\mathrm{I}}\rangle$ can be expressed as

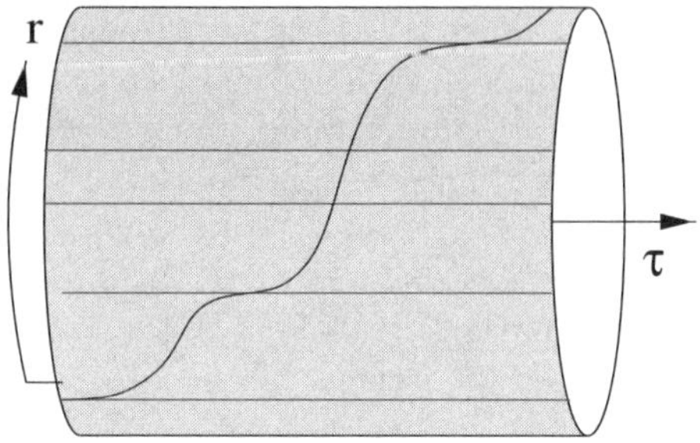

FIGURE 6.51. Schematic diagram representing a classical flux line configuration subject to a transverse magnetic field **A** and pinned by a random impurity potential. In executing a directed random walk, the flux line undergoes thermal fluctuations, and experiences a constant force from the magnetic field which acts in a direction perpendicular to the drift.

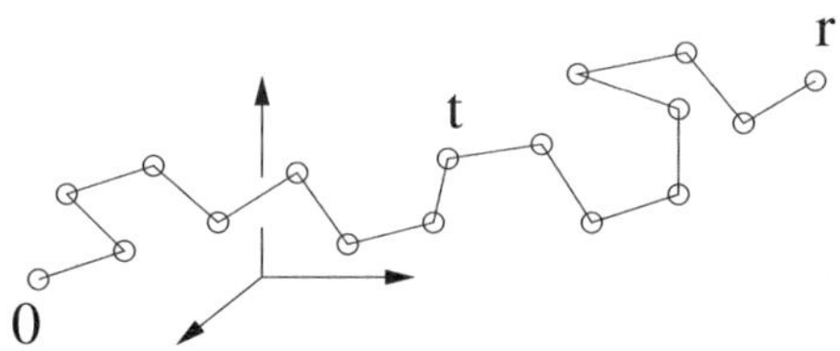

FIGURE 6.52. Schematic diagram showing a typical polymer configuration.

the functional integral

$$\mathcal{Z} = \int D\mathbf{r} e^{-(\int_0^T d\tau\, [m(d\mathbf{r}/d\tau)^2/2 - e\mathbf{A}/(mc)\cdot d\mathbf{r}/d\tau + V(\mathbf{r})])/\hbar}.$$

Regarding the variable τ as parameterising a length, the path integral can be interpreted as the statistical mechanics of a thermally fluctuating directed (i.e., always moving in the same longitudinal direction) flux line of length T, at temperature $\hbar$, subject to a transverse "magnetic field" $\mathbf{H}_\perp = \mathbf{A}e/mc$, and pinned by a quenched random washboard potential (see Fig. 6.51).

In the second case, the properties of a quantum particle subject to a random scalar potential has been investigated in connection with the statistical mechanics of a polymer chain with a contact interaction. In the continuum limit, the probability distribution $\mathcal{Z}(\mathbf{R}, t)$ of the end-to-end distance $\mathbf{R}$ of a self-interacting polymer chain of length t can be expressed in the form of a path integral [148, 149],

$$\mathcal{Z} = \int_{\mathbf{r}(0)=\mathbf{0}}^{\mathbf{r}(t)=\mathbf{R}} D\mathbf{r} e^{-\int_0^t d\tau\, m(d\mathbf{r}/d\tau)^2/2 - \int\int d\tau'\, d\tau''\, \gamma\delta^d(\mathbf{r}(\tau')-\mathbf{r}(\tau''))/2}.$$

where the contact interaction is specified by $\gamma \equiv 1/\pi\nu\tau$. The first term, known as the Wiener measure, determines the entropic contribution to the partition function, while the second term represents the repulsive local contact interaction between the monomers that make up the chain. Decoupling the interaction by means of a Hubbard–Stratonovich field, the distribution function can be identified with the ensemble average of the Feynman propagator of the Hamiltonian (5.29), $Z(\mathbf{r},t) = \langle\langle \mathbf{r}|e^{-t\hat{\mathcal{L}}_{\rm IS}}|\mathbf{0}\rangle\rangle_\Gamma$, where

$$\langle \mathbf{r}|e^{-t\widehat{H}}|\mathbf{0}\rangle = \int_{\mathbf{r}(0)=\mathbf{0}}^{\mathbf{r}(t)=\mathbf{R}} D\mathbf{r} e^{-\int_0^t d\tau\, [m(d\mathbf{r}/d\tau)^2/2 - i\Gamma(\mathbf{r}(\tau))]}$$

where Γ denotes a Gaussian δ-correlated impurity potential drawn from the distribution function (1.4).

Before subjecting both models to a field theoretic analysis, it is first necessary to identify the fundamental symmetries of the model Hamiltonians. In doing so, we will discover a connection between the generic class of non-Hermitian operators, and the chiral ensembles described above.

Symmetry Classification of Non-Hermitian Operators

In contrast to an Hermitian operator, the complex eigenvalues z_i of a general non-Hermitian operator are associated with right and left-hand eigenfunctions, i.e., $\widehat{H}|R_i\rangle = z_i|R_i\rangle$, and $L_i|\widehat{H} = \langle L_i|z_i$. Furthermore, although not a generic feature of a non-Hermitian operator, the right and left eigenfunctions typically form a complete (orthonormal) basis set,

$$\langle L_i \mid R_j\rangle = \delta_{ij}, \quad \sum_i |R_i\rangle\langle L_i| = \mathbf{1}. \tag{5.26}$$

This is indeed the case if there are no repeated eigenvalues, which will be true for a general operator such as the random or complex operators studied here.

As with the Hermitian counterparts, spectral properties of non-Hermitian operators can be conveniently expressed through the complex Green function

$$\hat{g}(z) = \frac{1}{z - \widehat{H}},$$

where $z = x + iy$ denotes the complex frequency argument. By inserting the resolution of identity (5.26), one obtains the spectral decomposition,

$$\hat{g}(z) = \sum_i |R_i\rangle \frac{1}{z - z_i} \langle L_i|.$$

Applied to the complex density of states (DoS),

$$\nu(z) \equiv \operatorname{tr} \delta(z - \widehat{H}) = \lim_{\eta\to 0} \frac{\eta^2}{\pi} \sum_i \frac{1}{(|z - z_i|^2 + \eta^2)^2},$$

one can establish the useful identity $\nu(z) = (1/\pi)\partial_{z^*} \operatorname{tr} \hat{g}(z)$ which shows that the random function $\operatorname{tr} g(z)$ is guaranteed to be *nonanalytic* everywhere in the complex z-plane where the eigenvalue density is nonvanishing. From this result, it follows that a perturbative analysis of random non-Hermitian operators can be ruled out: An expansion based on a purely diagrammatic scheme, $\hat{g}(z) = \sum_{n=0}^{\infty} \big(\hat{g}_0(z)V\big)^n \hat{g}_0(z)$, is inevitably analytic in z at any order [150].

Fortunately, such difficulties can be circumvented by expressing $\hat{g}(z)$ in terms of an auxiliary operator which is manifestly Hermitian [147, 150, 151]. In what proves to be an ingenious mapping, the Green function of the non-Hermitian operator is expressed as the off-diagonal element of the matrix Green function [151]

$$\hat{g}(z) = \widehat{G}^+_{21}(0), \quad \widehat{G}^+(\epsilon) = \frac{1}{\epsilon^+ - \widehat{\mathcal{H}}},$$

of an Hermitian Hamiltonian $\widehat{\mathcal{H}}$ which exhibits a chiral 2×2 block structure

$$\widehat{\mathcal{H}} = \begin{pmatrix} & \widehat{H} - z \\ \widehat{H}^\dagger - z^* & \end{pmatrix}. \tag{5.27}$$

I.e., to each non-Hermitian operator, an Hermitian counterpart of twice the rank can be defined.

The correspondence between the non-Hermitian and the Hermitian Hamiltonian allows for a classification of the former in terms of the symmetry properties of the latter. As usual, leaving aside a possible symplectic structure, according to the classification introduced in Section 5.1, the matrix construction places a general non-Hermitian Hamiltonian in the chiral symmetry class AIII, within which one can identify two subclasses of higher symmetry: CI, BDI. Yet, as we will shortly see, such a classification is not academic. Rather, it heavily constrains the form of the low-energy effective action of the model allowing for a simple phenomenology.

Applied to the problem of a diffusing particle subject to a random imaginary vector potential, the matrix Hamiltonian takes the form

$$\widehat{\mathcal{H}}_{\mathrm{IV}} = \left(\frac{\hat{p}^2}{2m} + \widetilde{V}(\mathbf{r}) - x\right)\sigma_1 + \left(y - \frac{e}{mc}\mathbf{A}\cdot\mathbf{p}\right)\sigma_2, \tag{5.28}$$

where $\widetilde{V}$ includes the diamagnetic contribution to the effective scalar potential. Therefore, according to the classification scheme introduced in Section 5.1, strictly at $y = 0$, $\widehat{\mathcal{H}}_{\mathrm{IV}}$ belongs to symmetry class BDI, while for $y \neq 0$ there is a crossover to the lower symmetry class AIII. From this result, we can deduce that the symmetry properties of a quantum particle subject to a random imaginary vector potential are isomorphic to the quantum properties of a sublattice Hamiltonian with real matrix elements [132].

Similarly, for the scalar potential, the corresponding matrix Hamiltonian

$$\widehat{\mathcal{H}}_{\rm IS} = \left(\frac{\hat{p}^2}{2m} - x\right)\sigma_1 + \big(y - \Gamma(\mathbf{r})\big)\sigma_2 \tag{5.29}$$

belongs to symmetry class CI independent of the magnitude of y. Put differently, the statistical properties of a quantum particle subject to a random scalar potential are isomorphic to the statistical properties of the zero energy quasi-particle states of a superconductor subject to a *real* random order parameter.

Field Theory of Non-Hermitian Operators

With this preparation, we are now in a position to construct an effective field theory. To do so, we might proceed from an *ab initio* approach implementing an ensemble averaging procedure on a generating functional. However, basing our considerations on the fundamental symmetries of the Hamiltonian, we can instead conjecture the form of the action describing the low-lying modes of the Hamiltonian. Therefore, according to the phenomenology of the statistical field theory, the low-energy effective action for the random imaginary vector potential are specified by the effective action

$$S_{\rm IV}[T] = -\int \left[\mathrm{str}\big(c_1 \partial T^{-1}\partial T + ic_2 y \sigma_3^{\rm PH}(T + T^{-1})\big) + c_3\big(\mathrm{str}(T^{-1}\partial T)\big)^2\right].$$

where according to the tabulated classification scheme of Zirnbauer [101], the supermatrices T belong to the manifold $\mathrm{GL}(2|2)/\,\mathrm{OSp}(2|2)$ corresponding symmetry class BDI [132]. For this model, the universal very low-energy properties of the effective action are described in detail in Ref. 147, while the higher energy diffusive properties of the action are described at the level of a perturbative one loop RG by Ref. 132.

Turning to the random imaginary scalar potential, a comparison with the dirty superconductor studied previously suggests that, at energies $y \ll 1/\tau$, the statistics are described by low-energy effective action

$$S_{\rm IS}[T] = -\int c_1\, \mathrm{str}(\partial T^{-1}\partial T),$$

where the supermatrices T belong to the group manifold $\mathrm{OSp}(2\,|\,2)$. Independent of y, this result suggests that the complex DoS is uniform at least over the interval $y \ll 1/\tau$ [152].

In applying the field theory above, it is necessary to exercise some caution. Our consideration of the effective action for the low-energy modes did not include a discussion of the source terms associated with the physical observables. In particular, we made no comment on whether the soft modes couple to the source terms. In fact, for non-Hermitian operators, the DoS is decoupled from the soft modes of the theory and is, instead, governed by fluctuations of the massive modes. Moreover, the localisation properties of "gauge invariant" combinations, $\langle L|\mathbf{r}|R\rangle$, also decouple. Again, a detailed discussion of these important issues lie beyond the scope of this review.

This completes our preliminary survey of the statistical properties of complex linear non-Hermitian operators. Our aim was to illustrate the manner in which coherence phenomena in random Schrödinger operators finds a counterpart in the behavior of random non-Hermitian operators. This last section also closes the chapter on novel symmetry classes and completes our survey of phase coherence phenomena in random media.

6 References

1. N.W. Ashcroft and N.D. Mermin, *Solid State Physics* (Saunders College, Philadelphia, 1976).
2. P.W. Anderson, *Basic Notions of Condensed Matter Physics* (Benjamin, Menlo Park, CA, 1984).
3. L.D. Landau and I.M. Lifshitz, *Physical Kinetics* (Pergamon Press, 1982).
4. For a review see L.P. Kouwenhoven et al., *Electron transport in quantum dots*, in "Mesoscopic Electron Transport," eds. L.L. Sohn, L.P. Kouwenhoven, and G. Schön (Kluwer, Dordrecht, 1997); C.M. Marcus et al., Chaos, Solitons & Fractals **8**, 1261 (1997), cond-mat/9703038.
5. M.J. Kelly, *Low-Dimensional Semiconductors* (Clarendon Press, Oxford, 1995).
6. C.W.J. Beenakker and H. van Houten, *Quantum transport in semiconducting nanostructures*, Solid State Physics, vol. 44 (Academic Press, New York, 1991), p. 1.
7. D.K. Ferry and S.M. Goodnick, *Transport in Nanostructures*, Cambridge Studies in Semi-Conductor Physics and Microelectronic Engineering, vol. 6 (Cambridge Univ. Press, 1997).
8. D. Mailly and M. Sanquer, *Sensitivity of quantum conductance fluctuations and of $1/f$ noise to time reversal symmetry*, J. Phys. France **2**, 357–364 (1992).
9. P.A. Lee and A.D. Stone, Phys. Rev. Lett. **55**, 1622 (1985).
10. B.L. Altshuler, JETP Lett. **41**, 649 (1985).
11. F.J. Dyson, J. Math. Phys. **3**, 140, 157, 166 (1962); F.J. Dyson and M.L. Mehta, J. Math. Phys. **4**, 701 (1963).

12. C.M. Marcus, private communication, unpublished.

13. S. Washburn, in "Mesoscopic Phenomena in Solids," eds. B.L. Altshuler, P.A. Lee and R.A. Webb (Elsevier, 1991); see also, S. Washburn and R.A. Webb, Adv. Phys. **35** (1986) 375; R.A. Webb, S. Washburn, C.P. Umbach, and R.B. Laibowitz, Phys. Rev. Lett **54**, 2696 (1985); R.A. Webb et al., in "Physics and Technology of Submicron Structures," eds. H. Heinrich, G. Bauer and F. Kuchar Springer Series in Solid State Science, vol. 83 (Springer, Berlin, 1988), p. 98.

14. D.E. Khmel′nitskii, Physica B+C **126**, 235 (1984).

15. L.P. Gor′kov, A.I. Larkin, and D.E. Khmel′nitskii, JETP Lett. **30**, 228 (1979).

16. B.L. Altshuler, A.G. Aronov and B.Z. Spivak, JETP Lett. **33**, 94 (1981).

17. D.Y. Sharvin and Y.V. Sharvin, JETP Lett. **34**, 272 (1981).

18. B.L. Altshuler and B.I. Shklovskii, JETP **64**, 127 (1986).

19. Y. Imry, Europhys. Lett. **1**, 249 (1986).

20. M.L. Mehta, *Random Matrices* (Academic Press, New York, 1991); T. Guhr, A. Mueller-Groeling and H.A. Weidenmüller, Phys. Rep. **299**, 189 (1998), cond-mat/9707301.

21. V.N. Prigodin, B.L. Altshuler, K.B. Efetov and S. Iida, Phys. Rev. Lett. **72**, 546 (1994).

22. P.W. Anderson, Phys. Rev. **109**, 1492 (1958).

23. J. Zittartz and J.S. Langer, Phys. Rev. **148**, 741 (1966).

24. B.I. Halperin and M. Lax, Phys. Rev. **148**, 722 (1966).

25. J.T. Edwards and D.J. Thouless, J. Phys. C **5**, 807 (1972).

26. D.J. Thouless, in "La matière mal condensée," eds. R. Balian, R. Maynard and G. Toulouse (North-Holland, Amsterdam, 1978), p. 1.

27. E. Abrahams, P.W. Anderson, D.C. Licciardello, and T.V. Ramakrishnan, Phys. Rev. Lett. **42**, 673 (1979).

28. C.M. Marcus, private communication.

29. L.P. Kouwenhoven, T.H. Oosterkamp, M.W.S. Danoesastro, M. Eto, D.G. Austing, T. Honda, and S. Tarucha, *Excitation spectra of circular few-electron quantum dots*, Science **278**, 1788 (1997), cond-mat/9708229.

30. L.P. Kouwenhoven and P.L. McEuen, *Single Electron Tunneling Through a Quantum Dot*, in "Nano-Science and Technology", ed. G. Timp (AIP Press, New York, 1996).

31. M.A. Kastner, Rev. Mod. Phys. **64**, 849 (1992).

32. L.S. Levitov and A.V. Shytov, cond-mat/9607136.

33. J. von Delft et al., Ann. Phys. **263**, 1 (1998).

34. B.L. Altshuler and A.G. Aronov, Sol. State. Commun. **39**, 115 (1979).

35. B.L. Altshuler and A.G. Aronov, in "Electron-Electron Interactions in Disordered Systems," eds. A.L. Efros and M. Pollak (North-Holland, Amsterdam, 1985), p. 1.

36. M.P. Sarachik and S.V. Kravchenko, *Novel phenomena in dilute electron systems in two dimensions*, Proc. Natl. Acad. Sci. USA **96**, 5900 (1999), cond-mat/9903292.

37. A.M. Finkelshtein, Sov. Sci. Rev. A Phys. **14**, 1 (1990).

38. G. Bergmann, Phys. Rep. **107**, 1 (1984).

39. P.A. Lee and T.V. Ramakrishnan, Rev. Mod. Phys. **57**, 287 (1985).

40. S. Chakravarty and A. Schmid, Phys. Rep. **140**, 193 (1986).

41. B.L. Altshuler and B.D. Simons, *Universalities: from Anderson Localization to Quantum Chaos*, a course of 10 lectures in the proceedings of the 1994 Summer Session LXI on "Mesoscopic Quantum Physics" at Les Houches, Nato ASI, eds. E. Akkermans, G. Montambaux, J.-L. Pichard and J. Zinn-Justin.

42. T.V. Ramakrishnan, in "Chance and Matter," eds. J. Souletie, J. Vannimenus and R. Stora (North-Holland, Amsterdam, 1987), p. 213.

43. B.L. Altshuler, A.G. Aronov, DE. Khmel′nitskii and A.I. Larkin, in "Quantum Theory of Solids," ed. I.M. Lifshits (Mir Publishers, Moscow, 1982), p. 130.

44. S.F. Edwards, Philos. Mag. **3**, 1020 (1958).

45. F.J. Wegner, Z. Phys. B **35**, 207 (1979).

46. K B. Efetov, Sov. Phys. JETP **82**, 872 (1982); *ibid.*, **83**, 833 (1982).

47. K.B. Efetov, Adv. Phys. **32**, 53 (1983); K. B. Efetov, *Supersymmetry in Disorder and Chaos* (Cambridge Univ. Press, New York, 1997).

48. J.J.M. Verbaarschot, H.A. Weidenmüller and M.R. Zirnbauer, Phys. Rep. **129**, 367 (1985).

49. S.F. Edwards and P.W. Anderson, J. Phys. F **5**, 965 (1975).

50. V.J. Emery, Phys. Rev. B **11**, 239 (1975).

51. J.J.M. Verbaarschot and M.R. Zirnbauer, J. Phys. A **17**, 1093 (1985).

52. M.R. Zirnbauer, cond-mat/9903338.

53. A. Kamenev and M. Mezard, cond-mat/9901110; cond-mat/9903001.

54. I.V. Yurkevitch and I.V. Lerner, cond-mat/9903025.

55. F.A. Berezin, *Introduction to Superanalysis* (Reidel, Dodrecht, 1987).

56. B.D. Simons, O. Agam and A.V. Andreev, J. Math. Phys. **38**, 1982 (1997).

57. B.D. Simons and B.L. Altshuler, Phys. Rev. Lett. **70**, 4063 (1993); *ibid.*, Phys. Rev. B **48**, 5422 (1993); B.D. Simons, P.A. Lee and B.L. Altshuler, Phys. Rev. Lett. **70**, 4122 (1993).

58. For a review, see M.C. Gutzwiller, *Chaos in Classical and Quantum Mechanics* (Springer, New York, 1990); F. Haake, *Signatures of Quantum Chaos* (Springer, Berlin, 1991).

59. D. Delande, in "Chaos and Quantum Physics," Les Houches Summer School Proceedings, vol. 52, eds. M.-J. Giannoni, A. Voros and J. Zinn-Justin (North-Holland, Amsterdam, 1991).

60. B.D. Simons, A. Hashimoto, M. Courtney, D. Kleppner and B.L. Altshuler, *New class of universal correlations in the spectra of hydrogen in a magnetic field*, Phys. Rev. Lett. **71**, 2899 (1993).

61. V.N. Prigodin, N. Taniguchi, A. Kudrolli, V. Kidambi and S. Shridar, *Spatial correlation in quantum chaotic systems with three-reversal symmetry: theory and experiment*, Phys. Rev. Lett. **75**, 2392 (1995).

62. E.P. Wigner, Ann. Math. **53**, 36 (1953).

63. C. E. Porter, ed. *Statistical Theories of Spectral Fluctuations* (Academic Press, New York, 1965).

64. L.P. Gor′kov and G.M. Eliashberg, Sov. Phys. JETP **21**, 940 (1965).

65. L. Schäfer and F.J. Wegner, Z. Phys. B **38**, 113 (1980).

66. K.B. Efetov, A.I. Larkin and D.E. Khmel′nitskii, Sov. Phys. JETP **52**, 568 (1980).

67. A. Houghton, A. Jevicky, R.D. Kenway and A. M. M. Pruisken, Phys. Rev. Lett. **45**, 394 (1980).

68. K. Junglich and R. Oppermann, Z. Phys. B **38**, 93 (1980).

69. S. Hikami, Phys. Rev. B **24**, 2671 (1981).

70. A.J. MaKane and M. Stone, Ann. Phys. (NY) **131**, 36 (1981).

71. O. Bohigas, M.J. Giannoni and C. Schmidt, Phys. Rev. Lett. **52**, 1 (1984); J. Physique Lett. **45**, L1615 (1984).

72. M.C. Gutzwiller, J. Math. Phys. **8**, 1979 (1967); **10**, 1004 (1969); **11**, 1791 (1970); **12**, 343 (1971).

73. M.V. Berry, Proc. Roy. Soc. London A **400**, 229 (1985).

74. M.V. Berry, in "Chaos and Quantum Physics", Les Houches Summer School Proceedings, vol. 52, eds. M.-J. Gianonni, A. Voros and J. Zinn-Justin (North-Holland, Amsterdam, 1991), p. 251.

75. P. Cvitanović and B. Eckhardt, J. Phys. A **24**, L237 (1991).

76. O. Agam, B.L. Altshuler and A.V. Andreev, Phys. Rev. Lett. **75**, 4389 (1995).

77. E. Bogomolny and J.P. Keating, Phys. Rev. Lett. **77**, 1472 (1996).

78. B.A. Muzykantskii and D.E. Khmel′nitskii, JETP Lett. **62**, 76 (1995)

79. A.V. Andreev, O. Agam, B.D. Simons and B.L. Altshuler, Phys. Rev. Lett. **76** 3947 (1996).

80. A. Altland, C.R. Offer and B.D. Simons, in "Supersymmetry and Trace Formulae: Chaos and Disorder," eds. I.V. Lerner, J.P. Keating and D.E. Khmel′nitskii (Plenum Press, 1999), p. 17.

81. M. R. Zirnbauer, in "Supersymmetry and Trace Formulae: Chaos and Disorder," eds. I.V. Lerner, J.P. Keating and D. E. Khmel′nitskii (Plenum Press, 1999), p. 153, chao-dyn/9812023.

82. P. Gaspard, G. Nicolis, A. Provata and S. Tasaki, Phys. Rev. E. **51**, 74 (1995).

83. D. Ruelle, Phys. Rev. Lett **56**, 405 (1986).

84. M. Pollicot, Ann. Math. **131**, 331 (1990).

85. G. Nicolis and C. Nicolis, Phys. Rev. A. **38**, 427 (1988).

86. H.H. Hasegawa and D.J. Driebe, Phys. Rev. E. **50**, 1781 (1994).

87. I.L. Aleiner and A.I. Larkin, Chaos, Solitons and Fractals **8**, 1179 (1997).

88. K. von Klitzing, G. Dorda and M. Pepper, Phys. Rev. Lett. **45**, 494 (1980); K. von Klitzing, Rev. Mod. Phys. **50**, 655 (1987).

89. A.M.M. Pruisken, Nuclear Phys. B **235**, 277 (1984); H. Levine, S. Libby and A.A.M. Pruisken, Nuclear Phys. B **240**, 30, 49, 71 (1985).

90. D.E. Khmel′nitskii, JETP Lett. **38**, 552 (1983).

91. H.P. Wei, D.C. Tsui, M.A. Paalanen and A.M.M. Pruisken, Phys. Rev. Lett. **61**, 1294 (1988).

92. M.R. Zirnbauer, hep-th/9905054.

93. M. Janssen, O. Viehwger, U. Fastenrath and J. Hajdu, *Introduction to the Theory of the Integer Quantum Hall Effect* (VCH Verlagsgesellschaft mbH, D-69451 Weinheim, Germany, 1994).

94. R.E. Prange and S.M. Girvin, eds., *The Quantum Hall Effect* (Springer-Verlag, New York, 1987).

95. L.V. Keldysh, Zh. Eksp. Teor. Fiz. **47**, 1515 (1964) [Sov. Phys. JETP **20**, 1018 (1965)].

96. M.L. Horbach and G. Schön, Ann. Phys. **2**, 51 (1993).

97. A. Kamenev and A.V. Andreev, cond-mat/9810191.

98. M. Gell-Mann and K.A. Brückner, Phys. Rev. **106**, 364 (1957); M. Gell-Mann, Phys. Rev. **106**, 369 (1957).

99. A.G. Aronov and Y.V. Sharvin, Rev. Mod. Phys. **59**, 755 (1987).

100. B.L. Altshuler, A.G. Aronov, M.E. Gershenson and Y.V. Sharvin, in "Physics Reviews", ed. I.M. Khalatnikov (Harwood Academic Publishers, Switzerland, 1987), p. 225.

101. M.R. Zirnbauer, J. Math. Phys. **37**, 4986 (1996).

102. S. Helgason, *Differential Geometry, Lie Groups and Symmetric Spaces*, (Academic Press, New York, 1978).

103. T. Nagao and S. Slevin, J. Math. Phys. **34**, 2075 (1993); *ibid.*, **34**, 2317 (1993).

104. J.J.M. Verbaaschot and I. Zahed, Phys. Rev. Lett. **70**, 3852 (1993); J.J.M. Verbaaschot, Phys. Rev. Lett. **72**, 2531 (1994).

105. A. Altland and M.R. Zirnbauer, Phys. Rev. Lett. **76**, 3420 (1996).

106. R. Oppermann, Nuclear Phys. B **280**, 753 (1987).

107. V.E. Kravtsov and R. Oppermann, Phys. Rev. B **43**, 10865 (1991).

108. A. Altland, J.P.D. Taras-Semchuk and B.D. Simons,*Field theory of mesoscopic f;uctuations in superconductor/normal-metal systems*, Adv. Phys. **49**, 321 (2000); cf. `http://www.tandf.co.uk`.

109. R. Bundschuh, C. Cassanello, D. Serban and M.R. Zirnbauer, Nucl. Phys. B **532**, 689 (1998), cond-mat/9806172.

110. P.W. Anderson, J. Phys. Chem. Solids **11**, 26 (1959).

111. G. Eilenberger, Z. Phys. **182**, 427 (1965); *ibid.*, **214**, 195 (1968).

112. K.D. Usadel, Phys. Rev. Lett. **25**, 507 (1970).

113. For a review, see *Nonequilibrium Superconductivity*, ed. V.L. Ginzburg (Nova Science Publications, 1988).

114. M.Yu. Kupriyanov and V.F. Lukichev, Zh. Eksp. Teor. Fiz. **94**, 139 (1988) [Sov. Phys. JETP **67**, 1163 (1988)].

115. L.P. Gor′kov and P. A. Kalugin, Pis′ma Zh. Eksp. Teor. Fiz. **41**, 208 (1985) [JETP Lett. **41**, 253 (1985)].

116. P.A. Lee, Phys. Rev. Lett. **71**, 1887 (1993).

117. A.V. Balatsky, A. Rosengren, and B.L. Altshuler, Phys. Rev. Lett. **73**, 720 (1994).

118. A.A. Nersesyan, A.M. Tsvelik, and F. Wenger, Phys. Rev. Lett. **72**, 2628 (1994); Nuclear Phys. B **438**, 561 (1994).

119. K. Ziegler, M.H. Hettler, and P.J. Hirschfeld, Phys. Rev. Lett. **77**, 3013 (1996).

120. T. Senthil et al., Phys. Rev. Lett. **81**, 4704 (1998); T. Senthil and M.P.A. Fisher, preprint cond-mat/9810238.

121. A. Altland, B. D. Simons and M. R. Zirnbauer, Phys. Rev. **B60**, 6893 (1999).

122. P.A. Lee and D.S. Fisher, Phys. Rev. Lett. **47**, 882 (1981); T. Sugiyama and N. Nagaosa, Phys. Rev. Lett. **70**, 1980 (1993); Y. Avishai et al., Phys. Rev. B **47**, 9561 (1993); D.Z. Liu et al., Phys. Rev. B **52**, 5858 (1995); D.N. Sheng and Z.Y. Weng, Phys. Rev. Lett. **75**, 2388 (1995); K. Yakubo and Y. Goto, Phys. Rev. B **54**, 13432 (1996); K. Yang and R.N. Bhatt, Phys. Rev. B **55**, R1922 (1997); M. Batsch et al., Physica B **249-251**, 792 (1998); X.C. Xie et al., Phys. Rev. Lett. **80**, 3563 (1998); A. Furusaki, Phys. Rev. Lett. 82, **604** (1999), cond-mat/9808059.

123. G. Gavazzi et al., Phys. Rev. B **47**, 15170 (1993); V. Kalmeyer et al., Phys. Rev. B **48**, 11095 (1993); S.-C. Zhang and D.P. Arovas, Phys. Rev. Lett. **72**, 1886 (1994); Y.B. Kim et al., Phys. Rev. B **52**, 16646 (1995).

124. D.K.K. Lee and J.T. Chalker, Phys. Rev. Lett. **72**, 1510 (1994).

125. A.G. Aronov et al., Phys. Rev. B **49**, 16609 (1994).

126. J. Miller and J. Wang, Phys. Rev. Lett. **76**, 1461 (1996).

127. B.I. Halperin, P.A. Lee and N. Read, Phys. Rev. B **47**, 7912 (1993).

128. S. Hikami et al., Nuclear Phys. B **408**, 2454 (1993); D.K.K. Lee, Phys. Rev. B **50**, 7743 (1994); K. Minakuchi and S. Hikami, Phys. Rev. B **53**, 10898 (1996); S. Hikami and K. Minakuchi, Phys. Rev. B **55**, 7155 (1997).

129. N. Nagaosa and P.A. Lee, Phys. Rev. Lett. **64**, 2450 (1990); X.-G. Wen and P.A. Lee, Phys. Rev. Lett. **76** 503 (1996).

130. E. Müller-Hartmann and E. Dagotto, Phys. Rev. B **54**, R6819 (1996).

131. R. Gade, Nuclear Phys. B **398**, 499 (1993); R. Gade and F. Wegner, Nuclear Phys. B**398**, 499 (1991).

132. T. Fukui, Nucl. Phys. B **562**, 477 (1999) , cond-mat/9902352.

133. M.R. Zirnbauer, J. Phys. A **29**, 7113 (1996).

134. This is an example of what in QCD is known as the phenomenon of "chiral symmetry breaking": Neglecting quark masses, the microscopic Yang–Mills action (corresponing to our ψ-action in the small ϵ-limit), is invariant under $G \times G$ where G is a symmetry group whose detailed structure depends on the flavour content of the theory, etc. (In our formalism, its rôle is played by $\mathrm{GL}(1 \mid 1)$.) Now, in the *effective* low energy QCD Lagrangian the microscopic chiral symmetry is spontaneously broken down to $G \times G \mapsto G$ in very much the same way as the symmetry group of our low energy Z-functional is broken down to $\mathrm{GL}(1 \mid 1)$ from $\mathrm{GL}(1 \mid 1) \times \mathrm{GL}(1 \mid 1)$.

135. A. Altland and B.D. Simons, J. Phys. A **32**, L353 (1999), cond-mat/9811134.

136. S. Hikami and A. Zee, Nuclear Phys. B **408**, 415 (1993); E. Brezin, S. Hikami and A. Zee, Nuclear Phys. B **464**, 411 (1996).

137. A.V. Andreev, B.D. Simons and N. Taniguchi, Nuclear Phys. B **432** [FS], 487 (1994).

138. For an odd number of sites, the block matrix elements of the Hamiltonian assume a *rectangular* form, the determinant vanishes and states at zero energy appear.

139. P.W. Brouwer et al., Phys. Rev. Lett. **81**, 862 (1998).

140. C.M. Mudry, P.W. Brouwer, and A. Furusaki, Phys. Rev. B **59**, 13221 (1999), cond-mat/9903026; P.W. Brouwer, C.M. Mudry and A. Furusaki, Phys. Rev. Lett **84**, 2913 (2000), cond-mat/9904200; Nucl. Phys. B **565**, 653 (2000), cond-mat/9904201.

141. C. Bernard and M. Golterman, Phys. Rev. D **49**, 486 (1994).

142. T. Guhr et al., Phys. Rep. **299**, 190 (1998).

143. M.B. Isichenko, Rev. Mod. Phys. **64**, 961 (1992).

144. V.V. Sokolov and V.G. Zelevinsky, Phys. Lett. B **202**, 10 (1988); F. Haake et al., Zeit. Phys. B **88**, 359 (1992); N. Lehmann et al., Nucl. Phys. A **582**, 223 (1995); Y.V. Fyodorov and H.-J. Sommers, J. Math. Phys. **38**, 1918 (1997).

145. H. Sompolinsky, A. Crisanti, and H.-J. Sommers, Phys. Rev. Lett. **61**, 259 (1988).

146. N. Hatano and D.R. Nelson, Phys. Rev. Lett. **77**, 570 (1996); Phys. Rev. B **56**, 8651 (1997).

147. K.B. Efetov, Phys. Rev. Lett. **79**, 491 (1997).

148. S.F. Edwards, Proc. Roy. Soc. **85**, 613 (1965).

149. H. Kleinert, *Path Integrals in Quantum Mechanics, Statistics and Polymer Physics* (World Scientific, Singapore, 1995).

150. H.-J. Sommers et al., Phys. Rev. Lett. **60**, 1895 (1988).

151. J.T. Chalker and J. Wang, Phys. Rev. Lett. **79**, 1797 (1997).

152. A.V. Izyumov and B.D. Simons, Europhys. Lett. **45**, 290 (1999).

7

QCD In Extreme Conditions

Frank Wilczek

ABSTRACT Recently we have made considerable progress in our understanding of the behavior of QCD in extreme conditions of high temperature or large baryon number density. Among the highlights are the prediction of a well-characterized true critical point, and the discovery that the ground state of three-flavor QCD at asymptotically high densities exhibits color-flavor locking. The critical point occurs at the unique temperature and density where a sharp distinction between an ionized plasma of quarks and gluons and the hadronic phase first appears. It appears to be accessible both to numerical and to laboratory experiments. Color-flavor locking provides a calculable, weak-coupling realization of confinement and chiral symmetry breaking. It also provides a microscopic realization of Han–Nambu charge assignments for quark quasiparticles, and of Yang–Mills theory for the physical vector mesons. Here I provide a self-contained introduction to these developments.

Introduction

In some ways, QCD is a mature subject. Its principles are precisely defined, and they have been extensively confirmed by experiment. QCD specifies unambiguous algorithms, capable of transmission to a Turing machine, that supply the answer to any physically meaningful question within its domain—any question, that is, about the strong interaction. I believe there is very little chance that the foundational equations of QCD will require significant revision in the foreseeable future. Indeed, as we shall soon review, these equations are deeply rooted in profound concepts of symmetry and local quantum field theory, which lead to them uniquely. So one cannot revise the equations without undermining these concepts.

Granting that the foundations are secure, we have the task—which is actually a wonderful opportunity—of building upon them. Due to the peculiarities of QCD, this is a particularly interesting and important challenge.

Interesting, because while the foundational equations are conceptually simple and mathematically beautiful, they seem at first sight to have nothing to do with reality. Notoriously, they refer exclusively to particles (quarks and gluons) that are not directly observed. Less spectacular, but more profound, is the fact that the equations exhibit a host of exact or approximate symmetries that are not apparent in the world. One finds that these symmetries are variously hidden: confined in the case of color, anomalous in the cases of scale invariance and axial baryon number, spontaneously broken in the case of chirality. It is fascinating to understand how a theory that superficially appears to be "too good for this world" actually manages to describe it accurately. Conversely it is pleasing to realize how the world is, in this profound and very specific way, simpler and more beautiful than it at first appears.

Important, because there are potential applications to which the microscopic theory has not yet rendered justice. An outstanding, historic, challenge is to derive the principles of nuclear physics. At present meaningful connections between the microscopic theory of QCD and the successful practical theory of atomic nuclei are few and tenuous, though in principle the former should comprehend the latter. This seems to be an intrinsically difficult problem, probably at least as difficult as computing the structure of complex molecules directly from QED. In both cases, the questions of interest revolve around small energy differences induced among valence structures, after saturation of much larger core forces. If one starts calculating from the basic equations, unfocussed, then small inaccuracies in the calculation of core parameters will blur the distinctions of interest.

A related but simpler class of problems is to calculate the spectrum of hadrons, their static properties, and a variety of operator matrix elements that are vital to the planning and interpretation of experiments. This is the QCD analogue of atomic physics. Steady progress has been made through numerical work, exploiting the full power of modern computing machines.

Other significant applications appear more accessible to analytic work, or to a combination of analytics and numerics.

The behavior of QCD at high temperature and low baryon number density is central to cosmology. Indeed, during the first few seconds of the Big Bang the matter content of the Universe was almost surely dominated by quark-gluon plasma. There are also ambitious, extensive programs planned to probe this regime experimentally.

The behavior of QCD at high baryon number density and (relatively) low temperature is central to extreme astrophysics—the description of neutron star interiors, neutron star collisions, and conditions near the core of collapsing stars (supernovae, hypernovae). Also, we might hope to find—and will find—insight into nuclear physics, coming down from the high-density side.

The special peculiarity of QCD, that its fundamental entities and abundant symmetries are well hidden in ordinary matter, lends elegance and focus to the discussion of its behavior in extreme conditions. Quarks, gluons, and the various symmetries will, in the right circumstances, come into their own. By tracing symmetries lost and found we will be able to distinguish sharply among different phases of hadronic matter, and to make some remarkably precise predictions about the transitions between them.

In these 5 lectures I hope to provide a reasonably self-contained introduction to the study of QCD in extreme conditions. The first lecture is a rapid tour of QCD itself. I've organized it as a cluster of related stories narrating how apparent symmetries of the fundamental equations are hidden by characteristic dynamical mechanisms. Lecture 2 and Lecture 3 are mainly devoted to QCD at high temperature, and Lecture 4 and Lecture 5 to QCD at high baryon number density. I have structured the lectures so that they head toward two climaxes: the prediction of a true critical point in real QCD, that ought to be accessible to numerical and laboratory experiments, in Lecture 3, and the prediction that at asymptotic densities QCD goes over into a color-flavor locked phase with remarkable properties including fully calculable realizations of confinement and chiral symmetry breaking, in Lecture 5. These are, I believe, remarkable results, and they bring us to frontiers of current research.

A wide variety of tools will be brought to bear, including three different renormalization groups (the usual "asymptotic freedom" version toward bare quarks and gluons at high virtuality, the usual "Kadanoff–Wilson–Fischer" version toward critical modes at a second-order phase transition, and the "Landa–Anderson" version toward quasiparticles near the Fermi surface), perturbative quantum field theory, effective field theory, instantons, lattice gauge theory, and BCS pairing theory. Of course I won't be able to give full-blown introductions to all these topics here; but I'll try to give coherent accounts of the concepts and results I actually need, and to supply appropriate standard references. In a few months, I hope, a comprehensive reference will be appearing.

Lecture 1 Symmetry and the Phenomena of QCD

1.1 Apparent and Actual Symmetries

Let me start with a slightly generalized and slightly idealized Lagrangian for QCD:

$$\mathcal{L} = -\frac{1}{4g^2}\,\mathrm{tr}\,G^{\mu\nu}G_{\mu\nu} + \sum_{j=1}^{f} \bar{\psi}_j(i\not{D})\psi_j \tag{1.1}$$

where

$$G_{\mu\nu} = \partial_\mu A_\nu - \partial_\nu A_\mu + i[A_\mu, A_\nu] \tag{1.2}$$

and

$$D = \partial_\mu + iA_\mu. \tag{1.3}$$

The A_μ are 3×3 traceless Hermitean matrices. Each spin-$\frac{1}{2}$ fermion quark field ψ_j carries a corresponding 3-component color index.

Equation (1.1) is both slightly generalized from real-world QCD, in that I've allowed for a variable number f of quarks; and slightly idealized, in that I've set all their masses to zero. Also, I've set the θ parameter to zero, once and for all. We will have much occasion to focus on particular numbers and masses of quarks in our considerations below, but (1.1) is a good simple starting point.

Our definition of the gauge potential differs from the more traditional one by

$$A_\mu^{\text{here}} = g A_\mu^{\text{usual}} \tag{1.4}$$

which is why the coupling constant does not appear in the covariant derivative. By writing the Lagrangian in this form we make it clear that $1/g$ is neither more nor less than a stiffness parameter. It informs us how big is the energetic cost to produce curvature in the gauge field.

The form of (1.1) is uniquely fixed by a few abstract postulates of a very general character. These are SU(3) gauge symmetry, together with the general principles of quantum field theory—special relativity, quantum mechanics, locality—and the criterion of renormalizability. It is renormalizability that forbids more complicated terms, such as an anomalous gluomagnetic moment term $\propto \bar{q}\sigma^{\mu\nu}G_{\mu\nu}q$.

Later I shall argue that it is very difficult to rigorously insure the existence of a quantum field theory that is not asymptotically free. Asymptotic freedom is a stronger requirement than renormalizability, and can only be achieved in theories containing non-Abelian gauge fields. Thus one can say, without absurdity, that even our postulates of gauge symmetry and renormalizability are gratuitous: both are required for the *existence* of a relativistic local quantum field theory.

The apparent symmetry of (1.1) is:

$$\mathcal{G}_{\text{apparent}} = \mathrm{SU}(3)^c \times \mathrm{SU}(f)_L \times \mathrm{SU}(f)_R \times \mathrm{U}(1)_B \times \mathrm{U}(1)_A \times \mathcal{R}^+_{\text{scale}}, \tag{1.5}$$

together of course with Poincaré invariance, P, C, and T. The factors, in turn, are local color symmetry, the freedom to freely rotate left-handed quarks among one another, the freedom to freely rotate right-handed quarks among one another, baryon number (= a common phase for all quark fields), axial baryon number (= equal and opposite phases for all left-handed and right-handed quark fields), and scale invariance.

The chiral $\mathrm{SU}(f)_L \times \mathrm{SU}(f)_R$ symmetries arise because the only interactions of the quarks, their minimal gauge couplings to gluons, are universal and helicity conserving. These chiral symmetries will be spoiled by nonzero quark mass terms, since such terms connect the two helicities. With quarks of nonzero but equal mass one would have only the diagonal (vector) $\mathrm{SU}(f)_{L+R}$ symmetry, while if the quarks have unequal nonzero masses this breaks up into a product of U(1)s. Thus the choice $m = 0$ can be stated as a postulate of enhanced symmetry. If the quarks masses are all nonzero, then P and T would be violated by a nonzero θ term, unless

$\theta = \pi$. For massless quarks, all values of θ are physically equivalent, so we lose nothing by fixing $\theta = 0$.

One could of course have written these chiral symmetries together with the two U(1) factors as $\mathrm{U}(f)_L \times \mathrm{U}(f)_R$, but the unwisdom of so doing will become apparent momentarily.

Finally the $\mathcal{R}^+_{\text{scale}}$ factor reflects that the only parameter in the theory, g, is dimensionless (in units with $\hbar = c = 1$, as usual). Thus the classical theory is invariant under a change in the unit used to measure length, or equivalently (inverse) mass. Indeed, the action $\int d^4x\mathcal{L}$ is invariant under the rescaling

$$x^\mu \to \lambda x^\mu; \quad A \to \lambda^{-1} A; \quad \psi \to \lambda^{-1}\psi. \tag{1.6}$$

The actual symmetry of QCD, and of the real world, is quite different from the apparent one. It is

$$\begin{aligned}\mathcal{G}_{\text{actual}} &= \mathrm{SU}(3)^c \times \mathrm{SU}(f)_L \times \mathrm{SU}(f)_R \times Z^f_A \times \mathrm{U}(1)_B \\ &\qquad\qquad\qquad\qquad \text{(asymptotic freedom, chiral anomaly)} \\ &\to \mathrm{SU}(3)^c \times \mathrm{SU}(f)_{L+R} \times \mathrm{U}(1)_B \\ &\qquad\qquad\qquad\qquad \text{(chiral condensation)} \\ &= \mathrm{SU}(f)_{L+R} \times \mathrm{U}(1)_B \text{ (confinement)}.\end{aligned} \tag{1.7}$$

Let me explain this cascade of symmetry reductions.

In the first line of (1.7), I've specified the subgroup of $\mathcal{G}_{\text{apparent}}$ which survives quantization. The $\mathcal{R}^+_{\text{scale}}$ of classical scale invariance is entirely lost, and the $\mathrm{U}(1)_A$ of axial baryon number is reduced to its discrete subgroup Z^f_A. Both fall victims to the need to regulate quantum fluctuations of highly virtual degrees of freedom, as I shall elaborate below. The breaking of scale invariance is associated with the running of the effective coupling, asymptotic freedom, and dimensional transmutation. The breaking of axial baryon number is associated with the triangle anomaly and instantons. Thus these symmetry removals arise from dynamical features of QCD that reflect its deep structure as a quantum field theory.

In the second line of (1.7), I've specified the subgroup of the symmetries of the quantized theory which are also symmetries of the ground state. This group is properly smaller, due to spontaneous symmetry breaking. That is, the stable solutions of the equations exhibit less symmetry than the equations themselves. Specifically, the ground state contains a condensate of quark-antiquark pairs of opposite handedness, which fills space-time uniformly. One cannot rotate the different helicities components independently while leaving the condensate invariant. The lightness of π mesons, and much of the detailed phenomenology of their interactions at low energies, can be understood as direct consequences of this spontaneous symmetry breaking.

In the third line of (1.7), I've acknowledged that local color gauge symmetry, which is so vital in formulating the theory, is actually not directly a property of any physical observable. Indeed, in constructing the Hilbert space of QCD, one must restrict oneself to gauge-invariant states. The auxiliary, extended Hilbert space that we use in perturbation theory does not have a positive-definite inner product. It's haunted by ghosts. Moreover, unlike the situation for QED, one discerns in the low-energy physics of QCD no obvious traces of gauge symmetry. Specifically, there are no long-range forces, nor do particles come in color multiplets. This is the essence of confinement, a tremendously important but amazingly elusive concept, as we shall discover repeatedly in these lectures. (If you look only at the world, not at a postulated micro-theory, what exactly does confinement mean? Don't fall into the trap of saying confinement means the unobservability of quarks—the quarks are a theoretical construct, not something you can observe (that's what you said!).)

Clearly, a major part of understanding QCD must involve understanding how its many apparent symmetries are lost, or realized in peculiar ways. The study of QCD in extreme conditions gives us fruitful new perspectives on these matters, for we can ask new, very sharp questions: Are symmetries restored, or lost, in phase transitions? Are they restored asymptotically?

Now I shall discuss each of the key dynamical phenomena: asymptotic freedom, confinement, chiral condensation, and chiral anomalies, in more detail.

1.2 Asymptotic Freedom

1.2.1 Running Coupling

Running of couplings is a general phenomenon in quantum field theory. Nominally empty space is full of virtual particle-antiparticle pairs of all types, and these have dynamical effects. Put another way, nominally empty space is a dynamical medium, and we can expect it to exhibit medium effects including dielectric and paramagnetic behavior, which amount (in a relativistic theory) to charge screening. In other words, the strength of the fields produced by a test charge will be modified by vacuum polarization, so that the *effective* value of its charge depends on the distance at which it is measured.

Asymptotic freedom is the special case of running couplings, in which the effective value of a charge measured to be finite at a given finite distance, decreases to zero when measured at very short distances. Thus asymptotic freedom involves antiscreening. Antiscreening is somewhat anti-intuitive, since it is the opposite of what we are accustomed to in elementary electrodynamics. Yet there is a fairly simple way to understand how it might be possible in a *relativistic, non-Abelian, gauge theory*:

- Because the theory is relativistic, magnetic forces are just as important as electric ones.
- Because it is a gauge theory, it contains vector mesons, with spin.

- Because it is non-Abelian these vector mesons carry charge, and their spins carry magnetic moments.

Now in electrodynamics we learn that spin response is paramagnetic—a spin (elementary magnetic dipole) tends to align with an imposed magnetic field, in such a way as to enhance the field. This, you'll realize if you think about it a moment, is antiscreening behavior. Thus there is a competition between normal electric screening (together with orbital diamagnetism) and antiscreening through spin paramagnetism. For virtual gluons, it turns out that spin paramagnetism is the dominant effect numerically.

When the effective coupling becomes weak, one can calculate the screening (or antiscreening) behavior in perturbation theory. In QCD one finds for the change in effective coupling

$$\frac{dg(\epsilon)}{d\ln\epsilon} = \beta_0 g^3 + \beta_1 g^5 + \cdots \tag{1.8}$$

with

$$\begin{aligned} \beta_0 &= \frac{-1}{16\pi^2}\left(11 - \frac{2}{3}f\right) \\ \beta_1 &= \left(\frac{1}{16\pi^2}\right)^2\left(102 - \frac{38}{3}f\right), \end{aligned} \tag{1.9}$$

where ϵ is the energy, or equivalently inverse length, scale at which the effective charge is defined. Thus if the $f \leq 16$ the effective coupling decreases to zero as the energy at which it is measured increases to infinity, which is asymptotic freedom. Taking the first term only, we see that the asymptotic behavior is approximately

$$\frac{1}{g(\epsilon)^2} = \frac{1}{g(1)^2} - \beta_0 \ln\epsilon. \tag{1.10}$$

1.2.2 Clouds, Jets, and Experiments

The asymptotic freedom of QCD can be exploited to simplify the calculation of many physical processes. To understand why that is so, consider the picture of quarks (or gluons) that it suggests. This is shown in Figure 7.1. The effective color charge of a quark, responsible for its strong interaction, accumulates over distance. We should think of it as a distributed cloud of induced color charge, all of the same sign, with no singular core. Now consider how this cloud is seen when probed at various wavelengths. If the wavelength is small, the effective color charge will be small, since only a small portion of the cloud is sampled. If the wavelength is large, the effective charge will be large.

One can also consider the complementary particle picture. If one suddenly imparts a large impulse to a quark (or gluon), liberating it from its cloud, the resulting object will have a small effective color charge, and will propagate almost freely, until it builds up a new cloud.

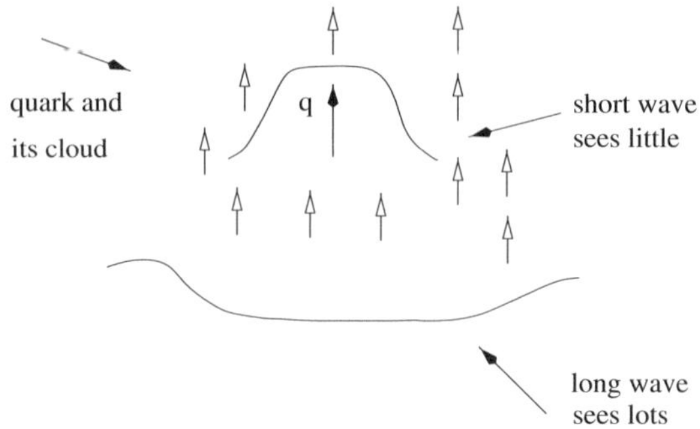

FIGURE 7.1. Distribution of the color charge in a cloud.

Of course the electric and weak charges of a quark are not shared by its color polarization cloud. These charges remain concentrated. Thus short-wavelength electromagnetic or weak probes can resolve pointlike quarks, whose 'strong' (color) interactions make only small corrections to free-particle propagation. More precisely, it is corrections that substantially change the energy-momentum of the color current which are guaranteed to be small. Big changes in the energy-momentum can only arise from radiation of gluons having short wavelength in the frame of the current, but such short-wavelength gluons see only a small effective color charge, as we've discussed.

So in physical process originating with the pointlike quarks directly produced by hard electroweak currents, as for instance in e^+e^- annihilation at high energy, one develops jets of rapidly-moving hadrons following the nominal paths of the original quarks. Hard gluon radiation processes do occasionally occur, of course, but at a small rate, calculable in perturbation theory. These hard gluons can (with small probability) induce jets of their own, which themselves occasionally radiate, and so forth. The "antenna pattern" of jets—including the relative probabilities for different topologies (numbers of jets), the energy and angular distributions for a given topology, and how all these quantities depend upon the total energy, can be calculated in exquisite detail. These predictions, which reflect in 1–1 fashion the structure of the fundamental interactions in the theory and the running of the coupling, have been extensively, and successfully, tested experimentally.

The science of exploiting asymptotic freedom to predict rates for a wide variety of hard processes is highly developed. Figure 7.2 displays results of comparisons between prediction and observation for a wide variety of experiments, in the form of determinations of the running coupling. Implicit, of course, is that within any given type of experiment, the theory gives a successful account of the functional form of dependence on event topology, energy distribution, angular distribution,. . . . The figure itself demonstrates

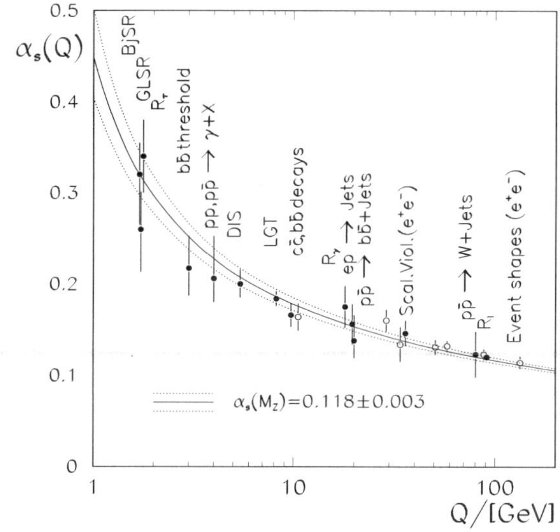

FIGURE 7.2. Running of the couplings from [1].

the success of (1.8), (1.9), suitably generalized to include the effect of quark masses, as a description of Nature.

1.2.3 Dimensional Transmutation and "Its From Bits"

Running of the coupling manifestly breaks the classical scale invariance $\mathcal{R}^+_{\text{scale}}$. It is an almost inevitable result of quantum field theory. Indeed, quantum field theory predicts as its first consequence the existence of a space-filling medium of virtual particles, whose polarization makes the effective coupling strength depend on the distance at which it is measured, or in other words causes the coupling to run. (There are very special classes of supersymmetric theories in which different types of virtual particles give cancelling effects, and the coupling does not run.)

Because its coupling runs, in QCD the analogue of Pauli's question:

> "Why is the value of the fine structure constant what it is?"—

receives a startling answer:

> "It's anything you like—at some scale or other."

We can simply declare it to be, say, 1/10, thereby defining the correct distance at which to measure it—i.e., the distance where it is 1/10! This is the phenomenon of dimensional transmutation. A dimensionless coupling constant has been transmuted into a length (or, equivalently, energy) scale.

In fact we can identify a scale explicitly, using (1.8):

$$M = \lim_{\epsilon \to \infty} \epsilon e^{u/\beta_0} u^{\beta_1/\beta_0^2} \tag{1.11}$$

where

$$u \equiv \frac{1}{g(\epsilon)^2}. \tag{1.12}$$

The form of (1.8) insures that the limit exists and is finite.

Due to its formal scale invariance, QCD with massless quarks appears naively—that is, classically—to be a one-parameter family of theories, distinguished from one another by the value of the coupling g. And none of these theories, it appears, defines a scale of distance. Through the magic of dimensional transmutation, QCD turns out instead to be a family of perfectly identical clones, each of which *does* define a distance scale. Indeed, the clones differ only in the units they employ to measure distance. This difference in units enters into comparisons of purely QCD quantities to quantities outside of QCD, such as the ratio of the proton mass to the electron mass. But it does not affect dimensionless quantities within QCD itself, such as ratios of hadronic masses or of isomer splittings.

Using $\hbar$ and c as units, and *no further inputs* the truncated version of QCD including just the u and d quarks, with their masses set to zero—what I call "QCD Lite"—accounts pretty accurately for the low-lying spectrum of nonstrange hadrons (demonstrably), nuclear physics (presumably), and much else besides. The only numerical inputs to QCD Lite are the number of colors—3 (binary 11)—and the number of quarks—2 (binary 10). Thus QCD Lite provides a remarkable realization of Wheeler's slogan "Getting Its From Bits"!

1.2.4 Limitations

The most straightforward applications of asymptotic freedom are to processes involving large energy scales only. This constraint pretty much limits one to inclusive processes induced by electroweak currents. Any external hadron introduces a small energy scale, namely its mass. By using clever tricks to isolate calculable subprocesses, one can calculate much more, as conveyed in Figure 7.2. However these tricks will take you only so far, and there is no easy way to exploit the small effective coupling at short distances to address truly low-energy phenomena, or to calculate the spectrum. If you try to calculate such things directly, you will encounter infrared divergences—so at least the theory is kind enough to warn you off. Similarly, large temperature or large chemical potential introduce large "typical" energies, but it is not trivial (and usually not even true) that this fact in itself will allow access, via asymptotic freedom, to interesting spectral or transport properties.

1.3 Confinement

1.3.1 Brute Facts and Crude Theory

An aura of mystery still seems to hover around the phenomenon of quark confinement. Historically, of course, it was a big surprise in world-modeling, and posed a major barrier both to the discovery and to the acceptance first of quarks, and then of modern QCD. And confinement is a genuinely profound and subtle dynamical phenomenon, as some of our later considerations will emphasize. Concerning the *fact* that QCD predicts confinement, however, there is no ambiguity. Our knowledge of the properties of the theory has moved far beyond abstract or hypothetical discussion of this point. Figure 7.3 exhibits the results of some direct calculations of the spectrum starting from the microscopic theory, with controlled errors, using the techniques of lattice gauge theory. There are neither massless flavor-singlet particles with long-range interactions, particles with quark or gluon quantum numbers, nor degenerate color multiplets. In fact the microscopic theory reproduces the observed spectrum extremely well, with no gratuitous additions.

Thus confinement is not a *practical* problem for modern QCD. Still, one would like to understand more precisely what it is, why it occurs and, particularly for these lectures, whether in can come undone in extreme conditions.

The simplest heuristic argument for confinement is due to Amati and Testa. It is based directly on (1.1). If the coefficient $1/g^2$ of the gauge curvature term is taken to zero, then upon varying the action with respect

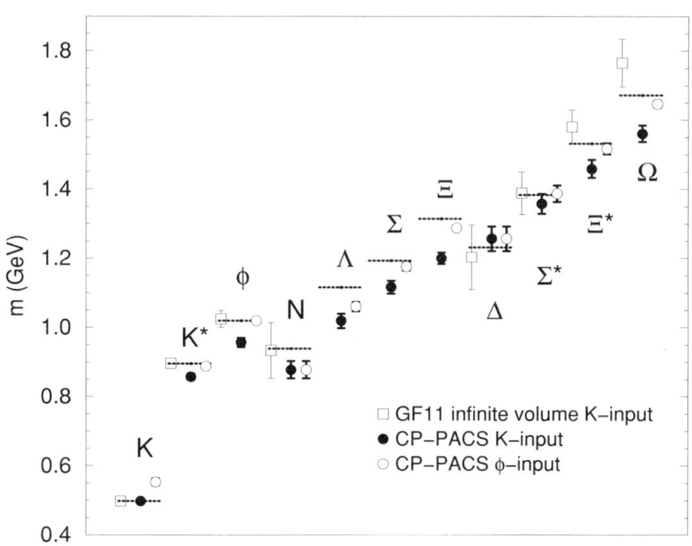

FIGURE 7.3. Spectrum from lattice gauge theory, from [2].

to the gauge potential A_μ we find that the color current vanishes, including its zero component, the color charge density—which is to say, color is confined. We can combine this idea with the running of the coupling, to argue (still more heuristically) that low-frequency modes, associated with large effective couplings, are confined, while high-frequency modes can be dynamically active. This is broadly consistent with the observed behavior in Nature, that quarks and gluons become visible in hard processes, but are not accessible to soft probes.

Unfortunately it is a very singular operation to throw away the highest-derivative terms in any differential equation, because they will always dominate for sufficiently abrupt variation. In high Reynolds number hydrodynamics, one addresses this difficulty with boundary layer theory. In QCD, the only method we know to make the simple Amati-Testa argument the starting point for a systematic approximation is by employing an artful discretization, lattice gauge theory, as I'll now discuss.

1.3.2 Lattice Gauge Theory Basics

The great virtue of the lattice version of QCD are that it provides an ultraviolet cutoff and that it allows a convenient strong-coupling expansion, while preserving a very large local gauge symmetry. Its drawbacks are that it destroys translation and rotation symmetry, that it has an awkward weak-coupling expansion, and that it mutilates the ultraviolet behavior of the continuum theory. But let's put these worries aside for the moment, and exploit the virtues. Also let's consider the pure glue theory. The extension to quarks will appear in Lecture 2.

The fundamental operation in gauge theory is parallel transport. The basic objects of the theory are 3×3 unitary matrices with unit determinant. They live on the oriented links of a cubic hyperlattice in four dimensions, and implement parallel transport. Thus the dynamical variable $U_{r,\hat\mu}$ is a matrix associated with the link starting from lattice point r and ending at $r+\hat\mu a$, where a is the lattice spacing. The matrix associated with the same link oriented in the opposite direction is the inverse, i.e. $U_{r+\hat\mu a,-\mu} = U^{-1}_{r,\hat\mu}$.

If there were an underlying continuum gauge potential A, the parallel transport would be

$$U_{r,\hat\mu} \text{ “=” } \overline{P} \exp i \int_r^{r+\hat\mu a} A_\mu \, dx^\mu , \tag{1.13}$$

where $\bar P$ denotes anti-path ordering. Indeed, this is the solution of the equation

$$\nabla_\mu U \equiv (\partial_\mu + i A_\mu) U = 0 \tag{1.14}$$

having the indicated endpoints. In line with this underlying structure, which we would like to recover in an appropriate limit, local gauge transformations $\Omega(r)$ should act on the lattice sites, and transform the U matrices according to

$$\widetilde{U}_{r,\mu} = \Omega(r) U_{r,\mu} \Omega(r + \hat{\mu}a)^{-1}. \tag{1.15}$$

The $\Omega(r)$ are, of course, 3×3 unitary matrices.

To form interaction terms invariant under (1.15) one must take the trace of a product of U matrices along a path of links forming a closed loop. The simplest possibility is simply the trace around a plaquette, e.g.,

$$\operatorname{tr} Pl_{r,\hat{x}\hat{y}} \equiv \operatorname{tr} U_{r,\hat{x}} U_{r+a\hat{x},\hat{y}} U_{r+a\hat{x}+a\hat{y},-\hat{x}} U_{r+a\hat{y},-\hat{y}}. \tag{1.16}$$

Putting (1.13) into (1.16) and expanding for small a, we find the first nontrivial term

$$\operatorname{tr} Pl_{r,\hat{x}\hat{y}} \to \operatorname{tr}\left(1 - \frac{a^4}{2} G^{xy} G_{xy}\right) + O(a^6). \tag{1.17}$$

The straightforward verification of (1.17) is quite arduous, but one can restrict its form *a priori* by exploiting gauge invariance, and then evaluate on a simple configuration such as $A_x = M(y - r_y)$. Note that a term $\propto a^2 G^{xy}$, which does appear in the plaquette product, vanishes when we take the trace.

Thus the simplest lattice gauge invariant action we can write, the Wilson action

$$S_W = \frac{1}{4g^2} \sum_{\text{plaquettes}} (3 - \operatorname{tr} Pl_{\square}) \tag{1.18}$$

reduces, formally, to the continuum action for very small lattice spacings. Hence we are invited to use it, and then attempt to justify the limit. (In the spirit of the Jesuit credo "It is more blessed to ask forgiveness than permission.")

To evaluate a correlation function of operators $\langle \mathcal{O}_1 \mathcal{O}_2 \cdots \rangle$, then, we must evaluate the integral

$$\langle \mathcal{O}_1 \mathcal{O}_2 \cdots \rangle = \frac{\int d\mathcal{M} \exp(-S_W) \mathcal{O}_1 \mathcal{O}_2 \cdots}{\int d\mathcal{M} \exp(-S_W)}, \tag{1.19}$$

where

$$d\mathcal{M} = \prod_{\text{links } r,\hat{\mu}} U_{r,\hat{\mu}}^{-1} dU_{r,\hat{\mu}} \tag{1.20}$$

is the product of invariant integrals over the gauge group.

1.3.3 Strong Coupling and Confinement

The lattice regularization permits one to formulate strong coupling perturbation theory in a simple, elegant way. When g gets large, we can simply expand e^{-S_W} in a power series in $1/g^2$. The result is to "bring down" activated plaquettes, one for each inverse power of $1/g^2$. When we integrate over a link incident on an activated plaquette, we encounter an extra power of U for that link.

To test for confinement, the traditional method is to study Wilson-Polyakov loops. These are simply the traces of products of U matrices around loops, similar to the terms that appear in the action. But now we want to consider large loops, instead of very small ones.

The motivation for this method is as follows. Suppose we put a very heavy quark into the system. This will stay at a fixed point in space, so its world-line will be simply a straight line in the $\hat{\tau}$ (Euclidian time) direction. The $\operatorname{tr} j \cdot A$ coupling of this color source will generate a product of U matrices along the links of its world-line. So to measure the potential between a heavy quark and a heavy antiquark at distance R we should measure the energy it takes to have a line like this and a similar line with U^{-1} matrices a distance R away. If we allow this configuration to persist for a long Euclidian time T, the cost should go as $e^{-V(R)T}$. Now to make the "measurement" clean we should imagine closing up the loop with short segments at the top and bottom. This corresponds to producing the quark-antiquark pair, letting them sit separated for a long time, and then annihilating them. With $R << T$, by taking the negative of the log, we will extract the potential. In a formula

$$V(R) = \lim_{T\to\infty} \frac{-1}{T} \ln \frac{\int d\mathcal{M} \exp(-S_W)\Pi}{\int d\mathcal{M} \exp(-S_W)}, \tag{1.21}$$

where Π is the trace of an ordered product of U matrices along the perimeter of a long rectangle with sides of length R and T, as shown in Figure 7.4(a).

With this background in hand, it becomes very easy to understand how confinement arises in strong coupling. The integral of a single U matrix over the group manifold, $\int U^{-1}\, dUU$, vanishes. (The change of variables $U \to gU$ leaves the measure invariant, so if $\int U^{-1}\, dUU = k$, then by changing variables we find $gk = k$ for any g, which of course means $k = 0$.) So does $\int U^{-1}\, dUU^{-1}$. Thus to find a nonzero contribution to (1.21) in strong coupling we must at least pull down plaquettes to share sides with each of the links in the Wilson loop.

But now you see there's a whole new set of interior links with single U matrices and vanishing group integrals. Clearly, to get a nonzero contribution we must tile the whole area spanned by the Wilson loop, as in Figure 7.4(b). This takes a number of plaquettes proportional to the area RT, at least. So the leading contribution to the Wilson loop in strong coupling goes as

$$V(R) \sim -\frac{1}{T} \ln \left(\frac{1}{g^2}\right)^{RT} \propto R. \tag{1.22}$$

The linear potential, of course, means that it is impossible to separate the color sources indefinitely, and so one has confinement.

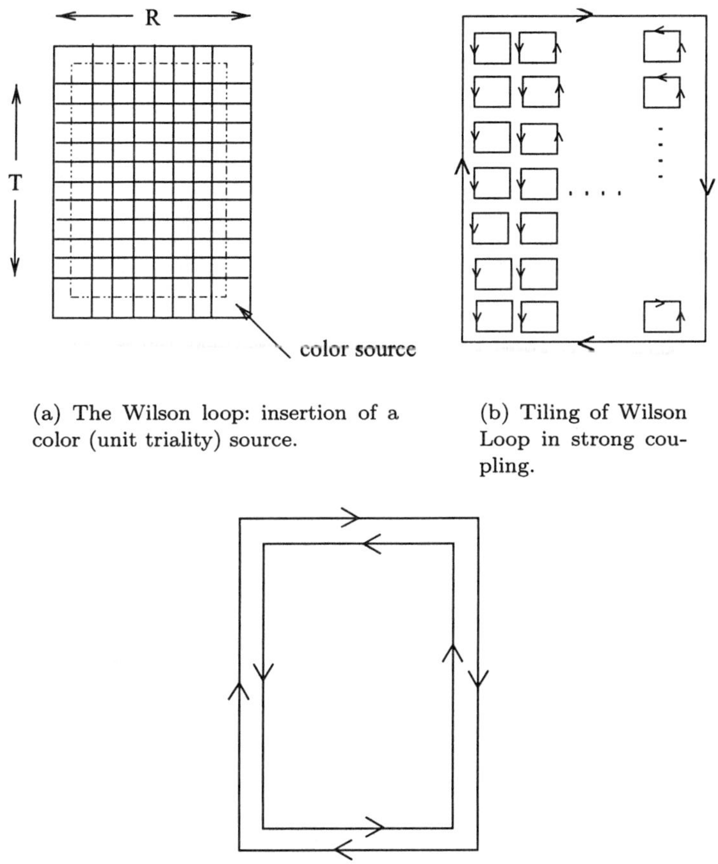

(a) The Wilson loop: insertion of a color (unit triality) source.

(b) Tiling of Wilson Loop in strong coupling.

(c) Perimeter law, in the theory with quarks.

FIGURE 7.4.

1.3.4 To the Continuum Limit

The strong coupling result forms the starting point for a convincing proof of confinement in (pure glue) QCD proper.

One first argues that the strong coupling perturbation theory has a finite radius of convergence. That can be done analytically. Then one investigates numerically whether there is a phase transition as a function of the coupling, as the coupling varies from strong to weak. It turns out there is a phase transition for U(1), but not for SU(3) (or SU(2)). When there is no phase transition, the theory remains in the same universality class, and its sharply defined qualitative properties cannot change.

Thus in the physically relevant SU(3) case:

- Since the strong coupling perturbation expansion converges, the lowest nontrivial order governs the asymptotic behavior of the Wilson loop, and exhibits confinement.
- Since the strong coupling theory is in the same universality class as the weak coupling theory, the weak coupling theory also exhibits confinement.
- Since asymptotic freedom implies that the weak coupling lattice theory reproduces the continuum theory, the continuum theory exhibits confinement.

Now we see that is fortunate, and reassuring for this circle of ideas, that there is a phase transition for U(1). Otherwise we'd have proved confinement in QED, which would be proving too much.

1.3.5 Foundational Remarks

To round out this discussion I would like to emphasize the deep connections among renormalizability, asymptotic freedom, and lattice gauge theory. To construct a relativistic quantum theory, one typically introduces at intermediate stages a cutoff, which spoils the locality or relativistic invariance of the theory. Then one attempts to remove the cutoff, while adjusting the defining parameters, in order to achieve a finite, cutoff-independent limiting theory. Renormalizable theories are those for which this can be done, order by order in a perturbation expansion around free field theory. That formulation of the problem of constructing a quantum field theory, while convenient for mathematical analysis, obviously begs the question whether this perturbation theory converges. For interesting quantum field theories, it rarely does.

A more straightforward procedure, conceptually, is to regulate the theory as a whole by discretizing it. This involves approximating space-time by a lattice, and spoils the continuous space-time symmetries of the theory. Then one attempts to remove dependence on the discretization, by refining it, while if necessary adjusting the defining parameters, to achieve a finite limiting theory that does not depend on the discretization, and therefore has a chance to respect the space-time symmetries. The redefinition of parameters is necessary, because in refining the discretization one is introducing new degrees of freedom. The earlier, coarser theory results from integrating out these degrees of freedom. If it is to represent the same physics it must incorporate their effects, for example in vacuum polarization. Operationally, one can demand that some observable(s) measured at scales well beyond the lattice spacing stays fixed as the discretization is refined. This fixes the free coupling(s). The question is then whether, having fixed the available parameters, the calculated values of *all* observables have finite limits.

This is very hard to prove, in general. The only case in which it is straightforward arises when the effects of integrating out the additional short-wavelength modes, that are introduced with each refinement of the lattice, can be captured accurately by a re-definition of the coupling parameter(s) already present in the theory. That, in turn, will occur in a straightforward way only if these modes are weakly coupled. For then perturbation theory will show us how to take the limit for the renormalizable couplings, while assuring us that naive power counting can be applied to argue away all nonrenormalizable ones. But of course the ultraviolet modes will be weakly coupled, if and only if the theory is asymptotically free.

Summarizing the argument, only those relativistic field theories which are asymptotically free can be argued in a straightforward way to exist. Furthermore, the only asymptotically free theories in four space-time dimensions involve non-Abelian gauge symmetry, with highly restricted matter content. So the axioms of gauge symmetry and renormalizability which we invoked to define QCD are, in a certain sense, redundant. They are implicit in the mere *existence* of nontrivial interacting quantum field theories.

1.4 Chiral Symmetry Breaking

1.4.1 Numerical and Laboratory Phenomena

The most direct evidence for chiral symmetry breaking in QCD comes form numerical simulation of the theory. One simply computes the expectation value

$$\langle \overline{q_{Li}} q^j_R \rangle \propto v \delta^j_i \neq 0 \tag{1.23}$$

in the ground state for the theory with massless quarks. This condensation, which breaks the chiral symmetry of the equations, is entirely analogous to the development of spontaneous magnetization in a ferromagnet. As in that case, for any finite sample (e.g., in any simulation) we must add an infinitesimal biasing field to stabilize a particular alignment.

In Eq. 1.23 I've chosen to align in the flavor diagonal direction, but in the absence of a biasing field any chirally rotated configuration, with $q_R \rightarrow U q_R$, will have the same energy but $\delta^i_j \rightarrow U^i_j$ in the expectation value, for any U $\in$ SU(f). There are several technical issues in the simulations that arise and must be addressed, but the numerical evidence that chiral symmetry is spontaneously broken is unambiguous and overwhelming, at least for $2 \leq f \leq 4$. I'll discuss this evidence in more detail in Lecture 2.

The historical path whereby spontaneous chiral symmetry was discovered as a property of Nature was of course quite different. Indeed, the discovery of chiral symmetry breaking in the strong interaction antedates by more than a decade the discovery of QCD as its microscopic theory.

The conceptual starting point for the historic development was the observation, coming into focus with the BCS theory of superconductivity, that if a symmetry is spontaneously broken there will be massless collective modes associated with this breakdown. (The "experimental" starting point was the Goldberger–Treiman relation; see below.) Quite generally, suppose the ground state of a physical system (e.g., a ferromagnet or the no-particle state of QCD at zero temperature) is characterized by the existence of a condensate

$$\langle \eta | \mathcal{M}^a_b | \eta \rangle = \eta^a_b \tag{1.24}$$

that violates a continuous symmetry of the underlying equations (e.g., rotational or chiral symmetry, respectively). Let the symmetry g of the underlying theory be implemented by the unitary operator $U(g)$, with $U(g)^{-1}MU(g) = \rho(g)M$. Then if $\rho(g)\eta \neq \eta$, which is the signature of symmetry breaking, the states

$$U(g)|\eta\rangle \equiv |\rho(g)\eta\rangle \tag{1.25}$$

will be physically distinct from the ground state, but energetically degenerate with it. By moving slowly within this manifold of states, as a function of space, we would expect to create states whose energy goes to zero as the wavelength of the variation goes to infinity. In a particle interpretation, the quanta of the field that creates such configurations will be massless. Furthermore, we have constructed a very specific realization of these quanta in terms of symmetry generators. This construction can be exploited to yield predictions for their properties. For example, if the broken symmetry is an internal symmetry, the quanta will be spin-0 particles.

Turning now specifically to the world of strong interactions, it is a striking fact that π mesons are spin-0 particles which are much lighter than all other hadrons. This suggests the possibility that they are associated with the spontaneous breakdown of an approximate internal symmetry. Their pseudoscalar character, and the fact that they form an isotriplet, suggests the breaking pattern

$$\mathrm{SU}(2)_L \times \mathrm{SU}(2)_R \to \mathrm{SU}(2)_{L+R}. \tag{1.26}$$

This very specific picture of pions as collective modes closely connected to broken chiral symmetry, besides explaining their quantum numbers and small mass, can be exploited to give many predictions about their low-energy behavior, as we shall discuss in Lecture 2. The phenomenological success of these predictions validates the hypothesis of spontaneously broken approximate chiral symmetry as a description of Nature.

Within QCD, this picture arises very naturally. If the u and d masses are small the basic equations of QCD will exhibit approximate chiral symmetry. And numerical work QCD spontaneously develops a symmetry-breaking condensate, as I mentioned. The general theoretical machinery for extracting predictions from spontaneous symmetry breaking remains valid and extremely valuable in modern QCD. Additionally, the specific form of intrinsic breaking in QCD, through small quark mass terms, has specific phenomenological consequences. I will spell out how all this works below in Lecture 2, when we discuss order parameters and effective Lagrangians. As we shall see, all these concepts take on additional twists, and become even more central, for QCD in extreme conditions.

1.4.2 Ironic Aside

Ironically, the first generation of developments in high-energy physics to be inspired by modern superconducitivity theory were inspired by BCS pairing theory in the limit that the gauge coupling—that is, electromagnetism, and hence the phenomenon of superconductivity—is neglected. In that limit it is the global symmetry of electron number that is violated by the formation of a Cooper pair condensate, and there is a massless collective mode. Spontaneous breaking of a global symmetry turns out to be the appropriate, and fruitful, idea for chiral symmetry breaking in the strong interaction.

There was an interval of several years before the second generation of developments, when the gauge coupling was reinstated. Only then did the primary phenomenon of superconductivity itself—the Meissner effect—enter the picture. Rechristened in its new context as the Higgs mechanism, it of course became central to modern electroweak interaction theory.

1.4.3 Pairing Heuristics

Just as for confinement, the *fact* of spontaneous chiral symmetry breaking in QCD is no longer negotiable. Still, just as for confinement, one would like to understand why and how it occurs, and whether there are circumstances in which it can come undone.

A heuristic model for chiral symmetry breaking was supplied by Nambu and Jona-Lasinio long before modern QCD. Amazingly, with some relabeling of the players the concepts they introduced still apply. Indeed, as we shall see, at high density they come to look better than ever. We shall be discussing pairing theory in great detail in Lecture 4, so this is just a foretaste.

Suppose one has an attractive four-fermion interaction

$$\mathcal{L}_{\text{int}} = g(\bar{\psi}\psi)(\bar{\psi}\psi). \tag{1.27}$$

Then one can imagine that it is energetically favorable to form a condensate

$$\langle\bar{\psi}\psi\rangle = v \neq 0, \tag{1.28}$$

since this condensate generates negative interaction energy. Indeed, if the condensate is so large that we can ignore quantum fluctuations, we shall have the condensation energy density

$$\Delta\mathcal{E} = -\Delta\mathcal{L} = -g\langle\bar{\psi}\psi\rangle^2. \tag{1.29}$$

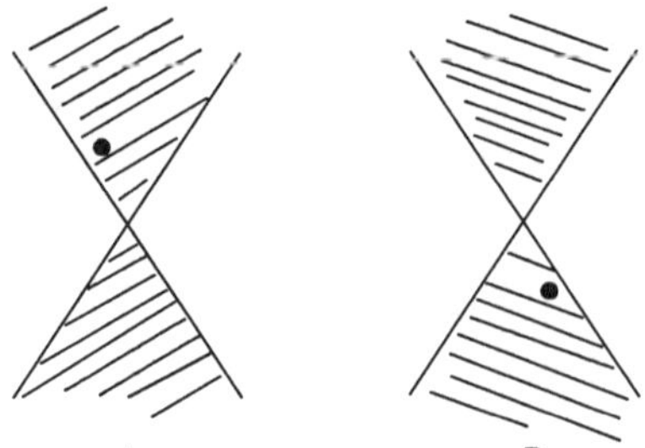

FIGURE 7.5. Pairing picture for chiral symmetry breaking. The dots indicate the position of modes that pair. In the limit of zero energy, you must go to the tip of the light cone.

To test this idea in the simplest crude way, write

$$(\bar{\psi}\psi)^2 = (\bar{\psi}\psi - v)^2 + 2v\bar{\psi}\psi - v^2 \tag{1.30}$$

and, in the interaction Lagrangian, discard the fluctuating first term (as is approximately valid at weak coupling). The other terms, when added to the standard kinetic energy term for massless fermions, generate the Lagrangian for free massive fermions. One can of course diagonalize this quadratic approximate Lagrangian and, by filling the negative energy sea, construct the appropriate zero fermion number density ground state (i.e., for a given value of the condensate). Then one can enforce consistency by calculating $\langle\bar{\psi}\psi\rangle$ in this state, and demanding that it is equal to the originally assumed value v. This consistency equation is called the gap equation, in honor of its ancestor in BCS theory. If the gap equation has a nontrivial solution, one will have lowered the energy by forming a condensate.

In QCD, the one-gluon exchange interaction is quite attractive in the quark-antiquark color singlet channel. This is hardly surprising, since by forming a singlet one cancels the charge and eliminates field energy. To make a scalar condensate in this channel, one must pair left-handed antiquarks with right-handed quarks. So there are simple heuristic reasons to anticipate the possibility of spontaneous chiral symmetry breaking in QCD.

Whether spontaneous chiral symmetry breaking actually occurs, however, is a delicate question, because it involves a competition. The interaction energy one gains by pairing up occupied particle and antiparticle modes must compete against the kinetic energy lost in occupying them. The kinetic energy can become arbitrarily small, but only for a density of states that likewise vanishes—the tip of the Lorentz cone as shown in Figure 7.5. So whether the interaction energy ever wins out, or not, is a delicate dynamical issue. For example, there is some evidence that as the number of flavors f grows the strength of chiral condensation (relative to the primary QCD scale) shrinks, and that for large enough f (6? 7?) it's gone. This comes about, presumably, because for larger f the effective coupling does not run so quickly to large values at small energy.

It's quite a different story at high density. In that case the density of states does not vanish, and spontaneous chiral symmetry breaking can appear at arbitrarily weak coupling, where we have excellent analytic control.

1.5 Chiral Anomalies and Instantons

1.5.1 The Historic Case

The original discovery of the chiral anomaly involved what in hindsight is a fairly complicated example of the phenomenon. It came about not through abstract investigation of mathematical models, but in the process of analyzing a very specific physical process, the decay $\pi^0 \to \gamma\gamma$. This is obviously an electromagnetic decay, so to treat it we must consider QCD coupled to QED. The combined theory, though it violates isospin, still appears to be invariant under axial I_3 symmetry. (I will work, for simplicity, in the limit of vanishing u and d quark masses. This can be shown to be a good approximation for estimating the $\pi\gamma\gamma$ vertex, though of course the actual π^0 mass must be inserted when we use this amplitude to calculate the decay rate.) If this were true, then the π^0 would still be accurately a collective Nambu–Goldstone mode associated with the spontaneous breaking of axial I_3 symmetry. The coupling $\pi^0 F^{\mu\nu}\widetilde{F}_{\mu\nu}$ would be forbidden, because at long wavelength such a Nambu-Goldstone mode is derivatively coupled to the corresponding current. Of course one can rewrite $\pi^0 F^{\mu\nu}\widetilde{F}_{\mu\nu} \to \frac{1}{2}\partial^\mu\pi^0 A^\nu\widetilde{F}_{\mu\nu}$ inside the Lagrangian, after an integration by parts. However the electromagnetic term is not part of the axial I_3 symmetry current, classically.

The brilliant result of Adler, Bell and Jackiw is that when one investigates the situation more deeply, using the full resources of quantum field theory, such a term does in fact occur. (Again, the original analysis antedates QCD, and therefore was couched in rather different language. Its details are fascinating and of considerable historical interest, but I will not discuss them here.) Moreover its coefficient can, given an underlying theory of the strong interaction, be calculated precisely. When the coefficient is calculated in a free quark model—with colored, fractionally charged quarks—one finds agreement between the predicted rate for $\pi^0 \to \gamma\gamma$ and experiment. Remarkably, the free-quark result remains valid in QCD.

The mechanism whereby the extra term is generated is quite subtle. It is most readily seen in perturbation theory, though a nonperturbative derivation is possible.

To regulate the triangle graph and other loop graphs with circulating fermions it is convenient to follow the procedure of Pauli and Villars. In this procedure, one introduces into the theory fictitious boson fields ϕ carrying all the same quantum numbers as the fermions (including spin $\frac{1}{2}$, and so flouting the spin-statistics theorem) but having a large mass M. Since the boson loops differ in sign from the fermion loops their contributions will cancel at very large virtual momentum, where the divergences arose. Thus one achieves finite integrals. Then one takes the limit $M \to \infty$, fixing a few low-energy amplitudes by setting them equal to their physical values. The basic result of renormalization theory is that in suitable (renormalizable) quantum field theories, having used a small number of low-energy amplitudes, to determine the couplings—which will be functions of the cutoff and a few physically determined parameters—all remaining physical amplitudes will approach a finite limit, order by order in perturbation theory. Of course the limiting amplitudes no longer contain contributions arising from the fictitious bosons as intermediate states, because those particles have been driven to infinite mass.

This is not the place to review the technicalities of renormalization theory. Fortunately, for present purposes we don't need their details. Our focus is simply the triangle graph for the two-photon matrix element of the divergence of the axial current. This graph, by naive power counting, is linearly convergent in the ultraviolet by power counting. In more detail, there are three fermion propagators, and gauge invariance pulls out two powers of momentum to make Fs from As. However, one will only be able to use gauge invariance if one has employed a gauge invariant regulator, like Pauli–Villars.

By the same power counting, the leading M dependence of the regulated integral, which arises from the Pauli–Villars boson loop, goes as $1/M$ for large M. This might seem to make it negligible. The large mass M, however, involves a large violation of chiral symmetry. Specifically, it implies a large coefficient $\propto M$ in the divergence of the axial current:

$$\partial_\mu(\bar{\phi}\gamma^\mu\gamma^5\phi) = 2M\bar{\phi}\gamma^5\phi. \tag{1.31}$$

(To have a regulated chiral symmetry that will behave sensibly as we remove the cutoff, the Pauli–Villars fields must transform in the same way as the quark fields they regulate.) This factor exactly cancels the convergence factor, leaving a finite residual contribution. The regulator, necessary to control the contribution of highly virtual quarks, spoils naive chiral symmetry. That is the essential mechanism of the chiral anomaly.

1.5.2 An Easier Version

The basic mechanism leading to anomalies is nicely illustrated, in a context where it is relatively easy to understand, by the process $h \to GG$ of Higgs particle decay into two gluons. Since this process is of independent and timely interest, I hope you'll forgive a brief diversion.

The coupling of the standard model Higgs boson to quarks is given as

$$\mathcal{L}_{\text{int}} = -2^{1/4}G_F^{1/2}h\sum m_i\bar{q}_iq_i, \tag{1.32}$$

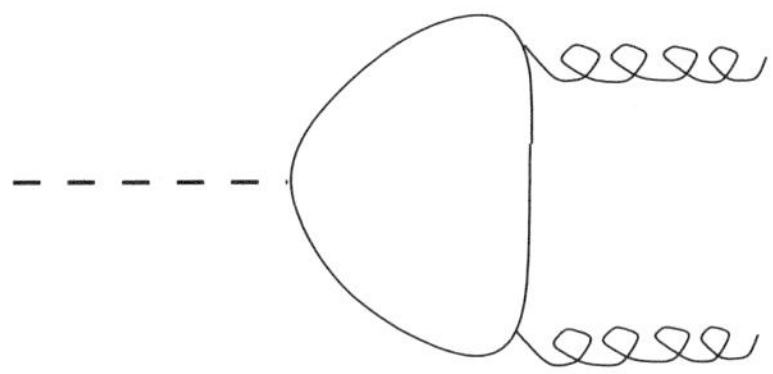

FIGURE 7.6. Triangle graph, crucial for anomalies and Higgs particle phenomenology.

where G_F is the Fermi coupling constant. The direct coupling to ordinary matter would seem to be extremely feeble, due to the very small masses of the u and d quarks. Ordinarily one would expect that the contribution from the heavy quarks would be suppressed, according to general "decoupling" theorems. Indeed, we'd be in pretty bad shape in physics if we always had to worry about big contributions to low-energy amplitudes from (potentially unknown) heavy particles. However there is an exception of sorts here, because the coupling grows with the mass. Thus the contribution to the dimension 5 induced interaction term

$$\mathcal{L}_{\text{induced}} = \kappa h \operatorname{tr} G^{\mu\nu} G_{\mu\nu} \tag{1.33}$$

arising through a heavy quark loop of circulating heavy quarks in the triangle graph of Figure 7.6, which by power counting one might expect to be inversely proportional to the mass of the quark, is instead unsuppressed. (Of course, this time the external legs are a Higgs particle and two gluons.) This is very similar to what we saw for the triangle anomaly, but now with real heavy particles rather than virtual regulators. One finds

$$\kappa = 2^{1/4} G_F^{1/2} \frac{g^2}{48\pi^2} \tag{1.34}$$

per heavy quark, in the large mass limit ($m_h \lesssim m_q$). For values $90\,\text{GeV} \lesssim m_h \lesssim 150\,\text{GeV}$ of the Higgs mass most interesting for ongoing searches this "anomalous" gluon coupling generates both the dominant mechanism for hadronic production of h particles, and a significant decay mode for them.

1.5.3 The Saga of Axial Baryon Number

For concreteness in this part I'll mostly take $f = 2$ and refer to the quarks as u and d. This simplifies the notation, loses nothing essential, and is close to reality.

The spontaneous breakdown of approximate chiral $\mathrm{SU}(2)_L \times \mathrm{SU}(2)_R \to \mathrm{SU}(2)_{L+R}$ is associated with an extensive, successful phenomenology. At the level of quarks in QCD, we understand it as the result of the development of a condensate $\langle \overline{u_L} u_R \rangle = \langle \overline{d_L} d_R \rangle$. This condensate also violates the axial baryon number symmetry under $(u_L, d_L) \to e^{i\alpha}(u_L, d_L)$,

$(u_R, d_R) \to e^{-i\alpha}(u_R, d_R)$, which is a symmetry of (1.1). One might expect, then, a light Nambu-Goldstone boson with the associated properties—a light, flavor-singlet pseudoscalar with highly constrained couplings. Alas, there is no such particle. Thereby hangs a tale.

The first observation is that there is an "anomalous" contribution to the divergence of the axial baryon number current, arising from the triangle graph of Figure 7.6. There's a new set of players—the axial baryon current instead of axial I_3, and gluons instead of photons—but they follow the same script. Thus we find

$$\partial_\mu j_B^{\mu 5} \equiv \partial_\mu(\bar{u}\gamma^\mu\gamma^5 u + \bar{d}\gamma^\mu\gamma^5 d) = \frac{g^2}{4\pi^2}\operatorname{tr}\widetilde{G}_{\mu\nu}G^{\mu\nu}, \tag{1.35}$$

where $\widetilde{G}_{\mu\nu} \equiv \frac{1}{2}\epsilon_{\mu\nu\alpha\beta}G^{\alpha\beta}$ is the dual field strength. The current $j_B^{\mu 5}$ is not conserved. Naive axial baryon number symmetry, generated by the spatial integral of j_B^{05}, is spoiled by an anomaly.

However, the existence of this anomaly does not in itself remove the problematic mode. For one has

$$\operatorname{tr} G^{\mu\nu}\widetilde{G}_{\mu\nu} = \partial_\mu K^\mu \tag{1.36}$$

where

$$K_\mu = \frac{1}{2}\epsilon_{\mu\alpha\beta\gamma}\operatorname{tr}\Big(A^\alpha\partial^\beta A^\gamma + \frac{2}{3}A^\alpha A^\beta A^\gamma\Big). \tag{1.37}$$

Thus it would appear that a modified symmetry, generated by the charge associated with the modified conserved current $j_\mu^{A5} - K_\mu$, is spontaneously broken, leaving us not much better off than before, still with the mistaken prediction of an extra light flavor singlet pseudoscalar.

Fortunately, though, (1.36) is itself problematic. The point is that K_μ is a gauge-dependent quantity, so that in principle it can be singular without implying the singularity of any physical observable.

We can be more specific about this in the context of a path integral treatment of the theory. In such a treatment we express quantum amplitudes as an integral of contributions from different classical field configurations. To have reasonable control of the functional integrals—to have a measure that is damped for large field strengths—we must consider the Euclidian form of the theory, rotating to imaginary values of the time. In this framework, consider the behavior of the gauge potentials A_μ at spatial infinity. For the integral of K_μ to acquire a nonvanishing surface term, and thereby violate the formal conservation law, requires that $A_\mu(x) \sim M_\mu(\hat{x})/|x|$. This is not allowed for generic forms of $M_\mu(\hat{x})$, since a nontrivial field strength $G \sim 1/r^2$ would result, leading to a logarithmically divergent action. But for special values of the boundary conditions $M_\mu(\hat{x})$ these can cancel, leaving behind a finite-action contribution to the functional integral that contributes to $\partial_\mu K^\mu$.

In the weak-coupling limit, we look for the field configurations that contribute to the amplitude of interest that have the smallest possible action.

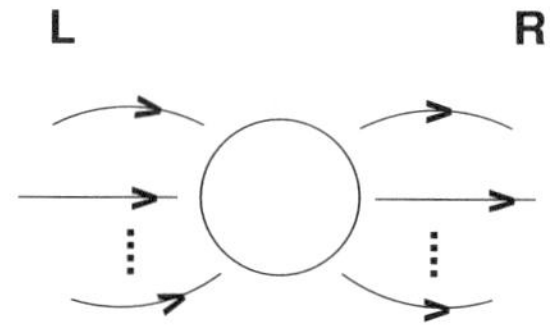

FIGURE 7.7. 'tHooft interaction. Chirality is violated by a knot of gluon energy.

For amplitudes that violate axial baryon number (nonvanishing $\int G^{\mu\nu}\widetilde{G}_{\mu\nu}$) these configurations are instantons, much-studied classical solutions of the Euclidean Yang–Mills equations.

A very important consistency check is that the integral $\int \mathrm{tr}\, G^{\mu\nu}\widetilde{G}_{\mu\nu}$, which according to the anomaly equation measure the violation of axial baryon number, must turn out to be quantized (that is, come in discrete units—it's a c-number integral). Indeed, since axial baryon number is quantized, changes in axial baryon number had also better be quantized. The facts we have discussed, that this integral can be thrown onto the surface at infinity and that its finiteness requires special conspiracies, suggest a connection to topology and the possibility of its quantization. These features can indeed be demonstrated, but I won't do that here.

In view of the easily proved inequality

$$\int \mathrm{tr}\, G^{\mu\nu}G_{\mu\nu} \geq \left| \int \mathrm{tr}\, G^{\mu\nu}\widetilde{G}_{\mu\nu} \right| \tag{1.38}$$

quantization of the right-hand side implies that the action of any configuration leading to axial baryon number violation is bounded below by a finite constant. Thus in the function integral such configurations are suppressed by a factor $e^{-\mathrm{const}/g^2}$. In particular, they vanish to all orders in perturbation theory!

Instantons violate axial baryon number, but none of the other chiral flavor symmetries. This allows us to visualize their effect quite simply, at a heuristic level. The flavor structure must contain a product of determinant factors

$$\mathcal{L}_{\mathrm{inst}} \propto \epsilon_{ij} q_L^i q_L^j \epsilon^{kl} \overline{q_{Rk}}\, \overline{q_{Rl}}. \tag{1.39}$$

For f flavors, we'd have f lefties in and f righties out. The structure of the color and spin indices, and the form-factor accompanying this interaction, are more tricky to work out. To do it one must solve for the fermion zero-modes in the presence of an instanton. The resulting effective nonlocal "t Hooft interaction" is represented pictorially in Figure 7.7: lots of fermions emerging from an extended gluon cloud.

Unfortunately, for reasons we have discussed before, the formal weak-coupling limit that leads us to focus on instantons is not under control for QCD in vacuum. So the arguments used in this section, while certainly suggestive, cannot be made quantitative in a convincing way. (Nevertheless, some reasonably successful phenomenology has been done by the saturation of the functional integral by superpositions of instantons and anti-instantons as a starting point.) Remarkably, under extreme conditions the theoretical situation for axial $U_A(1)$ breaking comes under much better control, with interesting consequences, as we'll see.

Lecture 2 High Temperature QCD: Asymptotic Properties

2.1 Significance of High Temperature QCD

In this lecture I shall be discussing the behavior of QCD (and some closely related models) at high temperature and zero baryon number density. Since that is a mouthful I'll just say high temperature.

The high temperature phase of QCD is of interest from many points of view. First of all, it is the answer to a fundamental question of obvious intrinsic interest: What happens to empty space, if you keep adding heat? Moreover, the high density phase of QCD was (almost certainly) the dominant form of matter during the earliest moments of the Big Bang. Moreover, it is a state of matter one can hope to approximate, and study systematically, in heavy ion collisions. Major efforts are being directed toward this goal and significant, encouraging results have already emerged. One can also simulate many aspects of the behavior with great flexibility and control, from first principles, using the techniques of lattice gauge theory. One can also make progress analytically. So there is a nice interplay among physical experiments, numerical experiments, and theory.

The fundamental theoretical result regarding the asymptotic high temperature phase is that it becomes quasi-free. That is, one can describe major features of this phase quantitatively by modeling it as a plasma of weakly interacting quarks and gluons. In this sense the fundamental degrees of freedom of the microscopic Lagrangian, ordinarily only indirectly and very fleetingly visible, become manifest (or at least, somewhat less fleetingly visible). Likewise the naive symmetry of the classical theory which, as we saw in Lecture 1, is vastly reduced in the familiar, low-temperature hadronic phase, gets restored asymptotically. In particular, chiral symmetry is restored, and confinement comes completely undone. Axial baryon number and scale symmetry, though never precisely restored, become increasingly accurate.

Since there are dramatic qualitative differences between the zero-temperature and the high-temperature phases, the question naturally arises whether there are sharp phase transitions separating them, and if so what is their nature. This turns out to be a rich and intricate story, whose answer depends in detail on the number of colors and light flavors. In the course of

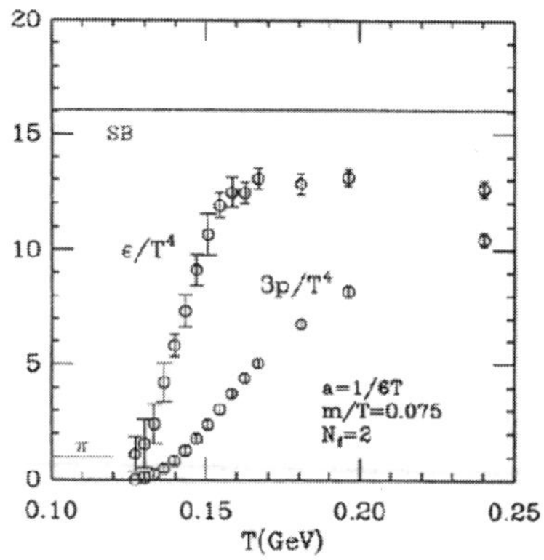

FIGURE 7.8. Energy density and pressure of 2-flavor QCD as a function of temperature.

addressing it, we shall have to refine and modify common, rough intuitions about chiral symmetry and (especially) confinement. After an involved but I think interesting and coherent story, building up from the study of various idealizations, we shall find that there is, plausibly, a true phase transition in real QCD that we can converge upon from several directions—experimentally, numerically, and analytically.

2.2 Numerical Indications for Quasi-Free Behavior

For technical reasons it has been difficult until recently to simulate QCD including dynamical quarks with realistically small masses. That situation is changing, but it will be a few years before accurate quantitative results for thermodynamic quantities for QCD with light dynamical quarks become available.

Fortunately, we can already learn a lot from the existing simulations using pure glue or glue plus moderately massive quarks.

Representative results for the temperature dependence of the energy density and pressure in the two flavor theory are shown in Figure 7.8. Clearly, there is a rapid crossover in the behavior, with dramatic rises in the energy density and pressure (even when normalized to T^4) over a small range of temperatures around 150 MeV.

A notable feature of the numerical results is that while the energy density (divided by T^4) ascends rapidly to something close to its asymptotic value, the pressure appears much more sluggish. Thus the behavior of the plasma, even in regard to this basic bulk property, differs significantly from a free gas of massless particles. It is a worthy challenge to compute the corrections to free behavior analytically in weak coupling. This is not entirely straightforward, due to the absence of magnetic screening in perturbation theory (concerning which, more below). For recent progress see [3].

In reality the only hadrons light on the scale of the computed crossover temperature are pions. Thus for temperatures significantly below this temperature (say $T \leq 120\,\text{MeV}$) one has a rather dilute gas of pions, with 3 massive degrees of freedom for the three possible charge states of these spinless particles. Asymptotically, on the other hand, one has a gas with three different flavors of quarks, each of which comes with two spins, three colors, and antiquarks. Also there are eight gluons, each with two helicities. Thus the number of degrees of freedom is $3 \times 2 \times 3 \times 2 + 8 \times 2 = 52$, of which all but the strange quarks are essentially massless. Evidently, the difference is gigantic! Remarkably, the change from one regime to the other appears to occur largely within a narrow range of temperatures around 150 MeV, amazingly low if regarded from the hadron side.

2.3 Ideas About Quark-Gluon Plasma

The physics of quark-gluon plasma is already a big subject with a vast literature. It will bloom further as the RHIC and ALICE programs gather data. Let me briefly sketch a few of the characteristic phenomena that have been discussed.

• A fundamental foundational result is the observation that distributions of particle energies in the final state are well described by thermal distributions corresponding to a freeze-out temperature around 120 MeV. This observation makes it extremely plausible, as had already been anticipated from theoretical work, that approximate thermal equilibrium (at least, kinetic equilibrium) is established—at higher temperatures, of course—in the initial fireball. That's very good news, both because it makes the theoretical analysis easier, and because it means that the collisions really are approximating the conditions which are of most fundamental interest.

• The most basic and profound prediction is what I have already mentioned, that one should have approximately the energy and pressure characteristic of the appropriate—large!—number of microscopic degrees of freedom. Qualitatively, this means among other things a steep rise in the specific heat, so that the rise of temperature with energy will slow markedly. Energy will go into particle production, not motion. In principle, the temperature is accessible either through measurement of the transverse momenta of hard leptons or photons emerging from the initial fireball. The entropy can be estimated from the final thermal particle distribution at freezeout, since the expansion and cooling should be roughly adiabatic until freezeout. Several more sophisticated flow diagnostics have been proposed to give handles on the full equation of state.

• Since strange quarks are expected to be much lighter than the lightest hadrons (K mesons) in which they are found, one can anticipate a significant rise in the relative multiplicities of strange (and antistrange) particles, relative to normal hadronic collisions. There is already a striking phenomenon of this kind seen at SPS, with a dramatic rise (as much as a factor 15) in Ω and $\bar{\Omega}$ production.

• Perhaps the single most striking experimental result to emerge so far from the study of heavy ion collisions is the suppression of J/ψ, relative to Drell–Yan background, in lead-lead collisions at the highest energies. When the ratio is plotted as a function of atomic weight and energy, a clear break in its behavior, relative to lower energies or lower atomic numbers, leaps to the eye. No hadron-based model of the collision process anticipated this break, and none has been successful in reproducing it. On the other hand an effect of just this sort was anticipated, based on simple qualitative arguments, to mark the onset of quark-gluon plasma behavior.
The basic point is that free gluons are very effective in dissociating J/ψ particles. In quark-gluon plasma there is an abundance of free gluons, while in the hadron phase there is a large mass gap for glue. Alternatively, we may say that in the plasma phase color screening prevents J/ψ binding. Unfortunately, while some effect of this kind is very plausible, and evidently does occur, it seems difficult to refine the heuristic argument into a really precise calculation.

• The best hope for a rigorous characterization of quark-gluon plasma behavior is probably comparison of experimental measurements to calculated predictions for quantities that can be addressed using (sophisticated extensions of) perturbative QCD. Among the most promising candidates are hard probes, such as high transverse momentum jets, high-mass dileptons, and energetic photons. Assuming thermal equilibrium—or a definite model of quasi-equilibrium—one can formulate reasonably precise expectations for the rates and distributions of these phenomena, with many cross-checks. Their use is analogous to the use of radiative probes in traditional plasma diagnostics. Another characteristic signature of quark-gluon plasma is softening of quark jet distributions due to their passage through the medium.

A more conjectural possibility, which has received much attention under the name "disordered chiral condensate" or DCC, is that the return to equilibrium as the fireball cools is marked by collective relaxation. Then one might see gross deviations from equipartition in the collective modes.

Specifically, as we shall discuss at length below, one expects that the large spontaneous breaking of chiral symmetry which occurs in the ground state comes undone at high temperatures. The transition from chiral symmetry breaking to chiral symmetry restoration is described by equations very similar to the equations that describe the loss of magnetization when one heats a magnet past its Curie temperature.

To make the analogy accurate we must envision the demagnetization taking place in the presence of a tiny external field, representing the small intrinsic breaking of chiral symmetry due to nonzero u and d quark masses. Now if, after being disordered at high temperature, the magnet is cooled rapidly there are two extreme possibilities for how it might relax back down to the ground state. According to one extreme picture, each spin separately and independently settles down to align with the external field. According to the other extreme picture, the spins first align with each other in large clumps, usually in some wrong direction, before the clumps relax collectively, as units, toward the correct alignment. In the latter case, one will have significant correlation phenomena, and the possibility of coherent radiation. In QCD, this will take the form of "pion lasing"—an abnormally large number of pions occupying a small region of phase space, with a highly non-Gaussian distribution of charged to neutral multiplicity, will emerge.

Because it is an intrinsically nonequilibrium phenomenon, the likelihood of DCC formation is hard to assess theoretically. It has been observed in some idealized numerical simulations.

Evidently there is a plenitude of signature phenomena in quark-gluon plasma that can and presumably will be explored in heavy ion collisions. We can look forward to a many-faceted dialogue between theory and experiment in coming years.

For the remainder of this lecture and in the next one, however, I will focus on the specific, narrower theoretical question of equilibrium phase transitions. This emphasis brings the advantage that the conceptual issues become well-posed and precise, and support a rich theory; on the other hand application of the results derived to the complex realities of heavy ion collisions is not straightforward.

2.4 Screening Versus Confinement

One should not assume, that because the quark-gluon plasma at high temperatures is conveniently described using very different degrees of freedom from those we use to describe the hadronic gas at low temperatures, there must be a sharp phase transition separating them. Indeed, ordinary plasmas are very different from gases of atoms (so different, that at Princeton they are studied on different campuses), but it is well understood that no strict phase transition separates them. The fraction of ionized atoms rises smoothly, though rather abruptly, from nearly (but not quite) zero at low temperatures to nearly unity at high temperatures.

With this cautionary example in mind, let us revisit the question of confinement. Previously we discussed the pure glue theory, and were able to give a precise definition of confinement in terms of the asymptotic behavior of Wilson loops. We were even able to understand in a very simple way why confinement is not at all a bizarre or mysterious behavior, but quite a reasonable possibility for a strong-coupling (or asymptotically free) gauge theory. Actually, it was lack of confinement that required some explaining away—a failure of the strong-coupling expansion, or a phase transition.

Now let's consider the theory with quarks.

The strong coupling expansion requires that we use the discretized lattice version of the theory. The basic idea of its extension to include quarks is quite simple, although there are great subtleties if one tries to do justice to chiral symmetries, and many algorithmic issues. These questions involve important, active areas of research. However they do not impact the basic issues of screening versus confinement, as discussed in this section.

To give the quarks dynamics, we need to supply a 'hopping' term. The sum over all links of

$$\Delta\mathcal{L}_{\text{hop}} = \psi(\hat{r})U_{r,\mu}\bar{\psi}(\hat{r}+\mu) \tag{2.1}$$

does the job, and reduces formally to the continuum action for a small. It has an evident gauge invariance, generalizing (1.15), whereby the ψ variables, which live on vertices, are simply multiplied by the corresponding Ωs.

Revisiting the question of tiling the Wilson loop, we see that now it is possible to get a nonzero contribution by propagating a single quark line around the perimeter, as shown in Figure 7.4(c). This is quite unlike the pure glue theory, where we were required to tile a whole area. The perimeter tiling corresponds to a potential which does not continue to grow at large distances, but rather saturates at a finite value. Physically, it corresponds to the production of a separated meson pair. The color sources, inserted by the two sides of the Wilson loop, can be saturated by a dynamical quark on one side, and a dynamical antiquark on the other. There is a finite energy to make the pair, but once it is made and combined with the sources into "mesons," the mesons have only short-range residual interactions, and the total energy does not grow with the distance.

There is a simple heuristic way to understand the difference between the two cases. There is an additive quantum number modulo 3, triality, characterizing color charges. It is one for quarks, minus one for antiquarks, and zero for gluons. If we write SU(3) indices on the fields, triality is simply the number of upper indices minus the number of lower ones. Because of the existence of the invariant epsilon symbol ϵ^{abc} triality can jump in units of three by color invariant processes, but not in units of one or two. In the pure glue theory all the dynamical fields have zero triality, so a source of unit triality cannot be screened. Furthermore the presence, or not, of unit triality can be determined by measurements made at great distances. We saw this in the strong coupling expansion. A triality source generated a "live" link that could be displaced by laying down plaquettes, but not cancelled. We have, therefore, a poor man's version of Gauss' law. If triality flux interferes with the correlations in the ground state, then as we separate source and antisource we will produce a finite change in vacuum energy per unit volume that extends over a growing volume, with confinement a conceivable outcome. By contrast, in the theory with dynamical quarks triality can be screened. In the absence of any strictly conserved quantity characterizing a source, it is difficult to imagine how its dynamical influence could extend to great distances. In fact, it would be hard to specify exactly what it is that is confined.

By the way, if the only dynamical quarks are extremely heavy ones then the area tiling can remain cheaper than the perimeter tiling up until very large values of the separation R. In this case one will have a linear interquark potential out to large R, supporting a spectrum of bound states up to an ionization threshold.

2.5 Models of Chiral Symmetry Breaking

To help ground our later discussions, I will now briefly discuss some basic elements of the phenomenology of chiral symmetry breaking in the observed strong interaction, and in QCD.

The circle of ideas around chiral symmetry breaking grew up around attempts to understand a remarkable formula discovered by Goldberger and Treiman. Their derivation of the formula made use of drastic and uncontrolled approximations, and is mainly of historical interest. The modern understanding starts from ideas introduced by Nambu and Gell-Mann and Levy, and developed with great ingenuity by many physicists. Their hypotheses are fully justified within QCD. Indeed, nowadays it is appropriate to start from QCD, and to interpret the necessary hypotheses within the microscopic theory.

Interpreted within QCD, the hypothesis of chiral symmetry breaking has two parts:

1. The u and d quark masses are small, so that the corresponding fundamental interaction terms $m_u \bar{u}u$ and $m_d \bar{d}d$ in the Lagrangian may be treated as perturbations.

   Thus we are invited to consider the properties of a zeroth-order theory with massless u and d quarks. In this limit, as we have discussed, there is an $\mathrm{SU}(2)_L \times \mathrm{SU}(2)_R$ chiral symmetry of the fundamental theory, rotating among the different helicities separately.

2. In the absence of u and d quark masses, the $\mathrm{SU}(2)_L \times \mathrm{SU}(2)_R$ chiral symmetry is spontaneously broken, down to the diagonal vector subgroup $\mathrm{SU}(2)_{L+R}$.

More precisely, the hypothesis is that a condensate

$$\langle \bar{u}u \rangle = \langle \bar{d}d \rangle = v \neq 0 \tag{2.2}$$

develops.

One can also consider extending these hypotheses to the s quark, but it is not entirely clear under what circumstances it is safe to treat $m_s \bar{s}s$ as a perturbation.

A consequence of these hypotheses is that one expects the existence of approximate Nambu–Goldstone bosons. If it were an exact symmetry that were spontaneously broken we would have exactly massless particles of this type; since there is some small intrinsic breaking, in addition to the larger spontaneous breaking, the corresponding Nambu-Goldstone particles acquire nonzero, but small, masses.

There are indeed particles within the observed hadron spectrum that are much lighter than any of their brethren, namely the π mesons. Furthermore the quantum numbers of the π mesons—$J^{PC} = 0^{-+}$, $\mathrm{SU}(2)_{L+R}$ (isospin) triplet—are what one requires for Nambu-Goldstone bosons arising from $\mathrm{SU}(2)_L \times \mathrm{SU}(2)_R \to \mathrm{SU}(2)_{L+R}$ breaking.

To see this, consider the physical origin of the Nambu-Goldstone bosons. They arise due to the possibility of obtaining low-energy field configurations by interpolating slowly, in space and time, among the energetically degenerate but inequivalent ground states one has due to spontaneous symmetry breaking. The inequivalent ground states are generated by three independent transformations of the type $(q, -q)$ in the Lie algebra of $\mathrm{SU}(2)_L \times \mathrm{SU}(2)_R$, which manifestly form an isotriplet of odd parity. Furthermore there is no preferred space-time direction in the condensate, so the quanta are spin 0.

Although it is fundamentally a phenomenon of the strong interaction, much of the interest of chiral symmetry derives from its connection with the weak interaction. Specifically, the currents that generate the approximate chiral symmetry of the strong interaction also appear in the weak interaction. The prototype application, the Goldberger–Treiman relation, exploits this connection. The pion decay $\pi^+ \to \mu^+\nu$ involves the hadronic matrix element of the axial vector current

$$\langle 0|A^5_\eta|\pi^+\rangle = F_\pi p_\eta \tag{2.3}$$

where p is the momentum.

Thus F_π is a directly measurable quantity. For the divergence of the axial current we find then

$$\langle 0|\partial^\eta A^5_\eta|\pi^+\rangle = F_\pi m_\pi^2. \tag{2.4}$$

We see here the connection between chiral symmetry and the mass of the pion: in the version of QCD with exact chiral symmetry the divergence would vanish, and so would the mass of the pion.

Now let us consider another matrix element that appears in describing another basic weak process, that is beta decay of the neutron. The nucleon matrix element of the axial current

$$\langle N|A^5_\eta|N\rangle = G_A \bar{u}\gamma_\eta\gamma_5 u \tag{2.5}$$

at small momentum transfer for nucleons nearly at rest. G_A is a quantity subject to strong-interaction corrections and it is therefore not the sort of

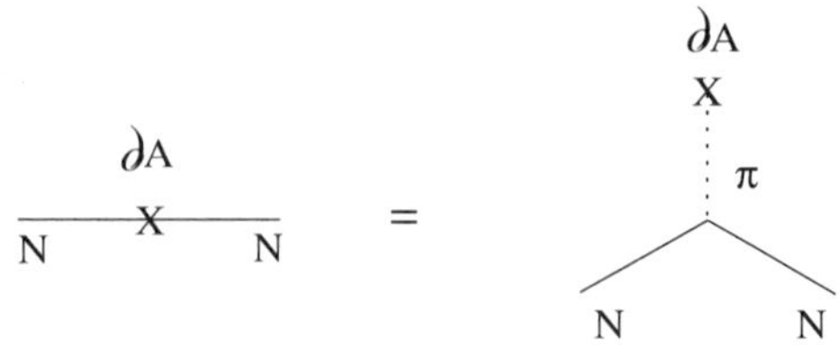

FIGURE 7.9. Saturation of the axial current divergence with the nearby π.

thing we can normally expect, in the absence of special insight, to calculate easily. It is measured to be about 1.2. Taking again the divergence, we have on the right-hand side $2MG_A\bar{u}\gamma_5 u$, not particularly small (beyond the kinematic suppression), whereas on the left-hand side we have the matrix element of a "small", chiral-symmetry breaking operator. However there is a contribution to this matrix element arising from the nucleon coupling to a π meson, which then communicates with the current divergence according to (2.4). The Feynman graph for this is shown in Figure 7.9. The factors of m_π cancel, and we find

$$g_Y F_\pi = 2m_N G_A, \tag{2.6}$$

which is the Goldberger–Treiman relation.

This logic of this derivation of the Goldberger-Treiman relation can be vastly generalized, to include matrix elements of low-momentum pions or currents between various states. It can be made systematic by using the technology of Ward identities. In this context one finds that in relating multi-current Green functions to multi-pion processes one must often evaluate current commutators. One obtains in this way a host of predictions for low-energy processes, which work remarkably well. (At high energies or momenta the saturation of axial currents with pions is no longer accurate.) The successful evaluation of such commutators, in agreement with experiment, using relations abstracted from free field theory (now justified in QCD), was a major step in the historical elucidation of the strong interaction, and in the revival of interest in quantum field theory in the late 1960s. When I reflect that this elaborate theoretical technology was developed before the microscopic theory, by working backward from nuggets of relevant data embedded within an overwhelming confusion of strong-interaction phenomena, I am lost in admiration.

Many of the results of Ward identity and current algebra gymnastics can be derived in an easier, more transparent way by writing down appropriate effective Lagrangians. If they embody the correct broken and unbroken symmetries, such Lagrangians will satisfy all the Ward identities that can be derived as consequences of these symmetries. Thus one can reproduce more transparently the valid results of the Green function analysis—and more. Generally the 'more'—relations derived from the effective Lagrangian away from the low-energy limit—will depend on nongeneric features of the effective Lagrangian, and should be ignored.

2.6 More Refined Numerical Experiments

With this background, we are now in a better position to discuss two more refined diagnostics of finite temperature QCD.

The first is the so-called Polyakov loop. It is basically half a Wilson loop. Let me be more precise. A standard result in path integral theory states that one can set up the partition function at finite temperature by passing to imaginary values of the time variable and requiring periodicity of the fields (antiperiodicity for fermions) under $\tau \to \tau + 1/T$. Now we can consider parallel transport around the circle of imaginary time:

$$\langle L \rangle = \left\langle \operatorname{Tr} \exp\left\{ i \int_0^\beta A_\tau^a \lambda^a \, d\tau \right\} \right\rangle. \tag{2.7}$$

By the same strong coupling argument as before, we anticipate that the expectation value of the Polyakov loop integral will vanish in a confining phase.

It is very pleasant that we can (following Polyakov) relate this anticipation to a symmetry principle. The action for the pure glue theory is left invariant if we multiply all the U matrices connecting (say) $\tau = 0$ to their temporal neighbor $\tau = a$ vertices by an element of the center of the group, since any plaquette product has either no such U or one going up and one down. Thus for SU(2) we can multiply by -10 (times the identity matrix), and we have the symmetry Z_2, whereas for SU(3) we can multiply by ω or ω^2, where $\omega = e^{2\pi i/3}$ is the cube root of unity, and we have the group Z_3. On the other hand the Polyakov loop is *not* left invariant under these operations, but rather multiplied by the corresponding numerical factor. If the expectation value of the loop vanishes, as is characteristic of the confined phase, then the confinement symmetry is valid. However if the expectation value of the Polyakov loop does not vanish the confinement symmetry is spontaneously broken.

The discrete "confinement symmetry" of the pure glue theory is *not* valid for the action including quarks. Indeed, we can think of it as multiplying different triality sectors containing N quarks, (modulo N for SU(N)) by different phases. "Virtual" quark world-lines winding around the imaginary time circle are not left invariant. Indeed, with the implementation above, terms in the action that hop quarks from $\tau = 0$ to $\tau = a$ are not invariant. By redefining $\psi(\tau = a)$ by a phase you can move the changes in the action to the next time slice, ..., but after winding around the circle you will just arrive back where you started.

This formal argument based on confinement symmetry agrees, of course, with our previous intuitive argument from consideration of conserved triality charge. In the pure glue theory there is a conserved flux that cannot

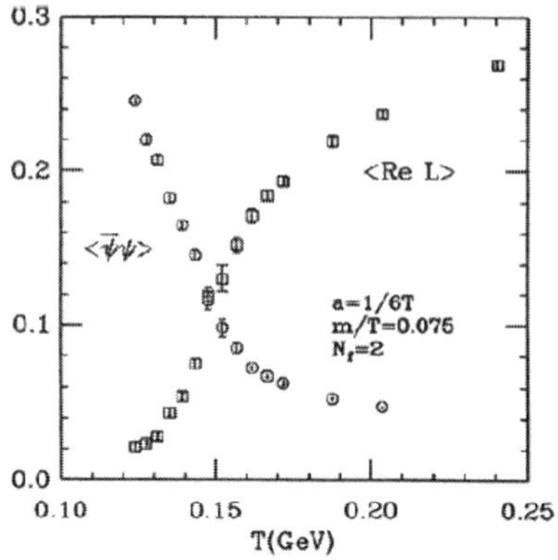

FIGURE 7.10. Behavior of the Polyakov loop $\langle L \rangle$, and the chiral condensate $\langle \bar{\psi}\psi \rangle$, in 2-Flavor QCD as functions of temperature.

be screened, an associated symmetry, and a strict criterion of confinement; in the theory with quarks all that structure is gone, and there is no strict definition of confinement to distinguish it from screening.

While there is no reason to expect the Polyakov loop strictly to vanish at the onset of deconfinement (since deconfinement is, as I've already belabored, an incoherent notion in this context), it remains a perfectly respectable observable. One might expect that if there is a crossover from a hadronic phase exhibiting pretty good confinement to a quark-gluon phase with very poor confinement the Polyakov loop should take a nose dive. This is indeed the behavior that shows up in the numerical simulation, as you see in Figure 7.10.

The second is the chiral condensate. While the notion of confinement gets fuzzy in the theory with quarks, the notion of chiral symmetry breaking is perfectly sharp (for massless quarks). And it exhibits very interesting dynamical behavior as a function of temperature. The simplest measure of chiral symmetry breaking is simply the expectation value $\langle \bar{\psi}\psi \rangle$. This quantity is perfectly accessible to lattice gauge theory. In Figure 7.10, you see that this expectation value does indeed take a dive. In the simulation it does not reach zero, because the quarks are not truly massless, but the possibility of a smooth decrease to zero at a finite value of T is certainly suggested. Studies with varying values of the quark masses, when extrapolated, further support this suggestion.

Lecture 3 High-Temperature QCD: Phase Transitions

3.1 Yoga of Phase Transitions and Order Parameters

Confinement, in the pure glue version of QCD (only), is a property we can associate with a definite symmetry, that is valid at low temperature but broken at high temperature. Chiral symmetry, in the versions of QCD with two or more massless quarks is, conversely, spontaneously broken at low temperatures (at least if the number of quarks is not too large) but restored at high temperatures. As emphasized by Landau, the presence or absence of a symmetry is a sharp, objective question, which in any given state of matter must have a yes or no answer. And if the answer is yes in one regime and no in another, passage from one regime to the other must be accompanied by a sharp phase transition. This situation is usually parameterized by some appropriate order parameter, that transforms nontrivially under the symmetry, and is zero on the unbroken side but nonzero on the broken side.

Phase transitions can occur without change of symmetry, or dynamical reasons—we shall see some simple examples below (where there is symmetry lurking just offstage). But by considering changes in symmetry, which *must* be associated with phase transitions, and behaviors of order parameters we will be able to say quite a lot, without doing any prohibitively difficult calculations.

3.1.1 Second Order Transitions

There are two broad classes of phase transitions, which have quite different qualitative properties near the transition point. First order transition are characterized by a finite discontinuity in the generic thermodynamic parameter—i.e. basically in anything except the free energy, which of course must be equal for the two phases at the transition point. Second-order transitions, on the other hand, are characterized by continuous but nonanalytic behavior of thermodynamic quantities.

In Nature second order transitions are less common than first order transitions, but they are especially interesting. Near first order transitions the two phases are simply "different," and are described by distinct expressions for the free energy (in terms of macroscopic variables). There is a wholesale reorganization of matter, even locally—there are jumps in intensive variables. Near second order transitions that is not the case. In a large but finite volume, a first order transition point will be marked by rare but sudden and drastic jumps from one phase to the other, going over in the infinite volume limit to hysteresis. In the same circumstances, a second order transition point will not exhibit any jumps, and the partition function will be a perfectly analytic function.

So how does the nonanalytic behavior arise? It can only arise from taking the infinite volume limit. This, in turn, implies that for a second order transition to occur there must be low-energy fluctuations of arbitrarily long wave length, since it is only such modes that can render the infinite volume limit subtle (otherwise the free energy must ultimately become simply additive in the volume, for large enough volumes). In terms of static quantities, there must be a diverging correlation length. In terms of particle physics, there must be massless particles. That is, if we quantized the modes under discussion, they would have massless quanta.

The hypothesis—or quasi-theorem, as motivated above—that nonanalytic behavior of thermodynamic quantities near a second order transition must arise from the dynamics of massless modes makes it possible, following Landau and Wilson, to make remarkably concrete and specific predictions about this behavior. The point is that it appears to be very difficult to construct consistent theories of massless particles, unless one considers small numbers of dimensions or large and exotic symmetry structures, that are inappropriate to the cases at hand. So if we specify the desired space dimension and symmetry we may find a unique theory of this kind, or none at all. Then the singular behavior of any possible second-order phase transitions with a given dimensionality and symmetry will be uniquely determined, independent of other details of the underlying microscopic theory. This is the hypothesis of *universality.*

Universality makes it possible to make rigorous predictions for the behavior of complicated physical systems—such as various versions of QCD—near second-order phase transitions by doing calculations in much simpler models.

Now let me give a few words of orientation about the formal aspects of such analyses, using the three dimensional Ising model as a prototype. We are interested in the singular behavior of thermodynamic functions near a possible second-order phase transition, where the magnetization decreases from a nonzero value to zero. The relevant low-energy, long-wavelength modes are gradual changes in the local average of the magnetization. Since we are working at long wavelengths it is appropriate to coarse grain, so the magnetization is described by a real three-dimensional scalar field $\phi(x)$. We are interested in the singularity of the partition function induced by fluctuations of $\phi(x)$. To find it, we need to construct the appropriate "universal" theory based on $\phi(x)$.

We want this theory to be describing fluctuations that are small in magnitude and long in wavelength, so we should use a Lagrangian with the smallest possible powers of $\phi(x)$ and of derivatives. Of course we need a quadratic term with two derivatives to get any nontrivial spatial behavior at all. To give this term its chance to shine, we will also need to put the mass equal to zero, since a mass term would always dominate the derivative term at long wavelength. Since we have $\phi \to -\phi$ symmetry, the next possibility is a ϕ^4 term. So our trial 'Lagrangian' (to be interpreted as H/T, the Hamiltonian divided by the temperature, in statistical mechanical language) is

$$\int d^3x\, \mathcal{L} = \int d^3x\, \left((\partial\phi)^2 + \lambda\phi^4\right). \tag{3.1}$$

Now if we count dimensions we see that for the Lagrangian to be dimensionless ϕ must have mass dimension $\frac{1}{2}$, and so λ must have mass dimension 1. According to naive dimensional analysis, therefore, we are not getting a scale-invariant theory.

We know, however, that interacting field theories contain another source of nontrivial scaling behavior. Because there are an infinite number of degrees of freedom, we must regulate the theory, and define a renormalized coupling at some finite momentum scale, which we fix to a physical value independent of the cutoff. Then in favorable cases we will get cutoff-independent answers, in terms of the renormalized coupling, as we take the cutoff to infinity. We can get a scale invariant theory from this set-up if the bare coupling $\lambda_b(\Lambda)$ one needs to insure a fixed renormalized coupling scales as $\lambda_b \propto \Lambda^{1/2}$. The proportionality constant is then our dimensionless parameter.

That reasoning is quite abstract, so it is informative to consider the situation also from a radically different perspective, due to Wilson and Fisher. We consider, formally, extending the theory to $4-\epsilon$ dimensions. It is of course problematic to construct a field theory in noninteger dimensions, but we can perfectly well continue the perturbation theory integrals, which ought to be adequate if the relevant coupling turns out to be small. Now in $4-\epsilon$ dimensions the same counting as above shows that the mass dimension of λ is ϵ. Thus as we rescale the momentum at which the coupling is defined there are two sources of variation. One is simple classical dimensional analysis, which tends to make the ϕ^4 more relevant at small momenta; the other is the running due to fluctuations (loops), familiar in its quantum version from QED or QCD, where it is interpreted as vacuum polarization due to virtual particles. Of course here we are concerned with classical fluctuations, but the equations are just the same. Since ϕ^4 in 4 dimensions is not asymptotically free, the two terms governing renormalization toward the infrared go as

$$\frac{d\tilde{\lambda}(p)}{dt} = \tilde{\lambda}\epsilon - b\tilde{\lambda}^2, \tag{3.2}$$

here $t = \ln(p_0/p)$, with p_0 some reference momentum, and $\tilde{\lambda} \equiv \lambda p^{-\epsilon}$, and the equation is valid for $p \ll \Lambda$ the cutoff and small λ and ϵ. b is a calculable *positive* number. So as $t \to \infty$, $\tilde{\lambda} \to \epsilon/b$, which is indeed small for small enough ϵ. We say there is a fixed-point coupling with this value. With Λ fixed, we approach a scale invariant theory for $p \ll \Lambda$. The funny dependence of the coupling λ (not $\tilde{\lambda}$!) extrapolated to momenta approaching the cutoff—what we would call the bare coupling—on the cutoff Λ itself is just what we anticipated before, on abstract grounds.

This construction makes it plausible that there can be a scale-invariant theory, but also makes it clear that this theory will not be easy to find in three dimensions, where the fixed-point coupling cannot be small. One approach, which works remarkably well, is to calculate around $4-\epsilon$ dimensions and extrapolate to $\epsilon = 1$. Another is to work directly in three dimensions, calculate to high orders in perturbation theory, and join on to the form at high orders, which is known from sophisticated semiclassical techniques. Finally, one can simply simulate the theory directly numerically. Any of these techniques would be prohibitively difficult to use in high temperature QCD directly, but thanks to universality we can get rigorous quantitative answers (to carefully selected questions, of course!) using much simpler models.

3.1.2 First Order Transitions

An important side-benefit of the analysis of how second-order transitions arise is that it alerts us to cases where this cannot occur. In the Ising model analysis, we found the scale invariant theory could arise when the mass parameter associated with the magnetization vanishes. Since we expect the effective mass parameter to be a function of temperature, $m^2 = m^2(T)$, it is reasonable to expect that this can happen at one particular value of the temperature. So (3.1) is a special case of the embedding set of Lagrangians

$$\int d^3x \, \mathcal{L} = \int d^3x \left((\partial\phi)^2 + m^2(T)\phi^2 + \lambda\phi^4\right) \tag{3.3}$$

describing the dynamics of the fluctuating magnetization not only exactly at, but also near, the critical transition temperature. This is reasonable from another point of view as well: when $m^2(T) < 0$ a nonzero expectation value for ϕ will be preferred, whereas for $m^2(T) > 0$ the expectation should vanish.

Now if we put the system in an external magnetic field, breaking the $\phi \to -\phi$ symmetry, then ϕ and ϕ^3 terms are allowed. Let's shift away the ϕ term. There will still be a T-dependent m^2, and it can go through zero. But now that will generally not give rise to a second-order transition, because in the presence of a small ϕ^3 term the expectation value will jump to a new minimum at a nonzero (positive) value of m^2. The special case where the cubic term vanishes simultaneously with m^2 can be accessed only if there is another control parameter available, in addition to the temperature. Then one has a so-called tricritical point. In the phase plane, the tricritical point appears as the terminus of a line along which there are weaker and weaker first-order transitions.

Even if the mean-field analysis allows a second-order transition, there will not be one if there is no suitable scale-invariant theory to represent the universality class. The mean field analysis ignored fluctuations, but as we have learned these are vital. In our discussion above, we saw that in order to construct the scale-invariant theory we needed to have a simple, finite limiting behavior of the effective coupling under renormalization group transformations toward the infrared. If there is no such limiting behavior, there cannot be a second-order transition. The physical interpretation of this outcome is simply that in such cases the fluctuations have grown out of control, resulting in a catastrophic rearrangement of the state—a first-order transition. Such an eventuality is, for obvious reasons, called a fluctuation driven first-order transition.

Since they are marked by finite discontinuities, first-order transitions are robust against small perturbations. Thus if we have a symmetry and an order parameter, whose change from a nonzero to a zero value forces the existence of a first-order transition according to either of the mechanisms I've just discussed, there will still be a first-order transition even if the symmetry is intrinsically slightly broken. There will be no strict order parameter, and thus we would not have been able to predict the necessity of a transition without referring to the nearby, unbroken variant of the theory.

Each and every one of the theoretical phenomena I have mentioned in these orienting sections plays a significant role in understanding the phase structure of QCD!

3.2 Application to Glue Theories

Let's recall the basic facts. When there are no quarks at all, then there is the possibility of a true confinement-deconfinement transition. As we have discussed, such a transition is characterized by the Polyakov order parameter

$$\langle L \rangle = \left\langle \mathrm{Tr}\exp\left\{ i \int_0^\beta A_\tau^a \lambda^a \, d\tau \right\} \right\rangle. \tag{3.4}$$

Here the expectation value is taken over the thermal ensemble of field configurations periodic in imaginary time τ with period $\beta = 1/T$, and λ is the representation matrix for the fundamental representation. This loop inserts quark quantum numbers into the ensemble. In the pure glue theory, the operator inserts a flux that cannot be screened, and alters the state by an irreducible amount out to infinity. This costs a finite energy per unit volume, and therefore infinite energy altogether. The expectation value of the loop would therefore be expected to vanish in the confined phase, while it acquires a nonzero value in the unconfined phase. L is multiplied by the appropriate complex root of unity when an element of the center of the gauge group is applied to the state. Thus symmetry under this discrete group (triality for color SU(3), diality for SU(2)) is broken in the unconfined phase expected to exist at high temperature.

Since there is a simple order parameter with well-defined symmetry properties, one can entertain the possibility of a second order transition. Indeed there does seem to be a second order transition for SU(2), in the universality class of the (inverted) 3d Ising model. 'Inverted' refers to the features that whereas in the Ising model the Z_2 symmetry is broken at low temperature, but restored at high temperature, the confinement Z_2 symmetry is valid at low temperature but broken at high temperature. This means the $m^2(T)$ goes through zero in the opposite direction, but of course one still has the same universal theory at the critical point and a simple correspondence away from it.

However for SU(3) the appropriate model is different. It is something called the 3-state Potts model. In the field-theoretic version of that model we must use a complex scalar field ϕ invariant under $\phi \rightarrow \omega\phi$, with ω the cube root of unity, to implement the symmetry. With such a field, cubic terms of the type

$$\Delta\mathcal{L} = \kappa(\phi^3 + \phi*^3) \tag{3.5}$$

are allowed. The existence of a cubic invariant implies that the transition will be first order.

Both these predictions prove to be true in large-scale numerical simulations of pure glue QCD.

3.3 Application to Chiral Transitions

Again, let's quickly recall the basics. With dynamical quarks there is no longer a confinement symmetry, but if we have f flavors of massless quarks there is an additional symmetry under chiral transformations in the group $\mathrm{SU}(f)_L \times \mathrm{SU}(f)_R \times \mathrm{U}(1)_B$ of independent special unitary rotations of the left- and right-handed fields, together with the overall vector baryon number symmetry. (The additional apparent axial baryon number symmetry, present at the classical level, is violated in quantum theory by the anomaly, as discussed earlier.) This chiral symmetry is believed on good grounds to break spontaneously down to vector $\mathrm{SU}(f) \times \mathrm{U}(1)$ at low temperatures; and to be restored at sufficiently high temperatures. Of course for $f = 1$ the chiral symmetry is vacuous; but for $f \geq 2$ there is a phase transition associated with restoration of chiral symmetry. Since there is a simple order parameter for this phase transition—namely, for example, the expectation value of the quark bilinear

$$M^i_j = \langle \bar{q}^i_L q_{Rj} \rangle \tag{3.6}$$

—one may again inquire concerning the possibility of a second order transition.

3.3.1 Formulation of Models

In order to describe a possible second-order transition quantitatively, we must try to find a tractable model in the same universality class. For the chiral order parameter (3.6) the relevant symmetries are independent unitary transformations of the left- and right-handed quark fields, under which

$$M \to U^\dagger M V. \tag{3.7}$$

These transformations generate an $\mathrm{SU}(f)_L \times \mathrm{SU}(f)_R \times \mathrm{U}(1)_V$ symmetry, after the anomaly in the axial baryon number current is taken into account. At the phase transition, the true symmetry is broken to $\mathrm{SU}(f)_{L+R} \times U(1)_V$.

To describe the critical behavior, it is sufficient to retain the degrees of freedom which develop long-range fluctuations at the critical point. It is natural to assume that these are associated with long-wavelength variations in the order parameter, whose magnitude is small and whose variations within the vacuum manifold therefore cost little energy near the transition. Thus the most plausible starting point for analyzing the critical behavior of a possible second-order phase transition in QCD is the Landau-Ginzburg free energy

$$\mathcal{F} = \mathrm{tr}\,\partial_i M^\dagger \partial_i M + \mu^2\,\mathrm{tr}\,M^\dagger M + \lambda_1\,\mathrm{tr}(M^\dagger M)^2 + \lambda_2(\mathrm{tr}\,M^\dagger M)^2. \tag{3.8}$$

Here μ^2 is the temperature-dependent renormalized $(\text{mass})^2$, which is negative below and positive above the critical point, while λ_1 and λ_2 parameterize the strength of the quartic couplings and are supposed to be smooth at the transition. The symmetry breaking pattern we want is $M \propto \mathbf{1}$ below the transition, which is what we shall find at the minimum of the potential, within a range of positive λ_1 and λ_2.

Actually (3.8) is not quite what we want. It has a full $\mathrm{U}(f) \times \mathrm{U}(f)$ symmetry, which breaks down to $\mathrm{U}(f)$. Thus it contains a massless Nambu-Goldstone boson for the axial baryon number symmetry, which is not present in the microscopic theory (QCD) we are trying to model.

For $f = 2$ one can solve this problem very neatly by implementing the symmetry in a slightly different way. Instead of general complex matrices, which in this case form a reducible representation of the chiral symmetry, we can restrict ourselves to unitary matrices with positive real determinant. This restriction on M is consistent with the transformation law (3.7) as long as U and V have equal phases, but not if the phases are unequal. Thus the unwanted axial U(1) symmetry is indeed removed. An important point is that for 2×2 matrices the condition of being a multiple of a unitary matrix is a linear condition, so that we can enforce it while remaining within the domain of renormalizable field theories. (Nothing like this is true for larger matrices.) It is very convenient to parameterize the 2×2 matrices in question in terms of four real parameters $(\sigma, \vec{\pi})$ and the Pauli matrices as

$$M = \sigma + i\vec{\pi}\cdot\vec{\tau}. \tag{3.9}$$

In this way, we arrive back at the original model of Gell-Mann and Levy.

After this pruning, the order parameter variables that have been retained have the quantum numbers of the scalar isoscalar density $\langle \bar{q}^i q_i \rangle$ and the pseudoscalar isovector densities $\langle \bar{q}^x \gamma_5 \vec{\tau} q \rangle$. Furthermore, the two *a priori* possible quartic couplings λ_1, λ_2 are not independent—a fact which proves to be very significant. In fact the model boils down to the theory of a four-component vector $\phi \equiv (\sigma, \vec{\pi})$ in internal space, that is to say the standard O(4) invariant $n = 4$ "Heisenberg magnet." For smaller numbers of components, this sort of model is a much-studied model for the critical behavior of magnets, with the vector of course representing the magnetization (or staggered magnetization).

For larger values of f the trick discussed above is no longer available. For $f = 3, 4$ one can break the unwanted U(1) symmetry by adding an additional determinantal interaction $\det M$. Beyond $f = 4$ that too is not entirely satisfactory, because it takes us outside the framework of renormalizable theories in four dimensions (which are suitable starting points for the construction of critical theories, using the ϵ expansion). However, as will soon appear, in the present context it is almost certainly academic anyway.

3.3.2 Fixed Points

Now we may search for second-order transitions within each model, by the standard method of looking for infrared stable fixed points of the renormalization group. The vector model has been studied in great depth, for arbitrary n. The existence of a fixed point has been established by detailed analysis of perturbation theory—taken to high orders and supplemented with estimates of the asymptotic behavior directly in three dimensions, and also *via* the ϵ expansion. The calculated critical exponents have been successfully compared with appropriate experiments, for $n \leq 3$. Thus there can be no serious doubt that there is a model for a second-order QCD chiral phase transition for two massless quarks, consistent with the symmetry of that theory.

On the other hand for $f \geq 3$ the structure of the renormalization group for the appropriate model is more complicated—there are then two independent couplings, and so to speak more ways to go wrong. In the lowest order of perturbation theory in the ϵ expansion, the calculations can be done quite simply. They indicate that there is a fixed point of the renormalization group, but that it is *not* infrared stable. In plain English, this means (if we can trust the ϵ expansion!) that, near the transition one would naively identify by taking μ^2 through zero in (3.8), M is actually subject to catastrophic fluctuations, that change the structure of the problem qualitatively. Thus the renormalization group fails to identify a consistent model for a second order transition, and indicates instead that the transition must be a first order transition. In important work Gausterer and Sanielevici have simulated the $f = 3$ matrix models directly, both with and without the determinantal interaction, to search for second-order transitions. None was found; the transition is always first order. These authors also studied the $f = 2$ model, and did find the expected second-order transition in that case. Thus the expectations drawn from simple ϵ expansion analysis appear to be vindicated. Direct simulation of the $n = 4$ magnet model, and comparison with $f = 2$ QCD using the translation dictionary discussed below, is also an attractive possibility.

Existing numerical calculations for QCD, are consistent with the prediction that the proposed models for the universality classes of possible QCD chiral phase transitions are appropriate, and that there is a big difference between the nature of the chiral transition in QCD for $f = 2$ and for larger f, with the former being second order and the latter first order.

3.4 Close Up on Two Flavors

If we accept that there is a second order transition for $f = 2$, we can derive many precise, testable consequences.

As we have just discussed the most plausible starting point for analyzing the critical behavior of a second-order phase transition in QCD is the Landau-Ginzburg free energy

$$F = \int d^3x \left[\frac{1}{2} \partial^i \phi^\alpha \partial_i \phi_\alpha + \frac{\mu^2}{2} \phi^\alpha \phi_\alpha + \frac{\lambda}{4} (\phi^\alpha \phi_\alpha)^2 \right], \tag{3.10}$$

of an $n = 4$ component scalar field. Here μ^2 is the temperature-dependent renormalized $(\text{mass})^2$, which is negative below and positive above the critical point, while λ is the strength of the quartic couplings and is supposed to be smooth at the transition. We neglect terms with higher powers of ϕ since $|\phi|$ is small near the transition. The symmetry breaking pattern we want is $M \propto \mathbf{1}$ (equivalently, $\langle \sigma \rangle \neq 0$; $\langle \vec{\pi} \rangle = 0$) below the transition which is indeed what we find at the minimum of the potential for positive λ. This model has been studied in depth for arbitrary η and spatial dimension d, and the existence of an infrared stable fixed point of the renormalization group for $\eta = 4$, $d = 3$ is firmly established. Hence, it is a model for a second order QCD chiral phase transition for two massless quarks.

When the free energy (3.10) is written in terms of σ and $\vec{\pi}$ it looks much like the original model of Gell-Mann and Levy. But there are two changes: there are no nucleon fields and only three (spatial) dimensions. These two changes reflect an important distinction. We are only proposing (3.10) as appropriate near the second order phase transition point. This is because it is only there that we can appeal to universality — the long-wavelength behavior of the σ and $\vec{\pi}$ fields is determined by the infrared fixed point of the renormalization group, and microscopic considerations are irrelevant to it. In Euclidean field theory at finite temperature, the integral over ω of zero temperature field theory is replaced by a sum over Matsubara frequencies ω_n given by $2n\pi T$ for bosons and $(2n+1)\pi T$ for fermions with n an integer. Hence, one is left with a Euclidian theory in three spatial dimensions with massless fields from the $n = 0$ terms in the boson sums and massive fields from the rest of the boson sums and the fermion sums. Hence, to discuss the massless modes of interest at the critical point, (3.10) is sufficient. We do not need to introduce nucleon fields or constituent quark fields.

3.4.1 Critical Exponents

The most important universal properties of the second order transition are the critical exponents, which we now define.

First, let us introduce the reduced temperature $t = (T - T_c)/T_c$. The exponents α, β, γ, η, and ν describe the singular behavior of the theory with strictly zero quark masses as $t \to 0$. For the specific heat one finds

$$C(T) \sim |t|^{-\alpha} + \text{less singular}. \tag{3.11}$$

The behavior of the order parameter defines β

$$\langle|\phi|\rangle \sim |t|^{\beta} \quad \text{for } t < 0. \tag{3.12}$$

η and ν describe the behavior of the correlation length ξ where

$$\begin{aligned} G_{\alpha\beta}(x) &\equiv \langle \phi(x)_\alpha \phi(0)_\beta \rangle - \langle \phi_\alpha \rangle \langle \phi_\beta \rangle \\ &\to \delta_{\alpha\beta} \frac{A}{|x|} \exp(-|x|/\xi) \quad \text{at large distances.} \end{aligned} \tag{3.13}$$

A is independent of $|x|$, but may depend on t. The correlation length exponent ν is defined by

$$\xi \sim |t|^{-\nu}. \tag{3.14}$$

Above T_c, where the correlation lengths are equal in the sigma and pion channels, the susceptibility exponent γ is defined by

$$\int d^3x G_{\alpha\beta}(x) \sim t^{-\gamma}. \tag{3.15}$$

The exponent η is defined through the behavior of the Fourier transform of the correlation function:

$$G_{\alpha\beta}(k \to 0) \sim k^{-2+\eta}. \tag{3.16}$$

The last exponent, δ is related to the behavior of the system in a small magnetic field H which explicitly breaks the O(4) symmetry. Let us first show that in a QCD context, H is proportional to a common quark mass $m_u = m_d \equiv m_q$. This common mass term may be represented by a 2×2 matrix $\mathcal{D}$ given by m_q times the identity matrix. We are now allowed to construct the free energy from invariants involving both $\mathcal{D}$ and $\mathcal{M}$. The lowest dimension term linear in $\mathcal{D}$ is just $\mathrm{tr}\, \mathcal{M}^\dagger \mathcal{D} = m_q \sigma$, which in magnet language is simply the coupling of the magnetization to an external field

$H \propto m_q$. In the presence of an external field, the order parameter is not zero at T_c. In fact,

$$\langle|\phi|\rangle(t=0, H \to 0) \sim H^{1/\delta}. \tag{3.17}$$

The six critical exponents defined above are related by four scaling relations. These are

$$\begin{aligned} \alpha &= 2 - d\nu \\ \beta &= \frac{\nu}{2}(d-2+\eta) \\ \gamma &= (2-\eta)\nu \\ \delta &= \frac{d+2-\eta}{d-2+\eta}. \end{aligned} \tag{3.18}$$

We therefore need values for η and ν for the four component magnet in $d = 3$. These were obtained in the remarkable work of Baker, Meiron and Nickel, who carried the perturbation theory to seven-loop order, and used information about the behavior of asymptotically large orders, and conformal mapping and Padé approximant techniques to obtain

$$\begin{aligned} \eta &= .03 \pm .01 \\ \nu &= .73 \pm .02. \end{aligned} \tag{3.19}$$

Using (3.18), the remaining exponents are $\alpha = -0.19 \pm .06$, $\beta = 0.38 \pm .01$, $\gamma = 1.44 \pm .04$ and $\delta = 4.82 \pm .05$. Since α is negative there is a cusp in the specific heat at T_c, rather than a divergence.

A more powerful result which includes (3.12) and (3.17) as special cases is the critical equation of state:

$$H = M|M|^{\delta-1}\kappa_1 g(\kappa_2 t|M|^{-1/\beta}) \tag{3.20}$$

in which here $H \equiv m$, $M \equiv \langle\bar{q}q\rangle = \langle|\phi|\rangle$, $t \equiv (T - T_c)/T_c$, g is a universal function, and κ_1 and κ_2 are nonuniversal constants. This equation of state could, of course, be compared directly with sufficiently accurate numerical simulations.

3.4.2 Tricritical Point

To this point I have been discussing a world with two massless quarks, and so we have implicitly been taking the strange quark mass to be infinite. On the other hand, I argued earlier that if the strange quark is massless, then the chiral phase transition is first order. Hence, as the strange quark mass is reduced from infinity to zero, at some point the phase transition must change from second order to first order. This point is called a tricritical point. There is numerical evidence that when the strange quark has its physical mass, and the two light quarks are taken strictly massless, the transition is second order, with the consequences discussed above. However this conclusion was controversial for many years, and even now may not be completely secure. In any case the physical strange quark mass is not drastically different from the critical value, so let's indulge our idle curiosity a little further (this will pay off shortly).

In a lattice simulation, the strange quark mass can be tuned to just the right value to reach the tricritical point. Let's discuss the critical exponents that would be observed in such a simulation.

Let us consider the effect of adding a massive but not infinitely massive strange quark to the two flavor theory. This will not introduce any new fields which become massless at T_c, and so the arguments leading to the free energy (3.10) are still valid. The only effect of the strange quark, then, is to renormalize the couplings. Renormalizing μ^2 simply shifts T_c, as does renormalizing λ unless λ becomes negative. In that case, one can no longer truncate the Landau-Ginzburg free energy at fourth order. After adding a sixth order term, and keeping track of small light quark masses, the free energy becomes

$$F = \int d^3x \left[\frac{1}{2}(\nabla\phi)^2 + \frac{\mu^2}{2}\phi^2 + \frac{\lambda}{4}(\phi^2)^2 + \frac{\kappa}{6}(\phi^2)^3 - H\sigma \right]. \tag{3.21}$$

While for positive λ, ϕ^2 increases continuously from zero as μ^2 goes through zero, for negative λ, ϕ^2 jumps discontinuously from zero to $|\lambda|/(2\kappa)$ when μ^2 goes through $\lambda^2/(4\kappa)$. Hence, the phase transition has become first order. Thus at the value of m_s where $\lambda = 0$, the phase transition changes continuously from second order to first order.

The singularities of thermodynamic functions near tricritical points, like the singularities near ordinary critical points, are universal. Hence, it is natural to propose that QCD with two massless flavors of quarks and with T near T_c and m_s near its tricritical value is in the universality class of the ϕ^6 Landau-Ginzburg model (3.21). This model has been studied extensively. Because the ϕ^6 interaction is strictly renormalizable in three dimensions, this model is much simpler to analyze than the ϕ^4 model of the ordinary critical point. No ϵ expansion is necessary, and the critical exponents all take their mean field values, up to calculable logarithmic corrections. Here I'll be content to show you the mean field tricritical exponents.

In mean field theory, the correlation function in momentum space is simply $G_{\alpha\beta}(k) = \delta_{\alpha\beta}(k^2 + \mu^2)^{-1}$. Since $\mu^2 \sim t$, this gives the exponents $\eta = 0$, $\gamma = 1$ and $\nu = \frac{1}{2}$. To calculate α and β, we minimize F for $H = \lambda = \nabla\phi = 0$, and find $\alpha = \frac{1}{2}$ and $\beta = 1/4$. To calculate δ, we minimize F for $t = \lambda = \nabla\phi = 0$ and find $\delta = 5$.

The result for the specific heat exponent α is particularly interesting, since it means that the specific heat diverges at the tricritical point, unlike at the ordinary critical point. This means that whereas for m_s large enough that the transition is second order the specific heat $C(T)$ has a cusp but is finite at $T = T_c$, as m_s is lowered to the tricritical value $C(T_c)$ should increase since at the tricritical point it diverges. This behavior should be seen in future lattice simulations.

Finally, at a tricritical point there is one more relevant operator than at a critical point, since two physical quantities (t and m_s) must be tuned to reach a tricritical point. Hence, a new exponent ϕ_t, the crossover exponent, is required. For $\lambda \neq 0$, tricritical behavior will be seen only for $|t| > t^*$, while for $|t| < t^*$, either ordinary critical behavior or first order behavior (depending on the sign of λ) results. t^* depends on λ according to

$$t^* \sim \lambda^{1/\phi_t}. \tag{3.22}$$

The mean field value of ϕ_t is obtained by minimizing the free energy F for $H = \nabla\phi = 0$, and is $\phi_t = \frac{1}{2}$. These mean field tricritical exponents, $\alpha = \frac{1}{2}$, $\beta = \frac{1}{4}$, $\gamma = 1$, $\delta = 5$, $\eta = 0$, $\nu = \frac{1}{2}$, and $\phi_t = \frac{1}{2}$ would describe the real world if m_s were smaller than it is, and will describe future lattice simulations with m_s chosen appropriately.

3.4.3 Robustness

If we now turn on small but nonzero u and d quark masses, the first-order line will persist, but the second-order transition will disappear, replaced by a crossover. And there will still be a true (tri)critical point, where the first-order line terminates! It will be in the universality class of a liquid-gas or Ising tricritical point, rather than the asymptotically free ϕ^6-type theory.

3.5 A Genuine Critical Point! (?)

In reality, of course, we cannot vary the strange quark mass. If the strange quark mass is, as seems to be indicated, so large that for massless u and d quarks we would have a second-order transition, then after taking into account the nonzero mass of these quarks we are left with only a crossover. Moreover since the mass of the physical pions is not so different from the critical temperature, we cannot expect that correlation length ever gets very long. Thus we are left with a crossover, and perhaps not a very dramatic one. This is a definite prediction, though perhaps a disappointing one for experimentalists, of the form "nothing striking occurs" (in the way of sharp phase transitions). But fortunately, as emphasized in recent work of Stephanov, Rajagopal, and Shuryak, that is not the end of the story.

Their basic insight is that although one cannot vary the strange quark mass experimentally, there is another control parameter that one can vary —the chemical potential. If the singularity in T at slightly unphysical values of m_s and zero μ continues into a singularity at the physical value of m_s for slightly costive μ, we will be back in business, with a true physical (tri)critical point, featuring diverging correlation lengths at a particular point in the μ, T plane.

3.5.1 Subnuclear Boiling

The same structure is also suggested by an entirely different line of thought. Consider now the behavior of two-flavor QCD as one varies the chemical potential μ at *zero* temperature. Models of the type we will be considering in the next two lectures suggest that in the case of two mass quarks there is a first-order transition from a state of broken chiral symmetry at $\mu = 0$ to a state with chiral symmetry restored at large μ. Since the transition is first order, it remains as a first-order transition in the presence of small symmetry-breaking perturbations (the quark masses), despite the absence of an order parameter. This idea is very much in the spirit of the phenomenologically successful MIT bag model, in which nucleons are pictured as droplets of chiral symmetry restored phase embedded in a broken symmetry vacuum. Indeed, if true, it comes very close to providing a microscopic justification of that model.

Physically, the suggested transition represents a sort of subnuclear boiling, to a phase in which quarks (or at least, as we shall see, a subset of them) are liberated to behave as nearly massless, weakly interacting particles.

If in fact there is such a transition, it becomes interesting to follow its fate as a function of temperature. At small temperatures, we shall still have a first-order transition at some value $\mu_c(T)$. We might expect that $\mu_c(T)$ decreases with T, since higher temperature should favor the higher-entropy 'boiled' state. This expectation can be convincingly justified on thermodynamic grounds. But we know that $\mu_c(T)$ cannot decrease to zero, since for $\mu = 0$ we have only a crossover. Again, the simplest possibility to reconcile the behaviors along the axes is to suppose that the line of first-order transitions terminates at a definite (μ_t, T_t). The strength of the first-order transition weakens as a function of temperature, until at T_t the strict distinction between hadron and quark phases has disappeared. This (μ_t, T_t) is the ripe fruit of our theoretical labors: the prediction of a true (tri)critical point in honest-to-god real world QCD, arrived at from qualitative, but deeply rooted, theoretical considerations!

3.5.2 Physical Signatures

Stephanov, Rajagopal and Shuryak (SRS) have made quite specific and detailed proposals for experimental signatures of the phase transition. A proper discussion of this would take us far afield, but a few comments are in order.

In a heavy ion collision one produces a baryon number poor fireball in the central region, but toward the fragmentation regions the baryon number density increases. As the different parts of the fireball expand and cool, each part traces out a different history in the (μ, T) plane. There will be critical fluctuations and long correlation lengths in a given region of phase space if—and only if—material in the region has passed close to $P = (\mu_t, T_t)$ during its history. For that class of events and regions of rapidity space, one can expect enhanced multiplicities of, and correlations among, low-momentum pions. These enhancements will have the distinctive feature of being nonmonotonic in the experimental control parameters, since one can miss the critical point on either of two sides.

The simplest observables to analyze are the event-by-event fluctuations of the mean transverse momentum of the charged particles in an event, p_T, and of the total charged multiplicity in an event, N. SRS calculate the magnitude of the effects of critical fluctuations on these and other observables, making predictions which, they hope, will allow experiments to find. As a necessary prelude, they analyze the contribution of noncritical thermodynamic fluctuations. They find that NA49 data (hep-ex/9904014) is consistent with the hypothesis that most of the event-by-event fluctuation observed in the data is thermodynamic in origin. This bodes well for the detectability of systematic changes in thermodynamic fluctuations near P.

As one example, consider the ratio of the width of the true event-by-event distributions of p_T to the width of the distribution in a sample of mixed events. SRS call this ratio $\sqrt{F}$. NA49 has measured $\sqrt{F} = 1.002 \pm 0.0002$, which is consistent with expectations for noncritical thermodynamic fluctuations. A detailed calculation suggests that critical fluctuations can increase $\sqrt{F}$ by 10–20%, fifty times the statistical error in the present measurement. There are other observables which are even more sensitive to critical effects. For example, a $\sqrt{F_{\text{soft}}}$ defined by using only the 10% softest pions in each event, might well be affected at the factor of two level.

NA49 data demonstrates very clearly that SPS collisions at $\sqrt{s} = 17\,\text{GeV}$ *do not* freeze out near the critical point. P has not yet been discovered. The nonmonotonic appearance and then disappearance (as $\sqrt{s}$ is varied) of any one of the signatures of the critical fluctuations mentioned above, or others, would be strong evidence for critical fluctuations. If nonmonotonic variation is seen in several of these observables, with the maxima in all signatures occurring at the same value of $\sqrt{s}$, it would turn strong evidence into an unambiguous discovery of the critical point. The quality of the present NA49 data, and the confidence with which we can use it to learn that collisions at $\sqrt{s} = 17\,\text{GeV}$ do not freeze out near the critical point make it plausible that critical behavior, if present, could be discerned experimentally. If and when the critical point P is discovered, it will appear prominently on the map of the phase diagram featured in any future textbook of QCD.

3.5.3 Wanted: Numerical Data

The web of general arguments I have outlined above seems to me to make it quite plausible that there is a true critical point in QCD. We can invoke the canonical yoga of second-order phase transitions to make precise, quantitative predictions for the singular behavior of thermodynamics near this point. Arguments of this sort, however, cannot address the nonuniversal but vital question of precisely where (μ_t, T_t) is. It is a great challenge to locate this point theoretically, and of course such knowledge, even if approximate, would greatly simplify the experimentalist search.

I believe it ought to be possible to locate the tricritical point numerically. For while it is notoriously difficult to deal with large chemical potentials at small temperature numerically, there are good reasons to be optimistic about high temperatures and relatively small chemical potentials, which is our concern here. Some combination of extrapolating from data taken at imaginary values of the chemical potential, and/or using the real part of the fermion determinant as the measure, may be adequate to sense the singularity, especially if we dial the strange quark mass to bring it close to $\mu = 0$ (and then extrapolate back to the physical strange quark mass).

If all these strands can be brought together, it will be a wonderful interweaving of theory, experiment, and numerics.

Lecture 4 High-Density QCD: Methods

The behavior of QCD at high density is intrinsically interesting, as the answer to the question: What happens to matter, if you keep squeezing it harder and harder? It is also directly relevant to the description of neutron star interiors, neutron star collisions, and events near the core of collapsing stars. Also, one might hope to obtain some insight into physics at "low" density—that is, ordinary nuclear density or just above—by approaching it from the high-density side.

4.1 Hopes, Doubts, and Fruition

Why might we anticipate QCD simplifies in the limit of high density? A crude answer is: "Asymptotic freedom meets the Fermi surface." One might argue, formally, that the only external mass scale characterizing the problem is the large chemical potential μ, so that if the effective coupling $\alpha_s(\mu)$ is small, as it will be for $\mu \gg \Lambda_{QCD}$, where $\Lambda_{QCD} \approx 200$ MeV is the primary QCD scale, then we have a weak coupling problem. More physically, one might argue that at large μ the relevant, low-energy degrees of freedom involve modes near the Fermi surface, which have large energy and momentum. An interaction between particles in these modes will either barely deflect them, or will involve a large momentum transfer. In the first case we don't care, while the second is governed by a small effective coupling.

These arguments are too quick, however. The formal argument is specious, if the perturbative expansion contains infrared divergences. And there are good reasons—two separate ones, in fact—to anticipate such divergences.

First, Fermi balls are generically unstable against the effect of attractive interactions, however weak, between pairs near the Fermi surface that carry equal and opposite momentum. This is the Cooper instability, which drives ordinary superconductivity in metals and the superfluidity of He3. It is possible to have an instability at arbitrarily weak coupling because occupied pair states can have very low energy, and they can all scatter into one another. Thus one is doing highly degenerate perturbation theory, and in such a situation even a very weak coupling can produce significant "nonperturbative" effects.

Second, nothing in our heuristic argument touches the gluons. To be sure the gluons will be subject to electric screening, but at zero frequency there is no magnetic screening, and infrared divergences do in fact arise, through exchange of soft magnetic gluons.

Fortunately, by persisting along this line of thought we find a path through the apparent difficulties. Several decades ago Bardeen, Cooper, and Schrieffer taught us, in the context of metallic superconductors, how the Cooper instability is resolved. We can easily adapt their methods to QCD. In electronic systems only rather subtle mechanisms can generate an attractive effective interaction near the Fermi surface, since the primary electron-electron interaction is Coulomb repulsion. In QCD, remarkably, it is much more straightforward. Even at the crudest level we find attraction. Indeed, two quarks, each a color triplet, can combine to form a single color antitriplet, thus reducing their total field energy.

The true ground state of the quarks is quite different from the naive Fermi balls. It is characterized by the formation of a coherent condensate, and the development of an energy gap. The condensation, which is energetically favorable, is inconsistent with a magnetic color field, and so such weak magnetic color fields are expelled. This is the color version of the famous Meissner effect in superconductivity, which is essentially identical to what is known as the Higgs phenomenon in particle physics. Magnetic screening of gluons, together with energy gaps for quark excitations, remove the potential sources of infrared divergences mentioned above. Thus we have good reasons to hope that a weak coupling—though, of course, nonperturbative—treatment of the high density state will be fully consistent and accurate.

The central result in recent developments is that this program can be carried to completion rigorously in QCD with sufficiently many (three or more) quark species. Thus the more refined, and fully adequate, answer to our earlier question is: "Asymptotic freedom meets the BCS groundstate." Together, these concepts can render the behavior of QCD at asymptotically high density calculable.

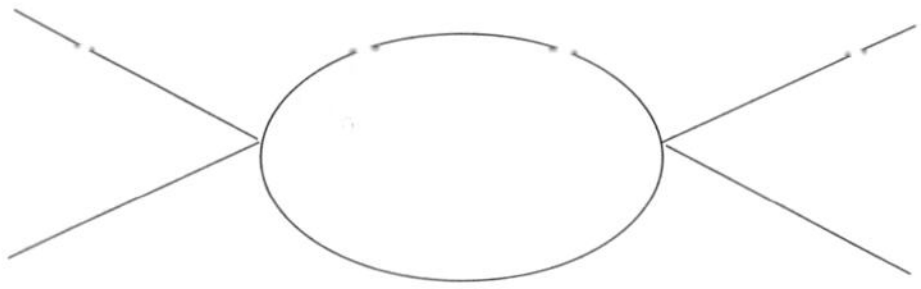

FIGURE 7.11. Graph contributing to the renormalization of four-fermion couplings.

4.2 Another Renormalization Group

To sculpt the problem, begin by assuming weak coupling, and focus on the quarks. Then the starting point is Fermi balls for all the quarks, and the low-energy excitations include states where some modes below the nominal Fermi surface are vacant and some modes above are occupied. The renormalization group, in a generalized sense, is a philosophy for dealing with problems involving nearly degenerate perturbation theory. In this approach, one attempts to map the original problem onto a problem with fewer degrees of freedom, by integrating out the effect of the higher-energy (or, in a relativistic theory, more virtual) modes. Then one finds a new formulation of the problem, in a smaller space, with new couplings. In favorable cases the reformulated problem is simpler than the original, and one can go ahead and solve it.

This account of the renormalization group might seem odd, at first sight, to high-energy physicists accustomed to using asymptotic freedom in QCD. That is because in traditional perturbative QCD one runs the procedure backward. When one integrates out highly virtual modes, one finds the theory becomes more strongly coupled. Simplicity arises when one asks questions that are somehow inclusive, so that to answer them one need not integrate out very much. It is then that the microscopic theory, which is ideally symmetric and constrained, applies directly. So one might say that the usual application of the renormalization group in QCD is fundamentally negative: it informs us how the fundamentally simple theory comes to look complicated at low energy, and helps us to identify situations where we can avoid the complexity.

Here, although we are still dealing with QCD, we are invoking quite a different renormalization group, one which conforms more closely to the Wilsonian paradigm. We consider the effect of integrating out modes whose energy is within the band $(\epsilon, \delta\epsilon)$ of the Fermi surface, on the modes of lower energy. This will renormalize the couplings of the remaining modes, due to graphs like those displayed in Figure 7.11. In addition the effect of higher-point interactions is suppressed, because the phase space for them shrinks, and it turns out that only four-fermion couplings survive unscathed (they are the marginal, as opposed to irrelevant, interactions). Indeed the most significant interactions are those involving particles or holes with equal and opposite three-momenta, since they can scatter through many intermediate states. For couplings g_η of this kind we find

$$\frac{dg_\eta}{d\ln\delta} = \kappa_\eta g_\eta^2. \tag{4.1}$$

Here η labels the color, flavor, angular momentum, ..., channel and in general we have a matrix equation—but let's keep it simple, so κ_η is a positive number. Then (4.1) is quite simple to integrate, and we have

$$\frac{1}{g_\eta(1)} - \frac{1}{g_\eta(\delta)} = \kappa_\eta \ln\delta. \tag{4.2}$$

Thus for $g_\eta(1)$ negative, corresponding to attraction, $|g_\eta(\delta)|$ will grow as $\delta \to 0$, and become singular when

$$\delta = e^{1/[\kappa_\eta g_\eta(1)]}. \tag{4.3}$$

Note that although the singularity occurs for arbitrarily weak attractive coupling, it is nonperturbative.

4.3 Pairing Theory

The renormalization group toward the Fermi surface helps us identify potential instabilities, but it does not indicate how they are resolved. The great achievement of BCS was to identify the form of the stable ground state the Cooper instability leads to. Their original calculation was variational, and that is still the most profound and informative approach, but simpler, operationally equivalent algorithms are now more commonly used. I will be very sketchy here, since this is textbook material.

The simplest and most beautiful results, luckily, occur in the version of QCD containing three quarks having equal masses. I say luckily, because this idealization applies to the real world, at densities so high that we can neglect the strange quark mass (yet not so high that we have to worry about charmed quarks). Until further notice, I'll be focusing on this case.

Most calculations to date have been based on model interaction Hamiltonians, that are motivated, but not strictly derived, from microscopic QCD. They are chosen as a compromise between realism and tractability. For concreteness I shall here follow, and consider

$$\begin{aligned} H &= \int d^3x\, \bar\psi(x)(\nabla\!\!\!/ - \mu\gamma_0)\psi(x) + H_I, \\ H_I &= K\sum_{\mu,A}\int d^3x\, \mathcal{F}\bar\psi(x)\gamma_\mu T^A\psi(x)\bar\psi(x)\gamma^\mu T^A\psi(x). \end{aligned} \tag{4.4}$$

Here the T^A are the color SU(3) generators, so the quantum numbers are those of one-gluon exchange. However instead of an honest gluon propagator we use an instantaneous contact interaction, modified by a form-factor

$\mathcal{F}$. $\mathcal{F}$ is taken to be a product of several momentum dependent factors $F(p)$, one for each leg, and to die off at large momentum. One convenient possibility is $F(p) = \left(\lambda^2/(p^2+\lambda^2)\right)^\nu$, where λ and ν can be varied to study sensitivity to the location and shape of the cutoff. The qualitative effect of the form-factor is to damp the spurious ultraviolet singularities introduced by H_I; microscopic QCD, of course, does have good ultraviolet behavior. One will tend to trust conclusions that do not depend sensitively on λ or ν. In practice, one finds that the crucial results—the form and magnitude of gaps—are rather forgiving.

Given the Hamiltonian, we can study the possibilities for symmetry breaking condensations. The most favorable condensation possibility so far identified is of the form

$$\begin{aligned}\langle q^{i\alpha}_{La}(p) q^{j\beta}_{Lb}(-p)\rangle &= -\langle q^{i\alpha}_{Ra}(p) q^{j\beta}_{Rb}(-p)\rangle \\ &= \epsilon^{ij}\left(\kappa_1(p^2)\delta^\alpha_a\delta^\beta_b + \kappa_2(p^2)\delta^\alpha_b\delta^\beta_a\right).\end{aligned} \tag{4.5}$$

Here we encounter the phenomenon of color-flavor locking. The ground state contains correlations whereby both color and flavor symmetry are spontaneously broken, but the diagonal subgroup, which applies both transformations simultaneously, remains valid. There are several good reasons to think that condensation of this form characterizes the true ground state, with lowest energy, at asymptotic densities. It corresponds to the most singular channel, in the renormalization group analysis discussed above. It produces a gap in all channels, and is perturbatively stable, so that it is certainly a convincing local minimum. It resembles the known order parameter for the B phase of superfluid He3. And it beats various more-or-less plausible competitors that have been investigated, by a wide margin.

Given the form of the condensate, one can fix the leading functional dependencies of $\kappa_1(p^2,\mu)$ and $\kappa_2(p^2,\mu)$ at weak coupling by a variational calculation. For present purposes, it is adequate to replace all possible contractions of the quark fields in (4.4) having the quantum numbers of (4.5) with their supposed expectation values, and diagonalize the quadratic part of the resulting Hamiltonian. The ground state is obtained, of course, by filling the lowest energy modes, up to the desired density. One then demands internal consistency, i.e. that the postulated expectation values are equal to the derived ones. Some tricky but basically straightforward algebra leads us to the result

$$\Delta_{1,8}(p^2) = F(p)^2\Delta_{1,8} \tag{4.6}$$

where Δ_1 and Δ_8 satisfy the coupled gap equations

$$\begin{aligned}\Delta_8 + \frac{1}{4}\Delta_1 &= \frac{16}{3}KG(\Delta_1) \\ \frac{1}{8}\Delta_1 &= \frac{16}{3}KG(\Delta_8)\end{aligned} \tag{4.7}$$

where we have defined

$$G(\Delta) = -\frac{1}{2}\sum_{\mathbf{k}}\left\{\frac{F(k)^4\Delta}{\sqrt{(k-\mu)^2+F(k)^4\Delta^2}}+\frac{F(k)^4\Delta}{\sqrt{(k+\mu)^2+F(k)^4\Delta^2}}\right\} \tag{4.8}$$

and

$$\begin{aligned}\kappa_1(p^2) &= \frac{1}{8K}\left(\Delta_8(p^2)+\frac{1}{8}\Delta_1(p^2)\right)\\ \kappa_2(p^2) &= \frac{3}{64K}\Delta_1(p^2).\end{aligned} \tag{4.9}$$

The Δ are defined so that $F(p)^2\Delta_{1,8}(p^2)$ are the gaps for singlet or octet excitations at three-momentum p. (4.7) must be solved numerically.

Finally, to obtain quantitative estimates of the gaps, we must normalize the parameters of our model Hamiltonian. One can do this very crudely by using the model Hamiltonian in the manner originally pioneered by Nambu and Jona-Lasinio, that is as the basis for a variational calculation of chiral symmetry breaking at zero density. The magnitude of this chiral condensate can then be fixed to experimental or numerical results. In this application we have no firm connection between the model and microscopic QCD, because there is no large momentum scale (or weak coupling parameter) in sight. Nevertheless a very large literature following this approach encourages us to hope that its results are not wildly wrong, quantitatively. Upon adopting this normalization procedure, one finds that gaps of order several tens of MeV near the Fermi surface are possible at moderate densities.

While this model treatment captures major features of the physics of color-flavor locking, with a little more work it is possible to do a much more rigorous calculation, and in particular to normalize directly to the known running of the coupling at large momentum. I'll sketch this now.

4.4 Taming the Magnetic Singularity

A proper discussion of the fully microscopic calculation is necessarily quite technical, and would be out of place here, but the spirit of the thing—and one of the most striking results—can be conveyed simply.

When retardation or relativistic effects are important a Hamiltonian treatment is no longer appropriate. One must pass to Lagrangian and graphical methods. (Theoretical challenge: is it possible to systematize these in a variational approach?) The gap equation appears as a self-consistency equation for the assumed condensation, shown graphically in Figure 7.12.

With a contact interaction, and throwing away manifestly spurious ultraviolet divergences, we obtain a gap equation of the type

$$\Delta \propto g^2 \int d\epsilon\, \frac{\Delta}{\sqrt{\epsilon^2+\Delta^2}}. \tag{4.10}$$

The phase space transverse to the Fermi surface cancels against a propagator, leaving the integral over the longitudinal distance ϵ to the Fermi surface. Note that the integral on the right diverges at small ϵ, so that as long as the proportionality constant is positive one will have nontrivial solutions for Δ, no matter how small is g. Indeed, one finds that for small g, $\Delta \sim e^{-\text{const}/g^2}$.

If we restore the gluon propagator, we will find a nontrivial angular integral, which diverges for forward scattering. That divergence will be killed, however, if the gluon acquires a mass $\propto g\Delta$ through the Meissner-Higgs mechanism. Thus we arrive at a gap equation of the type

$$\Delta \propto g^2 \int d\epsilon \, \frac{\Delta}{\sqrt{\epsilon^2 + \Delta^2}} \, dz \, \frac{\mu^2}{\mu^2 + (g\Delta)^2}. \tag{4.11}$$

Now one finds $\Delta \sim e^{-\text{const}/g}$!

A proper discussion of the microscopic gap equation is considerably more involved than this, but the conclusion that the gap goes exponentially in the inverse coupling (rather than its square) at weak coupling still emerges. It has the amusing consequence, that at asymptotically high densities the gap becomes arbitrarily large! This is because asymptotic freedom insures that it is the microscopic coupling $1/g(\mu)^2$ which vanishes logarithmically, so that $e^{-\text{const}/g(\mu)}$ does not shrink as fast as $1/\mu$. Since the "dimensional analysis" scale of the gap is set by μ, its linear growth wins out asymptotically.

Lecture 5 High-Density QCD: Color-Flavor Locking and Quark-Hadron Continuity

Color-flavor locking has many remarkable consequences. There is a gap for all colored excitations, including the gluons. This is, operationally, confinement. The photon picks up a gluonic component of just such a form as to ensure that all elementary excitations, including quarks, are integrally charged. Some of the gluons acquire nonzero, but integer-valued, electric charges. Baryon number is spontaneously broken, which renders the high-density material a superfluid.

If in addition the quarks are massless, then their chiral symmetry is spontaneously broken, by a new mechanism. The left-dynamics and the

FIGURE 7.12. Graphical form of the self-consistent equation for the condensate (gap equation).

right-dynamics separately lock to color; but since color allows only vector transformations, left is thereby locked to right.

You may notice several points of resemblance between the low-energy properties calculated for the high-density color-flavor locked phase and the ones you might expect at low density, based on semi-phenomenological considerations such as the MIT bag model, or experimental results in real-world QCD. The quarks play the role of low-lying baryons, the gluons play the role of the low-lying vector mesons, and the Nambu-Goldstone bosons of broken chiral symmetry play the role of the pseudoscalar octet. All the quantum numbers match, and the spectrum has gaps—or not—in all the right places. In addition we have baryon number superfluidity, which extends the expected pairing phenomena in nuclei. Overall, there is an uncanny match between all the universal, and several of the nonuniversal, features of the calculable high-density and the expected low-density phase. This leads us to suspect that there is no phase transition between them!

5.1 Gauge Symmetry (Non)Breaking

An aspect of (4.5) that might appear troubling at first sight, is its lack of gauge invariance. There are powerful general arguments that local gauge invariance cannot be broken. Indeed, local gauge invariance is really a tautology, stating the equality between redundant variables. Yet its 'breaking' is central to two of the most successful theories in physics, to BCS superconductivity theory and the standard model of electroweak interactions. In BCS theory we postulate a nonzero vacuum expectation value for the (electrically charged) Cooper pair field, and in the standard model we postulate a nonzero vacuum expectation value for the Higgs field, which violates both the weak isospin SU(2) and the weak hypercharge $U(1)$.

In each case, we should interpret the condensate as follows. We are working in a gauge theory at weak coupling. It is then very convenient to fix a gauge, because after we have done so—but not before!—the gauge potentials will make only small fluctuations around zero, which we will be able to take into account perturbatively. Of course at the end of any calculation we must restore the gauge symmetry, by averaging over the gauge fixing parameters (gauge unfixing). Only gauge-invariant results will survive this averaging. In a fixed gauge, however, one might capture important correlations, that characterize the ground state, by specifying the existence of nonzero condensates relative to that gauge choice. These condensates need not, and generally will not, break any symmetries.

For example, in the standard electroweak model one employs a nonzero vacuum expectation value for a Higgs doublet field $\langle \phi^a \rangle = v\delta^a_1$, which is not gauge invariant. One might be tempted to use the magnitude of its absolute square, which is gauge invariant, as an order parameter for the symmetry breaking, but $\langle \phi^\dagger \phi \rangle$ never vanishes, whether or not any symmetry is broken (and, of course, $\langle \phi^\dagger \phi \rangle$ breaks no symmetry). In fact there is no order parameter for the electroweak phase transition, and it has long been appreciated that one could, by allowing the SU(2) gauge couplings to become large, go over into a 'confined' regime while encountering no sharp phase transition. The most important gauge-invariant consequences one ordinarily infers from the condensate, of course, are the nonvanishing W and Z boson masses. This absence of massless bosons and long-range forces is the essence of confinement, or of the Meissner–Higgs effect. Evidently, when used with care, the notion of spontaneous gauge symmetry breaking can be an extremely convenient fiction—so it proves for (4.5).

5.2 Symmetry Accounting

The equations of our original model, QCD with three massless flavors, has the continuous symmetry group $\mathrm{SU}(3)^c \times \mathrm{SU}(3)_L \times \mathrm{SU}(3)_R \times \mathrm{U}(1)_B$. The Kronecker deltas that appear in the condensate (4.5) are invariant under neither color nor left-handed flavor nor right-handed flavor rotations separately. Only a global, diagonal SU(3) leaves the ground state invariant. Thus we have the symmetry breaking pattern

$$\mathrm{SU}(3)^c \times \mathrm{SU}(3)_L \times \mathrm{SU}(3)_R \times \mathrm{U}(1)_B \rightarrow \mathrm{SU}(3)_{c+L+R} \times Z_2. \tag{5.1}$$

5.2.1 Confinement

The breaking of local color symmetry implies that all the gluons acquire mass, according to the Meissner (or alternatively Higgs) effect. There are no long-range, $1/r$ interactions. There is no direct signature for the color degree of freedom—although, of course, in weak coupling one clearly perceives its avatars. It is veiled or, if you like, confined.

5.2.2 Chiral Symmetry Breaking

If we make a left-handed chiral rotation then we must compensate it by a color rotation, in order to leave the left-handed condensate invariant. Color rotations being vectorial, we must then in addition make a right-handed chiral rotation, in order to leave the right-handed condensate invariant. Thus chiral symmetry is spontaneously broken, by a new mechanism: although the left- and right-condensates are quite separate (and, before we include instantons—see below—not even phase coherent), because both are locked to color they are thereby locked to one another.

The spontaneous breaking of global chiral $\mathrm{SU}(3)_L \times \mathrm{SU}(3)_R$ brings with it an octet of pseudoscalar Nambu–Goldstone bosons, collective modes interpolating, in space-time, among the condensates related by the lost symmetry. These massless modes, as is familiar, are derivatively coupled, and therefore they do not generate singular long-range interactions.

5.2.3 Superfluidity and "Nuclear" Pairing

Less familiar, and perhaps disconcerting at first sight, is the loss of baryon number symmetry. This does not, however, portend proton decay, any more than does the nonvanishing condensate of helium atoms in superfluid He4. Given an isolated finite sample, the current divergence equation can be integrated over a surface surrounding the sample, and unambiguously indicates overall number conservation. To respect it, one should project onto states with a definite number of baryons, by integrating over states with different values of the condensate phase. This does not substantially alter the physics of the condensate, however, because the overlap between states of different phase is very small for a macroscopic sample. Roughly speaking, there is a finite mismatch per unit volume, so the overlap vanishes exponentially in the limit of infinite volume. The true meaning of the formal baryon number violation is that there are low-energy states with different distributions of baryon number, and easy transport among them. Indeed, the dynamics of the condensate is the dynamics of superfluidity: gradients in the Nambu–Goldstone mode are none other than the superfluid flow.

We know from experience that large nuclei exhibit strong even-odd effects, and an extensive phenomenology has been built up around the idea of pairing in nuclei. If electromagnetic Coulomb forces didn't spoil the fun, we could confidently expect that extended nuclear matter would exhibit the classic signatures of superfluidity. In our 3-flavor version the Coulomb forces do not come powerfully into play, since the charges of the quarks average out to electric neutrality. Furthermore, the tendency to superfluidity exhibited by ordinary nuclear matter should be enhanced by the additional channels operating coherently. So one should expect strong superfluidity at ordinary nuclear density, and it becomes less surprising that we find it at asymptotically large density too.

5.2.4 Global Order Parameters

I mentioned before that the Higgs mechanism as it operates in the electroweak sector of the standard model has no gauge-invariant signature. With color-flavor locking we're in better shape, because global as well as gauge symmetries are broken. Thus there are sharp differences between the color-flavor locked phase and the free phase. There must be phase transitions—as a function, say, of temperature—separating them.

In fact, it is a simple matter to extract gauge invariant order parameters from our primary, gauge variant condensate at weak coupling. For instance, to form a gauge invariant order parameter capturing chiral symmetry breaking we may take the product of the left-handed version of (4.5) with the right-handed version and saturate the color indices, to obtain

$$\begin{aligned}\langle q^{\alpha}_{La} q^{\beta}_{Lb} \bar{q}^{c}_{R\alpha} \bar{q}^{d}_{R\beta} \rangle &\sim \langle q^{\alpha}_{La} q^{\beta}_{Lb} \rangle \langle \bar{q}^{c}_{R\alpha} \bar{q}^{d}_{R\beta} \rangle \\ &\sim (\kappa_1^2 + \kappa_2^2)\delta^c_a \delta^d_b + 2\kappa_1\kappa_2 \delta^d_a \delta^c_b . \end{aligned} \tag{5.2}$$

Likewise we can take a product of three copies of the condensate and saturate the color indices, to obtain a gauge invariant order parameter for superfluidity. These secondary order parameters will survive gauge unfixing unscathed. Unlike the primary condensate from which they were derived, they are not just convenient fictions, but measurable realities.

5.2.5 A Subtlety: Axial Baryon Number

As it stands the chiral order parameter (5.2) is not quite the usual one, but roughly speaking its square. It leaves invariant an additional Z_2, under which the left-handed quark fields change sign. Actually this Z_2 is not a legitimate symmetry of the full theory, but suffers from an anomaly.

Since we can be working at weak coupling, we can be more specific. Our model Hamiltonian (4.4) was abstracted from one-gluon exchange, which is the main interaction among high-energy quarks in general, and so in particular for modes near our large Fermi surfaces. The instanton interaction is much less important, at least asymptotically, both because it is intrinsically smaller for energetic quarks, and because it involves six fermion fields, and hence (one can show) is irrelevant as one renormalizes toward the Fermi surface. However, it represents the leading contribution to axial baryon number violation. In particular, it is only $\mathrm{U}_A(1)$ violating interactions that fix the relative phase of our left- and right-handed condensates. So a model Hamiltonian that neglects them will have an additional symmetry that is not present in the full theory. After spontaneous breaking, which does occur in the axial baryon number channel, there will be a Nambu-Goldstone boson in the model theory, that in the full theory acquires an anomalously (pun intended) small mass. Similarly, in the full theory there will be a nonzero tertiary chiral condensate of the usual kind, bilinear in quark fields, but it will be parametrically smaller than (5.2).

5.3 Elementary Excitations

There are three sorts of elementary excitations. They are the modes produced directly by the fundamental quark and gluon fields, and the collective modes connected with spontaneous symmetry breaking.

The quark fields of course produce spin 1/2 fermions. Some of these are true long-lived quasiparticles, since there is nothing for them to decay into. They form an octet and a singlet under the residual diagonal SU(3). There is an energy gap for production of pairs above the ground state. Actually there are two gaps: a smaller one for the octet, and a larger one for the singlet.

The gluon fields produce an octet of spin 1 bosons. As previously mentioned, they acquire a mass by the Meissner–Higgs phenomenon. We have already discussed the Nambu-Goldstone bosons, too.

5.4 A Modified Photon

The notion of "confinement" I advertised earlier, phrased in terms of mass gaps and derivative interactions, might seem rather disembodied. So it is interesting to ask whether and how a more traditional and intuitive criterion of confinement—no fractionally charged excitations—is satisfied.

Before discussing electromagnetic charge we must identify the unbroken gauge symmetry, whose gauge boson defines the physical photon in our dense medium. The original electromagnetic gauge invariance is broken, but there is a combination of the original electromagnetic gauge symmetry and a color transformation which leaves the condensate invariant. Specifically, the original photon γ couples according to the matrix

$$\begin{pmatrix} \frac{2}{3} & 0 & 0 \\ 0 & -\frac{1}{3} & 0 \\ 0 & 0 & -\frac{1}{3} \end{pmatrix} \tag{5.3}$$

in flavor space, with strength e. There is a gluon G which couples to the matrix

$$\begin{pmatrix} -\frac{2}{3} & 0 & 0 \\ 0 & \frac{1}{3} & 0 \\ 0 & 0 & \frac{1}{3} \end{pmatrix} \tag{5.4}$$

in color space, with strength g. Then the combination

$$\tilde{\gamma} = \frac{g\gamma + eG}{\sqrt{e^2 + g^2}} \tag{5.5}$$

leaves the 'locking' Kronecker deltas in color-flavor space invariant. In our medium, it represents the physical photon. What happens here is similar to what occurs in the electroweak sector of the standard model, where both weak isospin and weak hypercharge are separately broken by the Higgs doublet, but a cunning combination remains unbroken, and defines electromagnetism.

5.4.1 Integer Charges!

Now with respect to $\tilde{\gamma}$ the electron charge is

$$\frac{-eg}{\sqrt{e^2 + g^2}}, \tag{5.6}$$

deriving of course solely from the γ piece of (5.5). The quarks have one flavor and one color index, so they pick up contributions from both pieces. In each sector we find the normalized charge unit $eg/\sqrt{e^2 + g^2}$, and it is multiplied by some choice from among $(\frac{2}{3}, -\frac{1}{3}, -\frac{1}{3})$ or $(-\frac{2}{3}, \frac{1}{3}, \frac{1}{3})$ respectively. The total, obviously, can be ± 1 or 0. Thus the excitations produced by the quark fields are integrally charged, in units of the electron charge.

Similarly the gluons have an upper color and a lower anti-color index, so that one faces similar choices, and reaches a similar conclusion. In particular, some of the gluons have become electrically charged. The pseudoscalar Nambu-Goldstone modes have an upper flavor and a lower anti-flavor index, and yet again the same conclusions follow. The superfluid mode, of course, is electrically neutral.

5.4.2 A Conceptual Jewel

It is fun to consider how a chunk of our color-flavor locked material would look. If the quarks were truly massless, then so would be Nambu–Goldstone bosons (at the level of pure QCD), and one might expect a rather unusual "bosonic metal," in which low-energy electromagnetic response is dominated by these modes. Actually electromagnetic radiative corrections lift the mass of the charged Nambu–Goldstone bosons, creating a gap for the charged channel. The same effect would be achieved by turning on a common nonzero quark mass. Thus the color-flavor locked material forms a transparent insulator. Altogether it would resemble a diamond, reflecting portions of incident light waves, but allowing finite portions through and out again!

5.5 *Quark-Hadron Continuity*

The universal features of the color-flavor locked state: confinement, chiral symmetry breaking down to vector SU(3), and superfluidity, are just what one would expect, based on standard phenomenological models and experience with real-world QCD at low density. Now we see that the low-lying spectrum likewise bears an uncanny resemblance to what one finds in the Particle Data Book (or rather what one would find, in a world of three degenerate quarks). It is hard to resist the inference that there is no phase transition separating them. Thus there need not be, and presumably is not, a sharp distinction between the low-density phase, where microscopic calculations are difficult but the convenient degrees of freedom are "obviously" hadrons, and the asymptotic high-density phase, where weak-coupling (but nonperturbative) calculations are possible, and the right degrees of freedom are elementary quarks and gluons plus collective modes associated with spontaneous symmetry breaking. We call this quark-hadron continuity. It might seem shocking that a quark can "be" a baryon, but remember that it is immersed in a sea of diquark condensate, wherein the distinction between one quark and three becomes negotiable.

5.6 *Remembrance of Things Past*

An entertaining aspect of the emergent structure is that two beautiful ideas from the pre-history of QCD, that were bypassed in its later development, have come very much back to center stage, now with microscopic validation. The quark-baryons of the color-flavor locked phase follow the charge assignments proposed by Han and Nambu. And the gluon-vector mesons derive from the Yang–Mills gauge principle—as originally proposed, for rho mesons!

5.7 More Quarks

For larger numbers of quarks, the story is qualitatively similar. Color symmetry is broken completely, and there is a gap in all quark channels, so the weak-coupling treatment is adequate. Color-flavor locking is so favorable that there seems to be a periodicity: if the number of quarks is a multiple of three, one finds condensation into 3×3 blocks, while if it is $4 + 3k$ or $5 + 3k$ one finds k color-flavor locking blocks together with special patterns characteristic of 4 or 5 flavors.

There is an amusing point here. QCD with a very large number of massless quarks, say 16, has an infrared fixed point at very weak coupling. Thus it should be quasi-free at zero density, forming a non-Abelian Coulomb phase, featuring conformal symmetry, no confinement, and no chiral symmetry breaking. To say the least, it does not much resemble real-world QCD. There are indications that this qualitative behavior may persist even for considerably fewer quarks (the critical number might be as small as 5 or 6). Nevertheless, at high density, we have discovered, these many-quark theories all support more-or-less normal-looking "nuclear matter"—including confinement and chiral symmetry breaking!

5.8 Fewer Quarks, and Reality

One can perform a similar analysis for two quark flavors. A new feature is that the instanton interaction now involves four rather than six quark legs, so it remains relevant as one renormalizes toward the Fermi surface. Either the one-gluon exchange or the instanton interaction, treated in the spirit above, favors condensation of the form

$$\langle q^{i\alpha}_{La}(p) q^{j\beta}_{Lb}(-p)\rangle = -\langle q^{i\alpha}_{Ra}(p) q^{j\beta}_{Lb}(-p)\rangle = \epsilon^{ij}\kappa(p^2)\epsilon^{\alpha\beta 3}\epsilon_{ab}. \tag{5.7}$$

Formally, (5.7) is quite closely related to (5.2), since $\epsilon^{\alpha\beta I}\epsilon_{abI} = 2(\delta^\alpha_a\delta^\beta_b - \delta^\alpha_b\delta^\beta_a)$. Their physical implications, however, are quite different.

To begin with, (5.7) does not lead to gaps in all quark channels. The quarks with color labels 1 and 2 acquire a gap, but quarks of the third color of quark are left untouched. Secondly, the color symmetry is not completely broken. A residual SU(2), acting among the first two colors, remains valid. For these reasons, perturbation theory about the ground state defined by (5.7) is *not* free of infrared divergences, and we do not have a fully reliable grip on the physics.

Nevertheless it is plausible that the qualitative features suggested by (5.7) are not grossly misleading. The residual SU(2) presumably produces confined glueballs of large mass, and assuming this occurs, the residual gapless quarks are weakly coupled.

Assuming for the moment that no further condensation occurs, for massless quarks we have the symmetry breaking pattern

$$\mathrm{SU}(3)^c \times \mathrm{SU}(2)_L \times \mathrm{SU}(2)_R \times \mathrm{U}(1)_B \\ \rightarrow \mathrm{SU}(2)^c \times \mathrm{SU}(2)_L \times \mathrm{SU}(2)_R \times \widetilde{\mathrm{U}}(1)_B. \quad (5.8)$$

Here the modified baryon number acts only on the third color of quarks. It is a combination of the original baryon number and a color generator, that are separately broken but when applied together leave the condensate invariant. Comparing to the zero-density ground state, one sees that color symmetry is reduced, chiral symmetry is restored, and baryon number is modified. Only the restoration of chiral symmetry is associated with a legitimate order parameter, and only it requires a sharp phase transition.

In the real world, with the u and d quarks light but not strictly massless, there is no rigorous argument that a phase transition is necessary. It is (barely) conceivable that one might extend quark-hadron continuity to this case. Due to medium modifications of baryon number and electromagnetic charge the third-color u and d quarks have the quantum numbers of nucleons. The idea that chiral symmetry is effectively restored in nuclear matter, however, seems problematic quantitatively. More plausible, perhaps, is that there is a first-order transition between nuclear matter and quark matter. This is suggested by some model calculations, and is the basis for an attractive interpretation of the MIT bag model, according to which baryons are droplets wherein chiral symmetry is restored.

5.8.1 Thresholds and Mismatches

In the real world there are two quarks, u and d, whose mass is much less than Λ_{QCD}, and one, s, whose mass is comparable to it. Two simple qualitative effects, that have major implications for the zero-temperature phase diagram, arise as consequences of this asymmetric spectrum. They are expected, whether one analyzes from the quark side or from the hadron side.

The first is that one can expect a threshold, in chemical potential (or pressure), for the appearance of any strangeness at all in the ground state. This will certainly hold true in the limit of large strange quark mass, and there is considerable evidence for it in the real world. This threshold is in addition to the threshold transitions at lower chemical potentials, from void to nuclear matter, and (presumably) from nuclear matter to two-flavor quark matter, as discussed above.

The second is that at equal chemical potential the Fermi surfaces of the different quarks will not match. This mismatch cuts off the Cooper instability in mixed channels. If the nominal gap is large compared to the

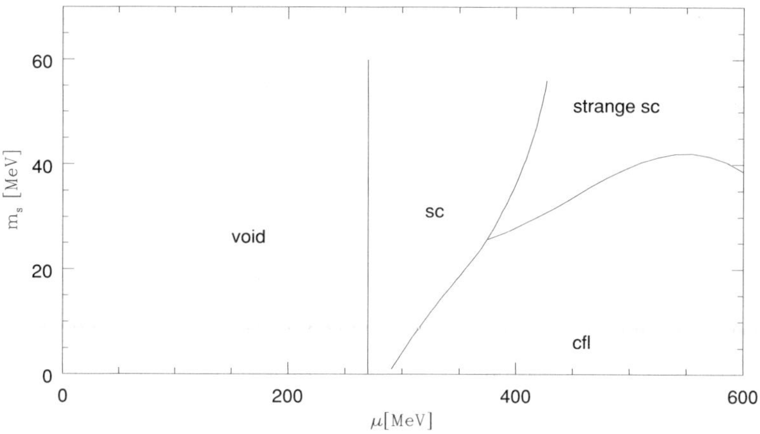

FIGURE 7.13. Phase diagram calculated in a schematic model.

mismatch, one can treat the mismatch as a perturbation. This will always be valid at asymptotically high densities, since the mismatch goes as m_s^2/μ, whereas the gap eventually grows with μ. If the nominal gap is small compared to the mismatch, condensation will not occur.

5.8.2 Assembling the Pieces

With these complications in mind, we can identify three major phases in the plane of chemical potential and strange quark mass, that reflect the simple microscopic physics we have surveyed above. (There might of course be additional "minor" phases—notably including normal nuclear matter!) There is 2-flavor quark matter, with restoration of chiral symmetry, and zero strangeness. Then there is a 2+1-flavor phase, in which the strange and nonstrange Fermi surfaces are badly mismatched, and one has independent dynamics for the corresponding low energy excitations. Here one expects strangeness to break spontaneously, by its own Fermi surface instability. Finally there is the color-flavor locked phase. A caricature version of the phase diagram in the (m_s, μ) plane, illustrating these features, is given in Figure 7.13.

5.8.3 Reality

The progress reported here, while remarkable, mainly concerns the asymptotic behavior of high-density QCD. Its extrapolation to practical densities is at present semi-quantitative at best. To do real justice to the potential applications, we need to learn how to do more accurate analytical and numerical work at moderate densities.

As regards analytical work, we can take heart from some recent progress on the equation of state at high temperature. Here there are extensive, interesting numerical results, which indicate that the behavior is quasi-free, but that there are very significant quantitative corrections to free quark-gluon plasma results, especially for the pressure. Thus it is plausible *a priori* that some weak-coupling, but nonperturbative, approach will be workable, and this seems to be proving out. An encouraging feature here is that the analytical techniques used for high temperature appear to be capable of extension to finite density without great difficulty.

Numerical work at finite density, unfortunately, is plagued by poor convergence. This arises because the functional integral is not positive definite configuration by configuration, so that importance sampling fails, and one is left looking for a small residual from much larger canceling quantities.

There are cases in which this problem does not arise. It does not arise for two colors. Although low-density hadronic matter is quite different in a two-color world than a three-color world—the baryons are bosons, so one does not get anything like a shell structure for nuclei—I see no reason to expect that the asymptotic, high-density phases should be markedly different. It would be quite interesting to see Fermi-surface behavior arising for two colors at high density (especially, for the ground state pressure), and even more interesting to see the effect of diquark condensations.

Another possibility, that I have been discussing with David Kaplan, is to engineer lattice gauge theories whose low-energy excitations

resemble those of finite density QCD near the Fermi surface, but which are embedded in a theory that is globally particle-hole symmetric, and so feature a positive-definite functional integral.

Aside from these tough quantitative issues, there are a number of directions in which the existing work should be expanded and generalized, that appear to be quite accessible. There is already a rich and important theory of the behavior of QCD at nonzero temperature and zero baryon number density. We should construct a unified picture of the phase structure as a function of both temperature and density; to make it fully illuminating, we should also allow at least the strange quark mass to vary. We should allow for the effect of electromagnetism (after all, this is largely what makes neutron stars what they are) and of rotation. We should consider other possibilities than a common chemical potential for all the quarks.

As physicists we should not, however, be satisfied with hoarding up formal, abstract knowledge. There are concrete experimental situations and astrophysical objects we must speak to. Hopefully, having mastered some of the basic vocabulary and grammar, we will soon be in a better position to participate in a two-way dialogue with Nature.

Background Material

General background on quantum field theory and QCD:

1. T.-P. Cheng and L.-F. Li, *Gauge Theory of Elementary Particles* (Oxford Univ. Press, London 1984).

2. M. Peskin and D. Schroder, *Introduction to Quantum Field Theory* (Addison-Wesley, Redwood City, CA, 1995).

Lattice gauge theory:

3. M. Creutz, *Quarks, Gluons, and Lattices* (Cambridge Univ. Press, Cambridge U.K., 1983).

Chiral symmetry and current algebra:

4. S. Weinberg, *Quantum Theory of Fields.* II, Chapter 19 (Cambridge Univ. Press, Cambridge U.K., 1996).

Renormalization group for critical phenomena:

5. D. Amit, *Field Theory, the Renormalization Group, and Critical Phenomena* (World Scientific, Singapore, 1984).

Renormalization group toward the Fermi surface:

6. J. Polchinski, hep-th/9210046;

7. R. Shankar, *Rev. Mod. Phys.* **66** (1993), 129.

Superconductivity:

8. J.R. Schrieffer, *Theory of Superconductivity* (Benjamin/Cummins, Reading, MA, 1984).

Tricritical points:

9. I. Lawrie and S. Sarbach, in *Phase Transitions and Critical Phenomena 9*, eds. C. Domb and J. Lebowitz (Academic Press, New York, 1984).

Quark-gluon plasma:

10. U. Heinz and M. Jacob, nucl-th/0002042.

Sources and Further Reading

The perspective on QCD adopted in Lecture 1, as a series of interweaving stories of symmetry and dynamics, is taken from my monograph *QCD* in preparation for Princeton University Press.

The material on high-temperature phase transitions in Lecture 2 and Lecture 3 is mostly taken from

1. F. Wilczek, *Internat. J. Modern Phys.* **A7** (1992), 3911.
2. K. Rajagopal and F. Wilczek, *Nuclear Phys.* **B399** (1993), 395.

The possibility of a true critical point was pointed out in

3. M. Stephanov, K. Rajagopal, and E. Shuryak *Phys. Rev. Lett.* **81** (1998), 4816.

Its possible experimental signatures are discussed in

4. M. Stephanov, in *Understanding Deconfinement in QCD*, eds. D. Blaschke, F. Karsch, and C.D. Roberts (World Scientific, Singapore, 1999), hep-ph/9906242.

For recent lattice results see

5. F. Karsch, Nucl. Phys. Proc. Suppl. **83** (2000), 14, hep-lat/9909006.

Lecture 4 and Lecture 5 are based on

6. F. Wilczek, Nucl.Phys. **A663-664** (2000), 257, hep-ph/9908480.

For further developments see especially

7. T. Schäfer, Nucl. Phys. **B575** (2000), 269, hep-ph/9909574.

These papers contain numerous additional references. Obviously, several topics discussed in the lectures are under very active development, and you should consult the web for up-to-date results.

References

1. M. Schmelling, *Status of the strong coupling constant*, plenary talk at the XXVIII International Conference on High Energy Physics, Warsaw, 1996, hep-ex/9701002.
2. R. Burkhalter, Nucl. Phys. Proc. Suppl. **73** (1999), 3, hep-lat/9810043.
3. J.O. Anderson, E. Braaten and M. Strickland, Phys. Rev. **D61** (2000), 014017, hep-ph/9905337; J.-P. Blaizot, E. Iancu and A. Rebjam, Phys. Rev. Lett. **83** (1999) 2906, hep-ph/9906340.